Study Guide & Solutions Book

ORGANIC CHEMISTRY: A SHORT COURSE

Eighth Edition

Harold Hart
Michigan State University

David J. Hart
Ohio State University

HOUGHTON MIFFLIN COMPANY   BOSTON
Dallas   Geneva, Illinois
Palo Alto   Princeton, New Jersey

Cover
"Midnight Variations," artist Kenneth Snelson's imagined atomic world. The image was created through use of a three-dimensional graphics computer.

ISBN: 0-395-57237-1

BCDEFGHIJ-H-9654321

CONTENTS

INTRODUCTION TO THE STUDENT

This study guide and solutions book was written to help you learn organic chemistry. The principles and facts of this subject are not easily learned by simply reading them, even repeatedly. Formulas, equations, and molecular structures are best mastered by *written* practice. To help you become thoroughly familiar with the material, we have included many problems within and at the end of each chapter in the text.

It is our experience that such questions are not put to their best use unless correct answers are also available. Indeed, answers alone are not enough. If you know how to work a problem and find that your answer agrees with the correct one, fine. But what if you work conscientiously, yet cannot solve the problem? You then give in to temptation, look up the answer, and encounter yet another dilemma—how in the world did the author get that answer? This solutions book has been written with this difficulty in mind. For many of the problems, all the reasoning involved in getting the correct answer is spelled out in detail. Many of the answers also include cross-references to the text. If you cannot solve a particular problem, these references will guide you to parts of the text that you should review.

Each chapter of the text is briefly summarized. Whenever pertinent, the chapter summary is followed by a list of all the new reactions and mechanisms encountered in that chapter. These lists should be especially helpful to you as you review for examinations.

When you study a new subject, it is always useful to know what is expected. To help you, we have included in this study guide a list of learning objectives for each chapter, that is, a list of what you should be able to do after you have read and studied that chapter. Your instructor may want to delete items from these lists of objectives or add to them. However, we believe that if you have mastered these objectives—and the problems should help you to do this—you should have no difficulty with examinations. Furthermore, you should be very well prepared for further courses that require this course as a prerequisite.

Near the end of this study guide you will find additional sections that may help you to study for the final examination in the course. The SUMMARY OF SYNTHETIC METHODS lists the important ways to synthesize each class of compounds discussed in the text. This is followed by the SUMMARY OF REACTION MECHANISMS. Both of these sections have references to appropriate portions of the text, in case you feel that further review is necessary. Finally, you will find two lists of sample test questions. The first deals with synthesis, and the second is a list of multiple-choice questions. Both of these sets should help you prepare for examinations.

In addition, we offer you a brief word of advice about how to learn the many reactions you will study during this course. First, learn the nomenclature systems thoroughly for each new class of compounds that is introduced. Then, rather than memorizing the particular examples of reactions given in the text, study reactions as being typical of a class of compounds. For example, if you are asked how compound A will react with compound B, proceed in the following way. First ask yourself, to what class of compounds does A belong? How does this class of compounds react with B (or with compounds of the general class to which B belongs)? Then proceed from the general reaction to the specific case at hand. This approach will probably help you to eliminate some of the memory work often associated with organic chemistry courses.

We urge you to study regularly, and hope that this study guide and solutions book will make it easier for you to do so.

Great effort has been expended to ensure the accuracy of the answers in this book. It is easy for errors to creep in, however, and we will be particularly grateful to anyone who will call them to our attention. Suggestions for improving the book will also be welcome. Send them to:

Harold Hart
Department of Chemistry
Michigan State University
East Lansing, Michigan 48824

David J. Hart
Department of Chemistry
The Ohio State University
Columbus, Ohio 43210

CHAPTER ONE: BONDING AND ISOMERISM

CHAPTER SUMMARY*

An **atom** consists of a nucleus surrounded by **electrons** arranged in **orbitals.** The electrons in the outer shell, or the **valence electrons**, are involved in bonding. **Ionic bonds** are formed by electron transfer from an **electropositive** atom to an **electronegative** atom. Atoms with similar electronegativities form **covalent bonds** by sharing electrons. A **single bond** is the sharing of one electron pair between two atoms.

Carbon, with four valence electrons, mainly forms covalent bonds. It usually forms four such bonds, and these may be with itself or with other atoms such as hydrogen, oxygen, nitrogen, chlorine, and sulfur. In pure covalent bonds, electrons are shared equally, but in **polar covalent bonds,** the electrons are displaced toward the more electronegative element. **Multiple bonds consist** of two or three electron pairs shared between atoms.

Structural (or constitutional) isomers are compounds with the same **molecular formulas** but different **structural formulas** (that is, different arrangements of the atoms in the molecule). **Isomerism** is especially important in organic chemistry because of the capacity of carbon atoms to be arranged in so many different ways: continuous chains, branched chains, and rings.

Structural formulas can be written so that every bond is shown, or in various abbreviated forms. For example, the formula for *n*-pentane (*n* stands for normal) can be written as

Some atoms, even in covalent compounds, carry a **formal charge,** defined as the number of valence electrons in the neutral atom minus the sum of the number of

*In the chapter summaries, terms whose meanings you should know appear in boldface type.

unshared electrons and half the number of shared electrons. **Resonance** occurs when we can write two or more structures for a molecule or ion with the same arrangement of atoms but different arrangements of the electrons. The correct structure of the molecule or ion is a **resonance hybrid** of the **contributing structures,** which are drawn with a double-headed arrow ($\leftrightarrow$) between them. Organic chemists use a curved arrow ($\frown$) to show the movement of an electron pair.

A **sigma (σ) bond** is formed between atoms by overlap of two atomic orbitals along the line that connects the atoms. Carbon uses sp^3-**hybridized orbitals** to form four such bonds. These bonds are directed from the carbon nucleus toward the corners of a tetrahedron. In **methane,** for example, the carbon is at the center and the four hydrogens are at the corners of a regular tetrahedron with H—C—H bond angles of 109.5°.

Carbon compounds can be classified according to their molecular framework as **acyclic** (not cyclic), **carbocyclic** (containing rings of carbon atoms), or **heterocyclic** (containing at least one ring atom that is not carbon). They may also be classified according to **functional group** (Table 1.5).

REACTION SUMMARY

$$2\ ROH\ +\ 2\ Na\ \longrightarrow\ 2\ RO^-\ Na^+\ +\ H_2$$

  alcohol   sodium   sodium alkoxide   hydrogen

LEARNING OBJECTIVES*

1. Know the meaning of: nucleus, electrons, protons, neutrons, atomic number, atomic weight, shells, orbitals, valence electrons, valence, kernel.

2. Know the meaning of: electropositive, electronegative, ionic and covalent bonds, radical, catenation, polar covalent bond, single and multiple bonds, nonbonding or unshared electron pair.

3. Know the meaning of: molecular formula, structural formula, structural (or constitutional) isomers, continuous and branched chain, formal charge, resonance, contributing structures, sigma (σ) bond, sp^3 hybrid orbitals, tetrahedral carbon.

*Although the objectives are often worded in the form of imperatives (that is, determine..., write..., draw...), these verbs are all to be preceded by the phrase "be able to...." This phrase has been omitted to avoid repetition.

4.   Know the meaning of: acyclic, carbocyclic, heterocyclic, functional group.

5.   Know the meaning of the following symbols:

   $\delta+$ $\delta-$ $\longrightarrow$ $\longleftrightarrow$ $\curvearrowright$

6.   Given a periodic table, determine the number of valence electrons of an element and write its electron-dot formula.

7.   Given two elements and a periodic table, tell which element is more electropositive or electronegative.

8.   Given the formula of a compound and a periodic table, classify the compound as ionic or covalent.

9.   Given an abbreviated structural formula of a compound, write its electron dot formula.

10.   Given a covalent bond, tell whether it is polar. If it is, predict the direction of bond polarity from the electronegativities of the atoms.

11.   Given a molecular formula, draw the structural formulas for all possible structural isomers.

12.   Given a structural formula abbreviated on one line of type, write the complete structure and clearly show the arrangement of atoms in the molecule.

13.   Given a line formula, such as , $\wedge\!\!\wedge\!\!\wedge$ (pentane), write the complete structure and clearly show the arrangement of atoms in the molecule. Tell how many hydrogens are attached to each carbon, what the molecular formula is, and what the functional groups are.

14.   Given a simple molecular formula, draw the electron-dot formula and determine whether each atom in the structure carries a formal charge.

15.   Draw electron-dot formulas that show all important contributors to a resonance hybrid.

16.   Predict the geometry of bonds around an atom, knowing the electron distribution in the orbitals.

17.   Draw in three dimensions with solid, wedged, and dashed bonds the tetrahedral bonding around sp^3-hybridized carbon atoms.

18.   Distinguish between acyclic, carbocyclic, and heterocyclic structures.

1. BONDING AND ISOMERISM

19.   Given a series of structural formulas, recognize compounds that belong to the same class (same functional group).

20.   Begin to recognize the important functional groups: alkene, alkyne, alcohol, ether, aldehyde, ketone, carboxylic acid, ester, amine, nitrile, amide, thiol, and thioether.

ANSWERS TO PROBLEMS

Problems Within the Chapter

1.1   Elements with fewer than four valence electrons tend to give them up and form positive ions: Al^{3+}, Li^+. Elements with more than four valence electrons tend to gain electrons to complete the valence shell, becoming negative ions: S^{2-}. Elements with half-filled valence shells give up electrons and form positive ions (H^+) or gain electrons and form negative ions (H^-).

1.2   Within any horizontal row in the periodic table, the most electropositive element appears farthest to the left. Na is more electropositive than Al and B is more electropositive than C. In a given column in the periodic table, the lower the element, the more electropositive it is. Al is more electropositive than B.

1.3   Within any horizontal row in the periodic table, the most electronegative element appears farthest to the right. F is more electronegative than O, O more than N. In a given column, the higher the element, the more electronegative it is. F is more electronegative than Cl.

1.4   Carbon is in Group IV and has a half-filled (or half-empty) valence shell. It is neither strongly electropositive nor strongly electronegative.

1.5   dichloromethane (methylene chloride)   trichloromethane (chloroform)

1.6   If the C—C bond length is 1.54 Å and the Cl—Cl bond length is 1.98 Å, we expect the C—Cl bond length to be about 1.76 Å: (1.54 + 1.98)/2. In fact, the C—Cl bond (1.75 Å) is longer than the C—C bond.

4

ANSWERS TO PROBLEMS

1.7

Propane

$$\begin{array}{ccc} H & H & H \\ | & | & | \\ H-C-C-C-H \\ | & | & | \\ H & H & H \end{array}$$

1.8 $N^{\delta+}$—$Cl^{\delta-}$ $S^{\delta+}$—$O^{\delta-}$. The first of these is a little risky to predict because the elements (N, Cl) are not in the same horizontal row. But because these elements are two groups apart and only one row apart, it is not too surprising that chlorine is more electronegative than nitrogen. The polarity of the S—O bond is easy to predict because both elements are in the same column, and the more electronegative atom appears nearest the top.

1.9 Both Cl and F are more electronegative than C.

$$\begin{array}{c} \delta- \\ F \\ | \delta+ \\ \delta- Cl-C-Cl \ \delta- \\ | \\ F \\ \delta- \end{array}$$

1.10 Both the C—O and H—O bonds are polar, and the oxygen is more electronegative than either carbon or hydrogen.

$$\begin{array}{c} H \\ | \leftrightarrow \\ H-C-O-H \\ | \quad \leftrightarrow \\ H \end{array}$$

1.11 H: C ::: N :   H : C ::: N :   H : C ::: N :

1.12 a. The carbon shown has 12 electrons around it, 4 more than is allowed.
      b. There are 20 valence electrons shown, whereas there should only be 16 (6 from each oxygen and 4 from the carbon).
      c. There is nothing wrong with this formula, but it does place a formal charge of -1 on the "left" oxygen and +1 on the "right" oxygen (see Sec. 1.11). This formula is one possible contributor to the resonance hybrid structure for carbon dioxide (see Sec. 1.12); it is less important than the structure with two carbon–oxygen double bonds because it takes energy to separate the + and - charges.

1.13 Methanal (formaldehyde), H_2CO. There are 12 valence electrons altogether (C = 4, H = 1, and O = 6). A double bond between C and O is necessary to put 8 electrons around each of these atoms.

5

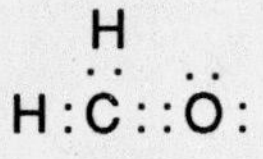

H
H:C::O: or

H
\
C=O
/
H

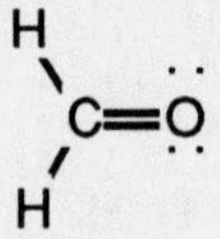

1.14   There are 10 valence electrons, 4 from C and 6 from O. An arrangement that puts 8 electrons around each atom is shown below. This structure puts a formal charge of -1 on C and +1 on O (see Sec. 1.11).

:C:::O: or   :C≡O:

1.15   If the carbon chain is linear, there are two possibilities:

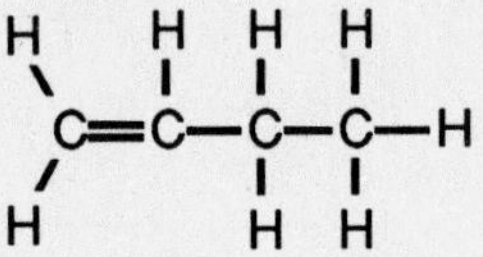

H   H H H
\ | | |
C=C—C—C—H   and   H—C—C=C—C—H
/ | |
H   H H

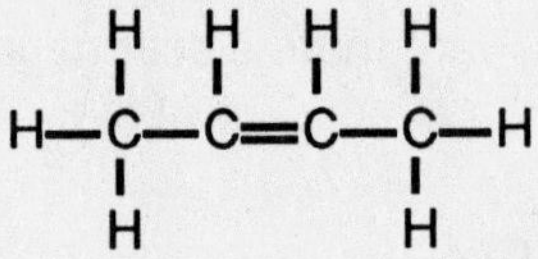

But the carbon chain can be branched, giving a third possibility:

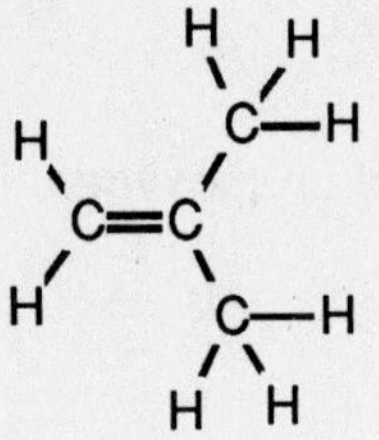

H H
\ /
C—H
H   /
\   /
C=C
/   \
H   C—H
/ \
H   H

1.16   a.   H   H
| /
H—C—N
| \
H   H

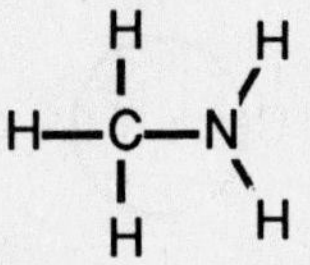

b.   H
|
H—C—O—H
|
H

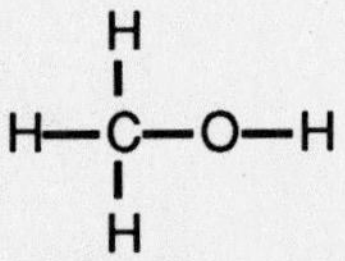

1.17   No, it does not. We cannot draw any structure for C_2H_5 that has four bonds to each carbon and one bond to each hydrogen.

1.18   First write the alcohols (compounds with an O–H group).

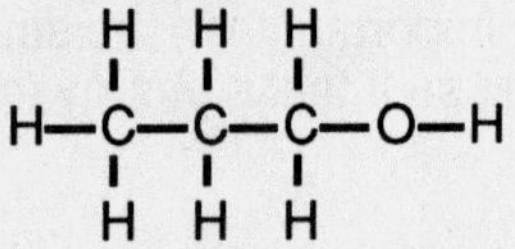

H H H
| | |
H—C—C—C—O—H   and   H—C—C—C—H
| | |
H H H

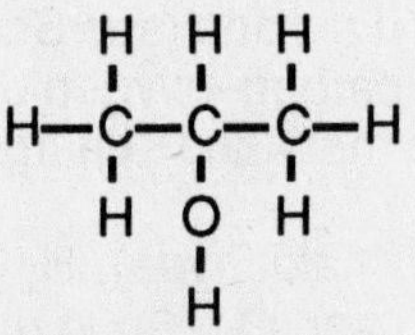

Then write the structures with a C–O–C bond (ethers).

6

```
    H       H  H
    |       |  |
H—C—O—C—C—H
    |       |  |
    H       H  H
```

There are no other possibilities. For example:

```
    H  H  H
    |  |  |
H—O—C—C—C—H          and
    |  |  |
    H  H  H
```

```
    H  H  H
    |  |  |
H—C—C—C—H
    |  |  |
    H  H  O
              |
              H
```

are the same as

```
        H  H  H
        |  |  |
   H—C—C—C—O—H
        |  |  |
        H  H  H
```

They all have the same bond connectivities and represent a single structure.
Similarly,

```
    H       H  H
    |       |  |
H—C—O—C—C—H          is the same as
    |       |  |
    H       H  H
```

```
    H  H       H
    |  |       |
H—C—C—O—C—H
    |  |       |
    H  H       H
```

1.19   From left to right: *n*-pentane, isopentane, and isopentane.

1.20

a.
```
     H   H
      \ /
  H—C
      \
  H—C—C—O—H
      /
  H—C      H
     / \
    H   H
```

b.
```
Cl          Cl
   \       /
    C = C
   /       \
Cl          Cl
```

1.21

```
  \       /
   \     /
    >—<
   /     \
  /       \
```

stands for the carbon skeleton

```
C          C
 \        /
  C — C
 /        \
C          C
```

The addition of the appropriate number of hydrogens on each carbon
completes the valence of 4.

1.22   ammonia

$$H—\overset{\overset{\displaystyle H}{|}}{\underset{\underset{\displaystyle H}{|}}{N}}\colon$$

formal charge on nitrogen = 5 - (2 + 3) = 0

ammonium ion

$$\left[H—\overset{\overset{\displaystyle H}{|}}{\underset{\underset{\displaystyle H}{|}}{N}}—H\right]^{+}$$

formal charge on nitrogen = 5 - (0 + 4) = +1

amide ion

$$\left[\colon\overset{\overset{\displaystyle H}{|}}{\underset{\underset{\displaystyle H}{|}}{N}}\colon\right]^{-}$$

formal charge on nitrogen = 5 - (4 + 2) = -1

The formal charge on hydrogen in all three cases is zero [1 - (0 + 1) = 0].

1.23   For the singly bonded oxygens, formal charge =
6 - (6 + 1) = -1.

For the doubly bonded oxygen, formal charge =
6 - (4 + 2) = 0.

$$\left[\colon\!\overset{\overset{\displaystyle \colon O\colon}{\|}}{\underset{}{O}}—C—\overset{}{O}\!\colon\right]^{2-}$$

For the carbon, formal charge = 4 - (0 + 4) = 0.

1.24   There are 24 valence electrons to use in bonding (6 from each oxygen, 5 from the nitrogen, and one more because of the negative charge). To arrange the atoms with eight valence electrons around each atom, we must have one nitrogen–oxygen double bond:

$$\left[\colon\!O\overset{\|}{—}N—O\colon \longleftrightarrow \colon O{=}N—O\colon \longleftrightarrow \colon O—N{=}O\colon\right]^{-}$$

The formal charge on nitrogen is 5 - (0 + 4) = +1.
The formal charge on singly bonded oxygens is 6 - (6 + 1) = −1.
The formal charge on doubly bonded oxygen is 6 - (4 + 2) = 0.
The net charge of the ion is −1 because each resonance structure has one positively charged nitrogen atom and two negatively charged oxygen atoms. In the resonance hybrid, the formal charge on the nitrogen is +1; on the

8

ANSWERS TO PROBLEMS

oxygens, the charge is –2/3 at each oxygen because each oxygen has a –1 charge in two of the three structures and zero charge in the third structure.

1.25   In tetrahedral methane, the H–C–H bond angle is 109.5°. In "planar" methane, this angle would be 90° and the bonding electrons would be closer together. Thus repulsion between electrons in different bonds would be greater in "planar" methane than in tetrahedral methane, and it would be less stable.

1.26   a. $CH_3CH_2CH_2O^- \ Na^+ \ + \ H_2$

b. $CH_3CHCH_2CH_3 \ + \ H_2$
$\quad\quad |$
$\quad\quad O^- \ Na^+$

1.27   a. C=C, alkene; O–H, hydroxyl group, alcohol
b. C=O, carbonyl group, ketone
c. C=C, alkene
d. C=C, alkene; C=O, ketone; O–H, alcohol

Additional Problems

1.28   The number of valence electrons is the same as the number of the group in the periodic table (Table 1.3) to which the element belongs.

a. $\cdot \overset{\cdot}{\underset{\cdot}{C}} \cdot$ b. $\overset{\cdot\cdot}{\underset{\cdot\cdot}{:F:}}$ c. $\cdot \overset{\cdot}{Si} \cdot$ d. $\cdot \overset{\cdot}{B}$ e. $\overset{\cdot\cdot}{\underset{\cdot\cdot}{:S:}}$ f. $\overset{\cdot}{:P} \cdot$

1.29   The bonds in sodium chloride are ionic; Cl is present as chloride ion (Cl^-); Cl^- reacts with Ag^+ to give AgCl, a white precipitate. The C–Cl bonds in CCl_4 are covalent; no Cl^- is present to react with Ag^+.

1.30   a. ionic   b. covalent   c. ionic   d. covalent
e. covalent   f. ionic   g. covalent   h. covalent

1.31

	Valence Electrons	Common Valence
a. O	6	2
b. H	1	1
c. Cl	7	1
d. N	5	3
e. S	6	2
f. C	4	4

Note that the *sum* of the number of valence electrons and the common valence is 8 in each case (except for H, where it is 2, the number of electrons in the completed first shell).

1.32 a.

```
      H
      |        ..
  H—C—Cl:
      |        ..
      H
```

b.

```
      H   H   H
      |   |   |
  H—C—C—C—H
      |   |   |
      H   H   H
```

c.

```
      H   H
      |   |        ..
  H—C—C—F:
      |   |        ..
      H   H
```

d.

```
      H      H
      |     /
  H—C—N:
      |     \
      H      H
```

e.

```
      H   H
      |   |        ..
  H—C—C—O—H
      |   |        ..
      H   H
```

f.

```
  H
   \        ..
    C=O
   /
  H
```

1.33 a.

```
  ..        ..
: Cl—Cl :
  ..        ..
```

Because the bond is between identical atoms, it is pure covalent (nonpolar).

b.

```
      H
      |  δ+    δ−
  H—C—F
      |
      H
```

Fluorine is more electronegative than carbon.

c.

```
  δ−   δ+   δ−
:O═══C═══O:
  ..            ..
```

The C=O bond is polar, and the oxygen is more electronegative than carbon.

d.

```
          ..
δ+ H—Br : δ−
          ..
```

The halogens are more electronegative than hydrogen.

e.

```
           δ−
            F
            |  F δ−
            | ⁄
δ− F—S—F δ−
         ⁄ |
        F  |
    δ−     F
           δ−
```

Fluorine is more electronegative than sulfur. Indeed, it is the most electronegative element. Note that the S has 12 electrons around it. Elements below the first full row sometimes have more than 8 valence electrons around them.

f.

$$H-\overset{\overset{\displaystyle H}{|}}{\underset{\underset{\displaystyle H}{|}}{C}}-H$$

Carbon and hydrogen have nearly identical electronegativities, and the bonds are essentially nonpolar.

g.

$$\overset{\delta-\quad\overset{\displaystyle\delta+}{}\quad\delta-}{:\overset{..}{O}=S=\overset{..}{O}:}$$

Oxygen is more electronegative than sulfur. Note that the S again has more than 8 valence electrons (see part e).

h.

$$H-\overset{\overset{\displaystyle H}{|}}{\underset{\underset{\displaystyle H}{|}}{\underset{\delta+}{C}}}-\overset{\delta-}{O}-\overset{\overset{\displaystyle H}{|}}{\underset{\underset{\displaystyle H}{|}}{\underset{\delta+}{C}}}-H$$

Oxygen is more electronegative than carbon, so the C–O bonds are polar covalent.

1.34 The O–H bond is polar, with the hydrogen $\delta+$. The reaction is analogous to the reaction of sodium with alcohols.

$$2 \quad \underset{H_3C}{\overset{\overset{\displaystyle O}{\parallel}}{C}}{\diagdown}_{OH} \quad + \quad 2\ Na \quad \longrightarrow \quad 2 \quad \underset{H_3C}{\overset{\overset{\displaystyle O}{\parallel}}{C}}{\diagdown}_{O^-\ Na^+} \quad + \quad H_2$$

1.35 a. C_3H_8 The only possible structure is $CH_3CH_2CH_3$.

b. C_3H_7Cl The chlorine can replace a hydrogen on an end carbon or on the middle carbon in the answer to a:

$CH_3CH_2CH_2Cl$ or $CH_3CH(Cl)CH_3$

c. $C_2H_4Cl_2$ The chlorines can either be attached to the same carbon or to different carbons:

CH_3CHCl_2 or $CH_2Cl–CH_2Cl$

d. $C_3H_6Br_2$ Be systematic. With one bromine on an end carbon there are three possibilities for the second bromine:

$$H-\overset{\overset{\displaystyle Br}{|}}{\underset{\underset{\displaystyle Br}{|}}{C}}-\overset{\overset{\displaystyle H}{|}}{\underset{\underset{\displaystyle H}{|}}{C}}-\overset{\overset{\displaystyle H}{|}}{\underset{\underset{\displaystyle H}{|}}{C}}-H \qquad\qquad H-\overset{\overset{\displaystyle H}{|}}{\underset{\underset{\displaystyle Br}{|}}{C}}-\overset{\overset{\displaystyle Br}{|}}{\underset{\underset{\displaystyle H}{|}}{C}}-\overset{\overset{\displaystyle H}{|}}{\underset{\underset{\displaystyle H}{|}}{C}}-H$$

11

$$\begin{array}{c} \text{H} \quad \text{H} \quad \text{Br} \\ | \quad | \quad | \\ \text{H}-\text{C}-\text{C}-\text{C}-\text{H} \\ | \quad | \quad | \\ \text{Br} \quad \text{H} \quad \text{H} \end{array}$$

If one bromine is on the middle carbon, the only new structure arises with the second bromine also on the middle carbon:

$$\begin{array}{c} \text{H} \quad \text{Br} \quad \text{H} \\ | \quad | \quad | \\ \text{H}-\text{C}-\text{C}-\text{C}-\text{H} \\ | \quad | \quad | \\ \text{H} \quad \text{Br} \quad \text{H} \end{array}$$

e. C_4H_9F The carbon chain may either be linear or branched. In each case there are two possible positions for the fluorine:

$CH_3CH_2CH_2CH_2F$ $CH_3CH(F)CH_2CH_3$
$(CH_3)_2CHCH_2F$ $(CH_3)_3CF$

f. $C_2H_2Cl_2$ The sum of the hydrogens and chlorines is 4, not 6. Therefore there must be a carbon–carbon double bond:

$CCl_2{=}CH_2$ or $CHCl{=}CHCl$
No carbocyclic structure is possible because there are only two carbon atoms.

g. C_3H_6 There must be a double bond, or with three carbons, a ring:

$CH_3CH{=}CH_2$ or

$$\begin{array}{c} \text{CH}_2 \\ \diagup \quad \diagdown \\ \text{H}_2\text{C}-\text{CH}_2 \end{array}$$

h. $C_4H_{10}O$ With an O–H bond, there are four possibilities:

$CH_3CH_2CH_2CH_2{-}OH$ $CH_3CH(OH)CH_2CH_3$
$(CH_3)_2CHCH_2{-}OH$ $(CH_3)_3C{-}OH$

There are also three possibilities in a C–O–C arrangement:

$CH_3{-}O{-}CH_2CH_2CH_3$ $CH_3CH_2{-}O{-}CH_2CH_3$
$(CH_3)_2CH{-}O{-}CH_3$

1.36 The problem can be approached systematically. Consider first a chain of six carbons, then a chain of five carbons with a one-carbon branch, and so on.

$$CH_3-CH_2-CH_2-CH_2-CH_2-CH_3 \qquad CH_3-CH_2-CH_2-\underset{\underset{\displaystyle CH_3}{|}}{CH}-CH_3$$

$$CH_3-CH_2-\underset{\underset{\displaystyle CH_3}{|}}{CH}-CH_2-CH_3 \qquad CH_3-CH_2-\underset{\overset{\displaystyle CH_3}{|}}{\underset{\underset{\displaystyle CH_3}{|}}{C}}-CH_3$$

$$CH_3-\underset{\underset{\displaystyle CH_3}{|}}{CH}-\underset{\underset{\displaystyle CH_3}{|}}{CH}-CH_3$$

1.37

a.
$$H-\overset{\overset{\displaystyle H}{|}}{\underset{\underset{\displaystyle H}{|}}{C}}-\overset{\overset{\displaystyle H}{|}}{\underset{\underset{\displaystyle H}{|}}{C}}-\overset{\overset{\displaystyle H}{|}}{\underset{\underset{\displaystyle H}{|}}{C}}-\overset{\overset{\displaystyle H}{|}}{\underset{\underset{\displaystyle H}{|}}{C}}-\overset{\overset{\displaystyle H}{|}}{\underset{\underset{\displaystyle H}{|}}{C}}-\overset{\overset{\displaystyle H}{|}}{\underset{\underset{\displaystyle H}{|}}{C}}-H$$

b.
$$H-\overset{\overset{\displaystyle H}{|}}{\underset{\underset{\displaystyle H}{|}}{C}}-\overset{\overset{\displaystyle H-C-H}{|}}{\underset{\underset{\displaystyle H-C-H}{|}}{C}}-\overset{\overset{\displaystyle H}{|}}{\underset{\underset{\displaystyle H}{|}}{C}}-\overset{\overset{\displaystyle H}{|}}{\underset{\underset{\displaystyle H}{|}}{C}}-H$$

c.
$$H-\overset{\overset{\displaystyle H}{|}}{\underset{\underset{\displaystyle H}{|}}{C}}-\overset{\overset{\displaystyle H}{|}}{\underset{\underset{\displaystyle O-H}{|}}{C}}-\overset{\overset{\displaystyle H}{|}}{\underset{\underset{\displaystyle H}{|}}{C}}-H$$

d.
$$H-\overset{\overset{\displaystyle H}{|}}{\underset{\underset{\displaystyle H}{|}}{C}}-\overset{\overset{\displaystyle H}{|}}{\underset{\underset{\displaystyle H}{|}}{C}}-S-\overset{\overset{\displaystyle H}{|}}{\underset{\underset{\displaystyle H}{|}}{C}}-\overset{\overset{\displaystyle H}{|}}{\underset{\underset{\displaystyle H}{|}}{C}}-H$$

e.
$$H-\overset{\overset{\displaystyle H}{|}}{\underset{\underset{\displaystyle Cl}{|}}{C}}-\overset{\overset{\displaystyle H}{|}}{\underset{\underset{\displaystyle O-H}{|}}{C}}-H$$

f.
$$\underset{\displaystyle H-C(H)(H)}{\overset{\displaystyle H-C(H)(H)}{N}}-\overset{\overset{\displaystyle H}{|}}{\underset{\underset{\displaystyle H}{|}}{C}}-\overset{\overset{\displaystyle H}{|}}{\underset{\underset{\displaystyle H}{|}}{C}}-H$$

1.38 a. CH_2=$CHCH_2CH_3$ b. 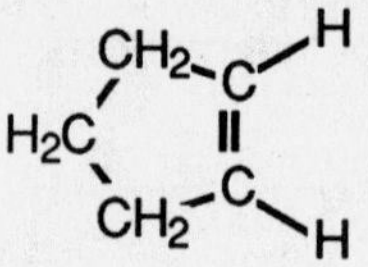 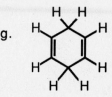 c. $(CH_3)_2CH-CH(CH_3)_2$

  d. 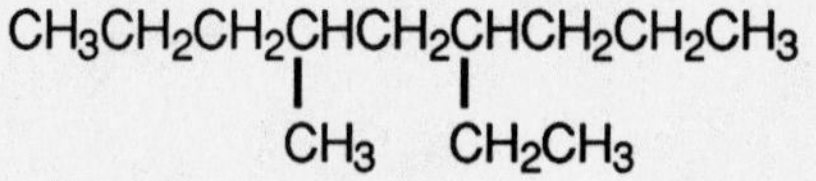 $CH_3CH_2CH_2CHCH_2CHCH_2CH_2CH_3$
       | |
       CH_3 CH_2CH_3

  e. $CH_3CH_2OCH_2CH_3$

  f. $(CH_3)_3C-C(CH_3)_3$ g.

  h. 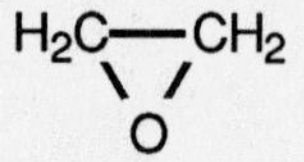 $H_2C\!-\!\!-\!\!CH_2$
         \   /
          O

1.39 a. 10 b. $C_{10}H_{18}O$ c. $(CH_3)_2C$=$CHCH_2CH_2C(CH_3)$=$CHCH_2OH$

1.40 First count the carbons, then the hydrogens, and finally the remaining atoms.

  a. $C_{16}H_{30}O$ b. C_6H_6 c. $C_{19}H_{28}O_2$
  d. $C_{10}H_{14}N_2$ e. $C_{10}H_{16}$ f. $C_5H_5N_5$

1.41 a. HONO   First determine the total number of valence electrons; H = 1,
                  O = 2 x 6 = 12, N = 5, for a total of 18. These must be
                  arranged in pairs so that the hydrogen has 2 and the other
                  atoms have 8 electrons around them.

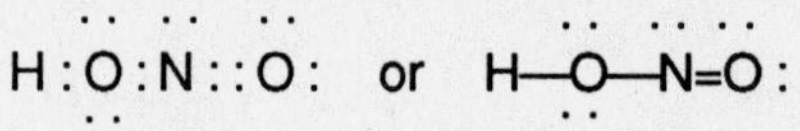

          H:O:N::O: or   H—O—N=O:

                  Using the formula for formal charge given in Sec. 1.11, it can
                  be determined that none of the atoms has a formal charge.

      b. $HONO_2$ There are 24 valence electrons. The nitrogen has a +1 formal
                  charge [5 – (0 + 4) = +1], and the singly bonded oxygen has a
                  –1 formal charge [6 – (6 + 1) = –1]. The whole molecule is
                  neutral.

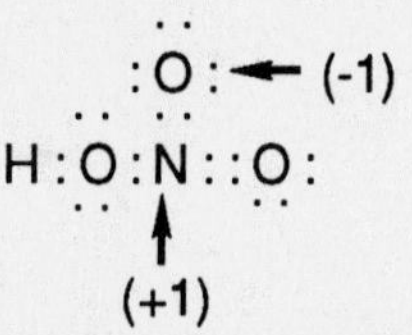

                  :O: ← (-1)

          H:O:N::O:
                  ↑
                (+1)

      c. H_2CO There are no formal charges.

$$\begin{array}{c} \text{H} \\ \cdot\cdot \quad \cdot\cdot \\ \text{H}:\text{C}::\text{O}: \end{array}$$

d. NH_4^+ The nitrogen has a +1 formal charge; see the answer to Prob. 1.22.

$$\begin{array}{c} \text{H} \quad \nearrow \; (+1)\\ \cdot\cdot \\ \text{H}:\text{N}:\text{H} \\ \cdot\cdot \\ \text{H} \end{array}$$

e. CN^- There are 10 valence electrons (C = 4, N = 5, plus 1 more because of the negative charge).
$^-:\text{C}:::\text{N}:$
The carbon has a –1 formal charge [4 – (2 + 3) = –1].

f. CO   There are 10 valence electrons.

$(-1)\longrightarrow:\text{C}:::\text{O}:\longleftarrow(+1)$

The carbon has a –1 formal charge, and the oxygen has a +1 formal charge. Carbon monoxide is isoelectronic with cyanide ion but has no net charge (–1 + 1 = 0).

g. SO_4^{2-} There are 32 valence electrons (6 each from the sulfur and oxygens, plus 2 more because of the double negative charge).

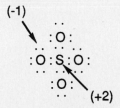

(-1)

(+2)

Each oxygen has a formal charge of –1 [6 – (6 + 1) = –1].
The sulfur has a formal charge of +2 [6 – (0 + 4) = +2].

h. BF_3 There 24 valence electrons (B = 3, F = 7). The structure is usually written with only 6 electrons around the boron. (See the top of page 16.)

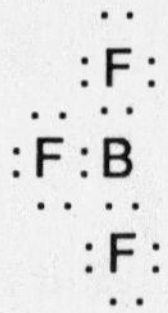

$$
\begin{array}{c}
\overset{\cdot\cdot}{:}F: \\
:F:B \\
\overset{\cdot\cdot}{:}F:
\end{array}
$$

In this case, there are no formal charges. This structure shows that BF_3 is a Lewis acid, which readily accepts an electron pair to complete the octet around the boron.

i. H_2O_2 There are 14 valence electrons and an O–O bond. There are no formal charges.

$$H : \overset{\cdot\cdot}{\underset{\cdot\cdot}{O}} : \overset{\cdot\cdot}{\underset{\cdot\cdot}{O}} : H$$

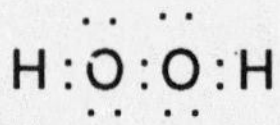

j. HCO_3^- There are 24 valence electrons involved. The hydrogen is attached to an oxygen, not to carbon.

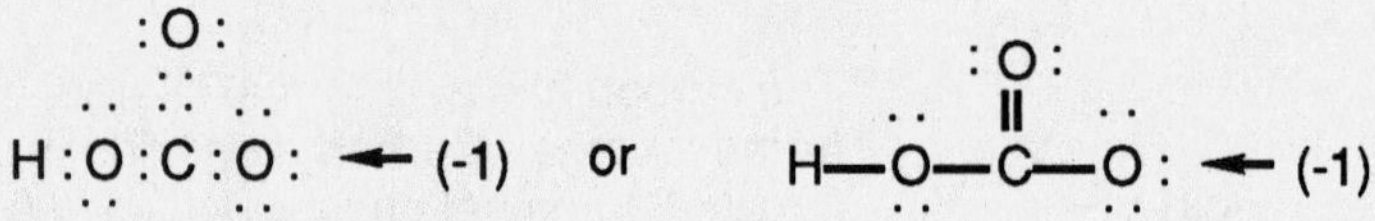

$$H : \overset{\cdot\cdot}{\underset{\cdot\cdot}{O}} : \overset{\cdot\cdot}{C} : \overset{:O:}{\underset{\cdot\cdot}{O}} : \leftarrow (-1) \quad \text{or} \quad H-\overset{\cdot\cdot}{\underset{\cdot\cdot}{O}}-\overset{:O:}{\overset{\|}{C}}-\overset{\cdot\cdot}{\underset{\cdot\cdot}{O}}: \leftarrow (-1)$$

The indicated oxygen has a formal charge: $6 - (6 + 1) = -1$. All other atoms are formally neutral.

1.42   This is a methyl carbocation, CH_3^+

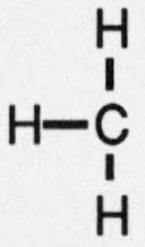

$$
\begin{array}{c}
H \\
| \\
H-C \\
| \\
H
\end{array}
\quad \text{formal charge} = 4 - (0 + 3) = +1
$$

This is a methyl free radical, $CH_3 \cdot$

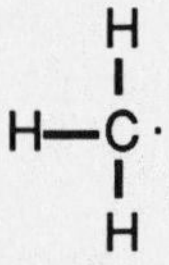

$$
\begin{array}{c}
H \\
| \\
H-C\cdot \\
| \\
H
\end{array}
\quad \text{formal charge} = 4 - (1 + 3) = 0
$$

This is a methyl carbanion, $CH_3 :^-$

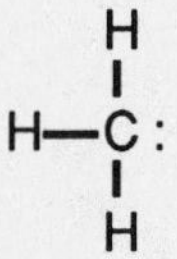

$$
\begin{array}{c}
H \\
| \\
H-C: \\
| \\
H
\end{array}
\quad \text{formal charge} = 4 - (2 + 3) = -1
$$

This is a methylene or carbene, CH_2 :

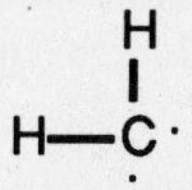

H
|
H—C ·   formal charge = $4 - (2 + 2) = 0$

All these fragments are extremely reactive. They may act as intermediates in organic reactions.

1.43   There are 18 valence electrons (6 from each of the oxygens, 5 from the nitrogen, and 1 from the negative charge).

$$\left[\ddot{:}\overset{..}{O}:\overset{..}{N}::\overset{..}{O}: \longleftrightarrow :\overset{..}{O}::\overset{..}{N}:\overset{..}{O}: \right]$$

or

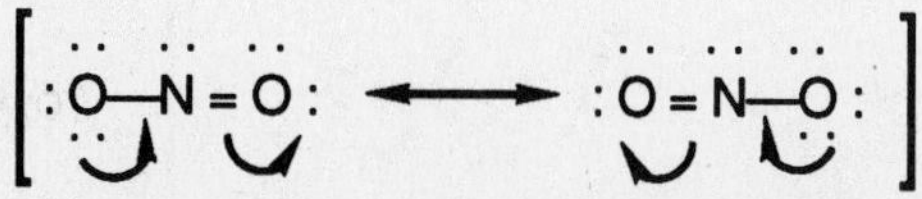

$$\left[:\overset{..}{O}-\overset{..}{N}=\overset{..}{O}: \longleftrightarrow :\overset{..}{O}=\overset{..}{N}-\overset{..}{O}: \right]$$

The negative charge in each contributor is on the singly bonded oxygen [6 − (6 + 1) = −1]. The other oxygen and the nitrogen have no formal charge. In the resonance hybrid, the negative charge is spread equally over the two oxygens; each is −1/2.

1.44   a.   There are 16 valence electrons $(3 \times 5 + 1)$. The formal charges on each nitrogen are shown below the structures.

$$\left[:\overset{..}{N}=\overset{..}{N}=\overset{..}{N}: \longleftrightarrow :N\equiv N-\overset{..}{N}: \right]$$

          −1 +1 −1   0 +1 −2

   b.   The singly bonded oxygen carries the negative charge in each contributor. In the resonance hybrid, this charge is spread equally over both oxygens.

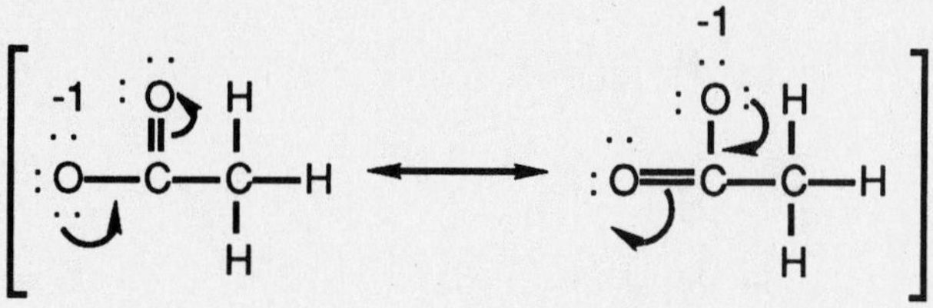

$$\left[\overset{-1\ :\overset{..}{O}}{:\overset{..}{O}-\overset{||}{C}-\overset{H}{\underset{H}{C}}-H} \longleftrightarrow \overset{-1\ :\overset{..}{O}}{:\overset{..}{O}=C-\overset{H}{\underset{H}{C}}-H} \right]$$

1.45 Each atom in both structures has a complete valence shell of electrons. There are no formal charges in the initial structure, but in the second structure the oxygen is formally negative and the nitrogen formally positive.

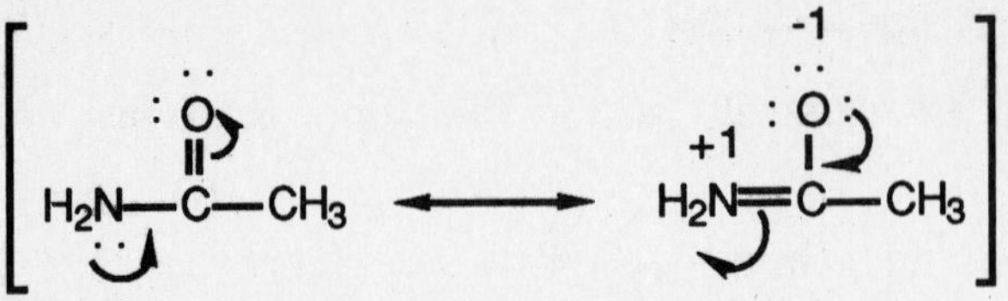

1.46 In the first structure, there are no formal charges. In the second structure, the oxygen is formally +1, and the ring carbon bearing the unshared electron pair is formally −1. (Do not forget to count the hydrogen that is attached to each ring carbon, except the one that is doubly bonded to oxygen.)

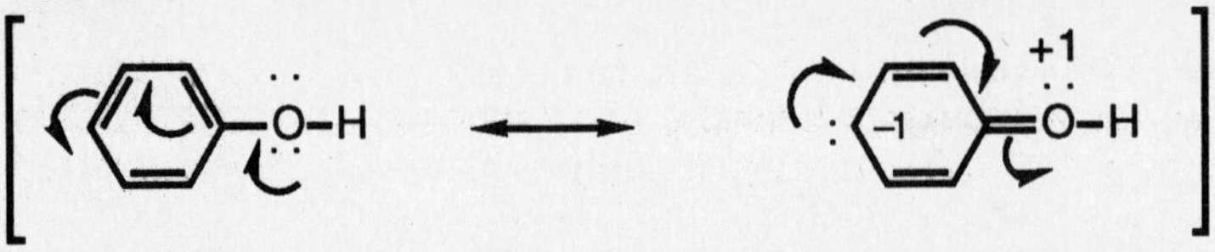

1.47

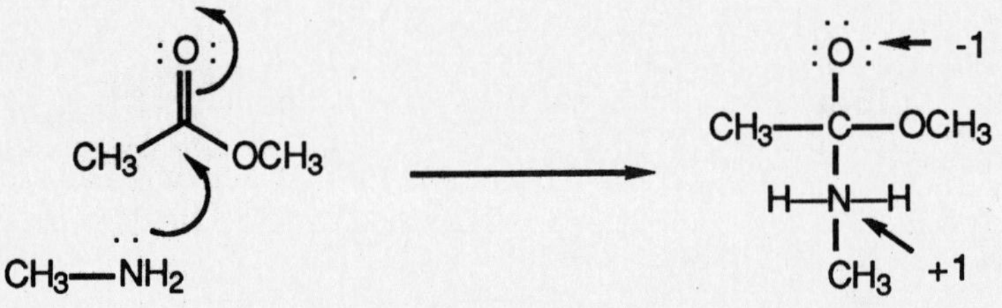

1.48 a. $CH_3O^- Na^+$

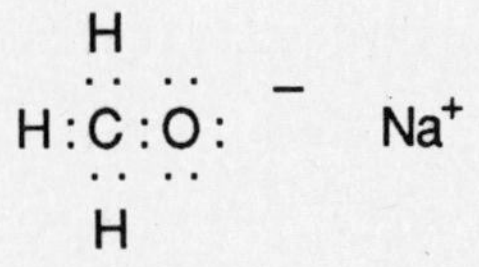

b. $NH_4^+ Cl^-$

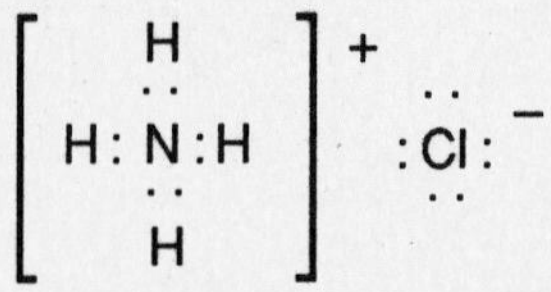

18

1.49   a. $CH_3CH_2\!-\!\overset{..}{\underset{..}{O}}\!-\!H$ b. $CH_3\!-\!\overset{\displaystyle :\overset{..}{O}:}{\overset{\|}{C}}\!-\!\overset{..}{\underset{..}{O}}\!-\!H$ c. $CH_3\!-\!\overset{\displaystyle H}{\underset{..}{N}}\!-\!CH_3$

   d. $CH_3\!-\!\overset{..}{\underset{..}{O}}\!-\!CH_2CH_2\!-\!\overset{..}{\underset{..}{O}}\!-\!H$

1.50   If the *s* and *p* orbitals were hybridized to *sp*³, two electrons would go into one of these orbitals and one electron would go into each of the remaining three orbitals.

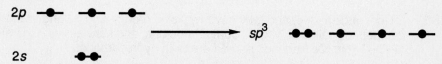

   The predicted geometry of ammonia would then be tetrahedral, with one hydrogen at each of three corners and the unshared pair at the fourth corner. In fact, ammonia has a pyramidal shape; a somewhat flattened tetrahedron. The H—N—H bond angle is 107° rather than 109°28'.

1.51   The ammonium ion is in fact isoelectronic with (has the same arrangement of electrons as) methane and consequently has the same geometry. Four *sp*³ orbitals of nitrogen each contain one electron. These orbitals then overlap with the 1*s* hydrogen orbitals, as in Figure 1.9.

1.52   The bonding is exactly as in methane. The geometry of silane is tetrahedral.

1.53

   The geometry is tetrahedral at carbon. It does not matter whether we draw the wedge bonds to the right or to the left of the carbon, or indeed "up" or "down."

1.54 Many correct answers are possible; a few are given here.

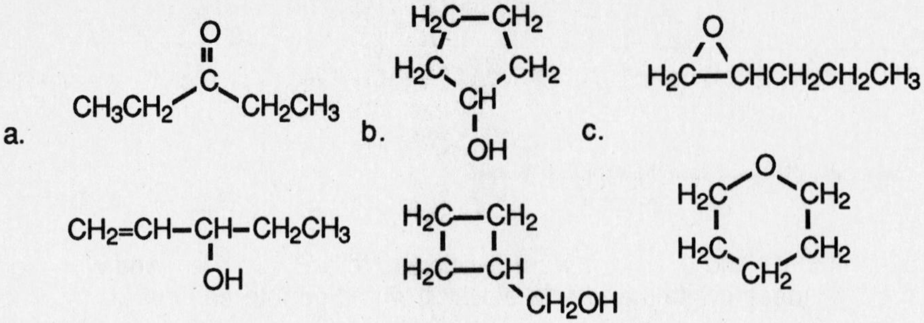

1.55 Compounds a, c, e, and g all have hydroxyl groups and belong to the class of compounds called alcohols. Compounds b, h, and i are ethers; they all have a C–O–C unit. Compounds d and f are both hydrocarbons.

1.56

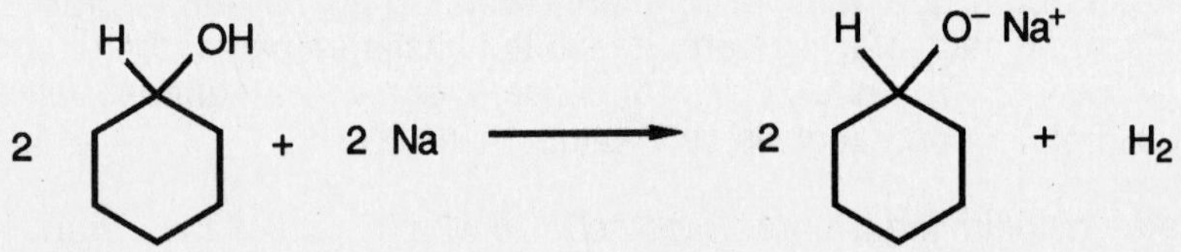

1.57 The more common functional groups are listed in Table 1.5. Often more than one answer is possible.

a. $CH_3CH_2CH_2CH_2OH$, $CH_3CH(OH)CH_2CH_3$, $(CH_3)_2CHCH_2OH$, and $(CH_3)_3COH$

b. $CH_3OCH_2CH_3$

c. $CH_3CH_2\underset{\underset{O}{\|}}{C}H$

d. $CH_3\overset{\overset{O}{\|}}{C}CH_2CH_3$

e. $CH_3CH_2CO_2H$

f. $H\overset{\overset{O}{\|}}{C}OCH_2CH_2CH_2CH_3$ $\quad H\overset{\overset{O}{\|}}{C}O\overset{\overset{CH_3}{|}}{C}HCH_2CH_3$ $\quad H\overset{\overset{O}{\|}}{C}OCH_2CH(CH_3)_2$

ANSWERS TO PROBLEMS

$$\underset{\substack{O \\ \| }}{} \text{HCOC(CH}_3)_3 \qquad \underset{\substack{O \\ \| }}{} \text{CH}_3\text{COCH}_2\text{CH}_2\text{CH}_3 \qquad \underset{\substack{O \\ \| }}{} \text{CH}_3\text{COCH(CH}_3)_2$$

$$\underset{\substack{O \\ \| }}{} \text{CH}_3\text{CH}_2\text{COCH}_2\text{CH}_3 \qquad \underset{\substack{O \\ \| }}{} \text{CH}_3\text{CH}_2\text{CH}_2\text{COCH}_3 \qquad \underset{\substack{O \\ \| }}{} \text{(CH}_3)_2\text{CHCOCH}_3$$

g. See part f.

h. $(CH_3)_3N$, $CH_3NHCH_2CH_3$, $CH_3CH_2CH_2NH_2$, and $(CH_3)_2CHNH_2$.

CHAPTER TWO: ALKANES AND CYCLOALKANES; CONFORMATIONAL AND GEOMETRIC ISOMERISM

CHAPTER SUMMARY

Hydrocarbons contain only carbon and hydrogen atoms. **Alkanes** are acyclic **saturated hydrocarbons** (contain only single bonds); **cycloalkanes** are similar but have carbon rings.

Alkanes have the general molecular formula C_nH_{2n+2}. The first four members of this **homologous series** are **methane, ethane, propane**, and **butane**; each member differs from the next by a $—CH_2—$, or **methylene group**. The **IUPAC** (International Union of Pure and Applied Chemistry) nomenclature system is used worldwide to name organic compounds. The IUPAC rules for naming alkanes are described in Secs. 2.4-2.6. **Alkyl groups**, alkanes minus one hydrogen atom, are named similarly except that the *-ane* ending is changed to *-yl*. The letter **R** stands for any alkyl group.

The two main natural sources of alkanes are **natural gas** and **petroleum**. Alkanes are insoluble in and less dense than water. Their boiling points increase with molecular weight and, for isomers, decrease with chain branching.

Conformations are different structures that are interconvertible by rotation about single bonds. For ethane (and in general), the **staggered** conformation is more stable than the eclipsed conformation (Figure 2.3).

The prefix *cyclo-* is used to name cycloalkanes. **Cyclopropane** is planar, but larger carbon rings are puckered. **Cyclohexane** exists mainly in a **chair conformation** with all bonds on adjacent carbons staggered. One bond on each carbon is **axial** (perpendicular to the mean carbon plane), the other is **equatorial** (roughly in that plane). The conformations can be interconverted by "flipping" the ring, which only requires bond rotation and occurs rapidly at room temperature for cyclohexane. Ring substituents usually prefer the less crowded, equatorial position.

Stereoisomers have the same order of atom attachments but different arrangements of the atoms in space. *Cis–trans* isomerism is one kind of stereoisomerism. For example, two substituents on a cycloalkane can be on either the same (*cis*) or opposite (*trans*) sides of the mean ring plane. Stereoisomers can be divided into two groups, **conformational isomers** (interconvertible by bond rotation) and **configurational isomers** (not interconvertible by bond rotation). *Cis–trans* **isomers** belong to the latter class.

22

MECHANISM SUMMARY

Alkanes are fuels; they burn in air if ignited. Complete combustion gives carbon dioxide and water; less complete combustion gives carbon monoxide or other less oxidized forms of carbon.

Alkanes react with **halogens** (chlorine or bromine) in a reaction initiated by heat or light. One or more hydrogens can be replaced by halogens. This **substitution reaction** occurs by a **free-radical chain mechanism**.

REACTION SUMMARY

Combustion

$$C_nH_{2n+2} \quad + \quad \left(\frac{3n + 1}{2} \right) O_2 \quad \longrightarrow \quad n\ CO_2 \quad + \quad (n + 1)\ H_2O$$

Halogenation (substitution)

$$R\!-\!H \ + \ X_2 \quad \xrightarrow{\text{heat or light}} \quad R\!-\!X \ + \ H\!-\!X \quad (X = Cl, Br)$$

MECHANISM SUMMARY

Halogenation

initiation
$$:\!X\!-\!X\!: \quad \longrightarrow \quad 2\ :\!X\!\cdot$$

propagation
$$R\!-\!H \ + \ :\!X\!\cdot \quad \longrightarrow \quad R\!\cdot \ + \ H\!-\!X\!:$$

$$R\!\cdot \quad + \quad :\!X\!-\!X\!: \quad \longrightarrow \quad R\!-\!X\!: \ + \ :\!X\!\cdot$$

termination
$$2\ :\!X\!\cdot \quad \longrightarrow \quad :\!X\!-\!X\!:$$

$$2\ R\!\cdot \quad \longrightarrow \quad R\!-\!R$$

$$R\!\cdot \quad + \quad :\!X\!\cdot \quad \longrightarrow \quad R\!-\!X\!:$$

2. ALKANES AND CYCLOALKANES

LEARNING OBJECTIVES

1. Know the meaning of: saturated hydrocarbon, alkane, cycloalkane, homologous series, methylene group.

2. Know the meaning of: conformation, staggered, eclipsed, "dash-wedge" projection, Newman projection, "sawhorse" projection, rotational isomers, rotamers.

3. Know the meaning of: chair conformation of cyclohexane, equatorial, axial, geometric or *cis–trans* isomerism, conformational and configurational isomerism.

4. Know the meaning of: substitution reaction, halogenation, chlorination, bromination, free-radical chain reaction, chain initiation, propagation, termination, combustion.

5. Given the IUPAC name of an alkane or cycloalkane, or a halogen-substituted alkane or cycloalkane, draw its structural formula.

6. Given the structural formula of an alkane or cycloalkane or a halogenated derivative, write the correct IUPAC name.

7. Know the common names of the alkyl groups, cycloalkyl groups, methylene halides, and haloforms.

8. Tell whether two hydrogens in a particular structure are identical or different from one another by determining whether they give the same or different products by monosubstitution with some group X.

9. Know the relationship between boiling points of alkanes and (a) their molecular weights and (b) the extent of chain branching.

10. Write all steps in the free-radical chain reaction between a halogen and an alkane, and identify the initiation, propagation, and termination steps.

11. Write a balanced equation for the complete combustion of an alkane or cycloalkane.

12. Draw, using dash-wedge, sawhorse, or Newman projection formulas, the important conformations of ethane, propane, butane, and various halogenated derivatives of these alkanes.

ANSWERS TO PROBLEMS

13.   Recognize, draw, and name *cis–trans* isomers of substituted cycloalkanes.

14.   Draw the chair conformation of cyclohexane, and show clearly the distinction between axial and equatorial bonds.

15.   Identify the more stable conformation of a monosubstituted cyclohexane; also, identify substituents as axial or equatorial when the structure is "flipped" from one chair conformation to another.

16.   Classify a pair of isomers as structural (constitutional) isomers or stereo-isomers and if the latter, as conformational or configurational (see Figure 2.5).

ANSWERS TO PROBLEMS

Problems Within the Chapter

2.1 $C_{20}H_{42}$; use the formula C_nH_{2n+2}, where $n = 20$.

2.2   The formulas in parts b and d fit the general formula C_nH_{2n+2} and are alkanes. C_8H_{16} (part a) has two fewer hydrogens than called for by the alkane formula and must be either an alkene or a cycloalkane. C_7H_{18} (part c) is an impossible molecular formula; it has too many hydrogens for the number of carbons.

2.3   a.   2-methylbutane (number the longest chain from left to right)
       b.   2-methylbutane (number the longest chain from right to left)

       The structures in parts a and b are identical, as the names show.

       c.   2,2-dimethylpropane

2.4   Bromochloromethane; the substituents are alphabetized when more than one is present.

2.5   a. $CH_3CH_2CH_2I$ b. $(CH_3)_2CHCl$ c. $(CH_3)_2CHCl$

       d. $(CH_3)_3C\!-\!I$ e. $(CH_3)_2CHCH_2Br$ f.   RF; the letter R stands for any alkyl group

2. ALKANES AND CYCLOALKANES

2.6   a. 2-fluoropropane   b. 4-chloro-2,2-dimethylpentane (*not* 2-chloro-4,4-dimethylpentane, which has higher numbers for the substituents)

2.7

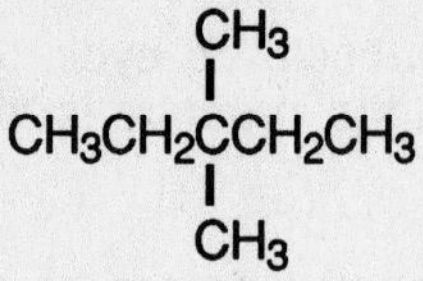

$$CH_3CH_2\overset{\displaystyle CH_3}{\underset{\displaystyle CH_3}{C}}CH_2CH_3$$

2.8

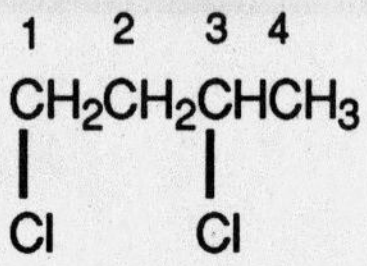

$$\overset{1}{C}H_2\overset{2}{C}H_2\overset{3}{C}H\overset{4}{C}H_3$$
$$\underset{Cl}{|}\qquad\underset{Cl}{|}$$

1,3-dichlorobutane

$$\overset{4}{C}H_2\overset{3}{C}H_2\overset{2}{C}H\overset{1}{C}H_3$$
$$\underset{^5CH_3}{|}\qquad\underset{CH_3}{|}$$

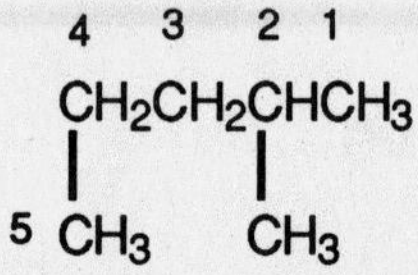

Correct name: 2-methylpentane; Carbon-5 is not a substituent. It is part of the longest chain.

2.9

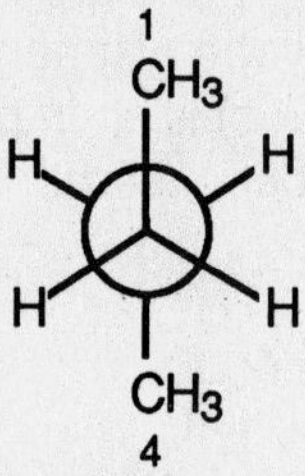

(anti)

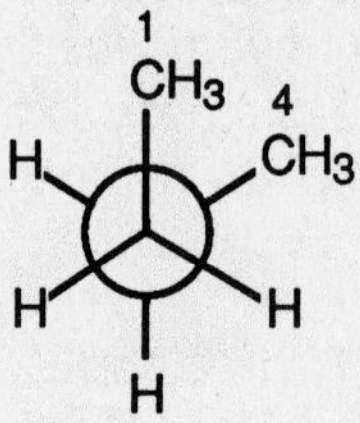

(gauche)

$$\overset{1}{C}H_3\overset{2}{C}H_2\overset{3}{C}H_2\overset{4}{C}H_3$$

Carbon-2 and carbon-3 in the chain each have two hydrogens and a methyl group attached. The staggered conformation at the left is the more stable. In the staggered conformation at the right, the two methyl substituents (large groups) come close to one another, a less stable arrangement than in the structure to the left. These conformations are referred to as *anti* and *gauche*, respectively.

2.10

a.

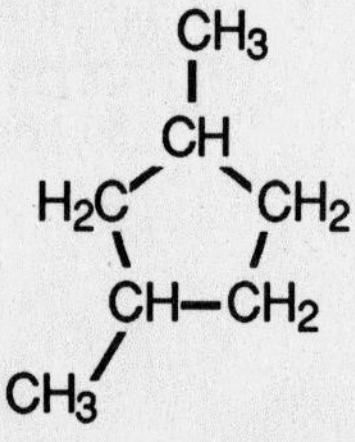

or

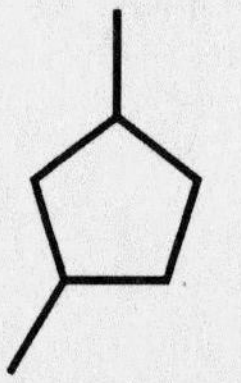

b.

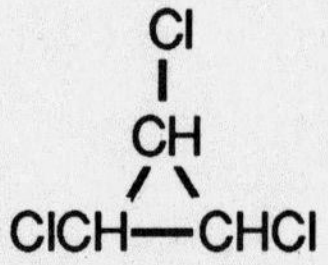

or

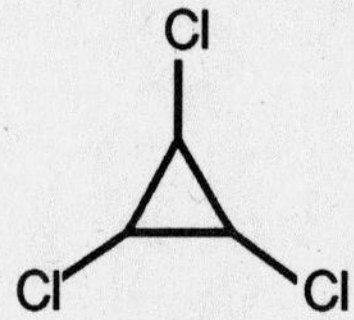

ANSWERS TO PROBLEMS

2.11 a. ethylcyclobutane b. 1,1-dichlorocyclopropane
c. 1-chloro-3-methylcyclobutane

2.12 Since each ring carbon is tetrahedral, the H—C—H plane and the C—C—C plane at any ring carbon are mutually perpendicular.

2.13 If you sight down the bond joining carbon-2 and carbon-3, you will see that the substituents on these carbons are eclipsed. The same is true for the bond between carbon-5 and carbon-6. Also, two of the hydrogens on carbon-1 and carbon-4, the "inside" hydrogens, come quite close to one another. All these factors destabilize boat cyclohexane compared to chair cyclohexane.

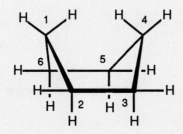

2.14 The *tert*-butyl group is much larger than a methyl group (eq. 2.3). Therefore the conformational preference is essentially 100% for an equatorial *tert*-butyl group:

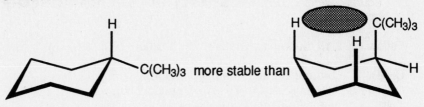

2.15

a.

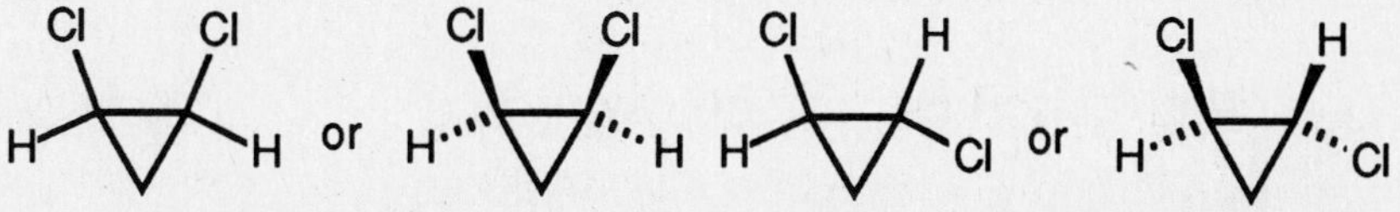

cis-1,2-dichlorocyclopropane   trans-1,2-dichlorocyclopropane

b.

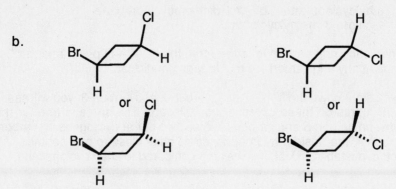

cis-1-bromo-3-chlorocyclobutane   trans-1-bromo-3-chlorocyclobutane

2.16   a. structural (constitutional) isomers
       b. configurational isomers (same bond pattern, but not interconvertible by bond rotations)
       c. conformational isomers

2.17   a. Formaldehyde (two C—O bonds) is more highly oxidized than methanol (one C—O bond).
       b. The carbons in methanol (CH_3OH) and dimethyl ether (CH_3OCH_3) are equally oxidized. Each carbon has one C—O bond and three C—H bonds.

2.18   Follow eq. 2.13, but replace chloro with bromo:

CH_3Br bromomethane (methyl bromide)
CH_2Br_2 dibromomethane (methylene bromide)
$CHBr_3$ tribromomethane (bromoform)
CBr_4 tetrabromomethane (carbon tetrabromide)

2.19 $CH_2CH_2CH_2CH_2CH_3$ $CH_3CHCH_2CH_2CH_3$ $CH_3CH_2CHCH_2CH_3$
         Br   Br   Br

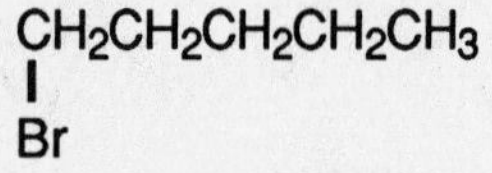

      1-bromopentane 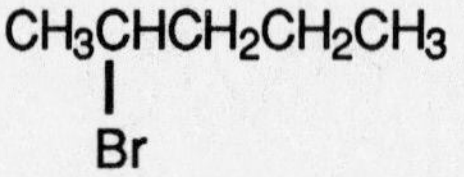   2-bromopentane 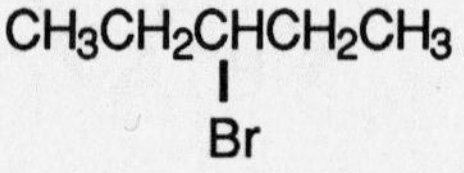   3-bromopentane

2.20   Four monochloro products can be obtained from octane, but only one monochloro product is obtained from cyclooctane.

CH₂CH₂CH₂CH₂CH₂CH₂CH₂CH₃   CH₃CHCH₂CH₂CH₂CH₂CH₂CH₃
| |
Cl   Cl

CH₃CH₂CHCH₂CH₂CH₂CH₂CH₃   CH₃CH₂CH₂CHCH₂CH₂CH₂CH₃
| |
Cl   Cl

2.21   Yes. All the hydrogens are equivalent, and monochlorination gives a single product.

$$
\begin{array}{c}
CH_3 \\
| \\
CH_3-C-CH_3 \\
| \\
CH_3
\end{array}
\quad
\xrightarrow[\text{heat or light}]{Cl_2}
\quad
\begin{array}{c}
CH_3 \\
| \\
CH_3-C-CH_2Cl \\
| \\
CH_3
\end{array}
$$

2.22   *initiation*   :Cl—Cl: $\xrightarrow{\text{heat or light}}$ 2 :Cl·

propagation   CH₄ + :Cl· $\longrightarrow$ CH₃· + H—Cl

CH₃· + :Cl—Cl: $\longrightarrow$ CH₃—Cl + :Cl·

termination   2 :Cl· $\longrightarrow$:Cl—Cl:

2 CH₃· $\longrightarrow$ CH₃—CH₃

CH₃· + :Cl· $\longrightarrow$ CH₃—Cl

2.23   The second termination step in the answer to Prob. 2.22 accounts for the formation of ethane in the chlorination of methane. Ethane can also be chlorinated, which explains the formation of small amounts of chloroethane in this reaction.

2. ALKANES AND CYCLOALKANES

Additional Problems

2.24   a. 3-methylpentane: First, note the root of the name (in this case, *pent*), and write down and number the carbon chain.

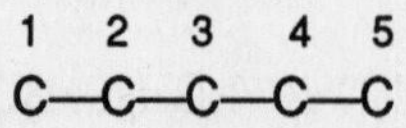

```
1   2   3   4   5
C—C—C—C—C
```

Next, locate the substituents (3-methyl).

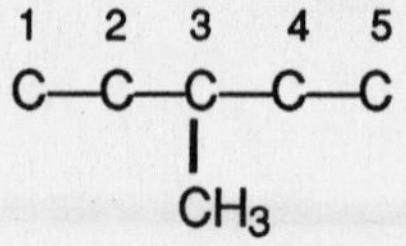

```
1   2   3   4   5
C—C—C—C—C
        |
       CH3
```

Finally, fill in the remaining hydrogens.

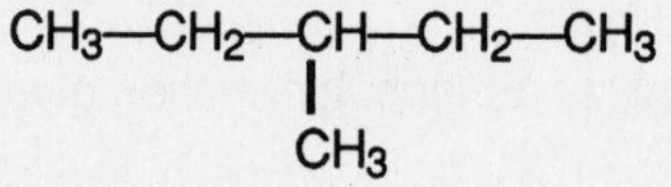

$$CH_3—CH_2—CH—CH_2—CH_3$$
$$|$$
$$CH_3$$

b. 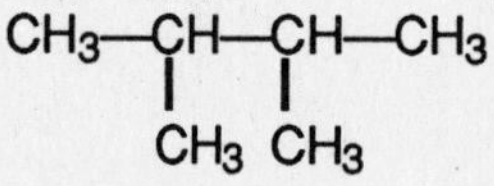 $CH_3—CH—CH—CH_3$
           $|$ $|$
          CH_3 CH_3

c. 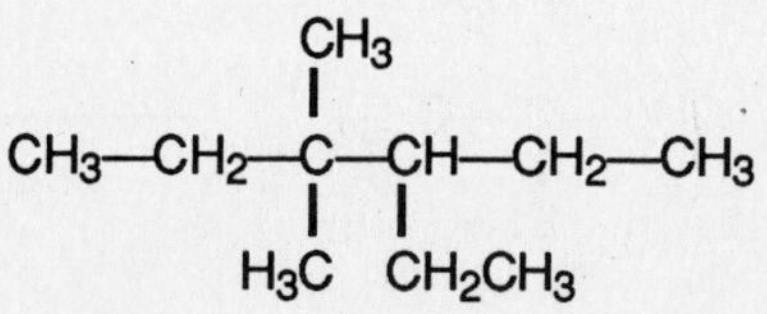 $CH_3—CH_2—C—CH—CH_2—CH_3$
                  CH_3 (above)
                H_3C CH_2CH_3

d. 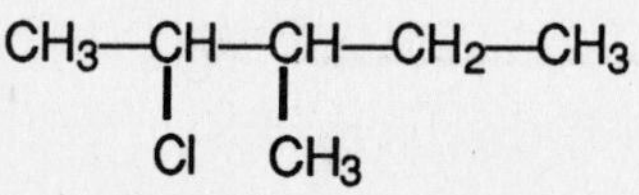 $CH_3—CH—CH—CH_2—CH_3$
            $|$ $|$
          Cl CH_3

e. 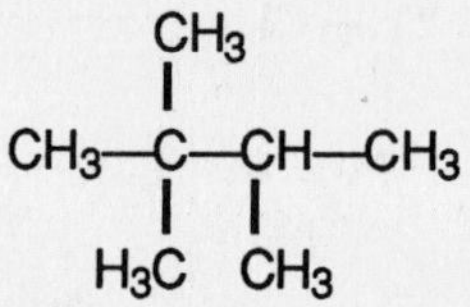 $CH_3—C—CH—CH_3$
              CH_3 (above)
         H_3C CH_3

f. 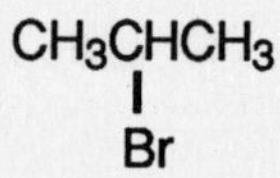 CH_3CHCH_3
              Br

g. 1,1-dichlorocyclopropane: The root *prop* indicates three carbons; the prefix *cyclo* designates that they form a ring.

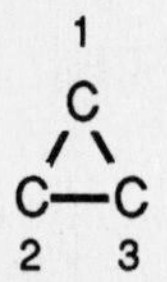

```
      1
      C
     / \
    C—C
    2   3
```

Next, the substituents are placed.

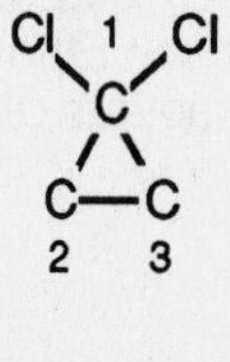

```
  Cl  1  Cl
    \  C  /
      /  \
     C——C
     2    3
```

30

Finally, the hydrogens are filled in.

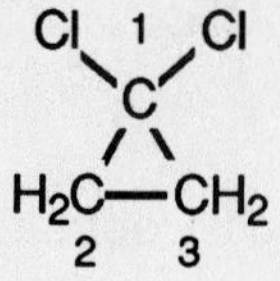

h. 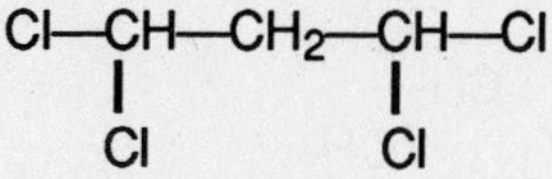 Cl—CH—CH₂—CH—Cl   or   CHCl₂CH₂CHCl₂   i.

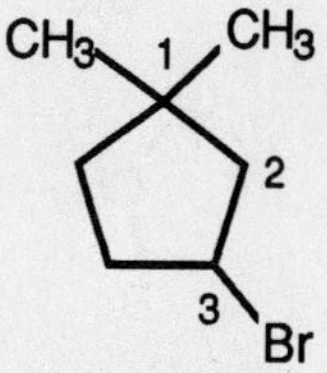

j. 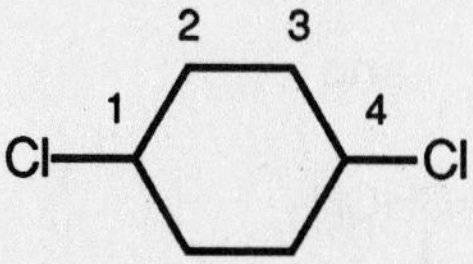 (Two isomers, *cis*- and *trans*-, are possible.)

2.25 a. CH₃CH₂CH₂CH₂CH₃   b.

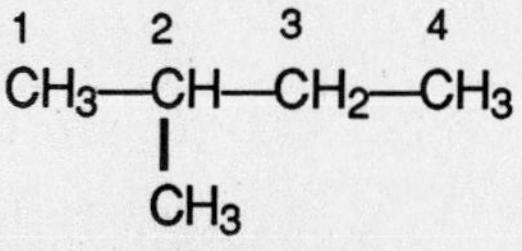

pentane

2-methylbutane

c.

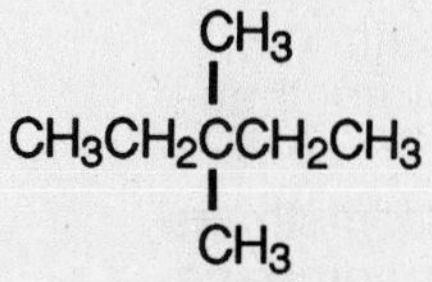

d.

3,3-dimethylpentane

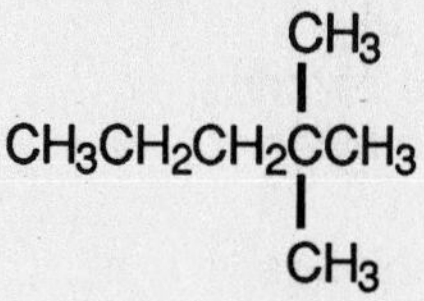

2,2-dimethylpentane

e.

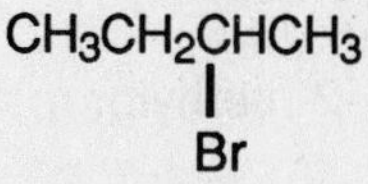

f.

2-bromobutane

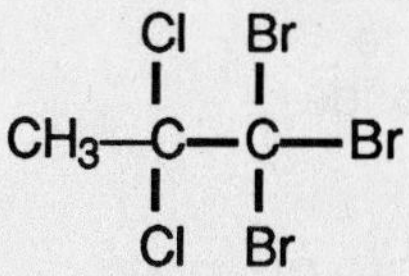

1,1,1-tribromo-2,2-dichloropropane

(The placement of commas and hyphens is important; this answer shows clearly how these punctuation marks are to be used.)

g.

$$CH_2CH_3$$
$$|$$
$$CH_3CH_2CCH_2CH_3$$
$$|$$
$$CH_2CH_3$$

3,3-diethylpentane

h.

$$\overset{2}{H_2C}\!\!-\!\!\overset{1}{CH_2}$$
$$|\quad\ |$$
$$Cl\quad Br$$

1-bromo-2-chloroethane

(The order of priority of the halo-gens follows alphabetical order.)

i.

$$CH_2\!\!-\!\!CH\!\!-\!\!CH\!\!-\!\!CH_3$$
$$|\quad\ \ |\quad\ \ |$$
$$Br\quad CH_3\ CH_3$$

1-bromo-2,3-dimethylbutane

j.

$$CH_2$$
$$H_2C\qquad CH_2$$
$$\backslash\qquad\ /$$
$$H_2C\!\!-\!\!CH_2$$

cyclopentane

k. CH_3I

iodomethane

l.

$$CH_3CHCH_3$$
$$|$$
$$Br$$

2-bromopropane

2.26

	Common	IUPAC
a.	methyl iodide	iodomethane
b.	ethyl chloride	chloroethane
c.	methylene chloride	dichloromethane
	(CH_2 = methylene)	
d.	bromoform	tribromomethane
e.	*n*-propyl chloride	1-chloropropane
f.	isopropyl bromide	2-bromopropane
g.	chloroform	trichloromethane
h.	cyclobutyl chloride	chlorocyclobutane
i.	*sec*-butyl iodide	2-iodobutane
j.	*tert*-butyl chloride	2-chloro-2-methylpropane

2.27   a. $CH_3CH_2CH_2CH_2CH_2CH_3$
The longest consecutive chain has six carbon atoms; the correct name is hexane.

b.

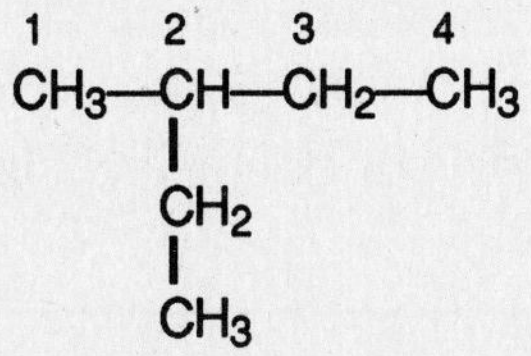

$$\overset{1}{CH_3}-\overset{2}{CH}-\overset{3}{CH_2}-\overset{4}{CH_3}$$
$$\underset{CH_3}{\overset{|}{CH_2}}$$

The longest chain was not selected when the compound was numbered. The correct numbering is

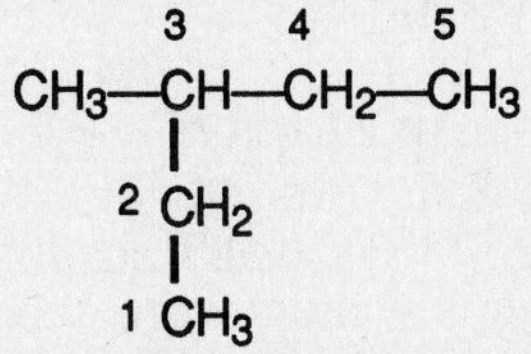

$$\overset{3}{CH_3}-\overset{4}{CH}-\overset{5}{CH_2}-CH_3$$
$$\overset{2}{CH_2}$$ 3-methylpentane
$$\overset{1}{CH_3}$$

c. The numbering started at the wrong end. The name should be 1,2-dichloropropane.

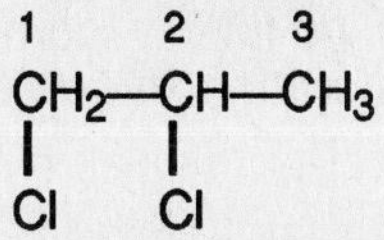

$$\overset{1}{CH_2}-\overset{2}{CH}-\overset{3}{CH_3}$$
$$\underset{Cl}{|}\quad\underset{Cl}{|}$$

d. The ring was numbered the wrong way to give the lowest substituent numbers. The correct name is 1,2-dimethylcyclobutane.

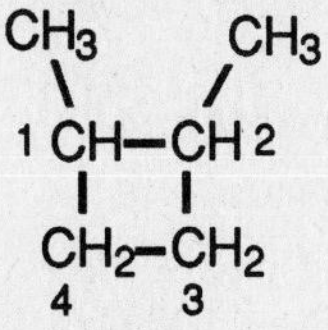

$$CH_3 \qquad CH_3$$
$$\overset{1}{CH}-\overset{2}{CH}$$
$$\overset{4}{CH_2}-\overset{3}{CH_2}$$

e. The longest chain was not selected. The correct name is 2-methyl-pentane.

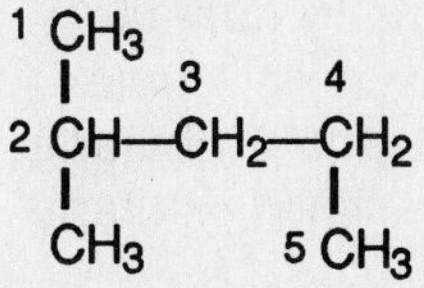

$$^1 CH_3$$
$$\overset{2}{CH}-\overset{3}{CH_2}-\overset{4}{CH_2}$$
$$CH_3 \qquad {}_5 CH_3$$

f. The correct name for the compound is 1-bromo-2-methylpropane (use the lower number).

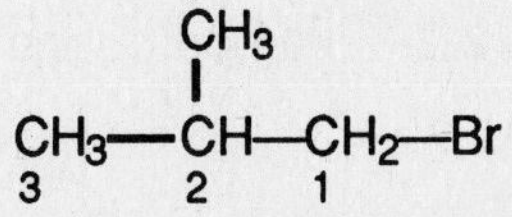

$$CH_3$$
$$\underset{3}{CH_3}-\underset{2}{CH}-\underset{1}{CH_2}-Br$$

2.28   The root of the name, heptadec, indicates a 17-carbon chain. The correct formula is

$$CH_3CHCH_2CH_2CH_2CH_2CH_2CH_2CH_2CH_2CH_2CH_2CH_2CH_2CH_2CH_3$$
$$|$$
$$CH_3$$

or $(CH_3)_2CH(CH_2)_{14}CH_3$

2.29   The molecular formulas are derived using C_nH_{2n+2}. They are $C_{102}H_{206}$, $C_{150}H_{302}$, $C_{198}H_{398}$, $C_{246}H_{494}$, and $C_{390}H_{782}$. Using approximate atomic weights of C = 12 and H = 1, the molecular weight of the C_{390} compound is 5462. However, using more precise atomic weights of C = 12.01 and H = 1.008, the molecular weight is 5472.2. (The small fractions in each atomic weight become significant when so many atoms are present in a single molecule.)

2.30   Approach each problem systematically. Start with the longest possible carbon chain and shorten it one carbon at a time until no further isomers are possible. To conserve space, the formulas below are written in condensed form, but you should write them out as expanded formulas.

a. $CH_3(CH_2)_2CH_3$ butane
      $(CH_3)_3CH$ 2-methylpropane

b. $CH_3CH_2CH_2CH_2Br$ 1-bromobutane
      $CH_3CHBrCH_2CH_3$ 2-bromobutane
      $(CH_3)_2CHCH_2Br$ 1-bromo-2-methylpropane
      $(CH_3)_3CBr$ 2-bromo-2-methylpropane

c. $CH_3(CH_2)_4CH_3$ hexane
      $CH_3CH(CH_3)CH_2CH_2CH_3$ 2-methylpentane
      $CH_3CH_2CH(CH_3)CH_2CH_3$ 3-methylpentane
      $CH_3CH(CH_3)CH(CH_3)CH_3$ 2,3-dimethylbutane
      $CH_3C(CH_3)_2CH_2CH_3$ 2,2-dimethylbutane

d. $CH_3CH_2CHBr_2$ 1,1-dibromopropane
      $CH_3CHBrCH_2Br$ 1,2-dibromopropane
      $CH_2BrCH_2CH_2Br$ 1,3-dibromopropane
      $CH_3CBr_2CH_3$ 2,2-dibromopropane

e.   The two hydrogens can be on either the same or different carbon atoms:

CH_2BrCCl_3	1-bromo-2,2,2-trichloroethane
$CH_2ClCBrCl_2$	1-bromo-1,1,2-trichloroethane
$CHBrClCHCl_2$	1-bromo-1,2,2-trichloroethane

f.

$CHBrClCH_2CH_3$	1-bromo-1-chloropropane
$CH_2BrCHClCH_3$	1-bromo-2-chloropropane
$CH_2BrCH_2CH_2Cl$	1-bromo-3-chloropropane
$CH_2ClCHBrCH_3$	2-bromo-1-chloropropane
$CH_3CBrClCH_3$	2-bromo-2-chloropropane

2.31   A cycloalkane has two fewer hydrogens than the corresponding alkane. Thus the general formula for a cycloalkane is C_nH_{2n}.

2.32   a. Start with a five-membered ring, then proceed to the four- and three-membered rings.

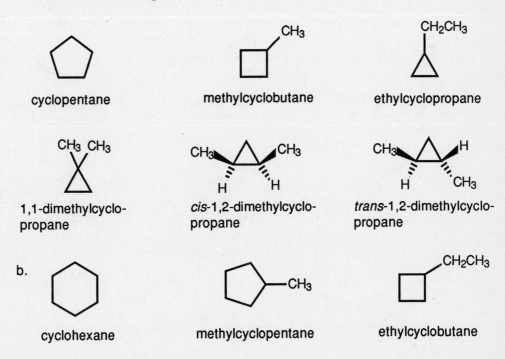

cyclopentane   methylcyclobutane   ethylcyclopropane

1,1-dimethylcyclo-propane   cis-1,2-dimethylcyclo-propane   trans-1,2-dimethylcyclo-propane

b.   cyclohexane   methylcyclopentane   ethylcyclobutane

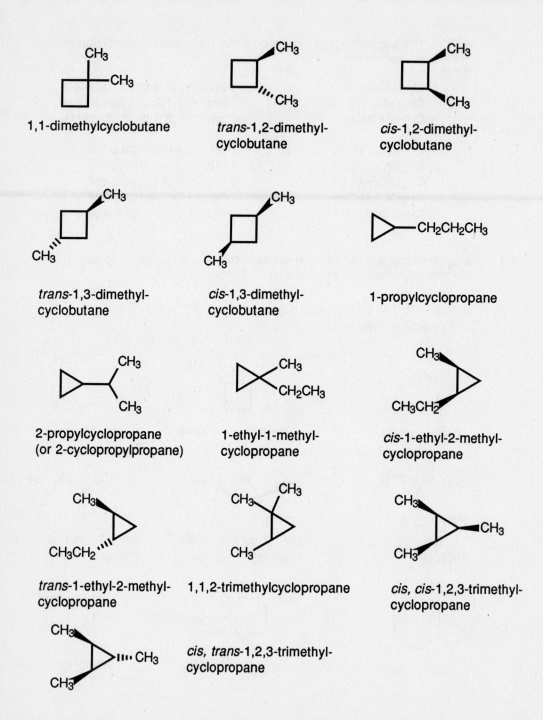

1,1-dimethylcyclobutane

trans-1,2-dimethyl-
cyclobutane

cis-1,2-dimethyl-
cyclobutane

trans-1,3-dimethyl-
cyclobutane

cis-1,3-dimethyl-
cyclobutane

1-propylcyclopropane

2-propylcyclopropane
(or 2-cyclopropylpropane)

1-ethyl-1-methyl-
cyclopropane

cis-1-ethyl-2-methyl-
cyclopropane

trans-1-ethyl-2-methyl-
cyclopropane

1,1,2-trimethylcyclopropane

cis, cis-1,2,3-trimethyl-
cyclopropane

cis, trans-1,2,3-trimethyl-
cyclopropane

2.33 The higher the molecular weight and the less branching of the chain, the higher the boiling point. On these grounds, the expected order from the lowest to highest boiling point should be e, d, c, a, b. The actual boiling points are as follows: e, 2-methylpentane (60°C); d, *n*-hexane (69°C); c, 3,3-dimethylpentane (86°C); a, 2-methylhexane (90°C); b, *n*-heptane (98.4°C).

2.34 The four conformations are as follows:

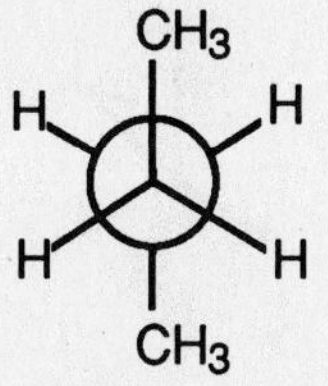

CH3
H — H
H — H
CH3

A: most stable staggered conformation

CH3
H — CH3
H — H
H

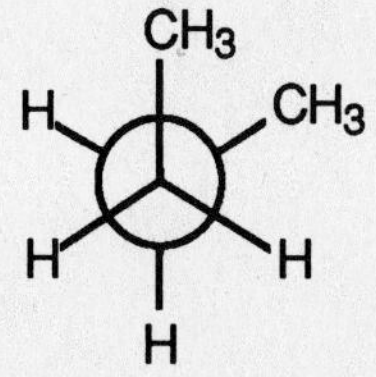

B: less stable staggered conformation (large groups are closer together)

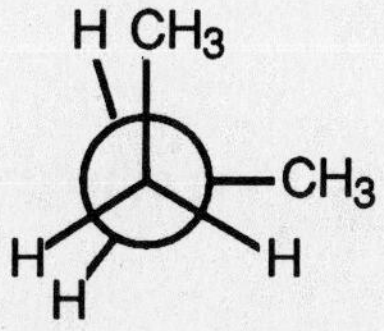

H CH3
—CH3
H
H
H

C: less stable than the staggered, but more stable than the eclipsed conformation with two methyls eclipsed

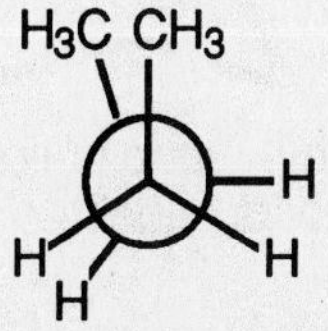

H3C CH3
—H
H
H
H

D: least stable of all four conformations

Staggered conformations are more stable than eclipsed conformations. Therefore A and B are more stable than C or D. Within each pair, CH3—CH3 interactions (for methyls on adjacent carbons) are avoided because of the large size of these groups.

2.35

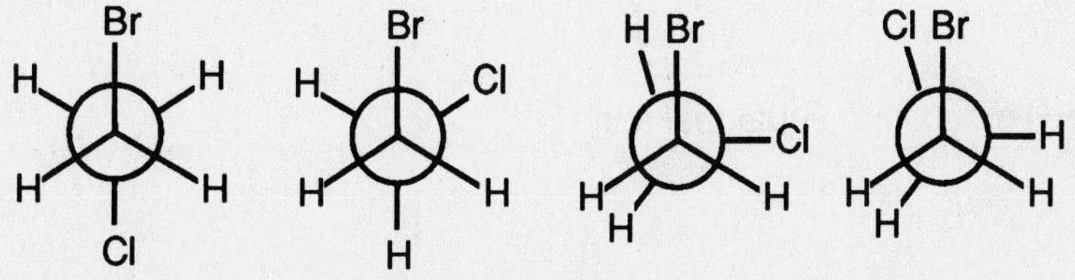

Br
H — H
H — H
Cl

Br
H — Cl
H — H
H

H Br
—Cl
H H
H

Cl Br
—H
H H
H

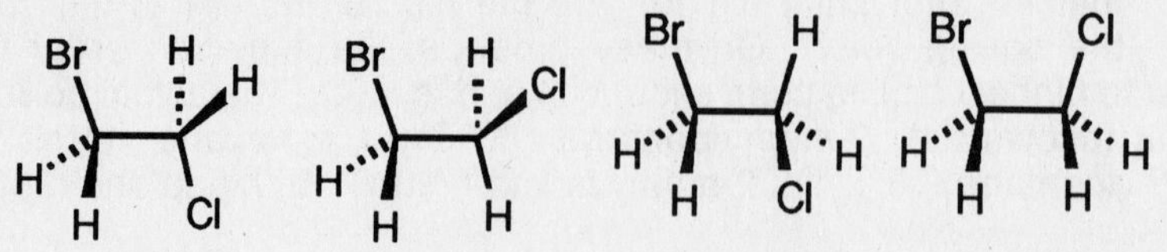

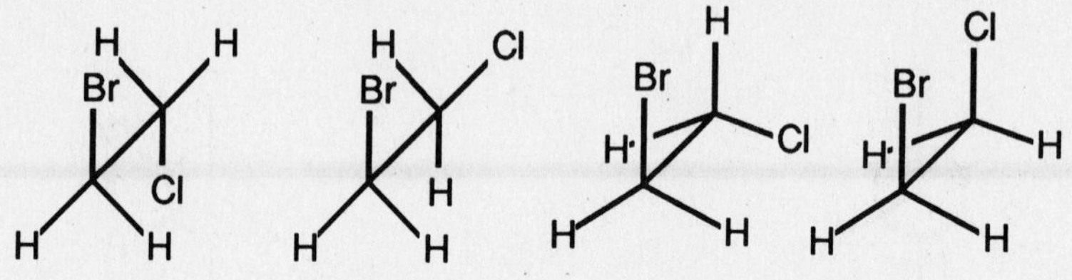

The stability decreases from left to right in the structures shown above.

2.36

a.

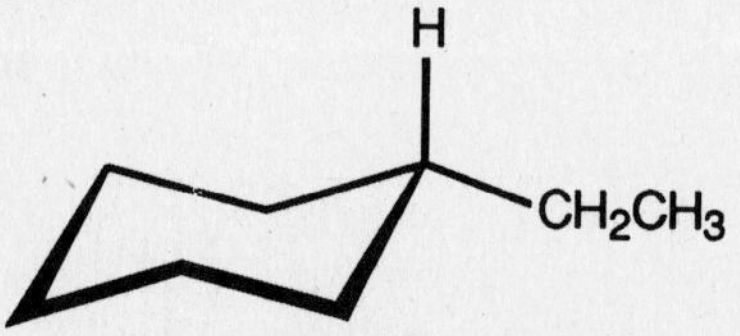

The ethyl group is equatorial. If we were to draw out the ethyl group, we would arrange its substituents in a staggered conformation.

b.

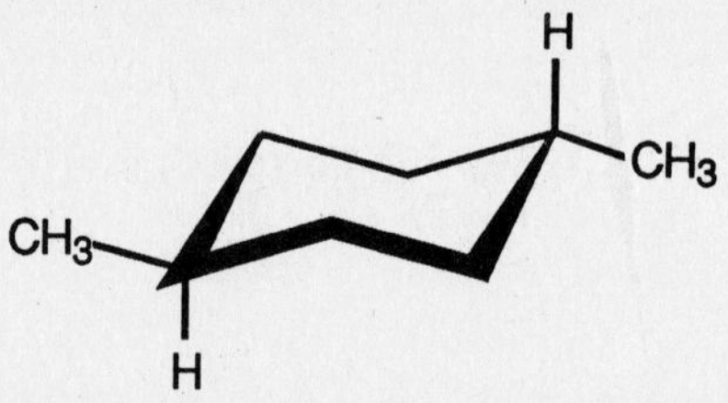

Both alkyl groups are equatorial and *trans*.

c.

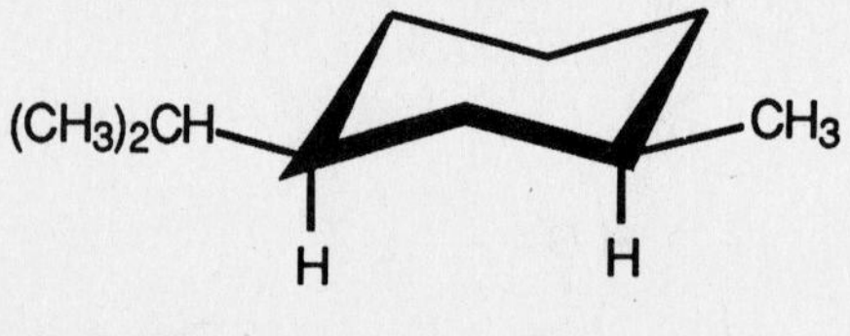

Both alkyl groups are equatorial and *cis*.

d.

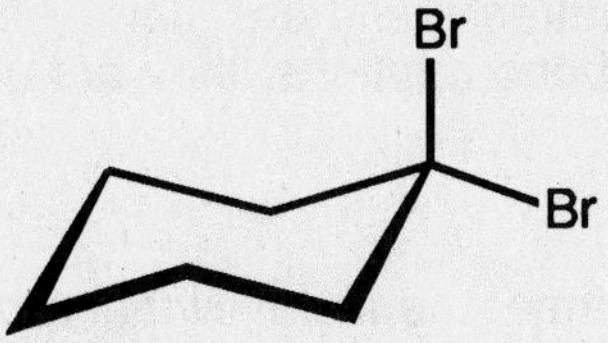

Br

Br

One bromine is axial; one is equatorial.

2.37   In each case, the formula at the left is named; the right-hand structure shows the other isomer.
a.   *cis*-1-chloro-3-methylcyclohexane
b.   *trans*-1,2-dichlorocyclohexane

2.38   *cis*-1,3-Dimethylcyclohexane can exist in a conformation in which both methyl substituents are equatorial:

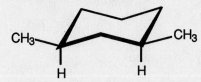

CH₃   CH₃

H   H

In the *trans* isomer, one methyl group must be axial. On the other hand, in 1,2- or 1,4-dimethylcyclohexane, only the *trans* isomer can have both methyls equatorial.

H   H

CH₃   CH₃
CH₃   CH₃

H   H

2.39   The *trans* isomer is very much more stable than the *cis* isomer because both *t*-butyl groups, which are very large, can be equatorial.

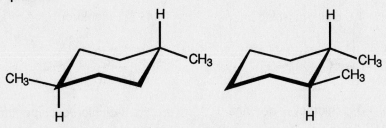

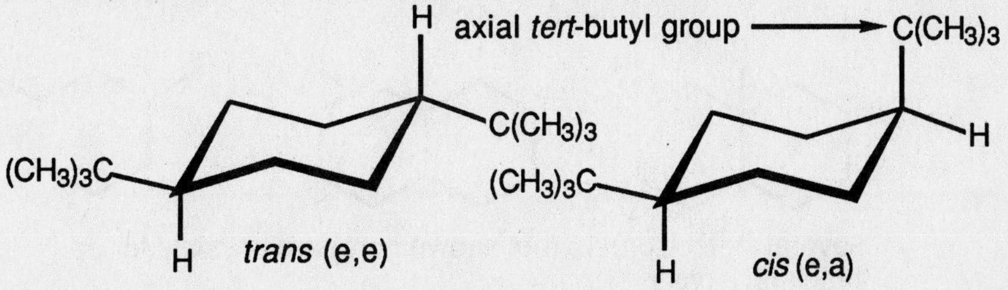

H   axial *tert*-butyl group ⟶ C(CH₃)₃

C(CH₃)₃

(CH₃)₃C   (CH₃)₃C   H

H   *trans* (e,e)   H   *cis* (e,a)

2. ALKANES AND CYCLOALKANES

2.40 a. All pairs have the same bond patterns and are stereoisomers. Since
          they are *not* interconvertible by bond rotations, they are configurational
          isomers.
      b. conformational isomers
      c. conformational isomers
      d. The bond patterns are *not* the same; the first Newman projection is for
          1,1-dichloropropane and the second is for 1,2-dichloropropane. Thus,
          these are structural (or constitutional) isomers.
      e. The structures are *identical* (both represent 2-methylbutane) and are not
          isomers at all.

2.41 We can obtain 1,1- or 1,2- or 1,3-dichlorocyclopentane. Of these, the last two
      can exist as *cis–trans* isomers.

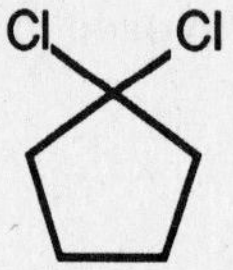

1,1-dichlorocyclopentane

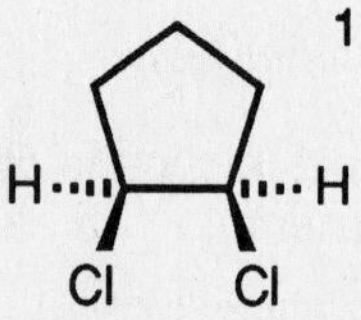

cis-1,2-dichlorocyclopentane

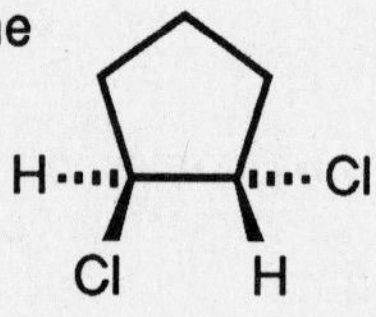

trans-1,2-dichlorocyclopentane

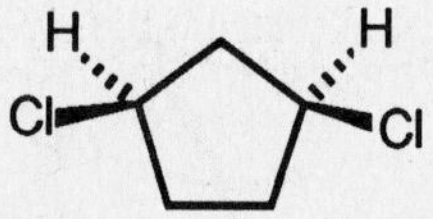

cis-1,3-dichlorocyclopentane

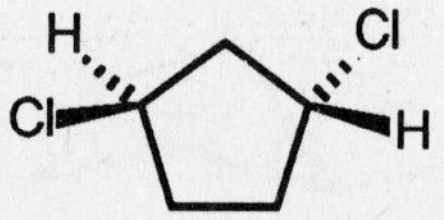

trans-1,3-dichlorocyclopentane

2.42 a. Three monochlorinated structural isomers are possible.

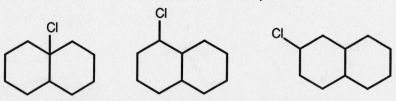

Several stereoisomers (not shown here) are possible for each
structural isomer.

ANSWERS TO PROBLEMS

b.   Three structural isomers

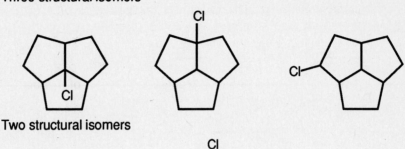

c.   Two structural isomers

2.43   a.   CH₃CH₂CH₂CH₂CH₃ + 8 O₂ ⟶ 5 CO₂ + 6 H₂O
                                                         carbon dioxide

b.   2 ⬠ + 15 O₂ ⟶ 10 CO₂ + 10 H₂O
                                         carbon dioxide

c.   CH₃CH₂CH₃ + Br₂ —heat or hv→ CH₃CH₂CH₂Br   and   CH₃CHCH₃
                                            1-bromopropane   Br + HBr
                                                                   2-bromopropane

d.   ⬠ + Cl₂ —heat or hv→ [cyclopentane with H and Cl] + HCl
                                      chlorocyclopentane

e.   CH₃CH₃ + 6 Cl₂ —heat or hv→ CCl₃CCl₃ + 6 HCl
                                         hexacloroethane

2.44   The four possible structures are
       CH₃CH₂CHCl₂   (1,1-dichloropropane)
       CH₃CH₂ClCH₂Cl (1,2-dichloropropane)
       CH₂ClCH₂CH₂Cl (1,3-dichloropropane)
       CH₃CCl₂CH₃   (2,2-dichloropropane)

2. ALKANES AND CYCLOALKANES

Only the last structure has all hydrogens equivalent and can be given only *one* trichloro compound. This structure must therefore be C:

$$CH_3CCl_2CH_3 \xrightarrow{Cl_2} CH_3CCl_2CH_2Cl$$

1,3-Dichloropropane has only two different "kinds" of hydrogens. It must be D:

$$CH_2ClCH_2CH_2Cl \xrightarrow{Cl_2} CH_2ClCHClCH_2Cl \quad \text{and} \quad CH_2ClCH_2CHCl_2$$

Next, A must be capable of giving 1,2,2-trichloropropane (the product from C). This is not possible for the 1,1 isomer since it already has two chlorines on carbon-1. Therefore, A must be 1,2-dichloropropane since it can give the 1,2,2-trichloro product (as well as 1,1,2- and 1,2,3-). By elimination, B is $CH_3CH_2CHCl_2$.

2.45 The equations follow the same pattern as in eqs. 2.16 through 2.21.

initiation $:\!\overset{..}{Cl}\!-\!\overset{..}{Cl}\!: \xrightarrow{\text{heat or light}} 2 \; :\!\overset{..}{Cl}\cdot$

propagation $CH_3CH_3 \;+\; :\!\overset{..}{Cl}\cdot \longrightarrow CH_3CH_2\cdot \;+\; H\!-\!Cl$

$CH_3CH_2\cdot \;+\; :\!\overset{..}{Cl}\!-\!\overset{..}{Cl}\!: \longrightarrow CH_3CH_2Cl \;+\; :\!\overset{..}{Cl}\cdot$

termination $2 \; :\!\overset{..}{Cl}\cdot \longrightarrow :\!\overset{..}{Cl}\!-\!\overset{..}{Cl}\!:$

$2 \;\; CH_3CH_2\cdot \longrightarrow CH_3CH_2CH_2CH_3$

$CH_3CH_2\cdot \;+\; :\!\overset{..}{Cl}\cdot \longrightarrow CH_3CH_2Cl$

The trace products resulting from the chain-termination steps would be butane and any chlorination products derived from it.

CHAPTER THREE: ALKENES AND ALKYNES

CHAPTER SUMMARY

Alkenes have a carbon-carbon bond and **alkynes** have a carbon–carbon triple bond. Nomenclature rules are given in Sec. 3.2.

Each carbon of a double bond is **trigonal**, or connected to only *three* other atoms. These atoms lie in a plane with bond angles of 120°. Ordinarily, rotation around double bonds is restricted. All six atoms of **ethylene** lie in a single plane. The C=C bond length is 1.34 Å, shorter than a C—C bond (1.54 Å). These facts can be explained by an orbital model with three sp^2 hybrid orbitals (one electron in each) and one *p* orbital perpendicular to these (containing the fourth electron). The double bond is formed by end-on overlap of sp^2 orbitals to form a σ bond and lateral overlap of aligned *p* orbitals to form a π **bond** (Figures 3.4 and 3.5). Since rotation around the double bond is restricted, *cis–trans* **isomerism** is possible if each carbon atom of the double bond has two different groups attached to it.

Alkenes react mainly by **addition**. Typical reagents that add to the double bond are halogens, hydrogen (metal catalyst required), water (acid catalyst required), and various acids. If either the alkene or the reagent is **symmetrical** (Table 3.2), only one product is possible. If *both* the alkene and reagent are unsymmetrical, however, two products are possible, in principle. In this case, **Markovnikov's rule** (Secs. 3.10 through 3.12) allows us to predict the product obtained.

Electrophilic additions occur by a two-step mechanism. In the first step, the **electrophile** adds in such a way as to form the most stable **carbocation** (the stability order is tertiary > secondary > primary). Then the carbocation combines with a **nucleophile** to give the product.

Conjugated dienes have alternating single and double bonds. They may undergo **1,2-** or **1,4-addition**. They also undergo cycloaddition reactions with alkenes (**Diels-Alder reaction**), a useful synthesis for six-membered rings.

Addition to double bonds may also occur by a **free-radical mechanism**. Polyethylene can be made in this way from the **monomer** ethylene.

Alkenes undergo a number of other reactions, such as **hydroboration**, permanganate oxidation, and **ozonolysis**.

3. ALKENES AND ALKYNES

Triple bonds are **linear** and the carbons are ***sp*-hybridized** (Figure 3.10). Alkynes, like alkenes, undergo addition reactions. A hydrogen connected to a triply bonded carbon is weakly acidic and can be removed by a very strong base such as **sodamide, NaNH$_2$**, to give acetylides.

REACTION SUMMARY

Additions to the Double Bond

$$\underset{/}{\overset{\backslash}{}}C=C\underset{\backslash}{\overset{/}{}} \; + \; X_2 \quad \xrightarrow{\text{(X = Cl or Br)}} \quad -\underset{X}{\overset{|}{C}}-\underset{X}{\overset{|}{C}}-$$

$$\underset{/}{\overset{\backslash}{}}C=C\underset{\backslash}{\overset{/}{}} \; + \; H_2 \quad \xrightarrow{\text{Pd, Pt, or Ni}} \quad -\underset{H}{\overset{|}{C}}-\underset{H}{\overset{|}{C}}-$$

$$\underset{/}{\overset{\backslash}{}}C=C\underset{\backslash}{\overset{/}{}} \; + \; H-OH \quad \xrightarrow{\;H^+\;} \quad -\underset{H}{\overset{|}{C}}-\underset{OH}{\overset{|}{C}}-$$

(an alcohol)

$$\underset{/}{\overset{\backslash}{}}C=C\underset{\backslash}{\overset{/}{}} \; + \; H-X \quad \longrightarrow \quad -\underset{H}{\overset{|}{C}}-\underset{X}{\overset{|}{C}}-$$

(X = —F, —Cl, —Br,
—I, —OSO$_3$H, —OCR)
                        ‖
                        O

Conjugated Dienes

$$C=C-C=C \quad \xrightarrow{\;X_2\;} \quad \underset{X}{\overset{|}{C}}-\underset{X}{\overset{|}{C}}-C=C \quad + \quad \underset{X}{\overset{|}{C}}-C=C-\underset{X}{\overset{|}{C}}$$

1,2-addition   1,4-addition

44

REACTION SUMMARY

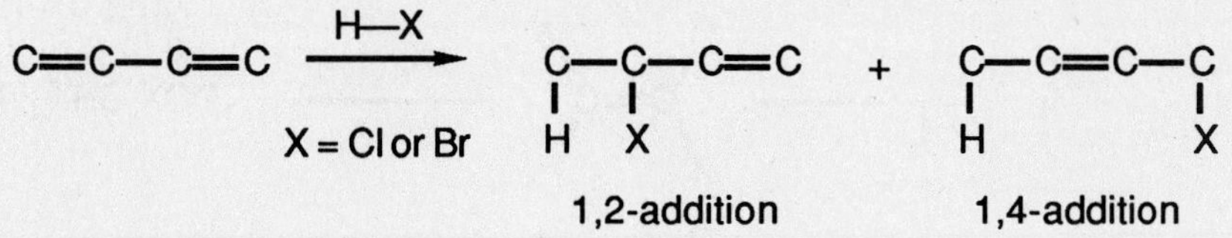

$$C=C-C=C \xrightarrow[\text{X = Cl or Br}]{\text{H—X}}$$

1,2-addition   +   1,4-addition

Cycloaddition (Diels–Alder)

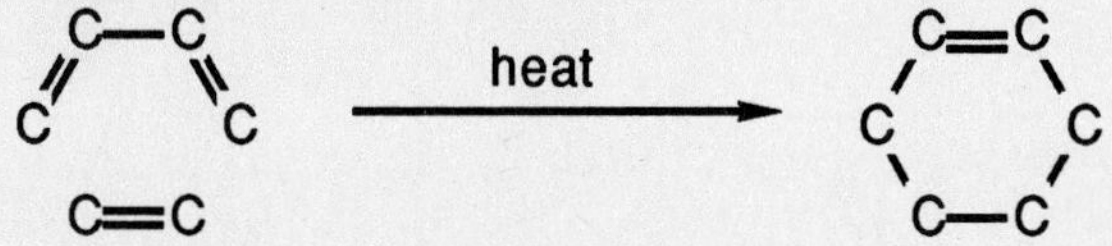

$$\xrightarrow{\text{heat}}$$

Polymerization of Ethylene

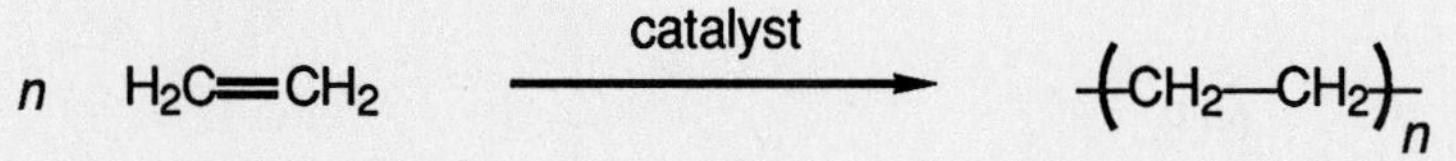

$$n \quad H_2C=CH_2 \xrightarrow{\text{catalyst}} \left(CH_2-CH_2\right)_n$$

Hydroboration–Oxidation

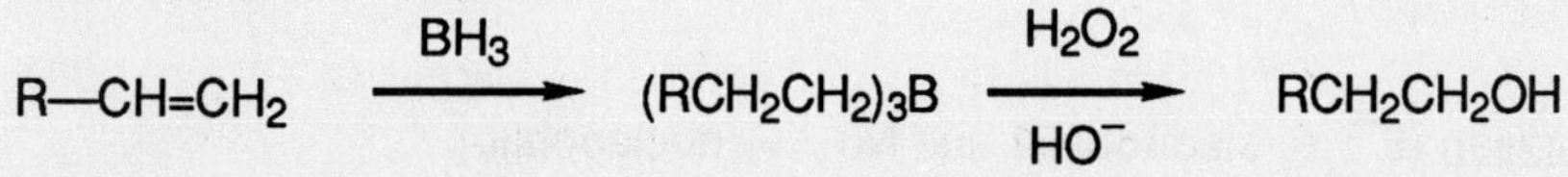

$$R-CH=CH_2 \xrightarrow{BH_3} (RCH_2CH_2)_3B \xrightarrow[HO^-]{H_2O_2} RCH_2CH_2OH$$

Permanganate Oxidation

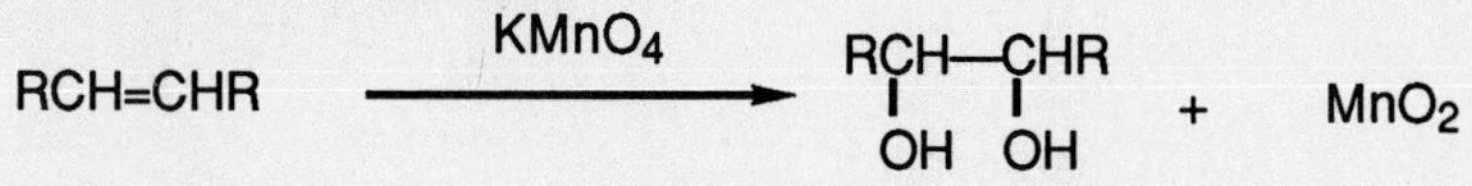

$$RCH=CHR \xrightarrow{KMnO_4} \underset{OH \quad OH}{RCH-CHR} + MnO_2$$

Ozonolysis

$$C=C \xrightarrow{O_3} C=O + O=C$$

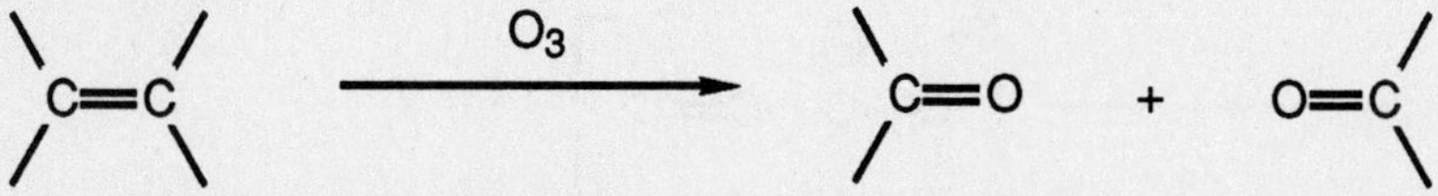

Additions to the Triple Bond

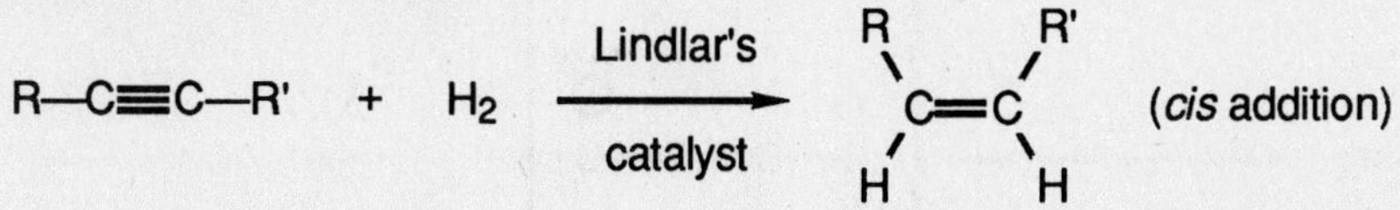

$$R-C\equiv C-R' + H_2 \xrightarrow[\text{catalyst}]{\text{Lindlar's}} \underset{H \quad H}{\overset{R \quad R'}{C=C}} \quad (\textit{cis addition})$$

3. ALKENES AND ALKYNES

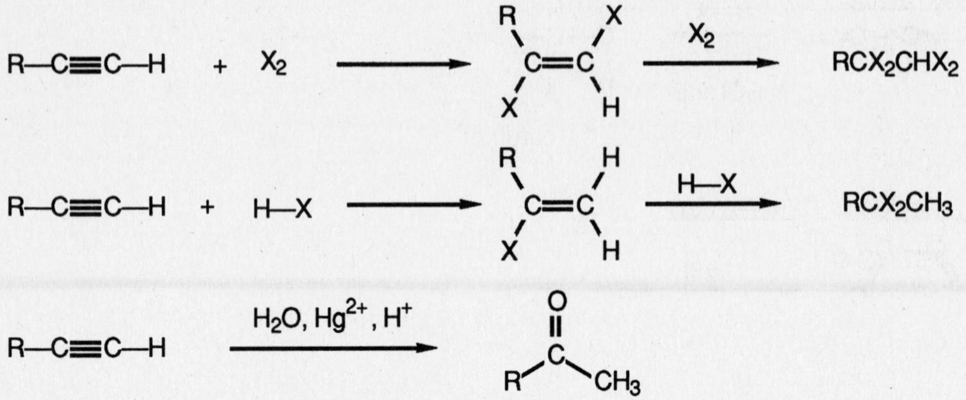

$$R-C\equiv C-H \quad + \quad X_2 \quad \longrightarrow \quad \underset{X}{\overset{R}{\underset{|}{C}}}=\underset{H}{\overset{X}{\underset{|}{C}}} \quad \xrightarrow{X_2} \quad RCX_2CHX_2$$

$$R-C\equiv C-H \quad + \quad H-X \quad \longrightarrow \quad \underset{X}{\overset{R}{\underset{|}{C}}}=\underset{H}{\overset{H}{\underset{|}{C}}} \quad \xrightarrow{H-X} \quad RCX_2CH_3$$

$$R-C\equiv C-H \quad \xrightarrow{H_2O,\ Hg^{2+},\ H^+} \quad \underset{R}{\overset{O}{\overset{\|}{C}}}\diagdown_{CH_3}$$

Alkyne Acidity

$$RC\equiv CH \quad + \quad NaNH_2 \quad \xrightarrow{NH_3} \quad RC\equiv C^- Na^+ \quad + \quad NH_3$$

MECHANISM SUMMARY

Electrophilic Addition (E^+ = electrophile and Nu : = nucleophile)

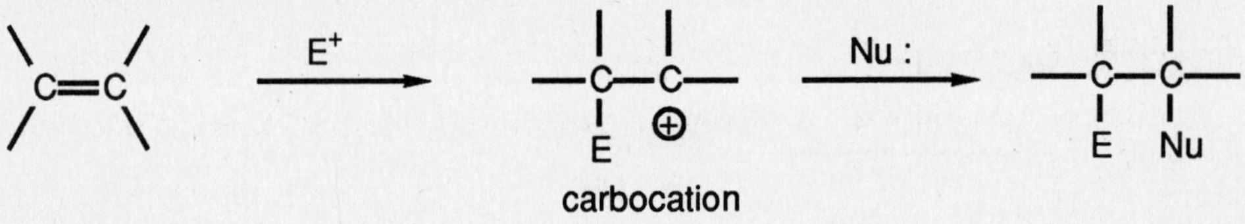

carbocation

1,4-Addition

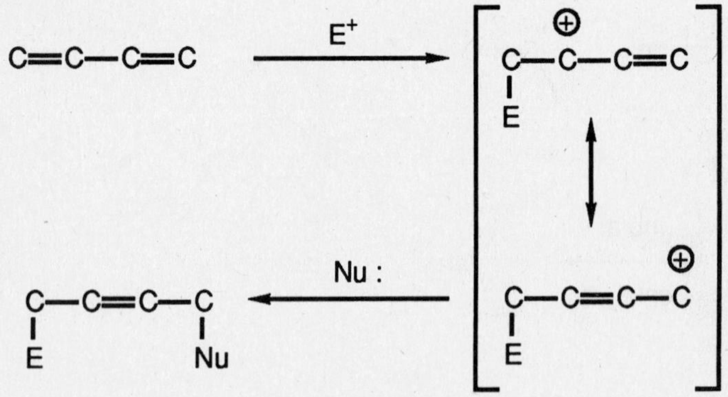

LEARNING OBJECTIVES

Free-Radical Polymerization of Ethylene

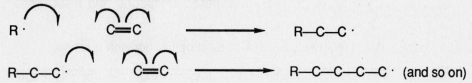

LEARNING OBJECTIVES

1. Know the meaning of: saturated and unsaturated; alkene, alkyne, and diene; conjugated, cumulated, and isolated double bonds; vinyl, allyl, and propargyl groups.

2. Know the meaning of: trigonal carbon, sp^2-hybridization, restricted rotation, σ and π bond, *cis* and *trans* double-bond isomers.

3. Know the meaning of: addition reaction, Markovnikov's rule, electrophile, nucleophile, symmetrical and unsymmetrical double bonds and reagents, carbocation.

4. Know the meaning of: 1,2-addition, 1,4-addition, cycloaddition, diene, dienophile, polymer, monomer, polymerization.

5. Know the meaning of: hydroboration, glycol, ozone, ozonolysis, combustion.

6. Given the structure of an acyclic or cyclic alkene, alkyne, diene, and so on, state the IUPAC name.

7. Given the IUPAC name of an alkene, alkyne, diene, and so on, write the structural formula.

8. Given the molecular formula of a hydrocarbon and the number of double bonds, triple bonds, or rings, draw the possible structures.

9. Given the name or abbreviated structure of an unsaturated compound, tell whether it can exist in *cis* and *trans* isomeric forms, and if so, how many. Draw them.

10. Given an alkene, alkyne, or diene, and one of the following reagents, draw the structure of the product (reagents: acids such as HCl, HBr, HI, and H_2SO_4; water in the presence of an acid catalyst; halogens such as Br_2 and Cl_2; hydrogen).

3. ALKENES AND ALKYNES

11. Given the structure or name of a compound that can be prepared by an addition reaction, deduce what unsaturated compound and what reagent react to form it.

12. Write the steps in the mechanism of an electrophilic addition reaction.

13. Given an unsymmetrical alkene and an unsymmetrical electrophilic reagent, give the structure of the predominant product (that is, apply Markovnikov's rule).

14. Given a conjugated diene and a reagent that adds to it, write the structures of the 1,2- and 1,4-addition products.

15. Given a diene and dienophile, write the structure of the resulting cycloaddition (Diels-Alder) adduct.

16. Given the structure of a cyclic compound that can be synthesized by the Diels-Alder reaction, deduce the structures of the required diene and dienophile.

17. Given an alkyne, write the structures of products obtained by adding one or two moles of a particular reagent to it.

18. Write the steps in the mechanism of ethylene polymerization catalyzed by a free radical.

19. Write the structure of the alcohol produced from the hydroboration–oxidation sequence when applied to a particular alkene.

20. Given an alkene or cycloalkene (or diene, and so on), write the structures of the expected ozonolysis products.

21. Given the structures of ozonolysis products, deduce the structure of the unsaturated hydrocarbon that produced them.

22. Draw orbital pictures for a double bond and a triple bond.

23. Draw conventional structures for the contributors to the resonance hybrid of an allyl cation.

24. Describe simple chemical tests that can distinguish an alkane from an alkene or alkyne.

25. Know the meaning of cracking, alkylation, isomerization, platforming, and octane number as applied to petroleum refining.

ANSWERS TO PROBLEMS

Problems Within the Chapter

3.1   The formula C_4H_6 corresponds to C_nH_{n-2}. The possibilities are one triple bond, two double bonds, one double bond and one ring, or two rings.

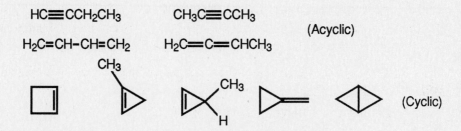

$$HC{\equiv}CCH_2CH_3 \qquad CH_3C{\equiv}CCH_3$$

(Acyclic)

$$H_2C{=}CH{-}CH{=}CH_2 \qquad H_2C{=}C{=}CHCH_3$$

(Cyclic)

3.2   Compounds b and c have alternating single and double bonds, and are conjugated. In a and d the double bonds are isolated.

3.3   a. $\overset{1}{C}lCH{=}\overset{2}{C}H\overset{3}{C}H_3$ 1-chloropropene

b. 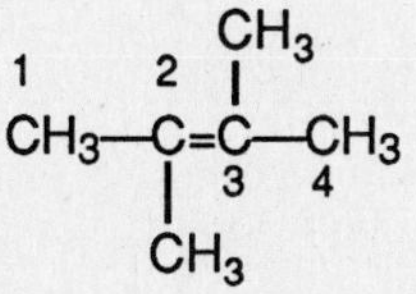   2,3-dimethyl-2-butene

c. 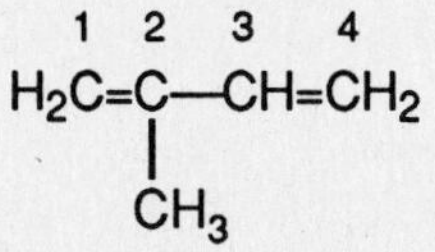   2-methyl-1,3-butadiene (also called isoprene)

d. 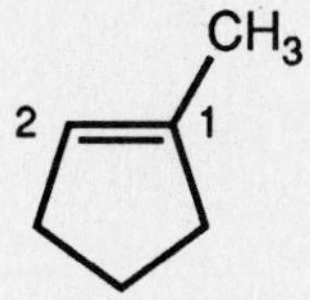   1-methylcyclopentene

3. ALKENES AND ALKYNES

e. $\overset{1}{H_2C}=\overset{2}{\underset{\underset{Cl}{|}}{C}}-\overset{3}{CH_3}$ 2-chloropropene

f. $\overset{1}{HC}\equiv\overset{2}{C}\overset{3}{C}H_2\overset{4}{C}H_2\overset{5}{C}H_2\overset{6}{C}H_3$ 1-hexyne

3.4   a.   First write out the five-carbon chain, with a double bond between carbon-2 and carbon-3:

$$\overset{1}{C}-\overset{2}{C}=\overset{3}{C}-\overset{4}{C}-\overset{5}{C}$$

Add the substituents:

$$\overset{1}{C}-\overset{2}{\underset{\underset{CH_3}{|}}{C}}=\overset{3}{C}-\overset{4}{\underset{\underset{CH_3}{|}}{C}}-\overset{5}{C}$$

Fill in the hydrogens:

$$CH_3-\underset{\underset{CH_3}{|}}{C}=CH-\underset{\underset{CH_3}{|}}{CH}-CH_3 \quad \text{or} \quad (CH_3)_2C=CHCH(CH_3)_2$$

b. $CH_3C\equiv CCH_2CH_2CH_3$

c.   [structure: cyclobutene ring with Br on each of the two double-bond carbons]

d. $H_2C=\underset{\underset{Cl}{|}}{C}-CH=CH_2$

3.5

a.   [cyclohexane with substituent]   or   [cyclohexane with CH=CH₂]   **vinyl group** $CH=CH_2$

b.   [cyclopropane]—$CH_2CH=CH_2$ **allyl group**

c. $HC\equiv CCH_2I$ **propargyl group**

3.6   Compounds a and d have only one possible structure because in each case one of the carbons of the double bond has two identical substituents:

ANSWERS TO PROBLEMS

a. H_2C=CHCH$_3$ b.

$$\underset{H}{\overset{CH_3CH_2}{\diagdown}}C=C\underset{H}{\overset{CH_2CH_3}{\diagup}}$$

and

$$\underset{H}{\overset{CH_3CH_2}{\diagdown}}C=C\underset{CH_2CH_3}{\overset{H}{\diagup}}$$

cis-3-hexene

trans-3-hexene

c.

$$\underset{H}{\overset{CH_3}{\diagdown}}C=C\underset{H}{\overset{CH_2CH_2CH_3}{\diagup}}$$

and

$$\underset{H}{\overset{CH_3}{\diagdown}}C=C\underset{CH_2CH_2CH_3}{\overset{H}{\diagup}}$$

d. $(CH_3)_2C$=CHCH$_3$

cis-2-hexene

trans-2-hexene

3.7   The electron pair in a σ bond lies directly between the nuclei it joins. In a π bond, the electron pair is further from the two nuclei that it joins. Therefore more energy is required to break a σ bond than a π bond.

3.8

a.

$$\underset{CH_3}{\overset{CH_3}{\diagdown}}C=CH_2 \ + \ H_2 \xrightarrow{catalyst} CH_3-\underset{CH_3}{\overset{H}{\underset{|}{\overset{|}{C}}}}-CH_3$$

b.

(cyclobutene with two CH$_3$) + H$_2$ $\xrightarrow{catalyst}$ (cyclobutane with CH$_3$, H, H, CH$_3$)

3.9   a.   One bromine atom adds to each doubly bonded carbon, and the double bond in the starting material becomes a single bond in the product:

H_2C=CHCH$_2$CH$_3$ + Br$_2$ ⟶ $H_2C-CHCH_2CH_3$ with Br, Br

b.

$$\underset{CH_3}{\overset{CH_3}{\diagdown}}C=CHCH_3 \ + \ Br_2 \ \longrightarrow \ CH_3-\underset{CH_3}{\overset{Br}{\underset{|}{\overset{|}{C}}}}-\overset{Br}{\underset{}{\overset{|}{C}}HCH_3$$

51

3.10   a. $CH_3CH=CHCH_3$ + H—OH $\xrightarrow{H^+}$ $CH_3CH_2CHCH_3$
                                                                             |
                                                                             OH

b. [cyclopentene structure] + H—OH $\xrightarrow{H^+}$ [cyclopentane structure]—OH

3.11   a. $CH_3CH=CHCH_3$ + H—I $\xrightarrow{H^+}$ $CH_3CH_2CHCH_3$
                                                                        |
                                                                        I

b. [cyclopentene structure] + H—Br $\xrightarrow{H^+}$ [cyclopentane structure]—Br

3.12   a. $H_2C=CHCH_2CH_3$
            ↑

This doubly bonded carbon has the most hydrogens.

            H—Cl
            ↑

This is the more electropositive part of the reagent.

                                    δ+ δ−
        $H_2C=CHCH_2CH_3$ + H—Cl ⟶ CH_3—$CHCH_2CH_3$
                                                          |
                                                          Cl

                                    δ+ δ−   OH
                                                                 |
b. CH_3—$C=CHCH_3$ + H—OH ⟶ CH_3—C—CH_2CH_3
              | |
              CH_3 CH_3

3.13 Since both carbons of the double bond have the same number of attached hydrogens (one), application of Markovnikov's rule is ambiguous. Both products are formed, and the reaction is not regiospecific.

$$CH_3CH=CHCH_2CH_3 \quad + \quad HCl \quad \longrightarrow \quad \begin{array}{c} CH_3CH_2CH(Cl)CH_2CH_3 \\ + \\ CH_3CH(Cl)CH_2CH_2CH_3 \end{array}$$

3.14 a. secondary b. primary c. tertiary

3.15 The order of stability is c > a > b.

3.16 eq. 3.17:

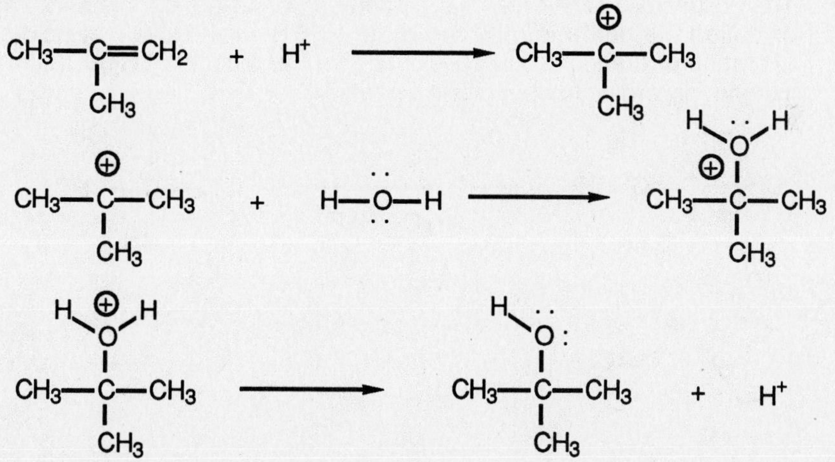

The intermediate carbocation is tertiary. If, in the first step, the proton had added to the other carbon of the double bond, the carbocation produced would have been primary:

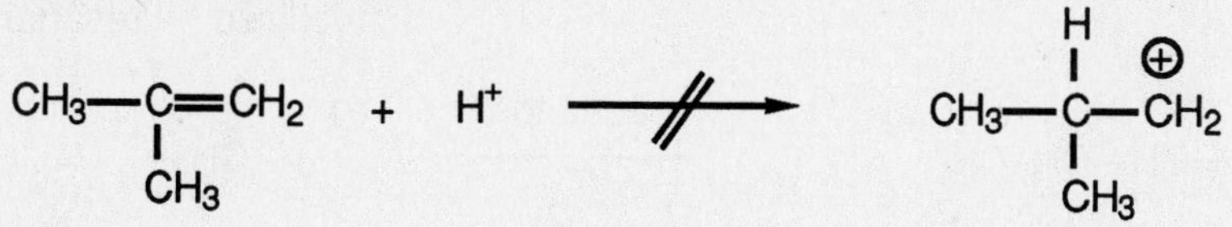

eq. 3.18:

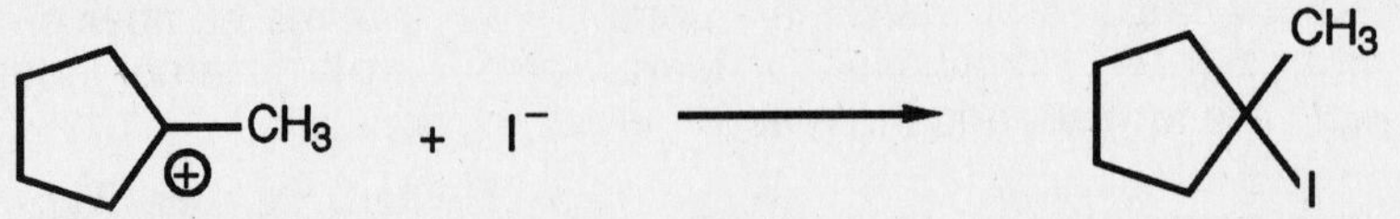

The intermediate carbocation is tertiary. Had the proton added to the methyl-bearing carbon, the intermediate carbocation would have been secondary:

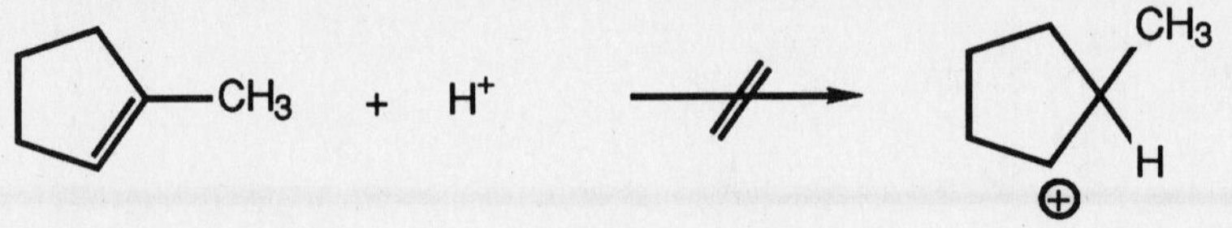

3.17 The boron adds to the *less* substituted carbon of the double bond, and in the oxidation the boron is replaced by an OH group. Note that acid-catalyzed hydration of the same alkene would give the alcohol $(CH_3)_2C(OH)CH_2CH_3$ instead, according to Markovnikov's rule.

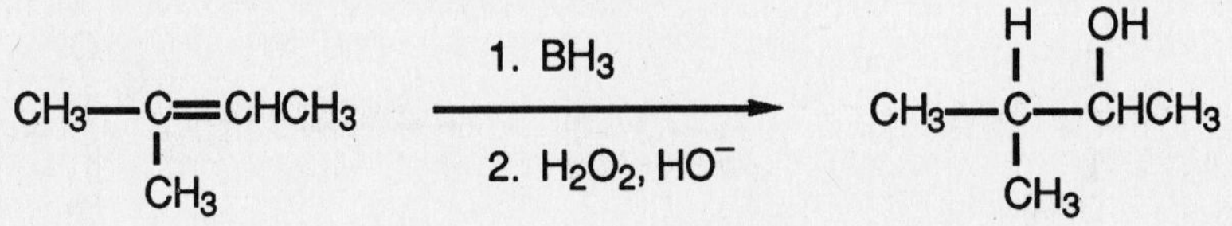

$$CH_3—\underset{\underset{CH_3}{|}}{C}=CHCH_3 \quad \xrightarrow[\text{2. } H_2O_2, \ HO^-]{\text{1. } BH_3} \quad CH_3—\underset{\underset{CH_3}{|}}{\overset{\overset{H}{|}}{C}}—\overset{\overset{OH}{|}}{C}HCH_3$$

3.18

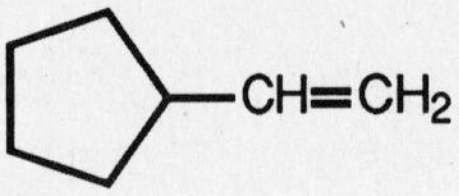

3.19 $\quad CH_2=CH—CH=CH_2 \quad + \quad H^+ \quad \xrightarrow[\text{C-1}]{\text{adds to}} \quad CH_3—\overset{\oplus}{C}H—CH=CH_2$

allylic carbocation, stabilized by resonance

$\quad CH_2=CH—CH=CH_2 \quad + \quad H^+ \quad \xrightarrow[\text{C-2}]{\text{adds to}} \quad \overset{\oplus}{C}H_2—CH_2—CH=CH_2$

primary carbocation, much less stable than the allylic carbocation

3.20

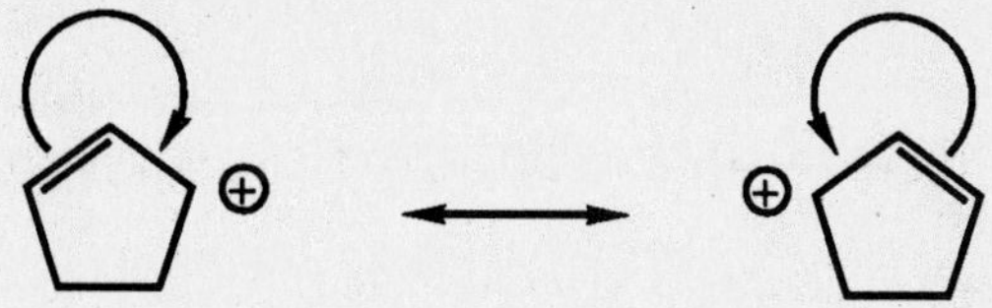

3.21

$$CH_2=CH{-}CH=CH_2 \xrightarrow{\ Br_2\ }$$

$$\begin{array}{c} CH_2{-}CH{-}CH=CH_2 \\ \;|\quad\;\; | \\ Br\quad Br \end{array}$$

1,2-addition

+

$$\begin{array}{c} CH_2{-}CH=CH{-}CH_2 \\ \;|\qquad\qquad\; | \\ Br\qquad\qquad Br \end{array}$$

1,4-addition

3.22 To find the structures of the diene and dienophile, break the cyclohexene ring just beyond the ring carbons that are connected to the double bond (the allylic carbons):

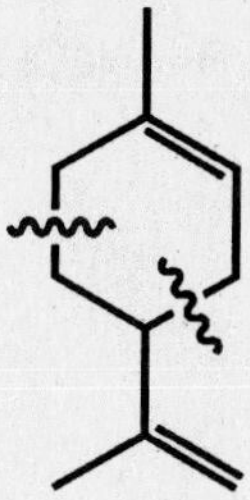

The equation is: $\xrightarrow{\text{heat}}$

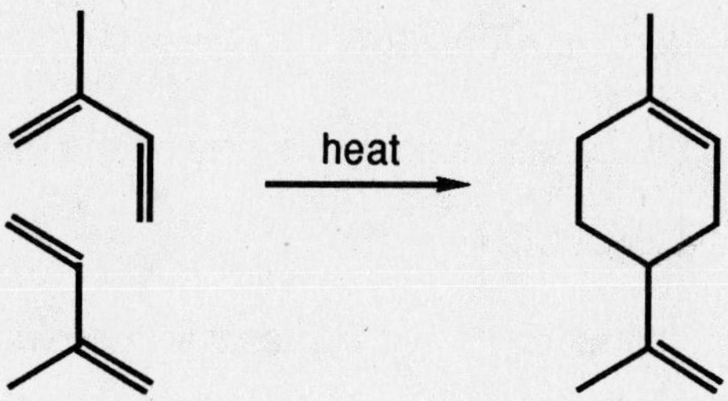

3.23

a.

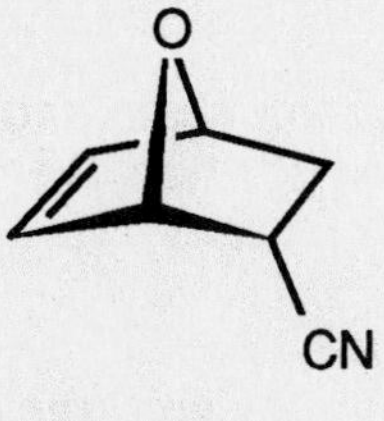

b.

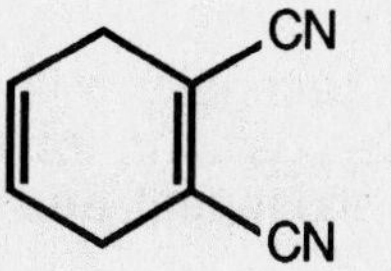

3.24 $CH_3CH{=\!=}CHCH_3$ + $KMnO_4$ $\longrightarrow$ $\begin{array}{c} CH_3CH{-}CHCH_3 \\ \;\;|\quad\;\; | \\ OH\;\; OH \end{array}$

3. ALKENES AND ALKYNES

3.25 $(CH_3)_2C=C(CH_3)_2$

3.26 a. $CH_3C \equiv CH \ + \ Cl_2 \longrightarrow CH_3CCl=CHCl$

b. $CH_3C \equiv CH \ + \ 2\ Cl_2 \longrightarrow CH_3CCl_2-CHCl_2$

c. $HC \equiv CCH_2CH_3 \ + \ HBr \longrightarrow CH_2=CBrCH_2CH_3$

$HC \equiv CCH_2CH_3 \ + \ 2\ HBr \longrightarrow CH_3CBr_2CH_2CH_3$

d. $HC \equiv CCH_2CH_3 \ + \ H-OH \xrightarrow[\ H^+\ ]{Hg^{2+}} CH_3\overset{\displaystyle O}{\overset{\|}{C}}CH_2CH_3$

3.27 Follow eq. 3.50 as a guideline:

$CH_3CH_2C \equiv CH \ + \ Na^+NH_2^- \longrightarrow CH_3CH_2C \equiv C^-\ Na^+ \ + \ NH_3$

3.28 $CH_3CH_2C \equiv C^-\ Na^+ \ + \ H-OH \longrightarrow CH_3CH_2C \equiv CH \ + \ NaOH$

3.29 2-Butyne has no hydrogens on the triple bond:

$CH_3C \equiv CCH_3$

Therefore, it does not react with sodium amide.

Additional Problems

3.30 In each case, start with the longest possible carbon chain and determine all possible positions for the double bond. Then shorten the chain by one carbon and repeat, and so on.

a. $CH_2=CHCH_2CH_3$ 1-butene
$CH_3CH=CHCH_3$ 2-butene (*cis* and *trans*)
$CH_2=C(CH_3)CH_3$ 2-methylpropene

b. $CH_2=CHCH_2CH_2CH_3$ 1-pentene
$CH_3CH=CHCH_2CH_3$ 2-pentene (*cis* and *trans*)

$CH_2=C(CH_3)CH_2CH_3$	2-methyl-1-butene
$CH_2=CHCH(CH_3)_2$	3-methyl-1-butene
$CH_3C(CH_3)=CHCH_3$	2-methyl-2-butene

c.
$CH_2=C=CHCH_2CH_3$	1,2-pentadiene
$CH_2=CH-CH=CHCH_3$	1,3-pentadiene
$CH_2=CHCH_2CH=CH_2$	1,4-pentadiene
$CH_3CH=C=CHCH_3$	2,3-pentadiene
$CH_2=C=C(CH_3)_2$	3-methyl-1,2-butadiene
$CH_2=C(CH_3)CH=CH_2$	2-methyl-1,3-butadiene

d.
$HC{\equiv}CCH_2CH_2CH_3$	1-pentyne
$CH_3C{\equiv}CCH_2CH_3$	2-pentyne
$HC{\equiv}CCH(CH_3)_2$	3-methyl-1-butyne

3.31 a. 2-pentene

b. 2-methyl-2-butene (number the chain from the end with the methyl substituent).

c. 1,2-dimethylcyclopentene (number the ring "through" the double bond).

d. 2-pentyne

e. 2-chloro-1,3-butadiene (number from the end with the chlorine substituent).

f. *trans*-2-hexene

g. *cis*-2-hexene

h. 3-(1-propyl)-2-hexene (number the longest chain containing the double bond).

i. 4-hexen-1-yne (number from the multiple bond nearest the end of the chain).

j. 1-penten-4-yne (the double bond receives the lowest numbers).

3.32 In general, write the longest carbon chain or ring and number it, then locate the double bond, place the substituents, and finally, write the correct number of hydrogens on each carbon atom.

a. $\overset{3}{CH_3}CH_2CH=CHCH_2CH_3$

b. $\underset{H_2C-CH_2}{\overset{HC=CH}{\vert\qquad\vert}}$

c. $Br-CH_2\overset{1}{CH}=\overset{3}{C}\overset{CH_3}{\diagup}_{\diagdown Br}$

d. $\overset{1}{HC}\equiv\overset{3}{C}-\underset{\underset{CH_3}{\vert}}{CH}-CH_2CH_3$

e. $\overset{1}{CH_2}=CHCH_2\overset{4}{CH}=CHCH_3$

f. $H_2C=CHBr$ $(CH_2=CH-$ is vinyl)

g. $H_2C=CHCH_2Cl$ $(CH_2=CHCH_2-$ is allyl)

h.

i.

Number "through" the
double bond

j.

Number "through" both
double bonds

3.33

a. $\overset{1}{CH_3}\overset{2}{CH}=\overset{3}{CH}\overset{4}{CH_3}$ 2-butene; use the lower of the
two numbers for the double bond.

b. $\overset{1}{CH_3}\overset{2}{C}\equiv\overset{3}{C}\overset{4}{CH_2}\overset{5}{CH_3}$ 2-pentyne; number the chain
from the other end.

c. $\overset{1}{H_2C}=\underset{\underset{\underset{3}{CH_2}\underset{4}{CH_3}}{\vert}}{\overset{2}{C}}-CH_3$ 2-methyl-1-butene; number the
longest chain.

d.   1-methylcyclopentene,   *not*

e. $\overset{1}{H_2C}=\underset{\underset{CH_3}{\vert}}{\overset{2}{C}}-\overset{3}{CH}=\overset{4}{CH_2}$ 2-methyl-1,3-butadiene; number to
give the substituent the lowest possible
number.

$$\overset{4}{CH_2}—\overset{3}{CH}=\overset{2}{CH}—\overset{1}{CH_3}$$

f. 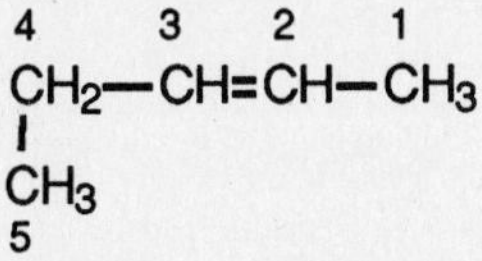 $CH_2—CH=CH—CH_3$ with CH_3 (5) below — 2-pentene; the "1-methyl" substituent lengthens the chain.

g. $\overset{1}{H_2C}=\overset{2}{CH}—\overset{3}{C}\equiv\overset{4}{CH}$ — 1-buten-3-yne; use the lower of the two numbers for the double bond.

h. $\overset{1}{H_2C}=\overset{2}{CH}—\overset{3}{C}\equiv\overset{4}{CH}$ — 1-buten-3-yne; name as an enyne, *not* an ynene.

3.34 a. The average values are 1.54Å , 1.34 Å, and 1.21 Å, respectively.

b. These single bonds are shorter than the usual 1.54Å because they are between $sp^2–sp^2$ (1.47 Å), $sp^2–sp$ (1.43 Å), and $sp–sp$ (1.37 Å) hybridized carbons. The more s character the orbitals have, the more closely the electrons are pulled in toward the nuclei and the shorter the bonds.

3.35 Review Sec. 3.5 if you have difficulty with this question.

a. Only one structure is possible since one of the doubly bonded carbons has two identical groups (hydrogens):

$H_2C=CHCH_2CH_2CH_3$

b.

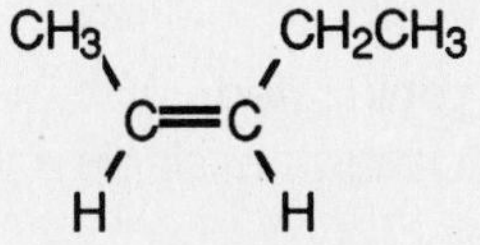

cis-2-pentene 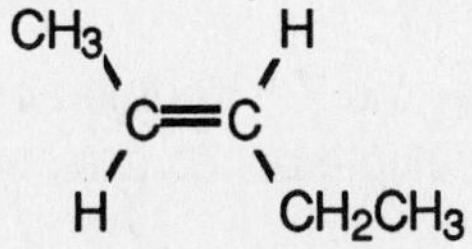   trans-2-pentene

c.

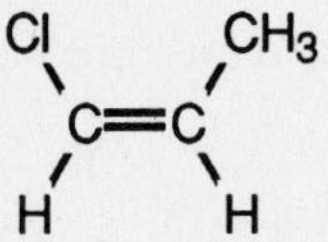

cis-1-chloropropene 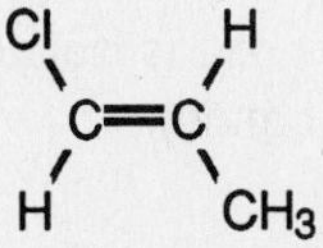   trans-1-chloropropene

59

3. ALKENES AND ALKYNES

d. Only one structure:

$H_2C=CHCH_2Cl$

e.

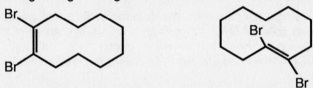

cis-1,3,5-hexatriene   trans-1,3,5-hexatriene

Only the central double bond has two different groups (a vinyl group and a hydrogen) attached to each carbon.

f. The ring is large enough to accommodate a *trans* double bond.

cis-1,2-dibromocyclodecene   trans-1,2-dibromocyclodecene

Normally, rings must be at least eight-membered to accommodate a *trans* double bond.

3.36

$$\overset{13}{HC}\equiv\overset{12}{C}-\overset{11}{C}\equiv\overset{10}{C}-\overset{9}{CH}=\overset{8}{C}=\overset{7}{CH}-\overset{6}{CH}=\overset{5}{CH}-\overset{4}{CH}=\overset{3}{CH}-\overset{2}{CH_2}-\overset{1}{\underset{\|}{C}}-OH$$

a. The 3-4, 5-6, and 7-8 double bonds are conjugated. Also, the 8-9, 10-11, and 12-13 multiple bonds are conjugated (alternate single and multiple bonds).

b. The 7-8 and 8-9 double bonds are cumulated.

c. Only the C=O bond is isolated (separated from the nearest multiple bond by two single bonds).

3.37 a. $CH_3CH{=}CHCH_3$ + Br_2 $\longrightarrow$ $CH_3CHBrCHBrCH_3$

2,3-dibromobutane

b. H_2C=$CHBr$ + Br_2 $\longrightarrow$ $CH_2BrCHBr_2$

1,1,2-tribromoethane

c. 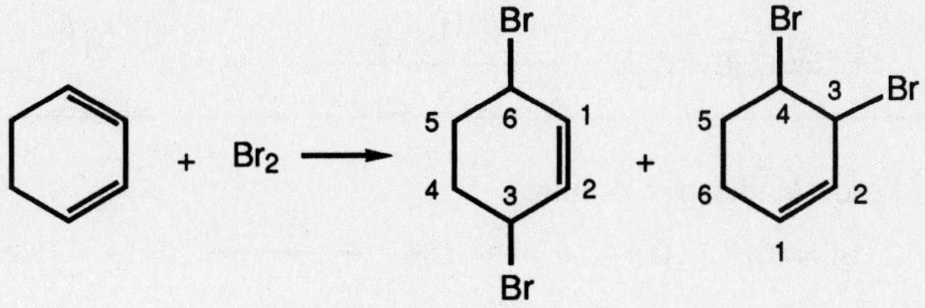 + Br_2 $\longrightarrow$ +

3,6-dibromocyclohexene 3,4-dibromocyclohexene

(1,4-addition)   (1,2-addition)

Compare with eqs. 3.28–3.30. The 1,4-addition product predominates.

d. 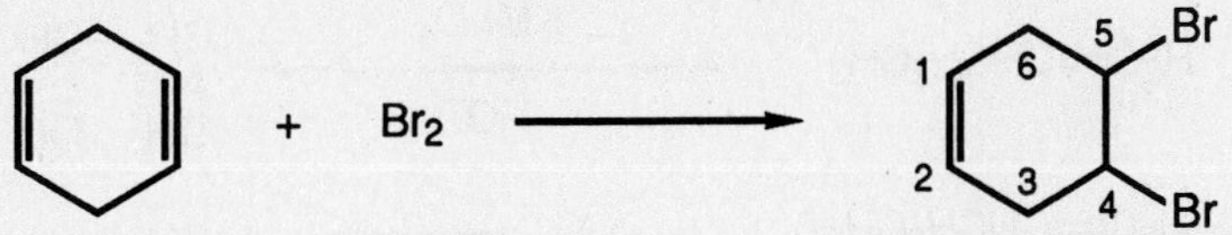 + Br_2 $\longrightarrow$

4,5-dibromocyclohexene

The double bonds are not conjugated, so only 1,2-addition is possible.

e. 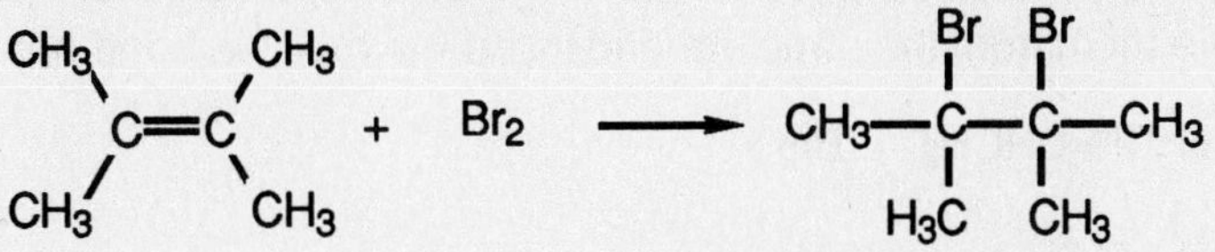 + Br_2 $\longrightarrow$

2,3-dibromo-2,3-dimethylbutane

3.38 a.  H_2C=$CHCH_2CH_3$ + Cl_2 $\longrightarrow$

b. Follow Markovnikov's rule:

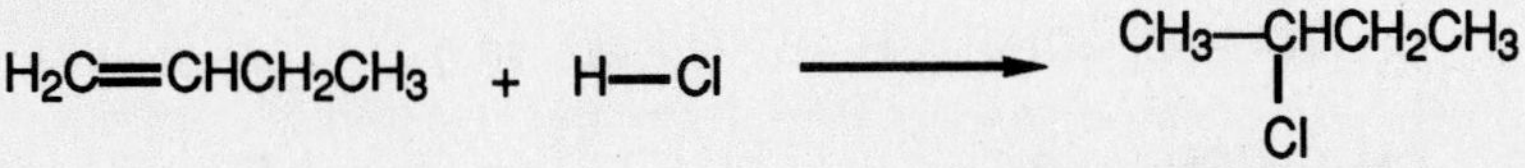

H_2C=$CHCH_2CH_3$ + H—Cl $\longrightarrow$

3. ALKENES AND ALKYNES

c. $H_2C=CHCH_2CH_3$ + H_2 $\xrightarrow{Pt}$ $CH_3-CH_2CH_2CH_3$

d. $H_2C=CHCH_2CH_3$ $\xrightarrow[\text{2. Zn, H}^+]{\text{1. O}_3}$ $O=CHCH_2CH_3$ + $H_2C=O$

e. Follow Markovnikov's rule:

$H_2C=CHCH_2CH_3$ + $H-OH$ $\xrightarrow{H^+}$ $CH_3-\underset{\underset{OH}{|}}{C}HCH_2CH_3$

f. $H_2C=CHCH_2CH_3$ $\xrightarrow[\text{2. H}_2\text{O}_2\text{, HO}^-]{\text{1. B}_2\text{H}_6}$ $HOCH_2-CH_2CH_2CH_3$

The product is the regioisomer of that obtained by acid-catalyzed addition of water (part e).

g. $H_2C=CHCH_2CH_3$ $\xrightarrow[\text{HO}^-]{\text{KMnO}_4}$ $\underset{\underset{OH}{|}}{C}H_2-\underset{\underset{OH}{|}}{C}HCH_2CH_3$

h. $H_2C=CHCH_2CH_3$ + $6 O_2$ $\longrightarrow$ $4 CO_2$ + $4 H_2O$

3.39 To work this kind of problem, try to locate (on adjacent carbons) the atoms or groups that must have come from the small molecule or reagent. Then remove them from the structure and insert the multiple bond appropriately.

a. $CH_3CH=CHCH_3$ + Br_2

b. $CH_2=C(CH_3)_2$ + HOH + H^+

c. $CH_2=CHCH_3$ + $H-OSO_3H$

d.

+ $H-Br$

e. $CH_2=CH-CH=CH_2$ + $H-Br$ (1,4-addition)

f. $CH_3C\equiv CCH_3$ + $2 Cl_2$

g.

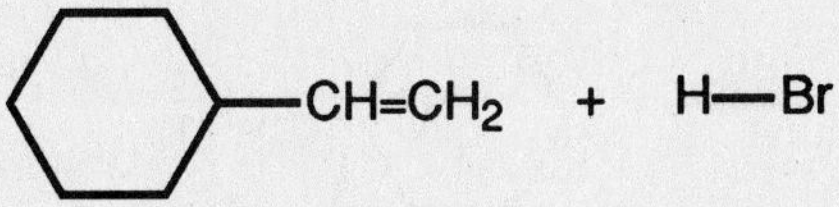

$$\text{(cyclohexyl)}-CH=CH_2 \quad + \quad H-Br$$

3.40 If the saturated hydrocarbon contained no rings, it would have the molecular formula $C_{15}H_{32}$. Since their are fewer hydrogens in $C_{15}H_{28}$, it must have two rings (two hydrogens are deleted per ring). Since caryophyllene absorbed 2 moles of H_2 ($C_{15}H_{24} + 2H_2 \rightarrow C_{15}H_{28}$), it must also have two double bonds or one triple bond. The structure of caryophyllene is shown below:

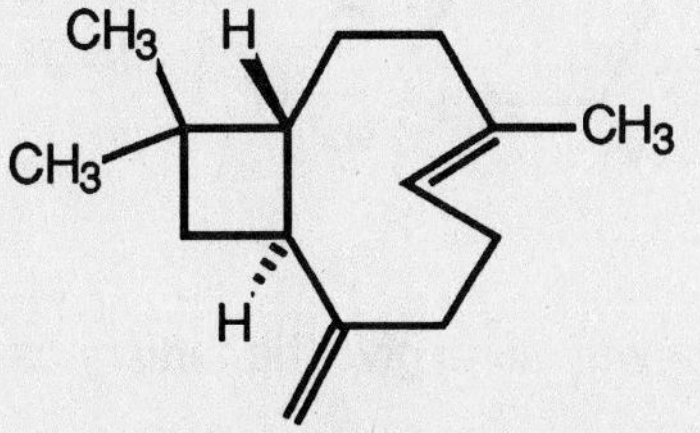

3.41 a. Electrophile; HCl can donate a proton to a nucleophile.
 b. Nucleophile; Cl^- can donate an electron pair to an electrophile.
 c. Electrophile; hydronium ion is a proton donor.
 d. Electrophile; the aluminum has only a sextet of electrons. It is electron-deficient and can accept an electron pair.
 e. Nucleophile; the negatively charged oxygen reacts with electrophiles.

3.42 Water can donate a proton to a nucleophile:

or an oxygen lone pair to an electrophile:

63

3.43

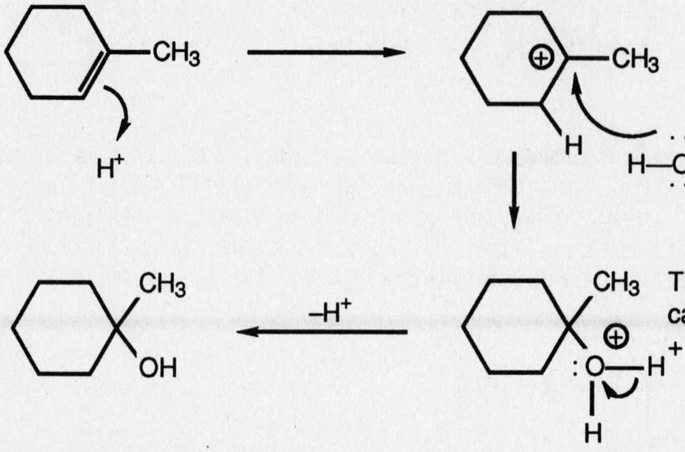

The oxygen carries a formal +1 charge.

In the first step, the proton adds in such a way as to give the tertiary carbocation intermediate.

3.44

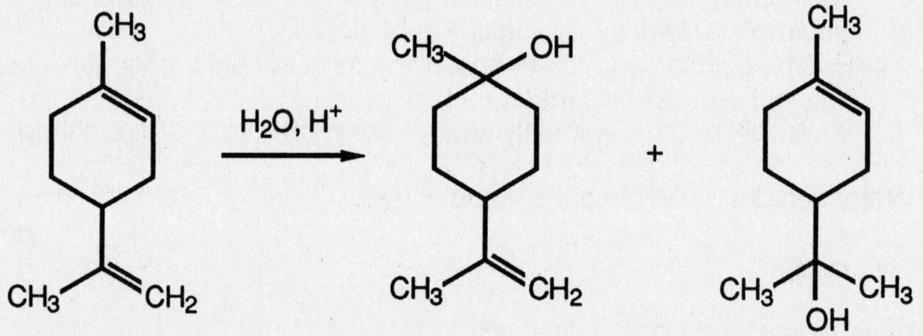

In each case, water adds according to Markovnikov's rule (via a tertiary carbonium ion). The diol is:

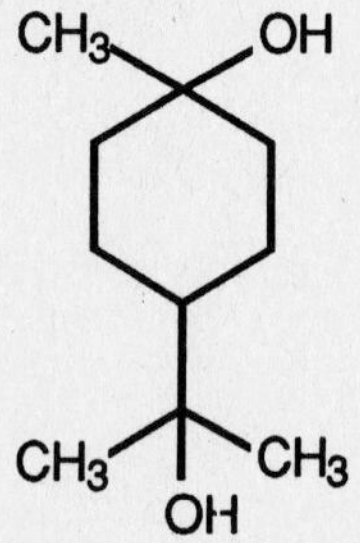

ANSWERS TO PROBLEMS

3.45 The ion has a symmetric structure.

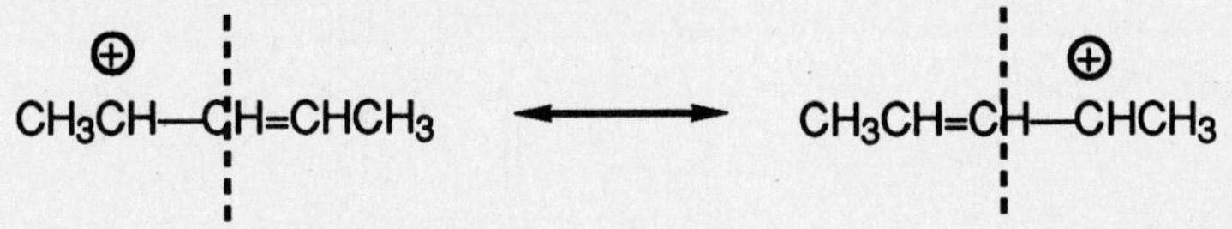

$$\overset{\oplus}{CH_3CH}-CH=CHCH_3 \quad \longleftrightarrow \quad CH_3CH=CH-\overset{\oplus}{CH}CH_3$$

3.46 $CH_3CH=CH-CH=CHCH_3 \xrightarrow{HBr}$

$$CH_3CH_2CH-CH=CHCH_3$$
$$\qquad\qquad\;|$$
$$\qquad\qquad Br$$

1,2-addition (4-bromo-2-hexene)

\+

$$CH_3CH_2CH=CH-CHCH_3$$
$$\qquad\qquad\qquad\;\;|$$
$$\qquad\qquad\qquad Br$$

1,4-addition (2-bromo-3-hexene)

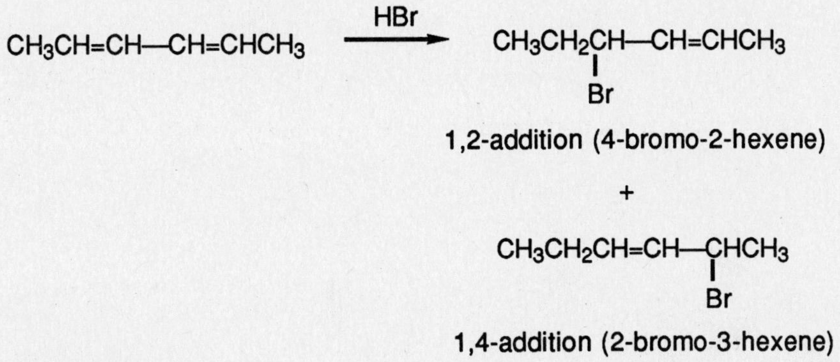

<u>Mechanism:</u>

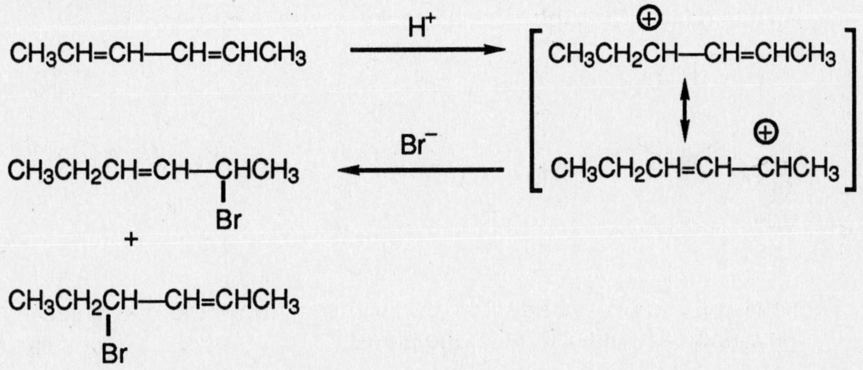

$$CH_3CH=CH-CH=CHCH_3 \xrightarrow{H^+} \left[\begin{array}{l} \overset{\oplus}{CH_3CH_2CH}-CH=CHCH_3 \\ \updownarrow \\ CH_3CH_2CH=CH-\overset{\oplus}{CH}CH_3 \end{array} \right]$$

$$CH_3CH_2CH=CH-CHCH_3 \xleftarrow{Br^-}$$
$$\qquad\qquad\qquad |$$
$$\qquad\qquad\qquad Br$$

\+

$$CH_3CH_2CH-CH=CHCH_3$$
$$\qquad\;\;|$$
$$\qquad Br$$

In the final step, the nucleophile Br⁻ can react with the allylic carbocation at either of the two positive carbons.

3.47

a.

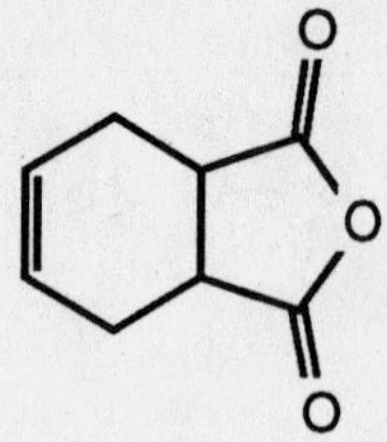

b.

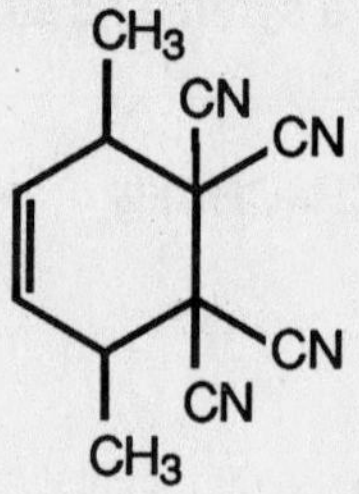

3.48 In each case, work backward from the cyclohexene double bond:

a.

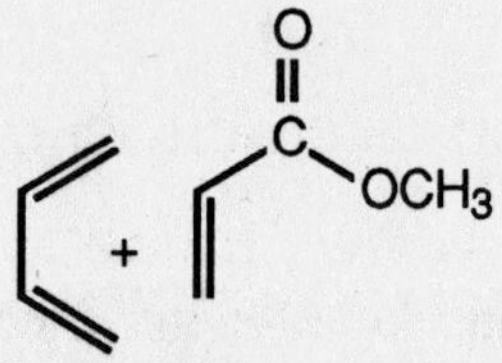

b.

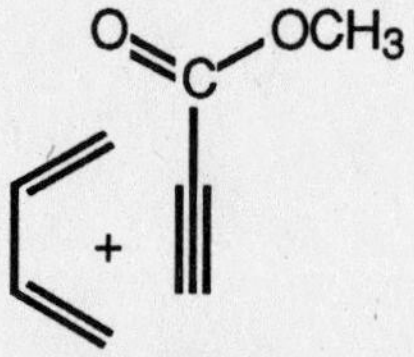

An alternative, although less satisfactory, answer for part b is

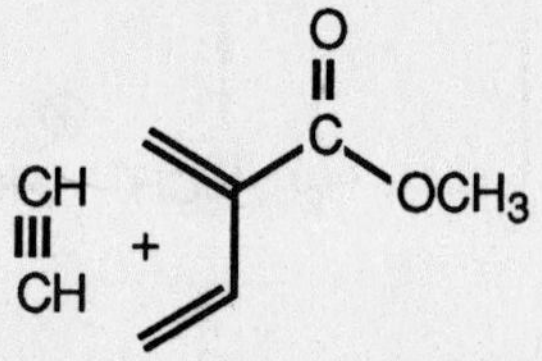

However, it is usually better to start with electron-withdrawing substituents on the dienophile rather than on the diene.

3.49 Each radical adds to a molecule of propene to give the more stable radical (secondary rather than primary):

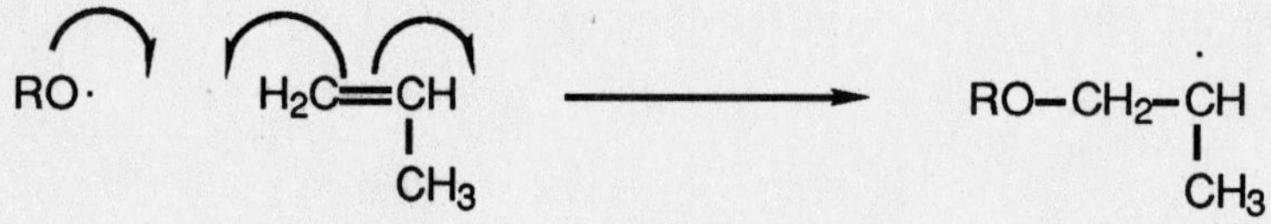

R—O—CH$_2$—CH   H$_2$C=CH $\longrightarrow$ ROCH$_2$CH—CH$_2$—CH
| | | |
CH$_3$ CH$_3$ CH$_3$ CH$_3$

and so on.

$+$CH$_2$—CH$+_n$ polypropylene
        |
       CH$_3$

3.50 a. 3 H$_2$C=C—CH$_2$CH$_3$ $\xrightarrow{\text{BH}_3}$ B$+$CH$_2$CHCH$_2$CH$_3$$)_3$
               | |
              CH$_3$ CH$_3$

H$_2$O$_2$ | NaOH

3 HOCH$_2$CHCH$_2$CH$_3$
            |
           CH$_3$

b. 3 $\xrightarrow{\text{BH}_3}$ $\left(\text{H}\cdots\cdots\cdots\right)_3$:B $\xrightarrow[\text{NaOH}]{\text{H}_2\text{O}_2}$ 3

In the first step, the boron and hydrogen add to the same face of the double bond, in a *cis* manner. In the second step, the OH takes the identical position occupied by the boron. Thus, in the product the H and OH groups are *cis*.

3.51

a.   =CH$_2$ + H—OH $\xrightarrow{\text{H}^+}$

Addition follows Markovnikov's rule, and the reaction proceeds via a tertiary carbocation intermediate.

3. ALKENES AND ALKYNES

b.

The boron adds to the less substituted, or CH_2, carbon. In the second step, the boron is replaced by the hydroxyl group.

3.52 Cyclohexene will rapidly decolorize a dilute solution of bromine in carbon tetrachloride (Sec. 3.9a) and will be oxidized by potassium permanganate, resulting in a color change from the purple of $KMnO_4$ to the brown solid MnO_2 (Sec. 3.16a). Cyclohexane, being saturated, does not react with either of these reagents.

3.53 The alkene that gave the particular aldehyde or ketone can be deduced by joining the two carbons attached to oxygens by a C=C double bond:

a. $CH_3CH_2CH=CHCH_2CH_3$ b. $(CH_3)_2C=CHCH_3$

c. $CH_2=CHCH(CH_3)_2$ d. cyclobutene

In the case of compound a, where *cis* and *trans* isomers are possible, either isomer gives the same ozonolysis products.

3.54 a. $CH_3C\equiv CCH_2CH_3 \ + \ 2 \ Cl_2 \longrightarrow$

b. $CH_3CH_2C\equiv CCH_2CH_3 \ + \ H_2 \xrightarrow{\text{Lindlar's catalyst}}$

This catalyst limits the addition to 1 mole of H_2, which adds in a *cis* manner.

c. $CH_3C\equiv CH \ + \ H_2O \xrightarrow[HgSO_4]{H^+}$

Compare with eq. 3.49, where R = CH_3.

d. $CH_3C\equiv CH \ + \ NaNH_2 \longrightarrow CH_3C\equiv C^- \ Na^+ \ + \ NH_3$

Compare with eq. 3.50, where R = CH_3.

ANSWERS TO PROBLEMS

3.55   a. 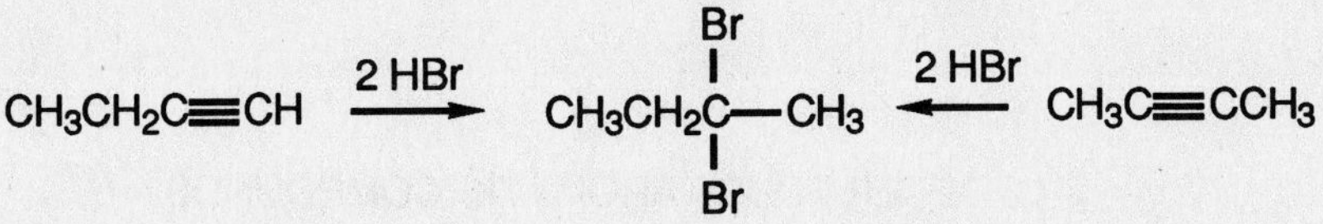 $CH_3CH_2C{\equiv}CH$ $\xrightarrow{\text{2 HBr}}$ $CH_3CH_2\overset{\displaystyle Br}{\underset{\displaystyle Br}{C}}\!-\!CH_3$ $\xleftarrow{\text{2 HBr}}$ $CH_3C{\equiv}CCH_3$

Either 1- or 2-butyne will add HBr in the manner shown.

b. 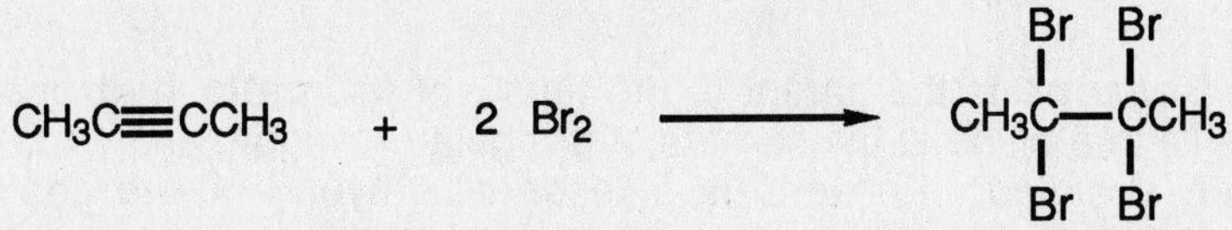 $CH_3C{\equiv}CCH_3$ + $2\ Br_2$ $\xrightarrow{\hspace{2cm}}$ $CH_3\overset{\displaystyle Br}{\underset{\displaystyle Br}{C}}\!-\!\overset{\displaystyle Br}{\underset{\displaystyle Br}{C}}CH_3$

The triple bond must be between C-2 and C-3 if the bromines are to be attached to those carbons in the product.

CHAPTER FOUR: AROMATIC COMPOUNDS

CHAPTER SUMMARY

Benzene is the parent of the family of **aromatic hydrocarbons.** Its six carbons lie in a plane at the corners of a regular hexagon, and each carbon has one hydrogen attached. Benzene is a resonance hybrid of two contributing Kekulé structures:

In orbital terms, each carbon is sp^2-hybridized. These orbitals form σ bonds to the hydrogen and the two neighboring carbons, all in the ring plane. A p orbital at each carbon is perpendicular to this plane, and the six electrons, one from each carbon, form an electron cloud of π orbitals that lie above and below the ring plane.

The bond angles in benzene are 120°. All C—C bond distances are equal (1.39 Å). The compound is more stable than either of the contributing Kekulé structures and has a **resonance** or **stabilization energy** of about 36 kcal/mol.

The nomenclature of benzene derivatives is described in Sec. 4.7. Common names and structures to be memorized include those of **toluene**, **styrene, phenol, aniline,** and **xylene.** Monosubstituted benzenes are named as benzene derivatives (bromobenzene, nitrobenzene, and so on). Disubstituted benzenes are named as *ortho*-(1,2-), *meta*- (1,3-), or *para*- (1,4-), depending on the relative positions of the substituents on the ring. Two important groups are **phenyl** (C_6H_5—) and **benzyl** ($C_6H_5CH_2$—).

Aromatic compounds react mainly by **electrophilic aromatic substitution**, in which one or more ring hydrogens are replaced by various electrophiles. Typical reactions are **chlorination, bromination, nitration, sulfonation, alkylation,** and **acylation** (the Friedel–Crafts reactions). The mechanism involves two steps: addition of the electrophile to a ring carbon, to produce an intermediate **benzenonium ion**, followed by proton loss to again achieve the (now substituted) aromatic system.

REACTION SUMMARY

Substituents already present on the ring affect the rate of further substitution and the position taken by the next substituent. Most groups are either *meta-directing* and *ring-deactivating* or *ortho,para-directing and ring-activating* (Table 4.1). An exception is the halogens, which are *ortho,para-directing* but *ring-deactivating*. These effects must be taken into account in devising syntheses of aromatic compounds.

Benzene is a major commercial chemical and is a source of styrene, phenol, other aromatics, acetone, and cyclohexane.

Polycyclic aromatic hydrocarbons, which are built of *fused* benzene rings, include **naphthalene, anthracene,** and **phenanthrene**. Some, such as benzo[a]pyrene, are carcinogens. Graphite, a common allotrope of carbon, consists of layers of planar, hexagonal rings separated by 3.4Å.

REACTION SUMMARY

Electrophilic Aromatic Substitution

Halogenation

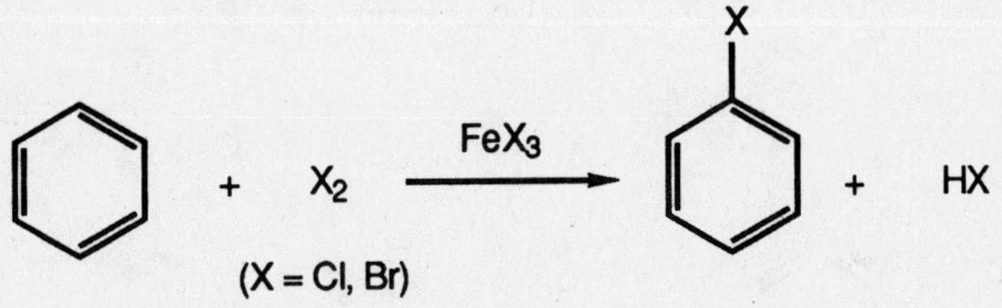

(X = Cl, Br)

Nitration

nitric acid   nitrobenzene

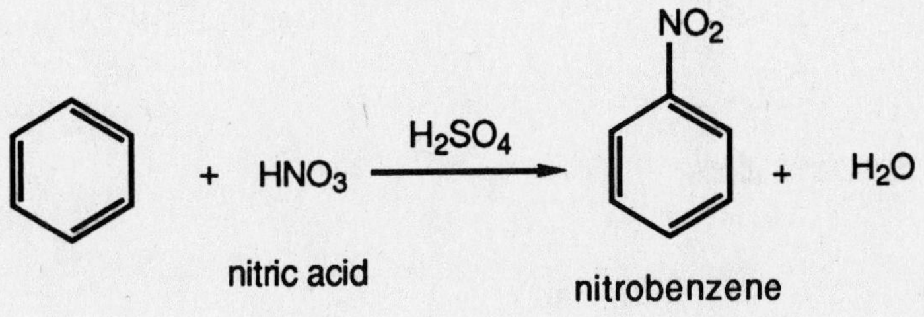

4. AROMATIC COMPOUNDS

Sulfonation

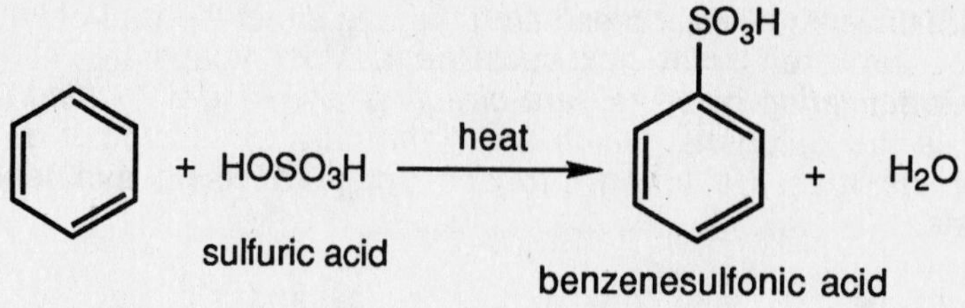

benzene + HOSO$_3$H $\xrightarrow{\text{heat}}$ benzenesulfonic acid (SO$_3$H) + H$_2$O

sulfuric acid

benzenesulfonic acid

Alkylation

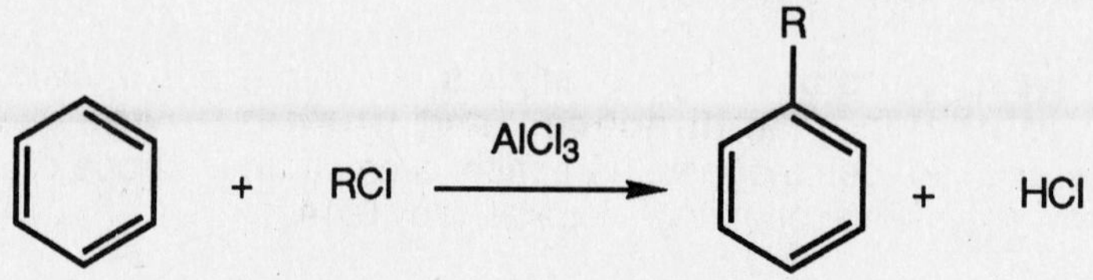

benzene + RCl $\xrightarrow{\text{AlCl}_3}$ R-substituted benzene + HCl

Friedel–Crafts reaction; R = an alkyl group

Alkylation

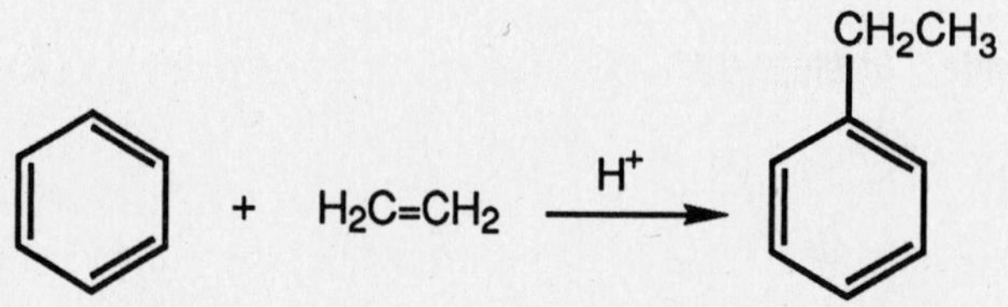

benzene + H$_2$C=CH$_2$ $\xrightarrow{\text{H}^+}$ ethylbenzene (CH$_2$CH$_3$)

Acylation

benzene + R–C(=O)Cl $\xrightarrow{\text{AlCl}_3}$ phenyl ketone (O=C–R) + HCl

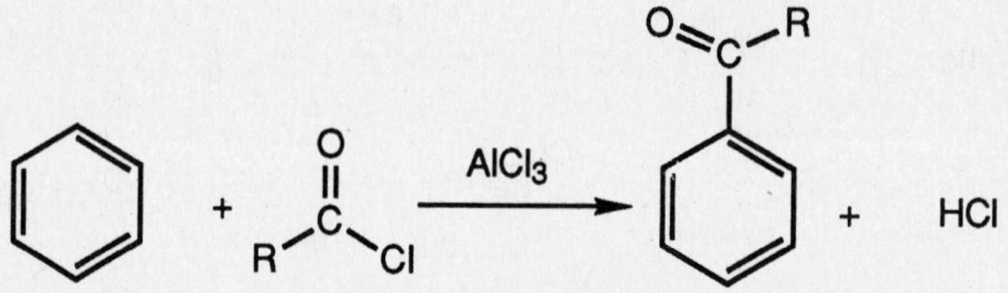

Catalytic Hydrogenation

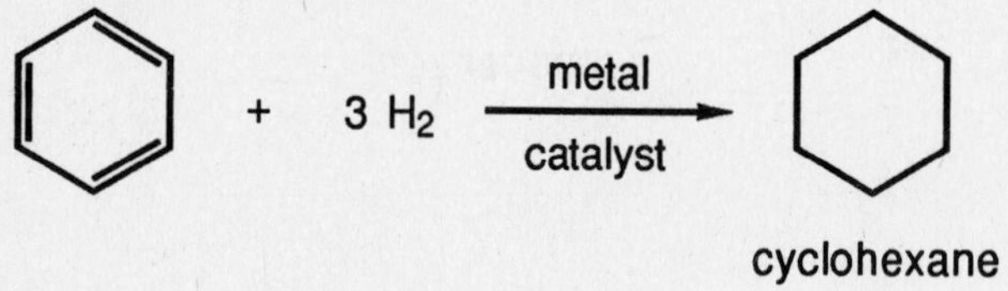

benzene + 3 H$_2$ $\xrightarrow[\text{catalyst}]{\text{metal}}$ cyclohexane

cyclohexane

ANSWERS TO PROBLEMS

MECHANISM SUMMARY

Electrophilic Aromatic Substitution

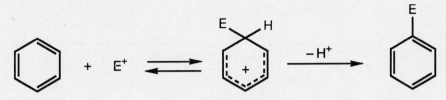

benzenonium ion

LEARNING OBJECTIVES

1. Know the meaning of: Kekulé structure, benzene resonance hybrid, resonance or stabilization energy.

2. Know the meaning of: *ortho, meta, para,* phenyl group (C_6H_5— or Ph—), benzyl group ($C_6H_5CH_2$—), aryl group (Ar—), benzene, toluene, styrene, phenol, aniline, xylene, arene.

3. Know the meaning of: electrophilic aromatic substitution, halogenation, nitration, sulfonation, alkylation, acylation, Friedel-Crafts reaction.

4. Know the meaning of: benzenonium ion, *ortho,para*-directing group, *meta*-directing group, ring-activating substituent, ring-deactivating substituent.

5. Know the meaning of: polycyclic aromatic hydrocarbon, naphthalene, anthracene, phenanthrene, carcinogenic, graphite.

6. Name and write the structures for aromatic compounds, especially mono-substituted and disubstituted benzenes and toluenes.

7. Given the reactants, write the structures of the main organic products of the common electrophilic aromatic substitution reactions (halogenation, nitration, sulfonation, alkylation, and acylation).

8. Write the steps in the mechanism for an electrophilic aromatic substitution reaction.

9. Draw the structures of the main contributors to the benzenonium ion resonance hybrid.

10.   Draw the structures of the main contributors to substituted benzenonium ions, and tell whether the substituent stabilizes or destabilizes the ion.

11.   Know which groups are *ortho,para*-directing, which are *meta*-directing, and explain why each group directs the way it does.

12.   Know which groups are ring-activating, ring-deactivating, and explain why each group effects the rate of electrophilic aromatic substitution as it does.

13.   Given two successive electrophilic aromatic substitution reactions, write the structure of the product, with substituents in the correct locations on the ring.

14.   Given a disubstituted or trisubstituted benzene, deduce the correct sequence in which to carry out electrophilic substitutions to give the product with the desired orientation.

ANSWERS TO PROBLEMS

Problems Within the Chapter

4.1   The formula C_6H_6 (or C_nH_{2n-6}) corresponds to any of the following possibilities: two triple bonds; one triple bond and two double bonds; one triple bond, one double bond, and one ring; one triple bond and two rings; four double bonds; three double bonds and one ring; two double bonds and two rings; one double bond and three rings; four rings. Obviously there are very many possibilities. One example from each of the above categories is shown below.

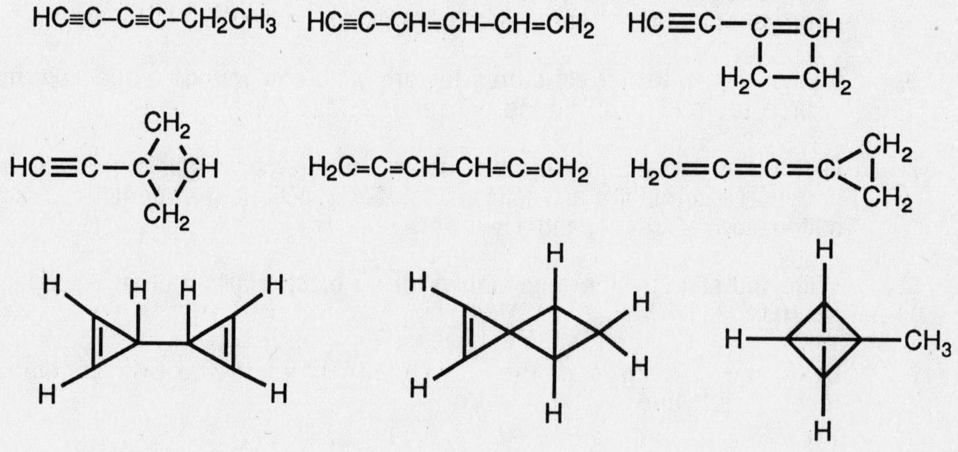

4.2 eq. 4.2:

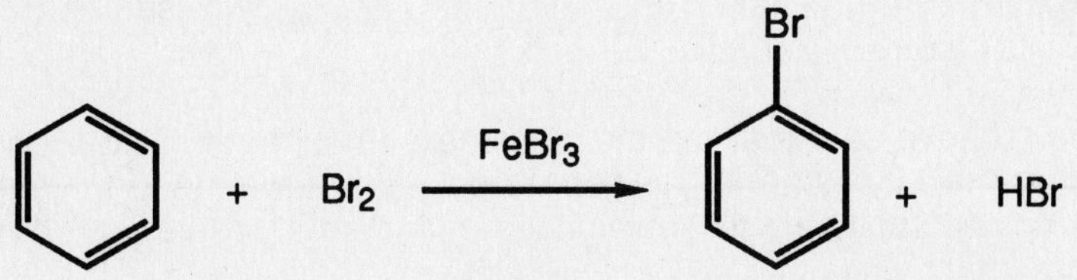

only one possible structure

eq. 4.4:

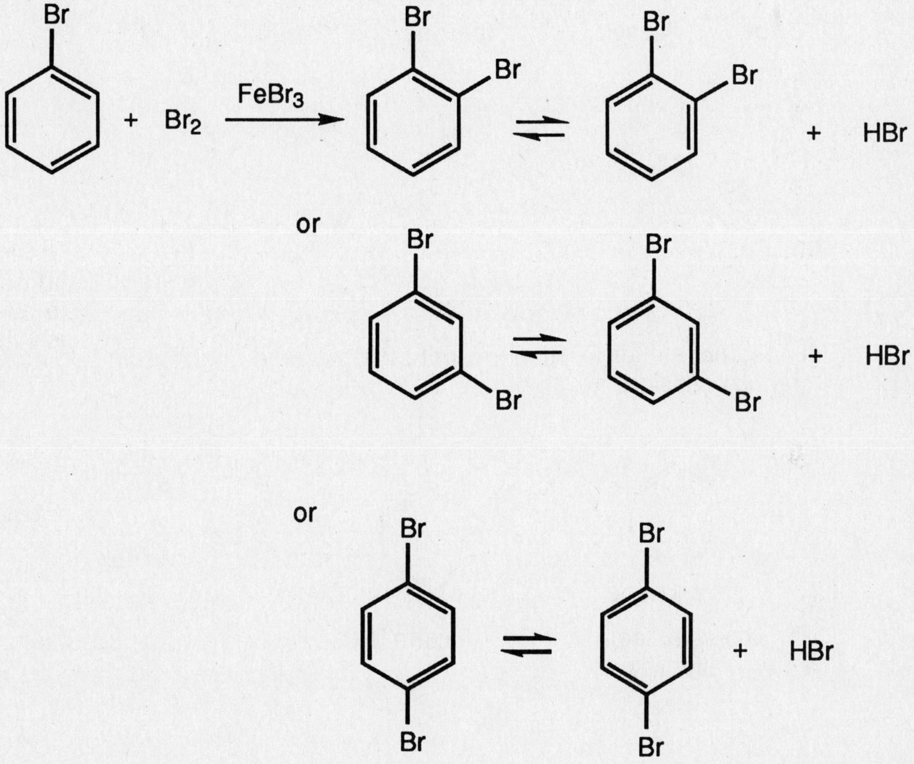

or

or

Yes, Kekulé's explanation accounts for one monobromobenzene and three dibromobenzenes, if the equilibrium between the two Kekulé structures for each dibromo isomer is taken into account. As you will see in Sec. 4.4, however, this picture is not quite correct.

4.3 Kekulé would have said that the structures are in such rapid equilibrium with one another that they cannot be separated. We know now, however, that there is only one such structure, which is not accurately represented by either Kekulé formula.

4.4

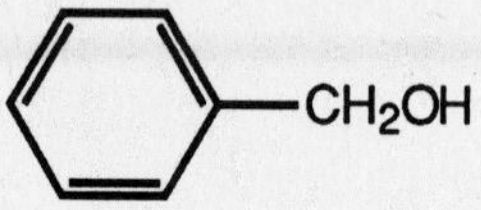

benzyl alcohol 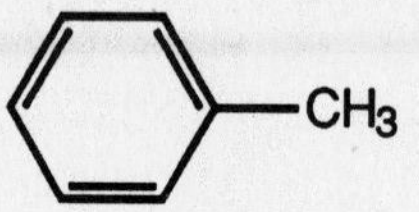   toluene 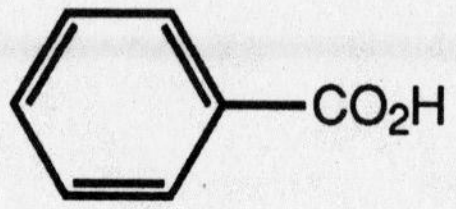   benzoic acid

4.5

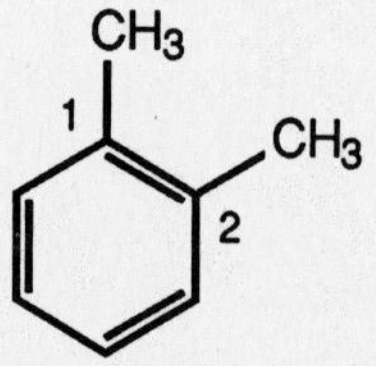

or

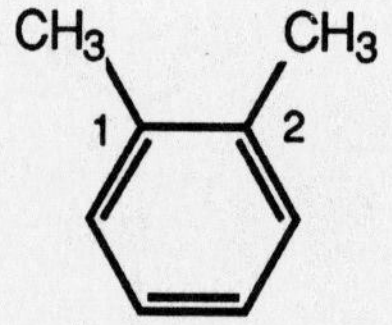

ortho-xylene

It does not matter whether we write the benzene ring standing on one corner or lying on a side.

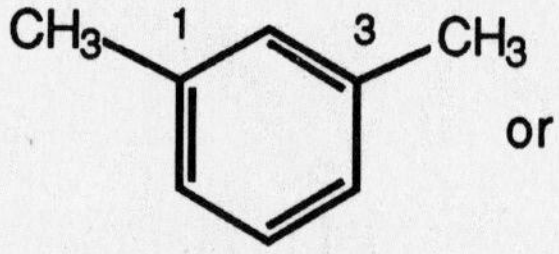

or 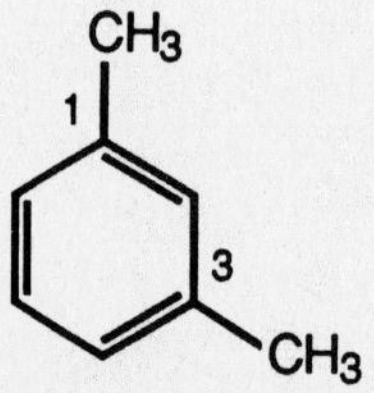   or

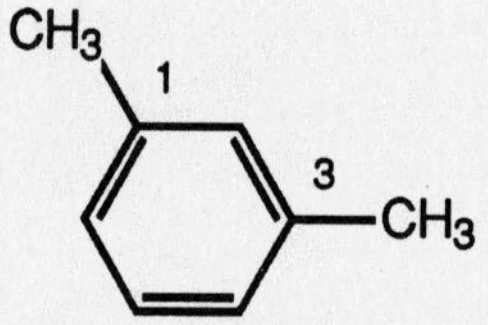

meta-xylene

The important feature of the structure is that the two methyl substituents are in a 1,3-relationship.

4.6   a.

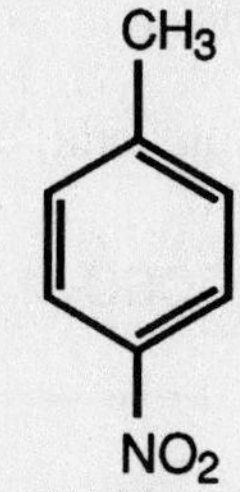

CH$_3$ / NO$_2$ (para)

b.

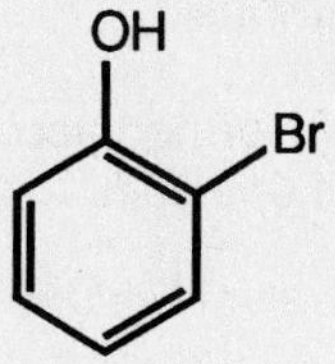

OH, Br (ortho)

c.

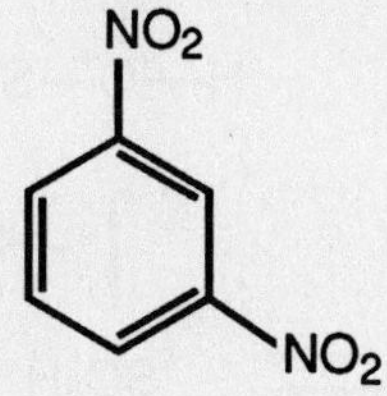

NO$_2$, NO$_2$ (meta)

d.

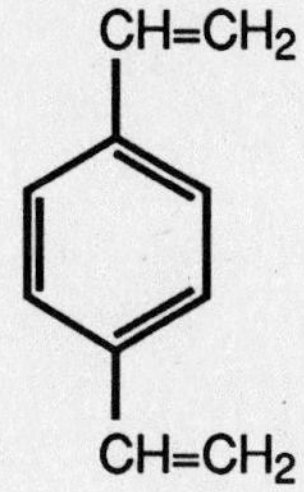

CH=CH$_2$ / CH=CH$_2$ (para)

4.7   a.

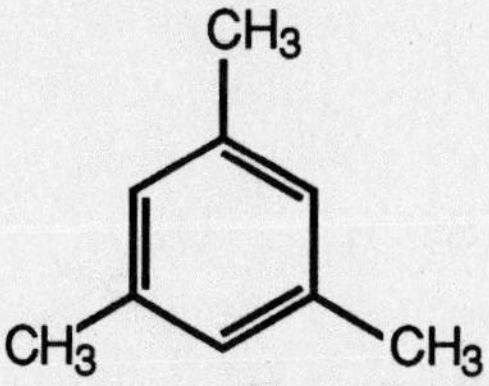

CH$_3$, CH$_3$, CH$_3$

b.

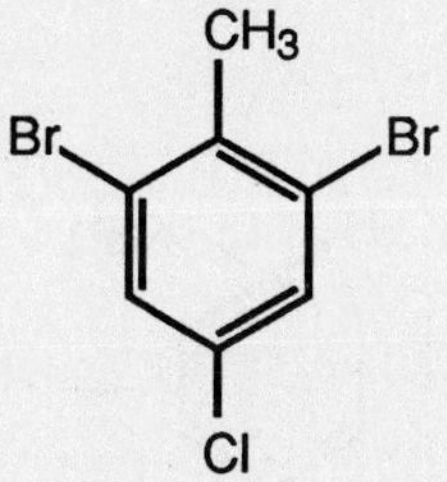

CH$_3$, Br, Br, Cl

4.8   a.

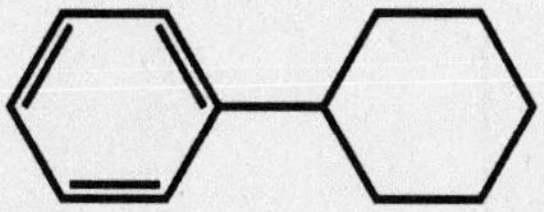

b.

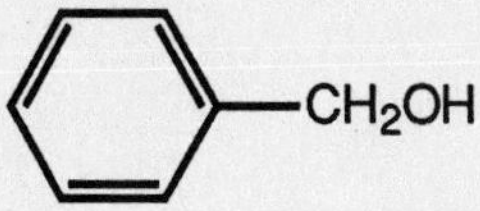

CH$_2$OH

c.

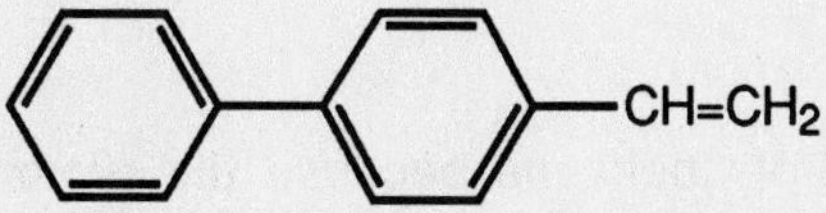

CH=CH$_2$

d.

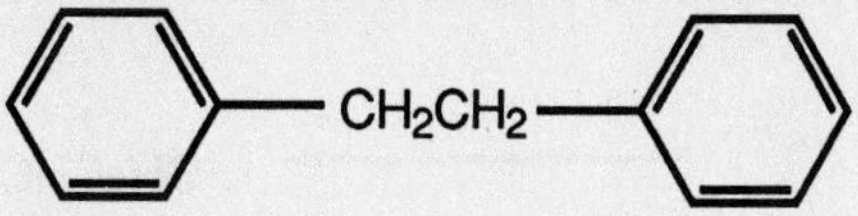

CH$_2$CH$_2$

77

4.9   a. phenylcyclopentane   b. *o*-benzylaniline

4.10   The electrophile is formed according to the following equilibria, beginning with the protonation of one sulfuric acid molecule by another:

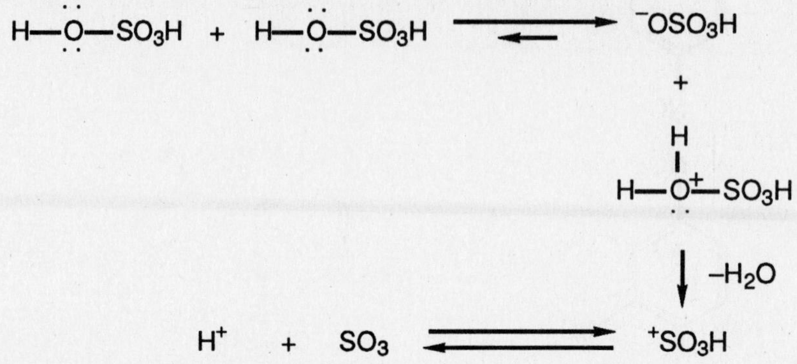

$$H-\ddot{O}-SO_3H \;+\; H-\ddot{O}-SO_3H \;\rightleftharpoons\; {}^-OSO_3H$$

$$+$$

$$\begin{array}{c} H \\ | \\ H-\overset{\pm}{\underset{\cdot\cdot}{O}}-SO_3H \end{array}$$

$$\downarrow -H_2O$$

$$H^+ \;+\; SO_3 \;\rightleftharpoons\; {}^+SO_3H$$

Using $^+SO_3H$ as the electrophile, we can write the sulfonation mechanism as follows:

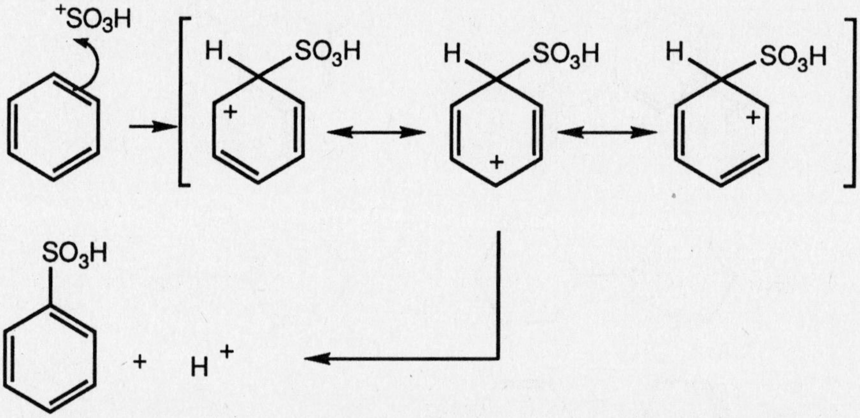

4.11   The product would be isopropylbenzene because the proton of the acid catalyst would add to propylene according to Markovnikov's rule to give the isopropyl cation:

$$CH_3CH{=}CH_2 \;+\; H^+ \;\longrightarrow\; CH_3\overset{+}{C}HCH_3$$

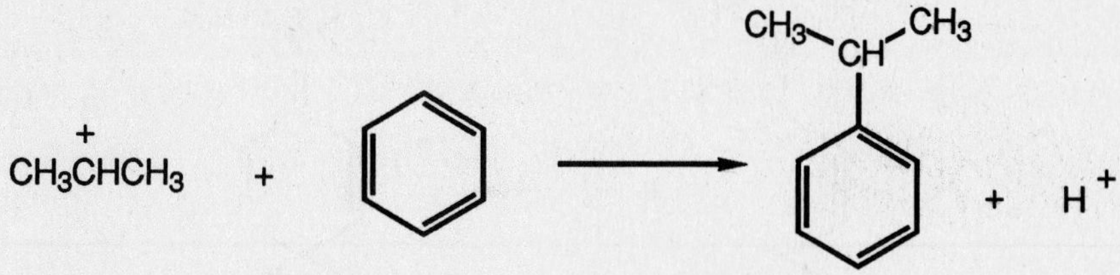

4.12   *ortho*

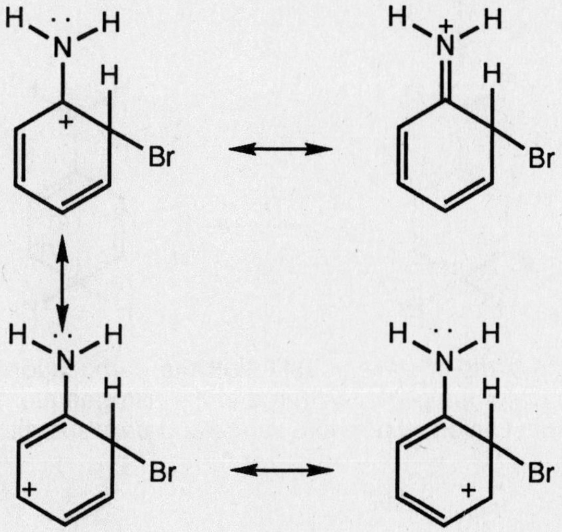

meta

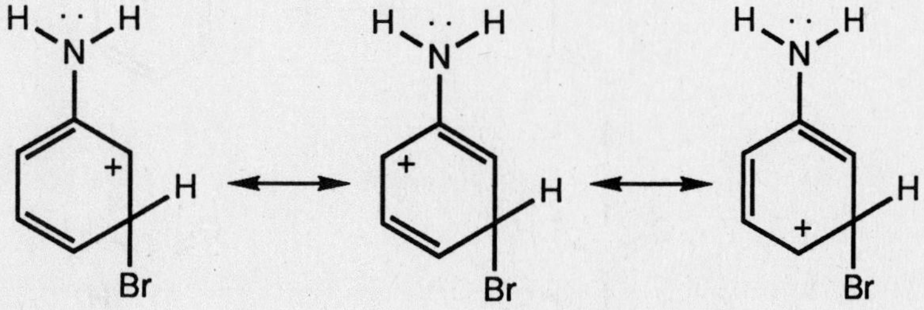

para

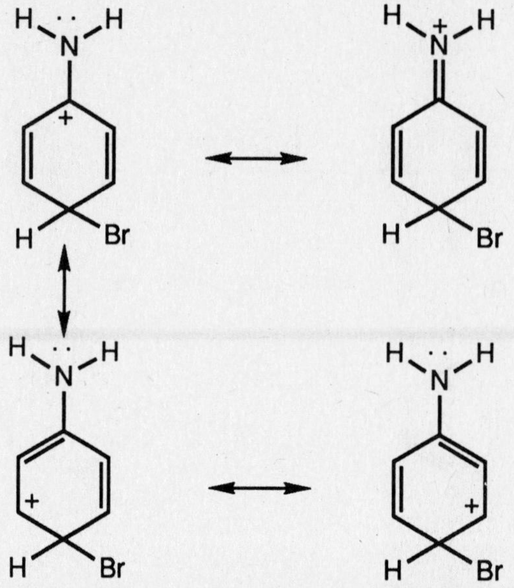

Note that in *ortho* or *para* substitution the carbocation can be stabilized by delocalization of the positive charge to the nitrogen atom. This is not possible for *meta* substitution. Therefore, *ortho* and *para* substitution are preferred.

4.13   *ortho*

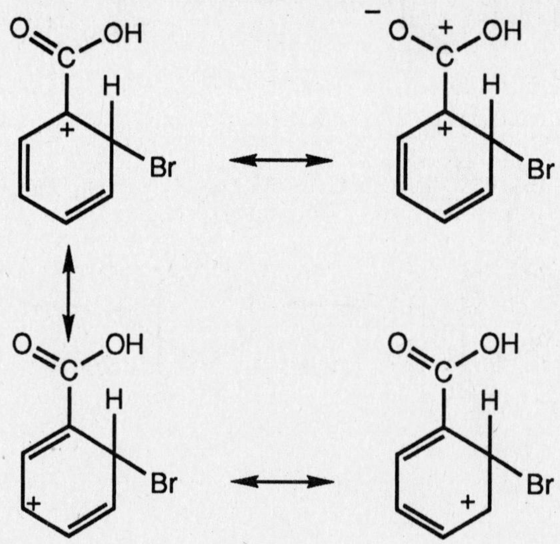

meta

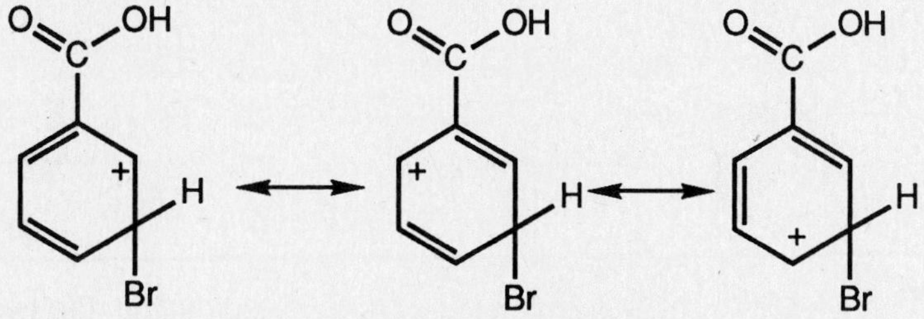

para

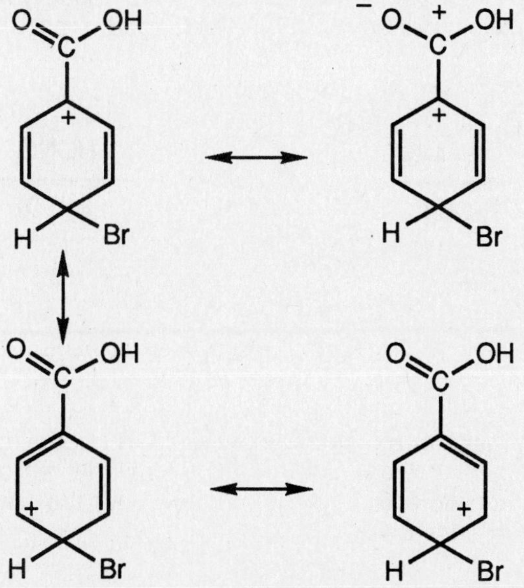

Note that in *ortho* or *para* substitution the positive charge of the benzenonium ion is adjacent to the partially positive carboxyl carbon. It is energetically unfavorable to have two adjacent like charges. In *meta* substitution, this unfavorable possibility does not exist. Thus, *meta* substitution predominates.

4.14   a.   Since the substituents are *meta* in the product, we must introduce the *meta*-directing substituent first:

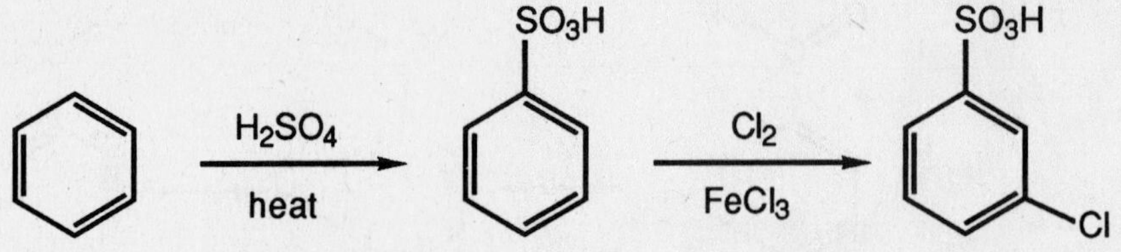

b. The methyl substituent is *ortho,para*-directing; the two isomers obtained in the final step would have to be separated. Usually the *para* isomer, in which the two substituents are furthest apart, predominates.

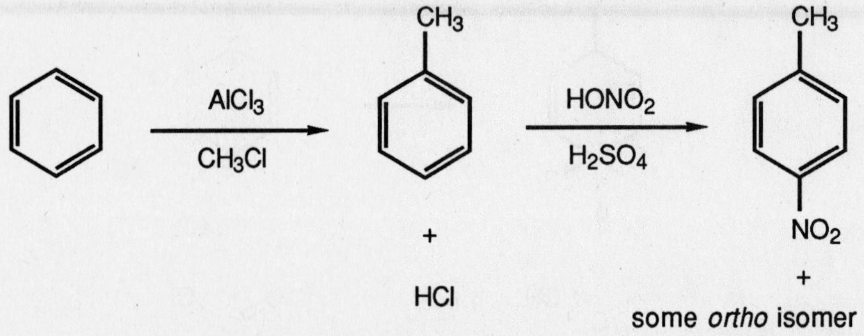

4.15 We could not make *m*-bromochlorobenzene in good yield this way because both Br and Cl are *ortho,para*-directing. Similarly, we could not prepare *p*-nitrobenzenesulfonic acid directly because both the nitro and sulfonic acid substituents are *meta*-directing.

4.16 Only two such contributors are possible:

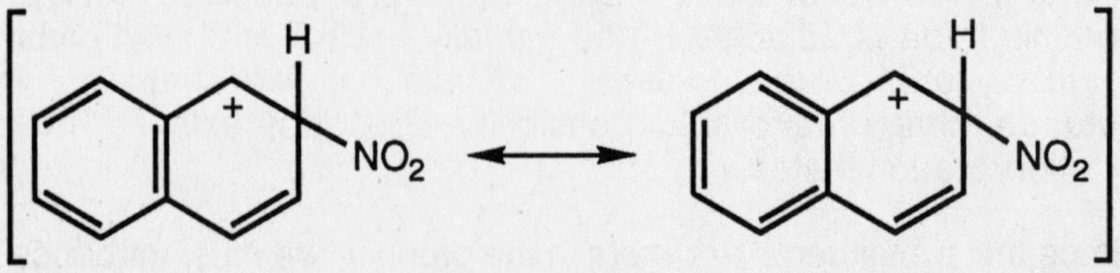

Any additional resonance contributors disrupt the benzenoid structure in the "left" ring. Since the intermediate carbocation for nitration of naphthalene at C-1 is more stable, substitution at that position is preferred.

4.17 The carbon-to-hydrogen ratios are: benzene = 1; naphthalene = 1.25; anthracene = 1.40; pyrene = 1.60. The percentage of carbon in a structure increases with the number of fused aromatic rings.

ANSWERS TO PROBLEMS

Additional Problems

4.18   a.

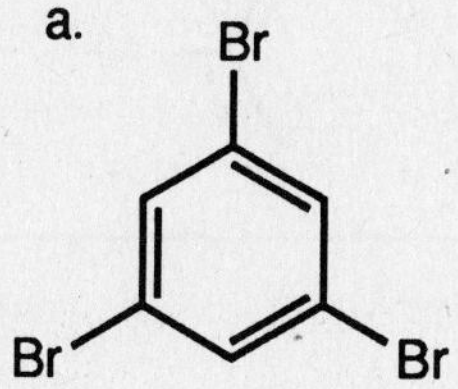

b.

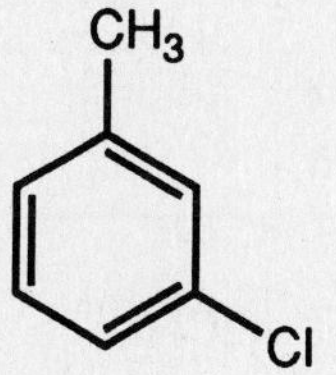

c.

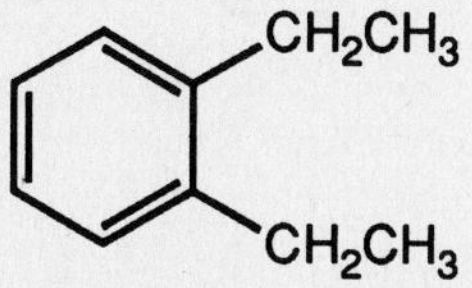

d.

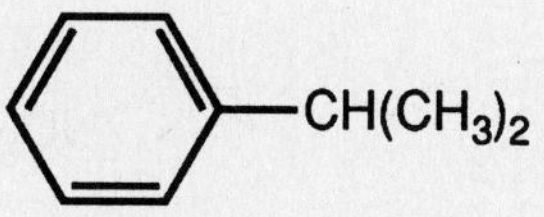

e.

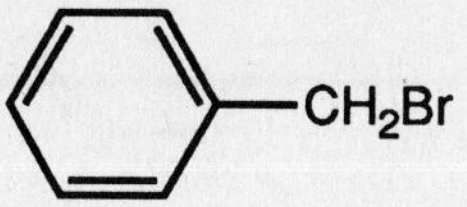

f.

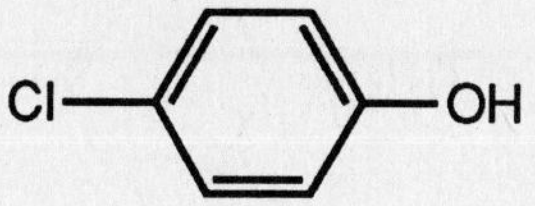

g.

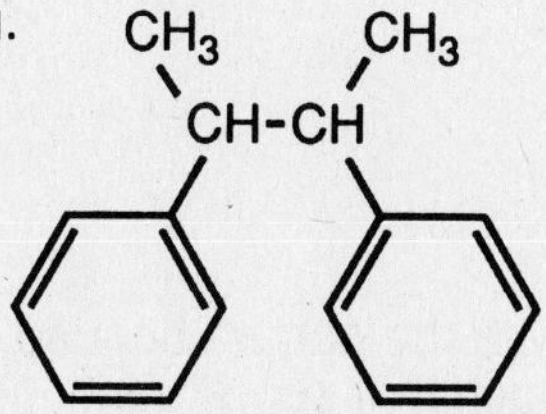

h.

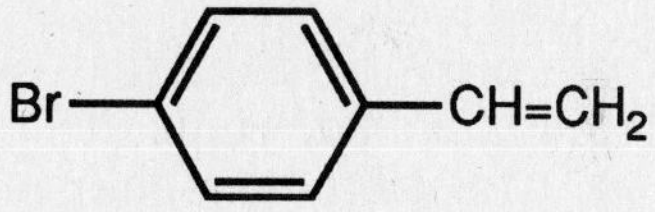

i.

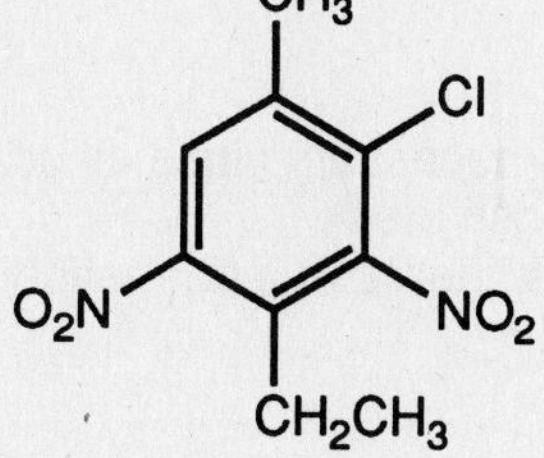

j.

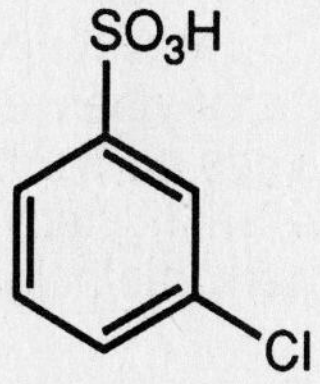

k.

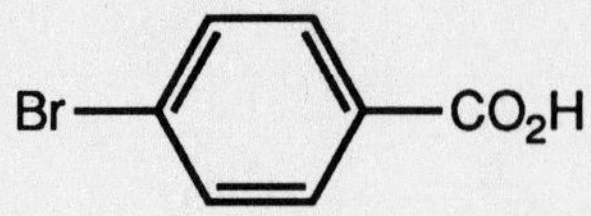

Br—⟨benzene ring⟩—CO_2H

l.

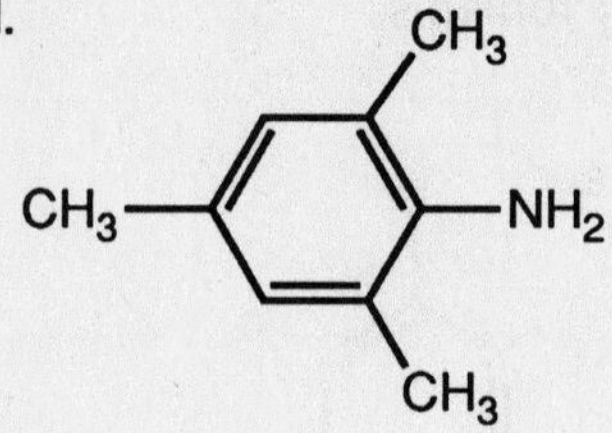

CH_3 / CH_3 / ⟨benzene ring⟩—NH_2 / CH_3

m.

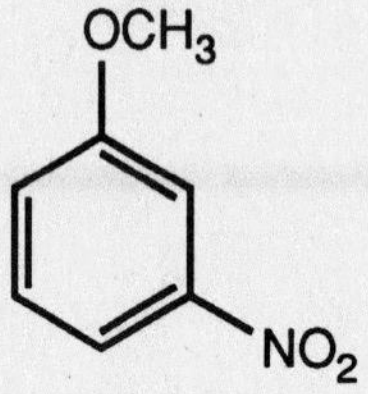

OCH_3 / ⟨benzene ring⟩ / NO_2

n.

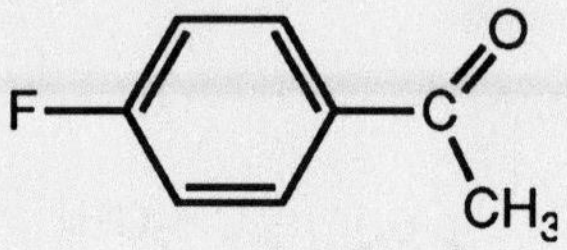

F—⟨benzene ring⟩—C(=O)—CH_3

o.

CH_3 / CH_3—⟨benzene ring⟩—CHO

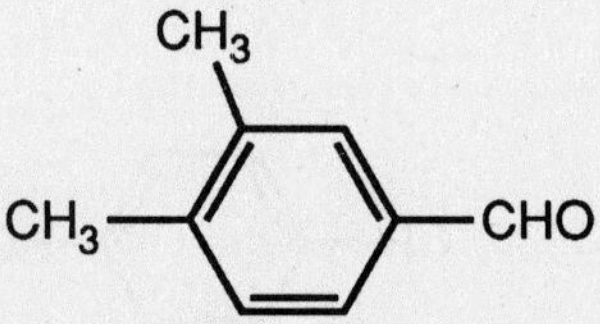

p.

Cl / ⟨benzene ring⟩—C(=O)—OH

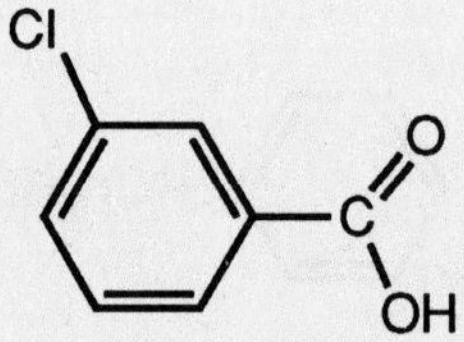

4.19   a.   propylbenzene (or 1-phenylpropane)
       b.   *m*-bromochlorobenzene (alphabetical order)
       c.   1,8-dibromonaphthalene
       d.   2,5-dichlorotoluene (start numbering with the carbon bonded to the
            methyl group and go around the ring such that substituents get the
            lowest possible numbers)
       e.   *p-tert*-butylphenol
       f.   *o*-chlorotoluene
       g.   3,5-dibromostyrene
       h.   hexamethylbenzene (no numbers are necessary since all possible
            positions on the benzene ring are substituted)
       i.   1-methyl-1-phenylcyclopropane (substituents in alphabetical order)
       j.   triphenylmethane

4.20   a.

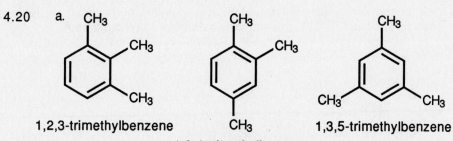

1,2,3-trimethylbenzene   1,3,5-trimethylbenzene

1,2,4-trimethylbenzene

b.

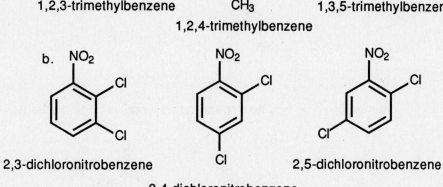

2,3-dichloronitrobenzene   2,5-dichloronitrobenzene

2,4-dichloronitrobenzene

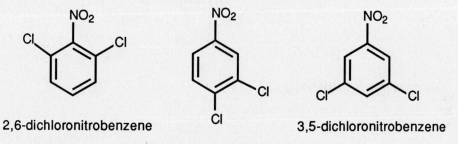

2,6-dichloronitrobenzene   3,5-dichloronitrobenzene

3,4-dichloronitrobenzene

4.21   The three possible structures are:

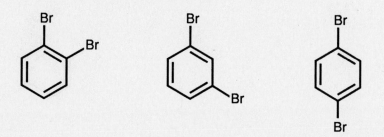

4. AROMATIC COMPOUNDS

Only in the *para* isomer are all four remaining hydrogens equivalent. Therefore, it can give only *one* mononitro derivative and must be A:

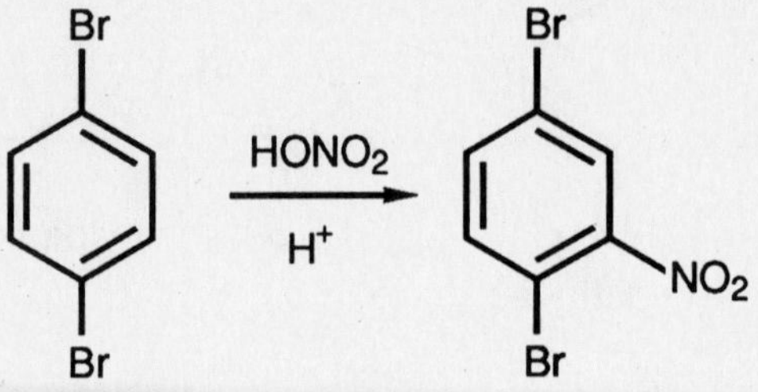

A (mp 87°C)

The *ortho* isomer can give only *two* mononitration products and must be B:

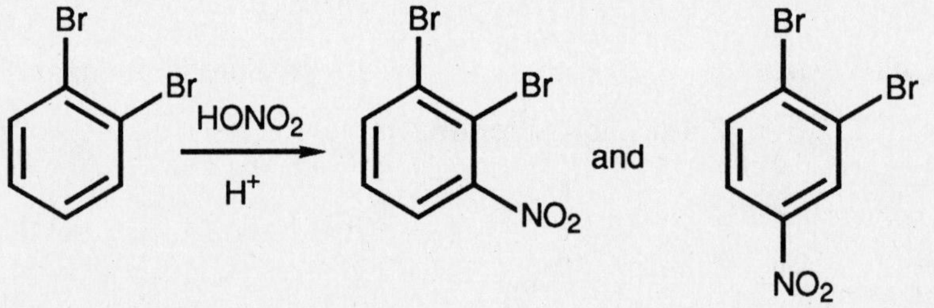

and

The *meta* isomer is therefore C:

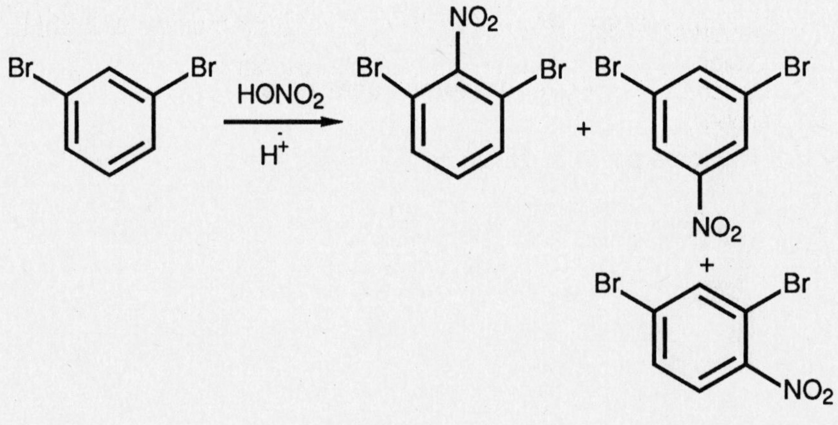

4.22 In each case six carbons are required for the benzene ring; the remaining carbons must be present as alkyl substituents:

a.

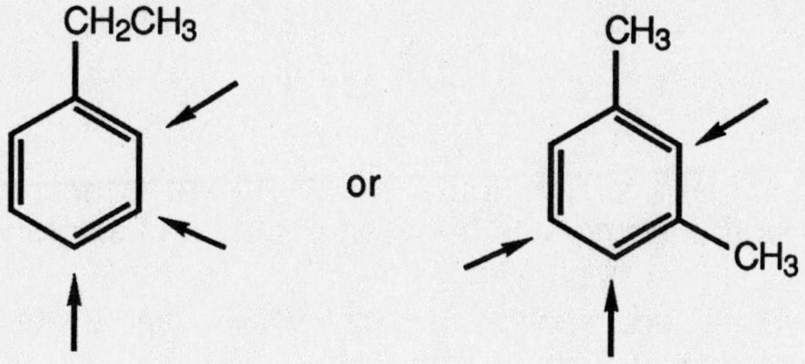

or

Each compound gives three monobromo derivatives, as shown by the arrows.

b.

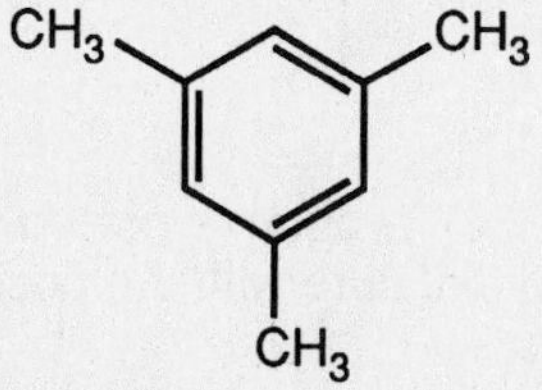

c.

The structure is symmetrical, and all three positions for aromatic substitution are equivalent.

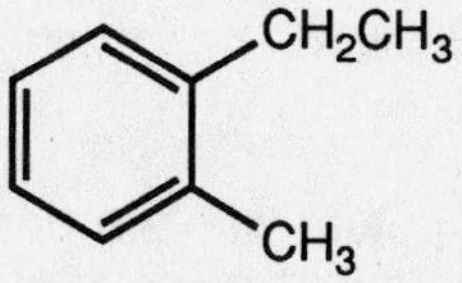

Substitution at each unoccupied ring position gives a different product.

4.23 The energy released on hydrogenating a carbon–carbon double bond is 26-30 kcal/mol (eq. 4.5). With four double bonds, we can calculate that 104-120 kcal/mol should be liberated when cyclooctatetraene is hydrogenated. The observed value (110 kcal/mol) falls within this range and suggests that cyclooctatetraene has no appreciable resonance energy. One reason is that cyclooctatetraene is not planar, and its tublike shape prevents overlap of the p orbitals around the ring.

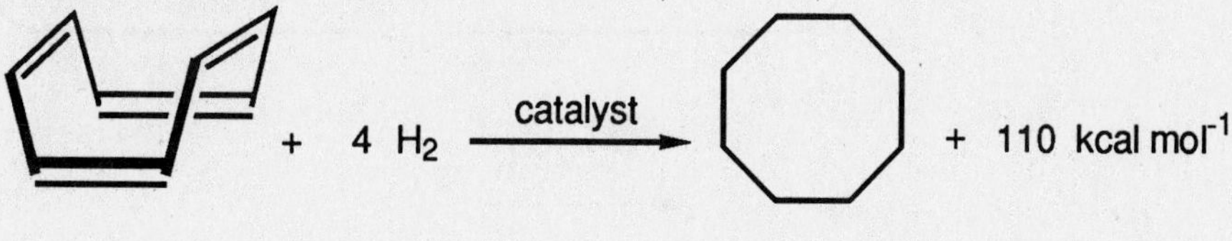

$+ \quad 4 \ H_2 \xrightarrow{\text{catalyst}}$

$+ \quad 110 \ \text{kcal mol}^{-1}$

4.24   The nitro group has two main contributing structures:

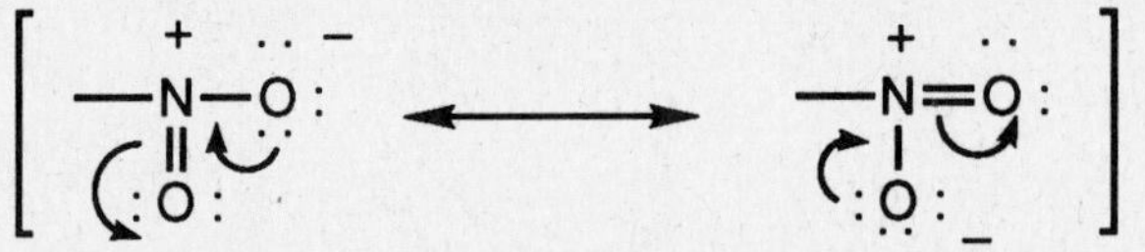

Since they are identical and contribute equally, there is only one type of nitrogen–oxygen bond, intermediate between double and single in length.

4.25 NO_2^+: There are 16 valence electrons available (N = 5, and 2 x O = 2 x 6 = 12 for a total of 17, but we must subtract 1 electron since the ion is positive).

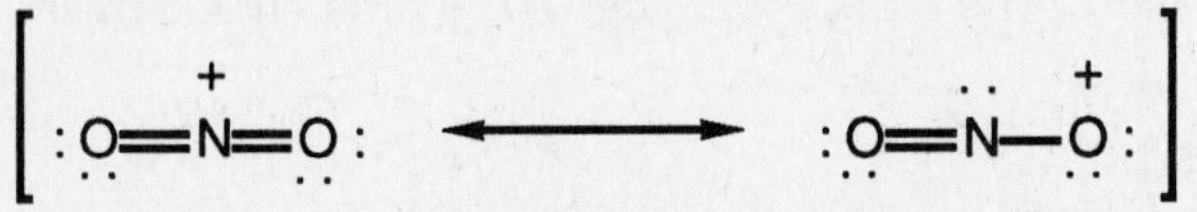

The structure with the positive charge on the nitrogen is preferred because each atom has an octet of electrons. In the structure with the positive charge on the oxygen, the oxygen atom has only six electrons around it. Note that in aromatic nitrations, it is the nitrogen atom of NO_2^+ that attacks and becomes attached to the aromatic ring.

4.26   a.   The electrophile, NO_2^+, is formed as in eq. 4.21. Then the mechanism follows the same steps as the solution to Example 4.2.

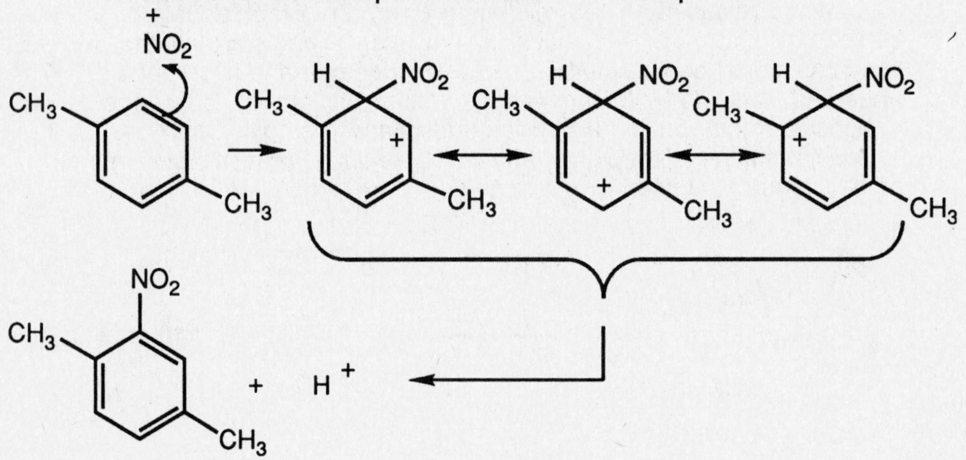

b. The electrophile is formed as in eq. 4.23:

$$(CH_3)_3CCl + AlCl_3 \longrightarrow (CH_3)_3C^+ + AlCl_4^-$$

Then

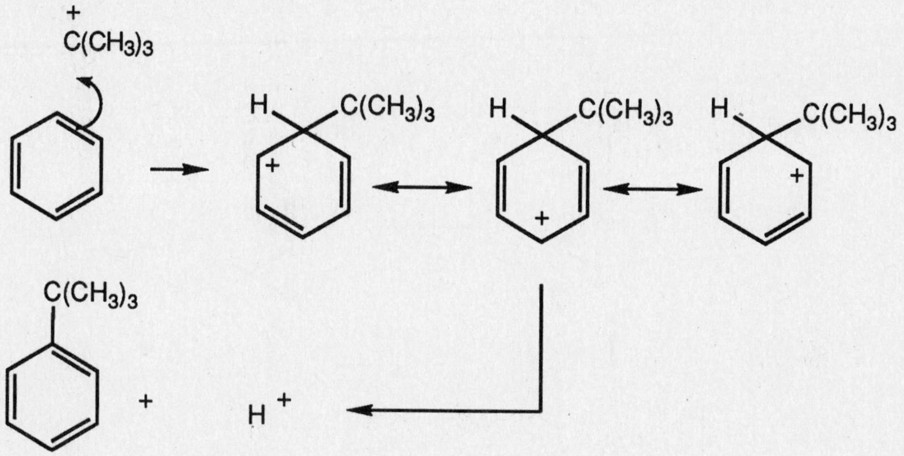

4.27   *ortho*

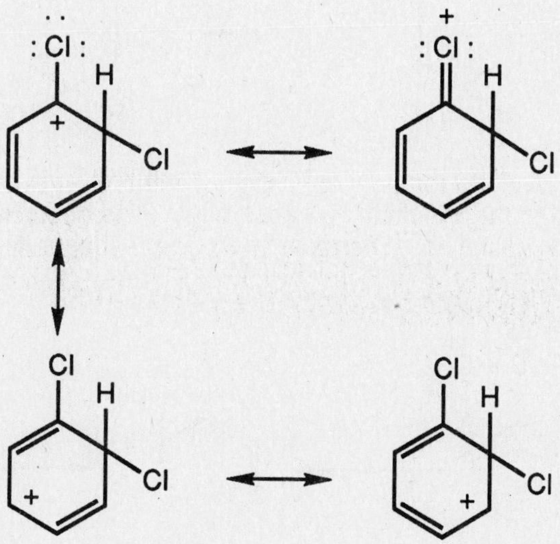

meta

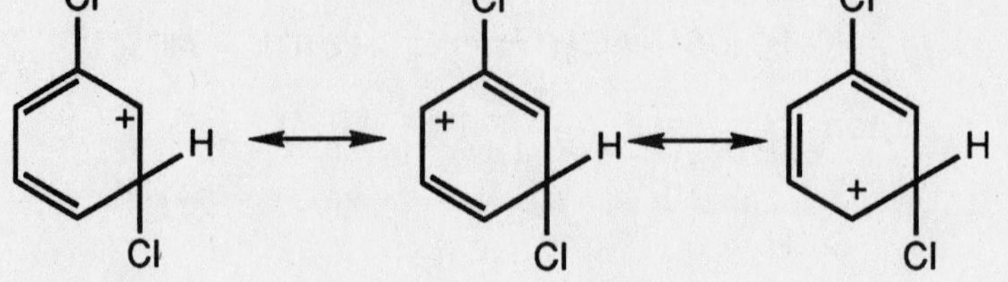

para

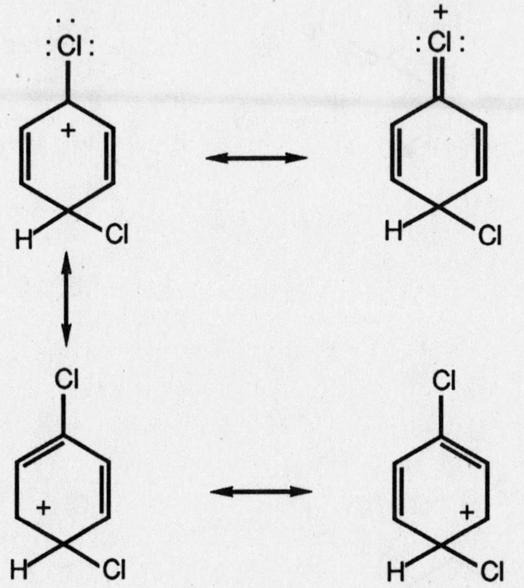

In the intermediate for *ortho* or *para* substitution, the positive charge can be delocalized to the initial chloro substituent. This delocalization is not possible with *meta* substitution. Therefore, *ortho,para*-substitution predominates.

4.28   *ortho*

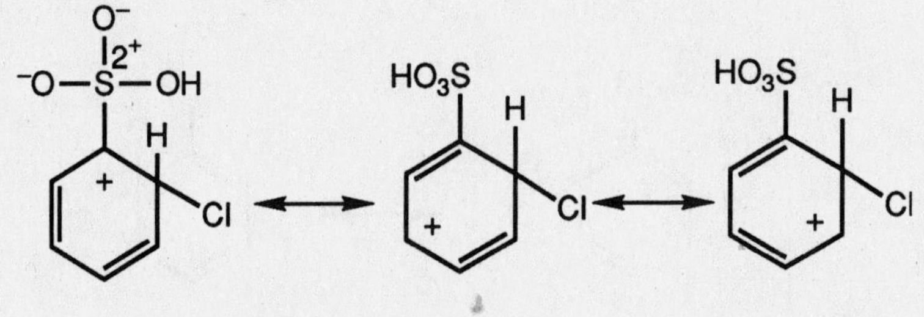

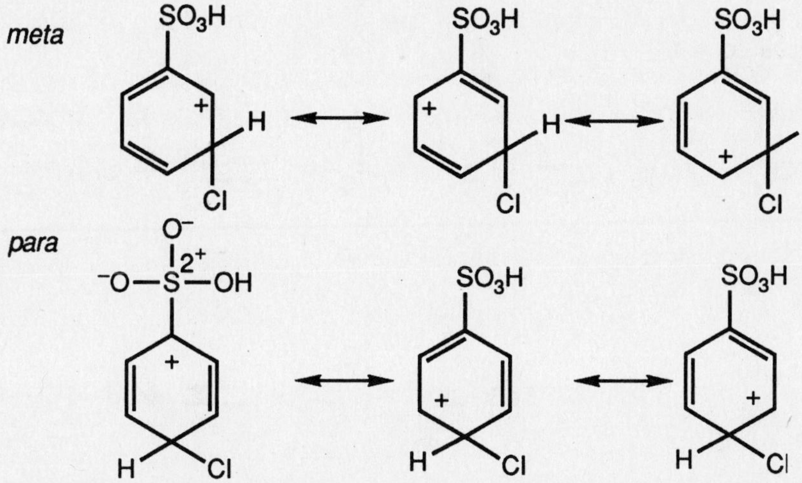

meta

para

The sulfur atom in the sulfonic acid group carries a partial positive charge because oxygen is more electronegative than sulfur. This partial charge is illustrated in the first contributor to the intermediates for *ortho* or *para* substitution. Note that two positive charges appear on adjacent atoms in these contributors, an unfavorable situation. No such contributor appears in the intermediate for *meta* substitution. Thus *meta* substitution predominates.

4.29   See Sec. 4.12 for a discussion of the orienting influence of substituents.

a.

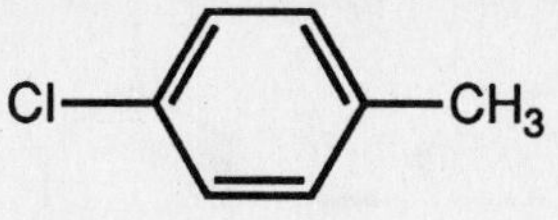

(and *ortho*)

b.

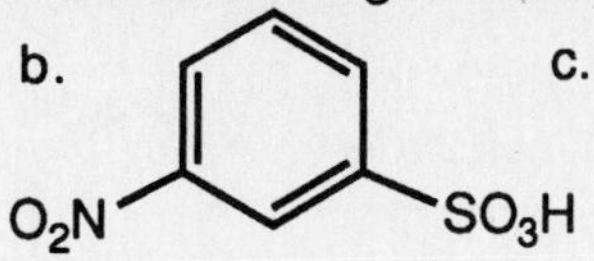

c.

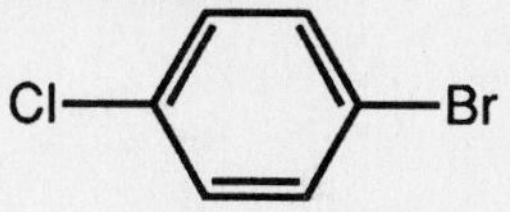

(and *ortho*)

d. Same as c.

e. Same as b.

f.

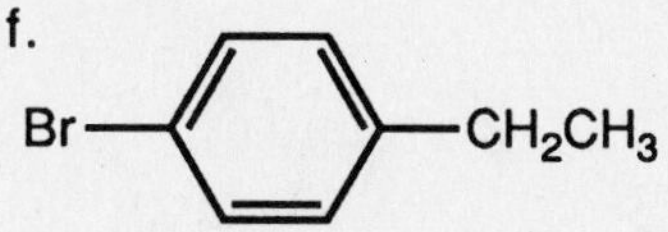

(and *ortho*)

g.

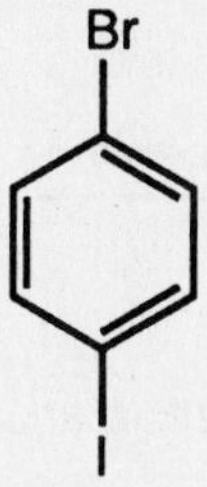

(and *ortho*)

h.

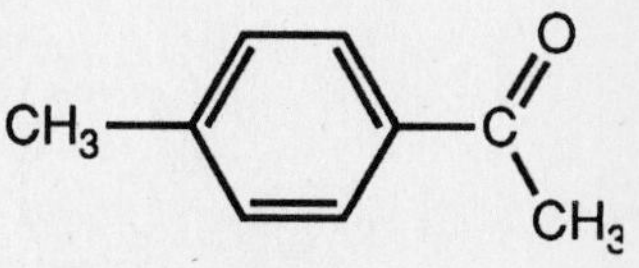

(and *ortho*)

4.30   If the halides were not identical, the following kind of exchange could occur (see eq. 4.16):

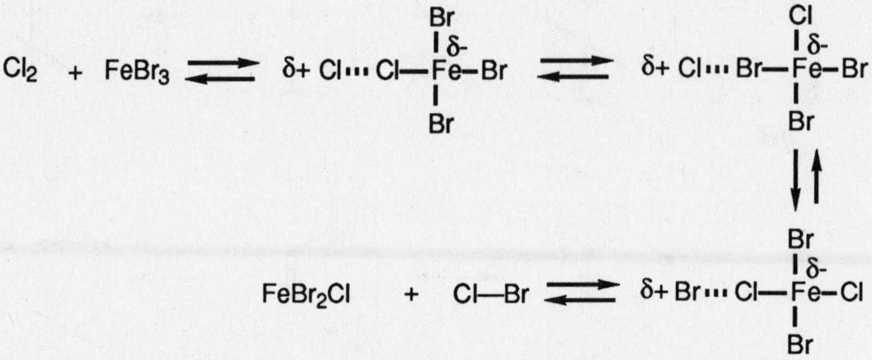

In this way, electrophilic bromine (Br^+) could be formed. Consequently, a mixture of chlorinated and brominated aromatic products would be obtained.

4.31   a.   Since the two substituents must end up in a *meta* relationship, the first one introduced into the benzene ring must be *meta*-directing. Therefore, nitrate first and then brominate.

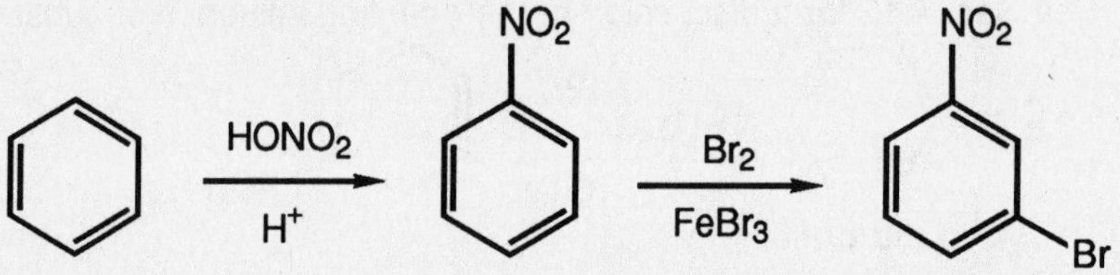

   b.   Alkyl groups are *ortho,para*-directing, but the —SO_3H group is *meta*-directing.

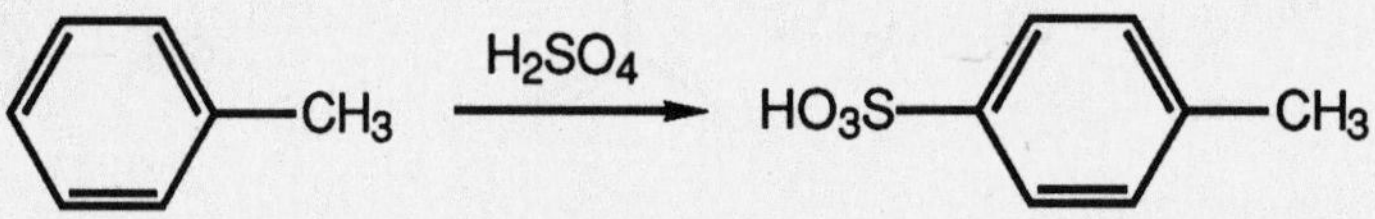

   c.   First make the ethylbenzene and then nitrate it:

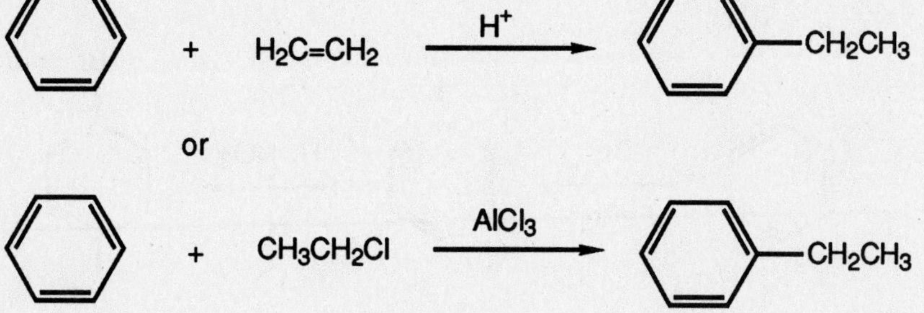

+ $H_2C=CH_2$ $\xrightarrow{H^+}$ —CH_2CH_3

or

+ CH_3CH_2Cl $\xrightarrow{AlCl_3}$ —CH_2CH_3

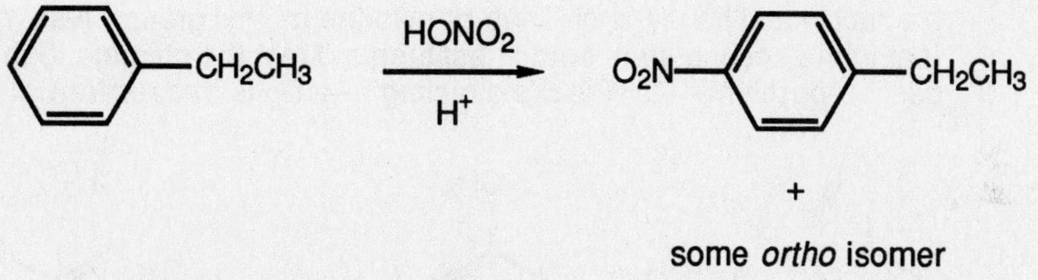

—CH_2CH_3 $\xrightarrow[H^+]{HONO_2}$ O_2N— —CH_2CH_3

+

some *ortho* isomer

d. Arenes can be converted to substituted cyclohexanes by catalytic hydrogenation.

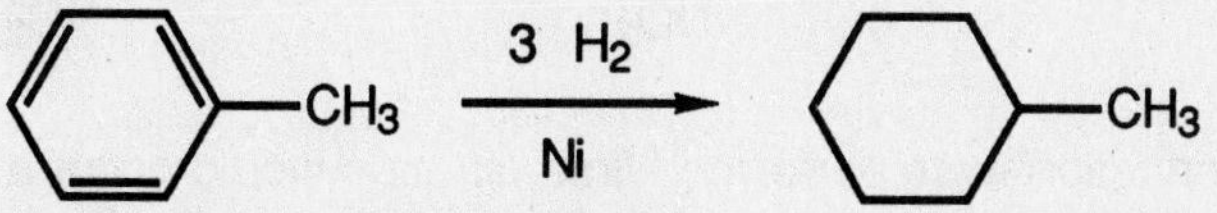

—CH_3 $\xrightarrow[Ni]{3\ H_2}$ —CH_3

e. The nitro substituent must be introduced first, to block the *para* position from the bromination.

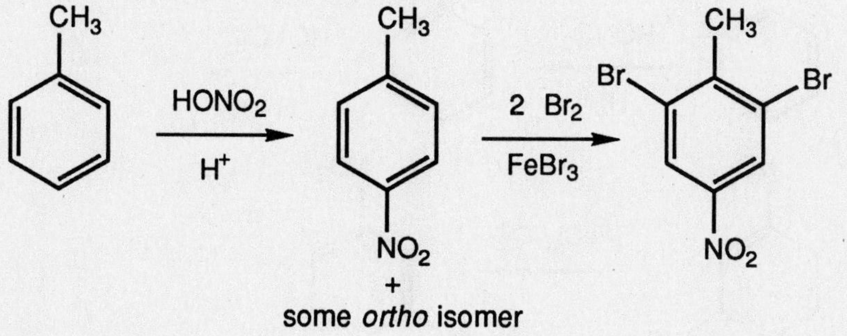

CH_3 $\xrightarrow[H^+]{HONO_2}$ CH_3 ... NO_2 $\xrightarrow[FeBr_3]{2\ Br_2}$ CH_3, Br, Br, NO_2

+

some *ortho* isomer

4. AROMATIC COMPOUNDS

f.   Compare with part a. The bromination must be performed first.

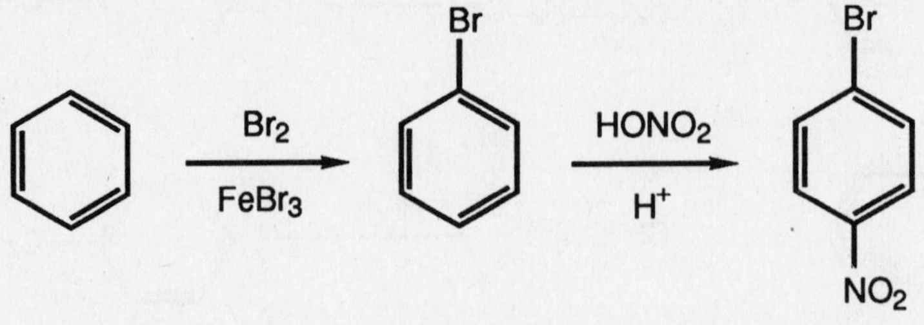

g.   If the chlorination were performed first, a considerable portion of product would have the chlorine *para* to the methyl group. Also, note that in the second step, both substituents direct the chlorine to the desired position (—CH3 is *o,p*-directing, —NO2 is *m*-directing).

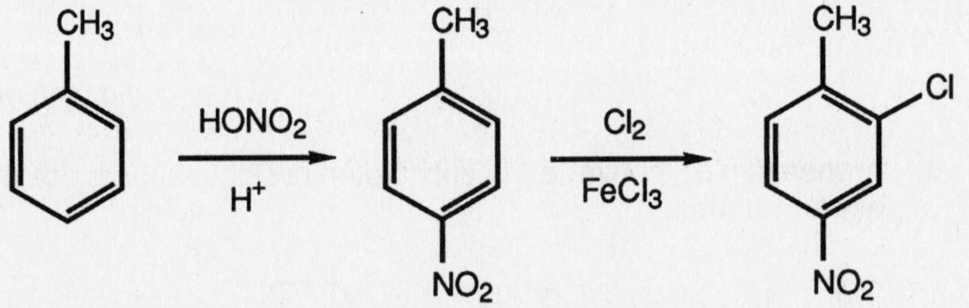

h.   If the chlorination were performed first, nitration would occur at the 2 and 4 positions. The nitro groups reinforce one another in the final chlorination.

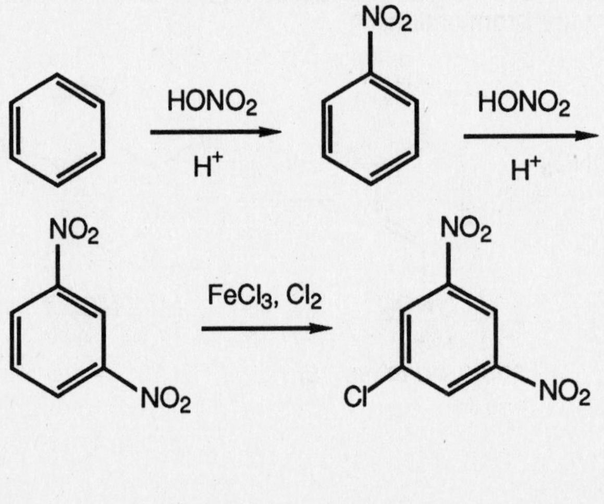

4.32 D_2SO_4 is a strong acid and a source of the electrophile D^+ (analogous to H^+ from H_2SO_4).

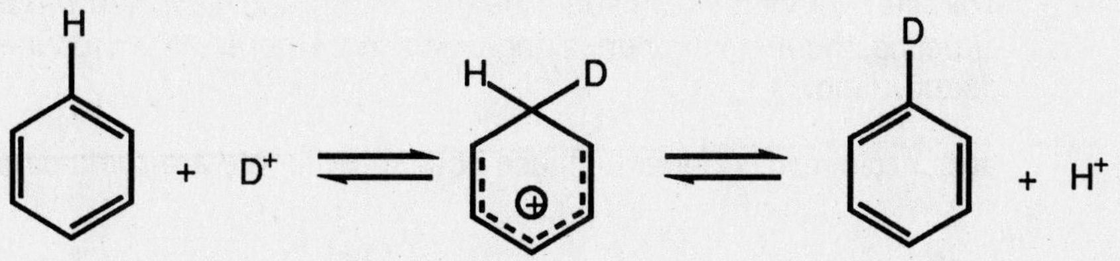

Loss of H^+ from the intermediate benzenonium ion results in replacement of H by D. With a large excess of D_2SO_4, these equilibria are shifted to the right, eventually resulting in fully deuterated benzene, C_6D_6.

4.33 a. *Ortho, para*-directing because of the unshared electron pairs on sulfur

—S̈CH₃

b. *Meta*-directing because of the positive charge on the nitrogen (electron-withdrawing)

—N⁺(CH₃)₃

c. *Ortho, para*-directing because of the unshared electron pairs on the oxygen

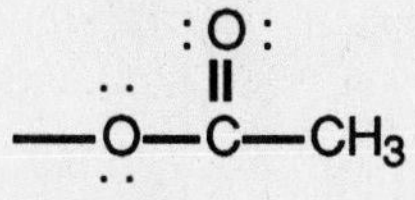

—Ö—C(=O:)—CH₃

d. *Meta*-directing because of the partial positive charge on the carbonyl carbon, due to contributors such as

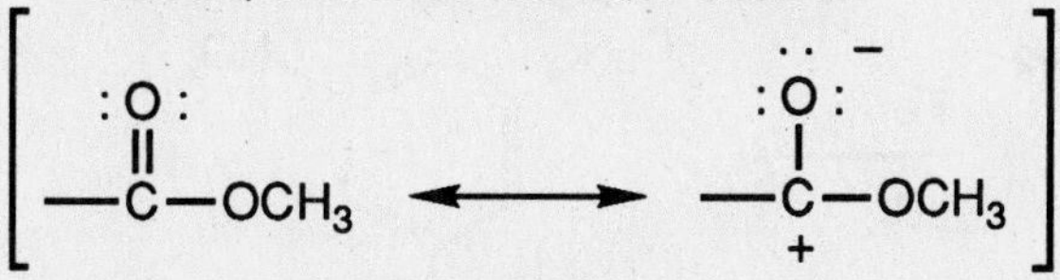

—C(=O:)—OCH₃ ⟷ —C⁺(—Ö:⁻)—OCH₃

4.34 Nitro groups are ring-deactivating. Thus, as we substitute nitro groups for hydrogens, we make the ring less and less reactive toward further electrophilic substitution and, therefore, must increase the severity of the reaction conditions.

4.35   a.   Anisole; the —OCH_3 group is ring-activating, whereas the —CO_2H group is ring-deactivating.

        b.   Toluene; although both substituents (—CH_3 and —Br) are *ortho,para*-directing, the methyl group is ring-activating, whereas the bromine is ring-deactivating.

4.36   3-Nitrobenzoic acid is better because both substituents are *meta*-directing:

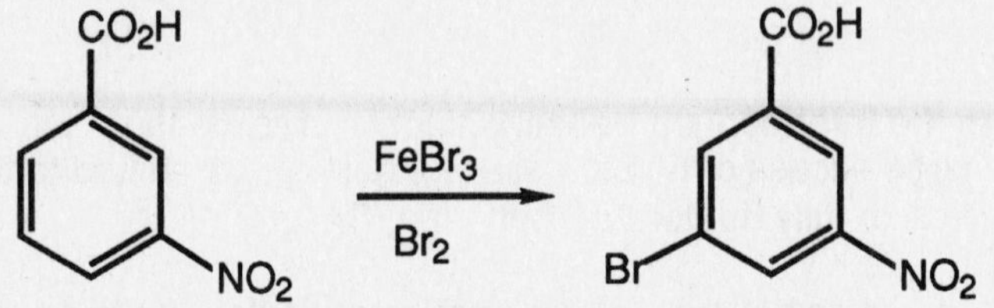

On the other hand, 3-bromobenzoic acid has an *ortho,para*-directing and a *meta*-directing substituent and on nitration would give a mixture of isomers:

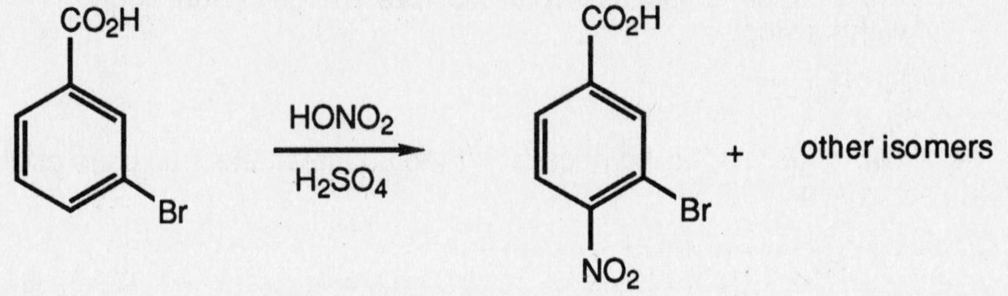

4.37   The *meta*-directing effect of both nitro groups reinforces substitution in the position shown.

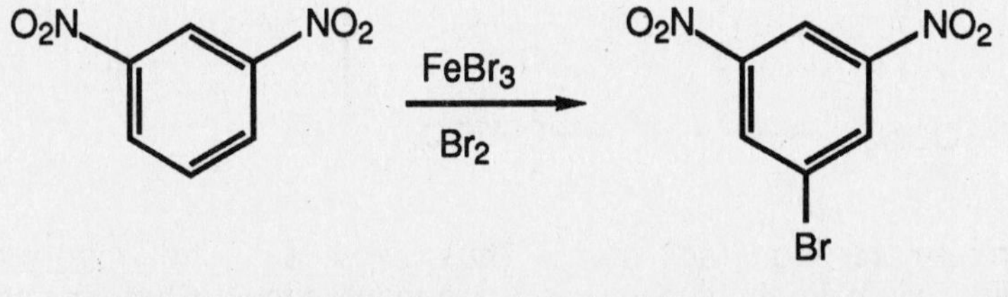

4.38   a.   Three different monosubstitution products are possible: at C-1 (equivalent to C-4, C-5, and C-8); at C-2 (equivalent to C-3, C-6, and C-7); and at C-9 (equivalent to C-10).

96

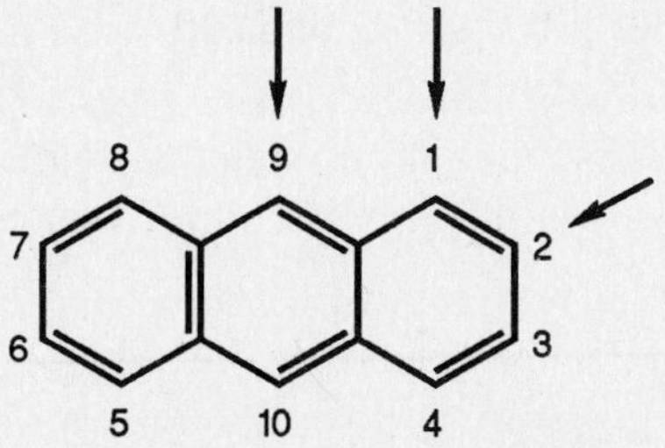

b. Five different monosubstitution products are possible: at C-1 (equivalent to C-8; at C-2 (equivalent to C-7); at C-3 (equivalent to C-6); at C-4 (equivalent to C-5); and at C-9 (equivalent to C-10).

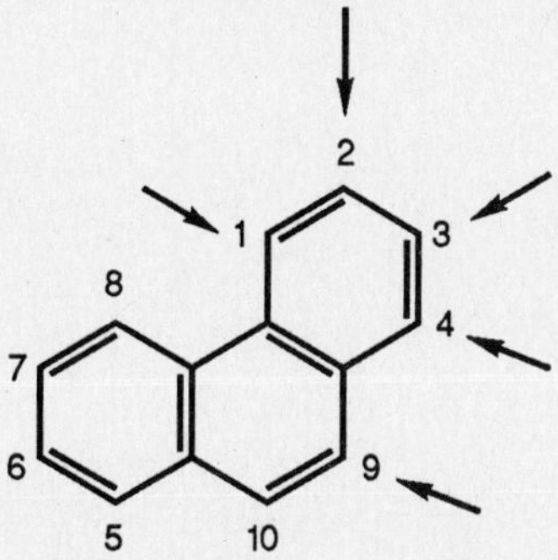

4.39 Nitration at C-9 is preferred over the other two possibilities (C-1 or C-2) because the intermediate benzenonium ion retains two benzenoid rings. Substitution at C-1 or C-2 gives an intermediate benzenonium ion with a naphthalene substructure, which has less resonance energy than two benzene rings.

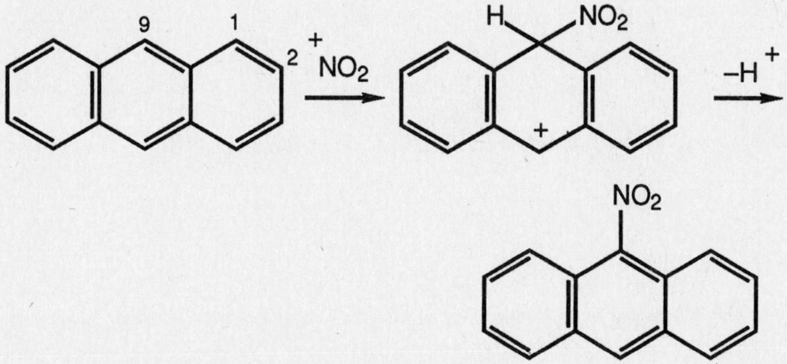

4.40   The carbon bonded to the electrophile (E) is sp^3-hybridized. The remaining carbons are sp^2-hybridized.

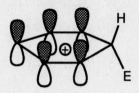

CHAPTER FIVE: STEREOISOMERISM

CHAPTER SUMMARY

Stereoisomers have the same atom connectivities but different arrangements of the atoms in space. They may be **chiral** or **achiral**. A stereoisomer is chiral if its mirror image is not identical or superimposable on the original molecule. It is achiral if the molecule and its mirror image are identical or superimposable. **Enantiomers** are a pair of molecules related as nonsuperimposable mirror images.

A carbon atom with four different groups attached is called an **asymmetric carbon** *or* **chiral center** *because it gives rise to stereoisomers (is* **stereogenic***)*. Any molecule with a **plane of symmetry** is achiral.

Configuration refers to the arrangement of groups attached to a chiral center. **Enantiomers** have opposite configurations. Configuration can be designated by the *R-S* **convention**. Groups attached to the chiral center are ranked in a priority order according to decreasing atomic number. When the chiral center is viewed from the side *opposite* the lowest-priority group, the center is said to be *R* if the other three groups, in decreasing priority order, form a *clockwise* array. If the three-group array is counterclockwise, the configuration is *S*. A similar convention (*E-Z*) has been applied to alkene *cis–trans* isomers.

Chiral molecules are **optically active**. They rotate a beam of plane-polarized light. They are dextrorotatory (+) or levorotatory (-), depending on whether they rotate the beam to the right or left, respectively. The rotations are measured with a **polarimeter** and are expressed as **specific rotations**, defined as

$$[\alpha]^{t}_{\lambda} = \frac{\alpha}{l \times c} \quad \text{(solvent)}$$

where α = observed rotation, l = length of sample tube in decimeters, c = concentration in g/mL, and the measurement conditions of temperature (*t*), wavelength of polarized light (λ), and solvent are given. Achiral molecules are **optically inactive.**

Pasteur showed that optical activity was related to molecular right- or left-handedness (chirality). Later, van't Hoff and LeBel proposed that the four valences of carbon are directed toward the corners of a tetrahedron. If the four attached groups

5. STEREOISOMERISM

are different, two arrangements are possible and are related as an object and its nonsuperimposable mirror image. Enantiomers differ *only* in chiral (or handed) properties, such as the *direction* of rotation of plane-polarized light. They have identical **achiral properties**, such as melting and boiling points.

Fischer projection formulas show three-dimensional structures in two dimensions. In such formulas, horizontal groups project toward the viewer, and vertical groups project away from the viewer. Fischer projections may be turned 180° (but not 90°) in the plane of the paper and still retain configuration. Also, while holding any one group fixed, the remaining three groups can be rotated either clockwise or counterclockwise without changing the configuration.

Diastereomers are stereoisomers that are not mirror images of one another. They may differ in all types of properties, and be chiral or achiral.

Compounds with *n* different chiral centers may exist in a maximum of 2^n forms. Of these, there will be $2^n/2$ pairs of enantiomers. Compounds from different enantiomeric pairs are diastereomers. If two (or more) of the chiral centers are identical, certain isomers will be achiral. A *meso* **form** is an optically inactive, achiral form of a compound with chiral centers. **Tartaric acid**, which has two identical chiral centers, exists in three forms, the *R,R* and *S,S* forms (a pair of enantiomers) and the achiral *meso* form.

Stereoisomers may be classified as either conformational or configurational, chiral or achiral, and enantiomers or diastereomers.

The stereochemistry of organic reactions depends on the nature of the reactants. Achiral molecules can react to give a chiral product. In such reactions, both enantiomers of the product will always be formed in equal amounts. When chiral molecules react with achiral reagents to create a new chiral center, diastereomers are formed in unequal amounts.

A **racemic form** is a 50:50 mixture of enantiomers. It is optically inactive. A racemic mixture of configurational isomers cannot be separated (resolved) by ordinary chemical means (distillation, crystallization, chromatography) unless the reagent is chiral. One way to separate a pair of enantiomers is to first convert them to diastereomers by reaction with a chiral reagent and then separate the diastereomers and regenerate the (now separated) enantiomers.

LEARNING OBJECTIVES

1. Know the meaning of: chiral, achiral, enantiomers, plane of symmetry, superimposable and nonsuperimposable mirror images, racemic mixture.

2.   Know the meaning of: asymmetric carbon atom, chiral center, *R-S* convention, priority order, *E-Z* convention, Fischer projection.

3.   Know the meaning of: diastereomer, *meso* compound, lactic acid, tartaric acid, resolution.

4.   Know the meaning of: plane-polarized light, polarimeter, optically active or optically inactive, observed rotation, specific rotation, dextrorotatory, levorotatory.

5.   Given the concentration of an optically active compound, length of the polarimeter tube, and observed rotation, calculate the specific rotation. Given any three of the four quantities mentioned, calculate the fourth.

6.   Given a structural formula, draw it in three dimensions and locate any plane of symmetry.

7.   Given the structure of a compound, determine if any chiral centers (asymmetric carbon atoms) are present.

8.   Given the structure or name of a compound, tell whether it is capable of optical activity.

9.   Know the rules for establishing priority orders of groups in the *R-S* convention.

10.   Given a compound with a chiral center, assign the priority order of groups attached to it.

11.   Given a chiral center in a molecule, assign the *R* or *S* configuration to it.

12.   Draw the three-dimensional formula of a molecule with a particular configuration, *R* or *S*.

13.   Given a pair of *cis–trans* isomers, assign the *E* or *Z* configuration.

14.   Draw the formula of an alkene with a particular configuration, *E* or *Z*.

15.   Given a structure with one or more chiral centers, draw a Fischer projection.

16.   Given Fischer projections of two isomers, tell their relationship (for example, same structure, enantiomers, diastereomers).

17.   Given a Fischer projection, assign *R* or *S* configuration to each chiral center.

5. STEREOISOMERISM

18.   Given a structure with more than one chiral center, tell how many stereo-
        isomers are possible and draw the structure of each. Tell what relationship
        the stereoisomers have to each other (for example, enantiomeric,
        diastereomeric).

19.   Tell whether or not a particular structure can exist as a *meso* form.

20.   Given a structure with two or more identical chiral centers, draw the structure of
        the *meso* form.

21.   Given a pair of stereoisomers, classify them as configurational or confor-
        mational, chiral or achiral, and enantiomers or diastereomers.

22.   Given a chemical reaction that gives a chiral product, tell the stereochemistry
        of the products.

ANSWERS TO PROBLEMS

Problems Within the Chapter

5.1   a. chiral   b. achiral   c. achiral   d. chiral
        e. achiral   f. chiral   g. chiral   h. achiral

        The achiral objects (teacup, football, tennis racket, and pencil) can be used
        with equal ease by right- or left-handed persons. Their mirror images are
        superimposable on the objects themselves. On the other hand, a golf club
        must be either left- or right-handed and is chiral; a shoe will fit a left or a right
        foot; a corkscrew may have a right- or left-handed spiral. These objects, as
        well as a portrait, have mirror images that are *not* identical with the objects
        themselves, and thus they are chiral.

5.2   The chiral centers are marked with an asterisk. Note that each chiral center has
        four different groups attached.

a.
$$CH_3CH_2 - \overset{\overset{\displaystyle H}{|}}{\underset{\underset{\displaystyle Cl}{|}}{C}}^{*} - CH_2CH_2CH_3$$

b.
$$CH_3 - \overset{\overset{\displaystyle H}{|}}{\underset{\underset{\displaystyle Cl}{|}}{C}}^{*} - \overset{\overset{\displaystyle H}{|}}{\underset{\underset{\displaystyle Cl}{|}}{C}}^{*} - CH_3$$

c.

d.
$$Br - \overset{\overset{\displaystyle H}{|}}{\underset{\underset{\displaystyle Cl}{|}}{C}}^{*} - CH_3$$

5.3 a.

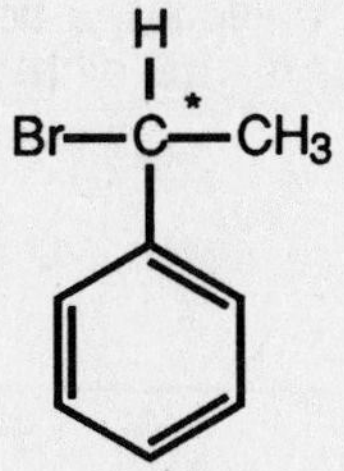

chiral

b.

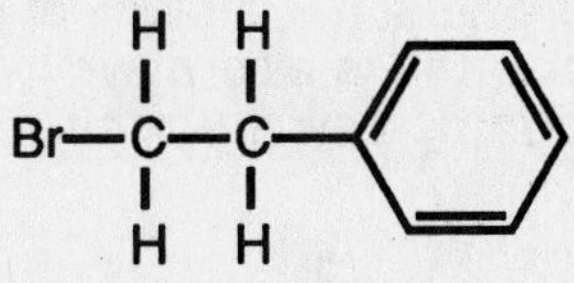

no chiral centers; achiral

5.4 Note that if the right structure is rotated 180° about the carbon–phenyl bond, the methyl and phenyl groups can be superimposed on those of the left structure, but the positions of the hydrogen and bromine will be interchanged.

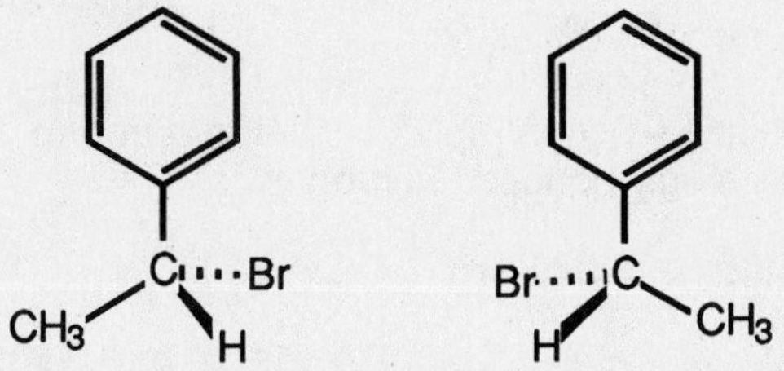

5.5 The planes of symmetry are (a) the three planes that pass through any pair of eclipsed hydrogens and (b) the perpendicular bisector of the C—C bond. Ethane in this conformation is achiral.

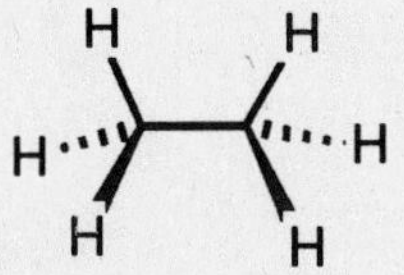

5.6 There are three planes of symmetry that pass through any pair of anti hydrogens. Ethane in this conformation is achiral.

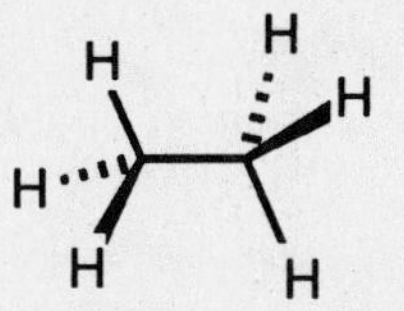

5.7 *Cis*-1,2-dichlorethene has a plane of symmetry that bisects the double bond. The molecular plane is also a symmetry plane. *Trans*-1,2-dichlorethene is planar. That plane is a symmetry plane.

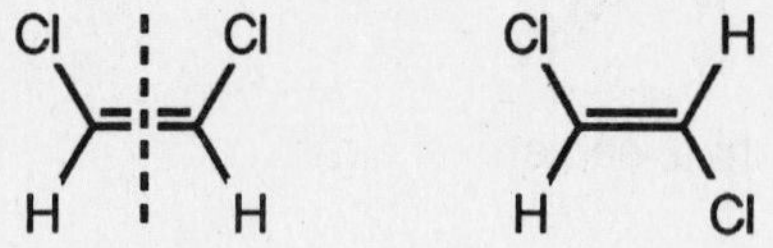

5.8 In each case, proceed from high to low priority.

    a.   —OH > —C(CH$_3$)$_3$ > —CH$_3$ > —H

    b.   —OCH$_3$ > —OH > —CH$_2$OH > —CH$_3$

    c.   —OH > —NHCH$_3$ > —CN > —CH$_2$NH$_2$

The carbon in the cyano group (—CN) is triply bonded to nitrogen, whereas the —CH$_2$NH$_2$ carbon is only singly bonded to nitrogen.

    d.   —CH(CH$_3$)$_2$ > —CH$_2$CH$_2$CH$_3$ > —CH$_2$CH$_3$ > —CH$_3$

5.9   a.   —C≡CH > —CH=CH$_2$

The acetylenic carbon (—C≡) is treated as though it is bonded to three carbons, while the olefinic carbon (—CH=) is treated as though it is bonded to two carbons and a hydrogen (see Sec. 5.4).

    b.

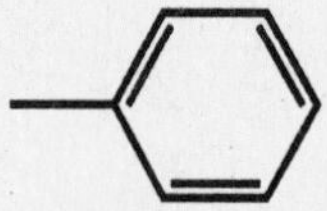

The phenyl carbon is treated as though it is bonded to three carbons.

    c.   —OH > —CH=O > —CH=CH$_2$ > —CH$_2$CH$_3$

5.10   a.   Priority order: OH > CH=O > CH$_3$ > H. Configuration is *S*.

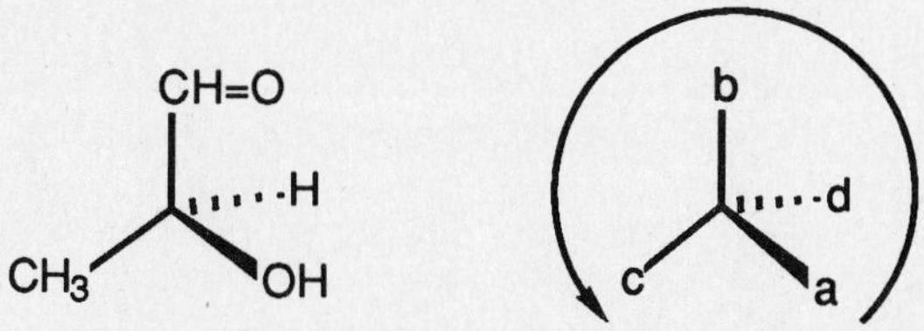

b. Priority order: NH_2 > —C_6H_5 > CH_3 > H. The configuration is *R*.

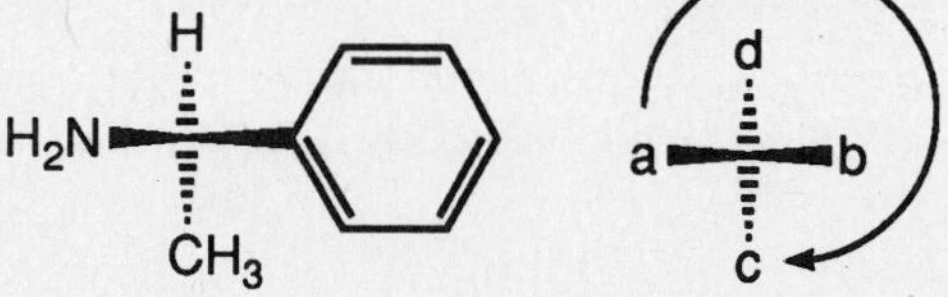

clockwise or *R*

5.11   a.   The priority of groups around the chiral center is as follows:

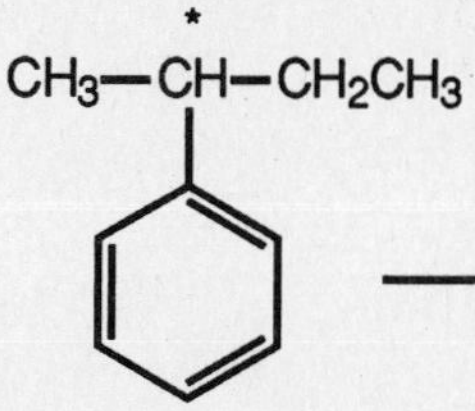

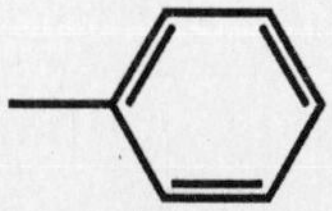

> —CH_2CH_3 > —CH_3 > —H

First draw the lowest priority group pointing away from you:

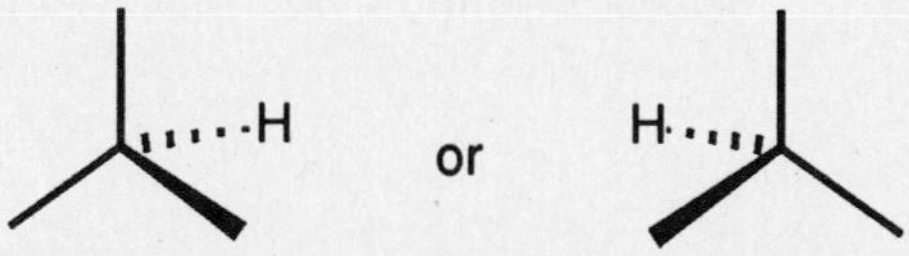

or

Then fill in the groups in priority order, clockwise (*R*),

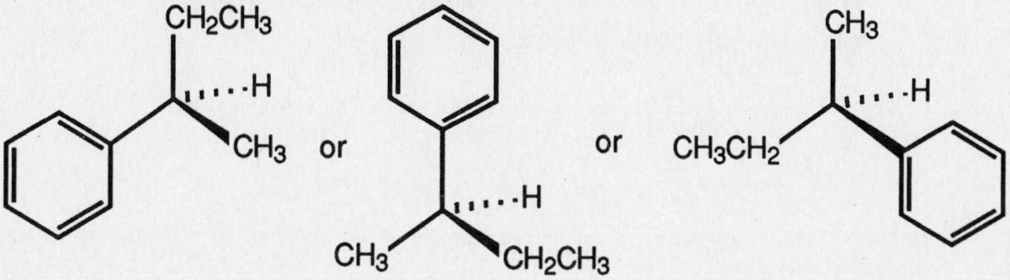

or   or

and similarly:

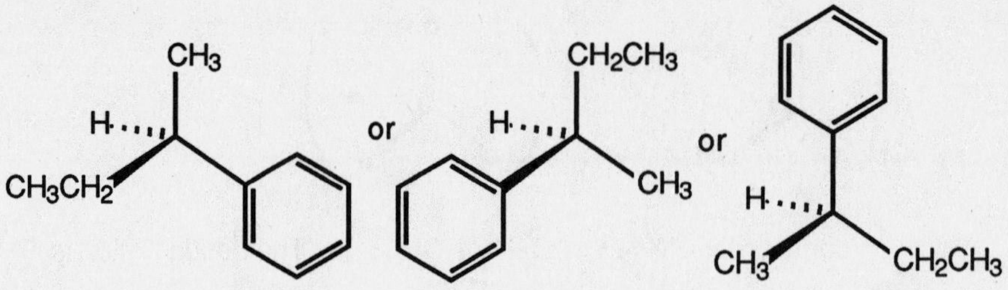

As you can see, there are many ways to write the correct answer. In subsequent problems, only one correct way will be shown. Work with models if you have difficulty.

b.   The priority around the chiral center is as follows:

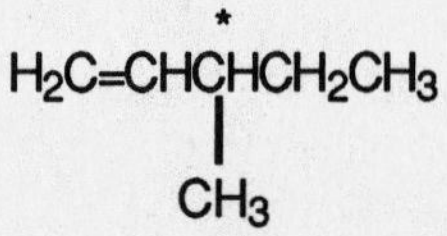

$$-CH=CH_2 \; > \; -CH_2CH_3 \; > \; -CH_3 \; > \; -H$$

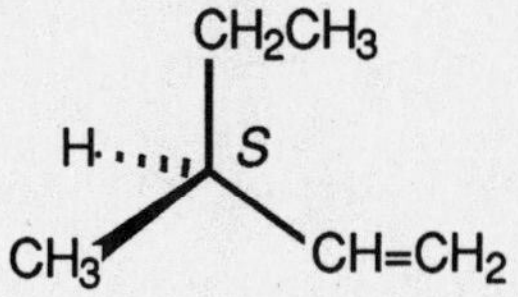

c.   The same rules apply for cyclic and acyclic compounds.

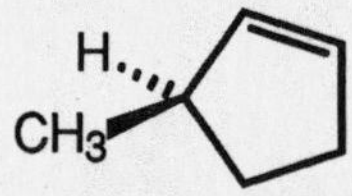

5.12   a.   The priority order at each doubly bonded carbon is $CH_3 > H$ and $CH_3CH_2 > H$. The configuration is Z.

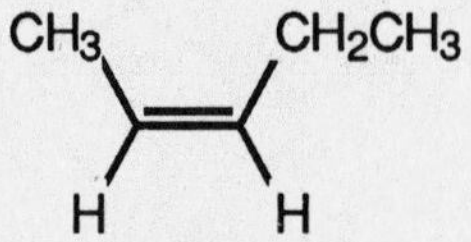

(Z)-2-pentene

ANSWERS TO PROBLEMS

b. The priority order is Br > Cl and F > H. The configuration is *Z*.

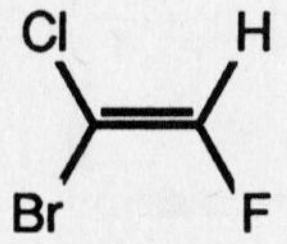

(*Z*)-1-bromo-1-choro-2-fluoroethene

5.13   a. The two highest priority groups, CH_3 and CH_3CH_2, are *entgegen*, or opposite.

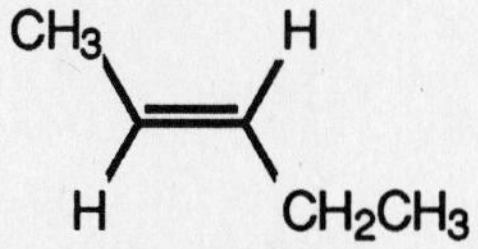

(*E*)-2-pentene

b. The priorities are $CH_2=CH—$ > H and $CH_3—$ > H. The two highest priority groups, $CH_2=CH—$ and $CH_3—$ are *zusammen* (together).

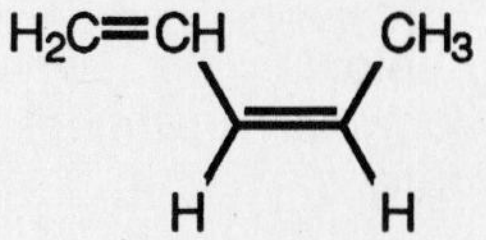

(*Z*)-1,3-pentadiene

5.14 $$[\alpha]_D^{20} = \frac{+0.66}{0.5 \times \dfrac{1.5}{50}} = 44° \quad \text{(ethanol)}$$

5.15

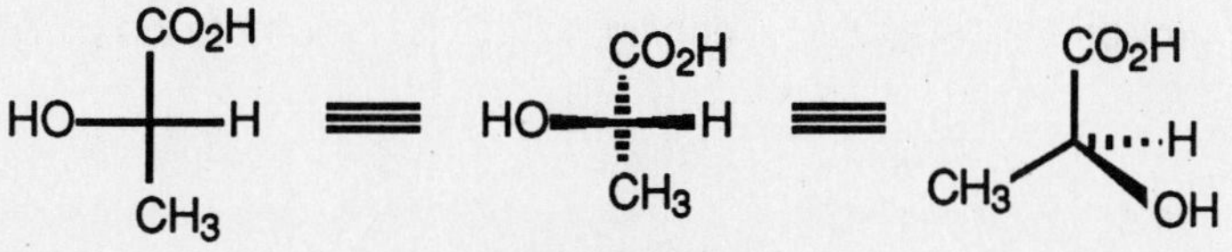

(*S*)-lactic acid

107

5.16

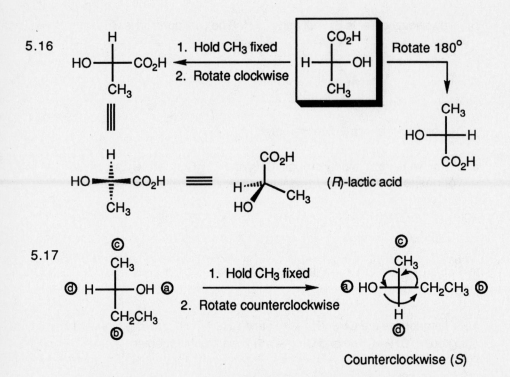

5.17

1. Hold CH_3 fixed

2. Rotate counterclockwise

Counterclockwise (S)

5.18   First consider C-2. The priorities are assigned below. The lowest priority group occupies a horizontal position. A counterclockwise arrangement of the remaining groups (a→b→c) indicates the R-configuration.

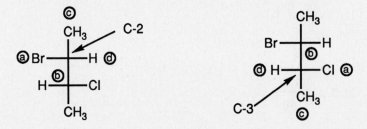

Next consider C-3. Once again, a counterclockwise arrangement of the remaining groups indicates the R-configuration (see footnote on p. 156).

ANSWERS TO PROBLEMS

5.19

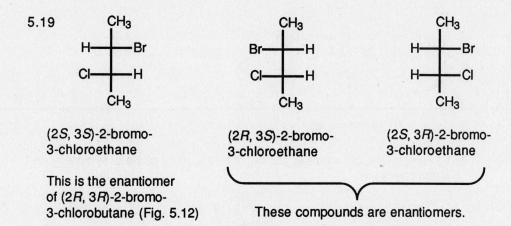

(2S, 3S)-2-bromo-
3-chloroethane

This is the enantiomer
of (2R, 3R)-2-bromo-
3-chlorobutane (Fig. 5.12)

(2R, 3S)-2-bromo-
3-chloroethane

(2S, 3R)-2-bromo-
3-chloroethane

These compounds are enantiomers.

5.20   The (2R,3R) and (2S,3S) isomers are a pair of enantiomers. Their specific
rotations will be equal in magnitude and opposite in sign.

The (2R,3R) and (2S,3R) isomers are a pair of diastereomers. Their specific
rotations will be unequal in magnitude and may or may not differ in sign.

5.21   There are *four* different chiral centers, marked below with asterisks. There are
therefore 2^4 = 16 possible isomers.

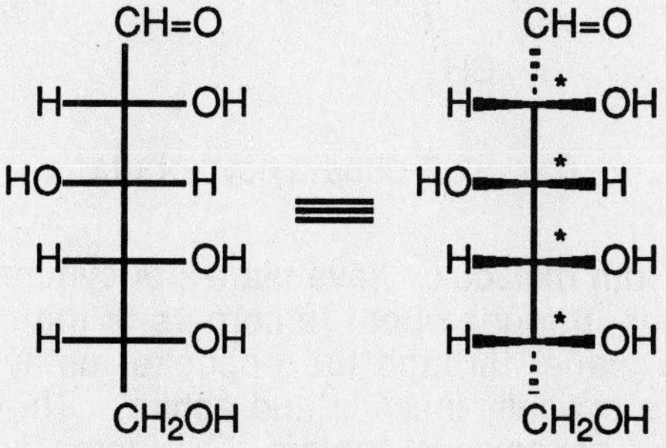

5.22   When the CH_2 groups are superimposed, the methyl groups of one mirror
image are superimposed on the hydrogens of the other mirror image. The
mirror images are nonsuperimposable and are therefore enantiomers.

109

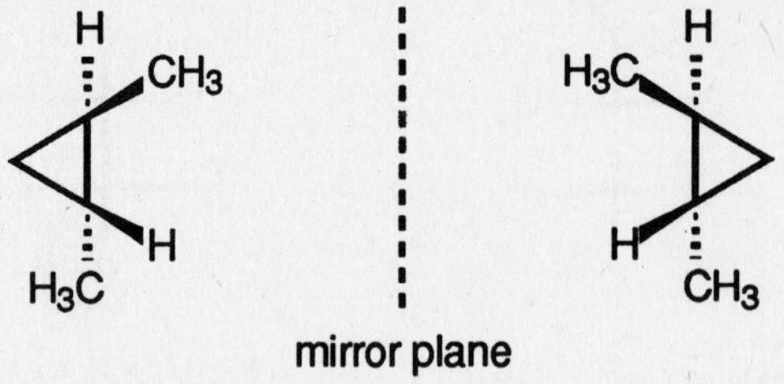

mirror plane

5.23   *Cis*-1,2-dimethylcyclopropane is achiral. It has a mirror plane of symmetry and is a *meso* compound.

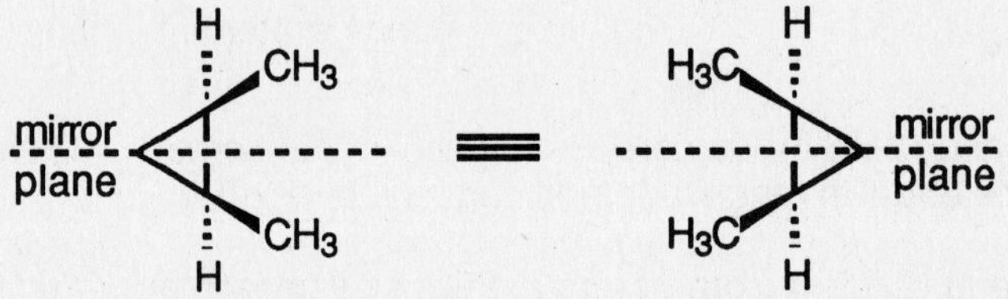

5.24

cis-1,3-dimethylcyclobutane   *trans*-1,3-dimethylcyclobutane

There are *no* chiral centers. Both molecules have planes of symmetry. The *cis* isomer has two such planes, through opposite corners of the ring. The *trans* isomer has one such plane, through the opposite methyl-bearing corners. Both compounds are optically inactive and achiral. They are not *meso* compounds because there are no chiral centers. To summarize, the two isomers are configurational, achiral diastereomers.

5.25

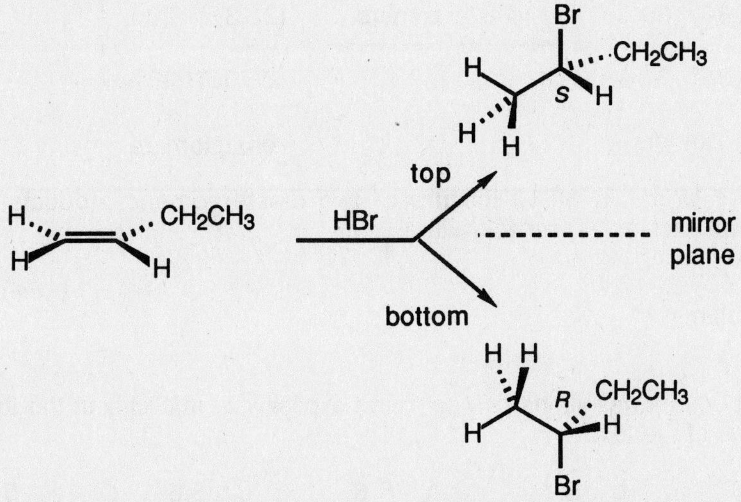

One-step addition of H—Br to the top face of the double bond gives (S)-2-bromobutane. Addition of H—Br to the bottom face gives (R)-2-bromobutane. Since 1-butene is achiral, the probability of addition to either face of the double bond is equal and the product will be racemic (an equal mixture of enantiomers).

5.26 If we draw the structure in three dimensions, we see that either hydrogen at C-2 can be replaced with equal probability:

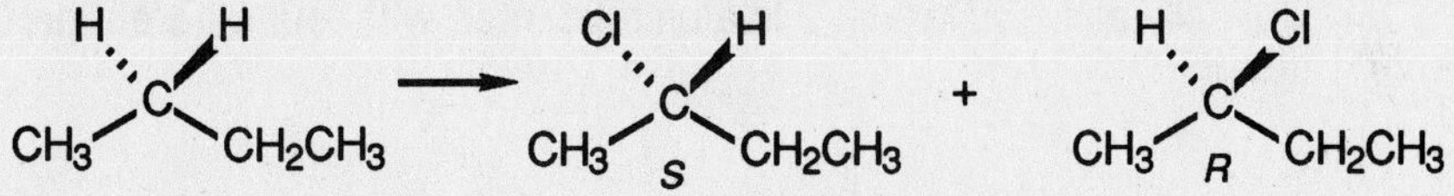

Thus a 50:50 mixture of the two enantiomers is obtained.

5.27 (S)-3-chloro-1-butene is the enantiomer of (R)-3-chloro-1-butene, and it will react with HBr to give the enantiomers of the products shown in eq. 5.7 [(2S, 3S)- and (2R, 3S)-2-bromo-3-chlorobutane], in the same 60:40 ratio. Therefore a racemic mixture of 3-chloro-1-butene will react with HBr to give the following mixture of 2-bromo-3-chlorobutanes:

5. STEREOISOMERISM

$$\left\{ \begin{array}{l} \text{(2R,3R)} \quad 30\% \\[1.5em] \text{(2S,3S)} \quad 30\% \end{array} \right. \quad \xleftrightarrow{\quad \text{diastereomers} \quad} \quad \left. \begin{array}{l} \text{(2S,3R)} \quad 20\% \\[1.5em] \text{(2R,3S)} \quad 20\% \end{array} \right\}$$

enantiomers   enantiomers

In other words, a 60:40 mixture of two diastereomeric products will be obtained, each as a racemic mixture.

Additional Problems

5.28   Each of these definitions can be found explicitly or implicitly in the following sections of the text:

a. 5.2	b. 5.2	c. 5.6	d. 5.6	e. 5.3
f. 5.3	g. 5.6, 5.7	h. 5.9	i. 5.10	j. 5.13

5.29   a.   Optically inactive. The molecule has several planes of symmetry, for example, the one that passes through the three carbon atoms.

   b.   Optically active. Carbon-2 is a chiral center, with four different groups attached.

   c.   Optically inactive. None of the carbon atoms has four different groups attached.

$$CH_3CH_2CHCH_2CH_2CH_3$$
$$|$$
$$CH_2CH_3$$

   d.   Optically active. The carbon marked with an asterisk is a chiral center:

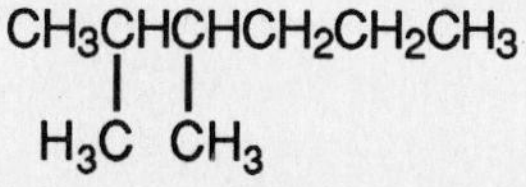

$$\overset{*}{CH_3CHCHCH_2CH_2CH_3}$$

$$\underset{H_3C \quad CH_3}{|\quad\ |}$$

e. The substance is optically inactive. It has a plane of symmetry perpendicu-
    lar to the five-membered ring, through carbon-1 and bisecting the bond
    between carbon-3 and carbon-4.

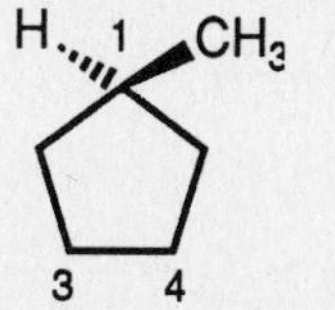

f. Optically active. The carbon marked with an asterisk is a chiral center, with
    four different groups attached. Even isotopes of the same element are
    sufficiently different to lead to optical activity.

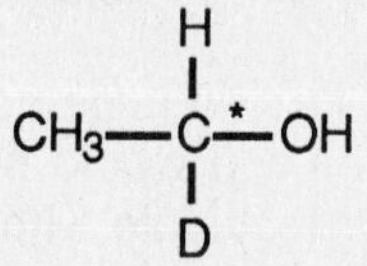

5.30 a.

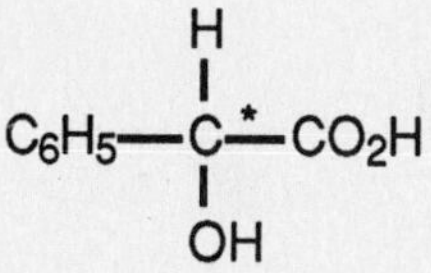

b.

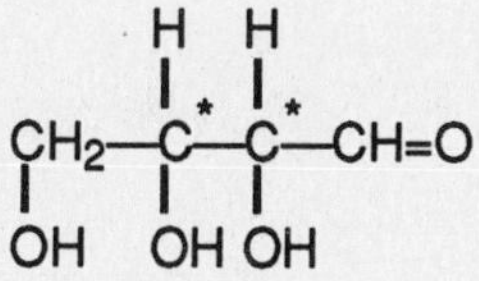

c.

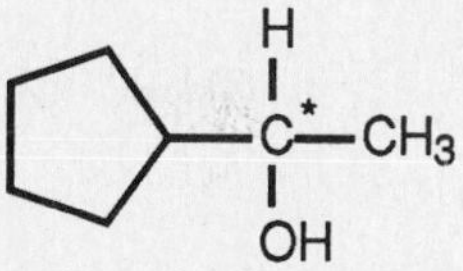

d.

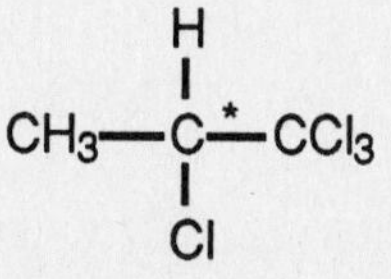

e.

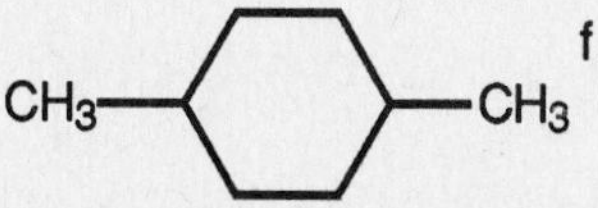

no chiral centers

f.

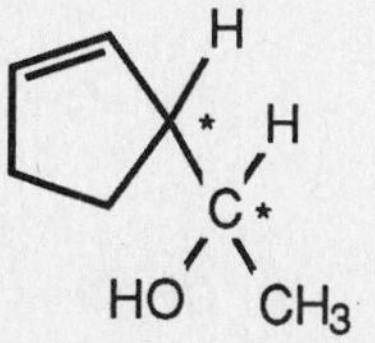

5.31 In each case the *observed* rotation would be doubled, but the specific rotation
      would remain constant. For example, if *c* is doubled, α will also double, but
      the fraction α/c in the formula for specific rotation will remain constant.

5.32
$$[\alpha]_D^{25} = \frac{1.33}{2.0 \ \times \ \dfrac{1}{100}} = 66.5^\circ \ \text{(water)}$$

5.33 a. identical   b. enantiomers

5.34 The following are examples. There may be other possibilities.

     a. $CH_3\overset{*}{C}HCH_2CH_3$ b. $CH_3\overset{*}{C}HCH_2CH_2CH_3$

                 $|$ $|$

              OH   Br

     c. $HOCH_2\overset{*}{C}HCH_2CH_3$ d. $CH_3\overset{*}{C}HCH{=}CH_2$

                  $|$ $|$

              OH CH_2CH_3

In each case the asymmetric carbon atom is marked with an asterisk.

5.35 All structures must contain only one double bond and no rings because if the monovalent chlorine were replaced by a hydrogen, we would have C_5H_{10}, corresponding to the general formula C_nH_{2n} (a molecule with one double bond or one ring). Since the chloride is described as unsaturated, there can be no ring present.

     a. $CH_2{=}CHCH_2CH_2CH_2Cl$ or $(CH_3)_2C{=}CHCH_2Cl$

     b. $CH_3CH{=}CCH_2Cl$ or $CH_3CH_2CH{=}CHCH_2Cl$ or $CH_3CH_2CH_2CH{=}CHCl$

                 $|$

               CH_3

     c. $CH_2{=}CH\overset{*}{C}HCH_2CH_3$ or $CH_2{=}CHCH_2\overset{*}{C}HCH_3$

                  $|$ $|$

              Cl   Cl

     d. $CH_3CH{=}CH\overset{*}{C}HCH_3$

                  $|$

              Cl

5.36 The rules for priority order are given in Sec. 5.4.

     a. $HO{-} > CH_3CH_2{-} > CH_3{-} > H{-}$
     b. $Cl{-} > C_6H_5{-} > CH_3{-} > H{-}$

c. HO— > —CH₂Cl > —CH₂OH > —CH₃
d. —CH=O > CH₂=CH— > CH₃CH₂CH₂— > CH₃CH₂—

5.37   a.

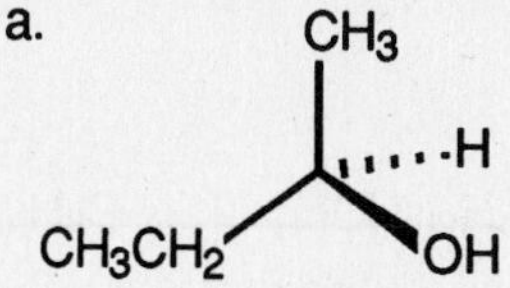

CH₃

⎪....·H

CH₃CH₂   OH

b.

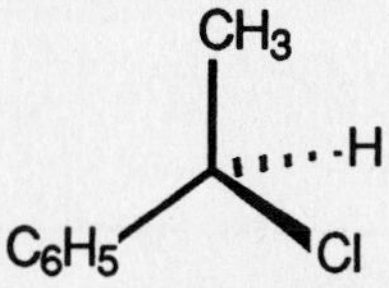

CH₃

⎪....·H

C₆H₅   Cl

c.

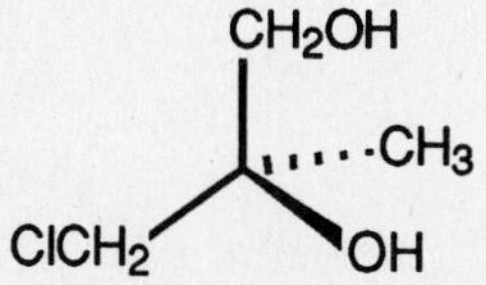

CH₂OH

⎪....·CH₃

ClCH₂   OH

d.

CH₂CH₂CH₃

⎪....·CH₂CH₃

CH₂=CH   CH=O

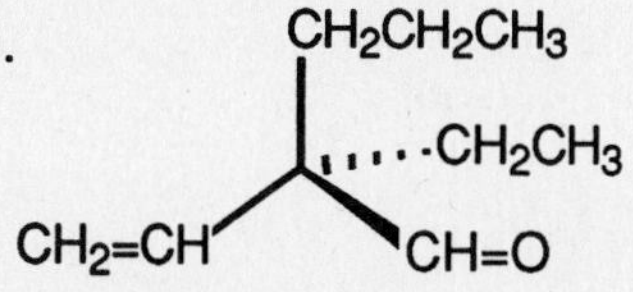

There are many ways to write these structures. They have been drawn here by putting the lowest priority group receding away from the viewer and the remaining three groups in a clockwise array, with the highest priority group at the lower right extending toward the viewer.

5.38   In each case, write down the groups in the proper priority order. Then view the chiral center from the face opposite the lowest priority group and determine whether the remaining array is clockwise (*R*) or counterclockwise (*S*). If you have difficulty, construct and examine molecular models.

a.

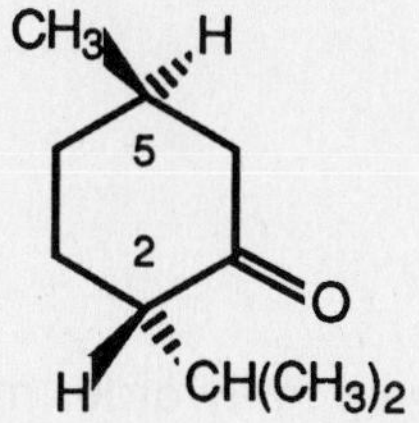

CH₃   H

5

2

O

H   CH(CH₃)₂

Carbon-5

$$\overset{O}{\overset{\|}{C}}$$

—CH₂C > —CH₂CH₂ > —CH₃ > —H

therefore *R*

Carbon-2

$$\overset{O}{\overset{\|}{C}}$$

—C > —CH(CH₃)₂ > —CH₂ > —H

therefore *S*

115

b.

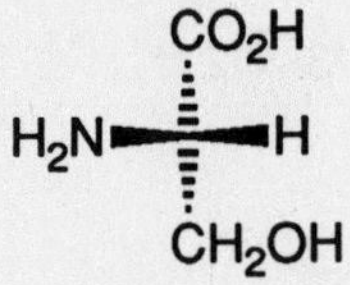

$$CO_2H$$

$$H_2N \text{---} H$$

$$CH_2OH$$

—NH₂ > —CO₂H > —CH₂OH > —H

therefore *S*

c.

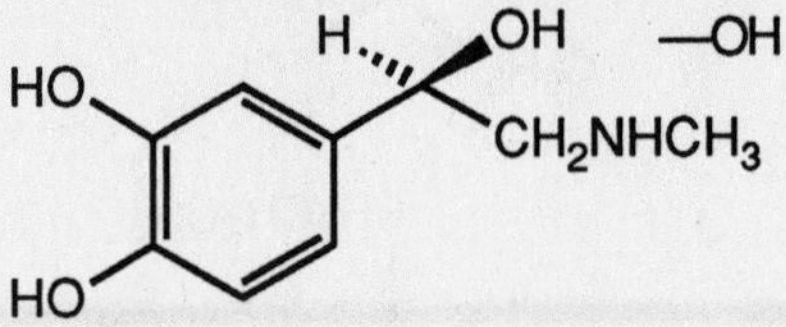

HO

H,,, OH

HO

CH₂NHCH₃

—OH > —CH₂NHCH₃ > —C₆H₃(OH)₂ > —H

therefore *R*

5.39

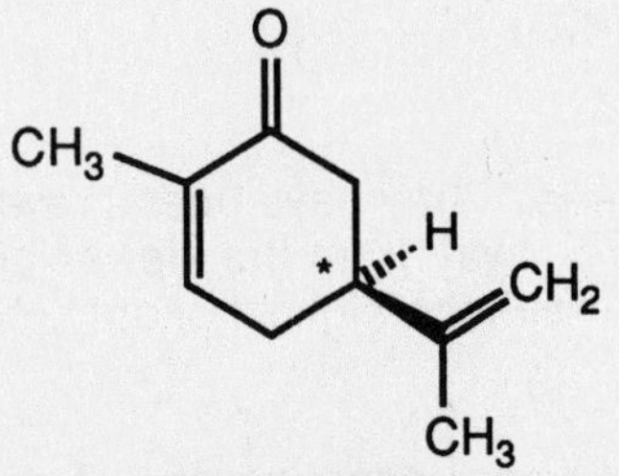

CH₃

O

*

H

CH₂

CH₃

(+)—carvone

The chiral center is marked with an asterisk. The priority order of groups at this center is

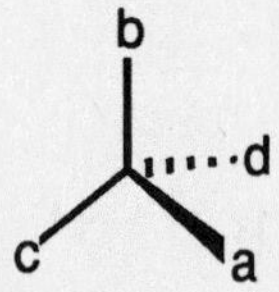

b

,,,,·d

c   a

and the configuration is S. A word about the priority order may be helpful. The

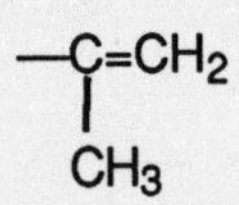

—C=CH₂

CH₃

group has three bonds from the attached carbon atom to the next atoms "out" and is therefore of the highest priority. The remaining groups both begin with —CH₂, so we must proceed further. One group is

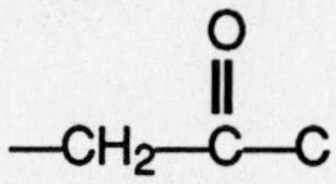

$$\overset{\displaystyle O}{\underset{\displaystyle }{\parallel}}$$

—CH$_2$—C—C

and the other is

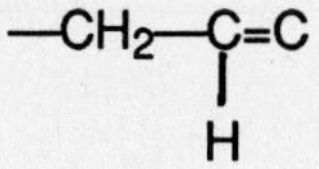

—CH$_2$—C=C
$\qquad\quad\ $|
$\qquad\quad$ H

Of these, the group with C=O has the higher priority because oxygen has a higher atomic number than carbon.

5.40   a.

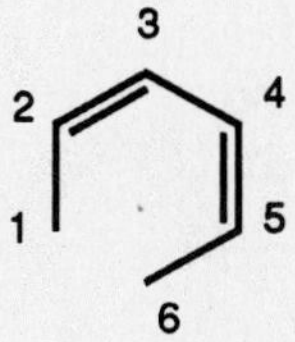

(Z,Z)-2,4-hexadiene
or more precisely,
(2Z,4Z)-2,4-hexadiene

If you have difficulty, draw out the full structure:

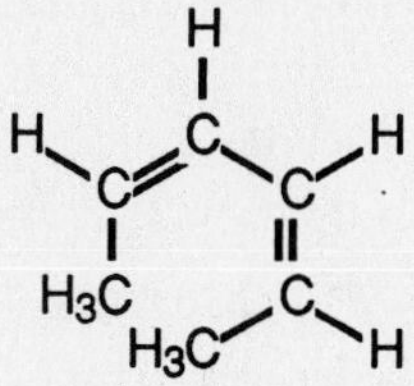

At the double bond between C-2 and C-3, the priority order is CH$_3$ > H and CH$_3$CH=CH— > H. The two high-priority groups, CH$_3$ and CH$_3$CH=CH—, are Z or *zusammen*. The same is true at the double bond between C-4 and C-5.

b.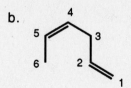

117

(Z)-1,4-hexadiene; There is no stereochemistry at the double bond joining C-1 and C-2 because both substituents at C-1 are identical (=CH$_2$).

c.

(E,E)-2,4-hexadiene

d.

There is no stereochemistry at either double bond. The IUPAC name is 1,5-hexadiene.

5.41 The structure has one double bond and one chiral center (C-4). Four isomers are possible (R or S at C-4 and E or Z at the double bond).

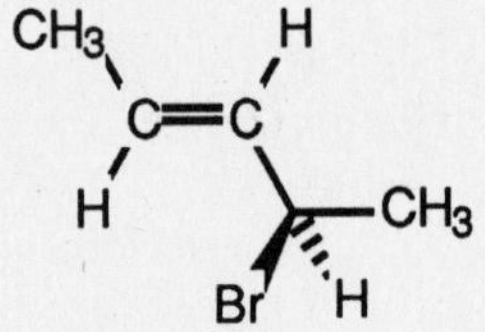

(R,E)

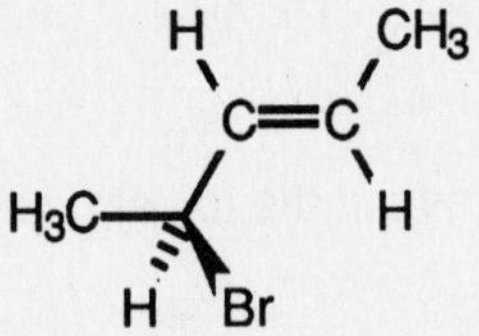

(S,E)

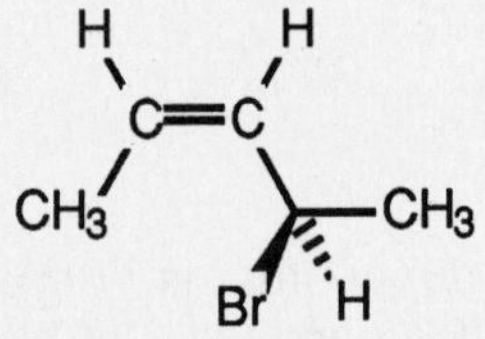

(R,Z)

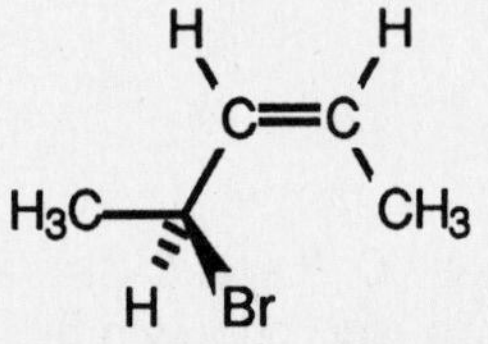

(S,Z)

The upper and lower sets form two pairs of enantiomers.

5.42 a.

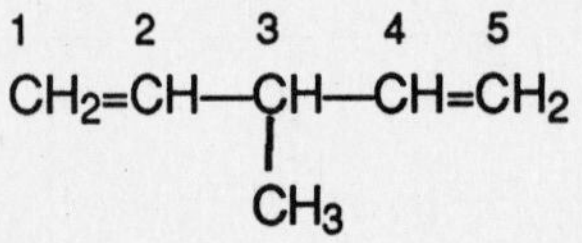

There are no chiral centers and no *cis–trans* possibilities at either double bond. Only one structure is possible.

b.

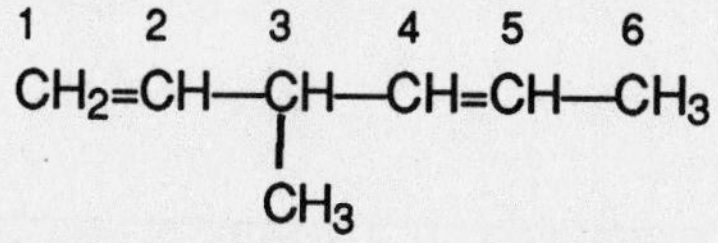

$$\underset{1}{CH_2}=\underset{2}{CH}-\underset{3}{CH}-\underset{4}{CH}=\underset{5}{CH}-\underset{6}{CH_3}$$

with CH_3 on C-3

Carbon-3 is a chiral center, and *cis–trans* isomers are possible at the double bond between C-4 and C-5. Therefore, four structures are possible (*R* or *S* at C-3, and *E* or *Z* at the double bond).

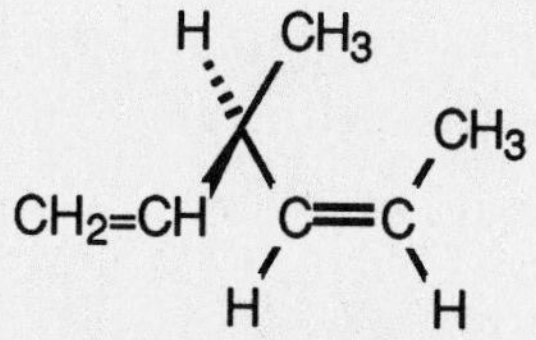

(*R,Z*)-3-methyl-1,4-hexadiene

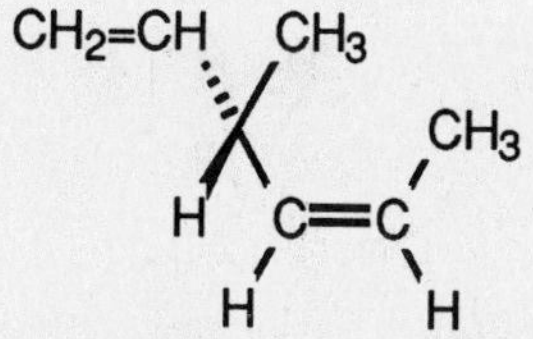

(*S,Z*)-3-methyl-1,4-hexadiene

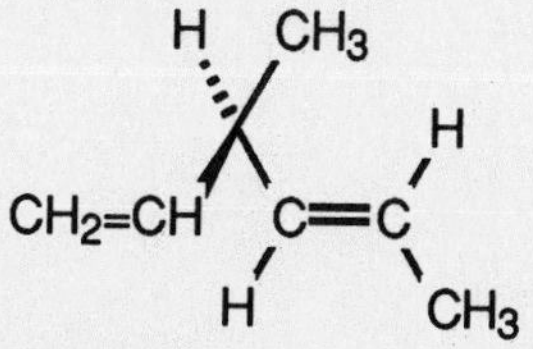

(*R,E*)-3-methyl-1,4-hexadiene

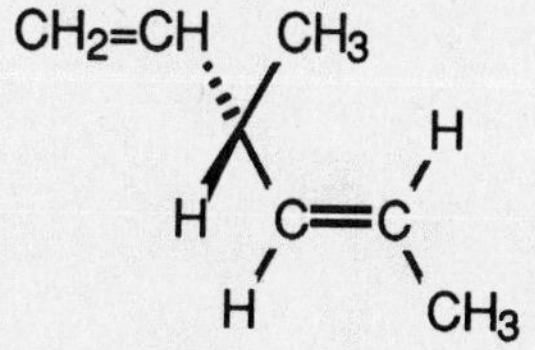

(*S,E*)-3-methyl-1,4-hexadiene

c.

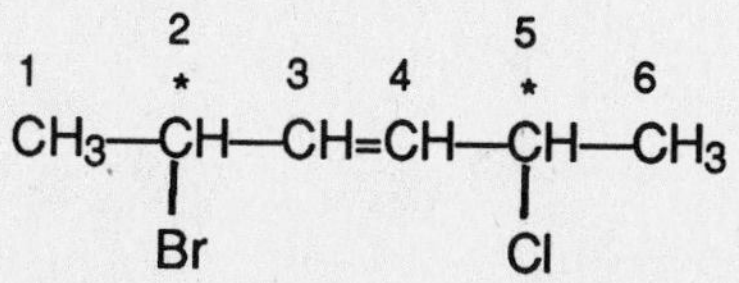

$$\underset{1}{CH_3}-\underset{2}{CH}-\underset{3}{CH}=\underset{4}{CH}-\underset{5}{CH}-\underset{6}{CH_3}$$

with Br on C-2 and Cl on C-5; C-2 and C-5 marked with asterisks

There are two chiral centers, marked with asterisks. Each can be either *R* or *S*. Also, the double bond joining C-3 and C-4 can be *E* or *Z*. Thus eight isomers are possible:

R, Z, R	*S, Z, R*
R, Z, S	*S, Z, S*
R, E, R	*S, E, R*
R, E, S	*S, E, S*

The first of these is shown below:

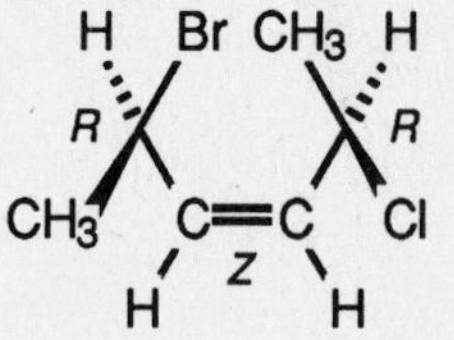

The other seven isomers can be drawn by interchanging one or more groups, using this structure as a guide. For example, the *S, Z, R* isomer is

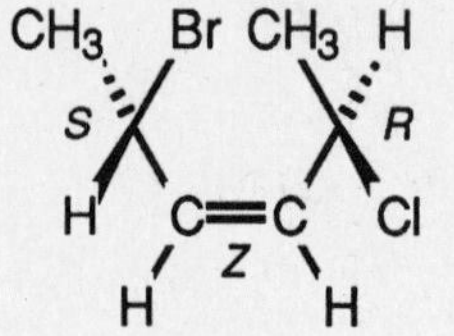

and so on.

d.

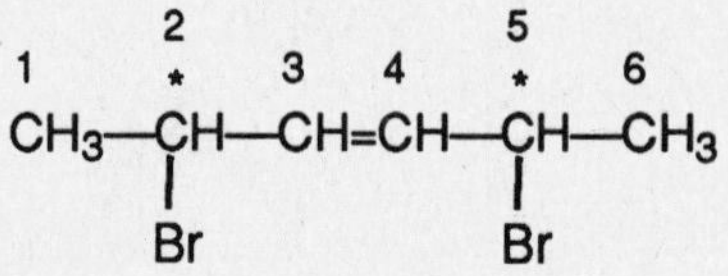

Compare with part c. In this case both chiral centers are identical. Therefore, two *meso* forms are possible, and the total number of isomers is reduced to six:

R, Z, R
R, Z, S   *(meso)*
R, E, R
R, E, S   *(meso)*
S, Z, S
S, E, S

There are two sets of enantiomers:

R, Z, R and *S, Z, S*
R, E, R and *S, E, S*

And there are two optically inactive, *meso* forms: *R, Z, S* and *R, E, S*. The R,Z,R isomer is shown below.

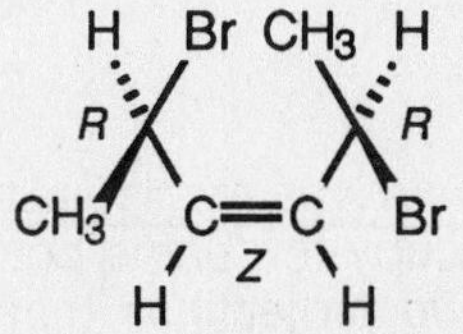

The other five structures can be derived from this one by interchanging groups. For example, the *R, Z, S meso* form is:

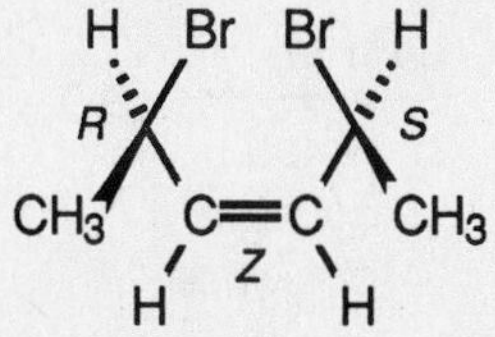

5.43   a.   The projections have the same configuration. There are several ways to work this problem. One involves assigning and comparing the absolute configuration (*R* or *S*) at each chiral carbon. For Fischer projection **A**, first assign a priority order to the four groups (OH > C$_2$H$_5$ > CH$_3$ > H). Next, hold the ethyl group fixed and rotate the remaining three groups clockwise (or counterclockwise) until the lowest priority group is at the top. This manipulation will not change the configuration at the chiral carbon (see Sec. 5.8). Then determine the arrangement of the three top priority groups (clockwise or counterclockwise). In this case they are clockwise, so the configuration is *R*.

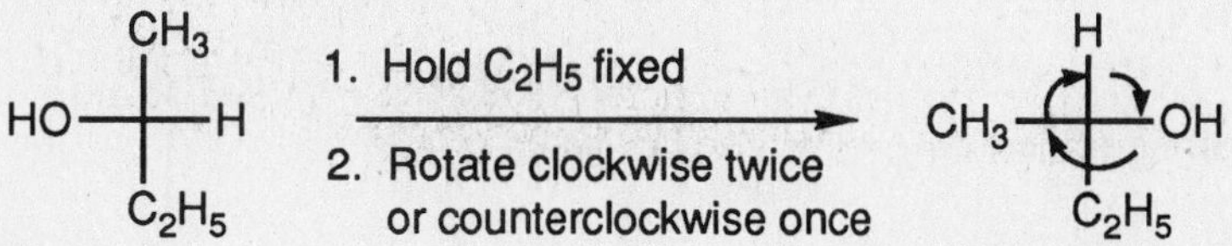

1. Hold C$_2$H$_5$ fixed

2. Rotate clockwise twice or counterclockwise once

(*R*)-2-butanol

The Fischer projection in part a already has the lowest priority group (H) at the bottom. The three top groups are arranged clockwise, so the configuration is also *R*:

OH
|
$CH_3 —\!\!\!\!— C_2H_5$ (R)-2-butanol
|
H

Another way to compare two Fischer projections involves manipulating one of them until two groups (adjacent or opposite) occupy the same positions as two groups on the other. Comparison of the two projections then reveals whether they have the same or opposite configurations. For example, the following manipulations of the Fischer projections in part a show it has the same configuration as structure **A**.

OH C_2H_5 CH_3
| 1. Fix H | 1. Fix OH |
CH_3——C_2H_5 ⟶   HO——CH_3 ⟶   HO——H
| 2. Rotate | 2. Rotate |
H   counter-   H   counter- C_2H_5
                clockwise   clockwise   **A**

b. identical
c. enantiomers
d. enantiomers

5.44   The Fischer projections in parts a and b have the lowest priority group in a vertical position (top or bottom), so we can use the method for assigning configuration discussed in Prob. 5.43.

a.   b.
Cl CH_3
| |
$CH_3 —\!\!\!\!— C_2H_5$ $Cl —\!\!\!\!— CH(CH_3)_2$
| |
H C_2H_5

R-configuration   R-configuration

c. In this case the lowest priority group is in a horizontal position. We can manipulate the structure such that this group occupies a vertical position, or observe that the arrangement of the three top priority groups is clockwise, signifying the S configuration (See Sec. 5.8).

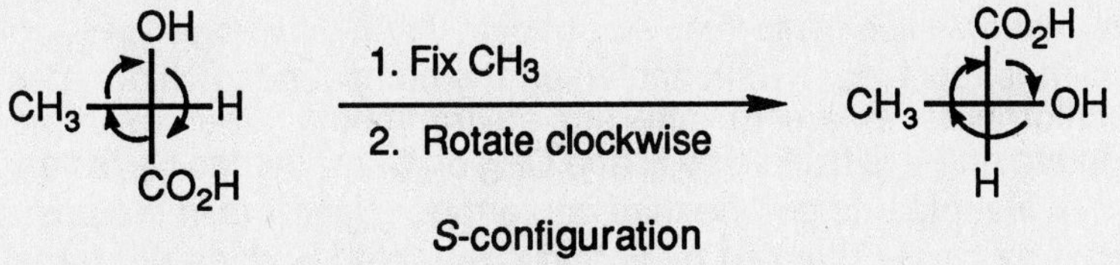

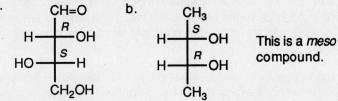

S-configuration

5.45 Use the method discussed in Prob. 5.44c.

a.

CH=O
H——OH   R
HO——H   S
CH₂OH

b.

CH₃
H——OH   S
H——OH   R
CH₃

This is a *meso* compound.

5.46 (2*R*, 3*S*, 4*R*, 5*R*).

5.47 For three different asymmetric carbons, $2^3 = 8$. The possibilities are as follows:

R—R—R   S—R—R
R—R—S   S—R—S
R—S—R   S—S—R
R—S—S   S—S—S

For four different asymmetric carbons, $2^4 = 16$. The possibilities are as follows:

R—R—R—R   R—S—R—R   S—R—R—R   S—S—R—R
R—R—R—S   R—S—R—S   S—R—R—S   S—S—R—S
R—R—S—R   R—S—S—R   S—R—S—R   S—S—S—R
R—R—S—S   R—S—S—S   S—R—S—S   S—S—S—S

5.48 Suppose we consider the dash-wedge formula for staggered conformations of one of the two enantiomers, say (*R*)-2-chlorobutane:

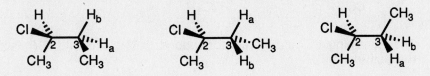

No matter which conformer we choose, the two hydrogens H_a and H_b (one of which is replaced by chlorine when the compound is chlorinated to give 2,3-dichlorobutane) are in different environments. For example, in the left conformer, H_a is flanked by H and CH_3 on C-2, whereas H_b is flanked by H and Cl. If we rotate to get the third conformer, H_b now is in the position occupied formerly by H_a (flanked by H and CH_3), but H_a *does not have the position occupied formerly by* H_b (it is flanked by CH_3 and Cl, not H and Cl). The positions of H_a and H_b cannot be interchanged. These hydrogens are said to be *diastereotopic* since if one or the other were replaced by some group that would make C-3 asymmetric, the products would be diastereomers (*not* enantiomers).

Since H_a and H_b are diastereotopic (and not enantiotopic), there is no reason why they should react identically. Thus the ratio of *meso* to racemic product need not be 1:1. The argument applies regardless of which isomer of 2-chlorobutane we consider and is therefore true of the racemic mixture as well.

5.49

This is the *meso* form. As drawn, this conformation has a center of symmetry, the midpoint of the central C—C bond. The plane of symmetry is readily seen if we rotate the "rear" carbon 180°:

The remaining two structures correspond to the S,S and R,R isomers, respectively:

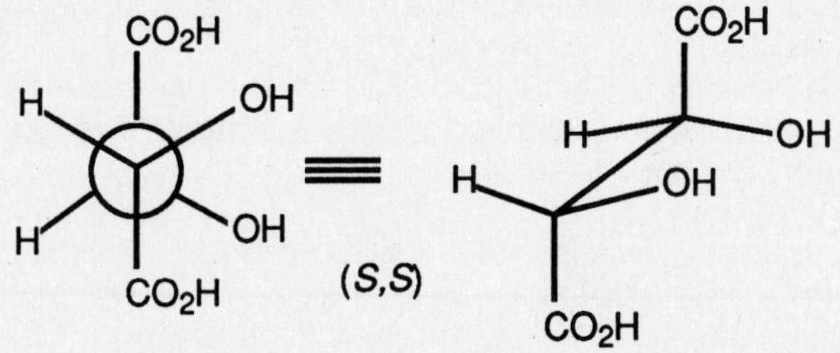

The priority order at each chiral center is OH > CO2H > CH(OH)CO2H > H.

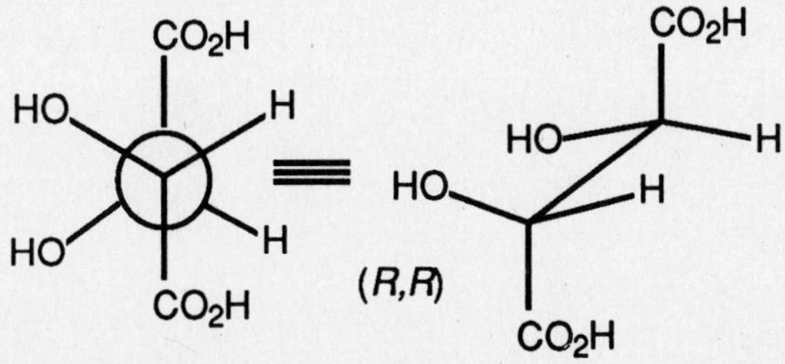

5.50

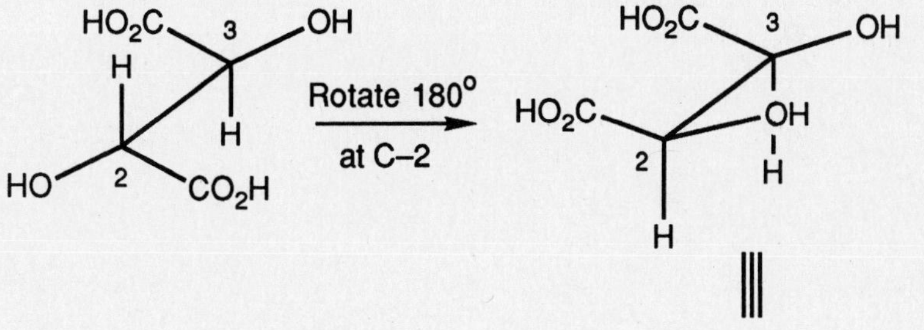

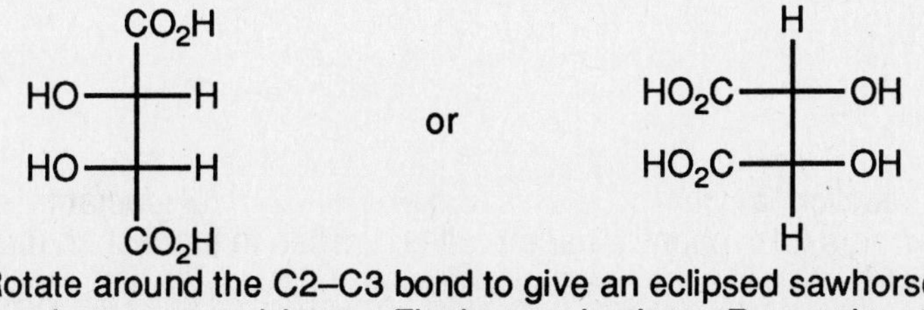

Rotate around the C2–C3 bond to give an eclipsed sawhorse projection. This can be converted into a Fischer projection. Remember that vertical lines project behind the plane of the page and horizontal lines project out of the page. There are several correct Fischer projections in this case. The plane of symmetry, that is easily seen in the projection shown above, shows that this is *meso*-tartaric acid.

5.51   The structures are conformational isomers. Both are achiral. They are diastereomers (stereoisomers but not mirror images).

5.52   The structures are conformational isomers. Both are chiral. They are enantiomers (they have nonsuperimposable mirror images). Although they are enantiomers, they interconvert readily through rotation about the C—C bond. They cannot, therefore, be separated from one another.

5.53   The compound has three different chiral centers, indicated in the formula below. There are therefore 2^3 or eight possible stereoisomers.

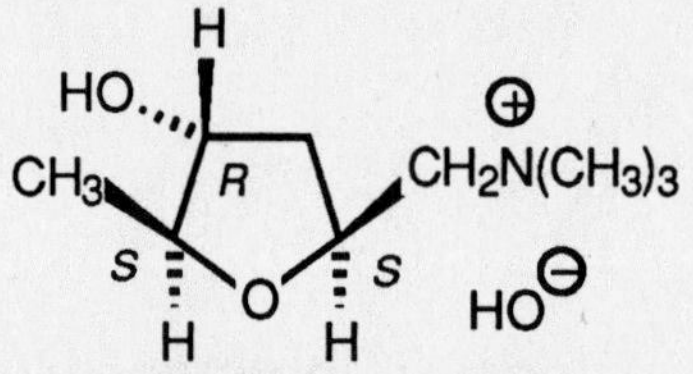

The configurational designation (*R* or *S*) at each center in the naturally occurring poison is shown.

5.54

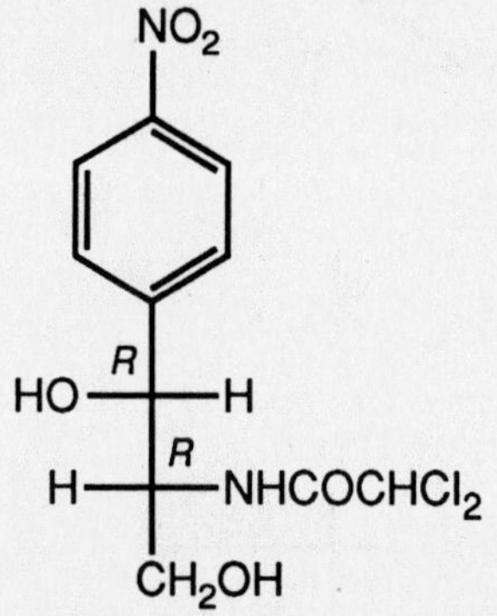

5.55   a. In this reaction, a *chiral* product is formed from *achiral* reactants, so both enantiomers (of 1-phenylethanol) will be formed in equal amounts (see Sec. 5.12).

b. In this reaction, a compound with one asymmetric center is converted to a compound with two asymmetric centers. The new asymmetric center can be either R or S and the latter is a *meso* compound.

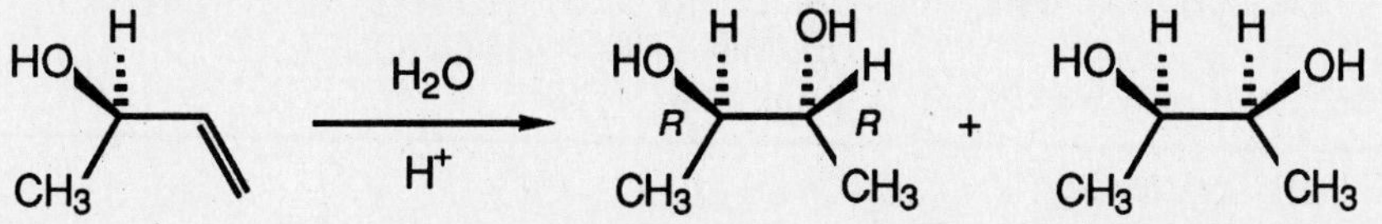

(2R,3R)-2,3-butandiol   *meso*-2,3-butandiol

5.56   The receptor site for (+)-carvone and (–)-carvone must be chiral such that the complex of the (+) and (–) isomers with the receptor are diasteromers. Therefore (+)-carvone and (–)-carvone have different achiral properties, including odor.

CHAPTER SIX: ORGANIC HALOGEN COMPOUNDS; SUBSTITUTION AND ELIMINATION REACTIONS

CHAPTER SUMMARY

Alkyl halides react with **nucleophiles,** reagents that can supply an electron pair to form a covalent bond, to give a product in which the nucleophile takes the place of the halogen. Table 6.1 gives sixteen examples of such **nucleophilic substitution reactions**, which can be used to convert alkyl halides to alcohols, ethers, esters, amines, thiols, alkyl cyanides, or acetylenes.

Nucleophilic substitution may occur by two mechanisms. The **S_N2 mechanism** is a one-step process. Its rate depends on the concentrations of substrate and nucleophile. If the halogen-bearing carbon is chiral, substitution occurs with inversion of configuration. The reaction is fastest for primary halides and slowest for tertiary halides.

The **S_N1 mechanism** is a two-step process. In the first step, the alkyl halide ionizes to a carbocation and a halide ion. In the second, fast step, the carbocation combines with the nucleophile. The overall rate is independent of nucleophile concentration. If the halogen-bearing carbon is chiral, substitution occurs with racemization. The reaction is fastest for tertiary halides and slowest for primary halides. The two mechanisms are compared in Table 6.2.

Elimination reactions often compete with substitution. They involve elimination of the halogen and a hydrogen from adjacent carbons to form an alkene. Like substitution, they occur by two main mechanisms. The **E2 mechanism** is a one-step process. The nucleophile acts as a base to remove the adjacent proton. The preferred form of the transition state is planar, with the hydrogen and the leaving group in an *anti* conformation.

The **E1 mechanism** has the same first step as the S_N1 mechanism. The resulting carbocation then loses a proton from a carbon atom adjacent to the positive carbon to form the alkene.

Several polyhalogen compounds have useful properties. Among them, carbon tetrachloride, chloroform, and methylene chloride are useful solvents. Other important polyhalogen compounds include **Halons** ($CBrClF_2$ and $CBrF_3$) used as fire extinguishers, tri- and tetrachloroethenes, used as dry-cleaning solvents, and **chloro-fluorocarbons** (CFCs) or **Freons** (CCl_3F, CCl_2F_2, $CClF_3$, $CHCl_2F$, $CHClF_2$), used as refrigerants, blowing agents, aerosol propellants, and solvents. **Teflon** is a polymer of tetrafluoroethene. It is used in nonstick coatings, Gore-Tex fabrics, insulators, and many other things. Certain **perfluorochemicals** dissolve high percentages of oxygen and can be used as artificial blood. Many halogen-containing compounds are important pesticides.

REACTION SUMMARY

Nucleophilic Substitution

$$Nu: \quad + \quad R{-}X \longrightarrow R{-}Nu^+ \quad + \quad X^-$$

$$Nu:^- \quad + \quad R{-}X \longrightarrow R{-}Nu \quad + \quad X^-$$

(See Table 6.1 for examples)

Elimination

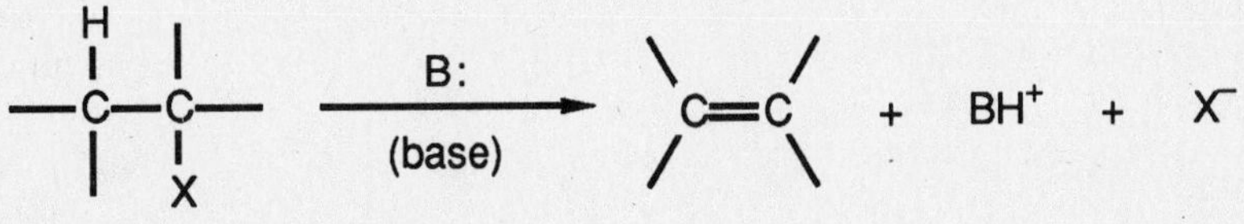

Preparation of Freons

$$CCl_4 \xrightarrow[\text{SbF}_5]{\text{HF}} CCl_3F, CCl_2F_2, \text{and so on}$$

Preparation of Teflon

$$F_2C{=}CF_2 \xrightarrow{\text{peroxide}} \left(CF_2CF_2\right)_n$$

6. ORGANIC HALOGEN COMPOUNDS; SUBSTITUTIONS AND ELIMINATIONS

MECHANISM SUMMARY

S$_N$2 (Bimolecular Nucleophilic Substitution)

Nu: + (nucleophile)

(substrate)

(leaving group)

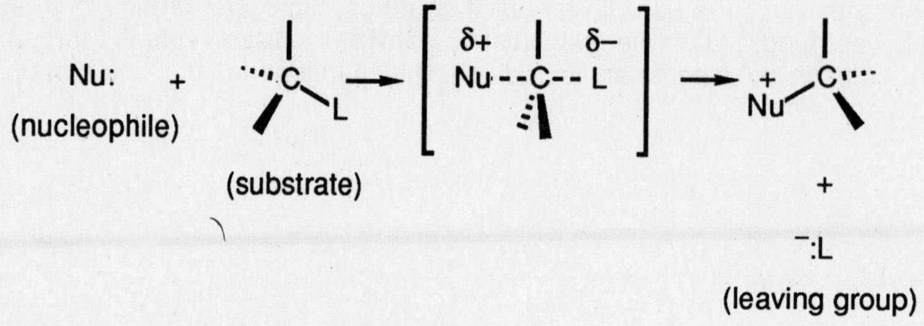

S$_N$1 (Unimolecular Nucleophilic Substitution)

(carbocation)

or

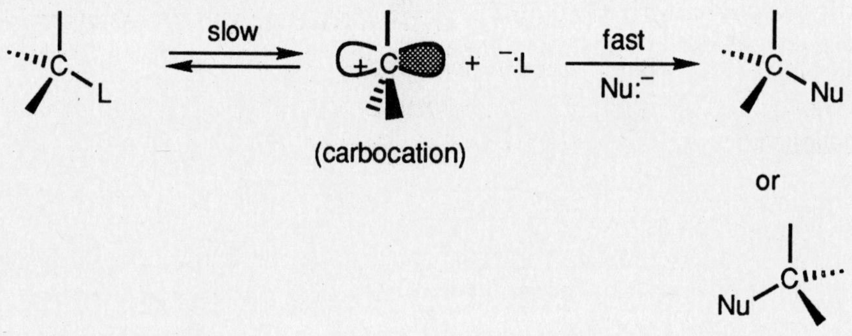

E2 (Bimolecular Elimination)

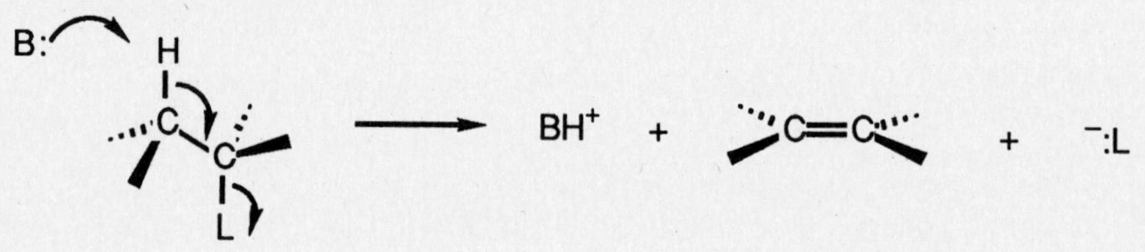

LEARNING OBJECTIVES

E1 (Unimolecular Elimination)

LEARNING OBJECTIVES

1. Know the meaning of: nucleophilic substitution reaction, nucleophile, substrate, leaving group.

2. Be familiar with the examples of nucleophilic substitution reactions listed in Table 6.1.

3. Know the meaning of: S_N2 mechanism, inversion of configuration, S_N1 mechanism, racemization, rate-determining step, E2 and E1 mechanisms.

4. Know the formulas of carbon tetrachloride, chloroform, methylene chloride, Freons, Halons, Teflon.

5. Given the name of an alkyl halide or a polyhalogen compound, write its structural formula.

6. Given the structural formula of an alkyl halide, write a correct name for it.

7. Write the equation for the reaction of an alkyl halide with any of the nucleophiles listed in Table 6.1. Recognize the class of organic compound to which the product belongs.

8. Given the structure of an alkyl halide, predict whether it is most likely to react with nucleophiles by an S_N1 or an S_N2 mechanism.

9. Given the structure of an alkyl halide and a nucleophile, write the equations that illustrate the formation of both the substitution and elimination products and be able to predict which path is likely to be favored.

10. Know the stereochemical outcome of S_N1 and S_N2 substitutions and E1 and E2 eliminations.

6. ORGANIC HALOGEN COMPOUNDS; SUBSTITUTIONS AND ELIMINATIONS

11. Given an alkyl halide with a particular stereochemistry, a nucleophile, and reaction conditions, predict the stereochemistry of the product of nucleophilic substitution.

12. Combine nucleophilic substitutions with previously studied reactions to devise a multistep synthesis of a given product.

ANSWERS TO PROBLEMS

Problems Within the Chapter

6.1 a. $NaOH + CH_3CH_2CH_2Br \longrightarrow CH_3CH_2CH_2OH + Na^+Br^-$

  (item 1, Table 6.1)

  b. $(CH_3CH_2)_3N + CH_3CH_2Br \longrightarrow (CH_3CH_2)_4N^+ \ Br^-$

  (item 10, Table 6.1)

  c.

  (item 11, Table 6.1)

6.2 a. $HS^- + CH_3CH_2CH_2CH_2Br \longrightarrow CH_3CH_2CH_2CH_2SH + Br^-$

  nucleophile   substrate   leaving group

  (item 11, Table 6.1)

  b. $HO^- + (CH_3)_2CHCH_2Br \longrightarrow (CH_3)_2CHCH_2OH + Br^-$

  nucleophile   substrate   leaving group

  (item 1, Table 6.1)

  c. $CH_3CH_2CH_2NH_2 + CH_3CH_2CH_2Br \longrightarrow (CH_3CH_2CH_2)_2\overset{+}{N}H_2$

  nucleophile   substrate $+ \ Br^-$ (leaving group)

  (item 8, Table 6.1) This reaction is followed by the acid–base equilibrium:

132

$$CH_3CH_2CH_2NH_2 \quad + \quad (CH_3CH_2CH_2)_2NH_2 \overset{+}{\rightleftharpoons} CH_3CH_2CH_2\overset{+}{N}H_3$$

$$+$$

$$(CH_3CH_2CH_2)_2NH$$

d. $(CH_3CH_2)_2S:$ + CH_3CH_2Br ⟶ $(CH_3CH_2)_3\overset{+}{S}:$ + Br^-

    nucleophile   substrate   leaving group

    (item 13, Table 6.1)

e. I^- + $H_2C{=}CHCH_2Br$ ⟶ $H_2C{=}CHCH_2I$ + Br^-

    nucleophile   substrate   leaving group

    (item 14, Table 6.1)

f.

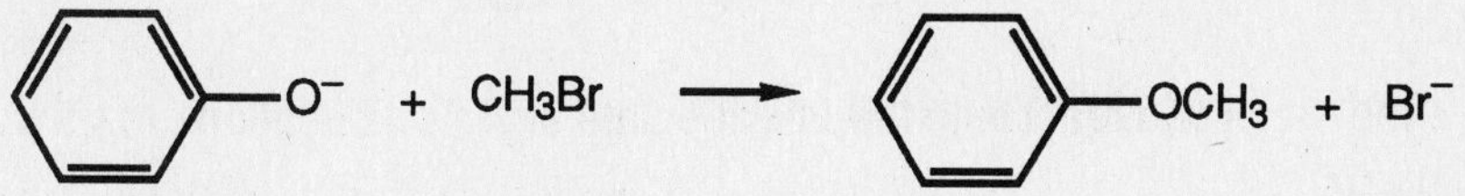

       ⬡—O^- + CH_3Br ⟶ ⬡—OCH_3 + Br^-

       nucleophile   substrate   leaving group

    rather than

      CH_3O^- +   ⬡—Br ↛ ⬡—OCH_3 + Br^-

    (item 2, Table 6.1; aryl halides do not undergo S_N2 displacement reactions)

6.3 CH_3
       |
    I—C—H
       |
    CH_2CH_3

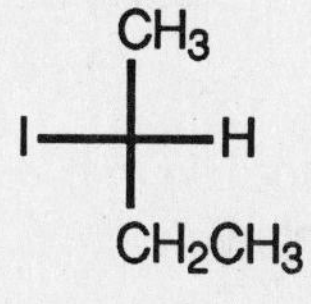

6.4 $CH_3CH{=}CHCH_2Br$ > $CH_2{=}C(CH_3)CH_2Br$ > $CH_2{=}CHCH(CH_3)Br$

    The more crowded the carbon where displacement occurs, the slower the reaction rate.

6.5 $CH_2=CHC(CH_3)_2Br$ will react faster than $CH_3CH_2C(CH_3)_2Br$ because ionization of the C—Br bond gives the more stable carbocation [tertiary (at one end) and allylic vs. tertiary].

$$H_2C=CH-\underset{\underset{CH_3}{|}}{\overset{\overset{CH_3}{|}}{C}}-Br \longrightarrow \left[\begin{array}{c} H_2C=CH-\underset{\underset{CH_3}{|}}{\overset{\overset{CH_3}{|}}{C}}+ \\ \updownarrow \\ \overset{+}{CH_2}CH=\underset{\underset{CH_3}{|}}{\overset{\overset{CH_3}{|}}{C}} \end{array} \right] \xrightarrow{CH_3OH} \begin{array}{c} H_2C=CH-\underset{\underset{CH_3}{|}}{\overset{\overset{CH_3}{|}}{C}}-OCH_3 \\ + \\ CH_3OCH_2CH=\underset{\underset{CH_3}{|}}{\overset{\overset{CH_3}{|}}{C}} \end{array}$$

The methanol can react with the intermediate allylic carbocation in two ways.

$$CH_3CH_2-\underset{\underset{CH_3}{|}}{\overset{\overset{CH_3}{|}}{C}}-Br \longrightarrow \left[CH_3CH_2-\underset{\underset{CH_3}{|}}{\overset{\overset{CH_3}{|}}{C}}+ \right] \xrightarrow{CH_3OH} CH_3CH_2-\underset{\underset{CH_3}{|}}{\overset{\overset{CH_3}{|}}{C}}-OCH_3$$

6.6   a. S_N2. The substrate is a secondary halide and may react by either S_N2 or S_N1. The nucleophile HS^- is a strong nucleophile, favoring S_N2.

   b. S_N1. The substrate is secondary and may react by either mechanism. The nucleophile (CH_3OH) is relatively weak and also polar, favoring the ionization mechanism.

   c. S_N2. The substrate is the same as in part b, but the nucleophile in this case is much stronger, favoring the displacement mechanism.

6.7   The reaction in eq. 6.4 involves a primary halide and strong nucleophile and thus proceeds by an S_N2 mechanism. The reaction in eq. 6.6 involves a tertiary halide and weak nucleophile, so it will proceed by an S_N1 mechanism. In eqs. 6.5 and 6.7, the substrate is a tertiary halide. Attack at the "rear" of the C—Br bond is sterically hindered, which slows down the rate of an S_N2 process drastically, allowing the competing elimination process to take over.

6.8   Five products are possible:

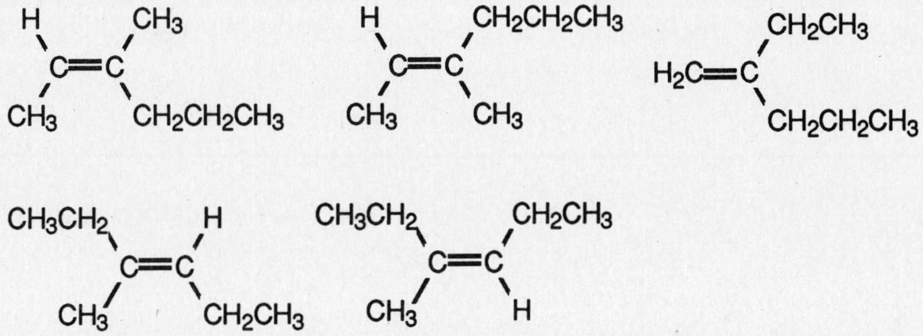

6.9

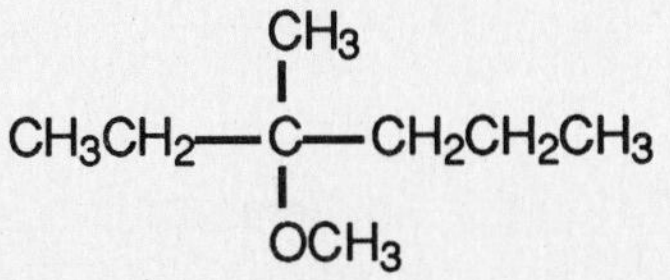

$$CH_3CH_2 - \overset{\overset{\displaystyle CH_3}{|}}{\underset{\underset{\displaystyle OCH_3}{|}}{C}} - CH_2CH_2CH_3$$

3-Bromo-3-methylhexane is a tertiary alkyl halide. It reacts with methanol, a weak nucleophile and polar solvent, to give an ether by an S_N1 mechanism.

Additional Problems

6.10   Each of these reactions involves displacement of a halogen by a nucleophile. Review Sec. 6.3 and Table 6.1.

a. $CH_3CH_2CH_2CH_2Br$ + NaI $\longrightarrow$ $CH_3CH_2CH_2CH_2I$ + NaBr

b. 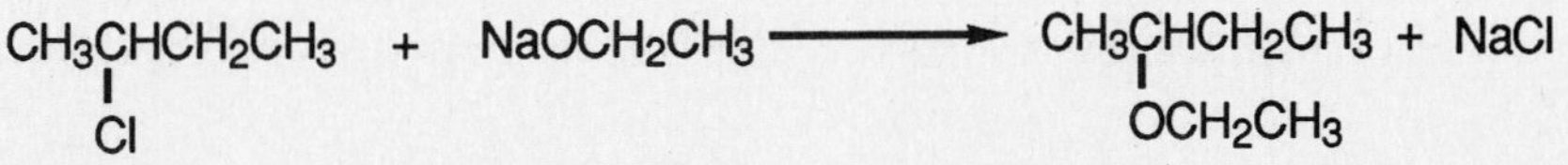 $CH_3\overset{\overset{\displaystyle }{}}{\underset{\underset{\displaystyle Cl}{|}}{C}}HCH_2CH_3$ + $NaOCH_2CH_3$ $\longrightarrow$ $CH_3\overset{}{\underset{\underset{\displaystyle OCH_2CH_3}{|}}{C}}HCH_2CH_3$ + NaCl

c. $(CH_3)_3CBr$ + H_2O $\longrightarrow$ $(CH_3)_3COH$ + HBr

The mechanism here is S_N1 (most of the other reactions in this problem occur by an S_N2 mechanism).

d. Cl—⟨benzene ring⟩—CH$_2$Cl ⟶ Cl—⟨benzene ring⟩—CH$_2$CN

+

NaCN

+

NaCl

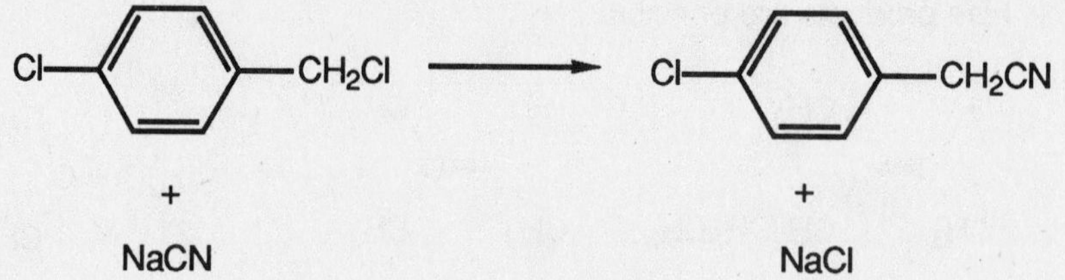

Substitution occurs only at the aliphatic (benzyl) carbon and not on the aromatic ring.

e. CH$_3$CH$_2$CH$_2$I + Na$^+$$^-$C≡CH ⟶ CH$_3CH_2CH_2$C≡CH

+

NaI

The use of acetylides as nucleophiles is a particularly important example of nucleophilic substitution because it results in a new carbon–carbon bond. Thus, larger organic molecules can be assembled from smaller ones using this method. The same is true for cyanide ion as a nucleophile (part d).

f. CH$_3$CHCH$_3$ + NaSH ⟶ CH$_3$CHCH$_3$ + NaCl
        | |
        Cl   SH

g. H$_2$C=CHCH$_2$Cl + 2 NH$_3$ ⟶ H$_2$C=CHCH$_2$NH$_2$

+

NH$_4$$^+Cl^-$

h. BrCH$_2$CH$_2$CH$_2$CH$_2$Br + 2 NaC≡N ⟶ N≡CCH$_2$CH$_2$CH$_2$CH$_2$C≡N

+

2 NaBr

Displacement occurs at both possible positions.

i. ⟨cyclohexane ring with CH$_3$ and Br⟩ + CH$_3$OH ⟶ ⟨cyclohexane ring with CH$_3$ and OCH$_3$⟩ + HBr

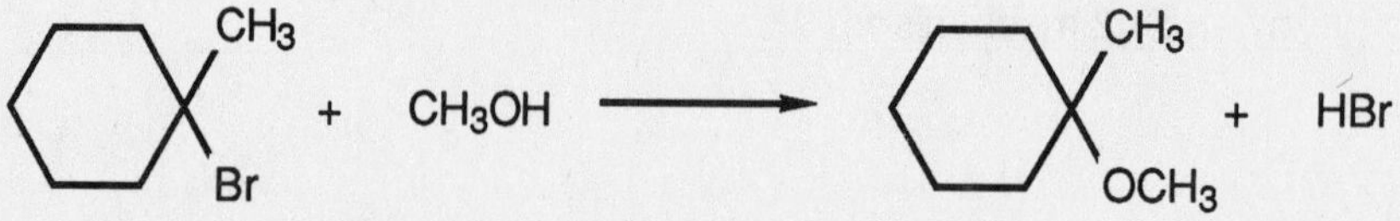

The starting halide is tertiary, and the mechanism is S$_N$1.

ANSWERS TO PROBLEMS

6.11   Use the equations in Table 6.1 as a guide.

    a. $CH_3CH_2CH_2Br + NH_3$ (item 7)

    b.. $CH_3CH_2I + CH_3CH_2S^-Na^+$ (item 12)

    c. $CH_3CH_2CH_2Br + HC\equiv C^-Na^+$ (item 16)

    d. $(CH_3)_2CHBr + (CH_3)_2CHO^-Na^+$ (item 2)

    e. 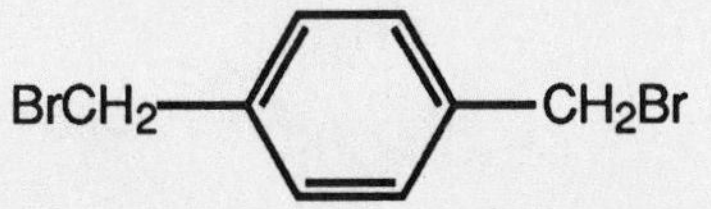 $BrCH_2$—⟨benzene ring⟩—CH_2Br + $Na^+{}^-C\equiv N$ (item 15)

    f. CH_3CH_2Br +   ⟨benzene ring⟩—O^-Na^+ (item 2)

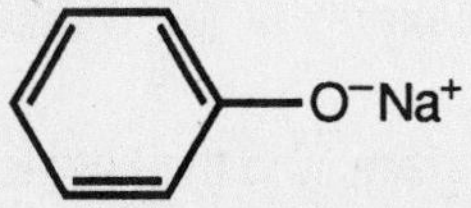

6.12   The configuration inverts if the reaction occurs by an S_N2 mechanism, but if the S_N1 mechanism prevails, considerable racemization occurs.

    a.   The nucleophile is methoxide ion, CH_3O^-. The alkyl halide is secondary, and the mechanism is S_N2.

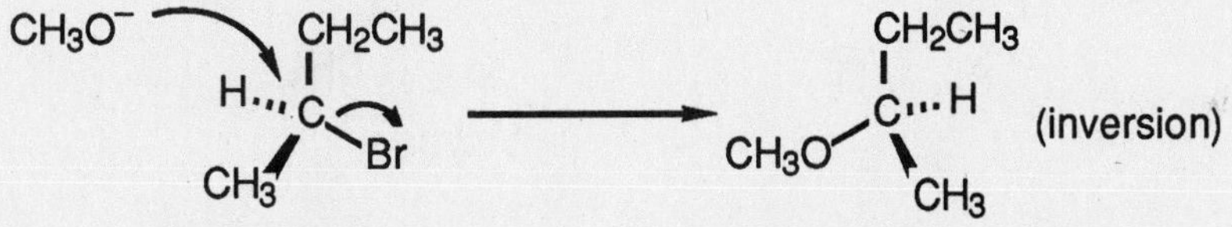

        (S)-2-bromobutane   (R)-2-methoxybutane

    b.   The alkyl halide is tertiary and the nucleophile is methanol (a weaker nucleophile than methoxide ion). The mechanism is S_N1, and the product is a mixture of R and S isomers.

137

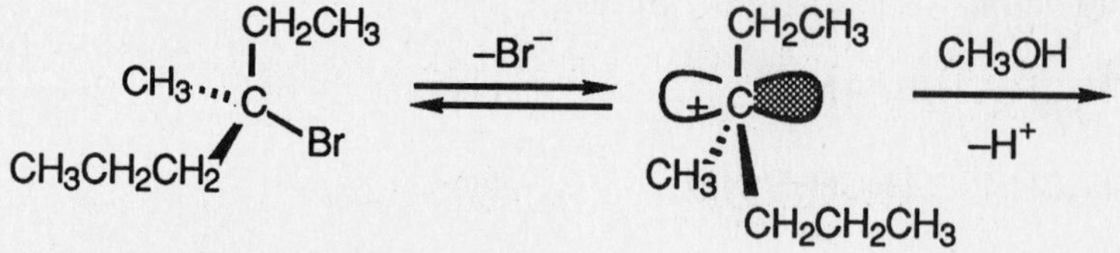

(R)-3-bromo-3-methylhexane

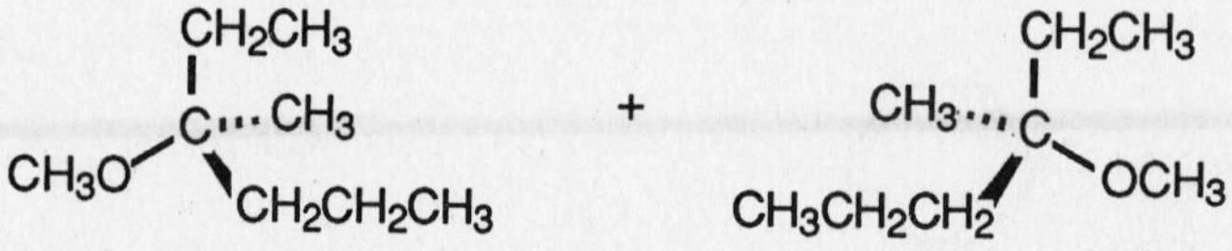

(S)-3-methoxy-3-methylhexane   (R)-3-methoxy-3-methylhexane

c. The alkyl halide is secondary, and the HS⁻ ion is a strong nucleophile. The mechanism is S_N2.

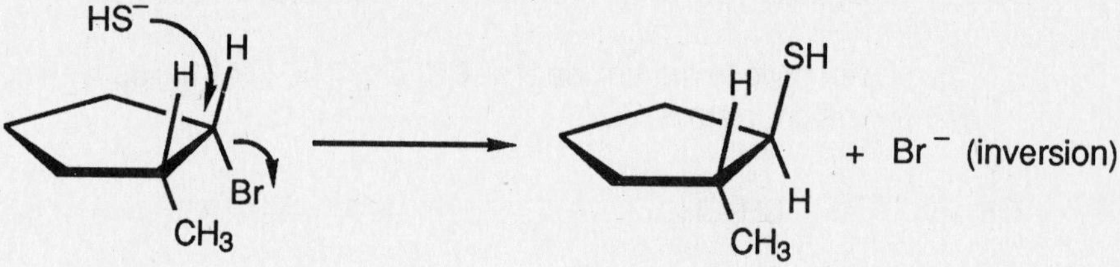

    cis   *trans*

6.13   a.

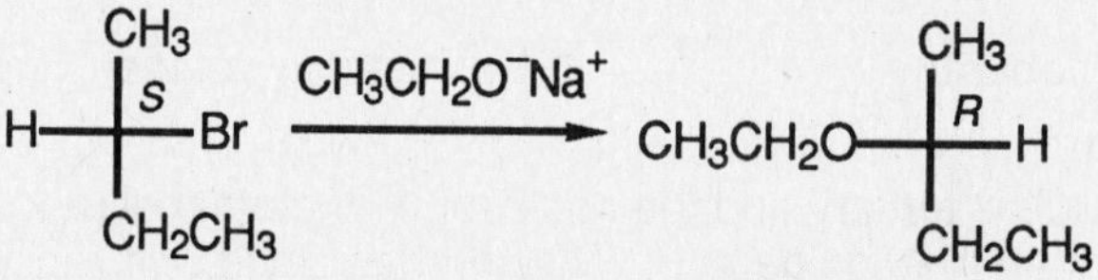

b.

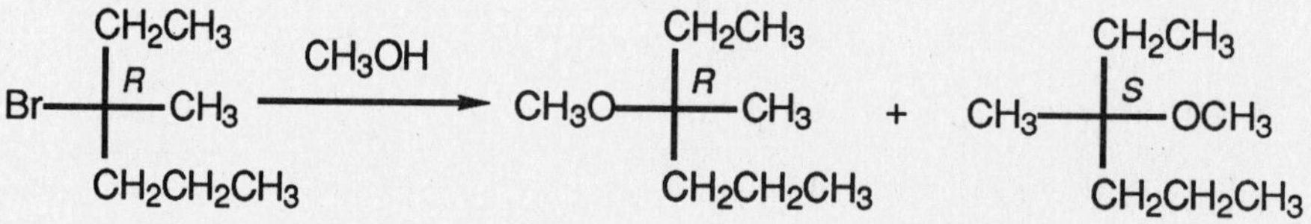

(R:S = 50:50)

6.14   a.   Sodium cyanide is a strong, anionic nucleophile. Thus the mechanism is S_N2 and the reactivity order of halides is primary > secondary > tertiary. Therefore,

$(CH_3)_2CHCH_2Br > CH_3CH(Br)CH_2CH_3 >> (CH_3)_3CBr$

b.   With 50% aqueous acetone, there is a weak nucleophile (H_2O) and a highly polar reaction medium favoring ionization, or the S_N1 mechanism. In this mechanism, the reactivity order of alkyl halides is tertiary > secondary > primary. Therefore,

$(CH_3)_3CBr > CH_3CH(Br)CH_2CH_3 >> (CH_3)_2CHCH_2Br$

6.15   An S_N2 displacement can occur. Since the leaving group and the nucleophile are identical (iodide ion), there is no change in the gross structure of the product. However, the configuration inverts every time a displacement occurs.

(R)-2-iodobutane   (S)-2-iodobutane

Since the enantiomer is produced, the optical rotation of the solution decreases. Eventually, as the concentration of the S enantiomer builds up, it too reacts with iodide ion to form some R isomer. Eventually an equilibrium (50:50) or racemic mixture is formed, and the solution is optically inactive.

6.16   The first step in the hydrolysis of any one of these halides is the ionization to a t-butyl cation:

$$(CH_3)_3C-X \xrightarrow{H_2O} (CH_3)_3C^+ + X^-$$

(X = Cl, Br, or I)

The product-determining step involves the partition of this intermediate between two paths—reaction with water and loss of a proton:

$$(CH_3)_3C-OH \xleftarrow{\quad H_2O \quad} (CH_3)_3C^+ \xrightarrow{\quad -H^+ \quad} H_2C=C(CH_3)_2$$

Since the halide ion is, to a first approximation, not involved in these steps, this partition occurs in the same ratio regardless of which alkyl halide is being hydrolyzed. This result provides experimental support for the S_N1 mechanism.

6.17　a.　The halide is tertiary, and the nucleophile is a relatively weak base. Hence the predominant mechanism is S_N1:

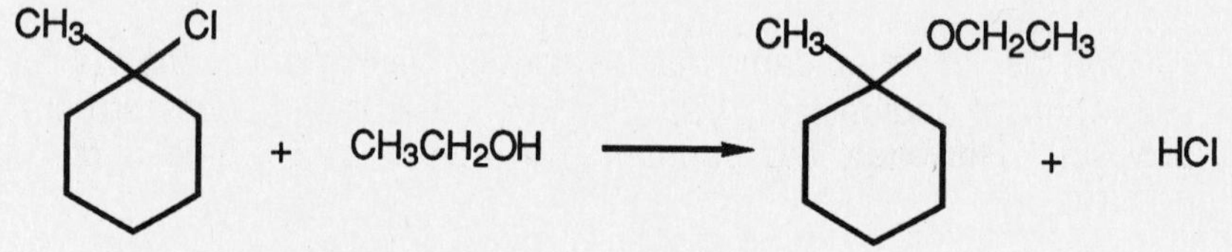

Some E1 reaction may occur in competition with S_N1, giving mainly the product with the double bond in the ring:

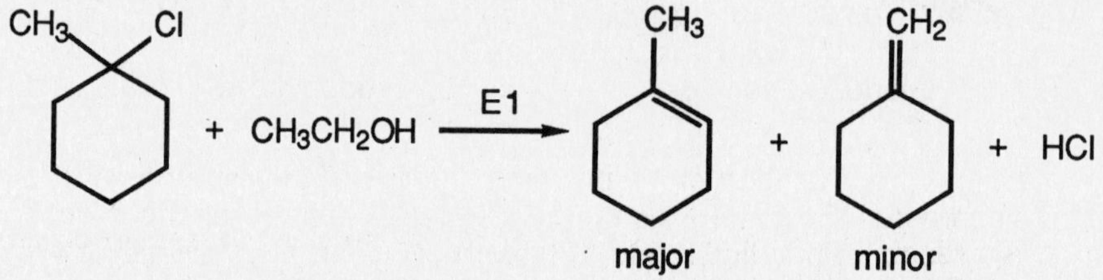

major　　　　minor

However, the main product will be the ether (S_N1).

　　b.　The nucleophile in this case is stronger, but the S_N2 process is not possible because the alkyl halide is tertiary. This nucleophile is also a strong base. Therefore, an E2 reaction will be preferred.

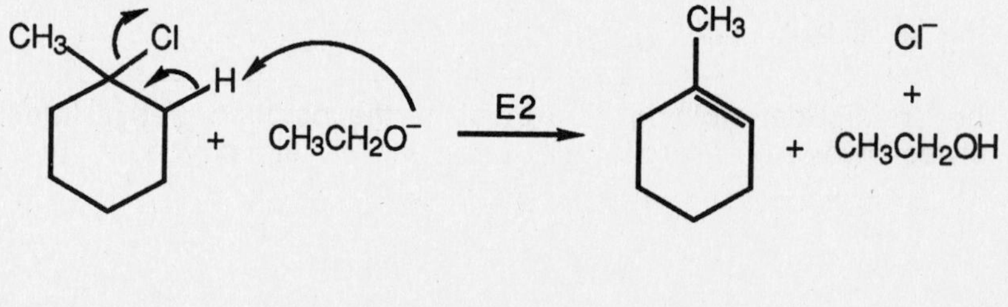

or

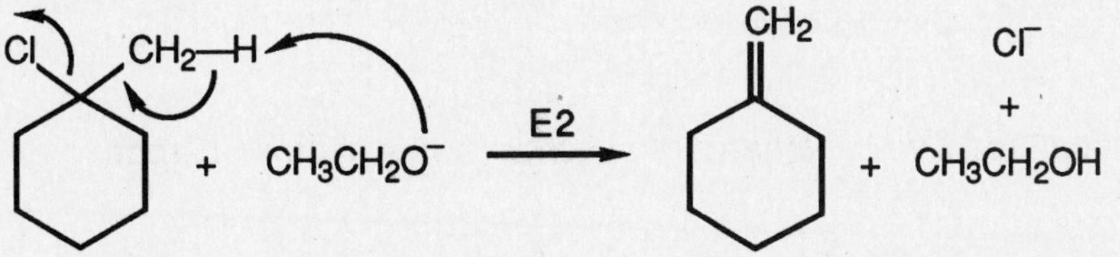

$$\text{E2}$$

The predominant product is 1-methylcyclohexene, the more stable of the two possible alkenes.

6.18

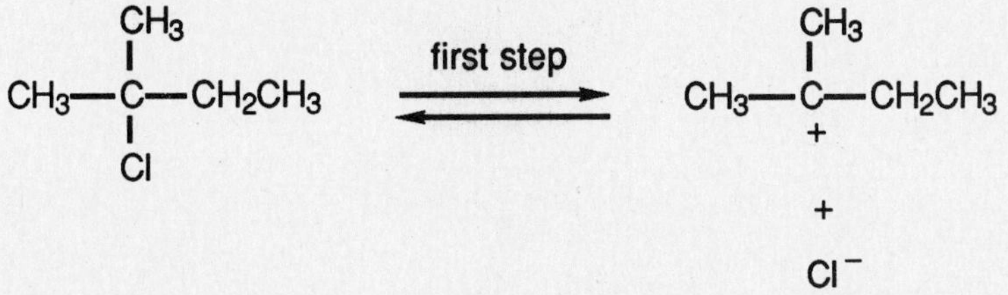

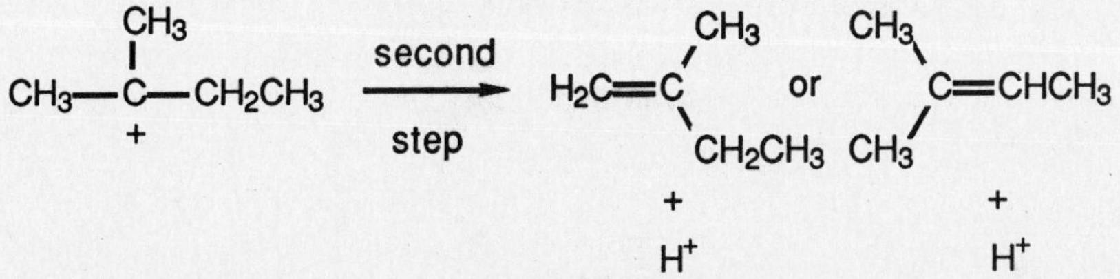

first step

second step

In the second step, the proton may be lost from either of the methyl carbons *or* from the methylene carbon, giving the two alkenes shown.

6.19   The first reaction involves a strong nucleophile (CH_3O^-), and the S_N2 mechanism is favored. Therefore, only one product is obtained. The second reaction involves a weak nucleophile (CH_3OH) that is also a fairly polar solvent, favoring the S_N1 mechanism:

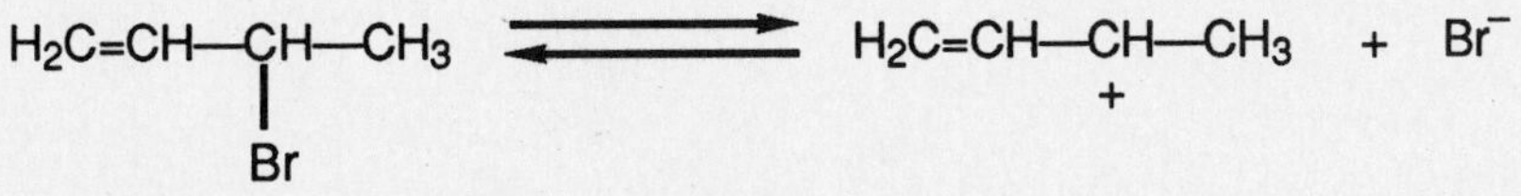

The carbocation is a resonance hybrid:

$$H_2C = CH - CH - CH_3 \overset{+}{\longleftrightarrow} \overset{+}{H_2C} - CH = CH - CH_3$$

It can react with methanol at either positively charged carbon, giving the two observed products.

6.20   a. $CH_3CH = CHCH_3 \xrightarrow{HBr} CH_3CHCH_2CH_3 \underset{S_N2}{\overset{NaOCH_3}{\longrightarrow}} CH_3CHCH_2CH_3$

with Br below the first product and OCH$_3$ below the second product.

b. $(CH_3)_2C = CHCH_3 \xrightarrow{HBr} CH_3CCH_2CH_3 \underset{S_N1}{\overset{CH_3OH}{\longrightarrow}} CH_3CCH_2CH_3$

with CH$_3$ above and Br below the first product, and CH$_3$ above and OCH$_3$ below the second product.

Here we use a weak nucleophile (base), CH_3OH, because the substrate is a tertiary bromide, and we want to favor the S_N1 mechanism. If we had used Na$^+$ $^-OCH_3$ instead, considerable elimination (E2) would have occurred. In part a, however, the alkyl halide was secondary, so the stronger nucleophile was required.

c. [benzene ring]–CH=CH$_2$ $\xrightarrow{HBr}$ [benzene ring]–CHCH$_3$ (with Br below) $\underset{S_N2}{\overset{NaCN}{\longrightarrow}}$

[benzene ring]–CHCH$_3$ (with CN below)

In the first step, addition occurs according to Markovnikov's rule. The second step occurs at the benzylic position by the S_N2 mechanism.

6.21   Work backward from the desired product to a reasonable starting material:

a.

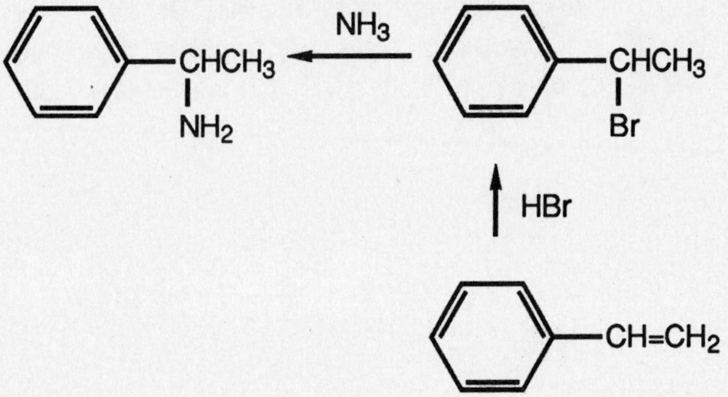

b. $CH_3CH_2CHCH_2CH_3$ $\xleftarrow{\text{NaSH}}$ $CH_3CH_2CHCH_2CH_3$
               | |
               SH   Br

                                                ↑ HBr

                                 $CH_3CH=CHCH_2CH_3$

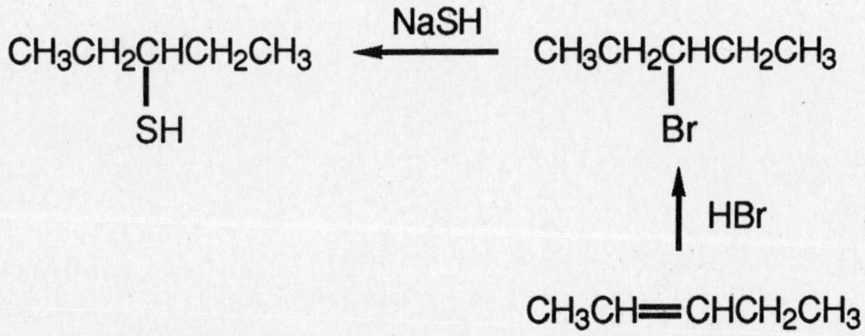

This synthesis will work, but it will *not* give a good yield because the first step can give not only 3-bromopentane but its regioisomer, 2-bromobutane (both products are obtained via secondary carbocation intermediates, which have nearly equal stabilities).

6.22 a. $2\ CH_3OH\ +\ 2\ Na\ \longrightarrow\ 2\ CH_3O^-\ Na^+\ +\ H_2$

   $CH_3O^-\ Na^+\ +\ CH_3CH_2Br\ \xrightarrow{\text{S}_\text{N}2}\ CH_3OCH_2CH_3\ +\ Na^+Br^-$

This two-step synthesis of ethers is called the Williamson synthesis (see Sec. 8.6).

   b. $(CH_3)_3CBr\ +\ CH_3OH\ \xrightarrow{\text{S}_\text{N}1}\ (CH_3)_3COCH_3\ +\ HBr$

We select this combination of reagents, *not* $(CH_3)_3COH + CH_3Br$, because methyl bromide, being primary, will not react by an S_N1 mechanism, and $(CH_3)_3COH$ is too weak a nucleophile to displace Br^- from CH_3Br in an S_N2 process. [$(CH_3)_3CO^-K^+$ would provide a strong enough nucleophile, and the reaction

$$(CH_3)_3CO^-K^+ \quad + \quad CH_3Br \quad \xrightarrow{\ S_N2\ } \quad (CH_3)_3COCH_3 \quad + \quad K^+Br^-$$

would provide an alternative synthesis of the desired product, but it uses an alkoxide rather than an alcohol as the problem specifies.]

6.23 a. $CH_2{=}CH{-}CH{=}CH_2 \ + \ HBr \ \xrightarrow{\ 1,4\ } \ CH_3{-}CH{=}CH{-}CH_2Br$

$$S_N2 \Big\downarrow \text{NaCN}$$

$$CH_3{-}CH{=}CH{-}CH_2C{\equiv}N$$

b. $CH_2{=}CH{-}CH{=}CH_2 \ + \ Br_2 \ \xrightarrow{\ 1,4\ } \ BrCH_2{-}CH{=}CH{-}CH_2Br$

$$S_N2 \Big\downarrow \text{2 NaCN}$$

$$N{\equiv}CCH_2{-}CH{=}CH{-}CH_2C{\equiv}N$$

6.24 This type of reaction sequence is a useful method for constructing C—C bonds.

a. $CH_3C{\equiv}CH \ + \ Na^+NH_2^- \ \xrightarrow{\ NH_3\ } \ CH_3C{\equiv}C^-Na^+$

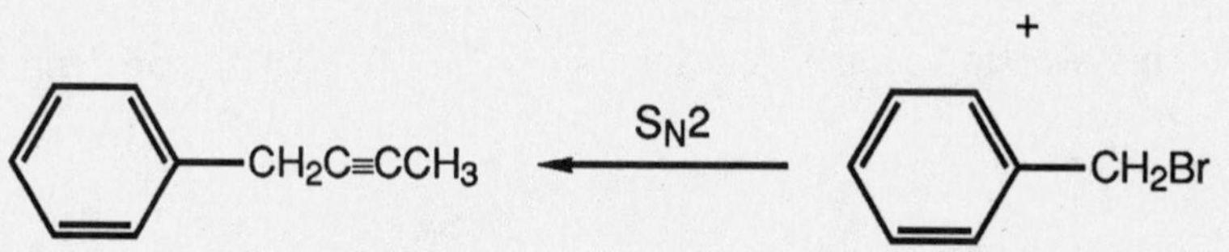

$$+$$

b. 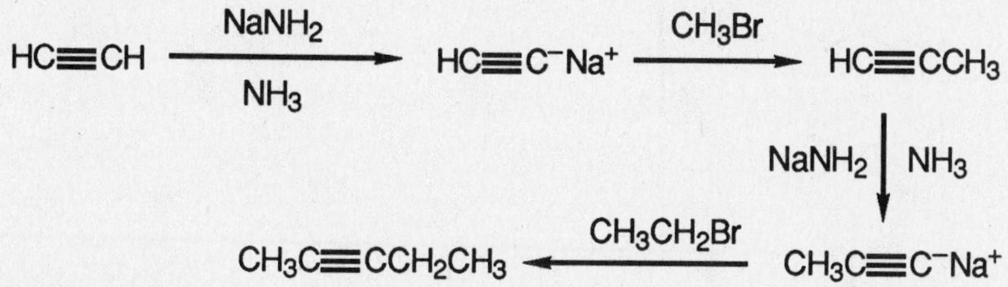 $HC\equiv CH$ $\xrightarrow[NH_3]{NaNH_2}$ $HC\equiv C^- Na^+$ $\xrightarrow{CH_3Br}$ $HC\equiv CCH_3$

$\downarrow NaNH_2 \mid NH_3$

$CH_3C\equiv CCH_2CH_3$ $\xleftarrow{CH_3CH_2Br}$ $CH_3C\equiv C^- Na^+$

The order in which the alkyl halides were used could be reversed, with the same overall result.

6.25 a.

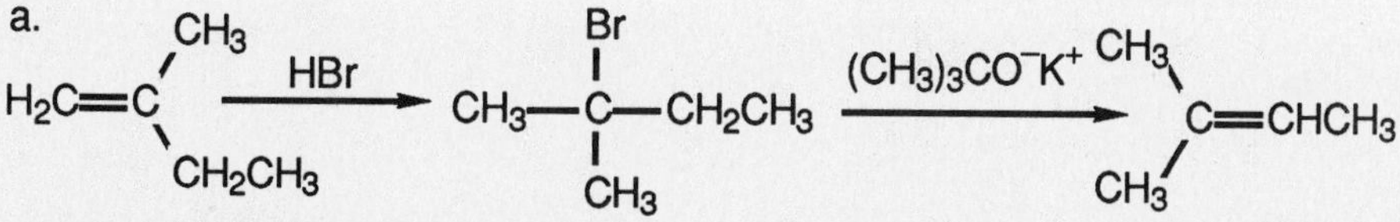

$H_2C=C \overset{CH_3}{\underset{CH_2CH_3}{}}$ $\xrightarrow{HBr}$ $CH_3-\overset{Br}{\underset{CH_3}{C}}-CH_2CH_3$ $\xrightarrow{(CH_3)_3CO^-K^+}$ $\overset{CH_3}{\underset{CH_3}{}}C=CHCH_3$

The first step follows Markovnikov's rule. The second step, an E2 elimination, gives mainly the desired trisubstituted alkene, although some of the starting alkene may also be formed.

b.

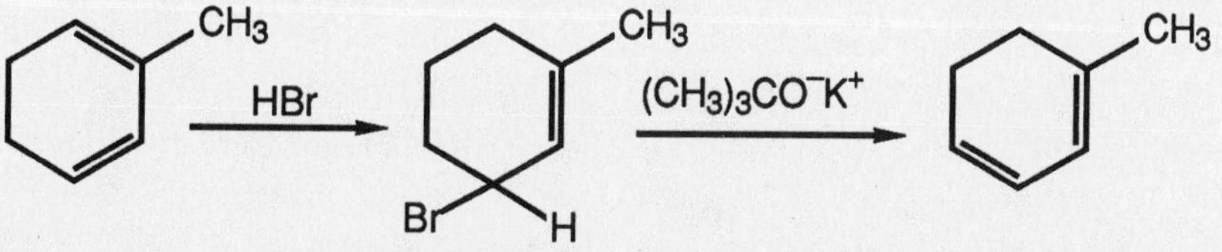

In the first step, proton addition occurs mainly at the carbon adjacent to the methyl group because the resulting allylic cation is tertiary (at one end):

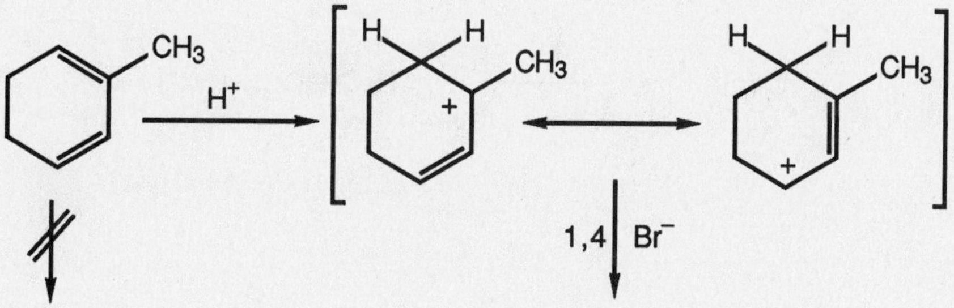

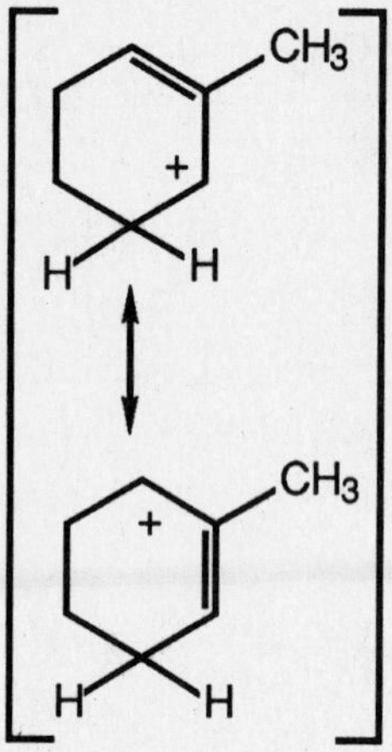

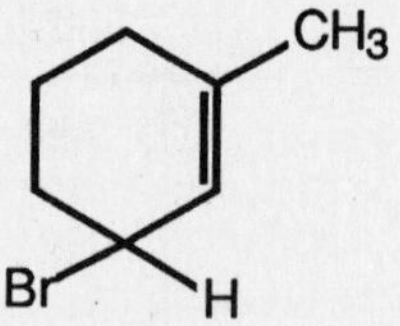

Elimination of HBr from this intermediate can yield the desired product.

6.26   a. $H_2C = CHCH_2Br$ $\xrightarrow{Na^+OH^-}$ $H_2C = CHCH_2OH$

$\downarrow$ H_2 | Pt

$CH_3CH_2CH_2OH$

b. $CH_3C \equiv CH$ $\xrightarrow[\text{NH}_3]{\text{NaNH}_2}$ $CH_3C \equiv C^- Na^+$

$\downarrow$ CH_3I

$CH_3C \equiv CCH_3$ $\xrightarrow[\text{Lindlar's catalyst}]{H_2}$ (product shown)

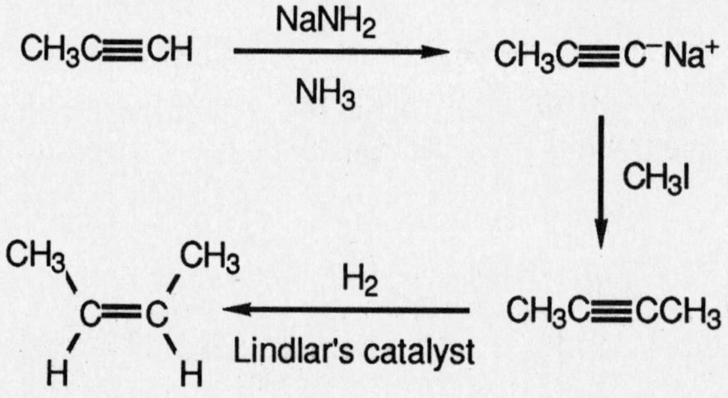

CHAPTER SEVEN: ALCOHOLS, PHENOLS, AND THIOLS

CHAPTER SUMMARY

The functional group of alcohols and phenols is the **hydroxyl group.** In alcohols, this group is connected to an aliphatic carbon, whereas in phenols, it is attached to an aromatic ring.

In the IUPAC system of nomenclature, the suffix for alcohols is *-ol.* Alcohols are classified as **primary, secondary**, or **tertiary** depending on whether one, two, or three organic groups are attached to the hydroxyl-bearing carbon. The nomenclature of alcohols and phenols is summarized in Secs. 7.2–7.4.

Alcohols and phenols form **hydrogen bonds.** These bonds account for the relatively high boiling points of these substances and the water solubility of lower members of the series.

Brønsted-Lowry and Lewis definitions of acids and bases are reviewed in Sec. 7.6. Alcohols are comparable in acidity to water, but phenols are 10^6 times more acidic. This increased acidity is due to charge delocalization (resonance) in phenoxide ions. Electron-withdrawing groups, such as —F and —NO_2, increase acidity, through either an **inductive** or a **resonance** effect, or both.

Alkoxides, the conjugate bases of alcohols, are prepared from alcohols by reaction with reactive metals or metal hydrides. They are used as organic bases. Because of the greater acidity of phenols, phenoxides can be obtained from phenols and aqueous base.

Alcohols and phenols are weak bases. They can be protonated on the oxygen by strong acids. This reaction is the first step in the acid-catalyzed dehydration of alcohols to alkenes and in the conversion of alcohols to alkyl halides by reaction with hydrogen halides. Alkyl halides can also be prepared from alcohols by reaction with **thionyl chloride** or **phosphorus halides**.

Primary alcohols can be oxidized to **carboxylic acids** using **Jones' reagent,** whereas secondary alcohols give **ketones**. Primary alcohols can be oxidized to aldehydes using **pyridinium chlorochromate** (PCC).

147

7. ALCOHOLS, PHENOLS, AND THIOLS

Glycols have two or more hydroxyl groups on adjacent carbons. **Ethylene glycol, glycerol,** and **sorbitol** are examples of glycols that are commercially important.

Three important industrial alcohols are **methanol, ethanol,** and **2-propanol.**

Phenols readily undergo aromatic substitution since the hydroxyl group is ring-activating and *ortho,para*-directing. Phenols are easily oxidized to **quinones.** Phenols with bulky *ortho* substituents are commercial antioxidants.

Examples of biologically important alcohols are **geraniol, farnesol,** and **cholesterol.**

The functional group of **thiols** is the **sulfhydryl group,** —SH. Thiols are also called **mercaptans** because of their reaction with mercury salts to form **mercaptides.** Thiols have intense, disagreeable odors. They are more acidic than alcohols and are easily oxidized to **disulfides.**

REACTION SUMMARY

Alkoxides from Alcohols

$$2 \text{ RO—H} + 2 \text{ Na} \longrightarrow 2 \text{ RO}^-\text{Na}^+ + \text{H}_2$$

$$\text{RO—H} + \text{NaH} \longrightarrow \text{RO}^-\text{Na}^+ + \text{H}_2$$

Phenoxides from Phenols

$$\text{Ar—OH} + \text{Na}^+\text{OH}^- \longrightarrow \text{ArO}^-\text{Na}^+ + \text{H}_2\text{O}$$

Dehydration of Alcohols

REACTION SUMMARY

Alkyl Halides from Alcohols

$$R\!-\!OH \ + \ HX \longrightarrow R\!-\!X \ + \ H_2O$$

$$R\!-\!OH \ + \ SOCl_2 \longrightarrow R\!-\!Cl \ + \ HCl \ + \ SO_2$$

$$3 \ R\!-\!OH \ + \ PX_3 \longrightarrow 3 \ R\!-\!X \ + \ H_3PO_3$$

Oxidation of Alcohols

$$RCH_2OH \xrightarrow{\ PCC\ } RCH\!=\!O$$

primary   aldehyde

$$RCH_2OH \xrightarrow[H^+]{CrO_3} RCO_2H$$

primary   carboxylic acid

$$R_2CHOH \xrightarrow[H^+]{CrO_3} R_2C\!=\!O$$

secondary   ketone

Aromatic Substitution in Phenols

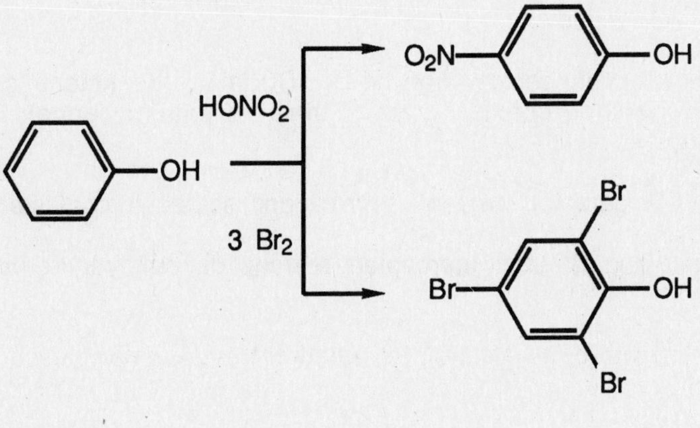

7. ALCOHOLS, PHENOLS, AND THIOLS

Oxidation of Phenols to Quinones

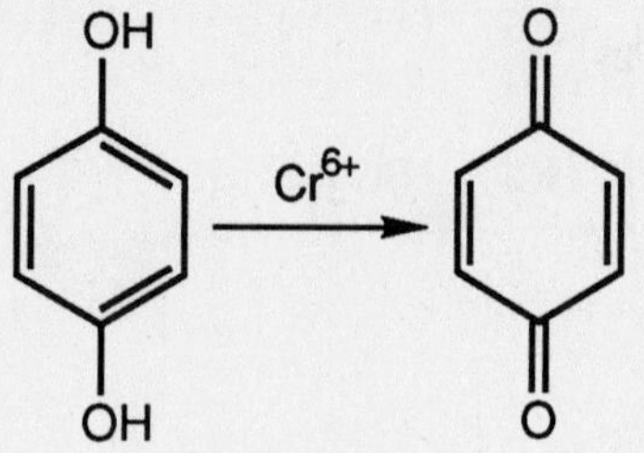

Thiols

$$RX + NaSH \longrightarrow RSH + Na^+X^- \text{ (preparation)}$$

$$RSH + NaOH \longrightarrow RS^-Na^+ + H_2O \text{ (acidity)}$$

$$2\ RSH \longrightarrow RS\text{—}SR \text{ (oxidation)}$$

LEARNING OBJECTIVES

1. Know the meaning of: alcohol, phenol, thiol, hydroxyl group, primary, secondary, and tertiary alcohol.

2. Know the meaning of: alkoxide, phenoxide, oxonium ion, alkyloxonium ion.

3. Know the meaning of: dehydration, thionyl chloride, phosphorus trichloride, phosphorus tribromide.

4. Know the meaning of: chromic anhydride, PCC, aldehyde, ketone, carboxylic acid, glycol, glycerol, sorbitol, glyceryl trinitrate (nitroglycerine), quinone, antioxidant.

5. Be familiar with: geraniol, farnesol, isoprene unit, squalene, cholesterol.

6. Know the meaning of: thiol, mercaptan, mercaptide, sulfhydryl group, disulfide.

7. Given the structure of an alcohol, tell whether it is a primary, secondary, or tertiary alcohol.

8. Given the IUPAC name of an alcohol or phenol, draw its structure.

9. Given the structure of an alcohol or phenol, assign it a correct name.

10. Explain the significance of hydrogen bonding of an alcohol or phenol with regard to solubility in water and boiling point.

11. Given a small group of compounds, including alcohols, phenols, and hydrocarbons, arrange them in order of water solubility, and construct a scheme for separating them based on acidity differences.

12. Given a group of compounds with similar molecular weights but differing potential for hydrogen bonding, arrange them in order of boiling point.

13. Draw the resonance contributors to phenoxide or substituted phenoxide ions, and discuss the acidity of the corresponding phenols.

14. Account for the acidity difference between alcohols and phenols.

15. Write equations for the reaction of a specific alcohol or phenol with sodium or sodium hydride or with an aqueous base (NaOH, KOH).

16. Write the structures for all possible dehydration products of a given alcohol, and predict which product should predominate.

17. Write the steps in the mechanism for the dehydration of a given alcohol. Given alcohols of different classes, tell which dehydration mechanism is most likely, and what the relative dehydration rates will be.

18. Write equations for the reaction of a given alcohol with HCl, HBr, or HI, with cold, concentrated H_2SO_4 or HNO_3, with PCl_3 or PBr_3, with thionyl chloride ($SOCl_2$), or with an oxidant such as chromic anhydride or PCC.

19. Write the steps in the mechanism for conversion of an alcohol to an alkyl halide.

20. Write equations for the reaction of phenol with dilute aqueous nitric acid and with bromine water.

21. Contrast the acidity of alcohols and thiols. Also contrast their reactivity toward oxidizing agents.

22. Write equations for the reaction of a given thiol with Hg^{2+}, base, or an oxidizing agent such as H_2O_2.

7. ALCOHOLS, PHENOLS, AND THIOLS

ANSWERS TO PROBLEMS

Problems Within the Chapter

7.1   a. Number from the hydroxyl-bearing carbon:

      3   2   1

$BrCH_2CH_2CH_2OH$ 3-bromo-1-propanol or just 3-bromopropanol

b. cyclopentanol

c. The alcohol takes precedence over the double bond.

     4 3 2 1

$H_2C=CHCH_2CH_2OH$ 3-buten-1-ol

7.2   a.
  1 2 3 4 5

$CH_3CHCH_2CH_2CH_3$ b. 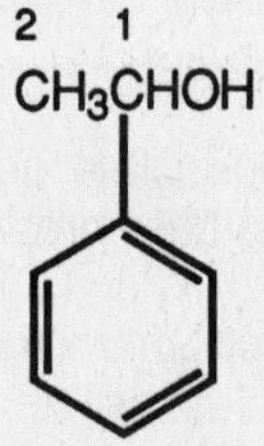
  2 1

CH_3CHOH c.
  1 2 3 4 5

$CH_3CHCH=CHCH_3$

    OH
             OH

7.3   It mixes two naming systems. Either *t*-butyl alcohol or 2-methyl-2-propanol is correct.

7.4   Methanol is usually grouped with the primary alcohols:

Primary: CH_3OH, CH_3CH_2OH, $CH_3CH_2CH_2OH$, $CH_3CH_2CH_2CH_2OH$, $(CH_3)_2CHCH_2OH$, $CH_2=CHCH_2OH$, $C_6H_5CH_2OH$ (all have a —CH_2OH group).

Secondary: $(CH_3)_2CHOH$, $CH_3CH(OH)CH_2CH_3$, cyclohexanol (all have a CHOH group).

Tertiary: $(CH_3)_3COH$

7.5   a. 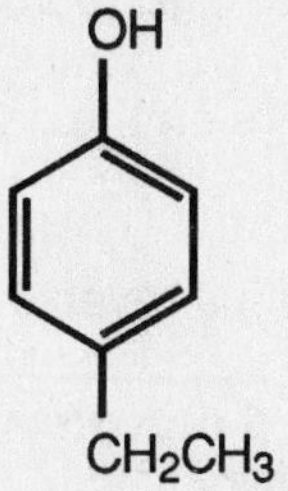   b. 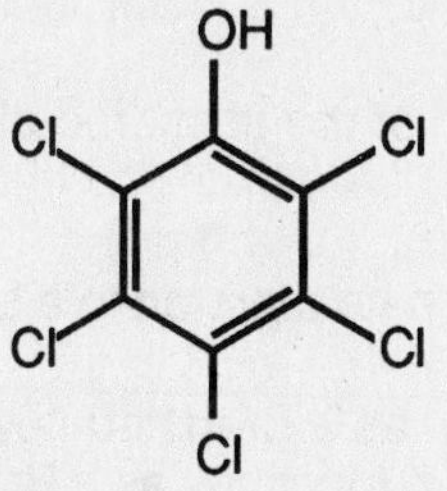   c.

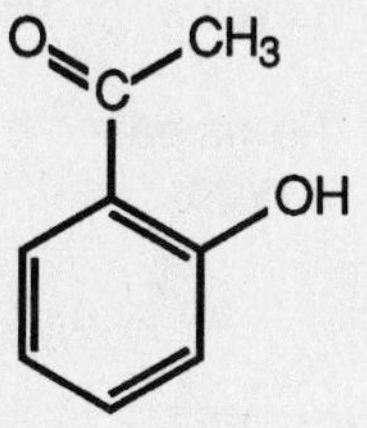

7.6   Another way to write eq. 7.6 is:

$$K_w[H_2O] = [H_3O^+][HO^-]$$

For the ionization of water (eq. 7.5), $[H_3O^+] = [HO^-]$.

Therefore $K_w[H_2O] = [H_3O^+]^2$ (or $[HO^-]^2$).

Substituting, (1.8×10^{-16}) $(55.5) = 10^{-14} = [H_3O^+]^2$ or $[H_3O^+] = 10^{-7}$.

7.7   Use eq. 7.7.

$$pK_a = -\log (1.0 \times 10^{-16}) = 16.0$$

7.8   Acetic acid has the lower pK_a and is the stronger acid.

7.9   a.   Lewis base; can donate its electron pair to an acid.

       b.   Lewis acid; the boron has only six valence electrons around it and can accept two more:

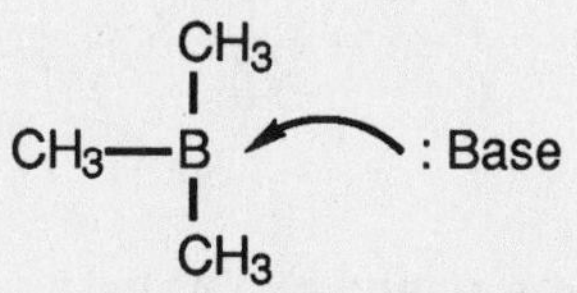

       c.   Lewis acid; can accept electron pairs to neutalize the positive charge.

       d.   Lewis base; because of the unshared electron pairs on the oxygen:

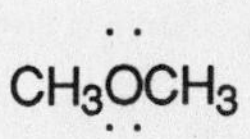

e. Lewis acid; this carbocation is isoelectronic with $(CH_3)_3B$ (part b).

f. Lewis base; because of the unshared electron pairs on the carbonyl oxygen.

7.10 NH_4^+ is an acid; it can donate a proton to a base, as in the reverse of eq. 7.10.

7.11 The amide anion functions as a Brønsted-Lowry base. It accepts a proton from a terminal acetylene, which functions as a Brønsted-Lowry acid.

7.12

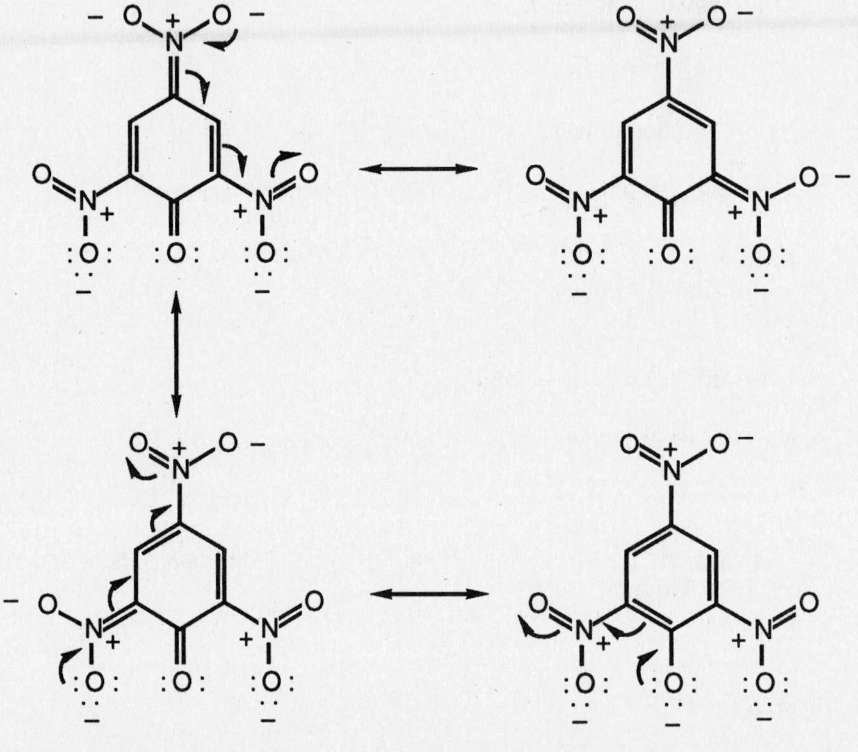

7.13 Both alcohols are weaker acids than the three phenols. Of the alcohols, 2-chloroethanol is the stronger acid because of the electronegativity of the chlorine substituent. Among the three phenols, acidity increases with increasing electronegativity of the *para* substituent: $CH_3 < H < Cl$.

$$CH_3CH_2OH \quad < \quad ClCH_2CH_2OH \quad <$$

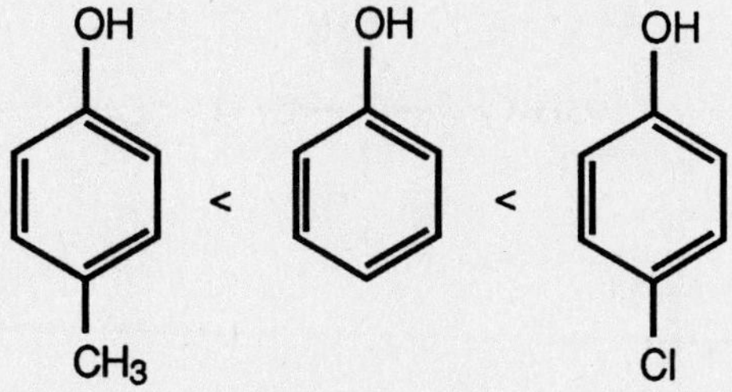

OH —⟨ring, CH3 para⟩— < OH —⟨ring⟩— < OH —⟨ring, Cl para⟩—

7.14 Follow the pattern of eq. 7.12:

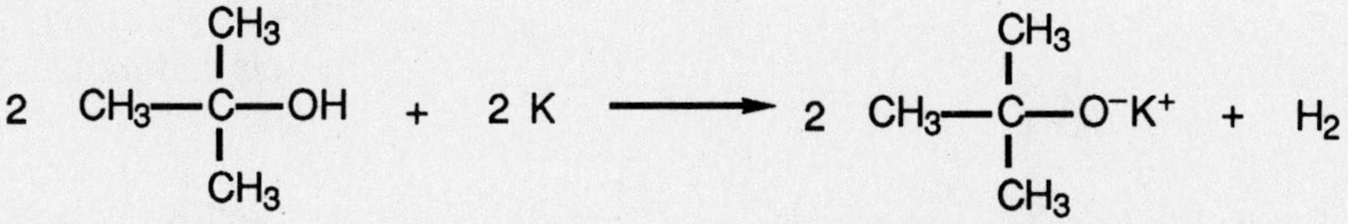

$$2 \quad CH_3-\underset{\underset{CH_3}{|}}{\overset{\overset{CH_3}{|}}{C}}-OH \quad + \quad 2\ K \quad \longrightarrow \quad 2 \quad CH_3-\underset{\underset{CH_3}{|}}{\overset{\overset{CH_3}{|}}{C}}-O^-K^+ \quad + \quad H_2$$

potassium *t*-butoxide

7.15 a. Follow the pattern of eq. 7.15:

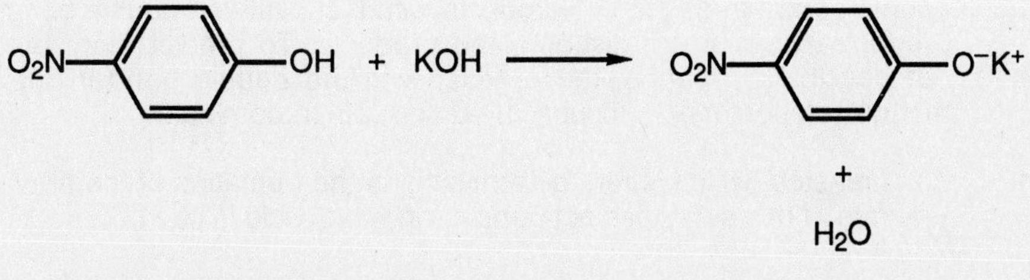

$$O_2N-\underset{}{\bigcirc}-OH \quad + \quad KOH \quad \longrightarrow \quad O_2N-\underset{}{\bigcirc}-O^-K^+$$

$$+$$

$$H_2O$$

b. Alcohols do not react with aqueous base. The equilibrium favors the starting material because hydroxide ion is a weaker base than alkoxide ion:

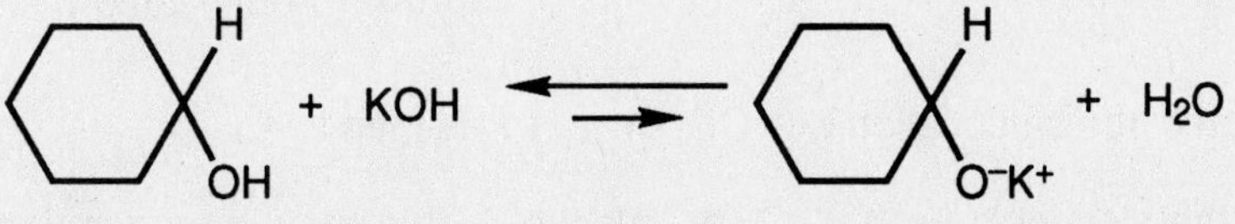

⟨cyclohexane with H and OH⟩ + KOH ⇌ ⟨cyclohexane with H and O⁻K⁺⟩ + H_2O

7.16 Write the structure of the alcohol, and consider products with a double bond between the hydroxyl-bearing carbon and each adjacent carbon that also has at least one hydrogen attached.

a.

$$CH_3CH_2-\underset{\underset{OH}{|}}{\overset{\overset{CH_3}{|}}{C}}-CH_2CH_2CH_3 \xrightarrow[\text{H}^+]{-H_2O} CH_3CH=\underset{CH_2CH_2CH_3}{\overset{CH_3}{C}}$$

(*cis* and *trans*)

+

$$CH_3CH_2-\underset{CH_2CH_2CH_3}{\overset{\overset{CH_2}{\|}}{C}} \quad + \quad \underset{CH_3CH_2}{\overset{CH_3}{C}}=CHCH_2CH_3$$

(*cis* and *trans*)

b.

The predominant product is generally the alkene with the most substituted double bond. In part a, the products with trisubstituted double bonds should predominate over the disubstituted product. To predict more precisely is not possible. In part b, the product with the double bond in the ring (1-methylcyclopentene) is trisubstituted and will predominate.

7.17 The first step, which is rate-determining, is the formation of the *t*-butyl cation. The rate of this step does not depend on which acid is used:

$$(CH_3)_3COH + H^+ \rightleftharpoons (CH_3)_3C-\overset{..}{\underset{\underset{H}{|}}{O}}{}^+ -H \xrightarrow[\text{slow step}]{S_N1} (CH_3)_3C^+ + H_2O$$

t-butyl cation

Reaction of the carbocation with Cl⁻, Br⁻, or I⁻ is then fast.

7.18 Unlike the alcohol in Prob. 7.17, the alcohol in this case is primary instead of tertiary. The rate-determining step is the S_N2 reaction:

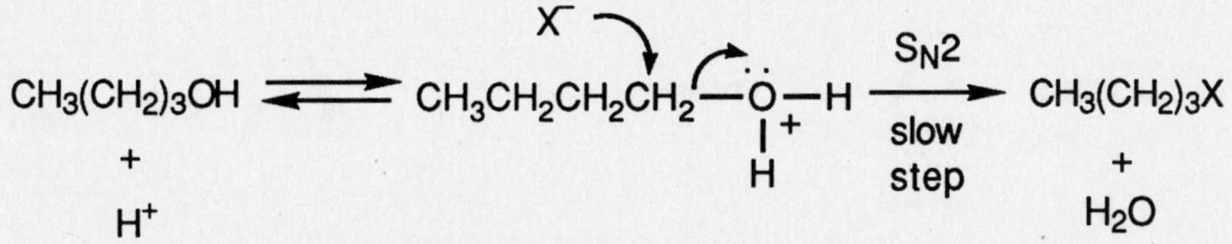

$$CH_3(CH_2)_3OH \rightleftharpoons CH_3CH_2CH_2CH_2-\overset{\overset{+}{|}}{\underset{H}{O}}-H \xrightarrow[\text{slow step}]{S_N2} CH_3(CH_2)_3X$$

$$+ \qquad\qquad\qquad\qquad\qquad\qquad\qquad + $$

$$H^+ \qquad\qquad\qquad\qquad\qquad\qquad\qquad\qquad H_2O$$

The rate of this step varies with the nucleophilicity of X^-; this order of nucleophilicity is $I^- > Br^- > Cl^-$.

7.19

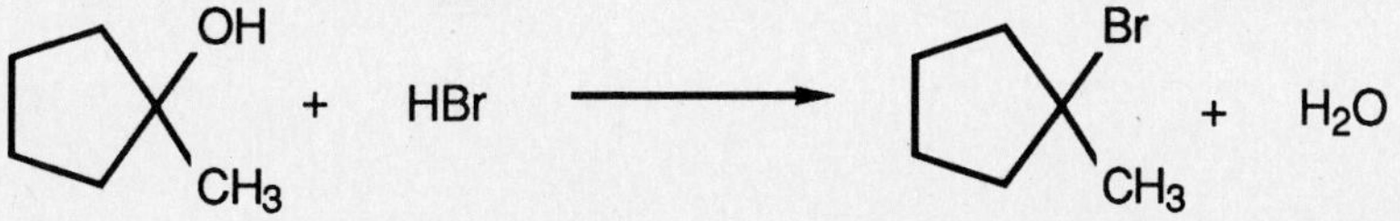

$$+ \quad HBr \longrightarrow$$

$$+ \quad H_2O$$

7.20 a. $CH_3(CH_2)_6CH_2OH + SOCl_2 \xrightarrow{\text{heat}} CH_3(CH_2)_6CH_2Cl + SO_2 + HCl$

b.

$$3\ CH_3\underset{\underset{OH}{|}}{C}HCH_3 + PBr_3 \longrightarrow 3\ CH_3\underset{\underset{Br}{|}}{C}HCH_3 + H_3PO_3$$

7.21 a.

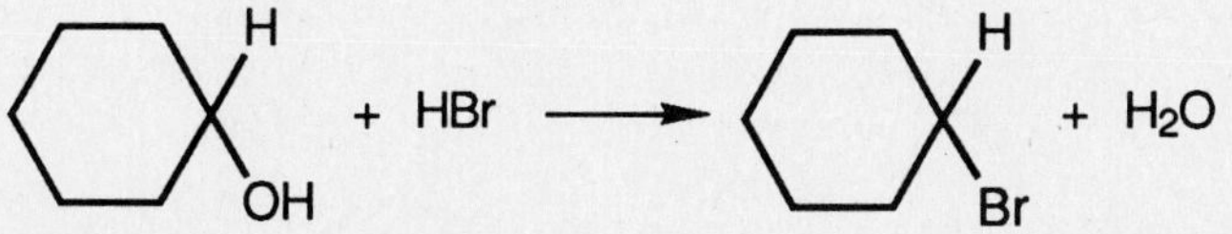

$$+ \quad HBr \longrightarrow$$

$$+ \quad H_2O \quad \text{(see eq. 7.25)}$$

Phenol does not react.

b.

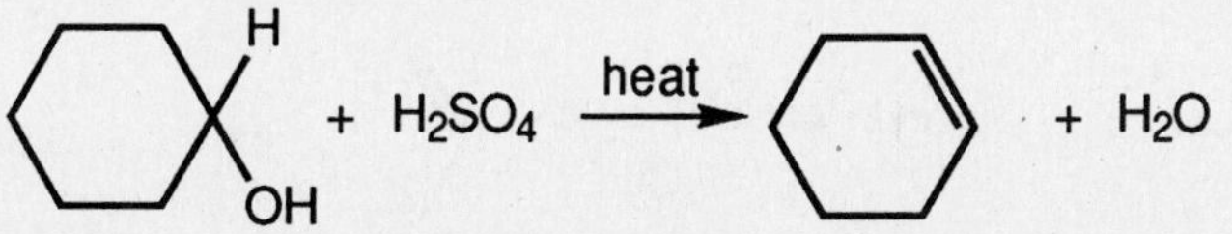

$$+ \quad H_2SO_4 \xrightarrow{\text{heat}}$$

$$+ \quad H_2O \quad \text{(compare with Prob. 7.16)}$$

Phenol may undergo electrophilic aromatic substitution:

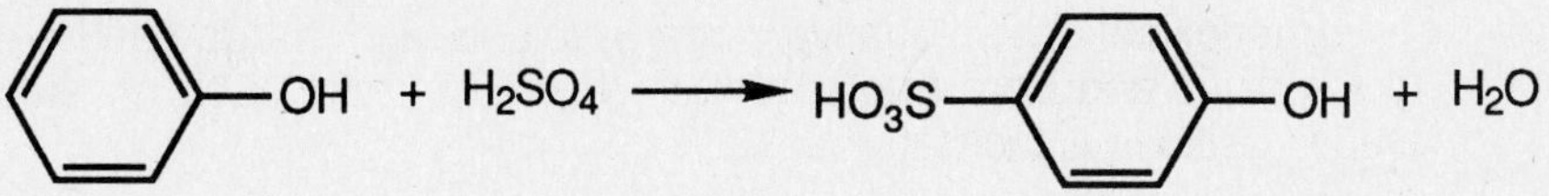

$$\text{--}OH + H_2SO_4 \longrightarrow HO_3S\text{--}\qquad\text{--}OH + H_2O$$

7. ALCOHOLS, PHENOLS, AND THIOLS

c.

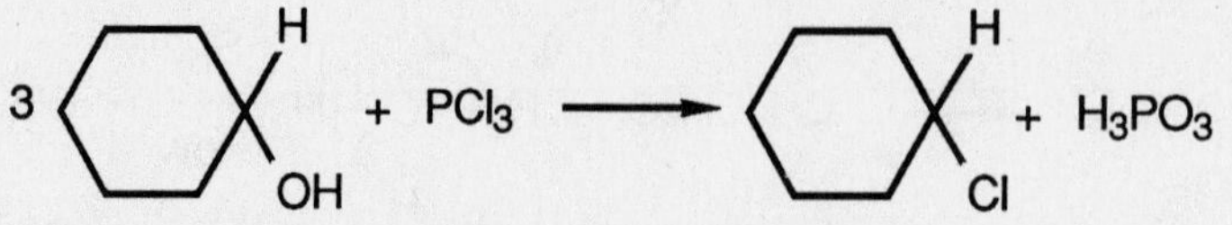

$$3 \text{ (cyclohexanol with H, OH)} + PCl_3 \longrightarrow \text{(cyclohexane with H, Cl)} + H_3PO_3 \quad \text{(compare with eq. 7.29)}$$

Phenol may react to form a phosphite ester:

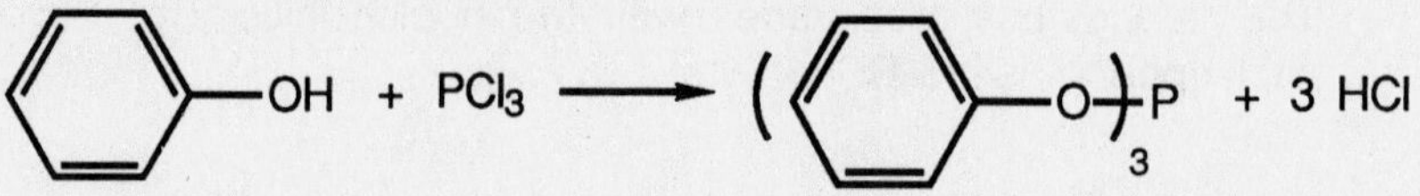

$$C_6H_5{-}OH + PCl_3 \longrightarrow (C_6H_5{-}O{-})_3 P + 3\ HCl$$

But the phenolic OH group cannot be replaced by Cl in this way.

7.22 a. The alcohol is secondary and gives a ketone.

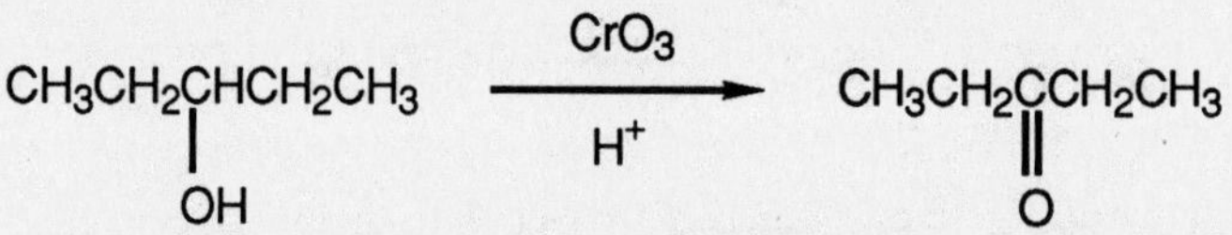

$$CH_3CH_2CHCH_2CH_3 \underset{H^+}{\overset{CrO_3}{\longrightarrow}} CH_3CH_2CCH_2CH_3$$
$$\qquad\quad |\qquad\qquad\qquad\qquad\qquad\ \|$$
$$\qquad\quad OH\qquad\qquad\qquad\qquad\qquad O$$

b. The alcohol is primary and gives a carboxylic acid.

$$C_6H_5CH_2CH_2CH_2CH_2OH \underset{H^+}{\overset{CrO_3}{\longrightarrow}} C_6H_5CH_2CH_2CH_2CO_2H$$

c. The alcohol is primary and gives an aldehyde.

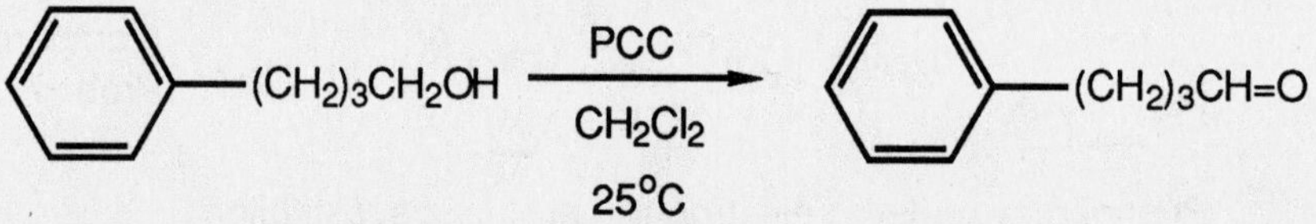

$$C_6H_5{-}(CH_2)_3CH_2OH \underset{CH_2Cl_2, 25^\circ C}{\overset{PCC}{\longrightarrow}} C_6H_5{-}(CH_2)_3CH{=}O$$

7.23 The phenoxide ion is negatively charged, and that charge can be delocalized to the *ortho* and *para* ring carbons. Therefore, attack by an electrophile at these positions is facilitated.

158

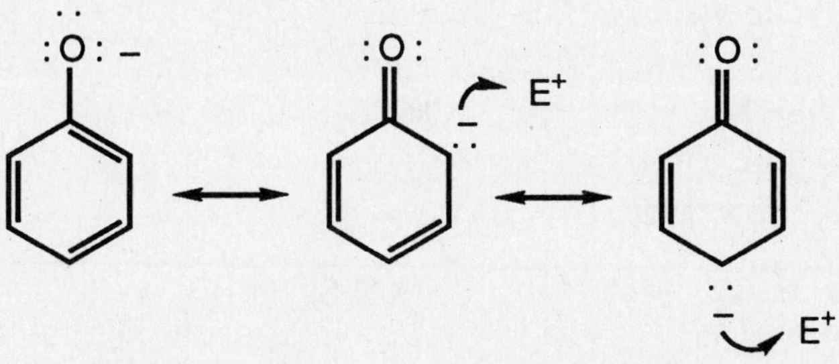

7.24 a. The hydroxyl group is more ring-activating than the methyl group. Thus substitution *ortho* to the hydroxyl group is preferred.

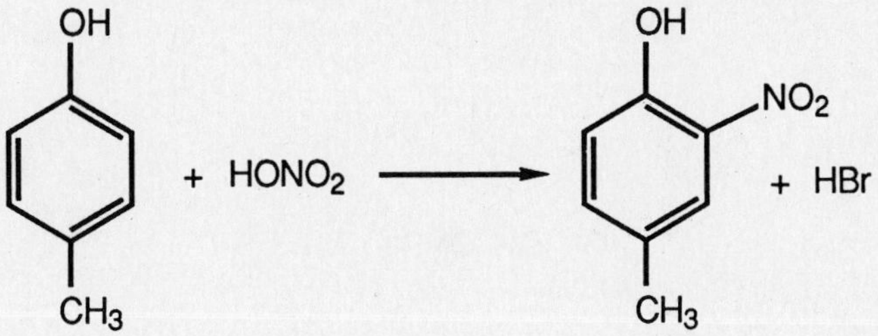

b. The hydroxyl group is ring-activating, whereas the chlorine is a ring-deactivating substituent. Therefore, substitution occurs *para* to the hydroxyl group. Substitution *ortho* to the hydroxyl group is less likely since the product would have three adjacent substituents, which would be quite crowded.

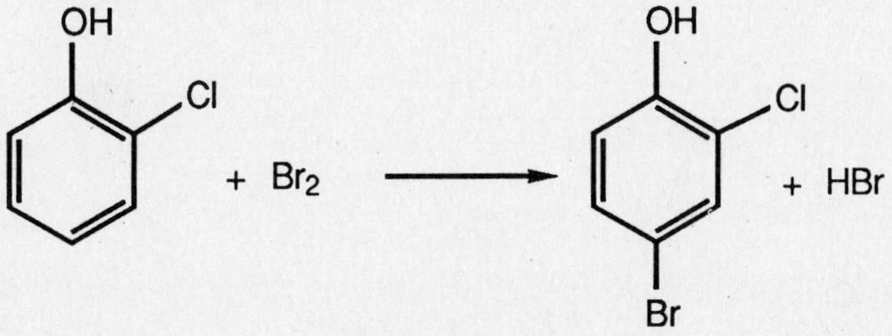

7.25 a. 1 2 3 4
$CH_3CHCH_2CH_3$
$\quad\quad|$
$\quad\;\;SH$

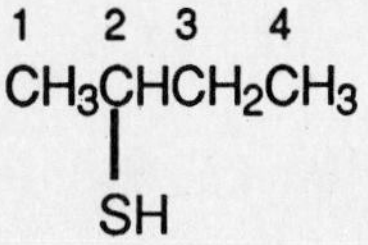

b. 1 2 3
CH_3CHCH_3
$\quad\quad|$
$\quad\;\;SH$

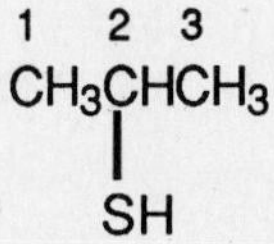

7.26 a. Follow eq. 7.44 as a guide.

$$CH_3CH_2SH + KOH \longrightarrow CH_3CH_2S^-K^+ + H_2O$$

b. Follow eq. 7.42 as a guide.

$$2\ CH_3CH_2SH + HgCl_2 \longrightarrow (CH_3CH_2S)_2Hg + 2\ HCl$$

Additional Problems

7.27

a.

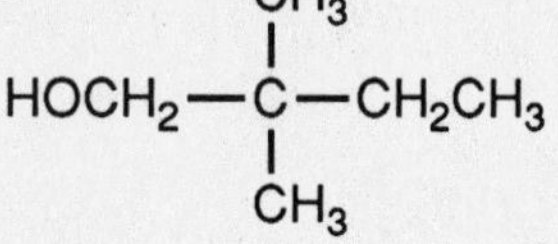

b.

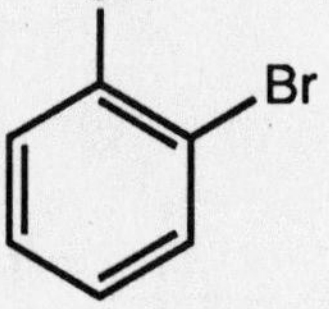

c.

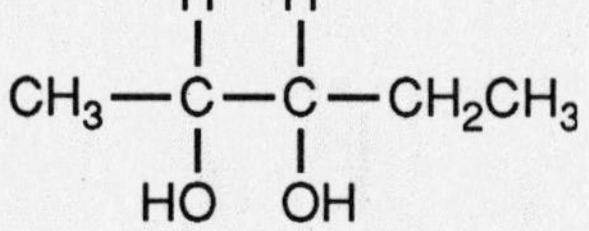

d.

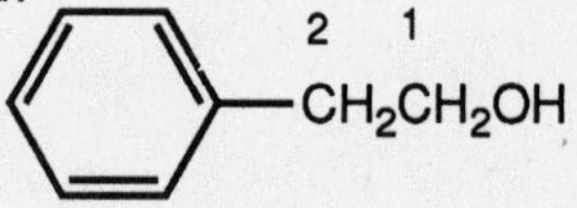

e. $CH_3CH_2ONO_2$

f.

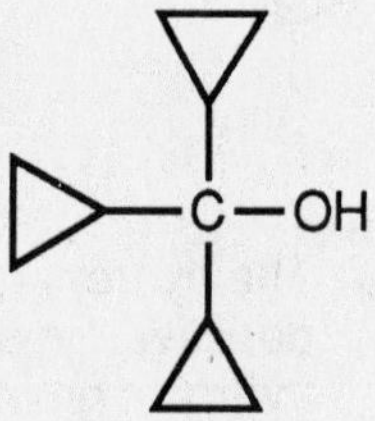

g. $CH_3CH_2O^-Na^+$

h.

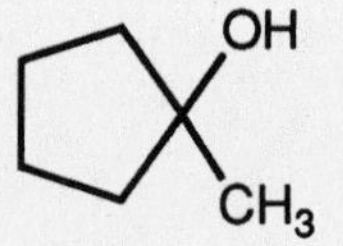

i.

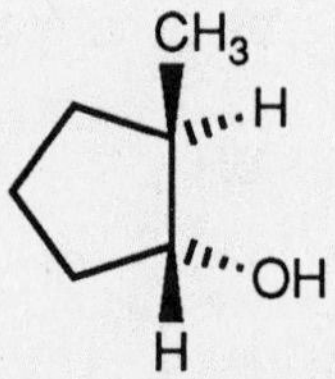

j.

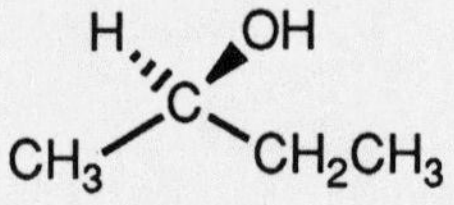

k.

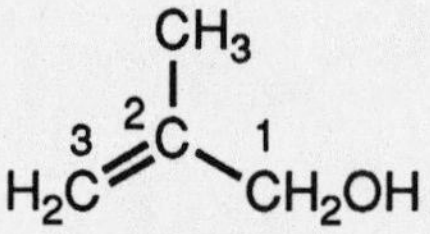

l.

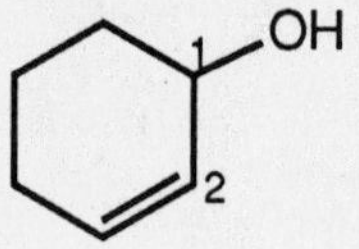

7.28 a. primary   d. primary   f. tertiary   h. tertiary
        i. secondary k. primary   l. secondary

160

7.29 a. 3,3-dimethyl-2-butanol
        c. 2,4-dichlorophenol
        e. *m*-bromophenol
        g. 2-buten-1-ol

        i.   1,2,3,4-butanetetraol
        k. *cis*-3-methylcyclobutanol
        m. 3-phenyl-2-propen-1-ol

        b. 3-bromo-2-methyl-2-butanol
        d. cyclopropanol
        f. diphenylmethanol
        h. 2-propanethiol (or isopropyl mercaptan)
        j   potassium *n*-propoxide
        l.   cyclopropanethiol
        n. 1-methyl-3-cyclopenten-1-ol

7.30 a. 3,3-Dimethyl-2-butanol; the hydroxyl should get the lower number.
        b. 2-Methyl-1-butanol; the longest chain was not selected.
        c. 2-Propen-1-ol (or allyl alcohol); the hydroxyl group should get the lower number.
        d. 3-Chlorocyclohexanol; number the ring from the hydroxyl-bearing carbon, in a direction that gives substituents the lowest possible numbers.
        e. 2,5-dibromophenol; give substituents the lowest possible numbers.
        f. *Sec*-butyl alcohol or 2-butanol; do not mix the common and IUPAC naming systems.

7.31 a. Ethyl chloride < 1-hexanol < ethanol. Both alcohols can hydrogen-bond with water and will be more soluble than the alkyl chloride. The lower-molecular-weight alcohol will be more soluble (it has a shorter hydrophobic carbon chain).
        b. 1-Pentanol < 1,5-pentanediol < 1,2,3,4,5-pentanepentaol. All three compounds have the same number of carbon atoms. Water solubility will therefore increase with increasing numbers of hydroxyl groups (that is, as the ratio of hydroxyl groups to carbon atoms increases).

7.32 a.

$$R-\overset{\cdot\cdot}{\underset{\cdot\cdot}{O}}-R + H^+ \;\rightleftharpoons\; R-\overset{\overset{\displaystyle H}{|}}{\underset{\cdot\cdot}{O}}\!^+\!-R$$

        b.

$$R-\overset{\displaystyle |}{\underset{\displaystyle R}{N}}-R + H^+ \;\rightleftharpoons\; R-\overset{\overset{\displaystyle H}{|}}{\underset{\displaystyle R}{N}}\!^+\!-R$$

        c.

$$\overset{\displaystyle R}{\underset{\displaystyle R}{{>}}}\!C{=}\overset{\cdot\cdot}{O}: \; + H^+ \;\rightleftharpoons\; \overset{\displaystyle R}{\underset{\displaystyle R}{{>}}}\!C{=}\overset{\cdot\cdot}{O}\!-H\;^+$$

7.33

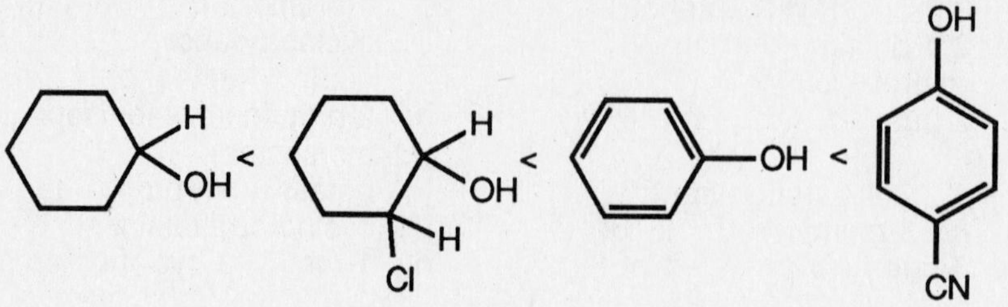

The two alcohols are less acidic than the two phenols. The electron-withdrawing chlorine substituent makes 2-chlorocyclohexanol a stronger acid than cyclohexanol. The electron-withdrawing cyano substituent makes *p*-cyanophenol a stronger acid than phenol. The negative charge in the *p*-cyanophenoxide ion can be delocalized to the nitrogen:

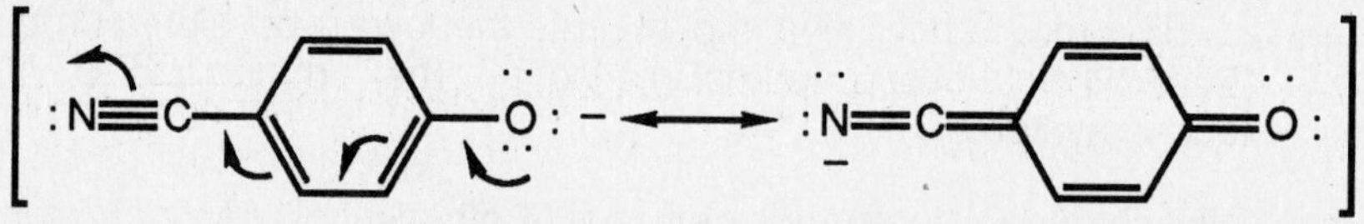

7.34 Since *t*-butyl alcohol is approximately 100 times *weaker* than ethanol as an acid (pK_a = 18 and pK_a = approximately 16, respectively), it follows that if we consider the conjugate bases, *t*-butoxide ion is a stronger base than ethoxide ion.

7.35 *p*-Methylphenol reacts with aqueous base, and the product is water soluble:

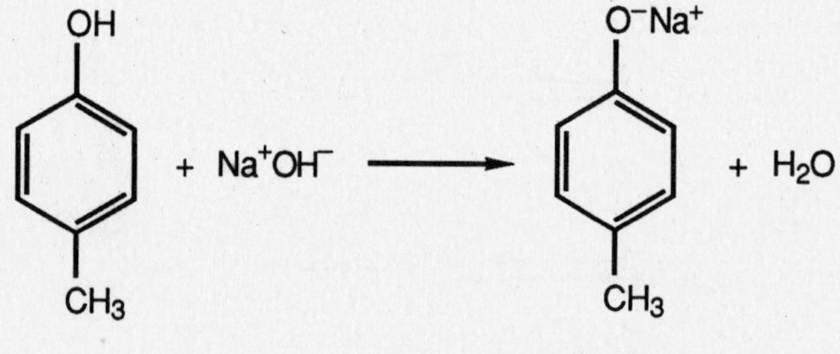

soluble in water layer

The cyclohexanol is not sufficiently acidic to react with sodium hydroxide. When the layers are separated, the cyclohexanol will be in the organic layer and the sodium *p*-methylphenoxide in the water layer. Acidification of the water layer allows the *p*-methylphenol to be recovered:

162

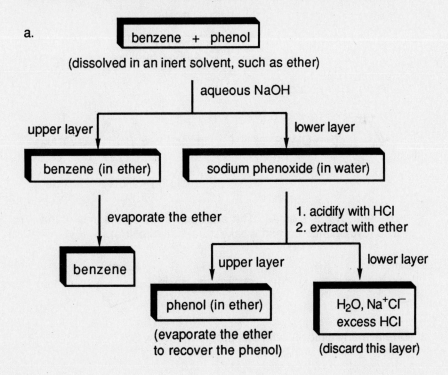

The *p*-methylphenol can then be extracted by an organic solvent.

7.36 In the laboratory, separations of this type are usually performed by dissolving the mixture in a low-boiling, inert organic solvent such as diethyl ether or methylene chloride. The solution is then extracted with an aqueous (neutral, acidic, or alkaline) solution that extracts one of the two components. The layers are then separated, and the organic layer is evaporated to recover the compound that was not extracted into the aqueous base. The aqueous layer is then treated in some way to recover the extracted compound.

a.

```
                    ┌─────────────────────┐
                    │  benzene  +  phenol │
                    └─────────────────────┘
          (dissolved in an inert solvent, such as ether)

                         aqueous NaOH

upper layer                              lower layer
  ┌────────────────────┐      ┌──────────────────────────────┐
  │ benzene (in ether) │      │ sodium phenoxide (in water)  │
  └────────────────────┘      └──────────────────────────────┘

     evaporate the ether            1. acidify with HCl
                                    2. extract with ether

      ┌──────────┐          upper layer          lower layer
      │ benzene  │     ┌──────────────────┐   ┌──────────────┐
      └──────────┘     │ phenol (in ether)│   │ H₂O, Na⁺Cl⁻  │
                       └──────────────────┘   │ excess HCl   │
                       (evaporate the ether   └──────────────┘
                        to recover the phenol) (discard this layer)
```

benzene + phenol

(dissolved in an inert solvent, such as ether)

aqueous NaOH

upper layer   lower layer

benzene (in ether)   sodium phenoxide (in water)

evaporate the ether   1. acidify with HCl
2. extract with ether

benzene   upper layer   lower layer

phenol (in ether) H_2O, Na^+Cl^- excess HCl

(evaporate the ether to recover the phenol)   (discard this layer)

b. The same procedure used in part a works here since phenol is extracted by base, whereas 1-hexanol, a much weaker acid, is not.

c. An ether solution of the two alcohols can be extracted with water. 1-Propanol is soluble, whereas 1-heptanol, with a much longer carbon chain, is not.

7.37   a. Use eq. 7.12 as a guide.

$$2 \ CH_3CHCH_2CH_3 \ + \ 2 \ K \ \longrightarrow \ 2 \ CH_3CHCH_2CH_3 \ + \ H_2$$
$$\overset{|}{OH} \qquad\qquad\qquad\qquad \overset{|}{O^-K^+}$$

potassium 2-butoxide

b. Use eq. 7.13 as a guide.

$$CH_3CHCH_3 \ + \ NaH \ \longrightarrow \ CH_3CHCH_3 \ + \ H_2$$
$$\overset{|}{OH} \qquad\qquad\qquad \overset{|}{O^-Na^+}$$

sodium isopropoxide
(or 2-propoxide)

c. Use eq. 7.15 as a guide.

$$Cl-\!\!\!\bigcirc\!\!\!-OH \ + \ NaOH \ \longrightarrow \ Cl-\!\!\!\bigcirc\!\!\!-O^-Na^+ \ + \ H_2O$$

sodium p-chlorophenoxide

d. Use eq. 7.14 as a guide. The equilibrium lies on the side of the weakest acid (cyclopentanol) and weakest base (hydroxide).

$$\bigcirc\!\!\!-OH \ + \ NaOH \ \underset{\longrightarrow}{\longleftarrow} \ \bigcirc\!\!\!-O^-Na^+ \ + \ H_2O$$

7.38   If you have any difficulty with this problem, review Sec. 7.9.

ANSWERS TO PROBLEMS

a.

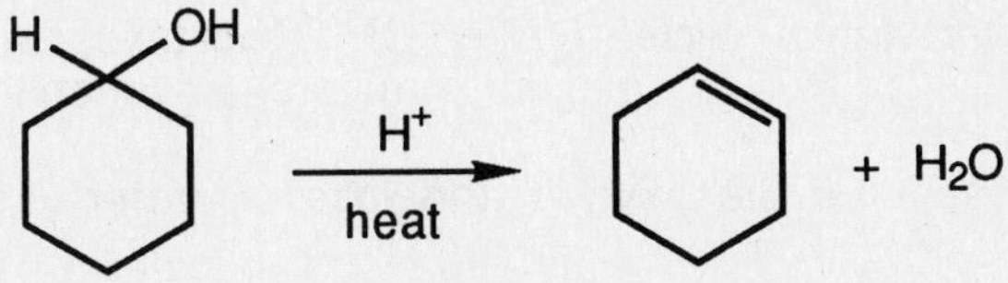

H, OH on cyclohexane ring $\xrightarrow[\text{heat}]{H^+}$ cyclohexene $+ H_2O$

b. $CH_3CHCH_2CH_3$ (with OH below) $\xrightarrow[\text{heat}]{H^+}$ $H_2C=CHCH_2CH_3$ $+ H_2O$

and

$CH_3CH=CHCH_3$ (*cis* and *trans*)

Of the three possible alkenes, *trans*-2-butene is the most stable and predominates.

c.

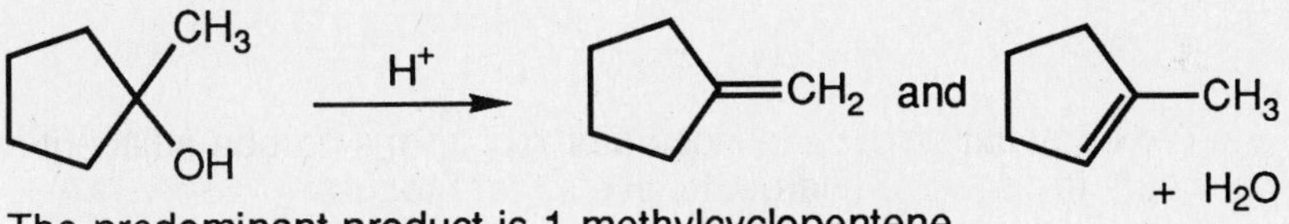

cyclopentane with CH₃ and OH $\xrightarrow{H^+}$ cyclopentane $=CH_2$ and cyclopentene $-CH_3$

$+ H_2O$

The predominant product is 1-methylcyclopentene.

d.

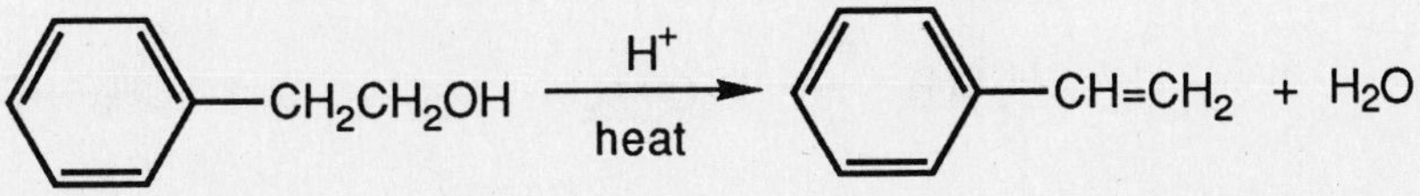

benzene ring $-CH_2CH_2OH$ $\xrightarrow[\text{heat}]{H^+}$ benzene ring $-CH=CH_2$ $+ H_2O$

7.39 In the reaction

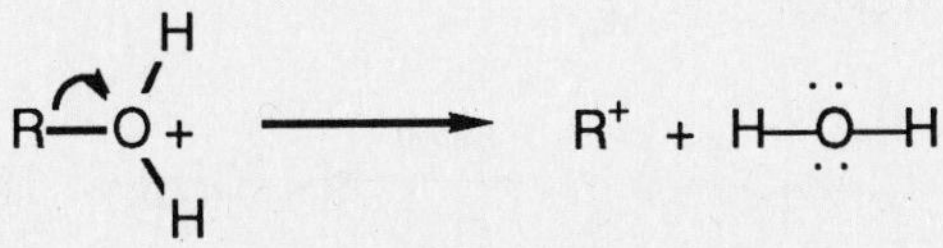

$R-O^+$ (with H's) $\longrightarrow$ $R^+ + H-\overset{..}{\underset{..}{O}}-H$

electrons flow toward the positive oxygen, and positive charge passes from the oxygen to carbon (R group).

In the reaction

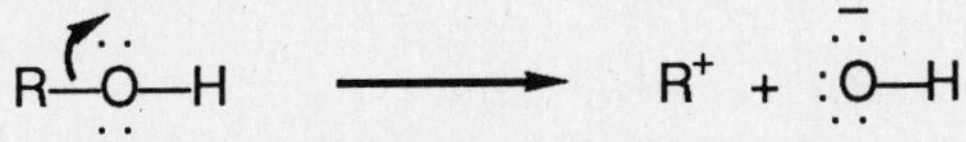

$R-\overset{..}{\underset{..}{O}}-H$ $\longrightarrow$ $R^+ + \overset{..}{\underset{..}{:}O}-H$

(a) The oxygen is not charged and therefore is less electron-demanding.
(b) Two oppositely charged species, R⁺ and OH⁻, must be separated.
   The second reaction thus requires much more energy than the first.

7.40 To begin, protonation of the oxygen and loss of water yield a tertiary carbocation:

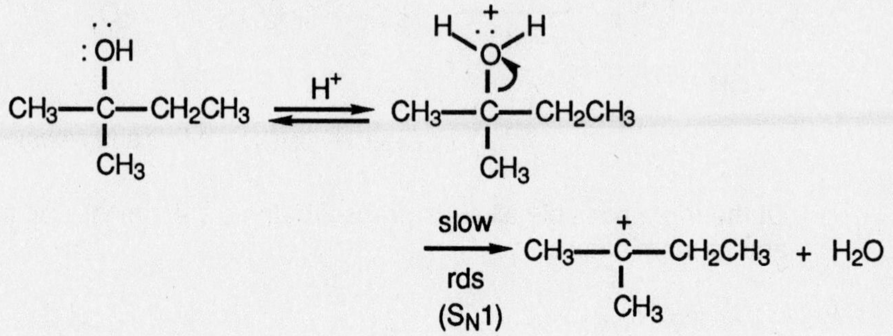

The carbocation can then lose a proton from a carbon adjacent to the one that bears the positive charge, to give either product:

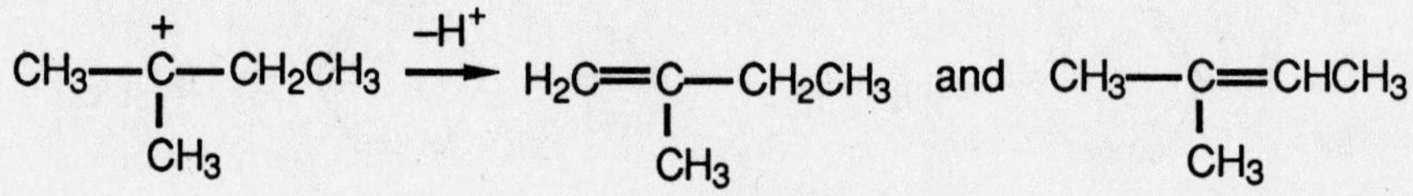

7.41 The by-product in eq. 7.26 is isobutylene, formed by an E1 process that competes with the main S_N1 reaction:

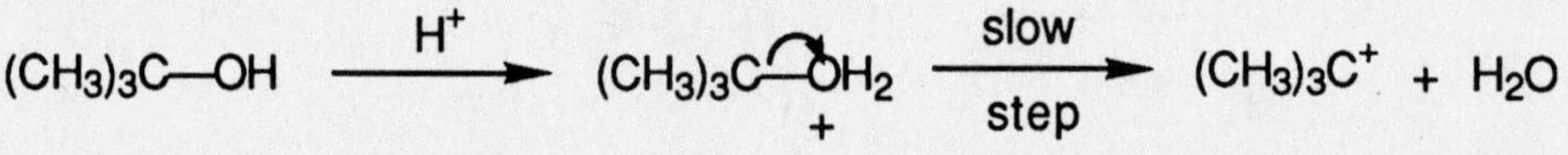

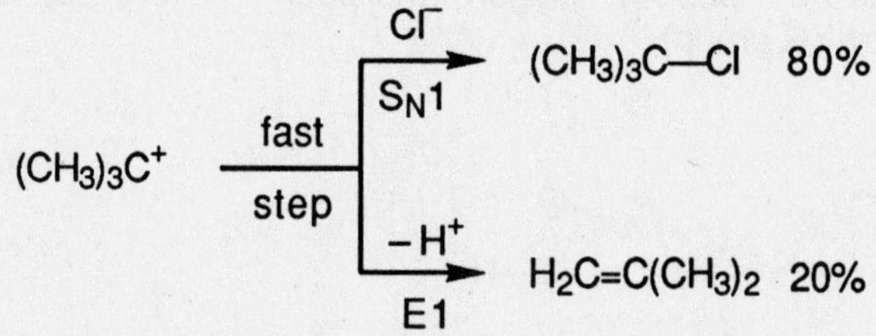

The reaction in eq. 7.27, on the other hand, involves a primary alcohol and proceeds by an S_N2 process:

$$CH_3CH_2CH_2CH_2\overset{+}{\underset{H}{-O-H}} \quad \xrightarrow{S_N2} \quad CH_3CH_2CH_2CH_2Cl \; + \; H_2O$$

Chloride ion is a *very* weak base; thus the E2 process cannot compete with the S_N2, and the yield of the substitution product is nearly 100%.

7.42   a.

$$CH_3\overset{\overset{\displaystyle CH_3}{|}}{\underset{\underset{\displaystyle OH}{|}}{C}}-CH_2CH_3 \quad HCl \longrightarrow \quad CH_3\overset{\overset{\displaystyle CH_3}{|}}{\underset{\underset{\displaystyle Cl}{|}}{C}}-CH_2CH_3 \; + \; H_2O$$

b. $2 \; CH_3(CH_2)_3CH_2OH \; + \; 2 \; Na \longrightarrow 2 \; CH_3(CH_2)_3CH_2O^-Na^+ \; + \; H_2$

sodium 1-pentoxide

c.

$$3 \quad \text{(cyclopentane ring)} \overset{H}{\underset{OH}{}} + \; PBr_3 \longrightarrow 3 \quad \text{(cyclopentane ring)} \overset{H}{\underset{Br}{}} + \; H_3PO_3$$

d.

$$\text{(phenyl)}\overset{\overset{\displaystyle H}{|}}{\underset{\underset{\displaystyle OH}{|}}{C}}-CH_3 \; + \; SOCl_2 \longrightarrow \text{(phenyl)}\overset{\overset{\displaystyle H}{|}}{\underset{\underset{\displaystyle Cl}{|}}{C}}-CH_3$$

$$+ \; SO_2 \; + \; HCl$$

e.

$$\text{(cyclopentane ring)}\overset{CH_3}{\underset{OH}{}} + \; HOSO_3H \quad \xrightarrow[-H_2O]{heat} \quad \text{(ring)}-CH_3 \; + \; \text{(ring)}=CH_2$$

(major)   (minor)

167

7. ALCOHOLS, PHENOLS, AND THIOLS

f. $HO-CH_2CH_2-OH + 2\ HONO_2 \longrightarrow O_2NO-CH_2CH_2-ONO_2$
$+$
$2\ H_2O$

g. $CH_3CH_2CH_2CH_2CH_2OH + NaOH \longrightarrow$ no reaction

h. $CH_3(CH_2)_6CH_2OH + HBr \xrightarrow{ZnBr_2} CH_3(CH_2)_6CH_2Br + H_2O$

i. $CH_3CHCH_2CH_2CH_3 \xrightarrow[H^+]{CrO_3} CH_3CCH_2CH_2CH_3$
    | ||
    OH   O

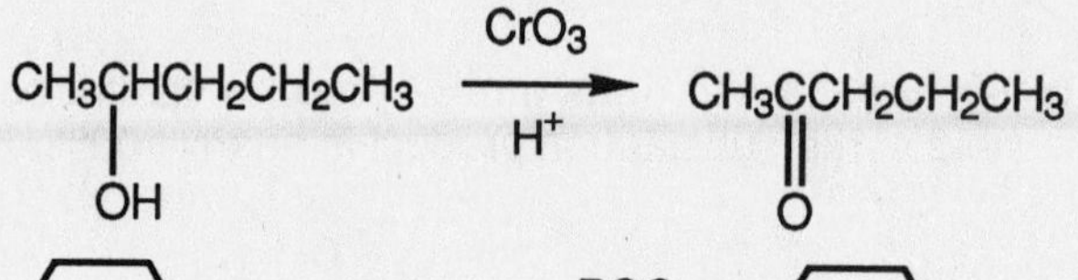

j. ⬡$-CH_2CH_2OH \xrightarrow{PCC}$ ⬡$-CH_2CH=O$

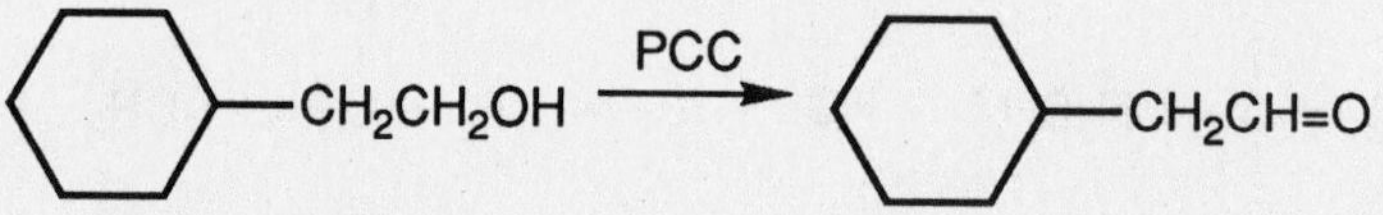

7.43 The mechanism involves protonation of the hydroxyl group and loss of water to form a carbocation:

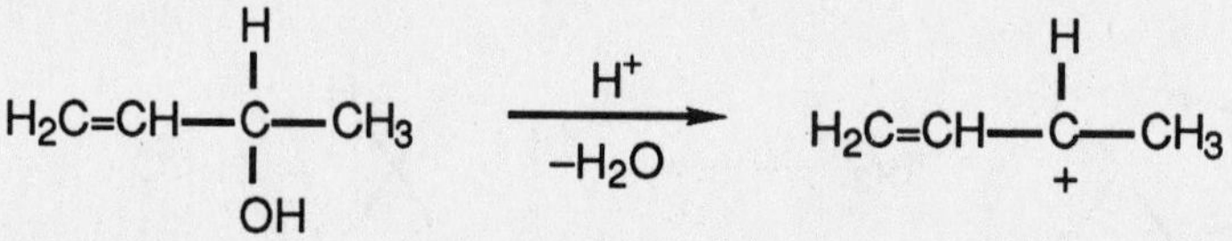

$$H_2C=CH-\underset{OH}{\overset{H}{C}}-CH_3 \xrightarrow[-H_2O]{H^+} H_2C=CH-\underset{+}{\overset{H}{C}}-CH_3$$

The carbocation is allylic and stabilized by resonance. This allylic ion can react with the nucleophile Cl^- at either end, giving the observed products:

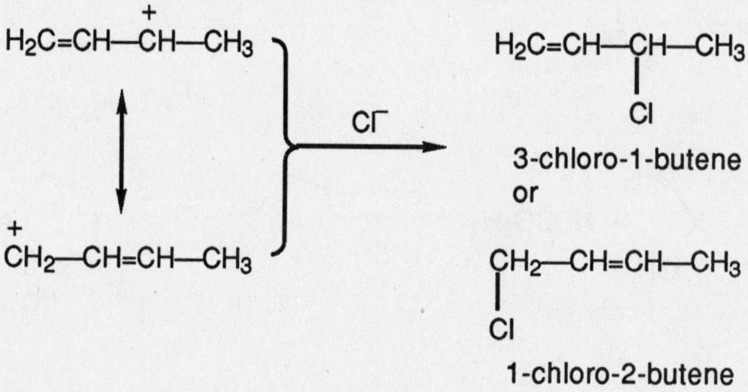

$H_2C=CH-\overset{+}{C}H-CH_3$ ⟷ $\overset{+}{C}H_2-CH=CH-CH_3$
$\xrightarrow{Cl^-}$

$H_2C=CH-\underset{Cl}{CH}-CH_3$
3-chloro-1-butene
or
$\underset{Cl}{CH_2}-CH=CH-CH_3$
1-chloro-2-butene

168

7.44   If necessary, review Secs. 3.9 through 3.12. 2-Butanol can be prepared by acid-catalyzed hydration of $CH_2=CHCH_2CH_3$ or $CH_3CH=CHCH_3$. *tert*-Butyl alcohol (2-methyl-2-propanol) can be prepared by acid-catalyzed hydration of $CH_2=C(CH_3)_2$. Neither $CH_3CH_2CH_2CH_2OH$ nor $(CH_3)_2CHCH_2OH$ can be obtained by this route [although they can be obtained by the (more expensive) hydroboration–oxidation sequence, Sec. 3.13].

7.45   In problems of this type, working backward from the final product sometimes helps. Ask yourself: What reactions that I have studied give this type of product? If there is more than one, see which one requires a precursor that is easily obtained from the compound given as the starting point for the synthesis.

a.

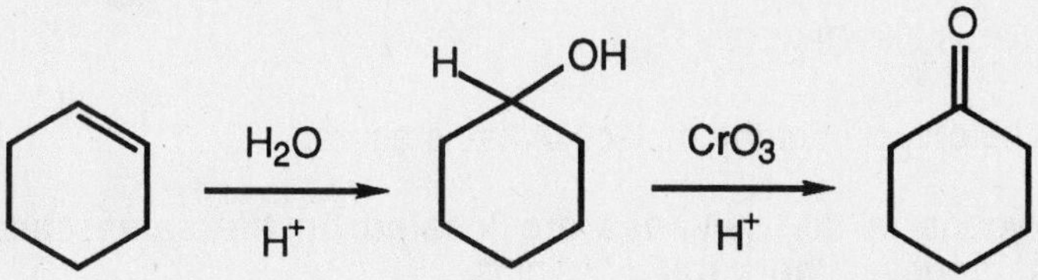

b. $CH_3CH_2CH_2CH_2Br$ $\xrightarrow[\text{H}_2\text{O}]{\text{NaOH}}$ $CH_3CH_2CH_2CH_2OH$ $\xrightarrow[\substack{\text{CH}_2\text{Cl}_2 \\ 25^\circ\text{C}}]{\text{PCC}}$

$CH_3CH_2CH_2CH=O$

c. $CH_3CH_2CH_2CH_2OH$ $\xrightarrow[\text{ZnBr}_2]{\text{HBr}}$ $CH_3CH_2CH_2CH_2Br$ $\xrightarrow{\text{Na}^+\text{SH}^-}$

$CH_3CH_2CH_2CH_2SH$

7. ALCOHOLS, PHENOLS, AND THIOLS

7.46 The secondary alcohol should be oxidized to a ketone:

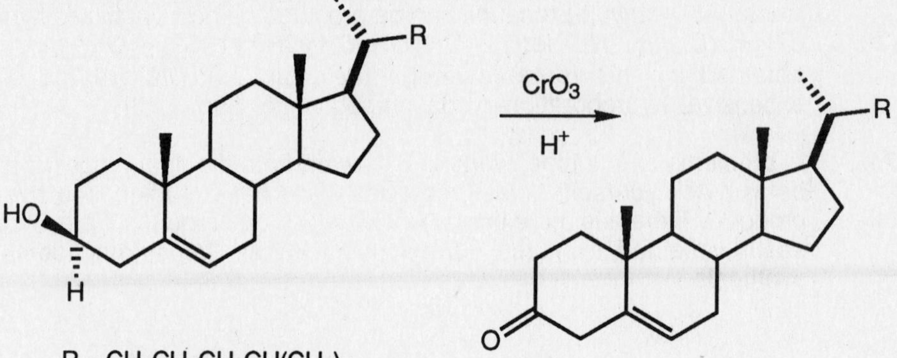

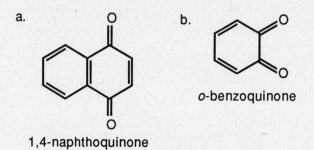

$$R = CH_2CH_2CH_2CH(CH_3)_2$$

That is the only part of the molecule that is altered.

7.47 The oxygens in the quinones are located on the same carbons as the hydroxyl groups in the starting phenols.

a.

1,4-naphthoquinone

b.

o-benzoquinone

7.48 The overall equation is as follows:

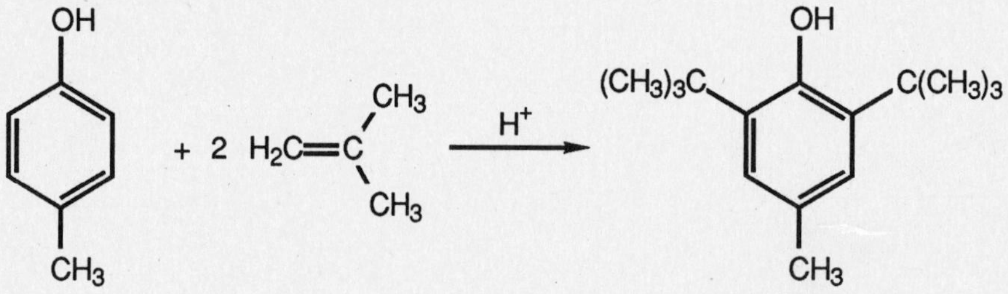

OH

$+ 2 \ H_2C\!\!=\!\!C \overset{\displaystyle CH_3}{\underset{\displaystyle CH_3}{\big\langle}}$ $\xrightarrow{H^+}$

CH_3

$(CH_3)_3C$ OH $C(CH_3)_3$

CH_3

BHT

The mechanism involves protonation of the 2-methylpropene according to Markovnikov's rule:

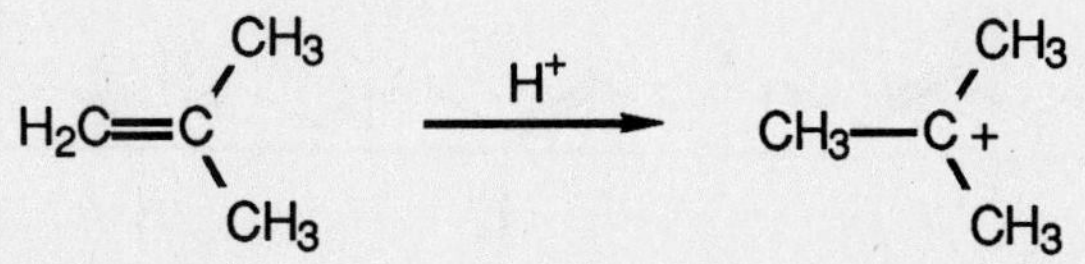

$$H_2C=C \begin{smallmatrix} CH_3 \\ \\ CH_3 \end{smallmatrix} \xrightarrow{\;H^+\;} CH_3-\overset{CH_3}{\underset{CH_3}{C}}+$$

The tertiary carbocation then acts as an electrophile toward the *p*-methylphenol:

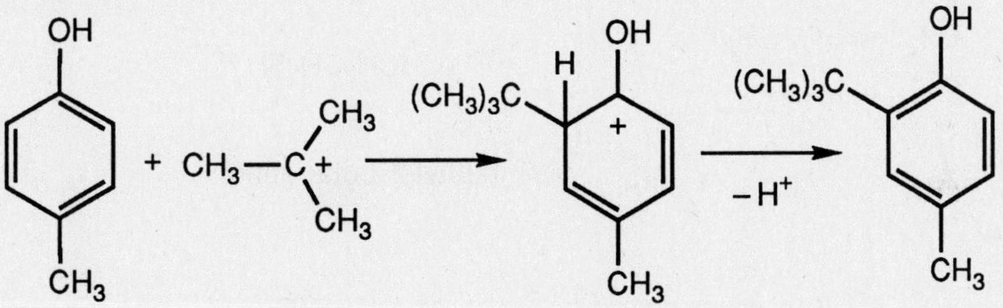

The steps are repeated to give the final product.

7.49   squalene

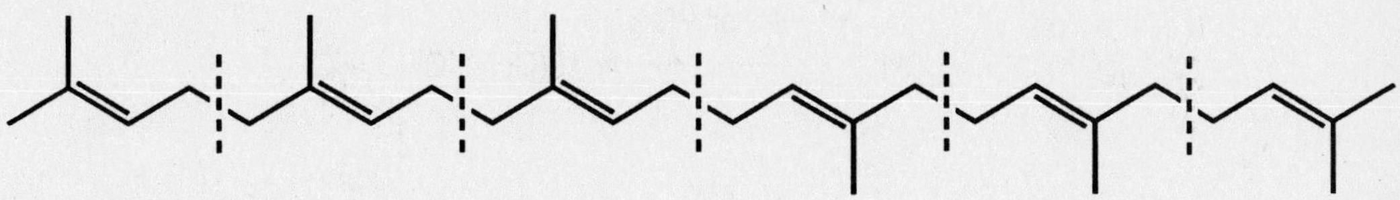

vitamin A

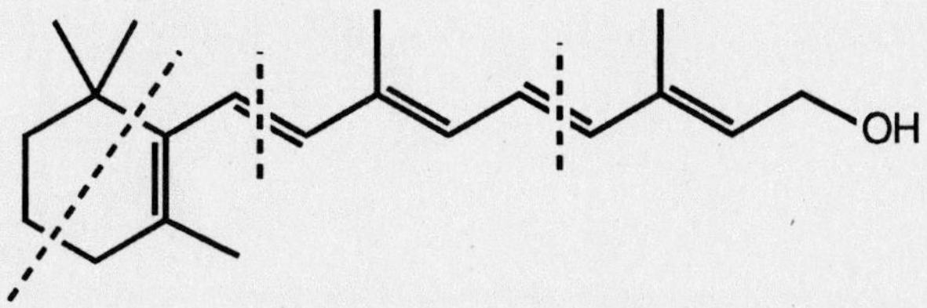

β-carotene

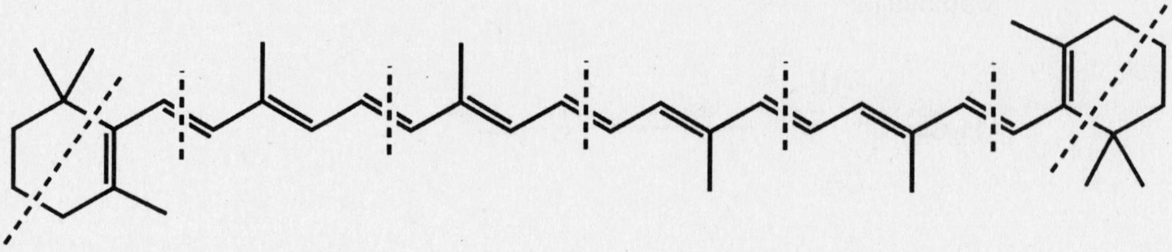

7.50

$$\underset{4}{CH_3}\underset{3}{CH}=\underset{2}{CH}\underset{1}{CH_2SH}$$

2-buten-1-thiol

$$\underset{CH_3}{\overset{CH_3}{>}}CHCH_2CH_2SH$$

3-methyl-1-butanethiol

7.51 $2\ CH_3SH \xrightarrow{H_2O_2} CH_3S\!-\!SCH_3$ (compare with eq. 7.45)

7.52 Work backward: from the disulfide to the thiol (eq. 7.45), to the alkyl halide (eq. 7.43), to the alcohol (eq. 7.25 or eq. 7.29).

$$(CH_3)_2CHCH_2CH_2OH \xrightarrow{HBr\ or\ PBr_3} (CH_3)_2CHCH_2CH_2Br$$

$$\downarrow NaSH$$

$$(CH_3)_2CHCH_2CH_2 \xleftarrow{H_2O_2} (CH_3)_2CHCH_2CH_2SH$$

$$\begin{array}{c} (CH_3)_2CHCH_2CH_2 \\ | \\ S \\ | \\ S \\ | \\ (CH_3)_2CHCH_2CH_2 \end{array}$$

CHAPTER EIGHT: ETHERS AND EPOXIDES

CHAPTER SUMMARY

Ethers have two organic groups, either alkyl or aryl, connected to a single oxygen atom (R—O—R'). In common names, the two organic groups are named and followed by the word *ether*, as in *ethyl methyl ether*, $CH_3CH_2OCH_3$. In the IUPAC system, the smaller **alkoxy group** is named as a substituent on the longer carbon chain. For the preceding example, the IUPAC name is *methoxyethane*.

Ethers have much lower boiling points than the alcohols with which they are isomeric because ethers cannot form intermolecular hydrogen bonds with themselves. They do, however, act as Lewis bases to form hydrogen bonds with compounds containing an —OH group (alcohols or water).

Ethers are excellent solvents for organic compounds. Their relative inertness makes them good solvents in which to carry out organic reactions.

Alkyl or aryl halides react with magnesium metal in **diethyl ether** or **tetrahydrofuran (THF)** to form **Grignard reagents, R—MgX.** Ethers stabilize these reagents by coordinating with the magnesium. Grignard reagents react with water, and the —MgX is replaced by —H, or if D_2O is used, by —D.

Diethyl ether is prepared commercially by intermolecular dehydration of ethanol with sulfuric acid. The **Williamson ether synthesis**, another route to ethers, involves preparation of an alkoxide from an alcohol and a reactive metal, followed by an S_N2 displacement between the alkoxide and an alkyl halide.

Ethers can be cleaved at the C—O bond by strong protonic (HBr) or Lewis (BBr_3) acids. The products are alcohols and/or alkyl halides.

Epoxides (oxiranes) are three-membered cyclic ethers. The simplest and commercially most important example is **ethylene oxide,** manufactured from ethylene, air, and a silver catalyst. In the laboratory, epoxides are most commonly prepared from alkenes and organic peracids.

Epoxides react with nucleophiles to give products in which the ring has opened. For example, acid-catalyzed hydration of ethylene oxide gives **ethylene glycol.** Other nucleophiles (such as alcohols) add similarly to epoxides, as do Grignard reagents. The latter is a useful two-carbon chain-lengthening reaction.

8. ETHERS AND EPOXIDES

Epoxy resins are polymers prepared by similar ring-opening reactions, using as raw materials the epoxide **epichlorhydrin** and the nucleophile **bisphenol-A**.

Cyclic ethers with larger rings than epoxides include **tetrahydrofuran (THF)**, **tetrahydropyran (THP)**, and **dioxane**. Large-ring cyclic polyethers, called **crown ethers**, can selectively bind metal ions, depending on the ring size.

REACTION SUMMARY

Grignard Reagents

$$R\text{—}X \ + \ Mg \ \xrightarrow{\text{ether or THF}} \ R\text{—}MgX \quad \text{(preparation)}$$

$$R\text{—}MgX \ + \ H\text{—}OH \longrightarrow RH \ + \ Mg(OH)X \ \Big\}$$
$$R\text{—}MgX \ + \ D\text{—}OD \longrightarrow RD \ + \ Mg(OD)X \ \Big\} \ \text{hydrolysis}$$

Ether Preparation

$$2 \ ROH \ \xrightarrow[140^{o}C]{H_2SO_4} \ ROR \ + \ H_2O$$

(best for primary alcohols; gives symmetrical ethers)

$$2 \ ROH \ + \ 2 \ Na \longrightarrow 2 \ RO^-Na^+ \ + \ H_2 \ \Big\}$$
$$RO^-Na^+ \ + \ R'X \longrightarrow ROR' \ + \ Na^+X^- \ \Big\} \ \begin{array}{l}\text{Williamson} \\ \text{synthesis}\end{array}$$

(best for R' = primary)

Ether Cleavage

$$R\text{—}O\text{—}R \ + \ HBr \longrightarrow RBr \ + \ ROH \ \xrightarrow{HBr} \ RBr \ + \ H_2O$$

REACTION SUMMARY

$$3\ R-O-R + 2\ BBr_3 \longrightarrow 6\ RBr + B_2O_3 \xrightarrow{3\ H_2O} 2\ H_3BO_3$$

Ethylene Oxide

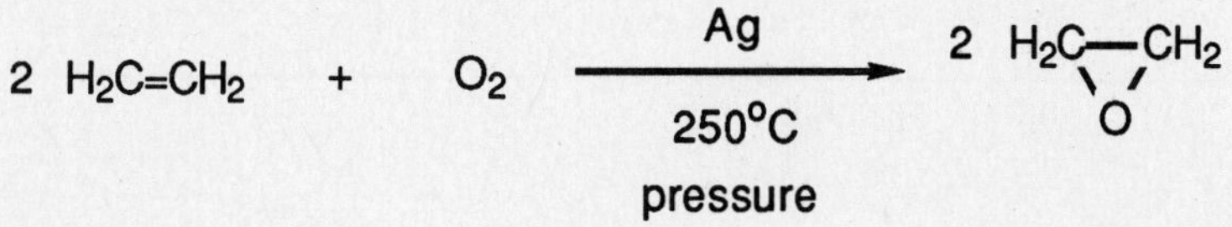

$$2\ H_2C=CH_2 + O_2 \xrightarrow[\substack{250^\circ C \\ pressure}]{Ag} 2\ H_2C-CH_2$$

Other Epoxides

alkene   peracid   epoxide   acid

Epoxide Ring Openings

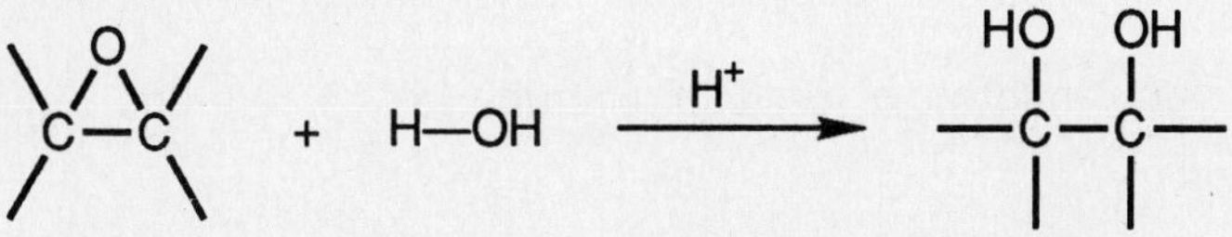

glycol

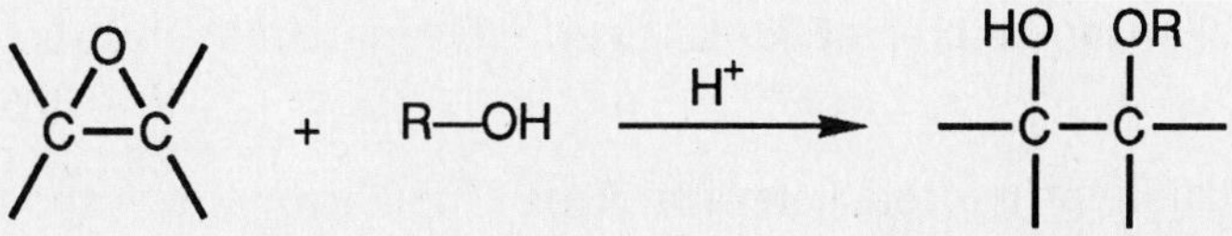

2-alkoxyalcohol

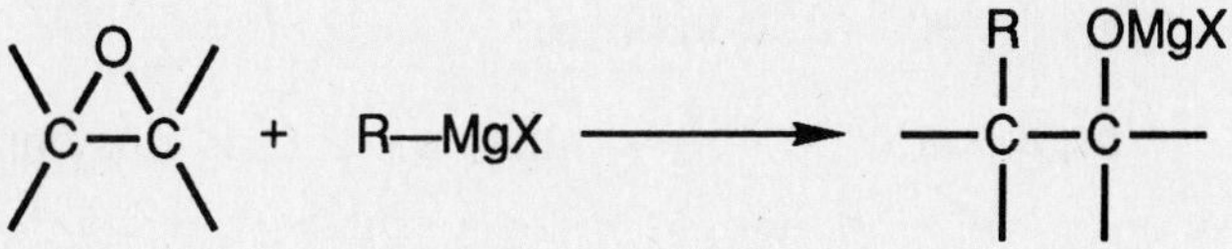

8. ETHERS AND EPOXIDES

$$\xrightarrow[\text{H}^+]{\text{H}_2\text{O}}$$

R OH
| |
—C—C—
| |

alcohol with two more
carbons than the
Grignard reagent

LEARNING OBJECTIVES

1. Know the meaning of: ether, alkoxy group, Grignard reagent, organometallic compound, ether cleavage.

2. Know the meaning of: epoxide, oxirane, organic peracid, nucleophilic addition to epoxides, diethylene glycol, epoxy resin, epichlorhydrin, bisphenol-A.

3. Know the meaning of: cyclic ether, tetrahydrofuran, furan, tetrahydropyran, dioxane, crown ethers.

4. Given the name of an ether or epoxide, write its structure, and vice versa.

5. Given the molecular formula, draw the structures of isomeric ethers and alcohols.

6. Compare the boiling points and solubilities in water of isomeric ethers and alcohols.

7. Write an equation for the preparation of a given Grignard reagent, and be able to name it.

8. Write an equation for the reaction of a given Grignard reagent and H_2O or D_2O.

9. Write an equation, using the appropriate Grignard reagent, for the preparation of a specific deuterium-labeled hydrocarbon.

10. Write equations for the preparation of a symmetrical and an unsymmetrical ether.

11. Write equations for the preparation of an ether using a Williamson synthesis.

12. Write the equation for the cleavage of an ether by a strong acid (HBr, HI, H_2SO_4) or a Lewis acid (BBr_3).

13.   Write the steps in the mechanism for cleavage of an ether.

14.   Write an equation for the preparation of an epoxide from the corresponding alkene.

15.   Write equations for the reaction of ethylene oxide or other epoxides with nucleophiles such as H^+ and H_2O, H^+ and alcohols, or a Grignard reagent.

16.   Write the steps in the mechanism for ring-opening reactions of ethylene oxide and other epoxides.

ANSWERS TO PROBLEMS

Problems Within the Chapter

8.1   a.   isopropyl methyl ether or 2-methoxypropane
      b.   phenyl *n*-propyl ether or 1-propoxybenzene or 1-phenoxypropane
      c.   1-methoxy-1-methylcyclopentane

8.2   a.   b. $CH_3CHCH_2CH_2CH_2CH_2CH_2CH_3$
                                        $\quad\ \ |$
                                        $\quad OCH_2CH_3$

8.3 $HOCH_2CH_2CH_2CH_2OH$ 1,4-butanediol

      $CH_3OCH_2CH_2CH_2OH$ 3-methoxy-1-propanol

      $CH_3OCH_2CH_2OCH_3$ 1,2-dimethoxyethane

      The compounds are listed in order of decreasing boiling point. The fewer hydroxyl groups, the fewer possibilities there are for intermolecular hydrogen bonding, and the lower the boiling point.

8.4   Yes. There are no acidic protons in the starting alkyl bromide.

8.5   We must first convert the alcohol to an alkyl halide before we can make the Grignard reagent.

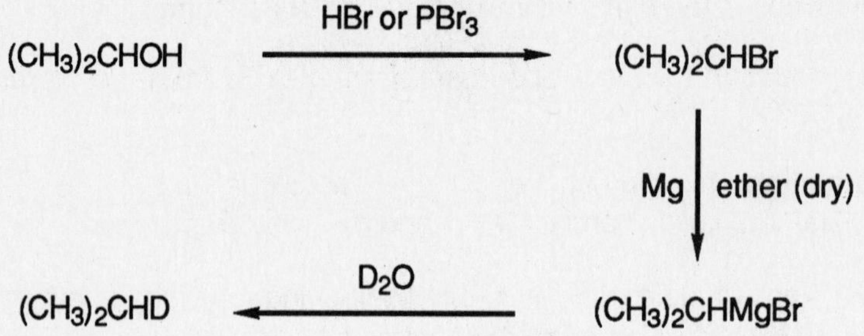

8. ETHERS AND EPOXIDES

$$(CH_3)_2CHOH \xrightarrow{\text{HBr or PBr}_3} (CH_3)_2CHBr$$

$$\downarrow \text{Mg} \quad \text{ether (dry)}$$

$$(CH_3)_2CHD \xleftarrow{\text{D}_2\text{O}} (CH_3)_2CHMgBr$$

8.6   Follow eq. 8.7 and then eq. 8.6 as guides.

$$CH_3CH_2CH_2Br \xrightarrow[\text{ether}]{\text{2 Li}} CH_3CH_2CH_2Li \xrightarrow{\text{D}_2\text{O}} CH_3CH_2CH_2D$$
$$+$$
$$LiBr$$

The resulting propane is labeled with one deuterium atom on one of the terminal carbon atoms.

8.7   The reaction occurs by an S_N2 mechanism. First a molecule of ethanol is protonated by the acid catalyst:

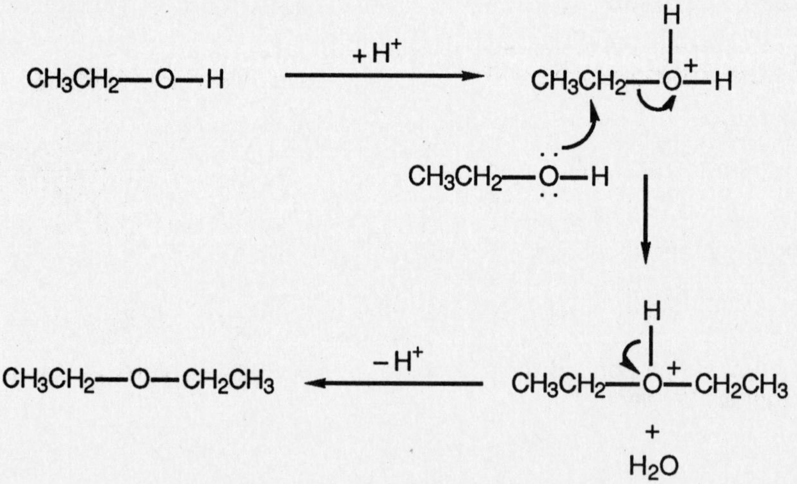

$$CH_3CH_2-O-H \xrightarrow{+H^+} CH_3CH_2-\overset{\overset{H}{|}}{O}{}^+-H$$

$$CH_3CH_2-\overset{..}{\underset{..}{O}}-H$$

$$\downarrow$$

$$CH_3CH_2-O-CH_2CH_3 \xleftarrow{-H^+} CH_3CH_2-\overset{\overset{H}{|}}{O}{}^+-CH_2CH_3$$
$$+$$
$$H_2O$$

Then a second molecule of ethanol displaces water, followed by loss of a proton to regenerate the acid catalyst. The reaction does not go by an S_N1

178

mechanism because the ethyl cation, which would be an intermediate, is primary and not easily formed.

8.8 Use eq. 8.8 as a model.

$$2 \ CH_3CH_2CH_2OH \xrightarrow[140°C]{H_2SO_4} CH_3CH_2CH_2OCH_2CH_2CH_3 + H_2O$$

8.9 First the double bond is protonated to give a *t*-butyl cation, which then reacts with methanol, a nucleophile, to give an oxonium ion. Loss of a proton gives *t*-butyl methyl ether. Notice that the acid is a catalyst; it is needed for a reaction to occur, but it is not consumed.

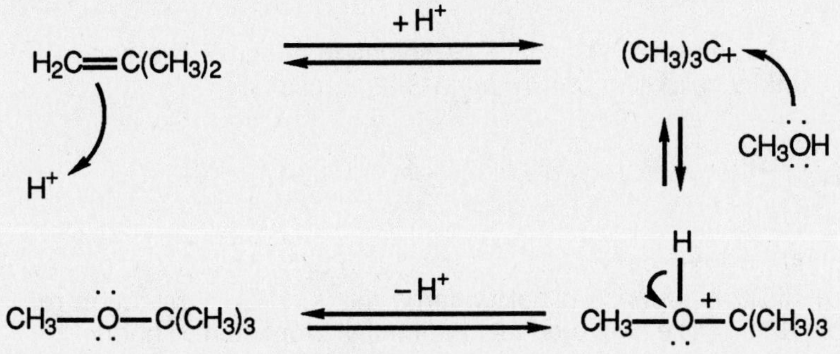

$$H_2C{=}C(CH_3)_2 \quad \xrightleftharpoons{+H^+} \quad (CH_3)_3C+$$

$$H^+ \qquad\qquad CH_3\overset{..}{O}H$$

$$CH_3{-}\overset{..}{\underset{..}{O}}{-}C(CH_3)_3 \quad \xrightleftharpoons{-H^+} \quad CH_3{-}\overset{\overset{H}{|}}{\underset{..}{O}}\overset{+}{-}C(CH_3)_3$$

8.10 a. $2 \ CH_3OH \ + \ 2 \ Na \longrightarrow 2 \ CH_3O^-Na^+ \ + \ H_2$

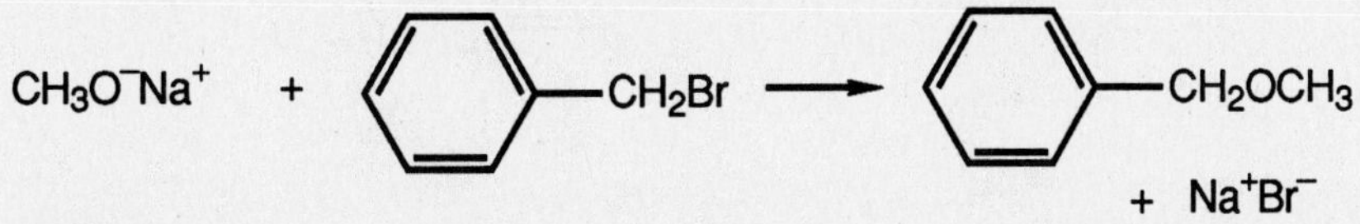

$$CH_3O^-Na^+ \ + \ \langle\bigcirc\rangle{-}CH_2Br \longrightarrow \langle\bigcirc\rangle{-}CH_2OCH_3$$
$$+ \ Na^+Br^-$$

or

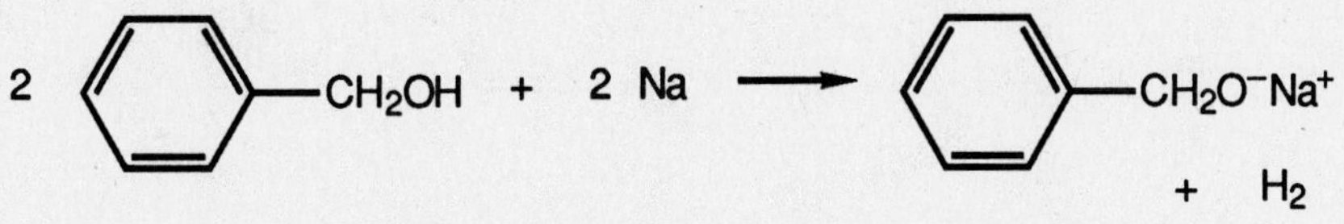

$$2 \ \langle\bigcirc\rangle{-}CH_2OH \ + \ 2 \ Na \longrightarrow \langle\bigcirc\rangle{-}CH_2O^-Na^+$$
$$+ \ H_2$$

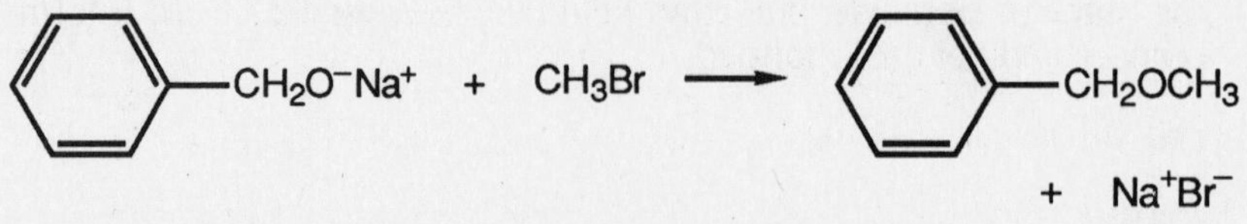

$$\text{C}_6\text{H}_5\text{—CH}_2\text{O}^-\text{Na}^+ \quad + \quad \text{CH}_3\text{Br} \longrightarrow \text{C}_6\text{H}_5\text{—CH}_2\text{OCH}_3$$
$$+ \quad \text{Na}^+\text{Br}^-$$

Although both methods will work, the first is probably preferable because S_N2 displacements occur quite easily for benzyl halides and because of the commercial availability of sodium methoxide.

b. $2\ (\text{CH}_3)_3\text{COH} \quad + \quad 2\ \text{K} \longrightarrow 2\ (\text{CH}_3)_3\text{CO}^-\text{K}^+ \quad + \quad \text{H}_2$

$(\text{CH}_3)_3\text{CO}^-\text{K}^+ \quad + \quad \text{CH}_3\text{Br} \longrightarrow (\text{CH}_3)_3\text{COCH}_3 \quad + \quad \text{K}^+\text{Br}^-$

We cannot use the alternative combination because the second step would fail; tertiary halides do *not* undergo S_N2 reactions:

$\text{CH}_3\text{O}^-\text{K}^+ \quad + \quad (\text{CH}_3)_3\text{CBr} \ \not\!\!\longrightarrow\ (\text{CH}_3)_3\text{COCH}_3 \quad + \quad \text{K}^+\text{Br}^-$

8.11 The intermediate oxonium ion reacts by an S_N2 mechanism (review Sec. 6.5). Aromatic compounds do not undergo this kind of substitution reaction (see Table 6.1), so the only products are methyl bromide and phenol.

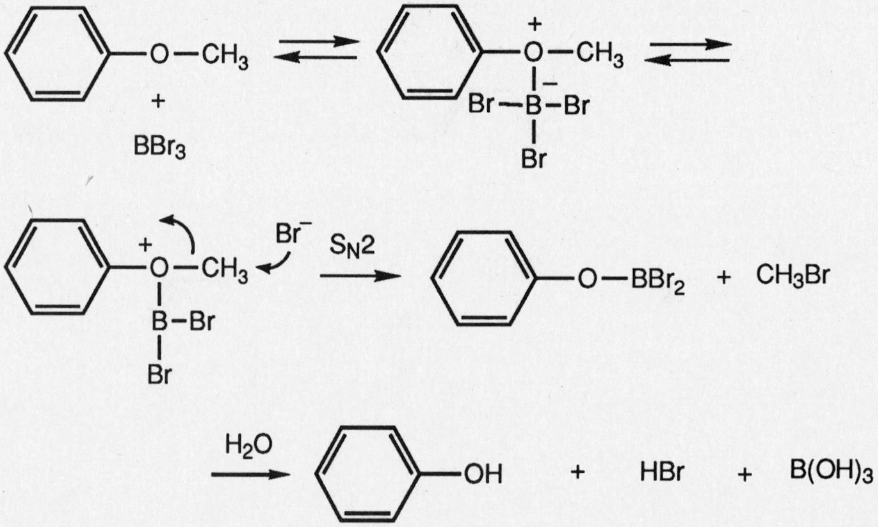

8.12 The reaction occurs by an S_N1 mechanism (review Sec. 6.6). The C—O bond to the *t*-butyl group cleaves to give an intermediate *t*-butyl cation that is much more stable than a phenyl cation.

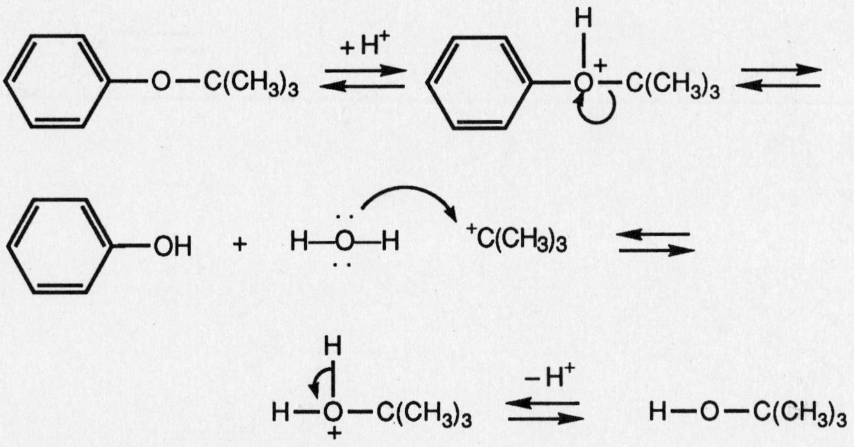

8.13 Follow eq. 8.18 as a guide.

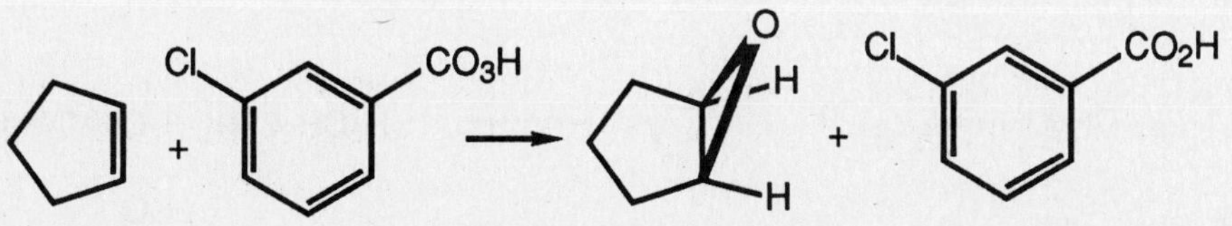

8.14 In the first step, the epoxide oxygen is protonated by the acid catalyst:

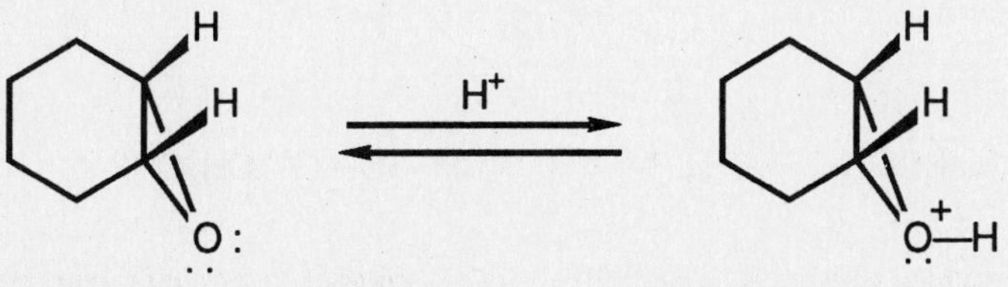

In the second step, water acts as a nucleophile in an S_N2 displacement:

The product is *trans*-1,2-cyclohexanediol:

8.15   Since two carbons, including the one bearing the hydroxyl group, come from the ethylene oxide (see eq. 8.21), we need a Grignard reagent with only three carbon atoms:

$$CH_3CH_2CH_2MgBr \quad + \quad \triangle \quad \longrightarrow \quad CH_3CH_2CH_2CH_2CH_2OMgBr$$

$$\downarrow H_2O$$

$$CH_3CH_2CH_2CH_2CH_2OH$$

Additional Problems

8.16 a. $CH_3CH_2CH_2OCH_2CH_2CH_3$ b. $(CH_3)_3C-O-CH_2CH_3$

c. $CH_3CH_2CHCH_2CH_2CH_3$ d. $H_2C=CHCH_2-O-CH_2CH=CH_2$
          |
          OCH_3

e.

Br—⟨benzene ring⟩—OCH$_2$CH$_3$

f.

(cyclopentane ring with OH, two H substituents (dashed wedges), and OCH$_2$CH$_3$)

g.   CH$_3$OCH$_2$CH$_2$OCH$_3$

h.   CH$_3$O—CH=CHCH$_3$

i.   CH$_3$CH—CH$_2$
              \O/

j.   CH$_3$O—⟨benzene ring⟩—OCH$_2$CH$_3$

8.17   a. diisopropyl ether
       c. propylene oxide
          (or methyloxirane)
       e. methoxycyclohexane
          (or cyclohexyl methyl ether)
       g. 2-ethoxypentane
       i. 1,2-epoxybutane
          (or ethyloxirane or 1-butene oxide)

       b. isobutyl methyl ether
       d. *p*-bromoanisole
          (or *p*-bromophenyl methyl ether)
       f. *t*-butyl phenyl ether
          (or 2-methyl-2-phenoxypropane)
       h. 2-methoxyethanol
       j. 3-methoxypropyne
          (or methyl propargyl ether)

8.18   Be systematic.

CH$_3$CH$_2$CH$_2$CH$_2$OH
1-butanol

CH$_3$CH$_2$CH(OH)CH$_3$
2-butanol

(CH$_3$)$_2$CHCH$_2$OH
2-methyl-1-propanol

(CH$_3$)$_3$COH
2-methyl-2-propanol

CH$_3$OCH$_2$CH$_2$CH$_3$
methyl *n*-propyl ether

CH$_3$OCH(CH$_3$)$_2$
methyl isopropyl ether

CH$_3$CH$_2$OCH$_2$CH$_3$
diethyl ether

8.19   The actual boiling points are as follows:

1-pentanol	CH$_3$CH$_2$CH$_2$CH$_2$CH$_2$OH	137°C
1,2-dimethoxyethane	CH$_3$OCH$_2$CH$_2$OCH$_3$	83°C
hexane	CH$_3$CH$_2$CH$_2$CH$_2$CH$_2$CH$_3$	69°C
ethyl *n*-propyl ether	CH$_3$CH$_2$OCH$_2$CH$_2$CH$_3$	64°C

1-Pentanol is the only one of these compounds capable of forming hydrogen bonds with itself. Thus it has the highest boiling point. Judging from the table in Sec. 8.3, we might expect hexane to have a slightly higher boiling point than a corresponding monoether, although the boiling points should be quite close. 1,2-Dimethoxyethane, with four polar C—O bonds, is expected to associate more than the monoether (with only two C—O bonds). Therefore we expect it to have a boiling point that is appreciably higher than that of ethyl *n*-propyl ether.

Regarding water solubility, 1,2-dimethoxyethane has two oxygens that can hydrogen-bond with water. The pentanol and the other ether have only one oxygen, and the hexane has none. We expect 1,2-dimethoxyethane to be the most soluble in water of these four compounds, and it is. In fact, the dimethoxyethane is completely soluble in water; 1-pentanol and ethyl *n*-propyl ether are only slightly soluble in water, and hexane is essentially insoluble in water.

8.20   a. $CH_3CH_2CH_2CH_2Br \xrightarrow[\text{ether}]{Mg} CH_3CH_2CH_2CH_2MgBr \xrightarrow{D_2O} CH_3CH_2CH_2CH_2D$

b. $CH_3OCH_2CH_2CH_2Br \xrightarrow[\text{ether}]{Mg} CH_3OCH_2CH_2CH_2MgBr \xrightarrow{D_2O} CH_3OCH_2CH_2CH_2D$

Note that the ether functional group can be tolerated in making a Grignard reagent (part b).

8.21

184

Friedel-Crafts alkylation of anisole would give *p-t*-butylanisole, the sterically least hindered product.

8.22 a. Since both alkyl groups are identical and primary, the dehydration route using sulfuric acid is preferred because it is least expensive and gives a good yield (see eq. 8.8).

$$CH_3CH_2CH_2CH_2OH \xrightarrow[\text{ether}]{H_2SO_4} CH_3CH_2CH_2CH_2OCH_2CH_2CH_2CH_3$$

b. The Williamson method is preferred. The sodium phenoxide can be prepared using NaOH instead of Na, because of the acidity of phenols:

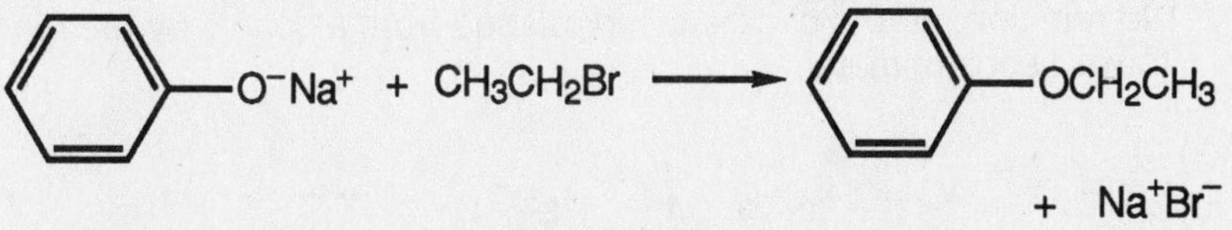

$$+ \ Na^+Br^-$$

c. The following Williamson ether synthesis is preferred. The alternate Williamson ether synthesis (the reaction between sodium ethoxide and *tert*-butyl iodide) would fail because dehydrohalogenation would be faster than substitution.

$$(CH_3)_3COH \xrightarrow{Na} (CH_3)_3CO^-Na^+ \xrightarrow{CH_3CH_2I} (CH_3)_3COCH_2CH_3$$

$$+ \qquad\qquad +$$
$$1/2 \ H_2 \qquad\qquad NaI$$

8.23 The second step fails because S_N2 displacements cannot be carried out on aryl halides.

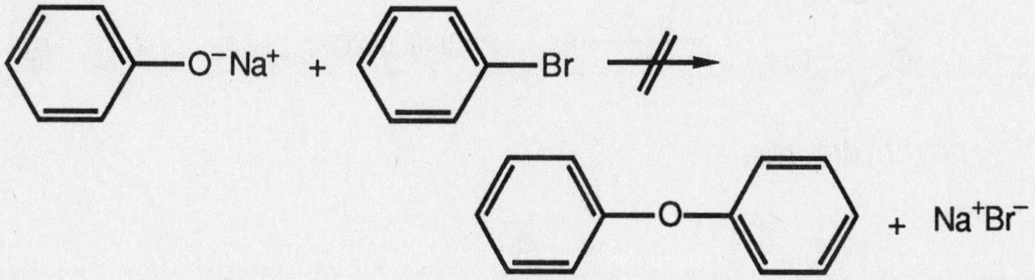

$$+ \ Na^+Br^-$$

8.24 The oxygen of the ether can be protonated, and the resulting highly polar dialkyloxonium ion is soluble in sulfuric acid:

$$R\text{—}\overset{..}{\underset{..}{O}}\text{—}R \;+\; H_2SO_4 \;\longrightarrow\; R\text{—}\overset{\displaystyle H}{\overset{|}{\underset{..}{O}^+}}\text{—}R \;+\; HSO_4^-$$

dialkyloxonium ion

Alkanes have no unshared electron pairs and are not protonated by sulfuric acid and thus remain insoluble in it.

8.25 a. No reaction; ethers (except for epoxides) are inert toward base.
b. $CH_3OCH_2CH_2CH_3 + 2\,HBr \rightarrow CH_3Br + CH_3CH_2CH_2Br + H_2O$
c. No reaction; ethers can be distinguished from alcohols by their inertness toward sodium metal.

d. $CH_3CH_2\text{—}\overset{..}{\underset{..}{O}}\text{—}CH_2CH_3 \;+\; H_2SO_4 \;\longrightarrow$

$$CH_3CH_2\text{—}\overset{\displaystyle H}{\overset{|}{\underset{..}{O}^+}}\text{—}CH_2CH_3 \;+\; HSO_4^-$$

The ether acts as a base and dissolves in the strong acid.

e.

$$\text{C}_6\text{H}_5\text{—}OCH_2CH_3 \xrightarrow{BBr_3} \text{C}_6\text{H}_5\text{—}OH + CH_3CH_2Br$$

Compare with eq. 8.15.

8.26

$$\overset{H_2C\text{—}CH_2}{\underset{H_2C\diagdown_{\;O}\diagup CH_2}{}} \xrightarrow{2\,HBr} BrCH_2CH_2CH_2CH_2Br \;+\; H_2O$$

tetrahydrofuran

8.27 $H_2C\!=\!CHCH_2CH_3$ $\xrightarrow[\text{(refer to eq. 8.18)}]{CH_3CO_3H}$ $H_2C\!-\!CHCH_2CH_3$ $\xrightarrow[\text{(refer to eq. 8.19)}]{H_2O,\ H^+}$

(the epoxide with O bridging)

$H_2C\!-\!CHCH_2CH_3$
$\quad\ |\quad\ |$
$\quad HO\quad OH$

8.28 a. $H_2C\!-\!CH_2$ + HBr $\longrightarrow$ $H_2C\!-\!CH_2$
$\qquad\ \ \diagdown O\diagup$ $\qquad\qquad\qquad\qquad\qquad\ |\quad\ |$
$\qquad\qquad\qquad\qquad\qquad\qquad\qquad\ \ HO\quad Br$

b. $H_2C\!-\!CH_2$ + 2 HBr $\longrightarrow$ $H_2C\!-\!CH_2$ + H_2O
$\qquad\ \ \diagdown O\diagup$ $\qquad\qquad\qquad\qquad\qquad\ \ |\quad\ |$
$\qquad\qquad\qquad\qquad\qquad\qquad\qquad\quad Br\quad Br$

The 2-bromoethanol formed in part (a) reacts as an alcohol with the second mole of HBr, to produce the dibromide.

c. $H_2C\!-\!CH_2$ + HO—⟨benzene ring⟩ $\xrightarrow{H^+}$ $HOCH_2CH_2O$—⟨benzene ring⟩
$\qquad\ \ \diagdown O\diagup$

2-phenoxyethanol

8.29   See eq. 8.20 for comparison.

$H_2C\!-\!CH_2$ + CH_3CH_2OH $\xrightarrow{H^+}$ $HOCH_2CH_2OCH_2CH_3$
$\ \diagdown O\diagup$

ethyl cellosolve

$H_2C\!-\!CH_2$ + $HOCH_2CH_2OCH_2CH_3$ $\xrightarrow{H^+}$
$\ \diagdown O\diagup$

$HOCH_2CH_2OCH_2CH_2OCH_2CH_3$

ethyl carbitol

8.30

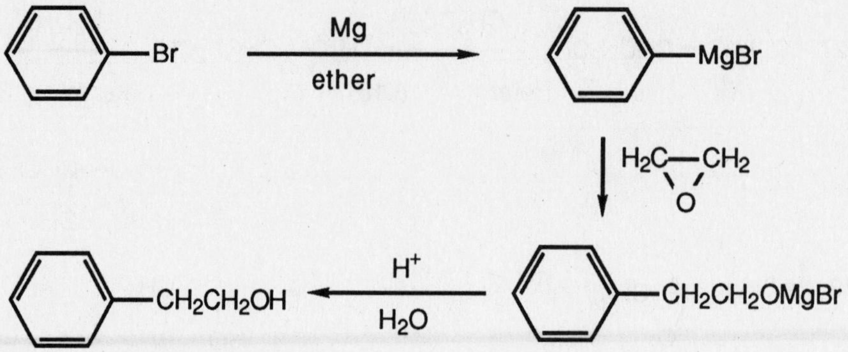

2-phenylethanol (oil of roses)

Compare with eqs. 8.2 and 8.21.

8.31 After protonation of the oxygen, the epoxide ring opens in an S_N1 manner to give the tertiary carbocation. This ion then reacts with a molecule of methanol (a weak nucleophile).

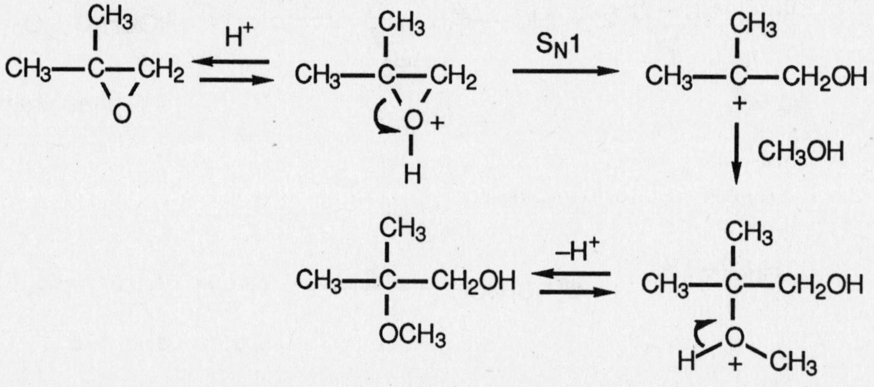

The regioisomer $(CH_3)_2C(OH)CH_2OCH_3$ is *not* formed because methanol is a weak nucleophile and S_N2 attack on the protonated oxirane at the primary carbon cannot compete with the fast S_N1 process.

8.32

ANSWERS TO PROBLEMS

8.33 $CH_3C{\equiv}CH$ + $NaNH_2$ $\xrightarrow{\text{(see eq. 3.50)}}$ $CH_3C{\equiv}C^-Na^+$ + NH_3

(see eq. 8.21) $\Big\downarrow$ $\begin{smallmatrix} H_2C{-}CH_2 \\ \diagdown O \diagup \end{smallmatrix}$

$CH_3C{\equiv}CCH_2CH_2OH$ $\xleftarrow[\text{H}_2\text{O}]{\text{H}^+}$ $CH_3C{\equiv}CCH_2CH_2O^-Na^+$

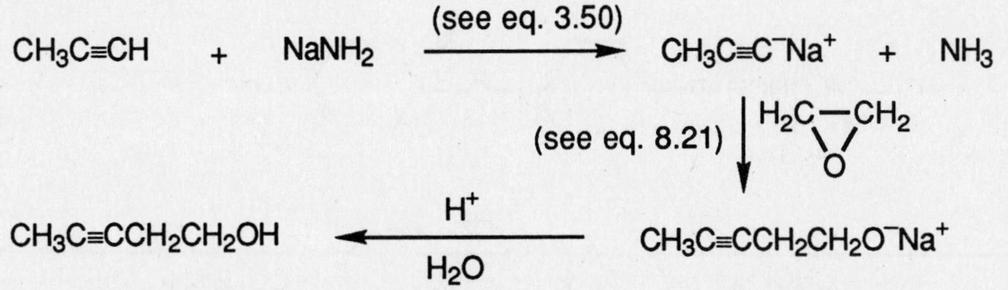

8.34   The first step gives 2-chloroethanol and occurs by an electrophilic addition mechanism (see Secs. 3.9 and 3.10) in which the hypochlorous acid behaves as $HO^{\delta-}{-}Cl^{\delta+}$:

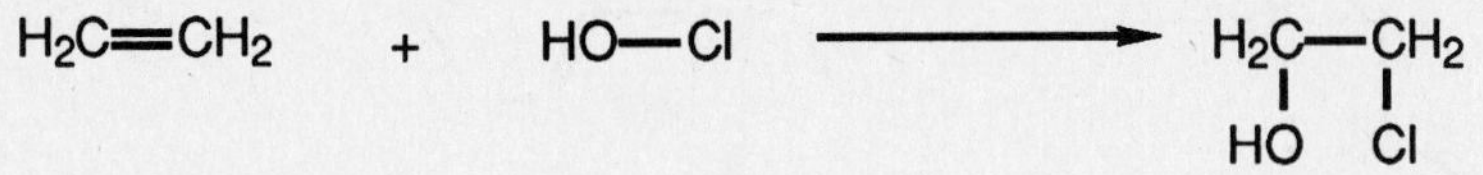

$H_2C{=}CH_2$ + $HO{-}Cl$ $\longrightarrow$ $\begin{smallmatrix} H_2C{-}CH_2 \\ | \quad\;\; | \\ HO \quad Cl \end{smallmatrix}$

The second step occurs by an intramolecular S_N2 displacement mechanism:

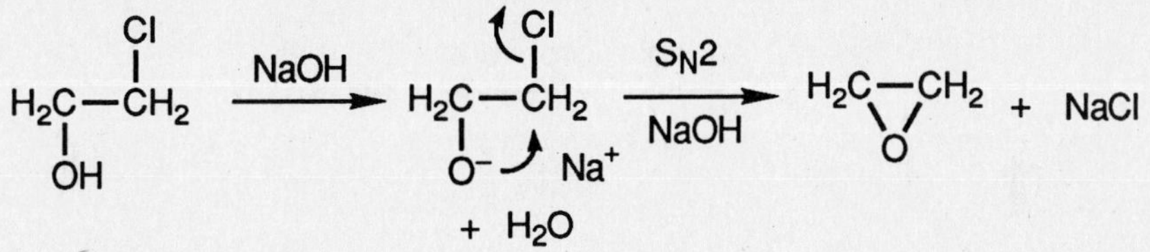

$\begin{smallmatrix} Cl \\ | \\ H_2C{-}CH_2 \\ | \\ OH \end{smallmatrix}$ $\xrightarrow{NaOH}$ $\begin{smallmatrix} Cl \\ | \\ H_2C{-}CH_2 \\ | \\ O^- \curvearrowright Na^+ \end{smallmatrix}$ $\xrightarrow[NaOH]{S_N2}$ $\begin{smallmatrix} H_2C{-}CH_2 \\ \diagdown O \diagup \end{smallmatrix}$ +   NaCl

$+\ H_2O$

8.35

189

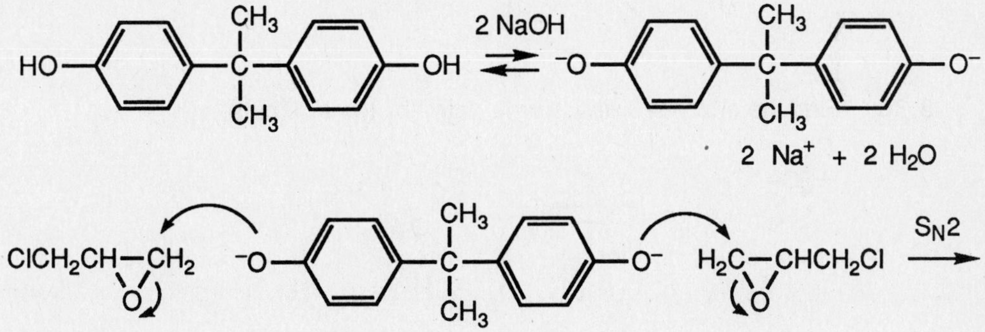

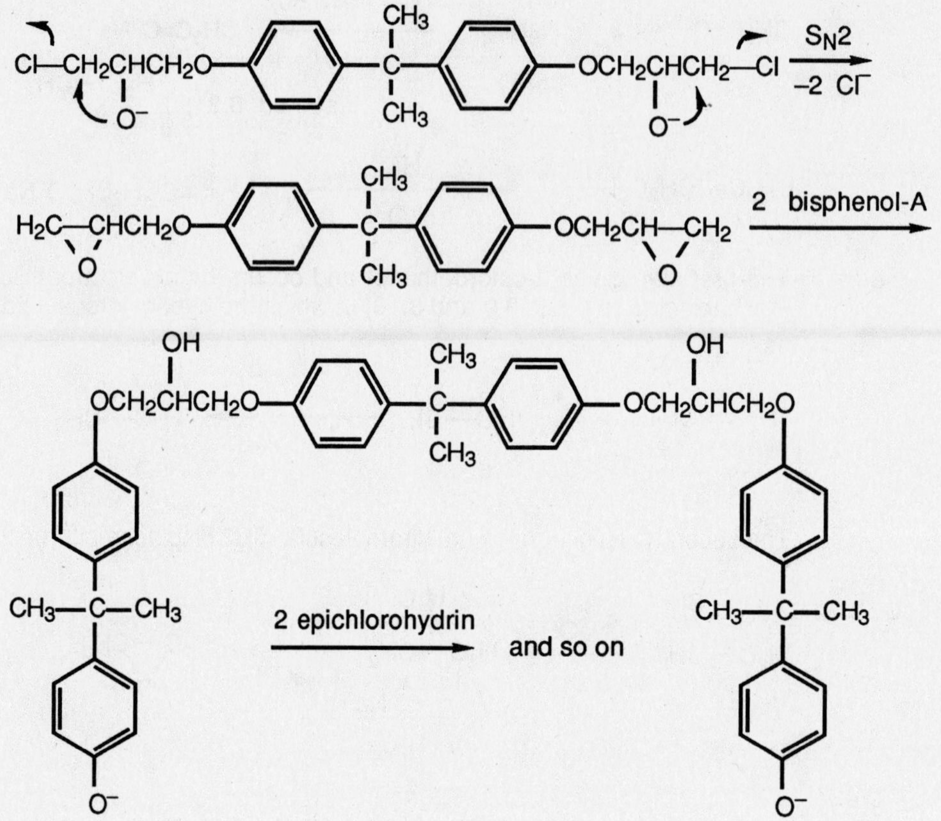

8.36   First the ethylene oxide is protonated by the acid catalyst:

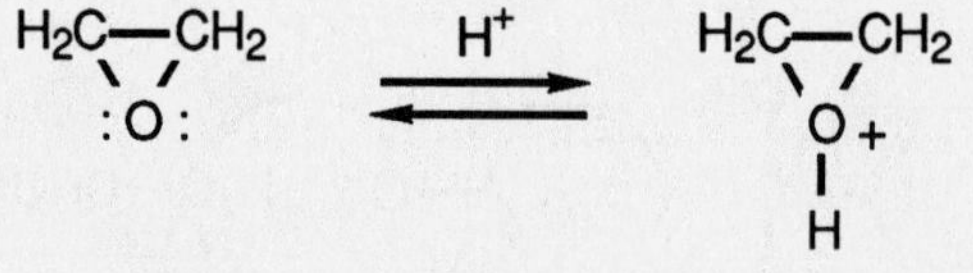

The alcohol or glycol then acts as nucleophile in an S_N2 displacement, which occurs quite easily because the epoxide ring opens in the process, thus relieving the strain associated with the small ring:

For eq. 8.20, R = —CH_3
or —CH_2CH_2OH

8.37   a.   Add a little of each compound to concentrated sulfuric acid, in separate test tubes. The ether is protonated and dissolves, whereas the hydrocarbon, being inert and less dense than sulfuric acid, simply floats on top.

     b.   Add a little bromine in carbon tetrachloride to each ether. The allyl phenyl ether, being unsaturated, quickly decolorizes the bromine, but the ethyl phenyl ether does not.

$$H_2C{=}CHCH_2OC_6H_5 \;+\; Br_2 \longrightarrow \underset{\underset{Br\;\;Br}{\textstyle|\;\;\;|}}{H_2C{-}CHCH_2OC_6H_5}$$

$$CH_3CH_2OC_6H_5 \;+\; Br_2 \longrightarrow \text{no reaction}$$

     c.   Add a small piece of sodium to each compound. The alcohol liberates a gas (hydrogen), whereas no gas bubbles are apparent in the ether.

$$\underset{\underset{OH}{\textstyle|}}{CH_3CHCH_2CH_3} \;+\; 2\,Na \longrightarrow \underset{\underset{O^-Na^+}{\textstyle|}}{CH_3CHCH_2CH_3} \;+\; H_2$$

$$CH_3OCH_2CH_2CH_3 \;+\; Na \longrightarrow \text{no reaction}$$

     d.   Add each compound to a little 10% aqueous sodium hydroxide. The phenol dissolves, whereas the ether is inert toward the base.

$$\text{C}_6\text{H}_5{-}\text{OCH}_3 \ + \ \text{Na}^+\text{OH}^- \longrightarrow \text{no reaction}$$

8.38 Since the product has only two carbons and the starting material has four carbons, two groups of two carbons must be separated by an ether oxygen:

$$-\text{C}-\text{C}-\text{O}-\text{C}-\text{C}-$$

The remaining two oxygens ($C_4H_{10}O_3$) must be at the ends of the chain. The desired structure is

$$\text{HO}-\text{CH}_2-\text{CH}_2-\text{O}-\text{CH}_2-\text{CH}_2-\text{OH}$$

and the equation for the reaction with HBr is

$$\text{HOCH}_2\text{CH}_2\text{OCH}_2\text{CH}_2\text{OH} \ + \ 4 \ \text{HBr} \longrightarrow \text{BrCH}_2\text{CH}_2\text{Br} \ + \ 3 \ \text{H}_2\text{O}$$

In this step the ether is cleaved, and the alcohol functions are also converted to alkyl halides.

8.39 The overall equation is

$$2 \ \text{HOCH}_2\text{CH}_2\text{OH} \ \xrightarrow[\text{heat}]{\text{H}^+} \ \text{(1,4-dioxane)} \ + \ 2 \ \text{H}_2\text{O}$$

The reaction is similar to the preparation of diethyl ether from ethanol (eq. 8.8).

Protonation of one hydroxyl group is followed by a nucleophilic (S_N2) displacement:

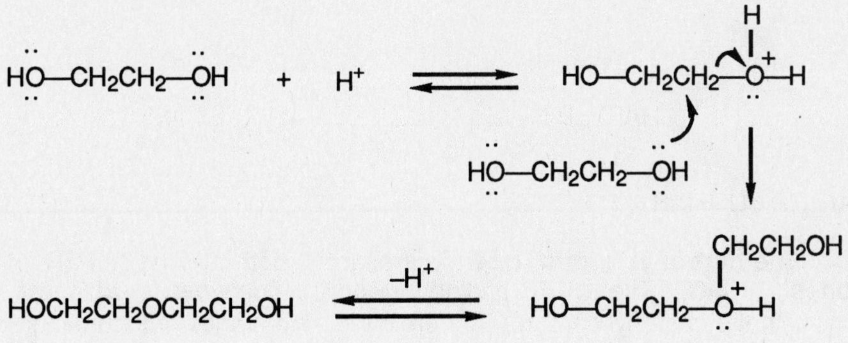

HO—CH₂CH₂—OH + H⁺ ⇌ HO—CH₂CH₂—O⁺—H

HO—CH₂CH₂—OH

HOCH₂CH₂OCH₂CH₂OH ⇌ (−H⁺) HO—CH₂CH₂—O⁺—H (CH₂CH₂OH)

The process is now repeated, except that the nucleophilic displacement is *intra*molecular:

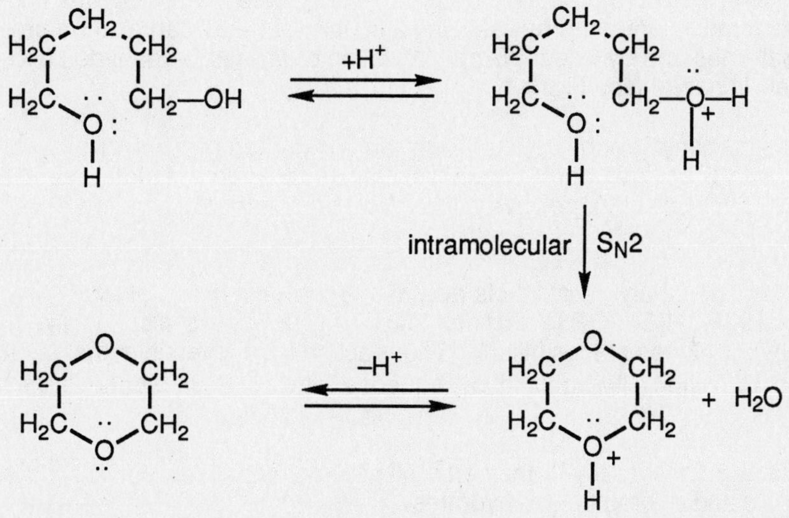

intramolecular | S_N2

−H⁺ + H₂O

+H⁺

CHAPTER NINE: ALDEHYDES AND KETONES

CHAPTER SUMMARY

The **carbonyl group, C=O**, is present in both **aldehydes** (RCH=O) and **ketones** ($R_2C=O$). The IUPAC ending for naming aldehydes is -*al*, and numbering begins with the carbonyl carbon. The ending for the names of ketones is -*one*, and the longest chain is numbered as usual. Common names are also widely used. Nomenclature is outlined in Sec. 9.2.

Formaldehyde, acetaldehyde, and **acetone** are important commercial chemicals, synthesized by special methods. In the laboratory, aldehydes and ketones are most commonly prepared by oxidizing alcohols, but they can also be prepared by hydrating alkynes and by Friedel-Crafts acylation of arenes. Aldehydes and ketones occur widely in nature (see Fig. 9.1).

The carbonyl group is planar, with the sp^2 carbon trigonal. The C=O bond is polarized, with C positive and O negative. Many carbonyl reactions are initiated by nucleophilic addition to the positive carbon and completed by addition of a proton to the oxygen.

With acid catalysis, alcohols add to the carbonyl group of aldehydes to give **hemiacetals** [RCH(OH)OR']. Further reaction with excess alcohol gives **acetals** [RCH(OR')$_2$]. Ketones react similarly. The reactions are reversible; that is, acetals or ketals can be readily hydrolyzed by aqueous acid to their alcohol and carbonyl components.

Water adds similarly to the carbonyl group of certain aldehydes (for example, formaldehyde and chloral) to give hydrates.

Grignard reagents add to carbonyl compounds. The products, after hydrolysis, are alcohols whose structure depends on that of the starting carbonyl compound . Formaldehyde gives *primary* alcohols, other aldehydes give *secondary* alcohols, and ketones give *tertiary* alcohols.

Hydrogen cyanide adds to carbonyl compounds as a carbon nucleophile to give **cyanohydrins** [R_2C(OH)CN].

Nitrogen nucleophiles add to the carbonyl group. Often, addition is followed by elimination of water to give a product with a $R_2C=NR$ group in place of the $R_2C=O$

group. For example, primary amines (R'NH$_2$) give **imines** (R$_2$C=NR'); **hydroxylamine** (NH$_2$OH) gives **oximes** (R$_2$C=NOH); and **hydrazine** (NH$_2$NH$_2$) gives **hydrazones** (R$_2$C=NNH$_2$).

Aldehydes and ketones are easily reduced to primary or secondary alcohols, respectively. Useful reagents for this purpose are **lithium aluminum hydride** (LiAlH$_4$) or **sodium borohydride** (NaBH$_4$).

Aldehydes are more easily oxidized than ketones. The **Tollens' silver mirror test** is positive for aldehydes and negative for ketones.

Aldehydes or ketones with an α-hydrogen exist as an equilibrium mixture of **keto** (H—C$_\alpha$—C=O) and **enol** (C$_\alpha$=C—OH) **tautomers**. The keto form usually predominates. An α-hydrogen is weakly acidic and can be removed by a base to produce a resonance-stabilized **enolate anion**. Deuterium exchange of α-hydrogens provides experimental evidence for enols as reaction intermediates.

In the **aldol condensation**, an enolate anion acts as a carbon nucleophile and adds to a carbonyl group to form a new carbon–carbon bond. Thus, the α-carbon of one aldehyde molecule becomes bonded to the carbonyl carbon of another aldehyde molecule to form an aldol (a 3-hydroxyaldehyde). In the **mixed aldol condensation,** the reactant with an α-hydrogen supplies the enolate anion, and the other reactant, usually without an α-hydrogen, supplies the carbonyl group to which the enolate ion adds. The aldol reaction is used commercially and also occurs in nature.

Quinones are cyclic conjugated diketones. They are colored compounds used as dyes. They also play important roles in reversible biological oxidation–reduction (electron-transfer) reactions.

REACTION SUMMARY

Preparation of Aldehydes and Ketones

(from primary alcohols)

9. ALDEHYDES AND KETONES

$$R-\overset{\overset{\displaystyle H}{|}}{\underset{\underset{\displaystyle R}{|}}{C}}-OH \quad \xrightarrow{Cr^{6+}} \quad \overset{\displaystyle O}{\overset{\|}{R-C-R}} \qquad \text{(from secondary alcohols)}$$

$$\text{benzene} \; + \; \overset{\displaystyle O}{\overset{\|}{R-C-Cl}} \quad \xrightarrow{AlCl_3} \quad \text{phenyl} - \overset{\displaystyle O}{\overset{\|}{C-R}} \qquad \text{(from arenes)}$$

$$R-C\equiv C-H \quad \xrightarrow[Hg^{2+}]{H_3O^+} \quad \overset{\displaystyle O}{\overset{\|}{R-C-CH_3}} \qquad \text{(from alkynes)}$$

Hemiacetals and Acetals

$$ROH \; + \; \overset{\displaystyle O}{\overset{\|}{R'-C-R''}} \quad \underset{}{\overset{H^+}{\rightleftharpoons}} \quad R'-\overset{\overset{\displaystyle OH}{|}}{\underset{\underset{\displaystyle OR}{|}}{C}}-R'' \quad \underset{H^+}{\overset{ROH}{\rightleftharpoons}} \quad R'-\overset{\overset{\displaystyle OR}{|}}{\underset{\underset{\displaystyle OR}{|}}{C}}-R'' \; + \; H_2O$$

Grignard Reagents

$$RMgX \; + \; H_2C=O \quad \longrightarrow \quad RCH_2OMgX \quad \xrightarrow{H_3O^+} \quad RCH_2OH$$

formaldehyde   primary alcohol

$$RMgX \; + \; R'HC=O \quad \longrightarrow \quad R'-\overset{\overset{\displaystyle H}{|}}{\underset{\underset{\displaystyle R}{|}}{C}}-OMgX \quad \xrightarrow{H_3O^+} \quad R'-\overset{\overset{\displaystyle H}{|}}{\underset{\underset{\displaystyle R}{|}}{C}}-OH$$

other aldehydes   secondary alcohol

$$RMgX \; + \; R'R''C=O \quad \longrightarrow \quad R''-\overset{\overset{\displaystyle R'}{|}}{\underset{\underset{\displaystyle R}{|}}{C}}-OMgX \quad \xrightarrow{H_3O^+} \quad R''-\overset{\overset{\displaystyle R'}{|}}{\underset{\underset{\displaystyle R}{|}}{C}}-OH$$

ketone   tertiary alcohol

REACTION SUMMARY

Hydrogen Cyanide

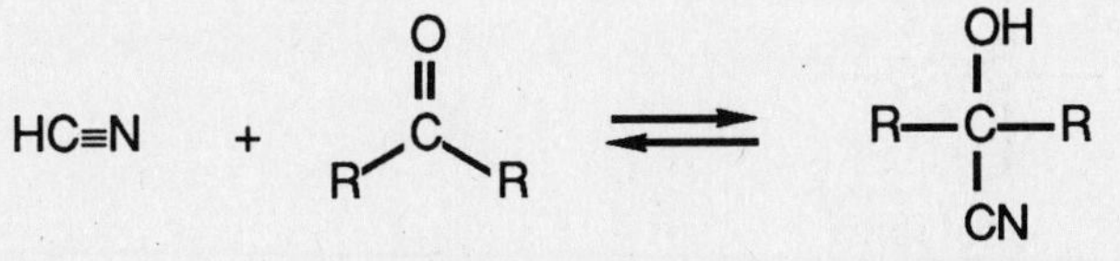

$HC{\equiv}N$ + [ketone/aldehyde: R—C(=O)—R] $\rightleftharpoons$ [R—C(OH)(CN)—R]

R = H, alkyl   cyanohydrin

Nitrogen Nucleophiles

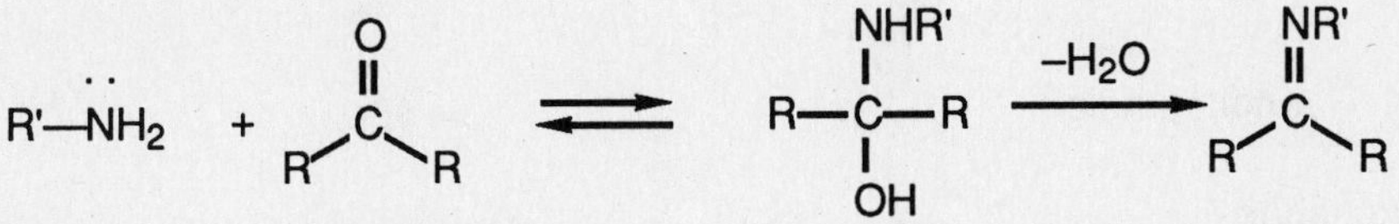

$R'{-}NH_2$ + [R—C(=O)—R] $\rightleftharpoons$ [R—C(NHR')(OH)—R] $\xrightarrow{-H_2O}$ [R—C(=NR')—R]

R = H, alkyl   imine

Reduction

[R—C(=O)—R] $\xrightarrow{\text{LiAlH}_4 \text{ or NaBH}_4}$ [R—C(OH)(H)—R]

Oxidation

$RCH{=}O$ + $2\ Ag(NH_3)_2^+$ + $3\ HO^-$ $\xrightarrow[\text{test}]{\text{Tollens'}}$

aldehyde

RCO_2^- + $2\ Ag^0\downarrow$ + $4\ NH_3\uparrow$ + $2\ H_2O$

silver mirror

$2\ RCH{=}O$ + O_2 $\longrightarrow$ $2\ RCO_2H$

9. ALDEHYDES AND KETONES

Tautomerism

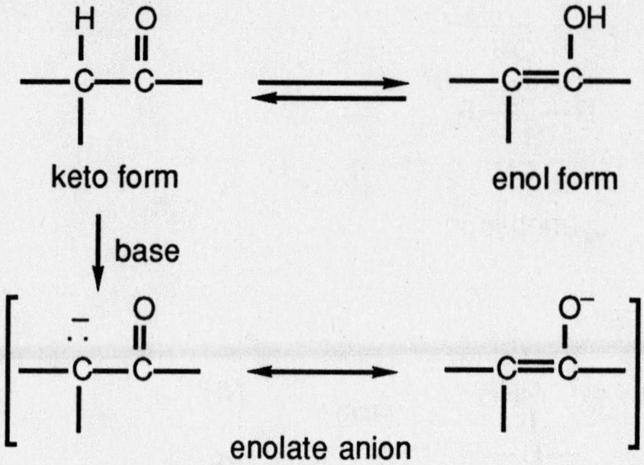

keto form   enol form

base

enolate anion

Deuterium Exchange

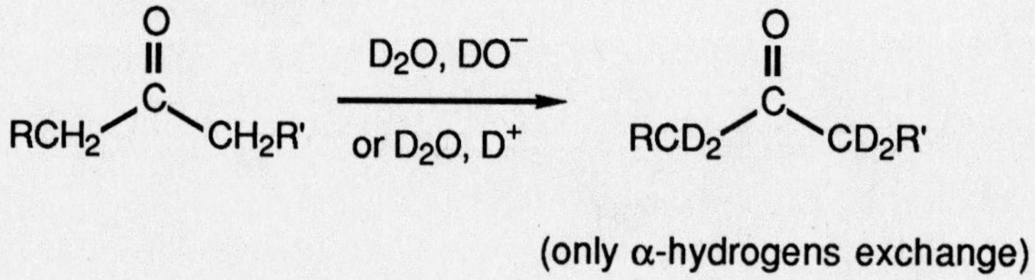

RCH_2—C(=O)—CH_2R' $\xrightarrow[\text{or } D_2O, D^+]{D_2O, DO^-}$ RCD_2—C(=O)—CD_2R'

(only α-hydrogens exchange)

Aldol Condensation

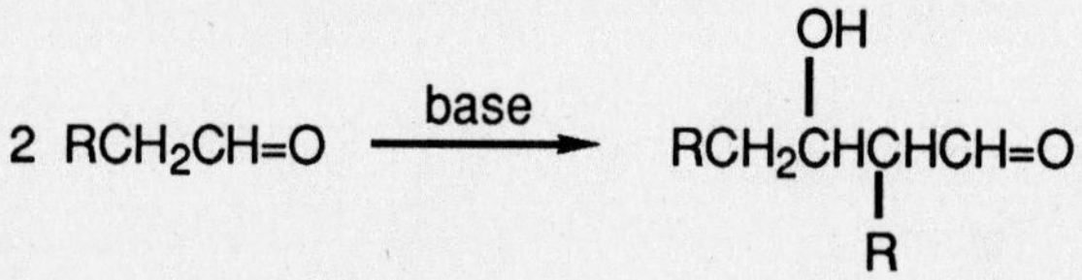

$$2\ RCH_2CH{=}O \xrightarrow{\text{base}} RCH_2CHCHCH{=}O$$

with OH on third carbon and R below

Mixed Aldol Condensation

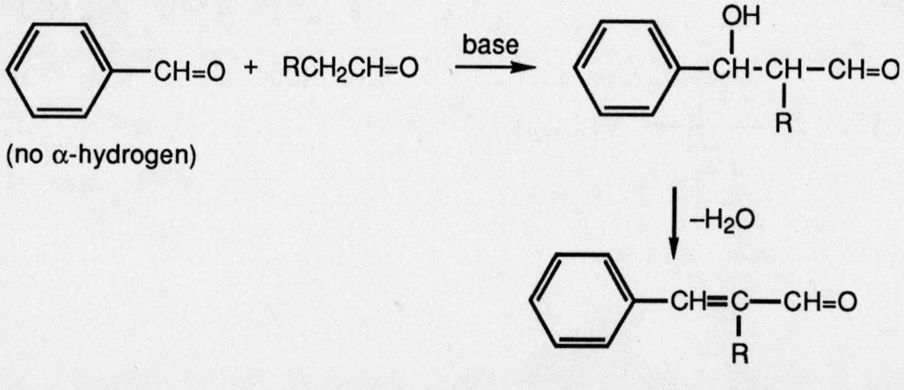

—CH=O + $RCH_2CH{=}O$ $\xrightarrow{\text{base}}$ —CH—CH—CH=O (with OH and R)

(no α-hydrogen)

$\downarrow$ $-H_2O$

—CH=C—CH=O (with R)

LEARNING OBJECTIVES

MECHANISM SUMMARY

Nucleophilic Addition

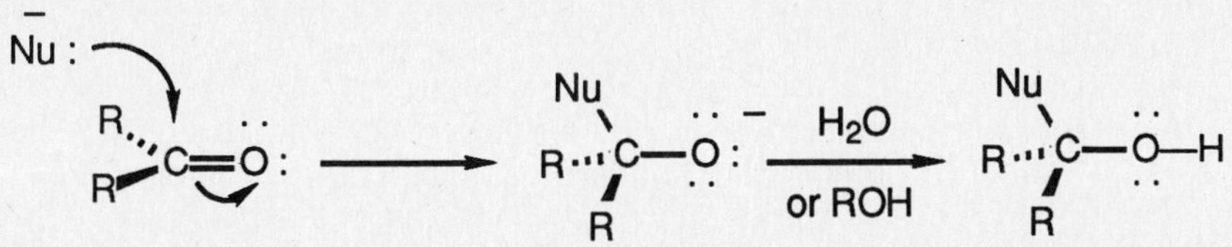

LEARNING OBJECTIVES

1. Know the meaning of: aldehyde, ketone, carbonyl group, formaldehyde, acetaldehyde, benzaldehyde, acetone, salicylaldehyde, acetophenone, benzophenone, carbaldehyde group.

2. Know the meaning of: nucleophilic addition, hemiacetal and acetal, aldehyde hydrate, cyanohydrin.

3. Know the meaning of: imine, hydroxylamine, oxime, hydrazine, hydrazone, phenylhydrazine, phenylhydrazone.

4. Know the meaning of: lithium aluminum hydride, sodium borohydride, Tollens' reagent, silver mirror test.

5. Know the meaning of: keto form, enol form, tautomers, tautomerism, enolate anion, α-hydrogen and α-carbon, aldol condensation, mixed aldol condensation.

6. Given the structure of an aldehyde or ketone, state its IUPAC name.

7. Given the IUPAC name of an aldehyde or ketone, write its structure.

8. Write the resonance contributors to the carbonyl group.

9. Given the structure or name of an aldehyde or ketone, write an equation for its reaction with the following nucleophiles: alcohol, cyanide ion, Grignard reagent or acetylide, hydroxylamine, hydrazine, phenylhydrazine, 2,4-dinitrophenylhydrazine, primary amine, lithium aluminum hydride, and sodium borohydride.

10. Explain the mechanism of acid catalysis of nucleophilic additions to the carbonyl group.

11. Write the steps in the mechanism of acetal formation and hydrolysis. Draw the structures of resonance contributors to intermediates in the mechanism.

12. Given a carbonyl compound and a Grignard reagent, write the structure of the alcohol that is formed when they react.

13. Given the structure of a primary, secondary, or tertiary alcohol, deduce what combination of aldehyde or ketone and Grignard reagent can be used for its synthesis.

14. Given the structure of an aldehyde or ketone, write the formula of the alcohol that is obtained from it by reduction.

15. Given the structure of an aldehyde, write the structure of the acid that is formed from it by oxidation.

16. Know which tests can distinguish an aldehyde from a ketone.

17. Given the structure of an aldehyde or ketone, write the structure of the corresponding enol and enolate anion.

18. Identify the α-hydrogens in an aldehyde or ketone, and be able to recognize that these hydrogens can be exchanged readily for deuterium.

19. Write the structure of the aldol formed by the self-condensation of an aldehyde of given structure.

20. Given two reacting carbonyl compounds, write the structure of the mixed aldol obtained from them.

21. Write the steps in the mechanism of the aldol condensation.

ANSWERS TO PROBLEMS

Problems Within the Chapter

9.1　a.　$CH_3CH_2CH_2CH_2CH=O$　　b.

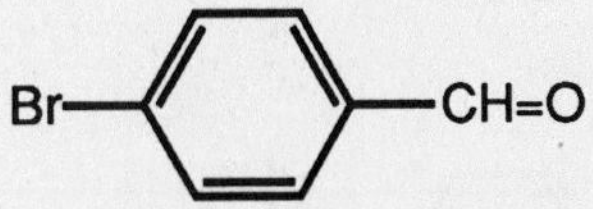

c.

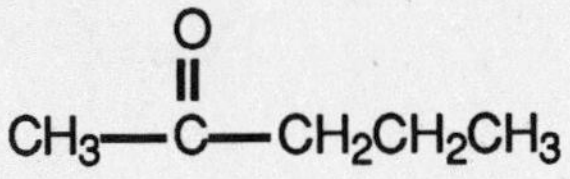

d.

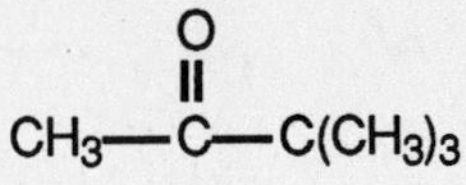

e.

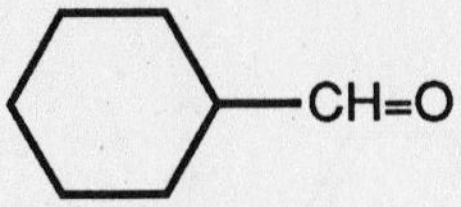

f.

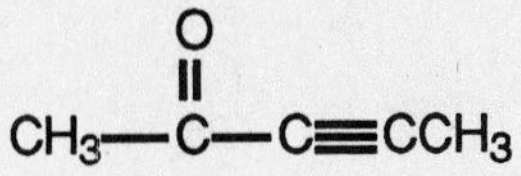

9.2　a.　3-methylbutanal
(no number is necessary for
the aldehyde function)

b.　2-butenal
(the number locates the double
bond between C-2 and C-3)

　　c.　cyclobutanone

d.　4-methyl-2-pentanone or
isobutyl methyl ketone

9.3

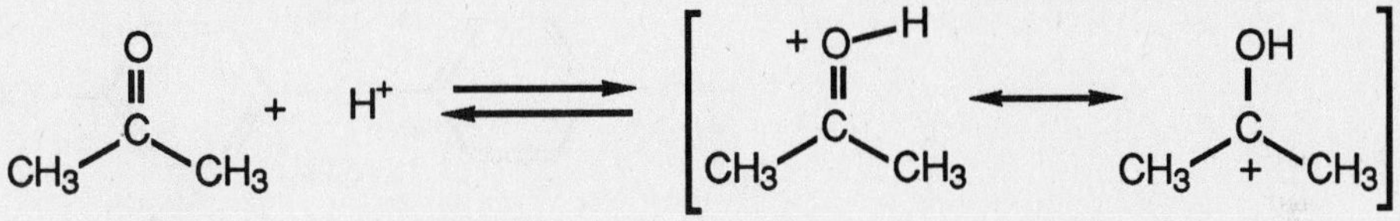

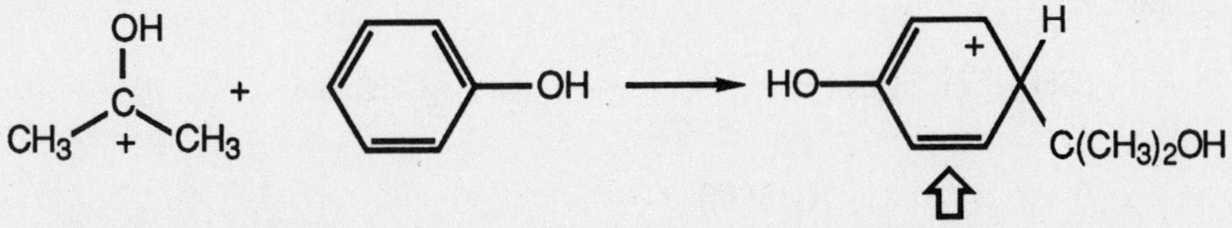

Can you write a resonance structure for this cation that places positive charge
on the oxygen atom ?

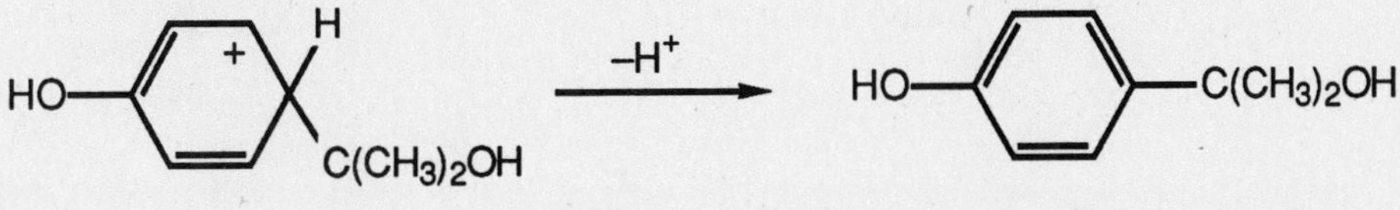

201

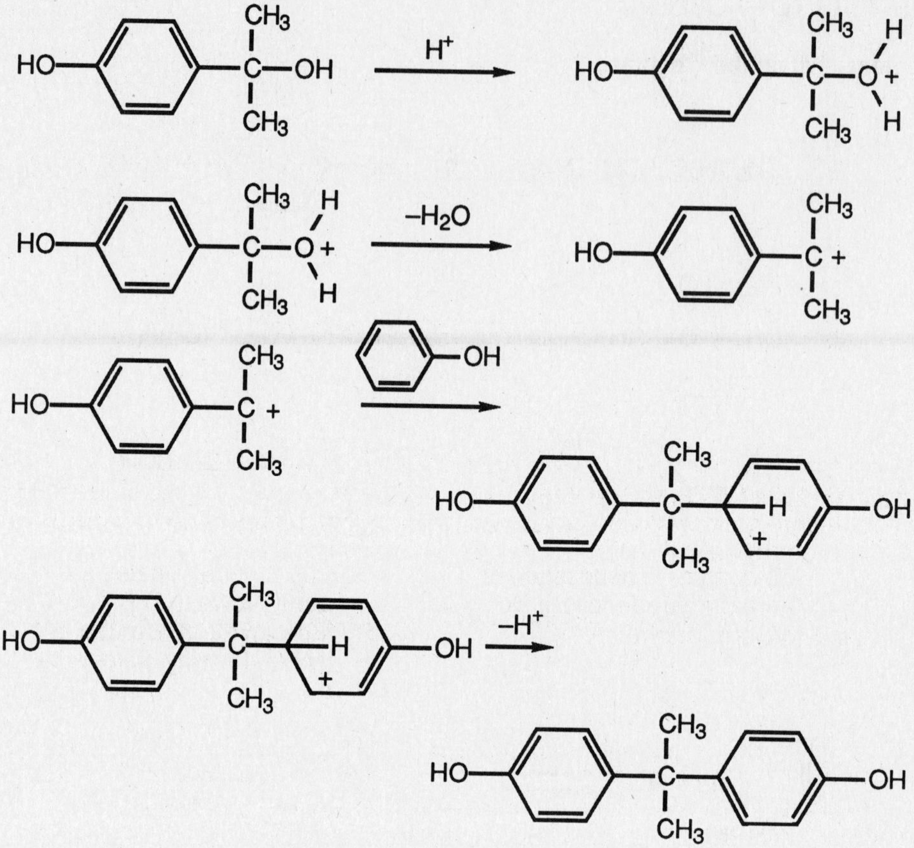

9.4   a. $(CH_3)_2CHCH_2OH$ $\xrightarrow[25°C]{PCC, CH_2Cl_2}$ $(CH_3)_2CHCH=O$

(see eq. 7.35)

b.

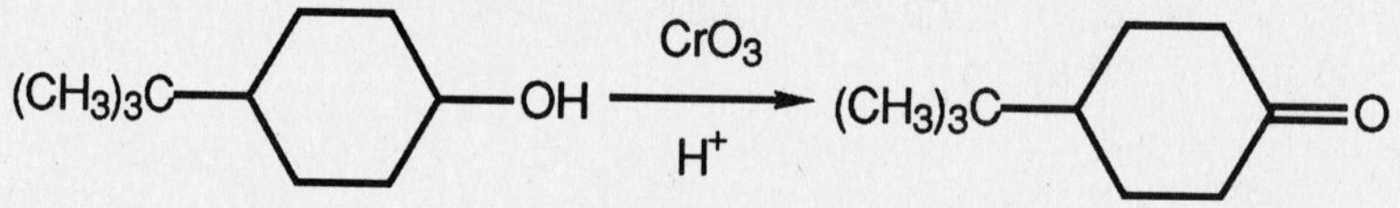

$(CH_3)_3C-$⬡$-OH$ $\xrightarrow[H^+]{CrO_3}$ $(CH_3)_3C-$⬡$=O$

(see eq. 7.33)

9.5

$$CH_3{-}\phi + CH_3{-}\underset{\displaystyle \overset{\displaystyle O}{\|}}{C}{-}Cl \xrightarrow{AlCl_3} CH_3{-}\phi{-}\underset{\displaystyle \overset{\displaystyle O}{\|}}{C}{-}CH_3$$

4-methylacetophenone
(methyl 4-methylphenyl ketone)

9.6

$$HC{\equiv}CCH_2CH_2CH_2CH_3 \xrightarrow[Hg^{2+}]{H_3O^+} CH_3{-}\underset{\displaystyle \underset{\displaystyle O}{\|}}{C}{-}CH_2CH_2CH_2CH_3$$

(see eq. 3.49)

9.7   a.   The carbonyl compound is more polar than the hydrocarbon but cannot form hydrogen bonds with itself. The compound with two hydroxyl groups hydrogen-bonds more effectively than the compound with just one hydroxyl group.

CH_3—⟨⟩—CH_3   <   CH=O—⟨⟩   <   CH_2OH—⟨⟩   <   OH—⟨⟩—OH

p-xylene
bp 138°C

benzaldehyde
bp 179°C

benzyl alcohol
bp 205°C

hydroquinone
bp 286°C

b.   Water solubility increases in the same order because of increasing possibilities for hydrogen bonding with water.

9.8 CH_3CH_2OH + $CH_3CH{=}O$ $\underset{}{\overset{H^+}{\rightleftharpoons}}$ $CH_3{-}\underset{\displaystyle \underset{H}{|}}{\overset{\displaystyle \overset{OCH_2CH_3}{|}}{C}}{-}OH$

hemiacetal

Mechanism:

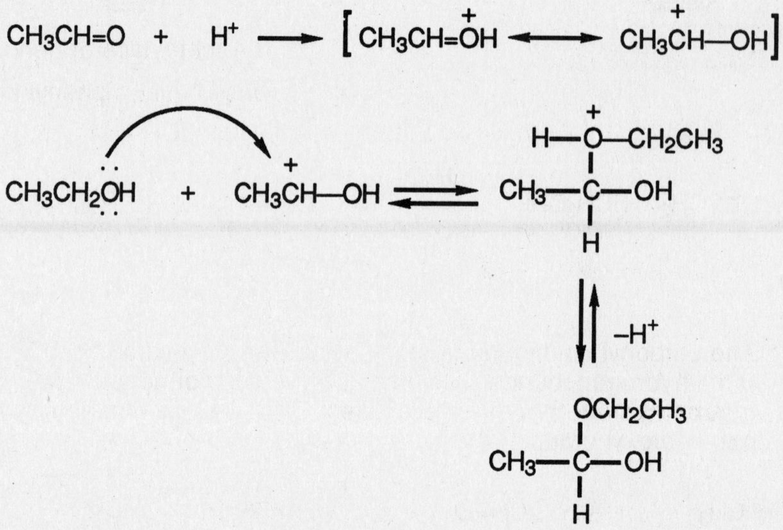

$$CH_3CH=O \ + \ H^+ \ \longrightarrow \ \left[CH_3CH=\overset{+}{O}H \ \longleftrightarrow \ CH_3\overset{+}{C}H-OH \right]$$

$$CH_3CH_2\ddot{O}H \ + \ CH_3\overset{+}{C}H-OH \ \rightleftharpoons \ CH_3-\underset{\underset{H}{|}}{\overset{\overset{+}{H-O-CH_2CH_3}}{C}}-OH$$

$$\Big\updownarrow \ -H^+$$

$$CH_3-\underset{\underset{H}{|}}{\overset{\overset{OCH_2CH_3}{|}}{C}}-OH$$

9.9

$$CH_3-\underset{\underset{H}{|}}{\overset{\overset{OCH_2CH_3}{|}}{C}}-OH \ + \ CH_3CH_2OH \ \xrightarrow{H^+} \ CH_3-\underset{\underset{H}{|}}{\overset{\overset{OCH_2CH_3}{|}}{C}}-OCH_2CH_3 \ + \ H_2O$$

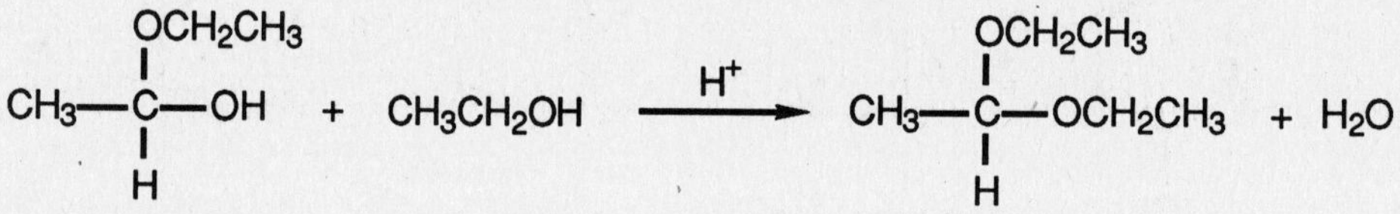

Mechanism:

$$CH_3-\underset{\underset{H}{|}}{\overset{\overset{OCH_2CH_3}{|}}{C}}-O-H \ \underset{H^+}{\rightleftharpoons} \ CH_3-\underset{\underset{H}{|}}{\overset{\overset{OCH_2CH_3}{|}}{C}}-\overset{+}{\underset{\underset{H}{|}}{O}}-H \ \underset{-H_2O}{\rightleftharpoons}$$

$$\left[CH_3-\underset{\underset{H}{|}}{\overset{+}{C}}-OCH_2CH_3 \ \longleftrightarrow \ CH_3-\underset{\underset{H}{|}}{C}=\overset{+}{O}CH_2CH_3 \right] \rightleftharpoons$$

$$CH_3CH_2\ddot{O}H$$

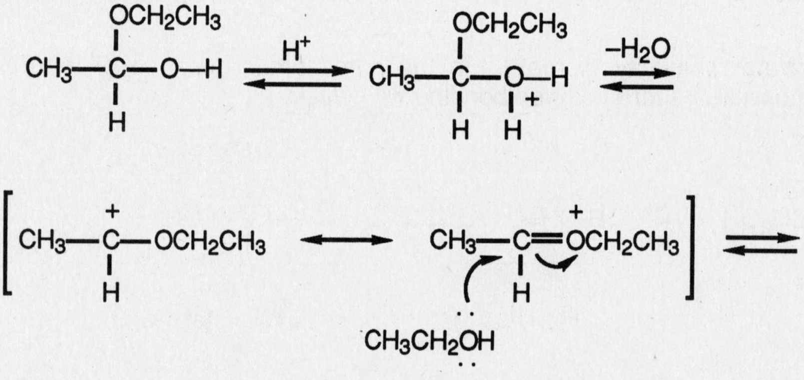

$$\text{H—}\overset{+}{\text{O}}\text{CH}_2\text{CH}_3 \qquad \text{O—CH}_2\text{CH}_3$$

$$\underset{\overset{|}{\text{H}}}{\text{CH}_3\text{—C—OCH}_2\text{CH}_3} \quad \underset{-\text{H}^+}{\rightleftharpoons} \quad \underset{\overset{|}{\text{H}}}{\text{CH}_3\text{—C—OCH}_2\text{CH}_3}$$

9.10 Protonation of acetone is followed by addition of ethylene glycol to the protonated carbonyl group. Loss of a proton gives the intermediate hemiacetal.

$$(\text{CH}_3)_2\text{C=O} \quad \overset{\text{H}^+}{\rightleftharpoons} \quad \left[(\text{CH}_3)_2\overset{+}{\text{C=OH}} \longleftrightarrow (\text{CH}_3)_2\overset{+}{\text{C}}\text{—OH} \right]$$

$$\rightleftharpoons$$

$$\text{HOCH}_2\text{CH}_2\text{OH}$$

$$\underset{(\text{CH}_3)_2\text{C—OCH}_2\text{CH}_2\text{OH}}{\overset{\text{OH}}{|}} \quad \underset{-\text{H}^+}{\rightleftharpoons} \quad \underset{\overset{|}{\text{H}}}{\overset{\overset{\text{OH}}{|}}{(\text{CH}_3)_2\text{C—}\overset{+}{\text{O}}\text{CH}_2\text{CH}_2\text{OH}}}$$

Protonation of the tertiary hydroxyl group and loss of water gives a resonance-stabilized carbocation. Nucleophilic attack of the remaining hydroxyl group on the carbocation followed by loss of a proton gives the ketal.

$$\underset{(\text{CH}_3)_2\text{C—OCH}_2\text{CH}_2\text{OH}}{\overset{\text{OH}}{|}} \quad \overset{\text{H}^+}{\rightleftharpoons} \quad \underset{(\text{CH}_3)_2\text{C—OCH}_2\text{CH}_2\text{OH}}{\overset{+\ \text{OH}_2}{|}}$$

$$\rightleftharpoons \quad -\text{H}_2\text{O}$$

Notice that each step is reversible. The reaction can be driven to the right by removing water as it is formed.

9.11   The proton first adds to one of the oxygens:

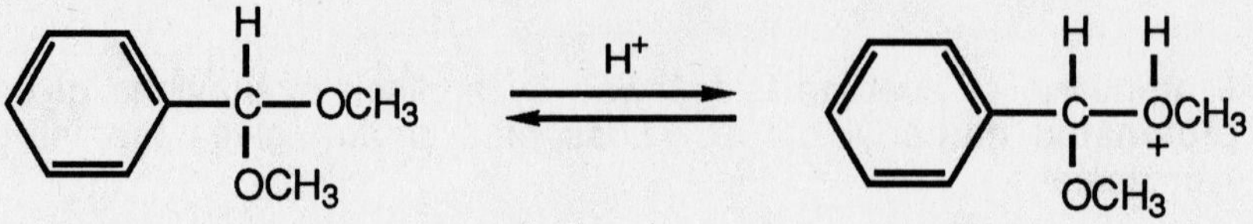

Loss of methanol gives a resonance-stabilized carbocation:

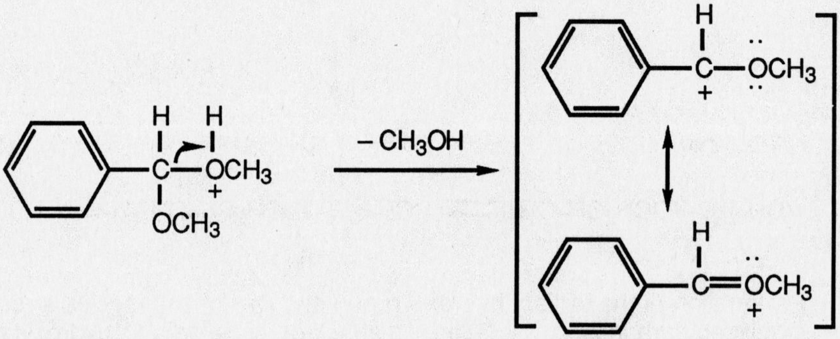

The carbocation reacts with water, which is a nucleophile and is present in large excess:

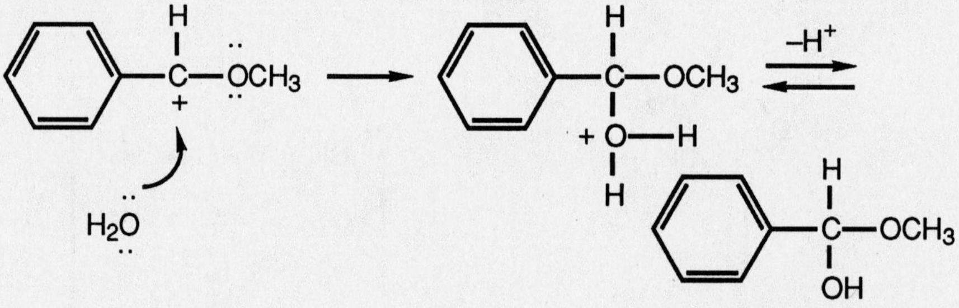

hemiacetal

The sequence is then repeated, beginning with protonation of the methoxyl oxygen of the hemiacetal:

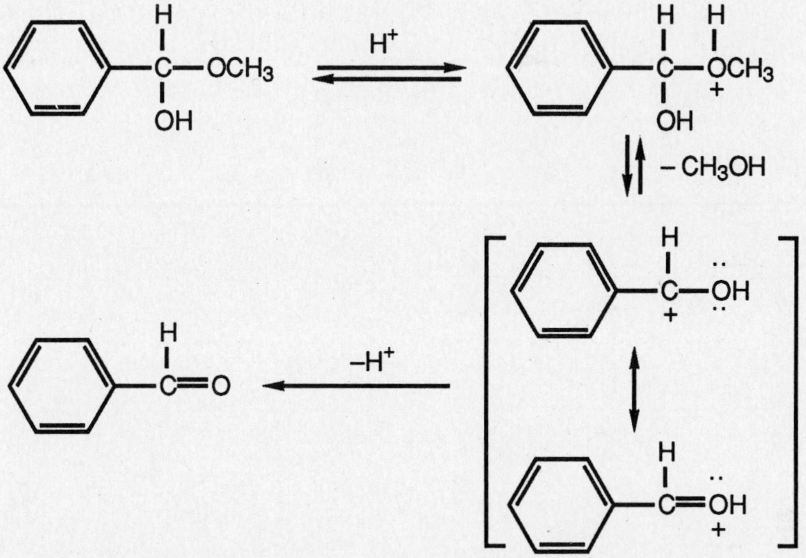

The whole process is driven forward because water is present in excess.

9.12   In the first step, one bromine is replaced by a hydroxyl group:

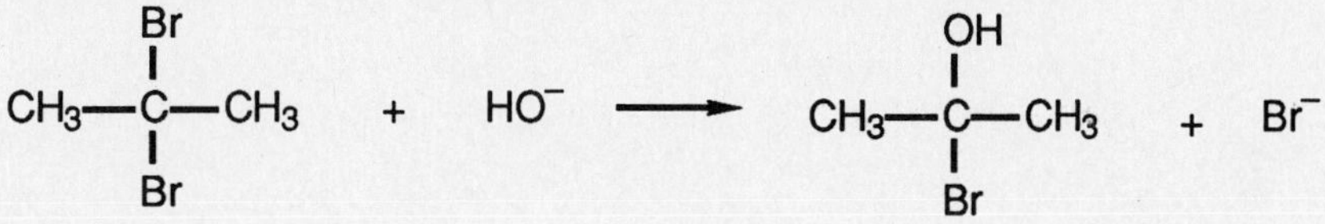

Loss of HBr gives acetone:

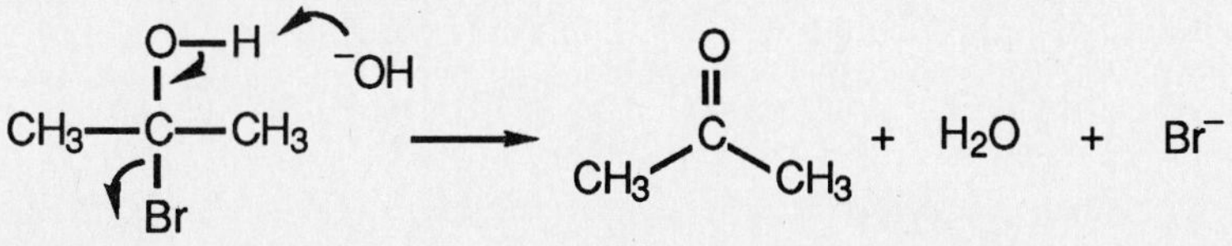

Even if both bromines were replaced by hydroxyl groups, the resulting diol (acetone hydrate) would lose water since acetone does not form a stable hydrate.

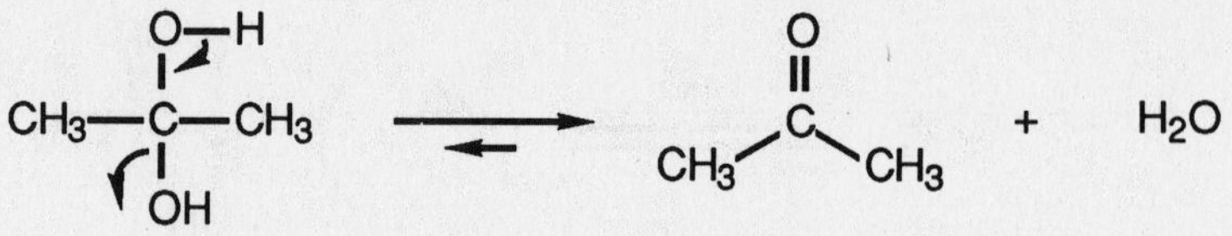

9.13 a. The alcohol is primary, so formaldehyde must be used as the carbonyl component:

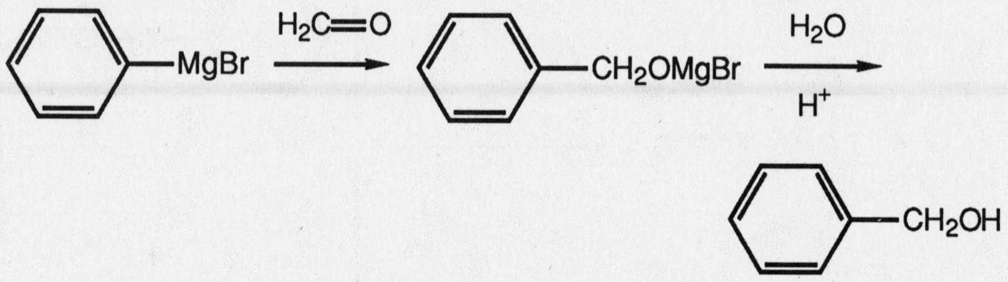

b. Only one R group attached to —C—OH comes from the Grignard reagent. The alcohol is tertiary, so the carbonyl component must be a ketone.

<u>Two possibilities:</u>

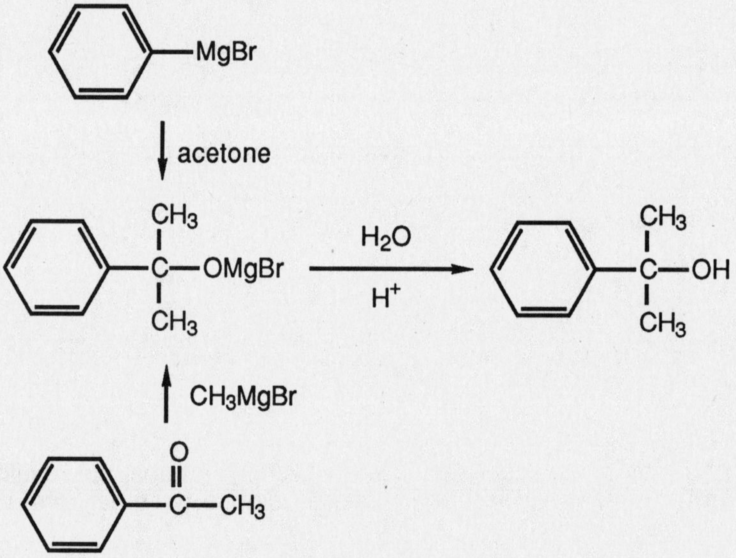

9.14   a. $CH_3CH=O$ +   HCN $\longrightarrow$ $CH_3-\overset{\overset{\displaystyle OH}{|}}{\underset{\underset{\displaystyle CN}{|}}{C}}-H$

b. 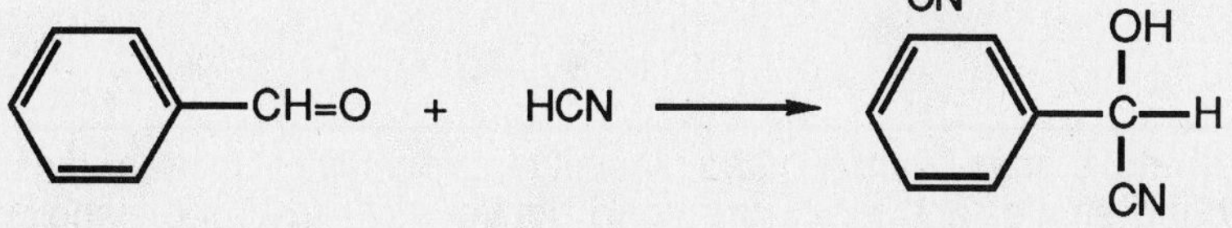   [benzene ring]—$CH=O$ +   HCN $\longrightarrow$ [benzene ring]—$\overset{\overset{\displaystyle OH}{|}}{\underset{\underset{\displaystyle CN}{|}}{C}}-H$

9.15   [benzene ring]—$CH=O$

+

H_2N—[benzene ring] $\longrightarrow$ [benzene ring]—$CH=N$—[benzene ring]   + H_2O

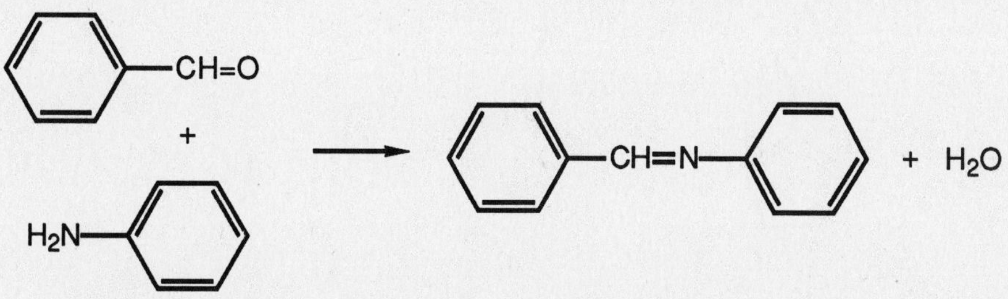

9.16   a. $CH_3CH_2CH=O$ + H_2N—OH $\longrightarrow$ $CH_3CH_2CH=N$—OH   + H_2O

b. 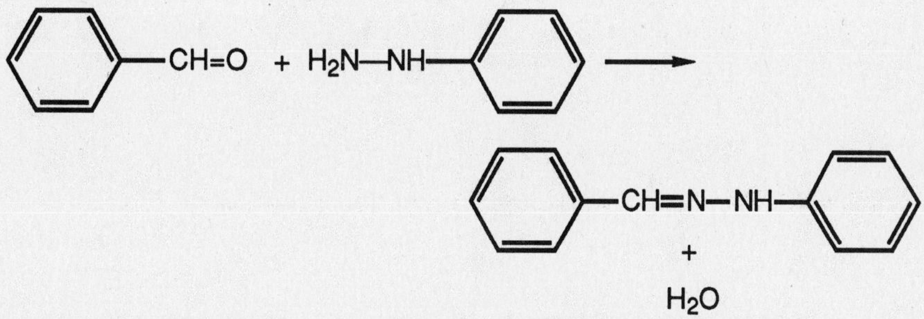   [benzene ring]—$CH=O$ + H_2N—NH—[benzene ring] $\longrightarrow$

[benzene ring]—$CH=N$—NH—[benzene ring]

+

H_2O

9.17   A metal hydride will reduce the C=O bond but not the C=C bond:

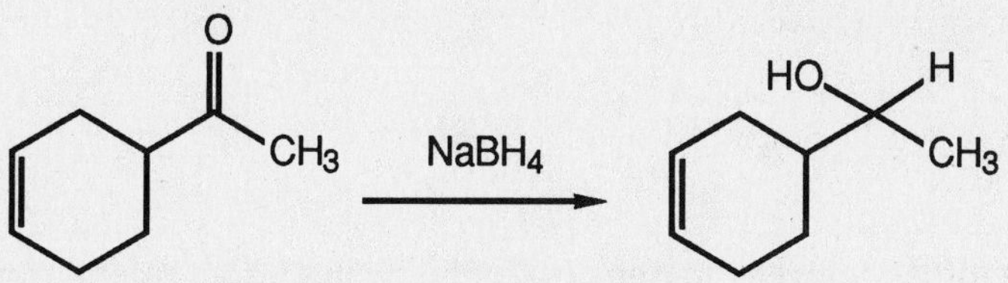

[cyclohexene with C(=O)CH₃ substituent] $\xrightarrow{\text{NaBH}_4}$ [cyclohexene with C(OH)(H)CH₃ substituent]

209

9. ALDEHYDES AND KETONES

9.18 Follow eq. 9.36 as a guide, replacing R with H

$$H_2C=O \ + \ 2\ Ag(NH_3)_2^+ \ + \ 3\ HO^- \longrightarrow$$

$$HCO_2^- \ + \ 2\ Ag\downarrow \ + \ 4\ NH_3\uparrow \ + \ 2\ H_2O$$

9.19 Remove an α-hydrogen and place it on the oxygen; make the carbon–oxygen bond single, and make the bond between the α-carbon and what was the carbonyl carbon double.

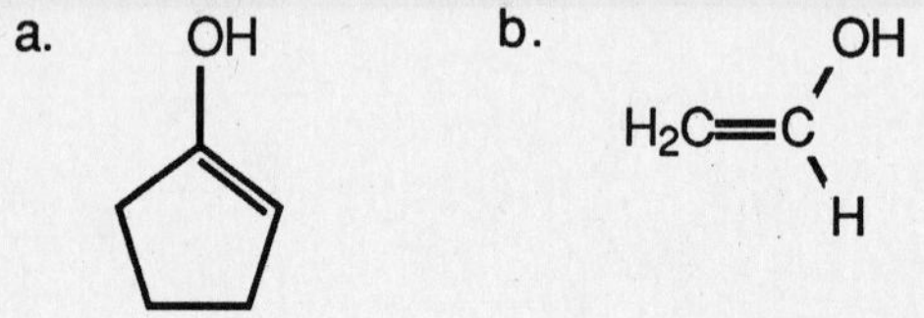

a.

b.

9.20 a.

b.

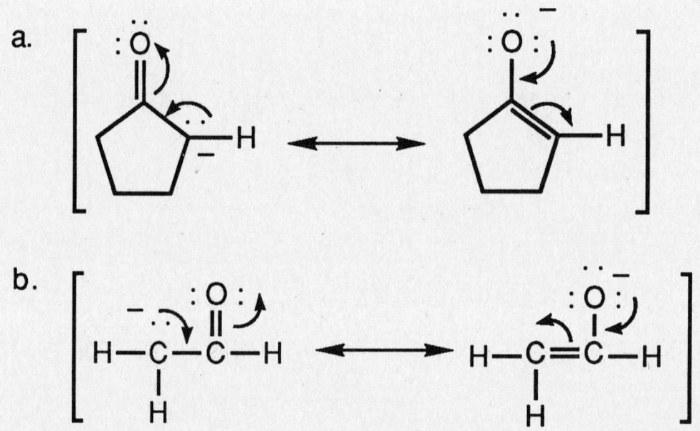

9.21 In each case, only the α-hydrogens can be readily exchanged.

a. There are three exchangeable hydrogens, indicated by the arrows.

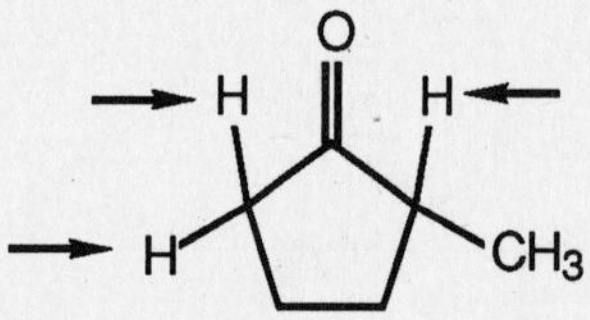

b. Only the three methyl hydrogens indicated by the arrow can be readily exchanged. The remaining methyl hydrogens are β, not α, with respect to the carbonyl group.

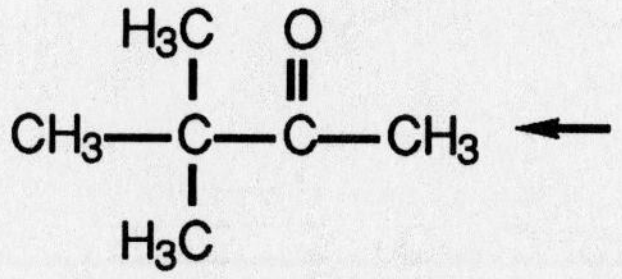

$$CH_3-\underset{\underset{H_3C}{|}}{\overset{\overset{H_3C}{|}}{C}}-\overset{\overset{O}{||}}{C}-CH_3 \longleftarrow$$

9.22 The acid catalyzes keto–enol tautomerization. The enol behaves as a nucleophile and chlorine as an electrophile.

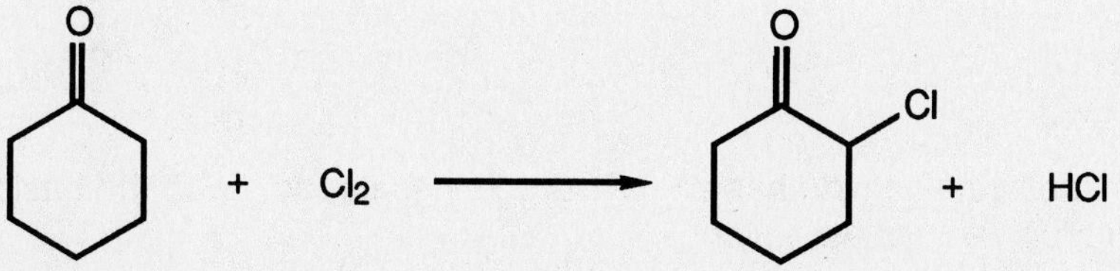

<u>Mechanism</u>

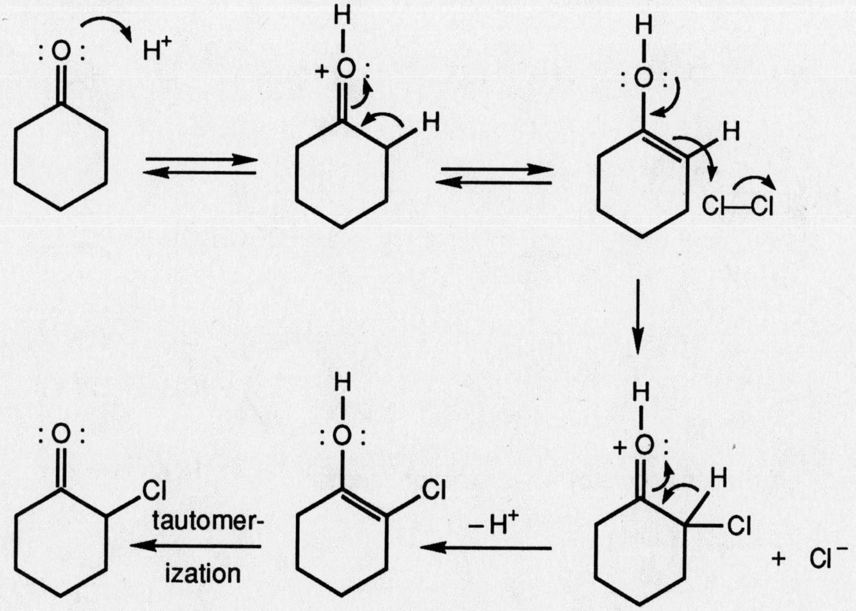

9.23 Follow eqs. 9.45-9.47 as a guide.

$$CH_3CH_2CH=O \quad + \quad HO^- \rightleftharpoons CH_3\overset{-}{C}HCH=O \quad + \quad H_2O$$

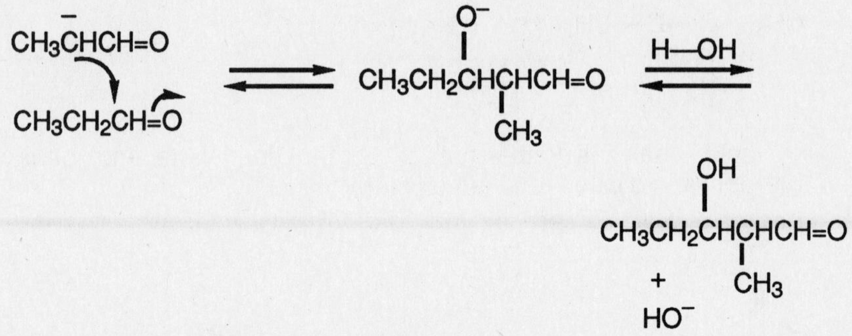

$$CH_3\overset{-}{C}HCH=O$$

$$CH_3CH_2CH\overset{\frown}{=}O$$

$$\rightleftharpoons \quad \underset{CH_3}{\overset{\overset{O^-}{|}}{CH_3CH_2CHCHCH=O}} \quad \overset{H-OH}{\rightleftharpoons}$$

$$\underset{\underset{HO^-}{+}}{\overset{\overset{OH}{|}}{CH_3CH_2CHCHCH=O}}$$
$$CH_3$$

9.24 Only the acetaldehyde has an α–hydrogen, so it reacts with the basic catalyst to produce an enolate anion:

$$CH_3CH=O \quad + \quad HO^- \rightleftharpoons \overset{-}{C}H_2CH=O \quad + \quad H_2O$$

The enolate anion then attacks the carbonyl group of benzaldehyde:

$$CH_3CH=O \quad + \quad HO^- \rightleftharpoons \overset{-}{C}H_2CH=O \quad + \quad H_2O$$

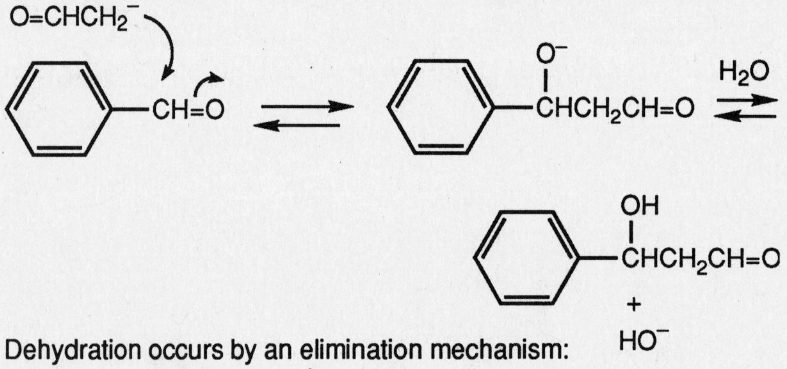

Dehydration occurs by an elimination mechanism:

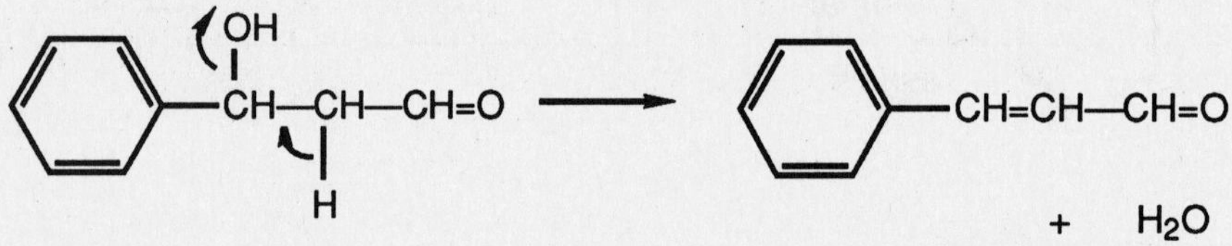

9.25   Propanal has the α-hydrogens. The aldol is

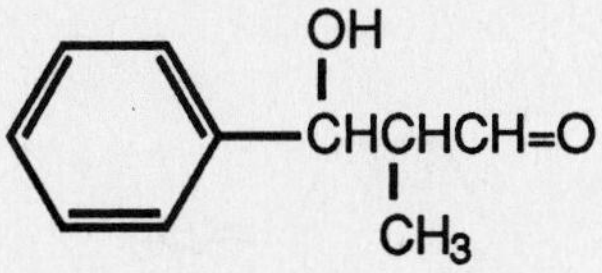

and its dehydration product is

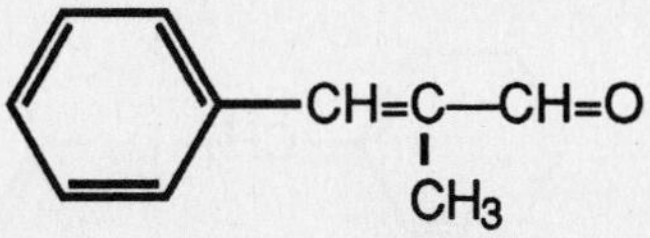

9.26   An aldol condensation–dehydration sequence can be used to construct the carbon skeleton.

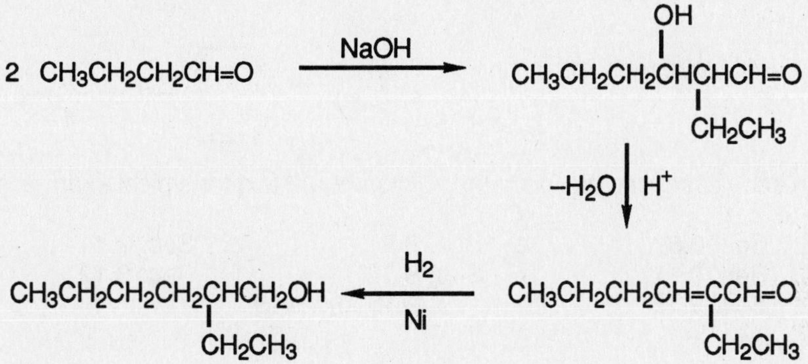

Catalytic hydrogenation of the double bonds completes the synthesis.

Additional Problems

9.27   a. 3-pentanone   b. hexanal
        c. benzophenone (diphenyl ketone)   d. p-bromobenzaldehyde
        e. cyclopentanone   f. 2,2-dimethylpropanal
        g. dicyclobutyl ketone   h. 3-penten-2-one
        i. bromoacetone (or bromopropanone)   j. 2,3-pentanedione

9.28   a.

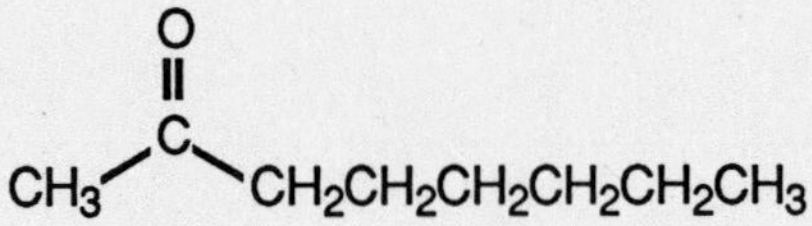

$$CH_3-\underset{\underset{O}{\parallel}}{C}-CH_2CH_2CH_2CH_2CH_2CH_3$$

b. $(CH_3)_2CHCH_2CH_2CH=O$

c.

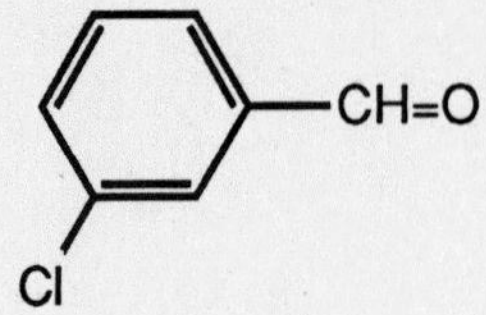

d.

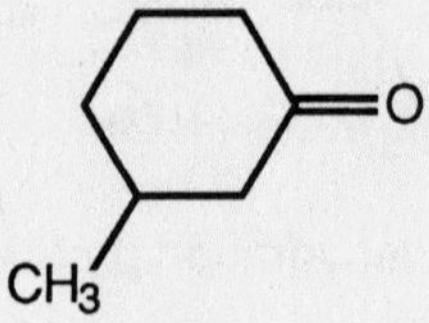

e. $CH_3CH=CHCHO$

f.

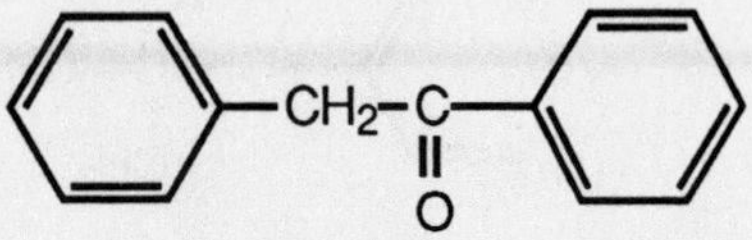

g.

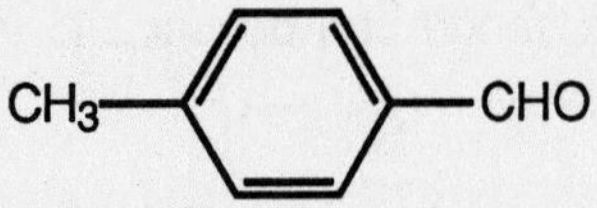

h.

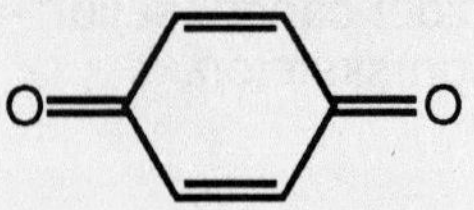

i. $CH_3(CH_2)_3CBr_2CH=O$

j.

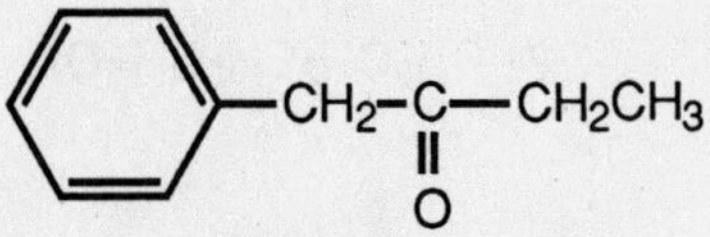

9.29   In each case, see the indicated section of the text for typical examples.

a.   Sec. 9.8   b.   Sec. 9.8   c.   Sec. 9.11
d.   Sec. 9.12   e.   Sec. 9.12   f.   Sec. 9.12
g.   Sec. 9.15   h.   Sec. 9.15 and 9.19

9.30   a.

$$\underset{\underset{OH}{|}}{CH_3CHCH_2CH_2CH_3} \xrightarrow[\underset{(see\ eq.\ 7.33)}{H^+}]{CrO_3} \underset{\underset{O}{\parallel}}{CH_3CCH_2CH_2CH_3}$$

b.

$$HC{\equiv}CCH_2CH_2CH_3 \xrightarrow[\underset{(see\ eq.\ 3.49)}{Hg^{2+}}]{H_3O^+} \underset{\underset{O}{\parallel}}{CH_3CCH_2CH_2CH_3}$$

214

9.31

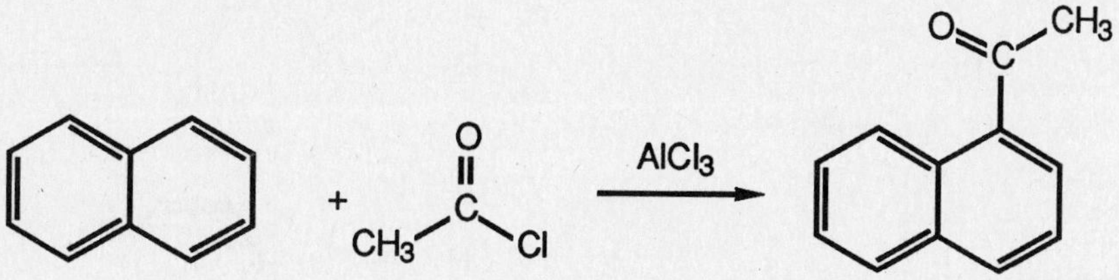

9.32 a.

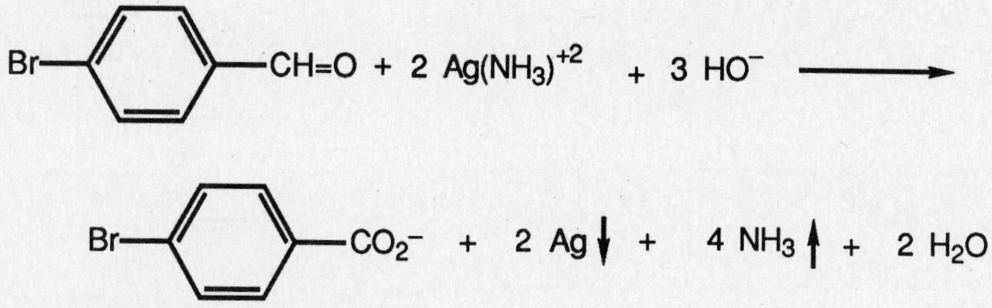

$$Br-C_6H_4-CH=O + 2\ Ag(NH_3)^{+2} + 3\ HO^- \longrightarrow$$

$$Br-C_6H_4-CO_2^- + 2\ Ag\downarrow + 4\ NH_3\uparrow + 2\ H_2O$$

p-bromobenzoate ion

b.

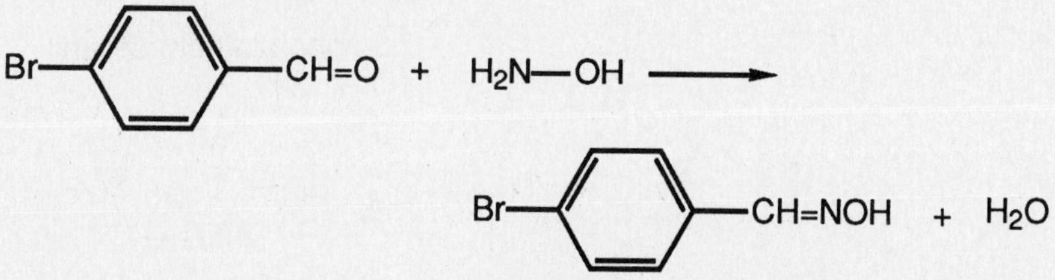

$$Br-C_6H_4-CH=O + H_2N-OH \longrightarrow$$

$$Br-C_6H_4-CH=NOH + H_2O$$

p-bromobenzaldoxime

c.

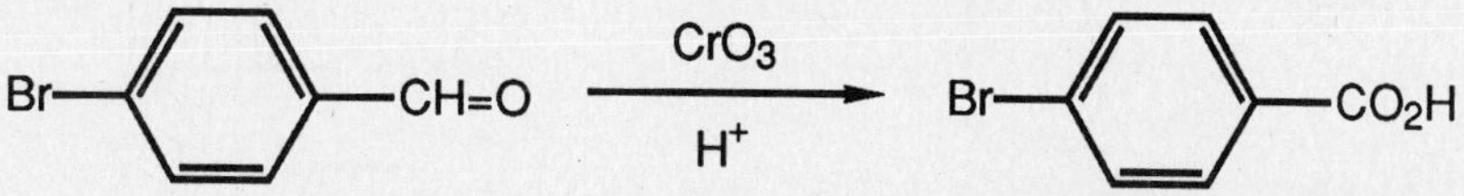

$$Br-C_6H_4-CH=O \xrightarrow[H^+]{CrO_3} Br-C_6H_4-CO_2H$$

p-bromobenzoic acid

d.

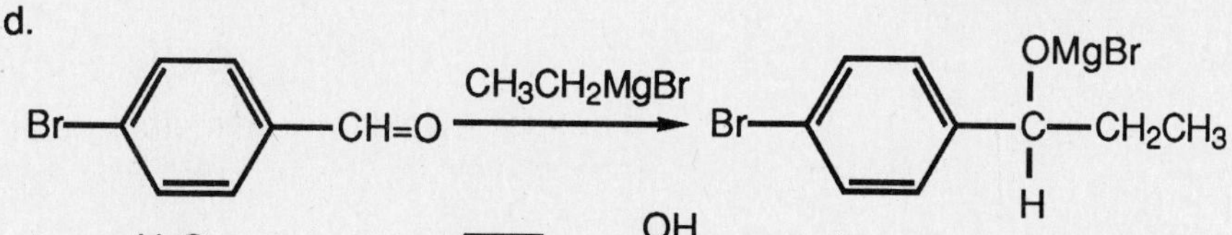

$$Br-C_6H_4-CH=O \xrightarrow{CH_3CH_2MgBr} Br-C_6H_4-\underset{\underset{H}{|}}{\overset{\overset{OMgBr}{|}}{C}}-CH_2CH_3$$

$$\xrightarrow[H^+]{H_2O} Br-C_6H_4-\underset{\underset{H}{|}}{\overset{\overset{OH}{|}}{C}}-CH_2CH_3$$

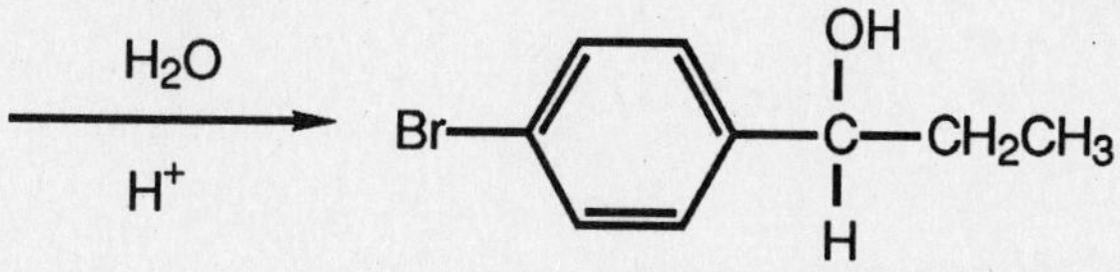

1-(*p*-bromophenyl)-
1-propanol

215

e.

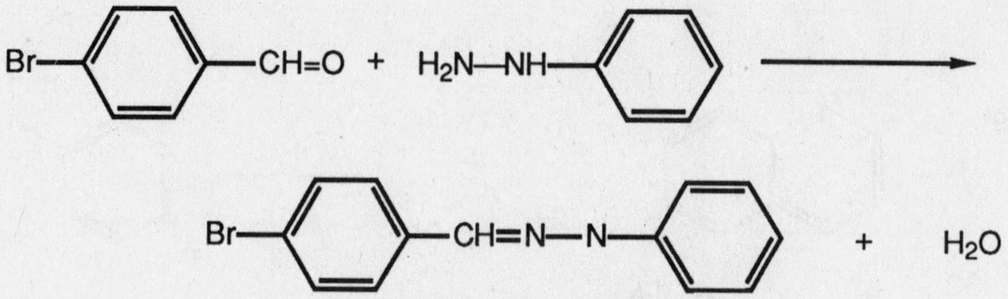

$$Br-\!\!\bigcirc\!\!-CH\!=\!O \ + \ H_2N-\!NH-\!\!\bigcirc \ \longrightarrow$$

$$Br-\!\!\bigcirc\!\!-CH\!=\!N-\!N-\!\!\bigcirc \ + \ H_2O$$

p-bromobenzaldehyde phenylhydrazone

f.

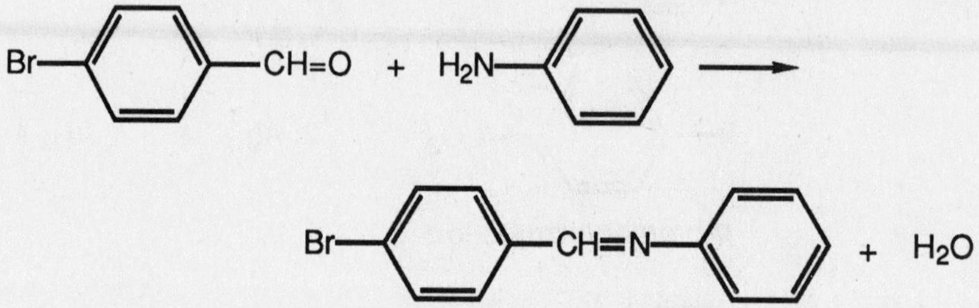

$$Br-\!\!\bigcirc\!\!-CH\!=\!O \ + \ H_2N-\!\!\bigcirc \ \longrightarrow$$

$$Br-\!\!\bigcirc\!\!-CH\!=\!N-\!\!\bigcirc \ + \ H_2O$$

p-bromobenzaldehyde anil

g.

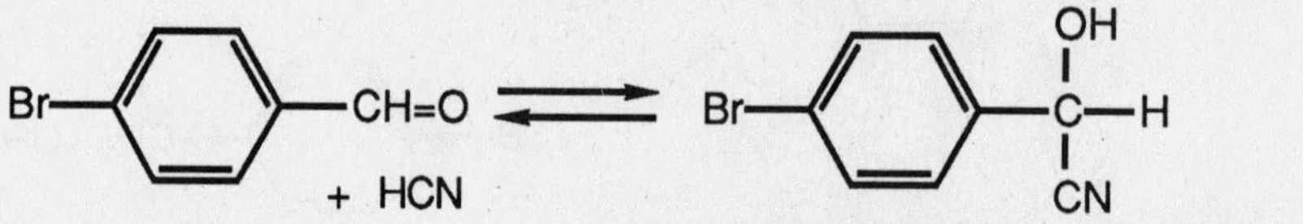

$$Br-\!\!\bigcirc\!\!-CH\!=\!O \ \rightleftharpoons \ Br-\!\!\bigcirc\!\!-\overset{\overset{\displaystyle OH}{|}}{\underset{\underset{\displaystyle CN}{|}}{C}}\!-H$$
$$+\ HCN$$

p-bromobenzaldehyde cyanohydrin

h.

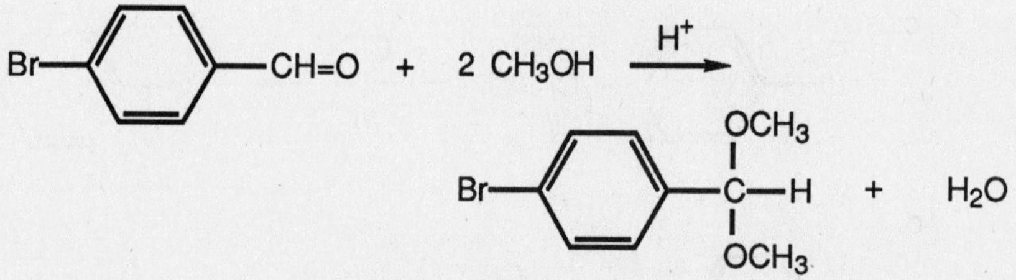

$$Br-\!\!\bigcirc\!\!-CH\!=\!O \ + \ 2\ CH_3OH \ \xrightarrow{H^+}$$

$$Br-\!\!\bigcirc\!\!-\overset{\overset{\displaystyle OCH_3}{|}}{\underset{\underset{\displaystyle OCH_3}{|}}{C}}\!-H \ + \ H_2O$$

p-bromobenzaldehyde dimethylacetal

i.

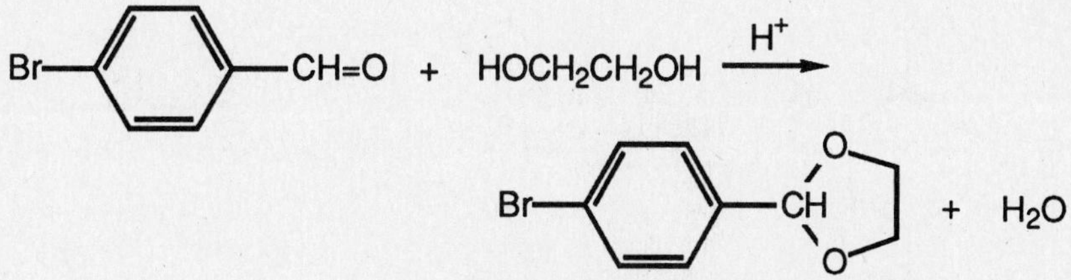

2-*p*-bromophenyl-1,3-dioxolane
(*p*-bromobenzaldehyde ethylene glycol acetal)

j.

p-bromobenzyl alcohol

9.33   a.   Use Tollens' reagent. The hexanal (an aldehyde) will react, whereas
            2-hexanone (a ketone) will not.
       b.   Again use Tollens' reagent. Alcohols (such as benzyl alcohol) do not
            react.
       c.   Both compounds are ketones, but 2-cyclopentenone has a carbon–
            carbon double bond and will be easily oxidized by potassium
            permanganate. The saturated ketone, cyclopentanone, will not react.

9.34   a.

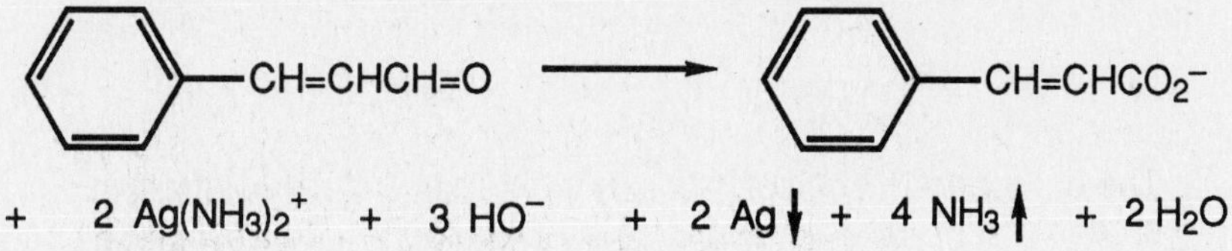

$+$ 2 $Ag(NH_3)_2^+$ $+$ 3 HO^- $+$ 2 Ag↓ $+$ 4 NH_3↑ $+$ 2 H_2O

       b.

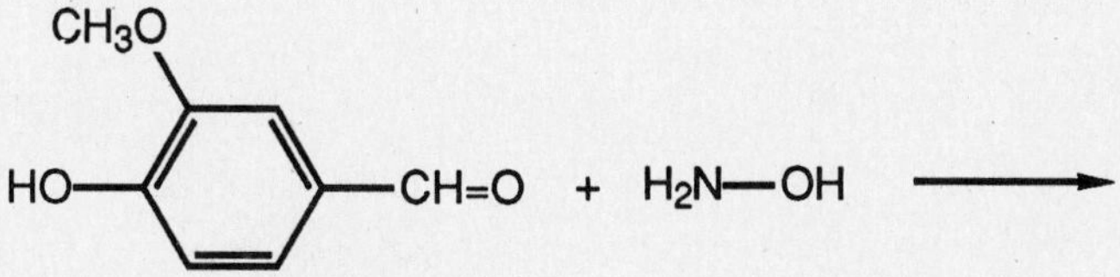

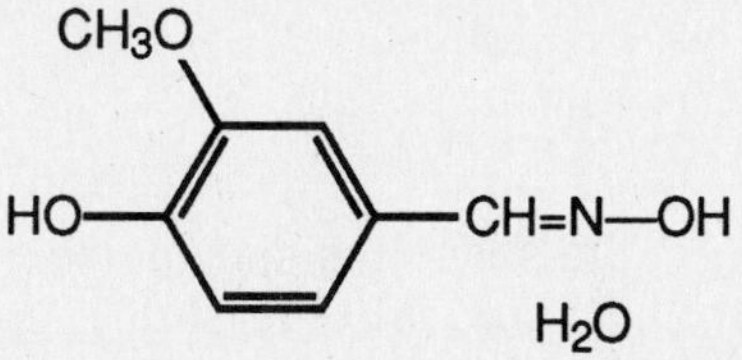

Only the aldehyde function
reacts with the reagent.

H_2O

217

c.

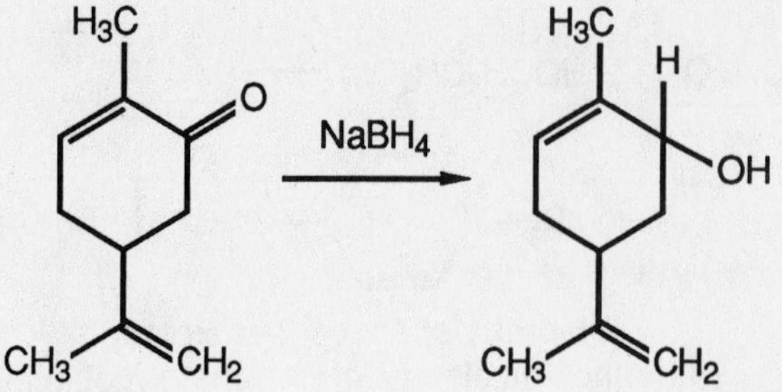

d.

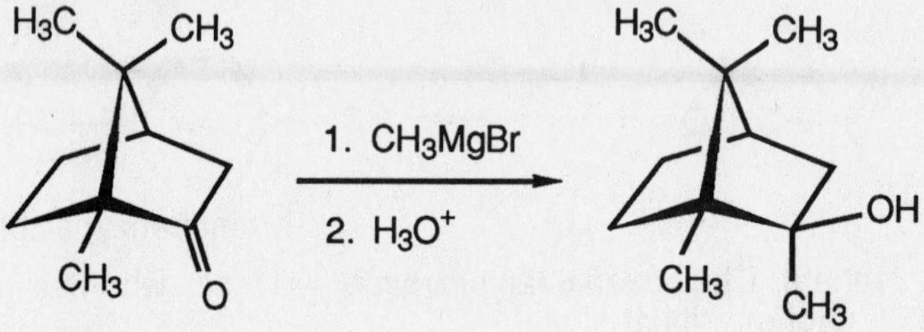

9.35 $CH_3CH_2CH_2CH_2CH_2CH_2CH=O$ $CH_3CH_2CH_2CCH_2CH_2CH_3$ $CH_3CHCCHCH_3$

bp 155°C   bp 144°C   bp 124°C

The compounds are isomers and have identical molecular weights. Each has a carbonyl group that, because of its polarity, can associate as follows as a consequence of intermolecular attraction between opposite charges:

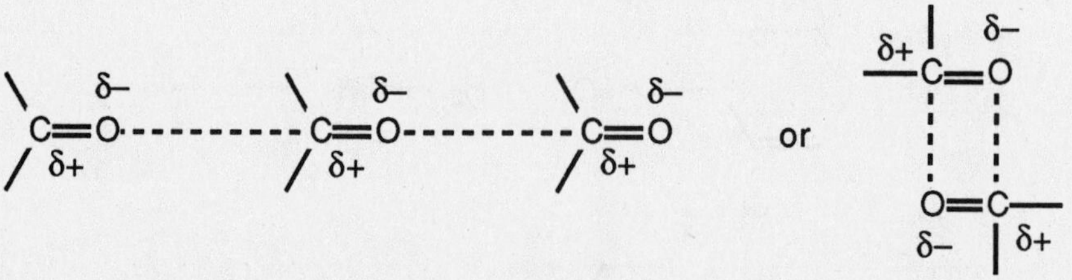

As we go from left to right in the series (as shown above), the carbonyl group is more and more hindered, or buried in the structure. Thus association is more difficult, and the boiling point decreases.

9.36   All parts of this problem involve the preparation or hydrolysis of hemiacetals or acetals (or the corresponding ketone derivatives). See Sec. 9.8.

a. $CH_3CH_2CH_2CH=O$ + 2 CH_3CH_2OH $\xrightarrow{H^+}$

$CH_3CH_2CH_2CH(OCH_2CH_3)_2$ + H_2O

b. $CH_3CH(OCH_3)_2$ $\xrightarrow[H_2O]{H^+}$ $CH_3CH=O$ + 2 CH_3OH

c.   In this case, the acetal is cyclic, and the product is a hydroxy aldehyde, which may exist in its cyclic hemiacetal form.

+ CH_3OH

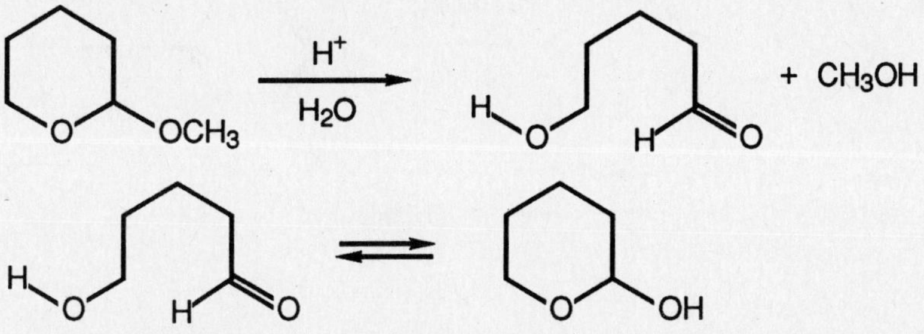

d. The starting material is a "double" acetal. It is completely hydrolyzed to two equivalents of the hydroxy aldehyde.

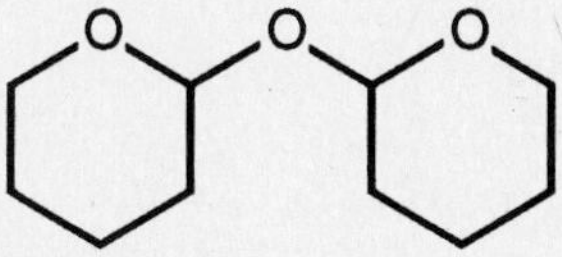

+ H_2O $\xrightarrow{H^+}$ 2 $HOCH_2CH_2CH_2CH_2CH=O$

The product, as in part c, may exist in its cyclic hemiacetal form. The reaction is analogous to the hydrolysis of certain carbohydrates (for example, disaccharides to monosaccharides; see Chapter 16). For practice, write the steps in the mechanism for this reaction.

e.   In this reaction, a hemiacetal is converted to an acetal.

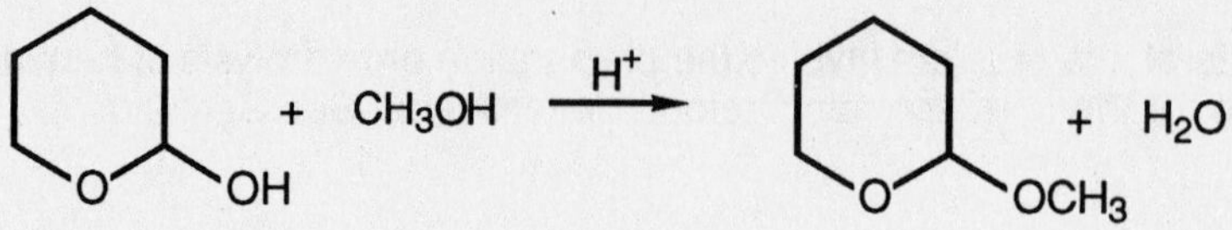

This is the reverse of the reaction in part c.

9.37 For guidance, review Sec. 9.10.

a. 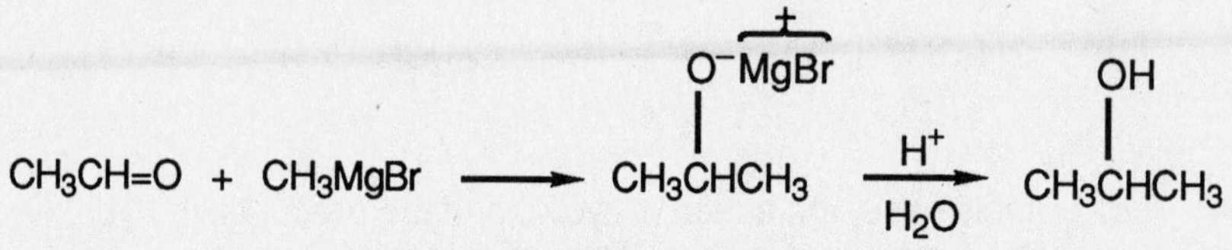 $CH_3CH=O + CH_3MgBr \longrightarrow$ CH_3CHCH_3 $\xrightarrow[H_2O]{H^+}$ CH_3CHCH_3 (with $O^- MgBr^+$ and OH groups)

b. 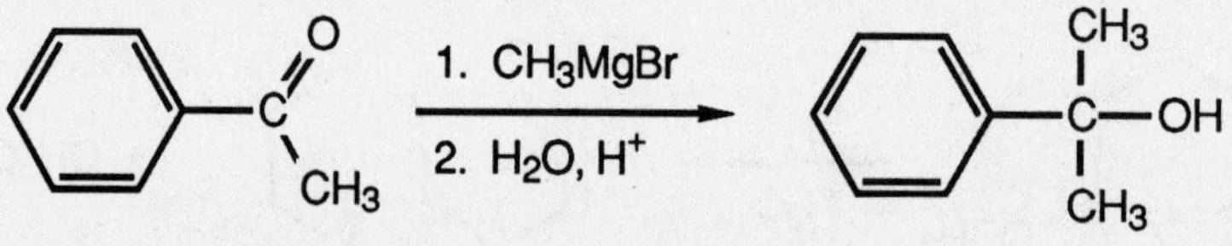 $\xrightarrow[\text{2. } H_2O, H^+]{\text{1. } CH_3MgBr}$

c. $H_2C=O$ $\xrightarrow[\text{2. } H_2O, H^+]{\text{1. } CH_3MgBr}$ CH_3CH_2OH

d. 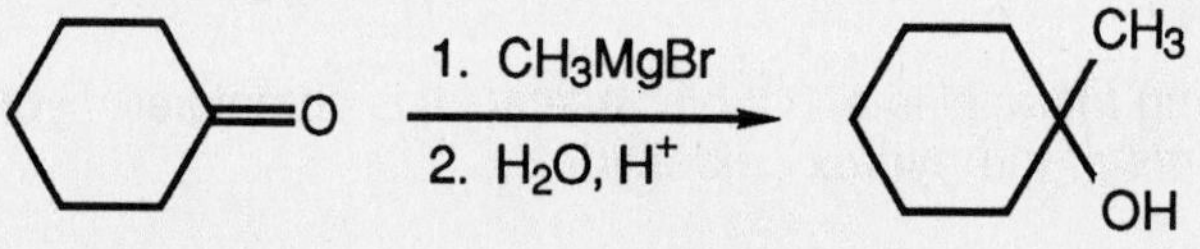 $\xrightarrow[\text{2. } H_2O, H^+]{\text{1. } CH_3MgBr}$

9.38 In each case write the structure of the alcohol:

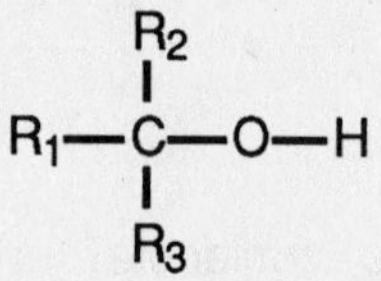

$R_1-\underset{\underset{R_3}{|}}{\overset{\overset{R_2}{|}}{C}}-O-H$

One of the R groups comes from the Grignard reagent. The rest of the molecule comes from the carbonyl compound. For example, if we select R_1 as the alkyl group to come from the Grignard reagent, then the carbonyl compound is R_2—C(=O)—R_3.

ANSWERS TO PROBLEMS

a.

$$CH_3CH_2CH_2CH_2 - \overset{\overset{\displaystyle H}{|}}{\underset{\underset{\displaystyle H}{|}}{C}} - O - H$$

$$CH_3CH_2CH_2CH_2MgX \; + \; \overset{H}{\underset{H}{\diagdown}}C{=}O \; \longrightarrow$$

$$CH_3CH_2CH_2CH_2 - \overset{\overset{\displaystyle H}{|}}{\underset{\underset{\displaystyle H}{|}}{C}} - O^-MgX \; \xrightarrow[H^+]{\overset{\pm}{\rightleftharpoons} \; H_2O} \; CH_3CH_2CH_2CH_2 - \overset{\overset{\displaystyle H}{|}}{\underset{\underset{\displaystyle H}{|}}{C}} - O - H$$

For the remaining cases, we will not write the equations but simply show how the initial reactants are derived.

b.

$$\boxed{CH_3CH_2} - \overset{\overset{\displaystyle H}{|}}{\underset{\underset{\displaystyle CH_2CH_3}{|}}{C}} - OH \qquad \text{from } CH_3CH_2MgX \; + \; CH_3CH_2CH{=}O$$

c.

$$\boxed{CH_3} - \overset{\overset{\displaystyle CH_3}{|}}{\underset{\underset{\displaystyle CH_2CH_3}{|}}{C}} - OH \qquad \text{from } CH_3MgX \; + \; CH_3\overset{\underset{\displaystyle \parallel}{O}}{C}CH_2CH_3$$

$$\boxed{CH_3CH_2} - \overset{\overset{\displaystyle CH_3}{|}}{\underset{\underset{\displaystyle CH_3}{|}}{C}} - OH \qquad \text{from } CH_3CH_2MgX \; + \; CH_3\overset{\underset{\displaystyle \parallel}{O}}{C}CH_3$$

Either of these combinations of reagents will work.

9. ALDEHYDES AND KETONES

d.

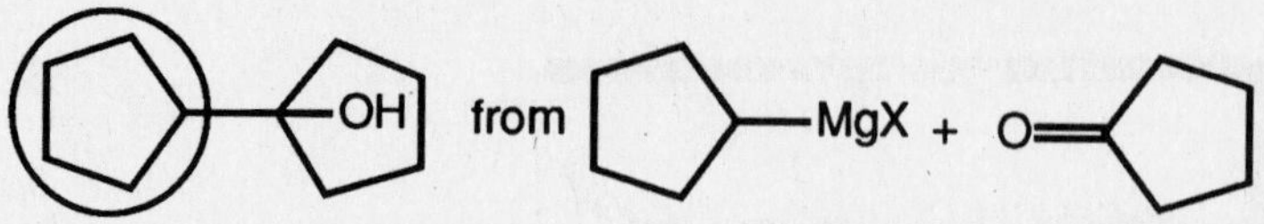

In this case, the "free-standing" R group is selected to come from the Grignard reagent.

e.

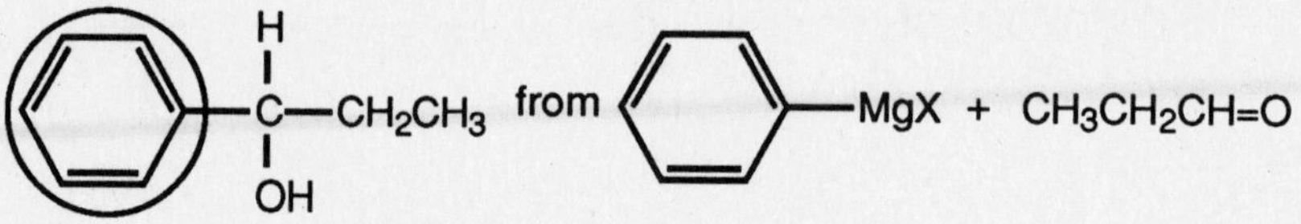

Either of these combinations of reagents will work.

f.

$$H_2C=CH-\underset{\underset{OH}{|}}{\overset{\overset{H}{|}}{C}}-CH_3 \qquad \text{from } H_2C=CHMgX \ + \ CH_3CH=O$$

$$H_2C=CH-\underset{\underset{OH}{|}}{\overset{\overset{H}{|}}{C}}-CH_3 \qquad \text{from } CH_3MgX \ + \ H_2C=CH-CH=O$$

Either of these combinations of reagents will work.

Vinyl Grignard reagents are known, although they are a bit more difficult to prepare than simple alkyl Grignard reagents. Either pair of reagents will work.

9.39   a.   The reaction is similar to that of a Grignard reagent with a ketone (see eq. 9.22). Also see eq. 9.24.

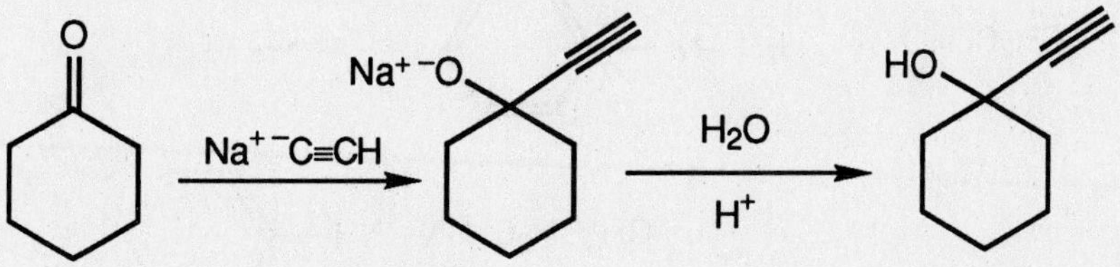

b.   See eq. 9.26 for guidance.

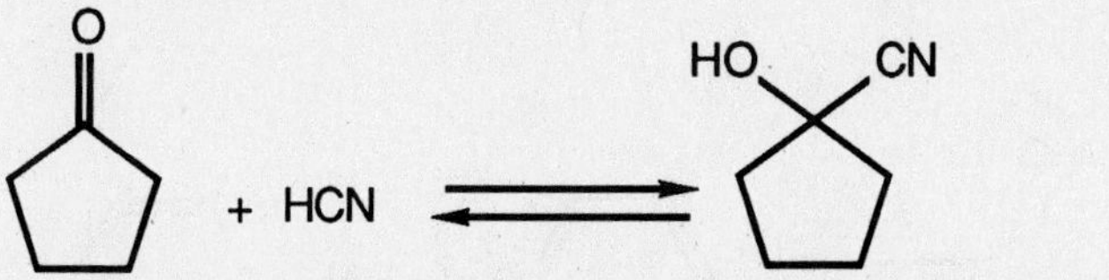

cyclopentanone cyanohydrin

c.   See Table 9.1 for guidance.

$$CH_3CCH_2CH_3 \;+\; NH_2OH \;\rightleftharpoons\; CH_3CCH_2CH_3 \;+\; H_2O$$
$$\underset{O}{\parallel} \qquad\qquad\qquad\qquad \underset{NOH}{\parallel}$$

d.   See Sec. 9.12 for examples.

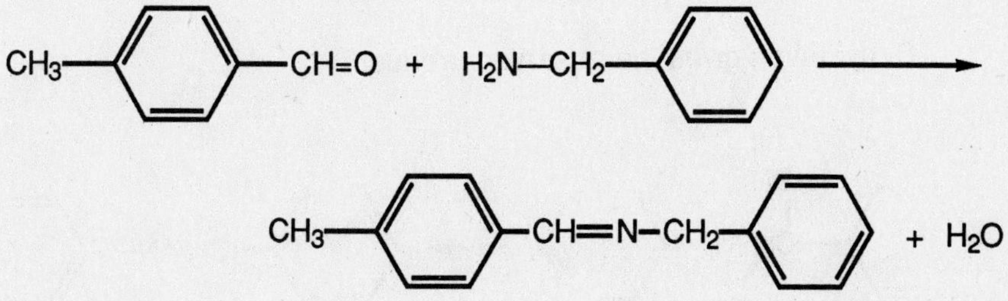

9. ALDEHYDES AND KETONES

e. See Table 9.1.

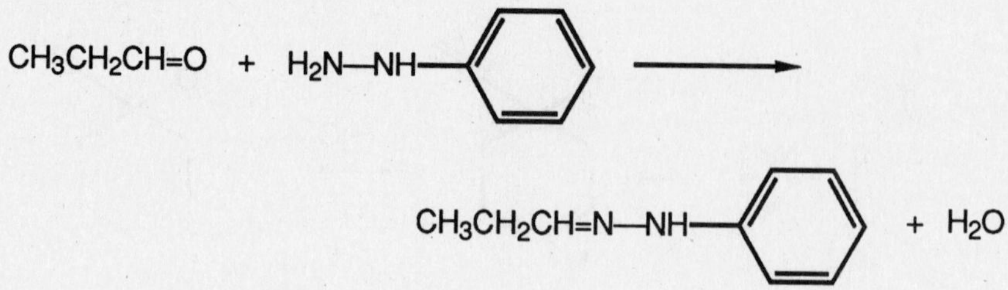

$$CH_3CH_2CH=O \ + \ H_2N-NH-\text{⬡} \longrightarrow$$

$$CH_3CH_2CH=N-NH-\text{⬡} \ + \ H_2O$$

9.40 a. The first step involves nucleophilic addition to the carbonyl group:

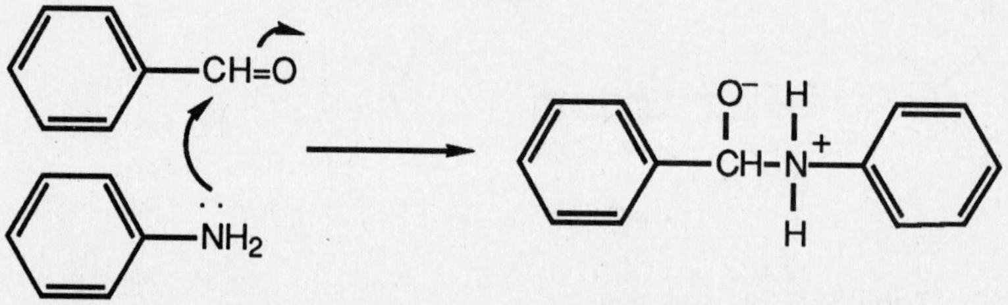

Proton transfer to the alkoxide part of the molecule is fast:

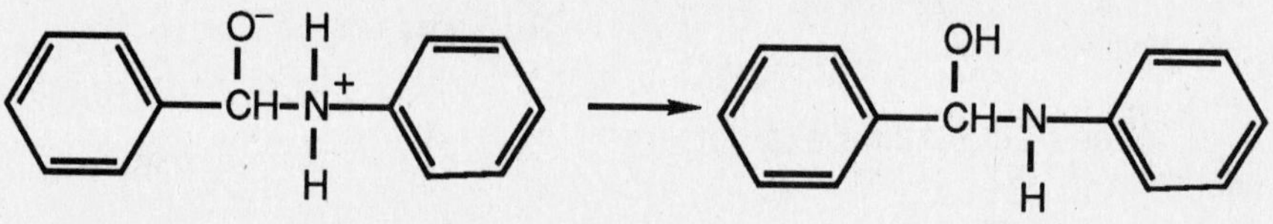

Dehydration gives the observed product:

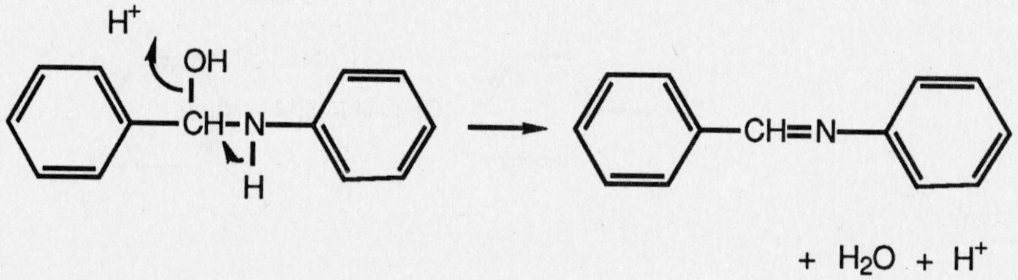

$$+ \ H_2O \ + \ H^+$$

b. The mechanism is similar to that in part a.

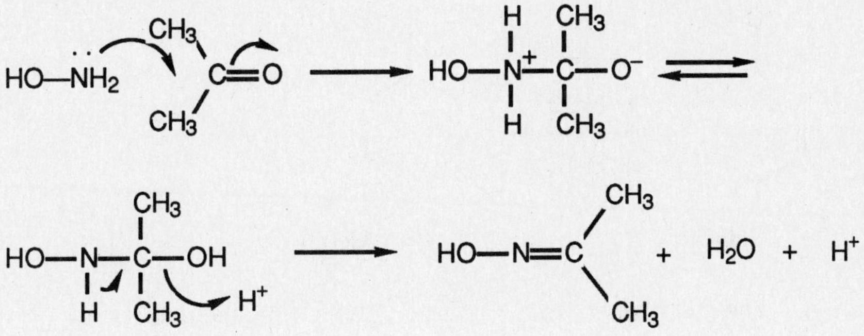

9.41   a.   See Sec. 9.13.

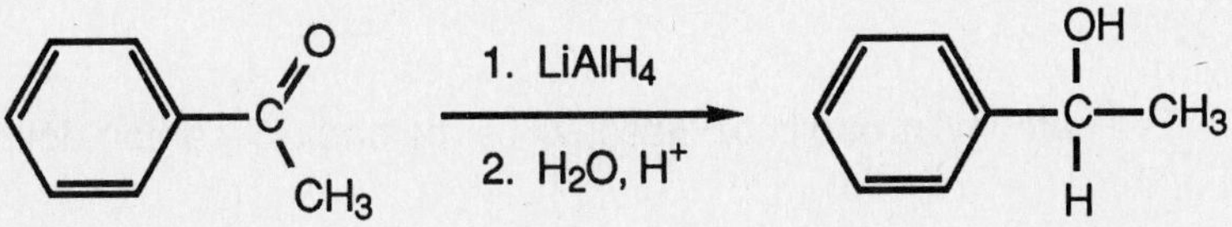

b.   See Sec. 9.17.

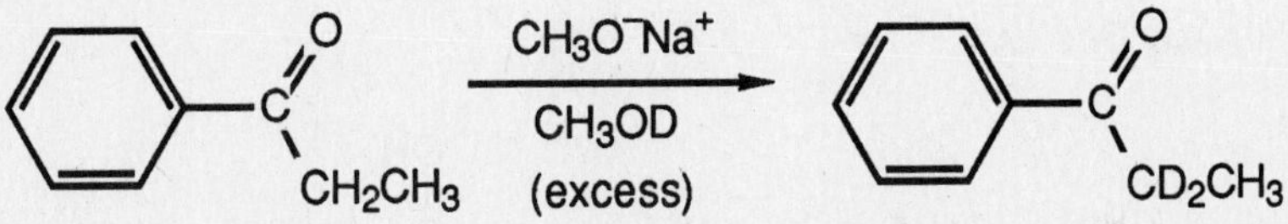

c.   See Sec. 9.13.

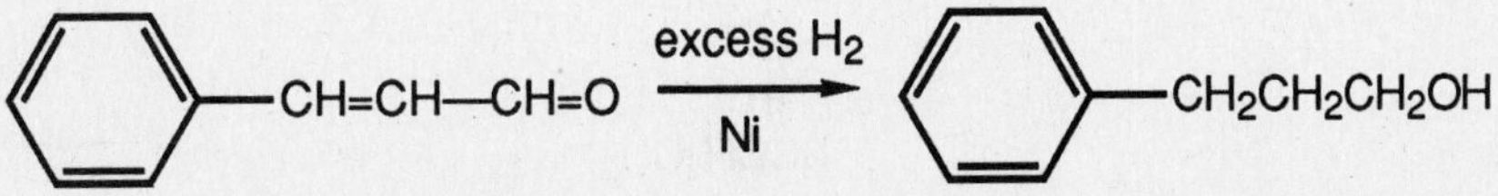

Usually the aromatic ring will not be reduced, although under certain reaction conditions even this is possible.

d.   The carbonyl group is reduced, but the carbon–carbon double bond and aromatic ring are not reduced.

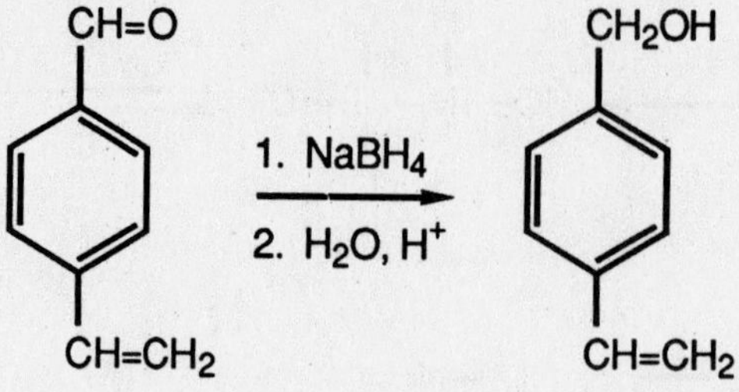

9. ALDEHYDES AND KETONES

e. The carbonyl group is oxidized, but the aromatic ring is not.

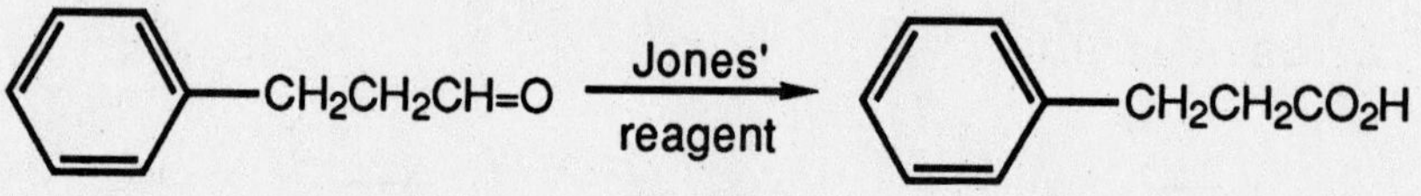

f. The carbonyl group is oxidized, while the carbon–carbon double bond is not.

$$CH_3CH=CHCH=O \xrightarrow{Ag_2O} CH_3CH=CHCO_2H$$

9.42 Review Sec. 9.15.

a.

There are two types of α-hydrogens in 2-butanone, and either may enolize:

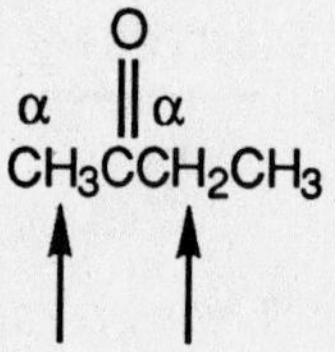

b.

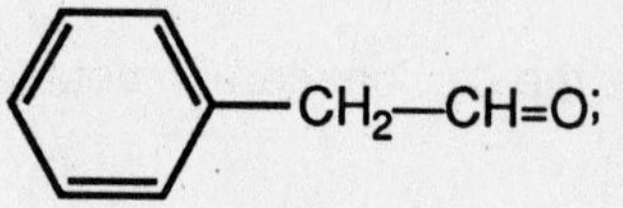

226

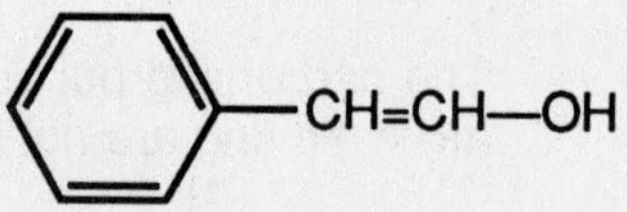

c.

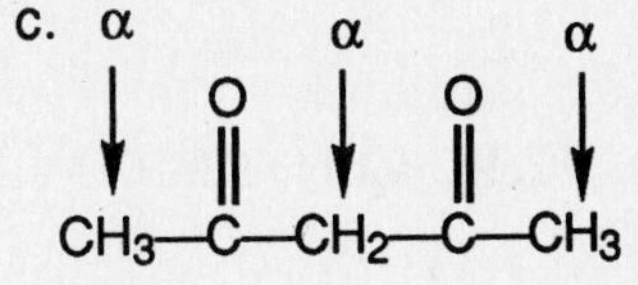

All the hydrogens are α to a carbonyl group.
The CH_2 hydrogens that are α to *two* carbonyl
groups are most likely to enolize.

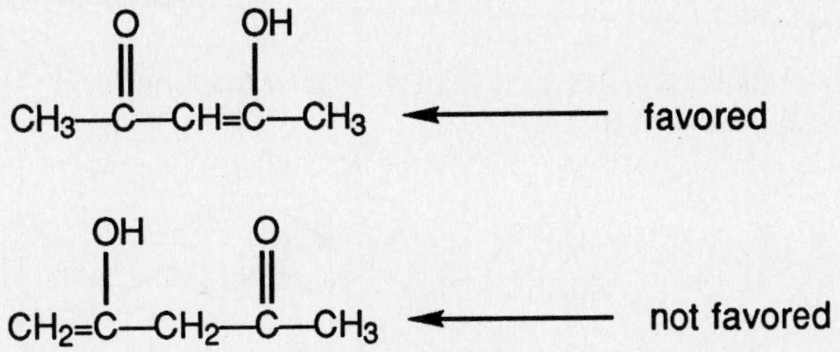

favored

not favored

9.43 Review Sec. 9.17. Only hydrogens α to a carbonyl group will be replaced by deuterium.

a.

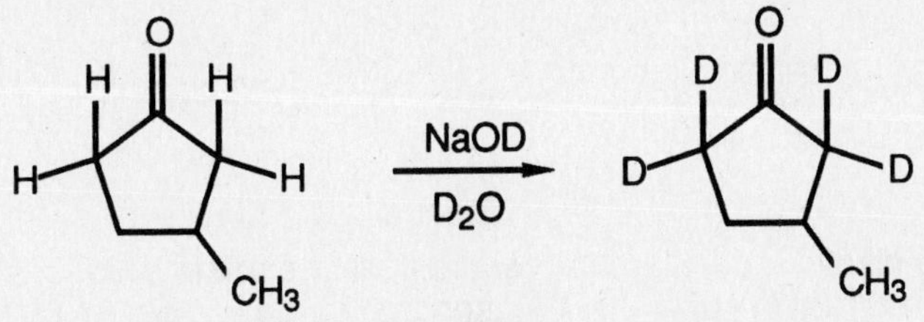

b.

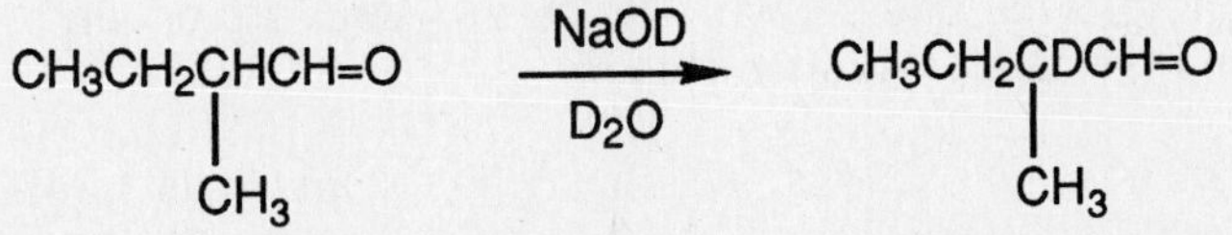

9.44 Follow eqs. 9.45-9.47. The steps in the mechanism are as follows:

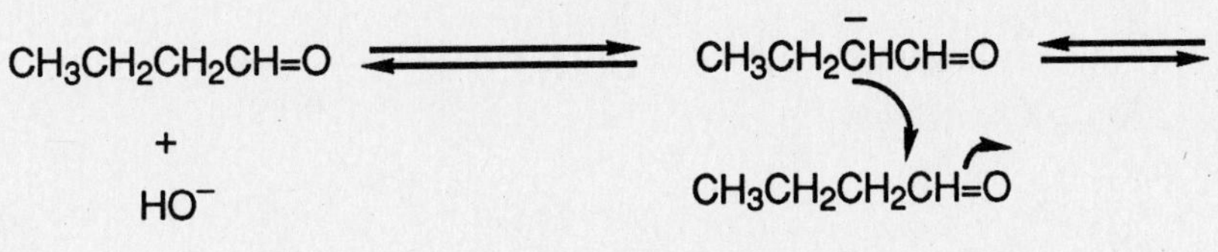

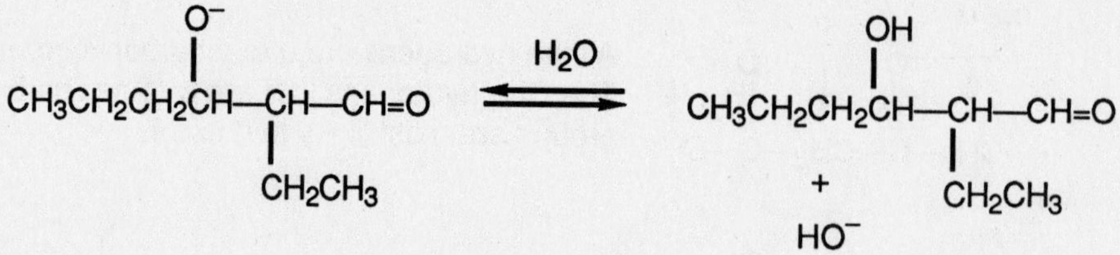

$$CH_3CH_2CH_2CH-\overset{\overset{\displaystyle O^-}{|}}{CH}-CH=O \overset{H_2O}{\underset{}{\rightleftharpoons}} CH_3CH_2CH_2CH-\overset{\overset{\displaystyle OH}{|}}{CH}-CH=O$$

9.45   The reaction occurs by an *intra*molecular aldol condensation, followed by dehydration of the resulting aldol:

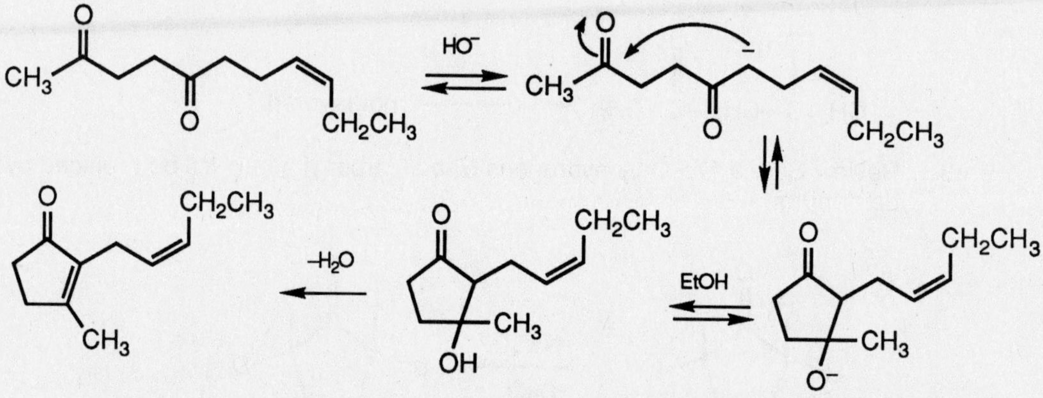

The starting diketone can form several different enolate anions. The one that can react most favorably with the second carbonyl group to form a five-membered ring is the one shown. All the equilibria are driven in the forward direction by the final dehydration step.

9.46   The product has 17 carbons, which suggests that it is formed from two benzaldehyde molecules (2 x 7 = 14 carbons) + one acetone molecule (3 carbons). The product forms by a double mixed aldol condensation:

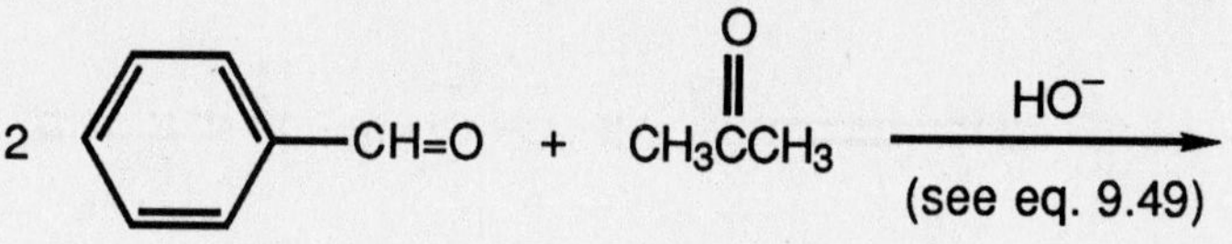

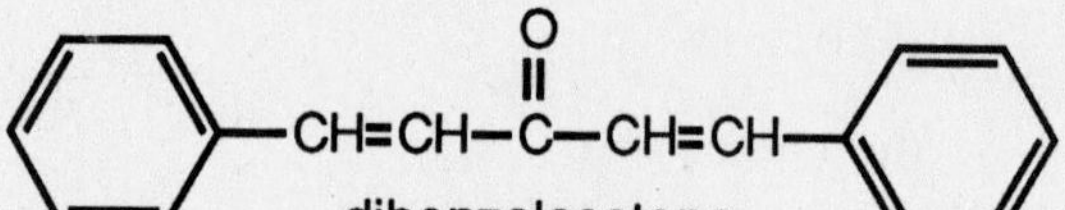

dibenzalacetone

The product is yellow because of the extended conjugated system of double bonds.

9.47   Step A: The reagents are ethylene glycol ($HOCH_2CH_2OH$) and H^+ (compare with eq. 9.15).

Step B: The reagent is chromic acid (compare with eq. 7.33).

Step C: The reagent is sodium acetylide, $HC{\equiv}C^-Na^+$ (compare with eq. 9.24).

Step D: The reagent is dilute acid, to hydrolyze the ketal (compare with eq. 9.17).

The carbonyl group in the six-membered ring must be "protected" so that in step C there will be only one carbonyl group (the one that is introduced into the five-membered ring) available for reaction with the sodium acetylide.

Although one might expect only Enovid to be formed, its double-bond isomer Norlutin is also formed, through an acid-catalyzed enolization (acid is the reagent used to hydrolyze the ketal):

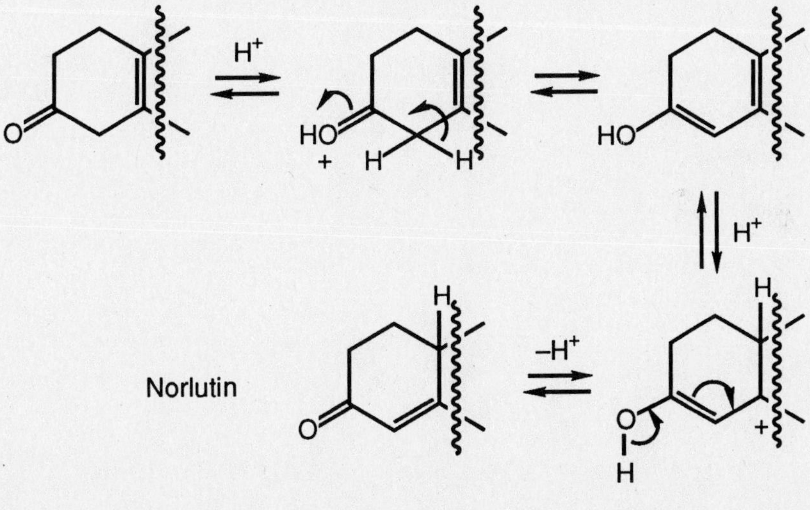

9.48   The carbonyl skeleton of lily aldehyde can be assembled by a mixed aldol condensation between propanal and the nonenolizable aldehyde, 4-tert-butylbenzaldehyde:

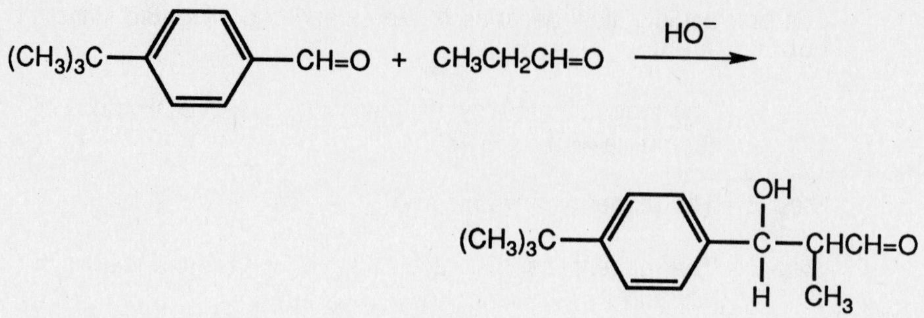

9. ALDEHYDES AND KETONES

$(CH_3)_3C$—⟨benzene ring⟩—$CH=O$ + $CH_3CH_2CH=O$ $\xrightarrow{HO^-}$

$(CH_3)_3C$—⟨benzene ring⟩—$\underset{H}{\overset{}{C}}$—$\underset{CH_3}{\overset{OH}{C}}HCH=O$

Dehydration and reduction of the carbon–carbon double bond completes the synthesis of lily aldehyde.

CHAPTER TEN: CARBOXYLIC ACIDS AND THEIR DERIVATIVES

CHAPTER SUMMARY

Carboxylic acids, the most important class of organic acids, contain the **carboxyl group,** —C(=O)—OH. The IUPAC ending for the names of these compounds is *–oic acid* but many common names (such as formic acid and acetic acid) are also used. An **acyl group,** R—C(=O)—, is named by changing the *–ic* ending of the corresponding acid to *-yl* [$CH_3C(=O)$— is acetyl].

The carboxyl group is polar and readily forms hydrogen bonds. A carboxylic acid dissociates to a **carboxylate anion** and a proton. In the carboxylate anion, the negative charge is delocalized equally over both oxygens. The pK_a's of simple carboxylic acids are about 4-5, but the acidity can be increased by electron-withdrawing substituents (such as chlorine) close to the carboxyl function.

Carboxylic acids react with bases to give salts. These are named by naming the cation first and then the carboxylate anion. The name of the anion is obtained by changing the *–ic* ending of the acid name to *–ate* (acet*ic* becomes acet*ate*).

Carboxylic acids are prepared by at least four methods: (1) by oxidation of primary alcohols or aldehydes, (2) by oxidation of an aromatic side chain, (3) from a Grignard reagent and carbon dioxide, or (4) by hydrolysis of a nitrile, $RC \equiv N$.

Carboxylic acid derivatives are compounds in which the carboxyl —OH group is replaced by other groups. Examples include **esters, acyl halides, anhydrides**, and **amides**.

Esters, RCO_2R', are named as salts are; the R' group is named first, followed by the name of the carboxylate group (for example, $CH_3CO_2CH_2CH_3$ is ethyl acetate). Esters can be prepared from an acid and an alcohol, with a mineral acid catalyst (**Fischer esterification**). The key step of the mechanism is nucleophilic attack by the alcohol on the protonated carbonyl group of the acid. Many esters are used as flavors and perfumes.

Saponification is the base-mediated hydrolysis of an ester, yielding its component carboxylate salt and alcohol. **Ammonolysis** of esters gives amides. Esters react with Grignard reagents to give tertiary alcohols. With lithium aluminum hydride, on the other hand, they are reduced to primary alcohols.

10. CARBOXYLIC ACIDS AND THEIR DERIVATIVES

Acid derivatives undergo nucleophilic substitution. The mechanism is as follows: the nucleophile adds to the trigonal carbonyl carbon to form a **tetrahedral intermediate,** which, through loss of a leaving group, becomes the trigonal product. The reaction can be regarded as an **acyl transfer**, the transfer of an acyl group from one nucleophile to another. The reactivity order of acid derivatives toward nucleophiles is acyl halides > anhydrides > esters > amides.

Acyl chlorides are prepared from acids and either $SOCl_2$ or PCl_5. They react rapidly with water to give acids, with alcohols to give esters, and with ammonia to give amides. **Acid anhydrides** react similarly but less rapidly. **Thioesters** are nature's acylating agents. They react with nucleophiles less rapidly than anhydrides but more rapidly than ordinary esters.

Amides can be prepared from ammonia and other acid derivatives. They can also be prepared by heating ammonium salts. They are named by replacing the *–ic* or *–oic* acid ending by *-amide*.

Because of resonance, the C—N bond in amides has considerable C=N character. Rotation about that bond is restricted, and the amide group is planar. Amides are polar, form hydrogen bonds, and have high boiling points considering their molecular weights.

Amides react slowly with nucleophiles (such as water and alcohols). They are reduced to amines by $LiAlH_4$. **Urea**, made from CO_2 and NH_3, is an important fertilizer.

Some reactions of acid derivatives are summarized in Table 10.5. **β-Keto esters** can be prepared by the **Claisen condensation**, a reaction analogous to the aldol condensation but involving **ester enolates** as the reactive intermediates.

REACTION SUMMARY

Acids

$$RCO_2H \rightleftharpoons RCO_2^- + H^+ \qquad \text{(ionization)}$$

$$RCO_2H + NaOH \longrightarrow RCO_2^- Na^+ + H_2O \quad \text{(salt formation)}$$

REACTION SUMMARY

Preparation of Acids

$$RCH_2OH \xrightarrow{Cr^{6+}} RCH{=}O \xrightarrow[\text{or } Ag^+]{Cr^{6+}} RCO_2H$$

$$ArCH_3 \xrightarrow[\text{or } O_2,\, Co^{3+}]{KMnO_4} ArCO_2H$$

$$RMgX + CO_2 \longrightarrow RCO_2MgX \xrightarrow{H_3O^+} RCO_2H$$

$$RC{\equiv}N + 2\, H_2O \xrightarrow{H^+ \text{ or } HO^-} RCO_2H + NH_3$$

Fischer Esterification

$$RCO_2H + R'OH \xrightarrow{H^+} RCO_2R' + H_2O$$

Saponification

$$RCO_2R' + NaOH \longrightarrow RCO_2^-\,Na^+ + R'OH$$

Ammonolysis of Esters

$$RCO_2R' + NH_3 \longrightarrow RCONH_2 + R'OH$$

Esters and Grignard Reagents

$$RCO_2R' \xrightarrow[\text{2. } H_3O^+]{\text{1. } 2\, R''MgX} R{-}\underset{\underset{R''}{|}}{\overset{\overset{R''}{|}}{C}}{-}OH + R'OH$$

Reduction of Esters

$$RCO_2R' + LiAlH_4 \longrightarrow RCH_2OH + R'OH$$

10. CARBOXYLIC ACIDS AND THEIR DERIVATIVES

Preparation of Acyl Chlorides

$$RCO_2H + SOCl_2 \longrightarrow RCOCl + HCl + SO_2$$

$$RCO_2H + PCl_5 \longrightarrow RCOCl + HCl + POCl_3$$

Reactions of Acyl Halides (or Anhydrides)

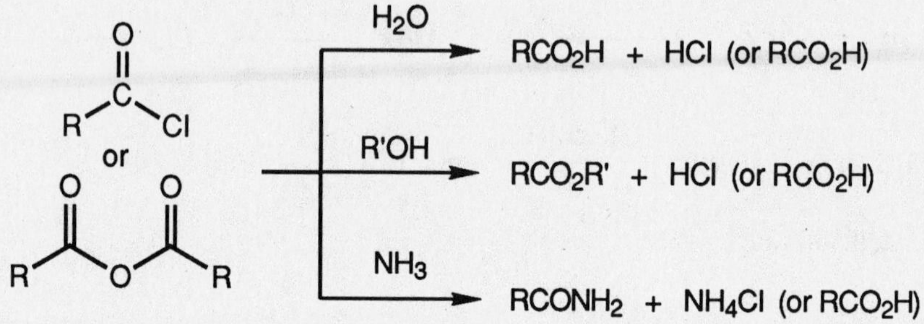

$$\xrightarrow{H_2O} RCO_2H + HCl \text{ (or } RCO_2H)$$

$$\xrightarrow{R'OH} RCO_2R' + HCl \text{ (or } RCO_2H)$$

$$\xrightarrow{NH_3} RCONH_2 + NH_4Cl \text{ (or } RCO_2H)$$

Amides from Ammonium Salts

$$RCO_2^- NH_4^+ \xrightarrow{\text{heat}} RCONH_2 + H_2O$$

Reactions of Amides

$$RCONH_2 + H_2O \xrightarrow{H^+ \text{ or } HO^-} RCO_2H + NH_3$$

$$RCONH_2 \xrightarrow{LiAlH_4} RCH_2NH_2$$

Urea

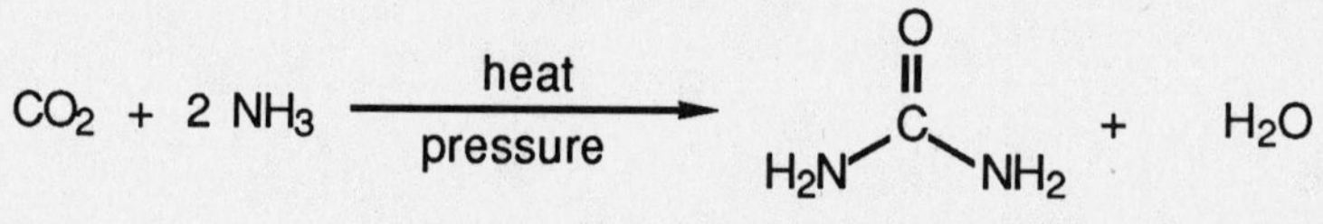

$$CO_2 + 2\,NH_3 \xrightarrow[\text{pressure}]{\text{heat}} H_2N{-}\overset{\displaystyle O}{\overset{\displaystyle \|}{C}}{-}NH_2 + H_2O$$

LEARNING OBJECTIVES

MECHANISM SUMMARY

Nucleophilic Addition–Elimination

LEARNING OBJECTIVES

1. Know the meaning of: carboxylic acid, carboxyl group, acyl group, carboxylate anion, acidity or ionization constant, inductive effect.

2. Know the meaning of: carboxylate salt, ester, acyl halide, acid anhydride, primary amide.

3. Know the meaning of: Fischer esterification, nucleophilic addition–elimination, tetrahedral intermediate, saponification, ammonolysis, acyl transfer.

4. Given the IUPAC name of a carboxylic acid, salt, ester, amide, acyl halide, or anhydride, write its structural formula, and given the structure, write the name.

5. Know the common names of the monocarboxylic acids listed in Table 10.1 and the dicarboxylic acids listed in Table 10.2.

6. Know the systems for designating carbons in a carboxylic acid chain by numbers (IUPAC) or by Greek letters (common).

7. Know how to name acyl groups and how to write a formula given a name that includes acyl group nomenclature.

8. Given the formula of a carboxylic acid, write an expression for its ionization constant, K_a.

9. Write the resonance structures of a carboxylate anion.

10. Given two or more carboxylic acids with closely related structures, rank them in order of increasing (or decreasing) acidities (pK_a's).

11. Tell whether a particular substituent will increase or decrease the acidity of a carboxylic acid.

10. CARBOXYLIC ACIDS AND THEIR DERIVATIVES

12.   Given a carboxylic acid and a base, write the equation for salt formation.

13.   Given a carboxylic acid, tell what aldehyde or primary alcohol is needed for its preparation by oxidation.

14.   Given an aromatic compound with alkyl substituents, tell what aromatic acid would be obtained from its oxidation.

15.   Given a carboxylic acid, write an equation for its synthesis by hydrolysis of a nitrile (cyanide) or by the Grignard method.

16.   Given an alcohol and an acid, write the equation for formation of the corresponding ester.

17.   Write the steps in the mechanism for the acid-catalyzed (Fischer) esterification of a given carboxylic acid with a given alcohol.

18.   Given the name or the structure of an ester, write the structure of the alcohol and acid from which it is derived.

19.   Write an equation for the reaction of a given ester with aqueous base (saponification).

20.   Write an equation for the reaction of a given ester with ammonia, a Grignard reagent, or lithium aluminum hydride.

21.   Given a particular acid halide, write an equation for its preparation from an acid.

22.   Write the equation for the reaction of a given acid halide or anhydride with a given nucleophile (especially with water, an alcohol, or ammonia).

23.   Write equations for the preparation of a given amide from an acyl halide, acid anhydride, or ammonium salt.

24.   Given a particular amide, write equations for its hydrolysis and reduction with lithium aluminum hydride.

25.   Given a particular product that can be prepared by any of the reactions in this chapter, deduce the structures of the reactants required for its preparation, and write the equation for the reaction.

26.   Write the Claisen condensation product from the reaction of an ester with an alkoxide or sodium hydride.

27.   Write the steps in the mechanism of the Claisen condensation.

ANSWERS TO PROBLEMS

Problems Within the Chapter

10.1   a. $CH_3CHCH_2CO_2H$ b. $(CH_3)_2CCO_2H$
                       | |
                      CL OH

c. $CH_3C \equiv CCO_2H$ d. $CH_3CHCH_2CH_2CH_2CO_2H$
                                                            |
                                                          $CH=O$

10.2   a.   3-phenylpropanoic acid
             b.   trichloroethanoic acid
             c.   4-hydroxy-2-butenoic acid
             d.   2,2-dimethylpropanoic acid

10.3   a.

b.

10.4   a.   cyclopropanecarboxylic acid
             b.   *p*-toluic acid (or *p*-methylbenzoic acid)

10.5   a.

b.

c.

d.

10.6 K_a is 1.8×10^{-5} for acetic acid and 1.5×10^{-3} for chloroacetic acid. K_a is larger for chloroacetic acid; it is the stronger acid. The ratio is:

$$\frac{1.5 \times 10^{-3}}{1.8 \times 10^{-5}} = 0.83 \times 10^2$$

In other words, chloroacetic acid is 83 times stronger than acetic acid.

10.7 K_a for benzoic acid is 6.6 x 10^{-5} or 0.66 x 10^{-4}. For o-, m-, and p-chlorobenzoic acids, K_a is 12.5, 1.6, and 1.0 x 10^{-4}, respectively. All three chloro acids are stronger than benzoic acid. However, the difference is greatest for the *ortho* isomer since in this isomer the chloro substituent is closest to the carboxyl group and exerts the maximum electron-withdrawing inductive effect. The effect decreases as the distance between the chloro substituent and the carboxyl group increases.

10.8 $BrCH_2CH_2CO_2H + K^+OH^- \rightarrow BrCH_2CH_2CO_2^-K^+ + H_2O$

10.9   First the alcohol must be converted to a halide. The halide can then be converted to the corresponding Grignard reagent, which can be reacted with carbon dioxide to provide the carboxylic acid.

10.10 It is not possible to conduct S_N2 displacements at sp^2-hybridized (aryl, vinyl) carbon atoms (see Table 6.1). The conversion can be accomplished via the Grignard reagent:

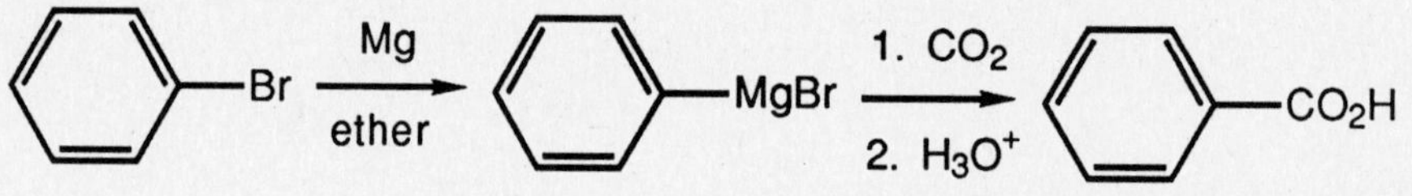

10.11

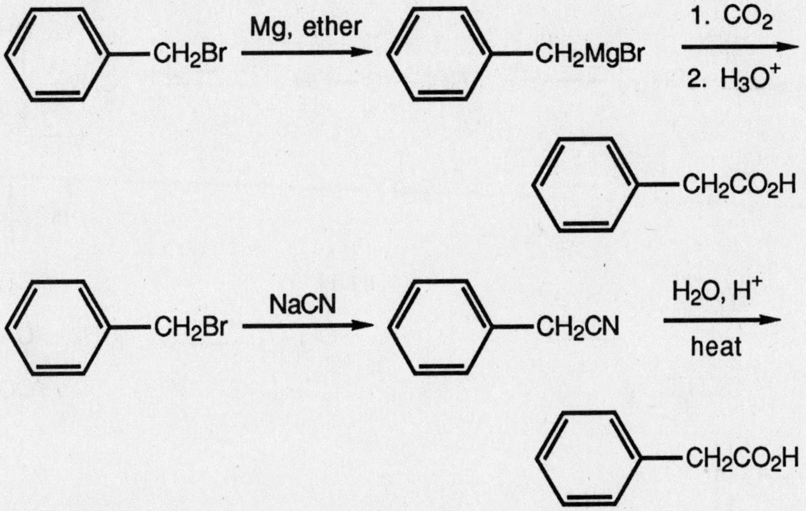

$$C_6H_5-CH_2Br \xrightarrow{\text{Mg, ether}} C_6H_5-CH_2MgBr \xrightarrow[\text{2. } H_3O^+]{\text{1. } CO_2}$$

$$C_6H_5-CH_2CO_2H$$

$$C_6H_5-CH_2Br \xrightarrow{\text{NaCN}} C_6H_5-CH_2CN \xrightarrow[\text{heat}]{H_2O, H^+}$$

$$C_6H_5-CH_2CO_2H$$

10.12 a. methyl methanoate   b. 1-propyl propanoate

10.13 a.

$$CH_3COCHCH_2CH_2CH_3$$
$$\overset{O}{\underset{CH_3}{\|}}$$

b.

$$CH_3CHCOCH_2CH_3$$
$$\overset{O}{\underset{CH_3}{\|}}$$

10.14 $CH_3CH_2CH_2CO_2H$ + $HOCH_2CH_2CH_3$ $\underset{}{\overset{H^+}{\rightleftarrows}}$

butanoic acid   propanol

$$CH_3CH_2CH_2CO_2CH_2CH_2CH_3 + H_2O$$

propyl butanoate

10.15

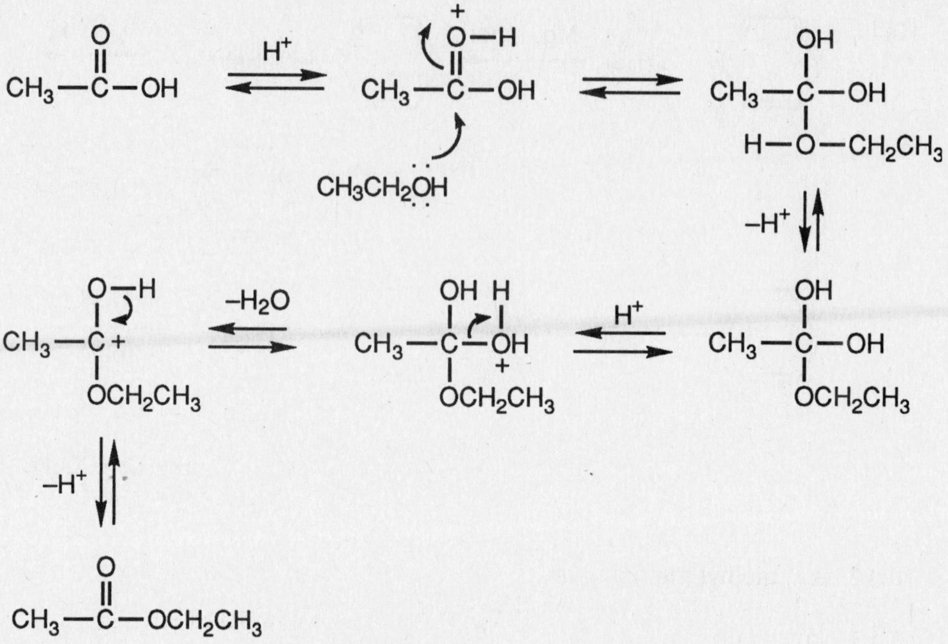

10.16

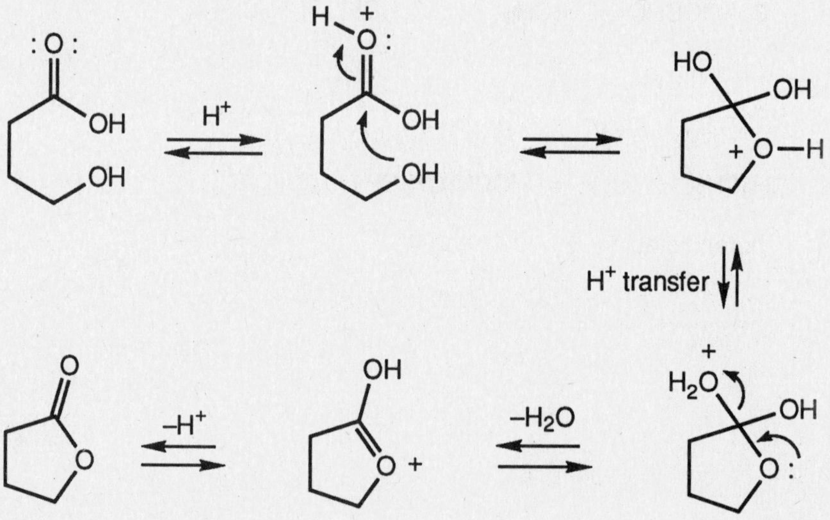

10.17

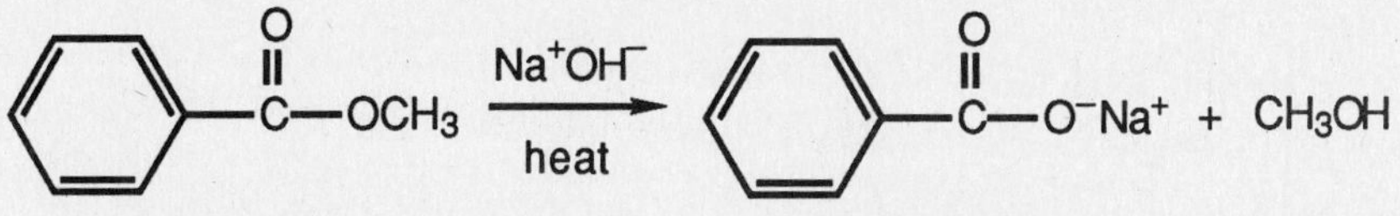

methyl benzoate → sodium benzoate + methanol

Na⁺OH⁻ / heat

10.18

NH₃ ⇌ ... H⁺ transfer ⇌

−CH₃O⁻ ⇌ ... −H⁺ ⇌

tautomerization ⇌

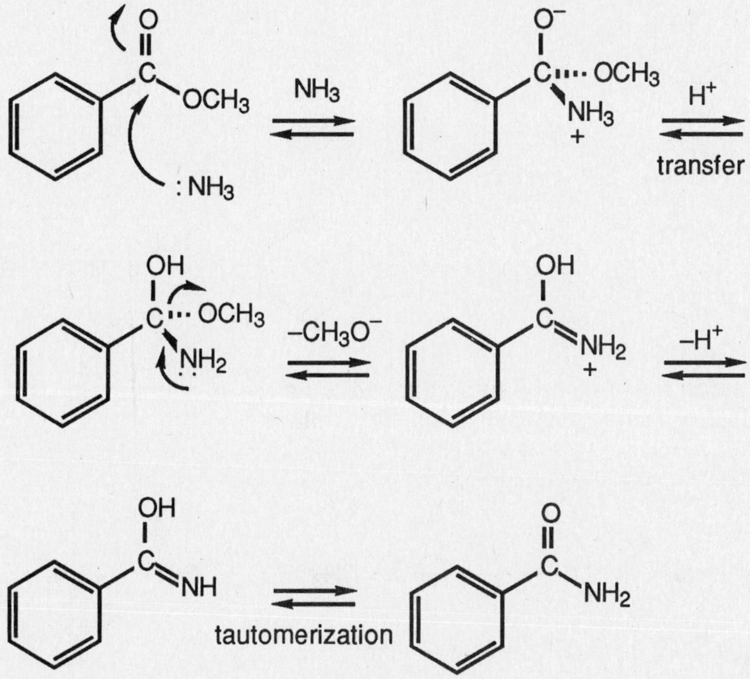

10.19 The Grignard reagent provides two of the three R groups attached to the hydroxyl-bearing carbon of the tertiary alcohol. The ester provides the third R group. So, from

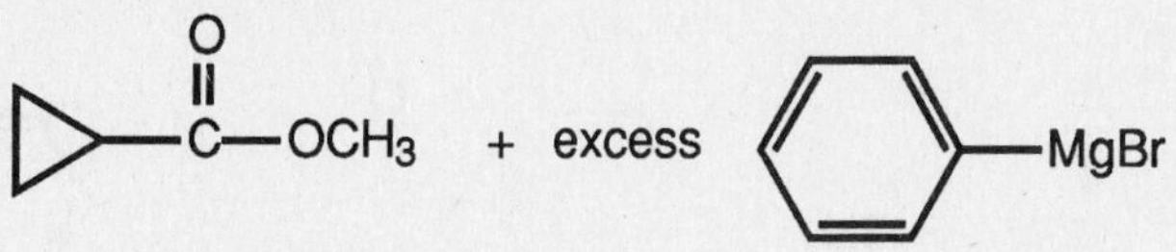

+ excess + —MgBr

we get the tertiary alcohol

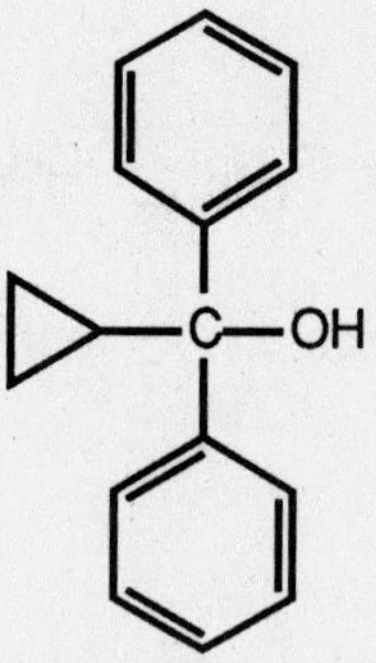

10.20

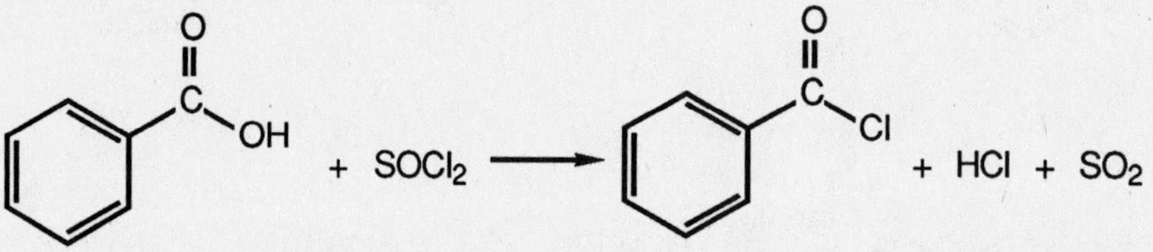

10.21 When acyl halides come in contact with the moist membranes of the nose, they hydrolyze, producing HCl, a severe irritant.

10.22 First prepare butanoyl chloride:

$$CH_3CH_2CH_2CO_2H \xrightarrow{SOCl_2} CH_3CH_2CH_2COCl$$

Then perform a Friedel-Crafts acylation:

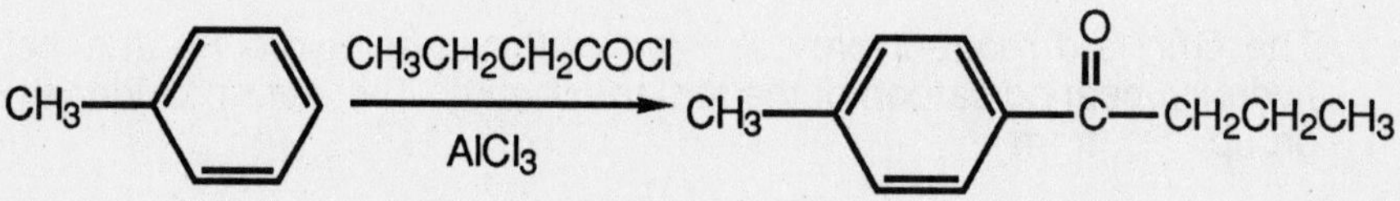

10.23 a.

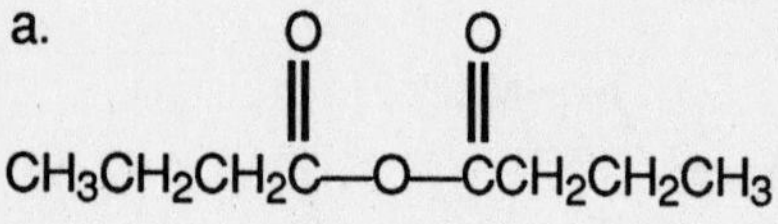

b.

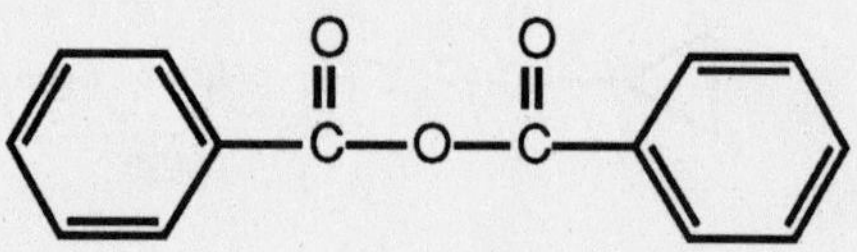

10.24

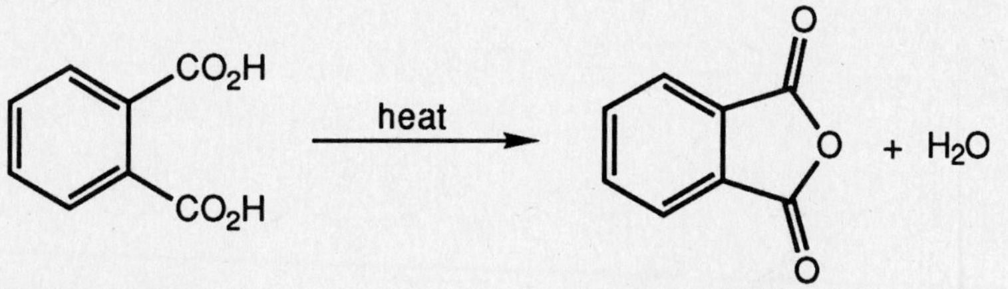

phthalic acid　　　　　　　　phthalic anhydride

10.25 No. The two carboxyl groups are *trans* to one another and cannot interact in an intramolecular fashion.

10.26 Use the middle part of eq. 10.35 as a guide.

$$CH_3-\overset{\overset{\displaystyle O}{\|}}{C}-O-\overset{\overset{\displaystyle O}{\|}}{C}-CH_3 \ + \ HOCH_2CH_2CH_2CH_3 \longrightarrow$$

$$CH_3-\overset{\overset{\displaystyle O}{\|}}{C}-O-CH_2CH_2CH_2CH_3 \ + \ CH_3CO_2H$$

10.27 Follow eq. 10.15.

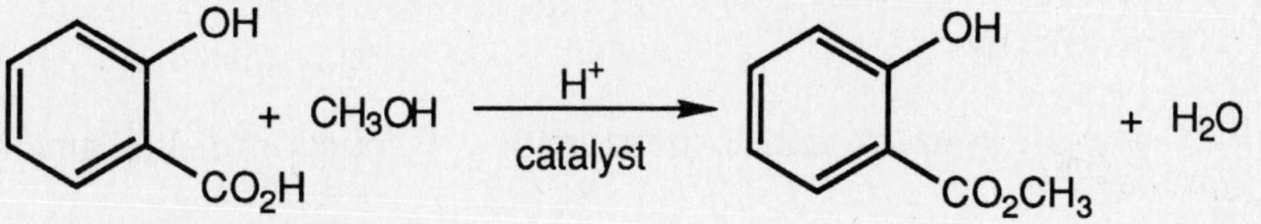

10.28 a. 2-methylpropanamide

b.

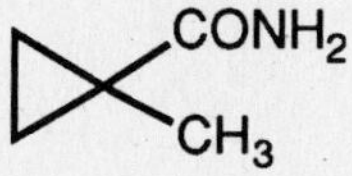

10.29 See Sec. 10.21:

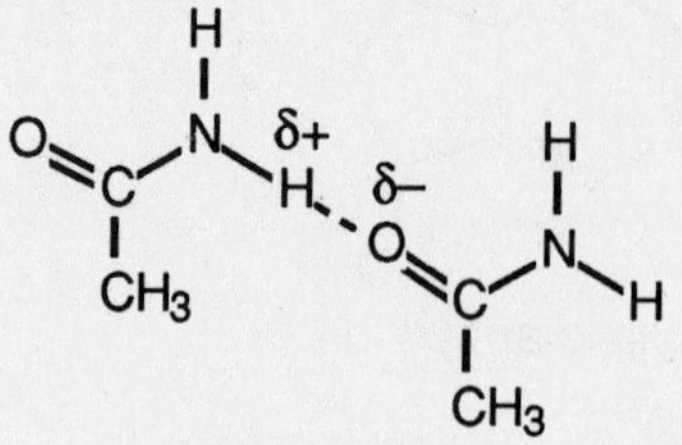

acetamide 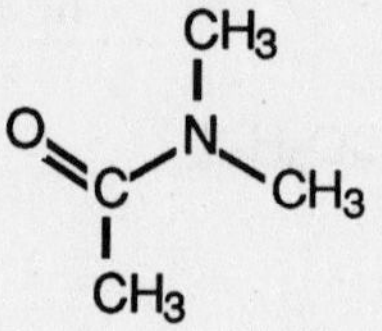   N,N-dimethylacetamide

The oxygen behaves as a Lewis base and the hydrogen behaves as a Lewis acid. N,N-dimethylacetamide has no acidic hydrogen, and hydrogen bonding cannot occur.

10.30 Follow eq. 10.38 , with R = CH₃.

$$CH_3-\overset{\overset{\textstyle O}{\|}}{C}-NH_2 \;+\; H_2O \;\xrightarrow{\;H^+ \text{ or } HO^-\;}\; CH_3-\overset{\overset{\textstyle O}{\|}}{C}-OH \;+\; NH_3$$

With acid catalysis, the products are $CH_3CO_2H + NH_4^+$.
With base catalysis, the products are $CH_3CO_2^- + NH_3$.

10.31 First convert benzoic acid to benzamide. Reduction of the amide gives benzylamine.

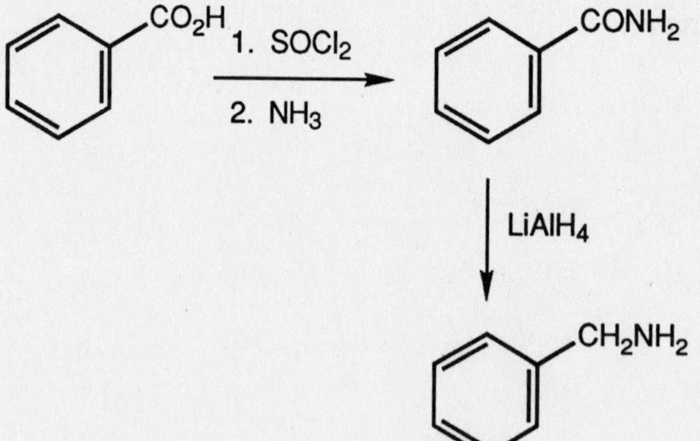

10.32

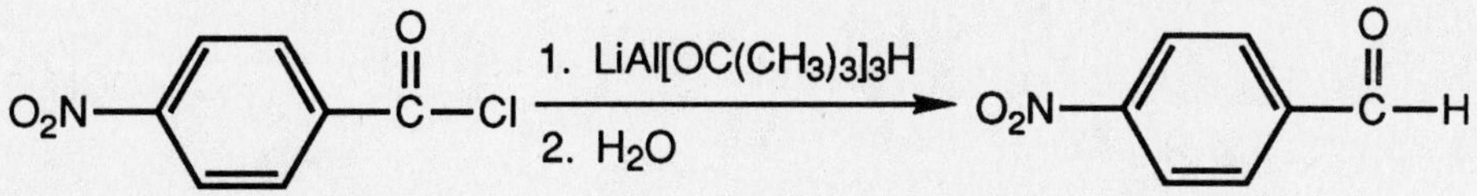

Mechanism:

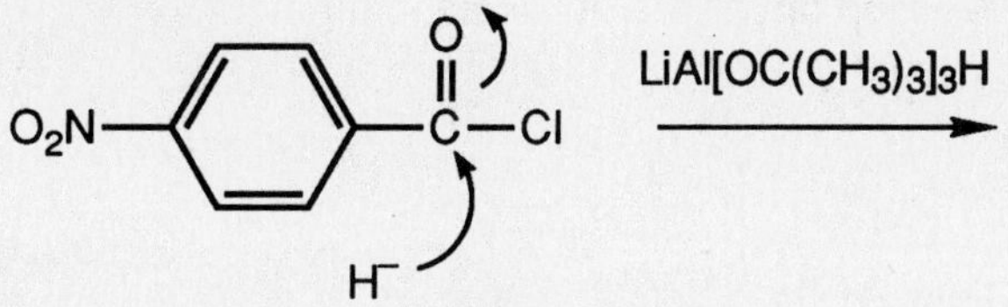

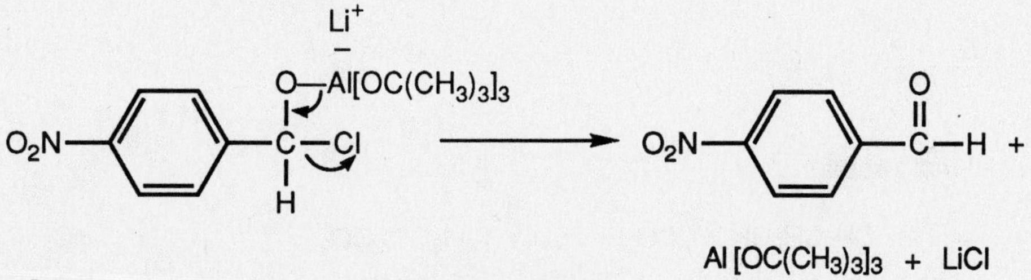

10.33 Step 1: Deprotonation gives an ester enolate.

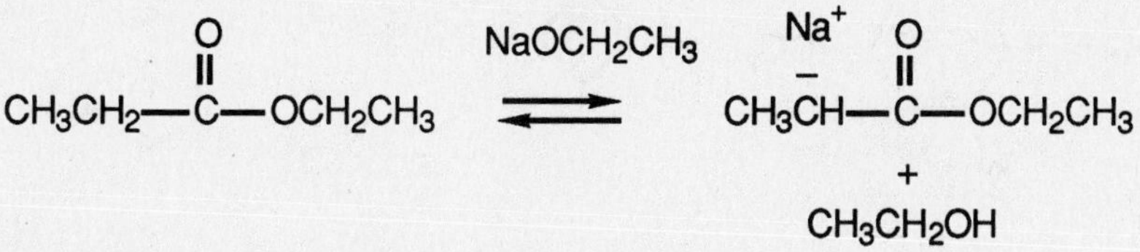

Step 2: Carbonyl addition gives a tetrahedral intermediate.

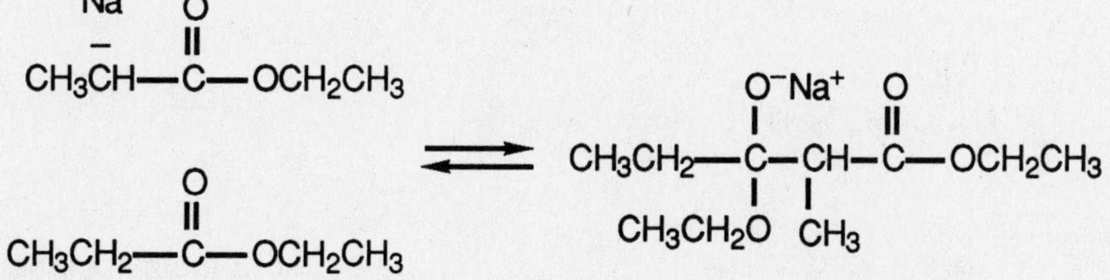

Step 3: The tetrahedral intermediate gives an ester plus ethoxide.

$$
\underset{\substack{\displaystyle |\\ CH_3CH_2O \quad CH_3}}{\overset{\substack{O^-Na^+\\ \displaystyle |}}{CH_3CH_2-C-CHCO_2CH_2CH_3}} \rightleftharpoons \underset{\substack{\displaystyle |\\ CH_3 \\ + \\ CH_3CH_2O^-Na^+}}{\overset{\substack{O \quad\quad O\\ \parallel \quad\quad \parallel}}{CH_3CH_2-C-CH-C-OCH_2CH_3}}
$$

Step 4: An acid-base reaction drives the equilibria to the right.

$$
\underset{\substack{\displaystyle |\\ CH_3 \quad +\\ NaOCH_2CH_3}}{\overset{\substack{O \quad\quad O\\ \parallel \quad\quad \parallel}}{CH_3CH_2-C-CH-C-OCH_2CH_3}} \longrightarrow \underset{\substack{\displaystyle |\\ CH_3 \quad +\\ CH_3CH_2OH}}{\overset{\substack{O \quad Na^+ \; O\\ \parallel \quad - \quad \parallel}}{CH_3CH_2-C-C-C-OCH_2CH_3}}
$$

Additional Problems

10.34   a. $CH_3CH_2\underset{\underset{CH_3}{|}}{C}HCH_2CO_2H$ b. $CH_3CH_2CCl_2CO_2H$ c. $CH_3CH_2\underset{\underset{OH}{|}}{C}HCH_2CH_2CO_2H$

d. $CH_3-\langle\!\!\!\bigcirc\!\!\!\rangle-CO_2H$ e. $\square-CO_2H$ f. (benzene ring with $-CO_2H$ and $-\overset{O}{\underset{\parallel}{C}}-CH_2CH_3$ substituents)

g. (benzene ring with $-CH_2CO_2H$)   h. (naphthalene with $-CO_2H$)   i. $H_2C=C(CH_3)CH(CH_3)CO_2H$

j. $CH_3\overset{O}{\underset{\parallel}{C}}CH_2CO_2H$ k. $HO_2CC(CH_3)_2CH_2CO_2H$ l. (lactone ring with CH_3 substituent)

10.35   a.   4-methylpentanoic acid   b.   3-bromo-2-methylbutanoic acid
       c.   o-nitrobenzoic acid   d.   2-phenylpropanoic acid

e. propenoic acid (or acrylic acid) f. cyclohexanecarboxylic acid
g. 2,2-difluoropropanoic acid h. 3-oxocyclopentanecarboxylic acid

i. 4-isobutyrylbenzoic acid j. 3-butynoic acid

10.36

$$CH_3CH_2-\overset{\overset{\displaystyle CH_3}{|}}{\underset{\underset{\displaystyle H}{|}}{C}}-\text{(benzene ring)}-\overset{\overset{\displaystyle CH_3}{|}}{\underset{\underset{\displaystyle H}{|}}{C}}-CO_2H$$

Ibuprofen

10.37 a. The molecular weights are identical (74). But acids hydrogen-bond more effectively than alcohols do.

$$CH_3CH_2CO_2H \;\; > \;\; CH_3CH_2CH_2CH_2OH$$

bp 141°C   bp 118°C

b. Chain branching generally lowers the boiling point. Thus for these isomeric acids the order is:

$$CH_3CH_2CH_2CH_2CO_2H \;\; > \;\; (CH_3)_3CCO_2H$$

bp 187°C   bp 164°C

10.38 The factors that affect acidity of carboxylic acids are discussed in Sec. 10.6.

a. $ClCH_2CO_2H$; both substituents, chlorine and bromine, are approximately the same distance from the carboxyl group, but chlorine is more electronegative than bromine.
b. o-Bromobenzoic acid; the bromine is closer to the carboxyl group and is an electron-withdrawing substituent. Compare the pK_a's of the corresponding chloro acids, given in Table 10.4.
c. CF_3CO_2H; fluorine is more electronegative than chlorine.
d. Benzoic acid; the methoxy group is an electron-releasing substituent when in the para position and may destabilize the anion because of the presence of structures such as:

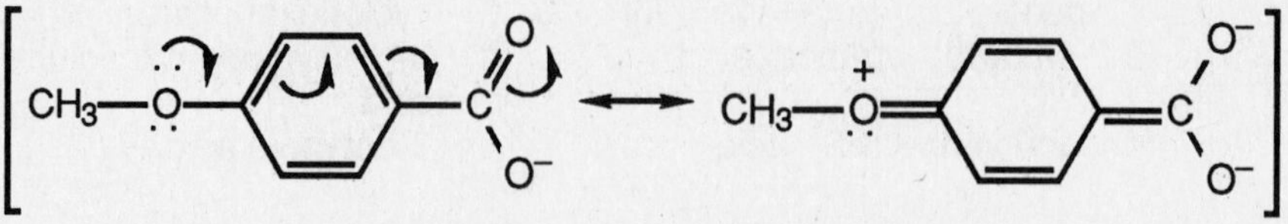

which bring two negative charges near one another.

e. $CH_3CHClCO_2H$; the chlorine, which is electron-withdrawing, is closer to the carboxyl group.

10.39 See Sec. 10.7 if you have any difficulty.

a. $ClCH_2CO_2H + K^+OH^- \longrightarrow ClCH_2CO_2^-K^+ + H_2O$

Salt formation occurs at room temperature. If the reagents are heated for some time, an S_N2 displacement on the primary chloride may also occur, giving the salt of hydroxyacetic acid, $HOCH_2CO_2^-K^+$.

b. $2\ CH_3(CH_2)_8CO_2H + Ca(OH)_2 \longrightarrow [CH_3(CH_2)_8CO_2^-]_2Ca^{2+} + 2\ H_2O$

10.40 a. $CH_3CH_2CH_2CH_2OH \xrightarrow{Na_2Cr_2O_7} CH_3CH_2CH_2CO_2H$

b. $CH_3CH_2CH_2OH \xrightarrow{HBr} CH_3CH_2CH_2Br$

$CH_3CH_2CH_2Br \xrightarrow{NaCN} CH_3CH_2CH_2CN \xrightarrow[H^+]{H_2O} CH_3CH_2CH_2CO_2H$

$CH_3CH_2CH_2Br \xrightarrow[\text{ether}]{Mg} CH_3CH_2CH_2MgBr \xrightarrow[\text{2. }H_2O,\ H^+]{\text{1. }CO_2} CH_3CH_2CH_2CO_2H$

c.

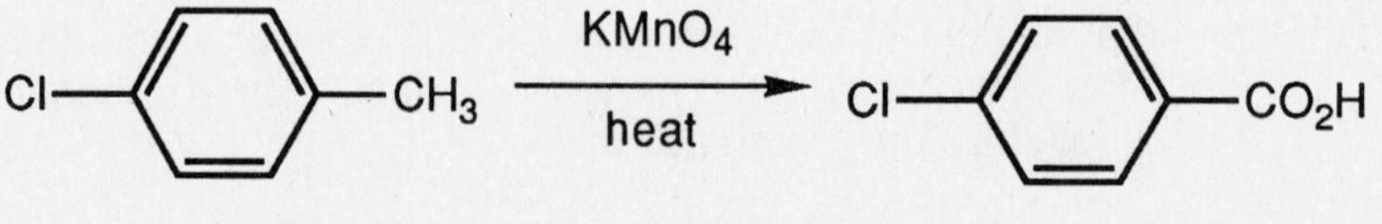

d.

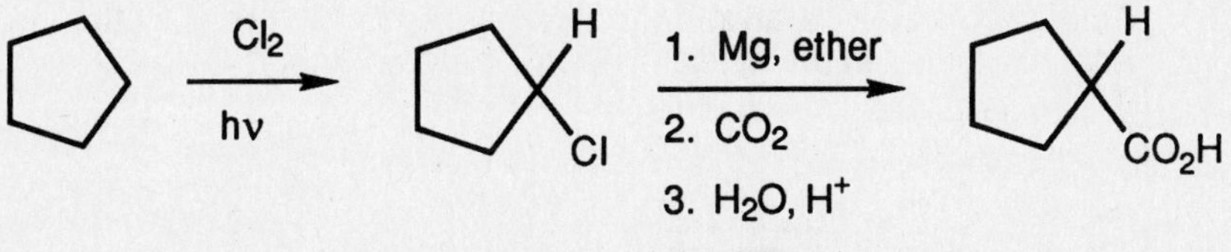

e.

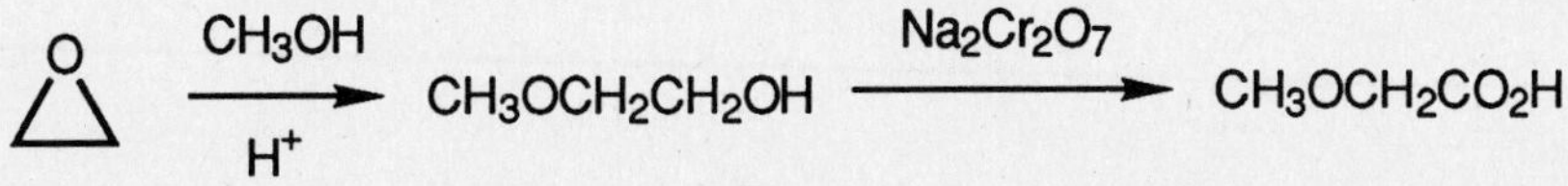

f.

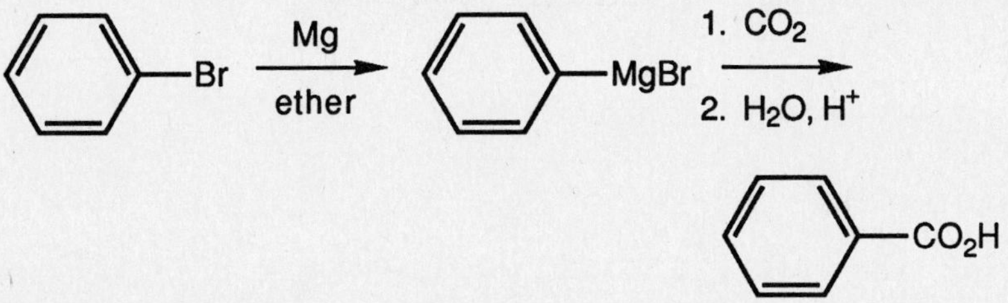

10.41 Catalysis by acid:

(protonated nitrile)

Nucleophilic attack by water on the protonated carbon–nitrogen triple bond, followed by proton transfers, gives the amide:

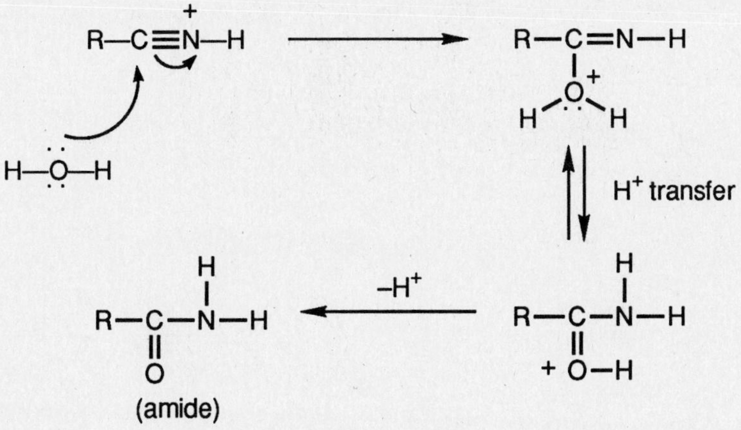

(amide)

Acid once again catalyzes the hydrolysis of the amide to the carboxylic acid and ammonia. In the acidic medium, the ammonia is immediately protonated.

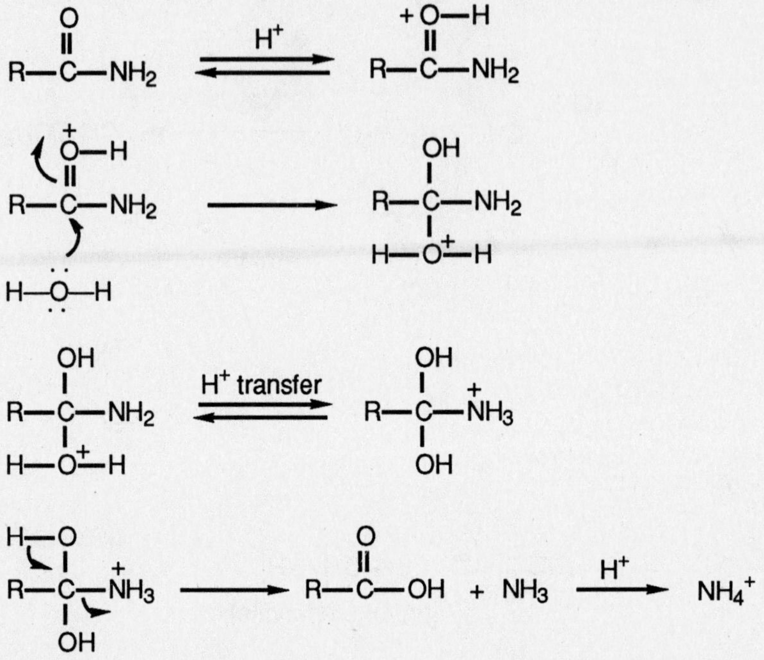

10.42 The nitrile route would involve an S_N2 displacement of bromide by cyanide, a highly unlikely step when the alkyl halide is tertiary. The Grignard route, on the other hand, works well for all alkyl halides, primary, secondary, and tertiary.

10.43 a. 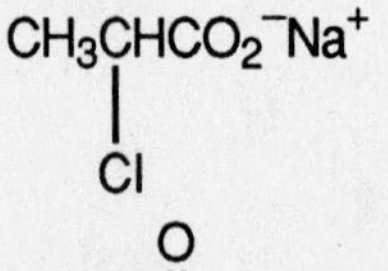 $CH_3CHCO_2^-Na^+$
   |
   Cl

b. $(CH_3CO_2^-)_2Ca^{2+}$

c.

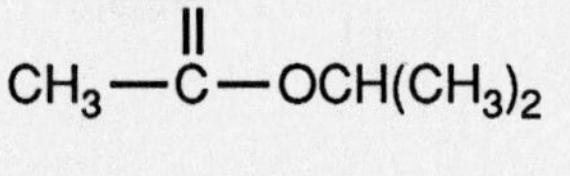

$$CH_3-\overset{\overset{\displaystyle O}{\|}}{C}-OCH(CH_3)_2$$

d.

$$H-\overset{\overset{\displaystyle O}{\|}}{C}-OCH_2CH_3$$

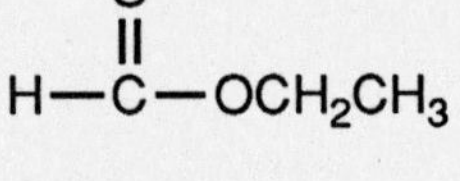

e.

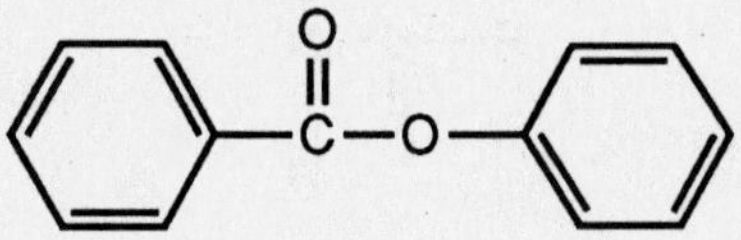

f.

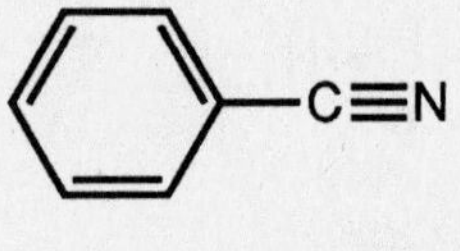

g.

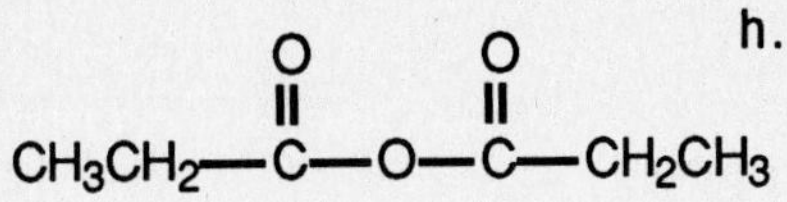

$$CH_3CH_2-\overset{\overset{\displaystyle O}{\|}}{C}-O-\overset{\overset{\displaystyle O}{\|}}{C}-CH_2CH_3$$

h.

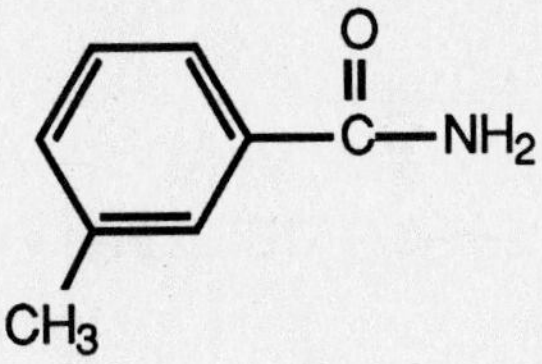

i.

$$ClCH_2CH_2CH_2-\overset{\overset{\displaystyle O}{\|}}{C}-Cl$$

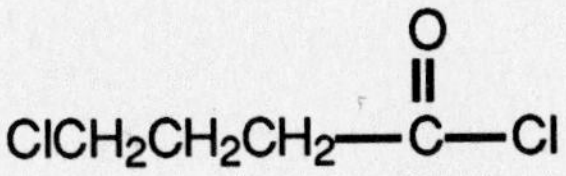

j.

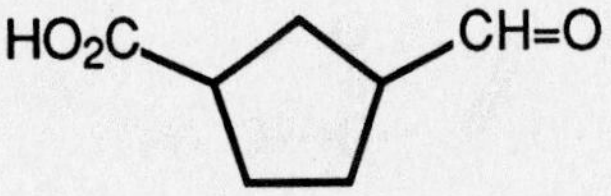

10.44
a. ammonium *p*-chlorobenzoate
b. calcium butanoate
c. phenyl 2-methylpropanoate
d. methyl trifluoroacetate
e. thioacetic acid
f. methyl thioacetate
g. formamide
h. cyclopropanecarboxylic anhydride

10.45 Compare with the answer to Prob. 10.15.

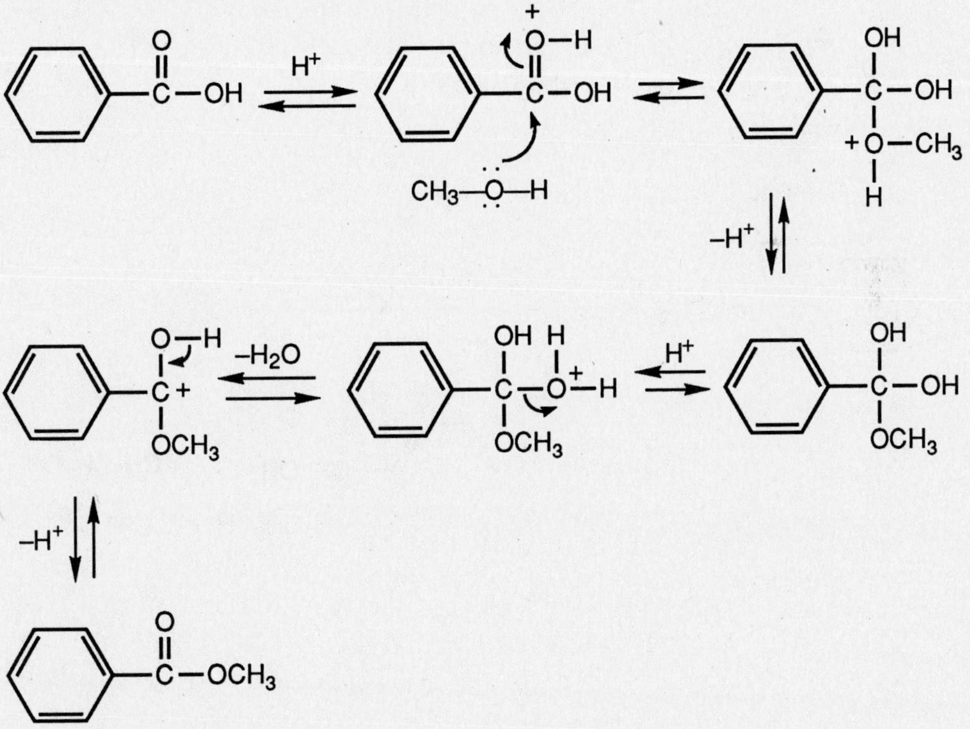

251

10.46 a. See eq. 10.20, where R = phenyl and R' = ethyl.

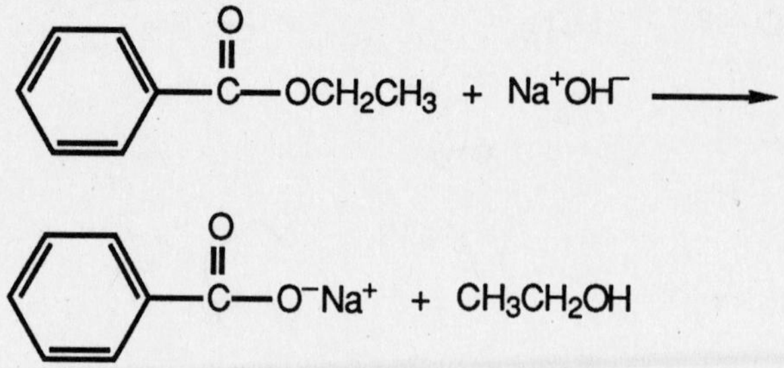

b. Compare with eq. 10.22.

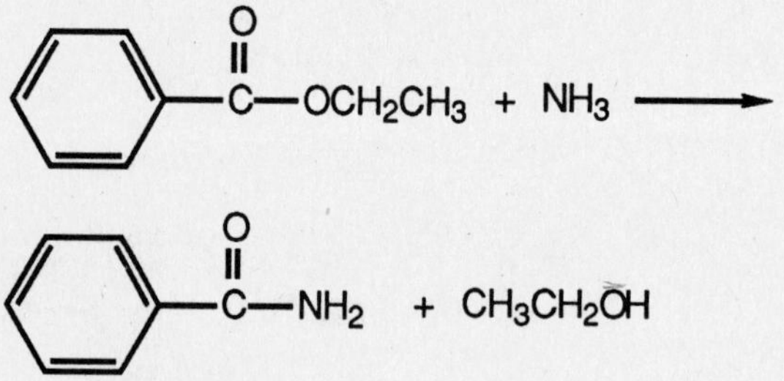

c. See Sec. 10.16.

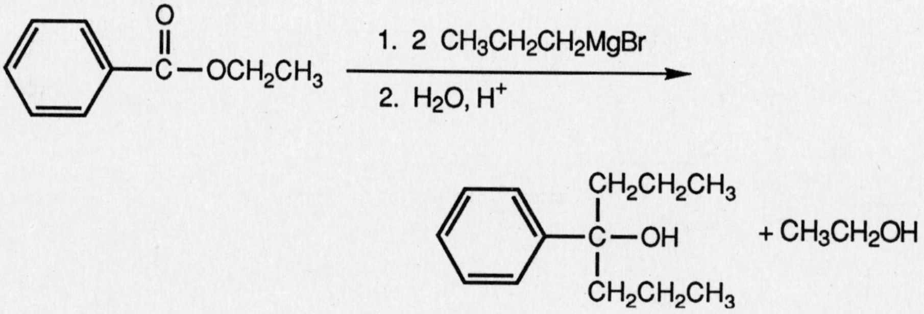

d. See Sec. 10.17.

$$\text{C}_6\text{H}_5-\overset{\overset{\displaystyle O}{\|}}{\text{C}}-\text{OCH}_2\text{CH}_3 \quad \xrightarrow[\text{2. H}_3\text{O}^+]{\text{1. LiAlH}_4}$$

$$\text{C}_6\text{H}_5-\text{CH}_2\text{OH} + \text{CH}_3\text{CH}_2\text{OH}$$

10.47 a.

$$\text{CH}_3\text{CH}_2-\overset{\overset{\displaystyle O}{\|}}{\text{C}}-\text{OCH}_3 \quad \rightleftharpoons \quad \text{CH}_3\text{CH}_2-\overset{\overset{\displaystyle O^-}{|}}{\underset{\underset{\displaystyle OH}{|}}{\text{C}}}-\text{OCH}_3$$

HO⁻

$$\text{CH}_3\text{CH}_2-\overset{\overset{\displaystyle O}{\|}}{\text{C}}-\text{O}^- + \text{CH}_3\text{OH} \quad \underset{\xleftarrow{\text{CH}_3\text{O}^-}}{\rightleftharpoons} \quad \text{CH}_3\text{CH}_2-\overset{\overset{\displaystyle O}{\|}}{\text{C}}-\text{OH} + \text{CH}_3\text{O}^-$$

b.

$$\text{CH}_3\text{CH}_2-\overset{\overset{\displaystyle O}{\|}}{\text{C}}-\text{OCH}_3 \quad \rightleftharpoons \quad \text{CH}_3\text{CH}_2-\overset{\overset{\displaystyle O^-}{|}}{\underset{\underset{\displaystyle +NH_3}{|}}{\text{C}}}-\text{OCH}_3$$

H₃N :

H⁺ transfer

$$\text{CH}_3\text{CH}_2-\overset{\overset{\displaystyle O}{\|}}{\text{C}}-\text{NH}_2 + \text{CH}_3\text{OH} \quad \rightleftharpoons \quad \text{CH}_3\text{CH}_2-\overset{\overset{\displaystyle O-H}{|}}{\underset{\underset{\displaystyle NH_2}{|}}{\text{C}}}-\text{OCH}_3$$

10.48 **a.** The carbonyl group in esters is less reactive than the carbonyl group of ketones because of the possibility of resonance in esters:

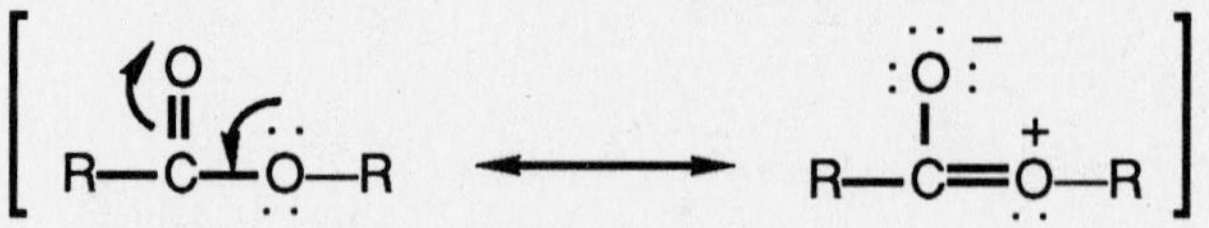

This delocalizes to the "ether" oxygen some of the positive charge usually associated with the carbonyl carbon atom:

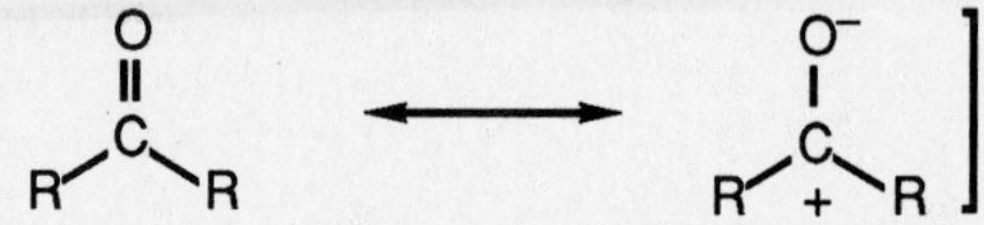

The carbonyl carbon in esters is therefore less susceptible to nucleophilic attack than is the carbonyl carbon of ketones.

b. In benzoyl chloride, the positive charge on the carbonyl carbon can be delocalized in the aromatic ring:

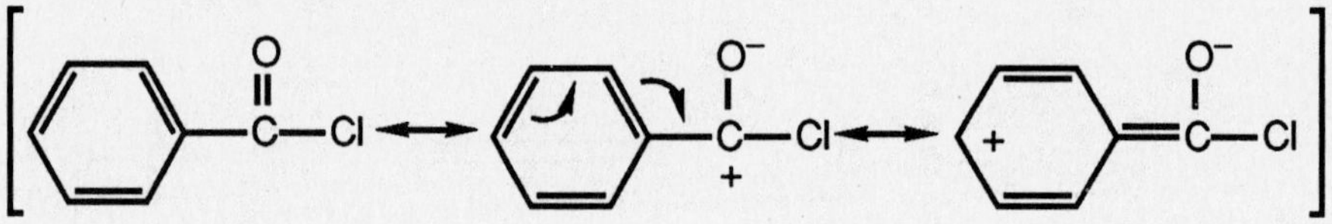

Such delocalization is not possible in cyclohexanecarbonyl chloride or any other aliphatic acid chloride. For this reason, aryl acid chlorides are usually less reactive toward nucleophiles than are aliphatic acid chlorides.

10.49 In each case, the two like organic groups attached to the hydroxyl-bearing carbon come from the Grignard reagent, and the third group comes from the acid part of the ester.

a. $CH_3CH_2CH_2MgBr$ +

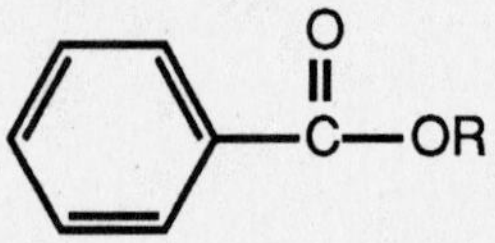

b.

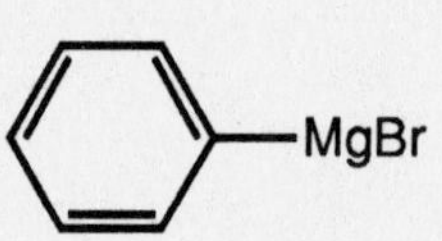

The identity of the R group in the ester does not affect the product structure. R is usually CH_3 or CH_3CH_2.

10.50

a. $CH_3-\overset{\overset{O}{\|}}{C}-Cl \ + \ H_2O \ \longrightarrow \ CH_3-\overset{\overset{O}{\|}}{C}-OH \ + \ HCl$

b.

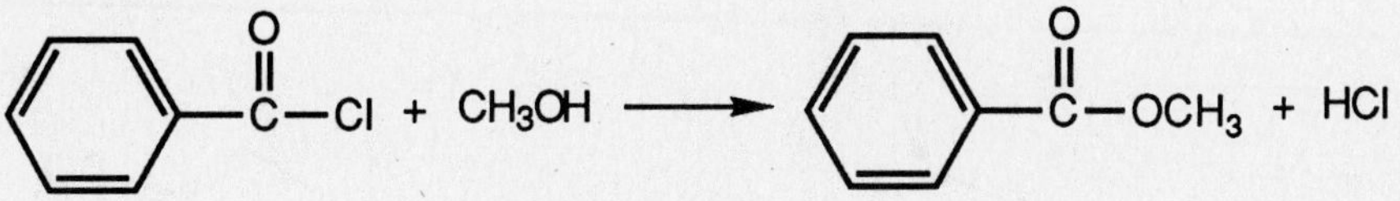

c. $CH_3-\overset{\overset{O}{\|}}{C}-O-\overset{\overset{O}{\|}}{C}-CH_3 \ + \ HOCH_2CH_2CH_2CH_2CH_3 \longrightarrow$

$CH_3-\overset{\overset{O}{\|}}{C}-OCH_2CH_2CH_2CH_2CH_3 \ + \ CH_3CO_2H$

d. $BrCH_2CH_2CH_2-\overset{\overset{O}{\|}}{C}-Br \ + \ 2\ NH_3 \longrightarrow BrCH_2CH_2CH_2-\overset{\overset{O}{\|}}{C}-NH_2$

$+ \ NH_4^+$

Note that the acyl bromide is much more reactive toward nucleophiles than is the alkyl bromide.

e. $CH_3CH_2CH_2CH_2-\overset{\overset{O}{\|}}{C}-OH \ + \ CH_3CH_2OH \ \overset{H^+}{\longrightarrow}$

$CH_3CH_2CH_2CH_2-\overset{\overset{O}{\|}}{C}-OCH_2CH_3 \ + \ H_2O$

f.

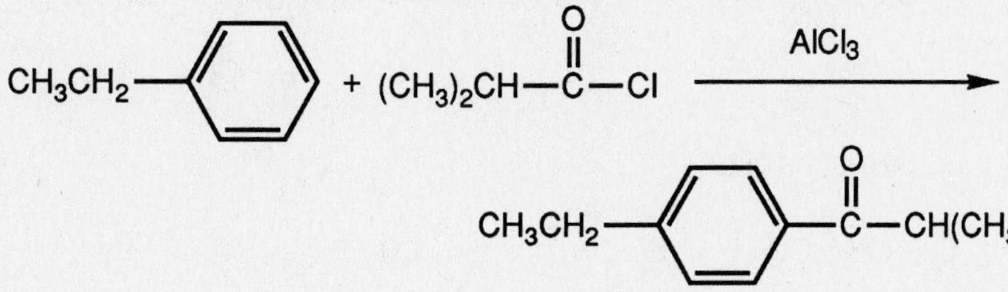

g.

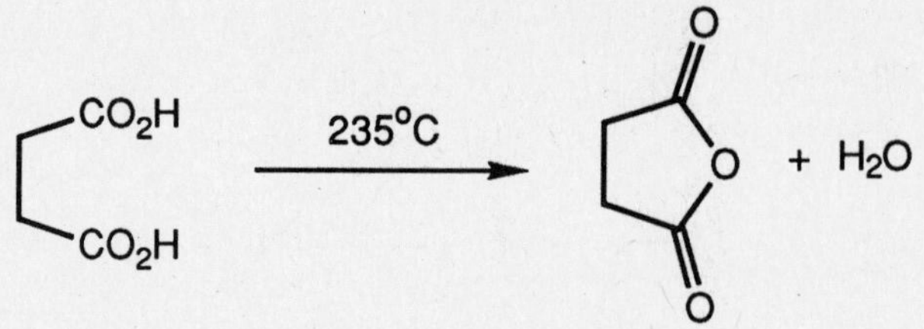

235°C → + H$_2$O

h.

CH$_3$OH (1 equiv) / H$^+$ 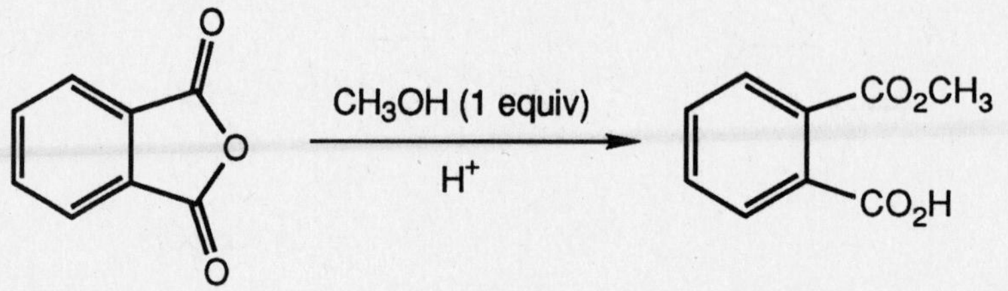 →

i.

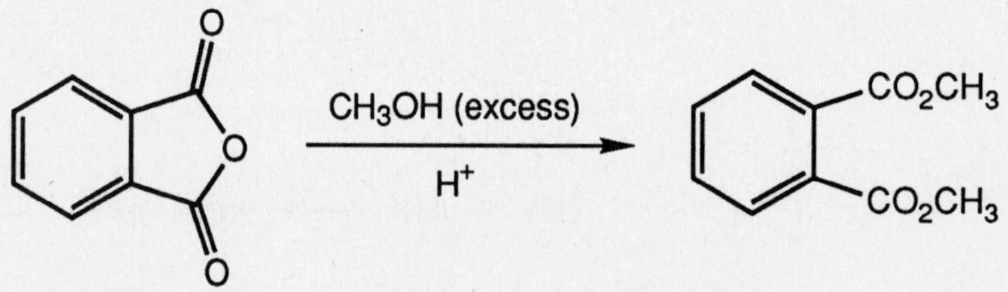

CH$_3$OH (excess) / H$^+$ →

j.

$$Cl-\overset{\overset{O}{\|}}{C}-\overset{\overset{O}{\|}}{C}-Cl \xrightarrow{NH_3} H_2N-\overset{\overset{O}{\|}}{C}-\overset{\overset{O}{\|}}{C}-NH_2$$

10.51 a. See eq. 10.30.

$$CH_3CH_2CH_2-\overset{\overset{O}{\|}}{C}-OH \ + \ PCl_5 \longrightarrow CH_3CH_2CH_2-\overset{\overset{O}{\|}}{C}-Cl$$

$$+ \ HCl \ + \ POCl_3$$

b. See eq. 10.29.

$$CH_3(CH_2)_8CO_2H \ + \ SOCl_2 \longrightarrow CH_3(CH_2)_8COCl \ + \ HCl \ + \ SO_2$$

c. See eq. 10.10.

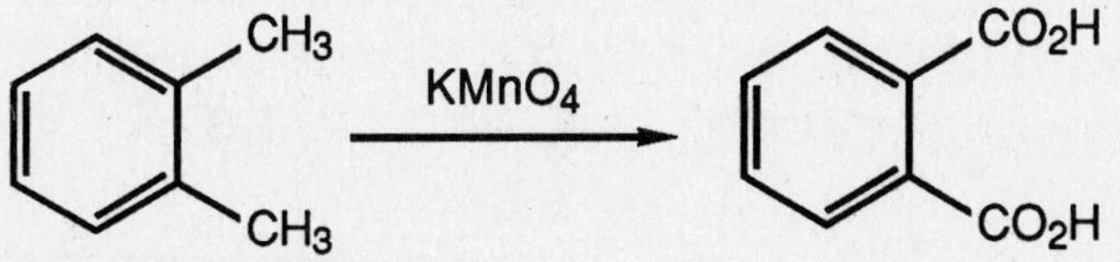

d. See eq. 10.37.

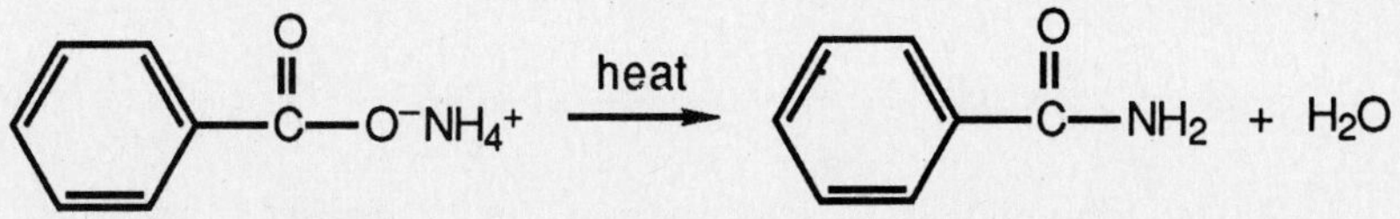

e. See eq. 10.39.

$$CH_3(CH_2)_5\text{—}\overset{\displaystyle O}{\overset{\|}{C}}\text{—}NH_2 \xrightarrow{LiAlH_4} CH_3(CH_2)_5CH_2NH_2$$

f. See eq. 10.25.

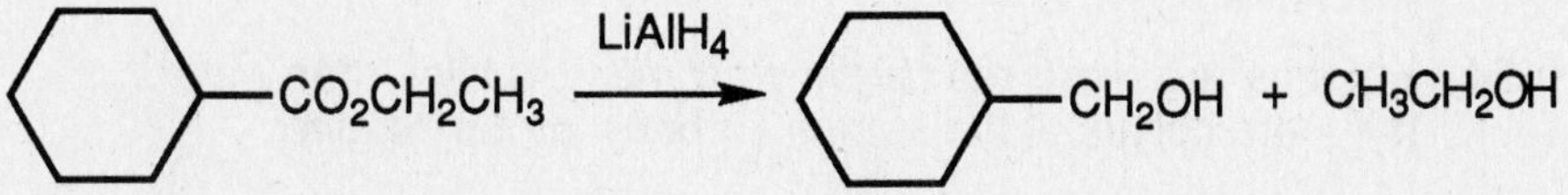

10.52 Review Sec. 10.21.

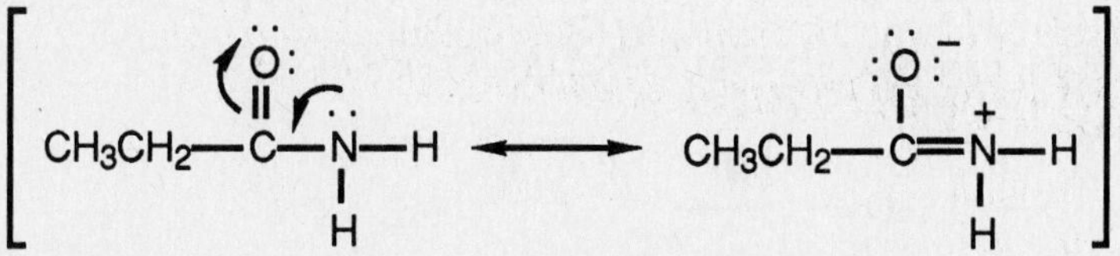

The six atoms that lie in one plane are

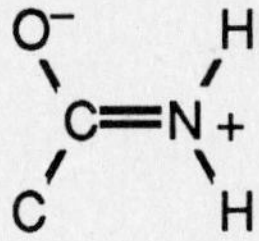

10.53 Consider the leaving group in each nucleophilic substitution:

$$R-\overset{\overset{\displaystyle O}{\|}}{C}-SR' + Nu^- : \longrightarrow R-\overset{\overset{\displaystyle O}{\|}}{C}-Nu + R'S^-$$

$$R-\overset{\overset{\displaystyle O}{\|}}{C}-O-\overset{\overset{\displaystyle O}{\|}}{C}-R' + Nu^- : \longrightarrow R-\overset{\overset{\displaystyle O}{\|}}{C}-Nu + {}^-O-\overset{\overset{\displaystyle O}{\|}}{C}-R'$$

$$R-\overset{\overset{\displaystyle O}{\|}}{C}-Cl + Nu^- : \longrightarrow R-\overset{\overset{\displaystyle O}{\|}}{C}-Nu + Cl^-$$

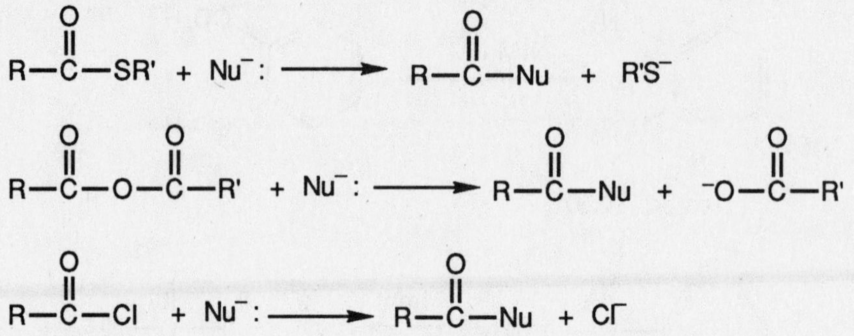

The K_a's of the conjugate acids increase in the order shown:

$RSH \cong 10^{-10}$; $RCO_2H \cong 10^{-5}$; HCl = strong acid. Therefore, the basicity of the anions decreases in the same order:

$R'S^- > RCO_2^- > Cl^-$

The weaker the basicity of the leaving group, the more facile the nucleophilic substitution. Therefore, the reactivity order is thioesters < anhydrides < acyl chlorides.

10.54 Ketones are more reactive toward nucleophiles than esters are. Reduction therefore occurs at the ketone carbonyl group, to give:

$CH_3CH(OH)CH_2CH_2CO_2CH_3$.

10.55 The method combines the formation of a cyanohydrin (Sec. 9.11) with the hydrolysis of a cyanide to an acid (Sec. 10.8d).

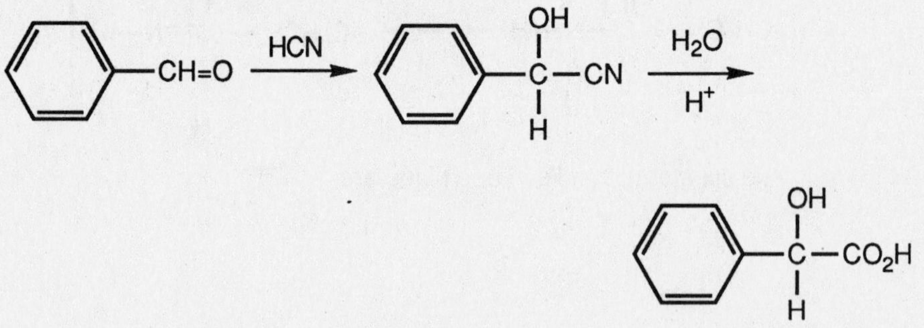

10.56 a. The compound is made of two isoprene units (heavy lines) linked together as shown by the dashed lines.

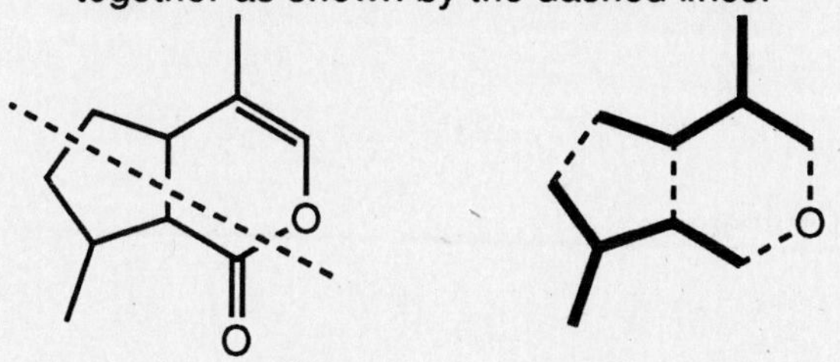

b. Nepatalactone has three chiral centers:

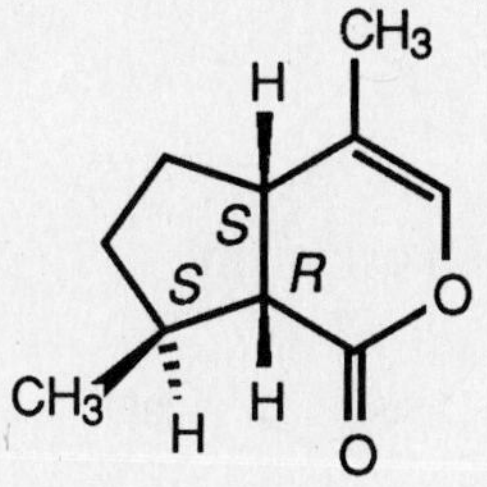

10.57 Since salicylic acid

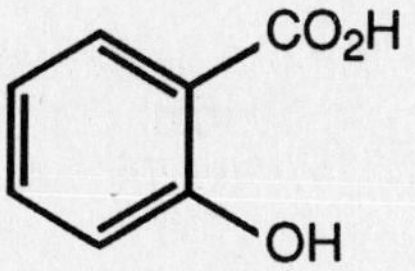

is an oxidation product, the starting formula should be

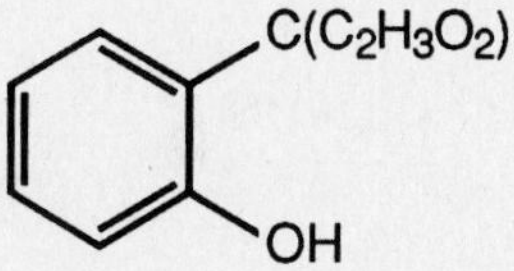

Since the side chain includes a carboxyl group (the compound is said to be a phenolic *acid*) this structure can be further refined:

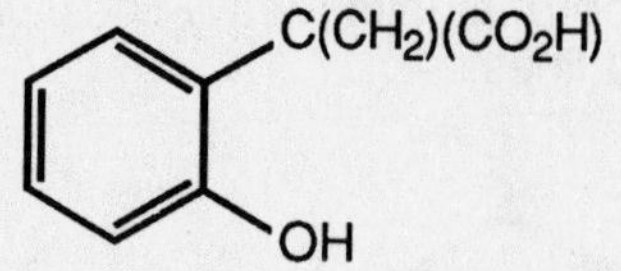

The two possible isomers are

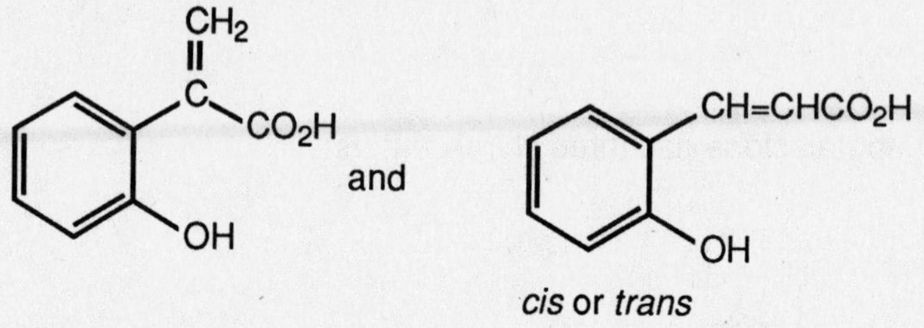

and

cis or trans

Only the second of these gives oxalic acid on oxidation:

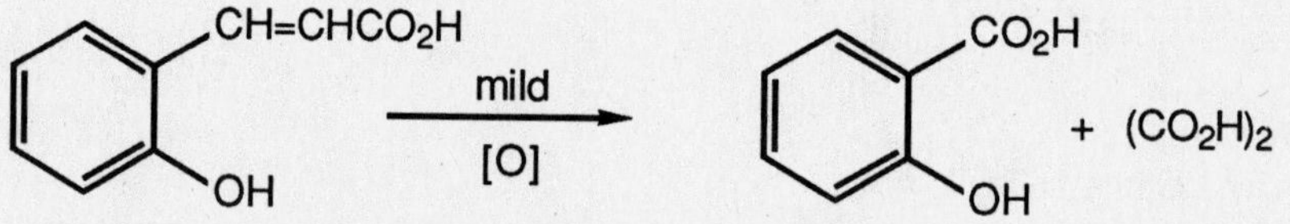

Therefore, the two isomeric forms are *cis*- and *trans*-*o*-hydroxycinnamic acid. Of these, the *cis* isomer easily loses water to form a lactone; the *trans* isomer cannot because the carboxyl and hydroxyl groups are too far apart.

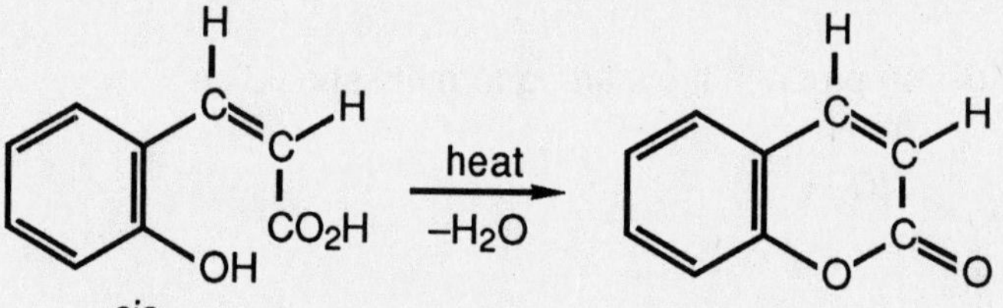

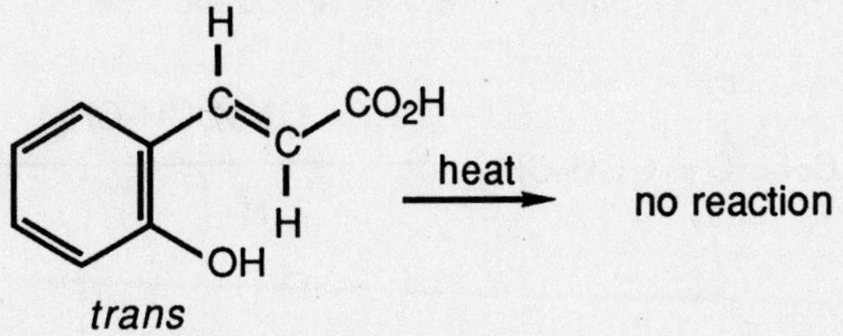

trans

heat ——→ no reaction

10.58 Protonation of the carboxyl oxygen gives the allylic cation shown:

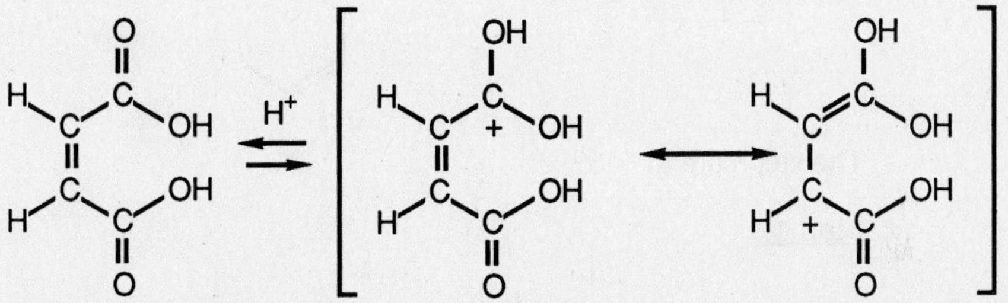

One of the resonance contributors to the cation has a single bond where the carbon–carbon double bond was. If rotation occurs around that single bond before proton loss, the product will be fumaric acid:

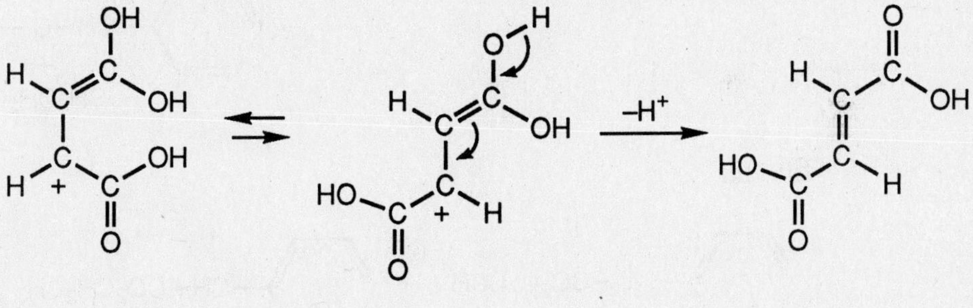

10.59 Use eqs. 10.42-10.44 as a guide. The overall equation is:

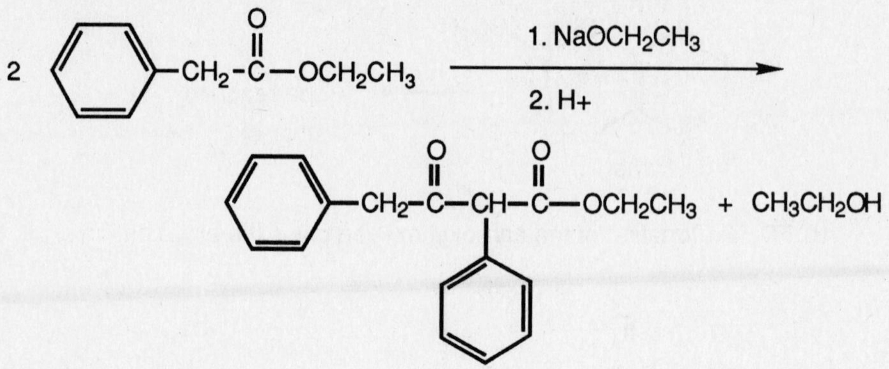

2 [benzene]-CH$_2$-C(=O)-OCH$_2$CH$_3$ $\xrightarrow[\text{2. H+}]{\text{1. NaOCH}_2\text{CH}_3}$

[benzene]-CH$_2$-C(=O)-CH-C(=O)-OCH$_2$CH$_3$ + CH$_3$CH$_2$OH

(with phenyl substituent on CH)

The steps are as follows:

Step 1:

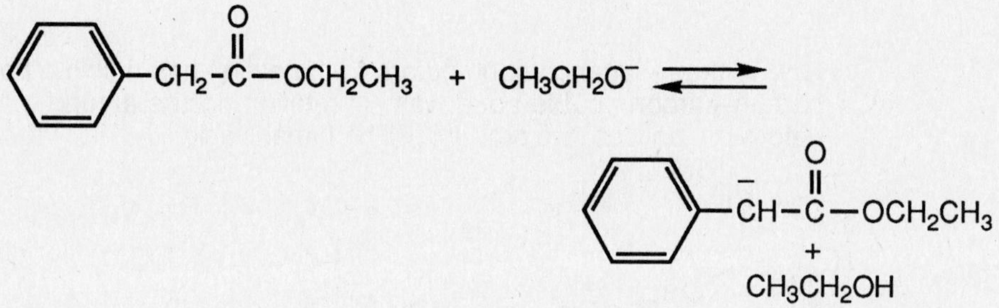

[benzene]-CH$_2$-C(=O)-OCH$_2$CH$_3$ + CH$_3$CH$_2$O$^-$ $\rightleftharpoons$

[benzene]-CH$^-$-C(=O)-OCH$_2$CH$_3$ + CH$_3$CH$_2$OH

Step 2:

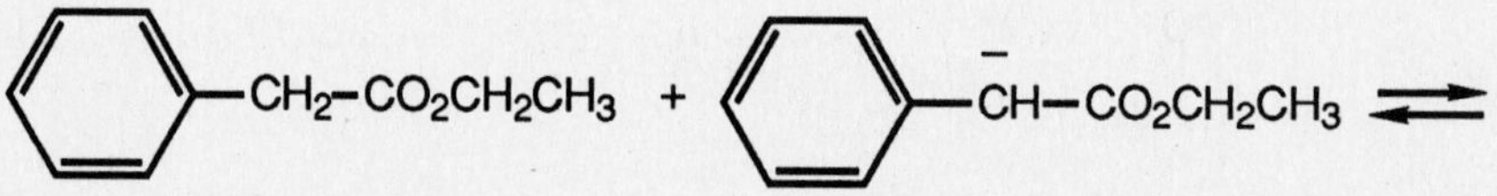

[benzene]-CH$_2$-CO$_2$CH$_2$CH$_3$ + [benzene]-CH$^-$-CO$_2$CH$_2$CH$_3$ $\rightleftharpoons$

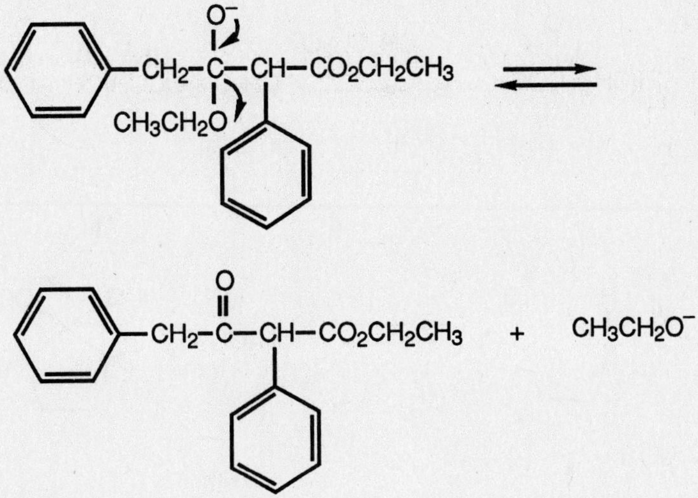

$$\text{—CH}_2\text{—}\overset{\overset{\displaystyle O^-}{\|}}{\underset{\underset{\displaystyle CH_3CH_2O}{}}{C}}\text{—CH—CO}_2\text{CH}_2\text{CH}_3 \quad \rightleftharpoons$$

$$\text{—CH}_2\text{—}\overset{\overset{\displaystyle O}{\|}}{C}\text{—CH—CO}_2\text{CH}_2\text{CH}_3 \quad + \quad CH_3CH_2O^-$$

Step 3:

$$\text{—CH}_2\text{—}\overset{\overset{\displaystyle O}{\|}}{C}\text{—CH—}\overset{\overset{\displaystyle O}{\|}}{C}\text{—OCH}_2\text{CH}_3 + CH_3CH_2O^- \longrightarrow$$

$$\text{—CH}_2\text{—}\overset{\overset{\displaystyle O}{\|}}{C}\text{=}\overset{-}{C}\text{=}\overset{\overset{\displaystyle O}{\|}}{C}\text{—OCH}_2\text{CH}_3 + CH_3CH_2OH$$

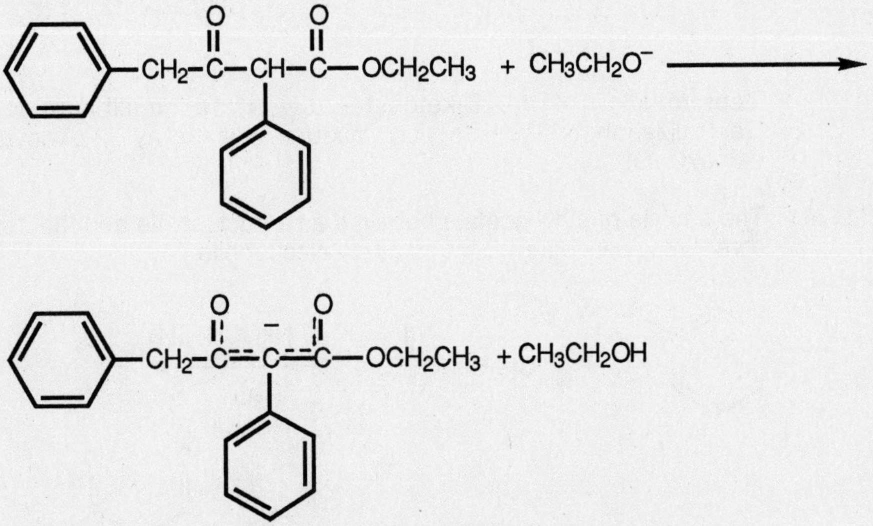

The reaction is neutralized with acid to obtain the final product.

10.60

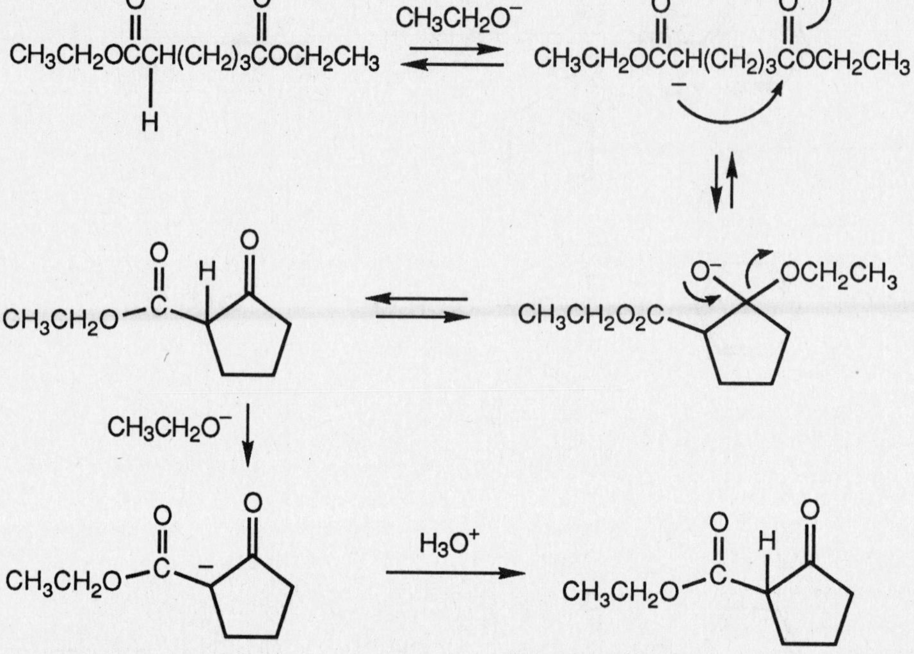

Deprotonation of the β-ketoester drives the equilibrium to the right. Neutralization of the reaction mixture gives ethyl 2-oxocyclopentane-carboxylate.

11.61 The enolate of ethyl acetate behaves as a nucleophile and the nonenolizable ester, ethyl benzoate, behaves as an electrophile.

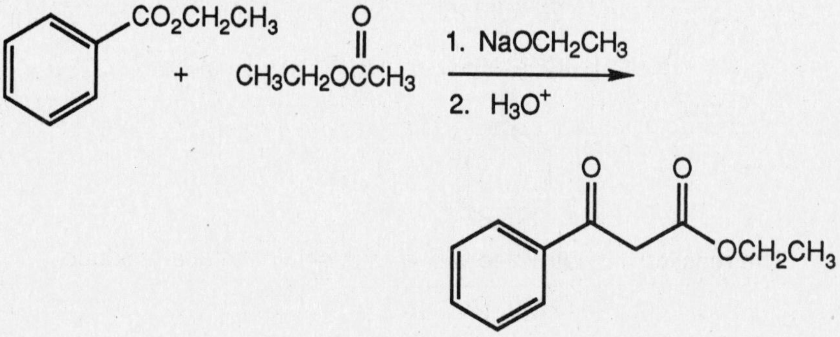

CHAPTER ELEVEN: AMINES AND RELATED NITROGEN COMPOUNDS

CHAPTER SUMMARY

Amines are organic derivatives of ammonia. They may be **primary, secondary,** or **tertiary**, depending on whether one, two, or three organic groups are attached to the nitrogen. The nitrogen is sp^3-hybridized and pyramidal, nearly tetrahedral.

The **amino group** is —NH_2. Amines are named according to the Chemical Abstracts (CA) system by adding the suffix *-amine* to the names of the alkyl groups attached to the nitrogen. Amines can also be named using the IUPAC system in which the amino group is named as a substituent. Aromatic amines are named as derivatives of aniline or of the aromatic ring system.

Amines form intermolecular N—H···N bonds. Their boiling points are higher than those of alkanes but lower than those of alcohols with comparable molecular weights. Lower members of the series are water soluble because of N···H—O bonding.

Amines can be prepared by S_N2 alkylation of ammonia or 1° and 2° amines. Aromatic amines are made by reduction of the corresponding nitro compounds. Amides, nitriles, and imines can also be reduced to amines.

Amines are weak bases. Alkylamines and ammonia are of comparable basicity, but aromatic amines are much weaker as a result of delocalization of the unshared electron pair on nitrogen to the *ortho* and *para* carbons of the aromatic ring.

Amides are much weaker bases than amines because of delocalization of the unshared electron pair on nitrogen to the adjacent carbonyl oxygen. Amides are stronger Brønsted acids than amines because of the partial positive charge on the amide nitrogen and resonance in the **amidate anion.**

Amines react with strong acids to form **amine salts**. The pK_a's of amine salts are related to the base strength of the corresponding amines. Alkylammonium salts have pK_a's of 9-10, while arylammonium salts have pK_a's of 4-5. The fact that these salts are usually water soluble can be taken advantage of in separating amines from neutral or acidic contaminants. Chiral amines can be used to resolve enantiomeric acids, through the formation of diastereomeric salts.

11. AMINES AND RELATED NITROGEN COMPOUNDS

Primary and secondary amines react with acid derivatives to form amides. Amides made commercially this way include **acetanilide** and **N,N-diethyl-**m**-toluamide** (the insect repellent Off). Primary and secondary amines react with arylsulfonyl chlorides to give sulfonamides such as the sulfa drugs **sulfanilamide** and **sulfadiazine.**

Tertiary amines react with alkyl halides to form **quarternary ammonium salts**. An example of this type of salt that has important biological properties is **choline** (2-hydroxyethyltrimethylammonium ion).

Primary aromatic amines react with nitrous acid to give **aryldiazonium ions**, ArN_2^+, which are useful intermediates in synthesis of aromatic compounds. The process by which they are formed is called **diazotization**. The nitrogen in these ions can readily be replaced by various nucleophiles (OH, Cl, Br, I, CN). Diazonium ions couple with reactive aromatics, such as amines or phenols, to form **azo compounds**, which are useful as dyes.

REACTION SUMMARY

Alkylation of Ammonia and Amines

$$R-X + 2 NH_3 \longrightarrow R-NH_2 + NH_4^+X^-$$

Reduction Routes to Amines

REACTION SUMMARY

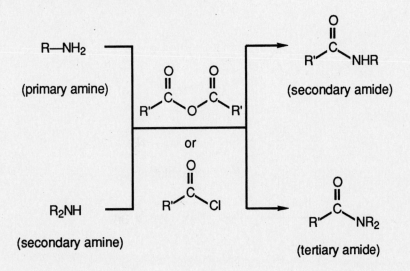

Amine Basicity

$$R-NH_2 + H-OH \longrightarrow R-\overset{+}{N}H_3 + {}^-OH$$

$$R-NH_2 + H-Cl \longrightarrow R-\overset{+}{N}H_3 + Cl^-$$

Acylation of Primary and Secondary Amines

11. AMINES AND RELATED NITROGEN COMPOUNDS

Hinsberg Test (Sulfonamides)

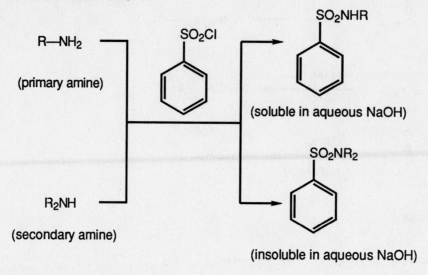

R—NH$_2$

(primary amine)

R$_2$NH

(secondary amine)

SO$_2$Cl

SO$_2$NHR

(soluble in aqueous NaOH)

SO$_2$NR$_2$

(insoluble in aqueous NaOH)

Quaternary Ammonium Salts

$$R_3N \ + \ R'X \longrightarrow R_3\overset{+}{N}-R' \ X^-$$

Aryldiazonium Salts

$$ArNH_2 \ + \ HONO \ \xrightarrow{\ HX\ } \ ArN_2^+ \ X^-$$

(aryldiazonium ion)

$$ArN_2^+ \ + \ H_2O \ \xrightarrow{\ heat\ } \ ArOH \ + \ N_2 \ + \ H^+$$

(phenols)

$$ArN_2^+ \ + \ HX \ \xrightarrow{\ Cu_2X_2\ } \ ArX \quad (X = Cl, Br)$$

MECHANISM SUMMARY

$ArN_2^+ + KI \longrightarrow ArI$

$ArN_2^+ + KCN \xrightarrow{Cu_2(CN)_2} ArCN$

$ArN_2^+ + HBF_4 \longrightarrow ArF$

$ArN_2^+ + H_3PO_2 \longrightarrow ArH$

Diazo Coupling

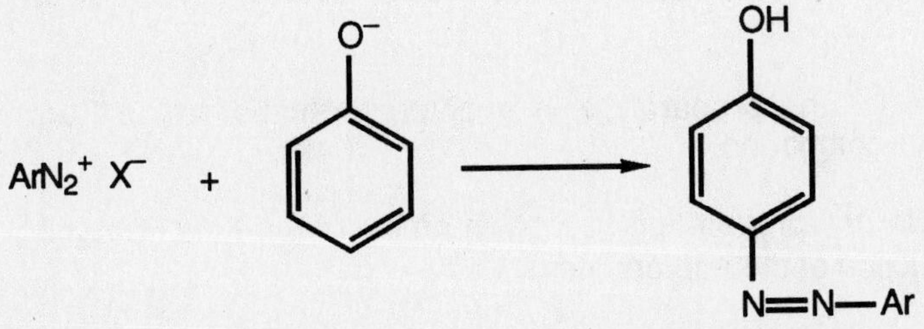

$ArN_2^+ \ X^-$ +

(azo compound)

MECHANISM SUMMARY

Diazotization

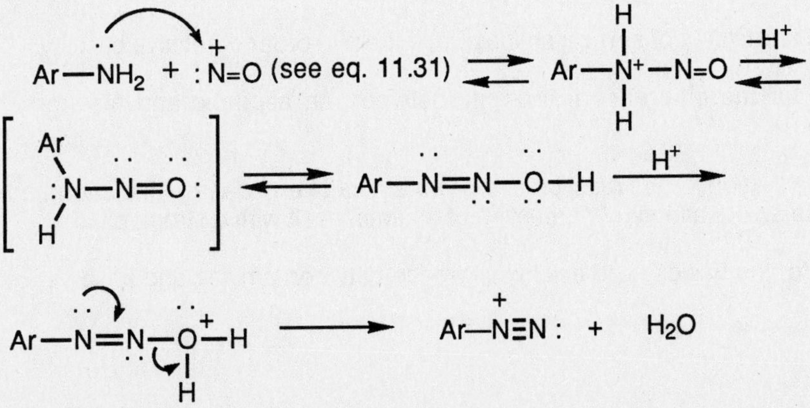

11. AMINES AND RELATED NITROGEN COMPOUNDS

LEARNING OBJECTIVES

1. Know the meaning of: primary, secondary, and tertiary amine; amino group, aniline, amine salt, quaternary ammonium salt.

2. Know the meaning of: nitrous acid, diazonium ion, diazotization, azo coupling.

3. Given the structure of an amine, identify it as primary, secondary, or tertiary.

4. Given the structure of an amine, name it. Also, given the name of an amine, write its structural formula.

5. Explain the effect of hydrogen bonding on the boiling points of amines and their solubility in water.

6. Write an equation for the reaction between ammonia or an amine of any class and an alkyl halide.

7. Write an equation for the preparation of a given aromatic amine from the corresponding nitro compound.

8. Write an equation for the preparation of a given amine of the type RCH_2NH_2 or $ArCH_2NH_2$ by reduction of the appropriate nitrile.

9. Write an equation for the preparation of a secondary amine from a ketone, primary amine, and sodium cyanoborohydride.

10. Write an equation for the dissociation of an amine in water.

11. Write an expression for K_a of any amine salt.

12. Draw the important contributors to the resonance hybrid for an aromatic amine.

13. Given the structures of several amines, rank them in order of relative basicity.

14. Account for the difference in basicity between an aliphatic and an aromatic amine.

15. Write an equation for the reaction of a given amine of any class with a strong acid. Also, write an equation for the reaction of an amine salt with a strong base.

16. Account for the basicity and acidity difference between amines and amides.

17. Explain, with the aid of equations, how you can separate an amine from a mixture containing neutral and/or acidic compounds.

18. Explain how chiral amines can be used to resolve a mixture of enantiomeric acids.

19. Write an equation for the reaction of a given primary or secondary amine with an acid anhydride or acid halide.

20. Explain the Hinsberg test for distinguishing between primary, secondary, and tertiary amines.

21 Write the steps in the mechanism for acylation of a primary or secondary amine.

22. Write an equation for the diazotization of a given primary aromatic amine.

23. Write the equations for the reaction of an aromatic diazonium salt with: aqueous base; $HX + Cu_2X_2$ (X = Cl, Br); $KCN + Cu_2(CN)_2$; HBF_4; and H_3PO_2.

24. Write an equation for the coupling of an aromatic diazonium salt with a phenol or aromatic amine.

ANSWERS TO PROBLEMS

Problems Within the Chapter

11.1 a. primary   b. secondary
      c. primary   d. tertiary

11.2 N,N-dimethyl-3-pentanamine

11.3 a. t-butylamine or 2-methyl-2-propanamine
      b. 2-aminoethanol
      c. p-nitroaniline

11.4 a. $(CH_3CH_2CH_2)_2NH$ b. $CH_3CH_2CH(NH_2)CH_2CH_2CH_3$

c.

$$CH_3 \quad CH_3$$

(benzene ring with substituents)

$$CH_3 \quad CH_3$$
$$CH_3 \quad -NH_2$$
$$CH_3 \quad CH_3$$

d.

$$CH_3-CH-CH_2CH_3$$
$$\quad\quad |$$
$$\quad N(CH_3)_2$$

11.5 Trimethylamine has no hydrogens on the nitrogen: $(CH_3)_3N$. Thus intermolecular hydrogen bonding is not possible. In contrast, intermolecular hydrogen bonding is possible for $CH_3CH_2CH_2NH_2$, and this raises its boiling point considerably above that of its tertiary isomer.

11.6 a. $\quad CH_3CH_2CH_2Br \;+\; 2\;\overset{..}{N}H_3 \;\longrightarrow\; CH_3CH_2CH_2NH_2 \;+\; NH_4{}^+Br^-$

b. $\quad CH_3CH_2I \;+\; 2\;(CH_3CH_2)_2\overset{..}{N}H \;\longrightarrow\; (CH_3CH_2)_3N \;+\; (CH_3CH_2)_2\overset{+}{N}H_2\; I^-$

c. $\quad (CH_3)_3N: \;+\; CH_3I \;\longrightarrow\; (CH_3)_4N^+\; I^-$

d.

$CH_3CH_2CH_2NH_2\; +$ (benzene ring with CH_2Br) $\longrightarrow$ (benzene ring with $\overset{+}{CH_2NH_2}CH_2CH_2CH_3\; Br^-$)

$\Big\downarrow CH_3CH_2CH_2NH_2$

(benzene ring with $CH_2NHCH_2CH_2CH_3$)

$CH_3CH_2CH_2\overset{+}{N}H_3\; Br^-\; +$

11.7

2 (benzene ring with NH_2) $+\; CH_3CH_2Br \;\longrightarrow$ (benzene ring with $NHCH_2CH_3$) $+$ (benzene ring with $\overset{+}{N}H_3\; Br^-$)

11.8 Nitration of toluene twice gives mainly the 2,4-dinitro product. Reduction of the nitro groups completes the synthesis.

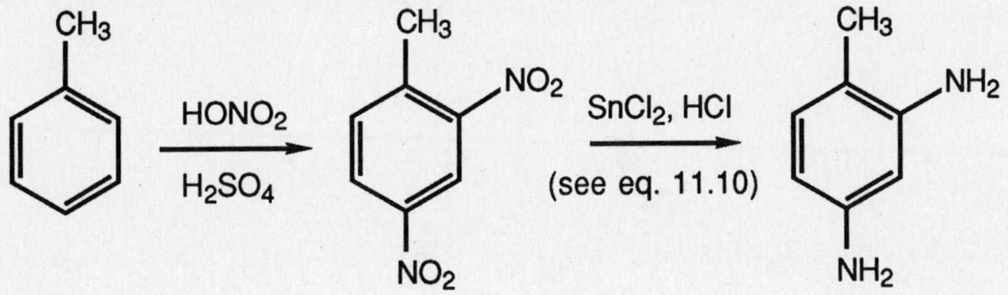

11.9 See eq. 11.11.

11.10 See eq. 11.12.

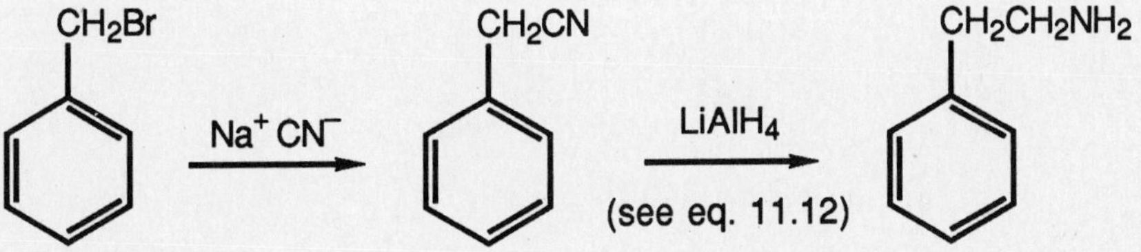

11.11 See eq. 11.13.

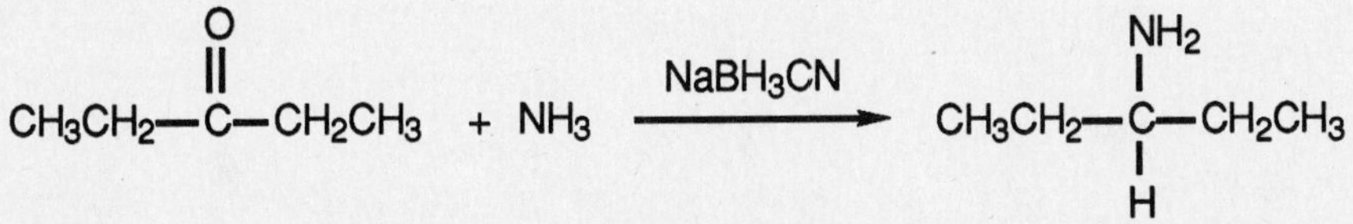

11.12 See eq. 11.14.

$$(CH_3)_3N + H_2O \rightleftharpoons (CH_3)_3\overset{+}{N}H \ ^-OH$$

11.13 $ClCH_2CH_2NH_2$ is a weaker base than $CH_3CH_2NH_2$. The chlorine substituent is electron-withdrawing compared to hydrogen and will destabilize the protonated base because of the repulsion between the positive charge on nitrogen and the partial positive charge on C-2 due to the C—Cl bond moment:

$$\overset{\delta-}{Cl}\!\!-\!\!\overset{\delta+}{CH_2}CH_2\overset{+}{N}H_3 \quad \text{compared to} \quad CH_3CH_2\overset{+}{N}H_3$$

Ethylamine is therefore easier to protonate (more basic) than 2-chloroethyl-amine.

11.14 Alkyl groups are electron-donating and stabilize the positively charged ammonium ion relative to the amine. Therefore N,N-dimethylaniline is a stronger base than N-methylaniline, which is a stronger base than aniline. The electron-withdrawing chlorine in p-chloroaniline destabilizes the positively charged ammonium ion. Therefore p-chloroaniline is a weaker base than aniline:

weaker base ⟶ stronger base

11.15 The order of the substituents by increasing electron-donating ability is $-NO_2 <$ —H < —CH_3. Therefore, the basicities will increase in that order:

11.16 Amides are less basic than amines, and aromatic amines are less basic than aliphatic amines. The order is therefore:

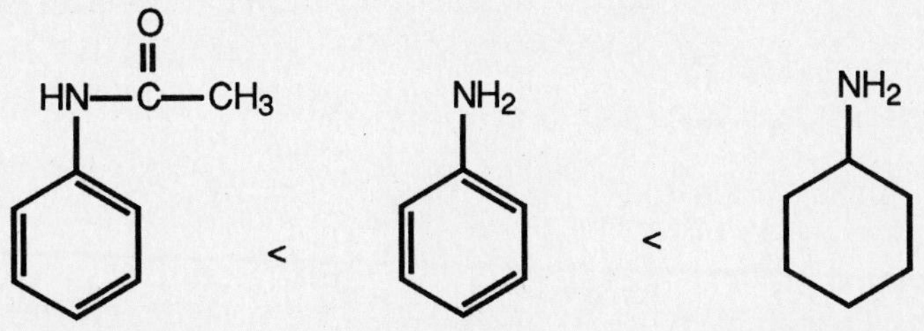

acetanilide   <   aniline   <   cyclohexylamine

The acidity increases in the reverse direction.

11.17

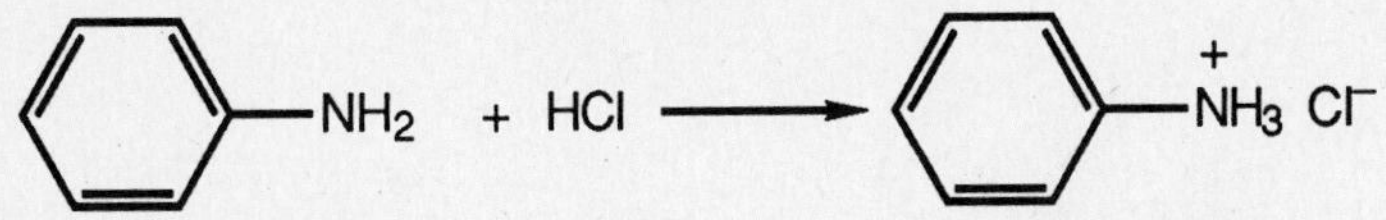

anilinium chloride

11.18

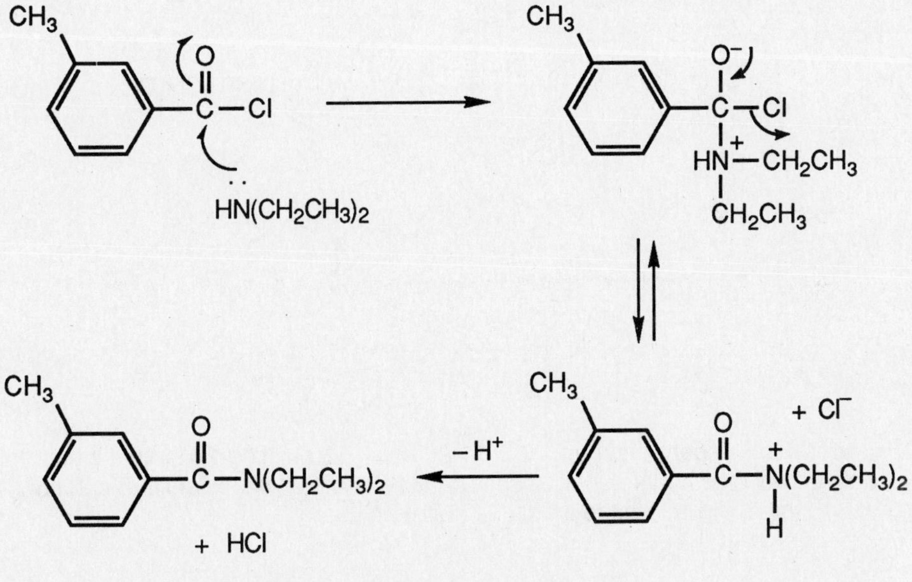

The sodium hydroxide combines with the HCl to form sodium chloride and water. Otherwise, the HCl would protonate the diethylamine and prevent it from functioning as a nucleophile.

11.19 This is an example of a nucleophilic acyl substitution or acyl transfer (see Sec. 10.18).

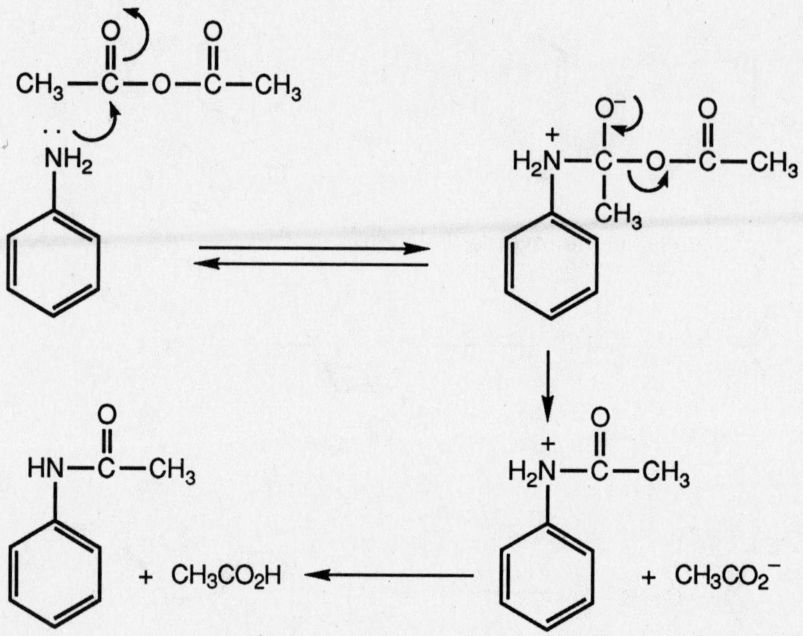

11.20

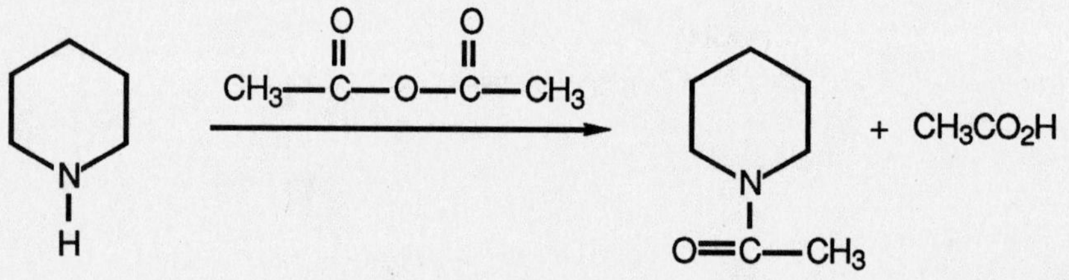

11.21 Perform a Hinsberg test (see eqs. 11.26 and 11.27). The 1° amine, *p*-methyl-aniline, will give a base-soluble sulfonamide, while N-methylaniline, a second-ary amine, will give a base-insoluble sulfonamide.

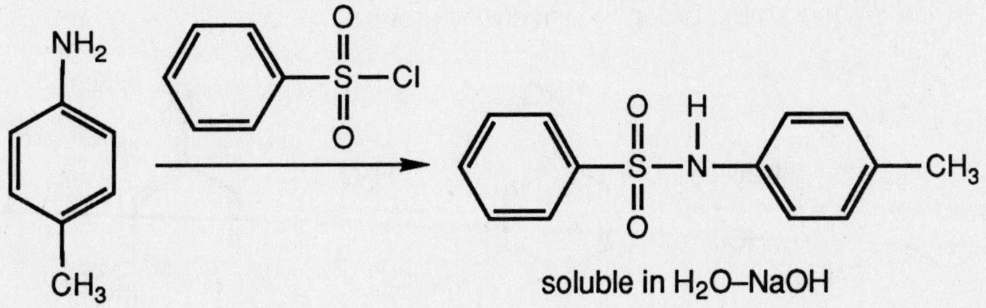

soluble in H_2O–NaOH

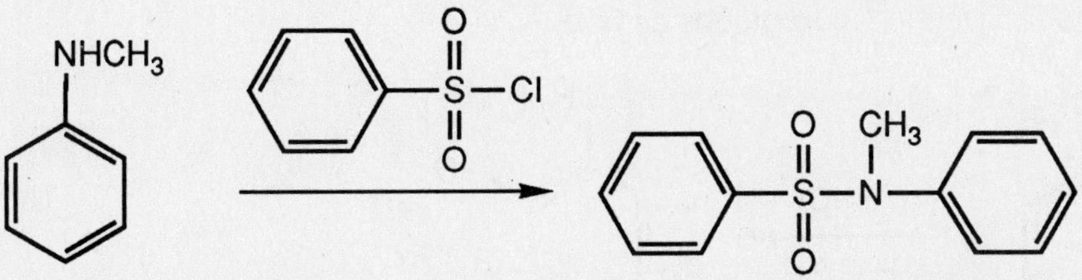

no acidic hydrogen

insoluble in H_2O–NaOH

11.22 a. Use the *meta*-directing nitro group to establish the proper relationship between the two substituents:

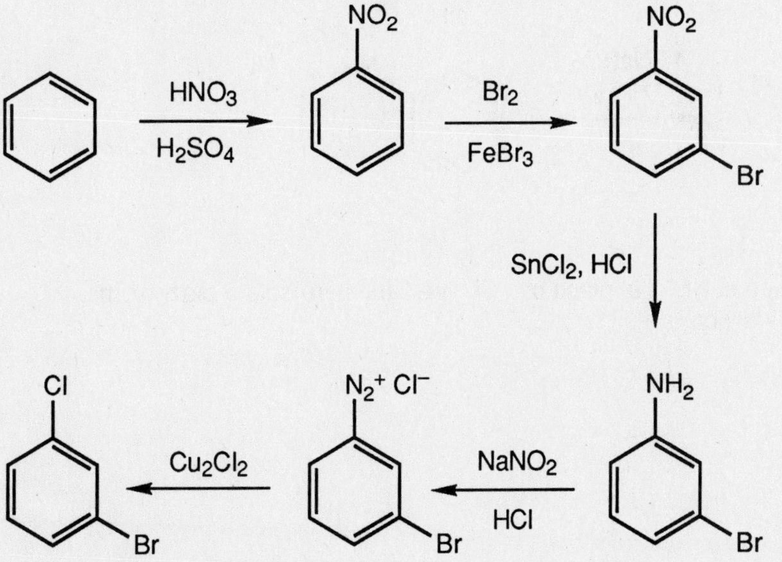

b. Convert the amino group to a hydroxyl group.

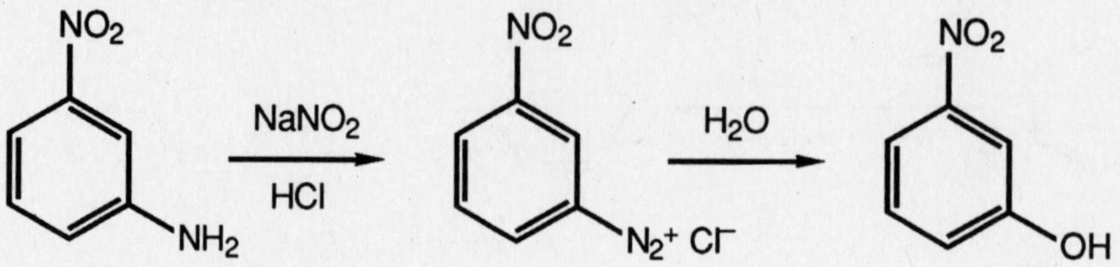

c. Introduce the fluoro groups as follows:

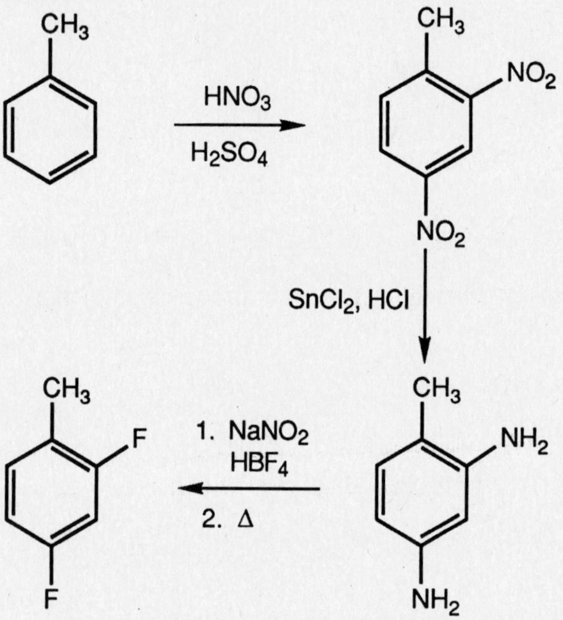

The use of HBF_4 in place of HCl gives an intermediate diazonium tetrafluoroborate.

d. Use the amino group to introduce the bromines and then replace the amino group by a hydrogen.

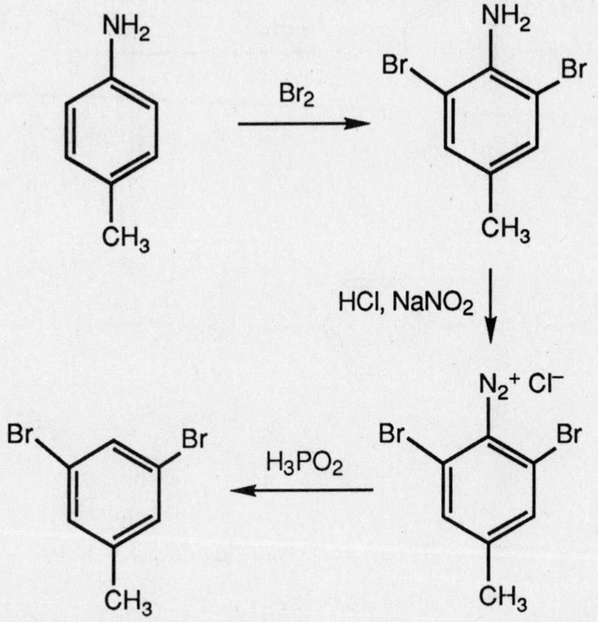

11.23

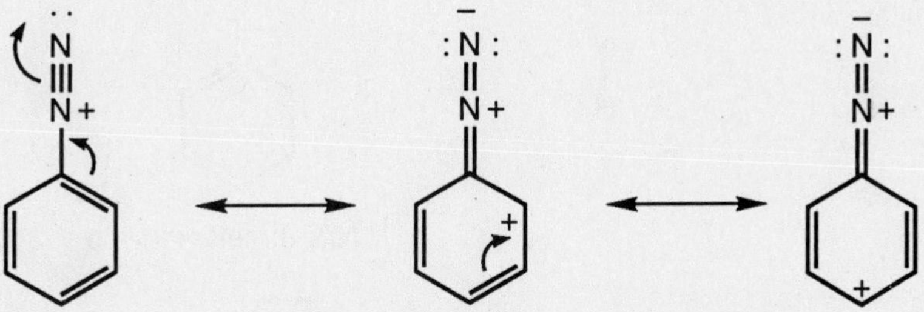

11.24

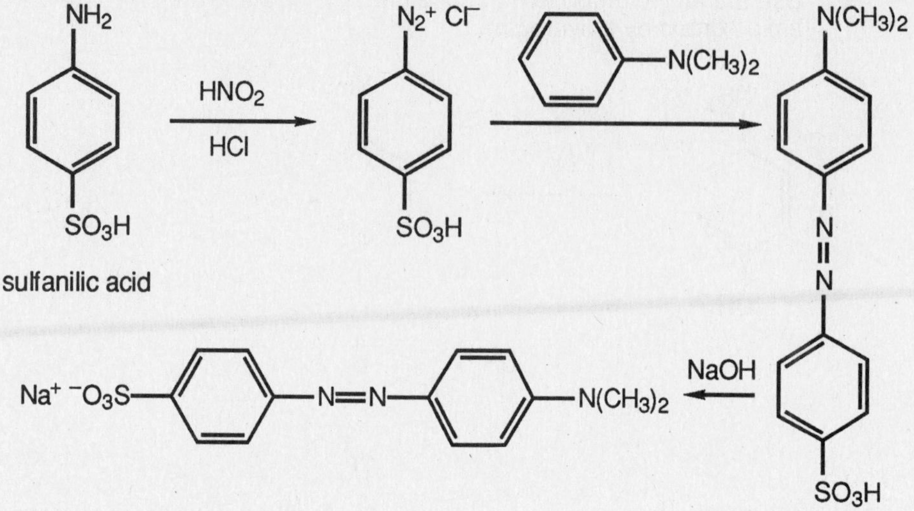

sulfanilic acid

Additional Problems

11.25 Many correct answers are possible, but only one example is given in each case.

a. CH_3NH_2

methylamine

b.

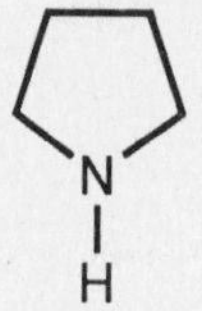

pyrrolidine

c.

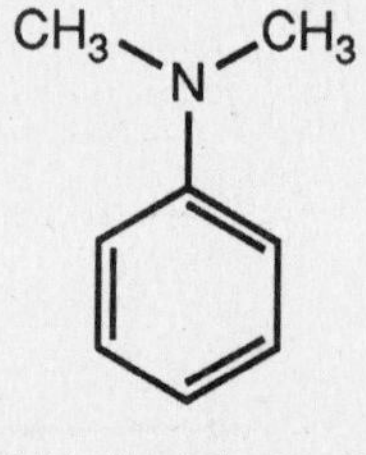

N,N-dimethylaniline

d.

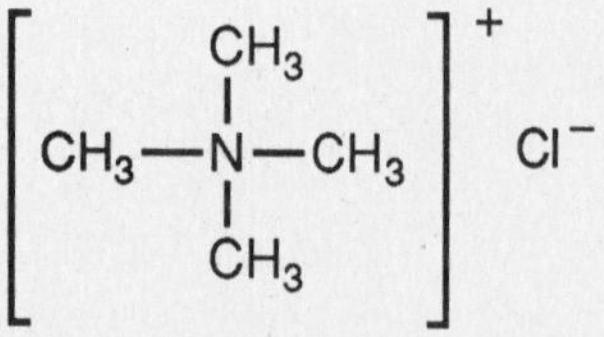

tetramethylammonium chloride

e.

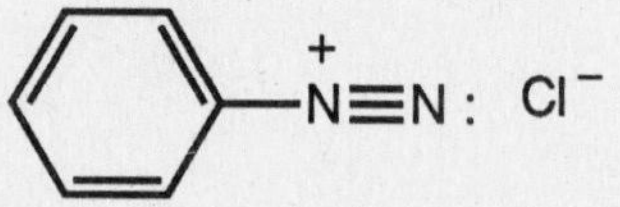

benzenediazonium chloride

f.

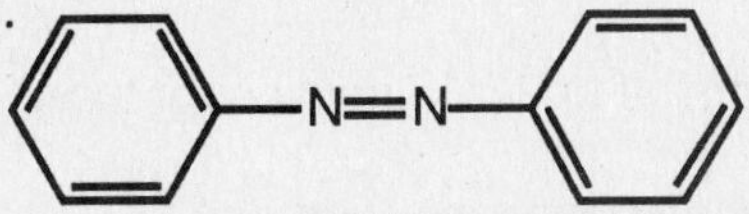

azobenzene

g.

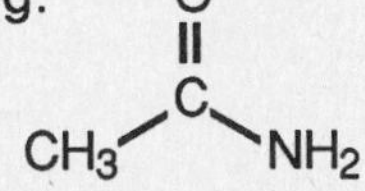

$$CH_3 - \overset{\overset{\displaystyle O}{\|}}{C} - NH_2$$

acetamide

h.

$$\text{—}SO_2NH_2$$

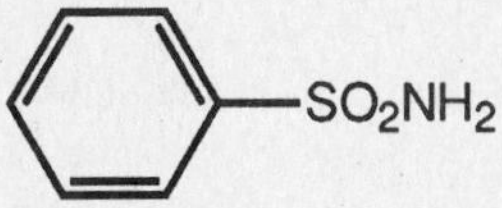

benzenesulfonamide

11.26 a. 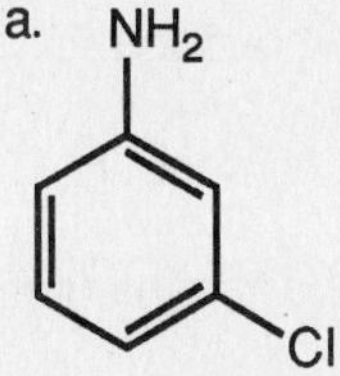 NH_2

Cl

b. $CH_3CHCH_2CH_3$
    $\quad\quad |$
    $\quad\quad NH_2$

c. $CH_3CHCH_2CH_2CH_2CH_3$
    $\quad\quad |$
    $\quad\quad NH_2$

d. $CH_3CH_2CH_2N(CH_3)_2$

e. 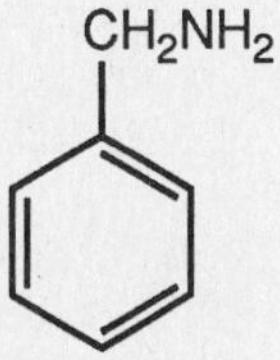 CH_2NH_2

f. $CH_3CH - CH_2$
    $\quad\quad |\quad\quad |$
    $\quad\quad H_2N\quad NH_2$

g. 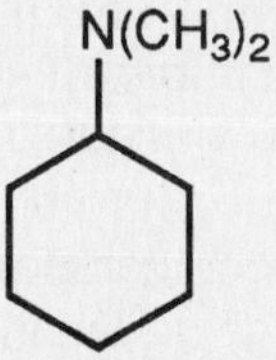 $N(CH_3)_2$

h. $(CH_3CH_2)_4N^+\ Br^-$

i. 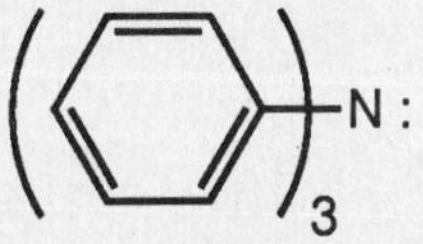 $\left(\underset{}{} \right)_3 \!\!\!- N:$

j. 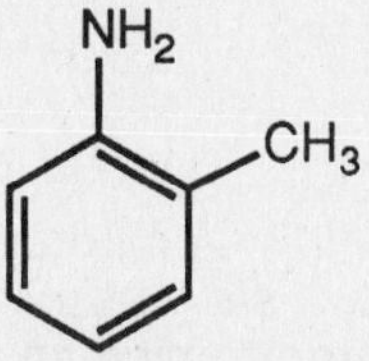 NH_2
    CH_3

k. $(CH_3)_3C - NH_2$

l. $(CH_3CH_2)_2CHN(CH_3)_2$

11.27 a. *p*-bromoaniline

b. methylpropylamine
    (or N-methylpropanamine)

c. diethylmethylamine

d. tetramethylammonium chloride

e. 4-amino-2-butanol

f. 4-aminocyclohexanone

g. *p*-bromobenzene–
    diazonium chloride

h. N-methyl-*p*-toluidine

i. aminocyclopentane
    (or cyclopentylamine)

j. 1,6-diaminohexane

11.28 CH$_3$CH$_2$CH$_2$CH$_2$NH$_2$ butylamine or 1-butanamine (primary)

CH$_3$CH$_2$CH(NH$_2$)CH$_3$ 2-butylamine or 2-butanamine (primary)

(CH$_3$)$_2$CHCH$_2$NH$_2$ 2-methylpropanamine (primary)

(CH$_3$)$_3$CNH$_2$ 2-methyl-2-propanamine (primary)

CH$_3$CH$_2$CH$_2$NHCH$_3$ N-methylpropanamine (secondary)

(CH$_3$)$_2$CHNHCH$_3$ N-methyl-2-propanamine (secondary)

(CH$_3$CH$_2$)NH   diethylamine or N-ethylethanamine (secondary)

(CH$_3$)$_2$NCH$_2$CH$_3$ ethyldimethylamine or N,N-dimethylethanamine (tertiary)

11.29 See the answer to Prob. 11.5. Since intermolecular hydrogen bonding is not possible for trimethylamine, the difference between its boiling point and that of the structurally similar isobutane is relatively small. On the other hand, the possibilities for intermolecular hydrogen bonding in propylamine enhance the difference between its boiling point and that of the structurally similar butane.

The other factor is the polarity of the C$^{\delta+}$—N$^{\delta-}$ bond compared with the C—C bond. Trimethylamine is more polar than isobutane and tends to associate:

This association accounts for the modest (~13°C) difference in the boiling points of these two compounds. Undoubtedly, such association contributes *part* of the boiling point difference between propylamine and butane, the rest of the difference being due to hydrogen bonding.

11.30 1,2-Diaminoethane, with two amino groups, has more possibilities for inter-molecular hydrogen bonding than does propylamine, for example.

$$—N\cdots\cdots H—N—CH_2CH_2—N—H\cdots\cdots N—$$
$$\quad\quad\quad\quad\quad\quad\;\; H\quad\quad\quad\quad H$$

11.31 The boiling point order is:

pentane < methyl *n*-propyl ether < 1-aminobutane < 1-butanol

O—H···O bonds are more effective than N—H···N bonds, which explains the order of the last two compounds. No hydrogen bonding is possible in the first two compounds, but C—O bonds are polar, giving the ether a higher boiling point than the alkane. The actual boiling points are pentane, 36°C; methyl *n*-propyl ether, 39°C; 1-aminobutane, 78°C; 1-butanol, 118°C.

11.32 a. Alkylate aniline twice with ethyl bromide or ethyl iodide.

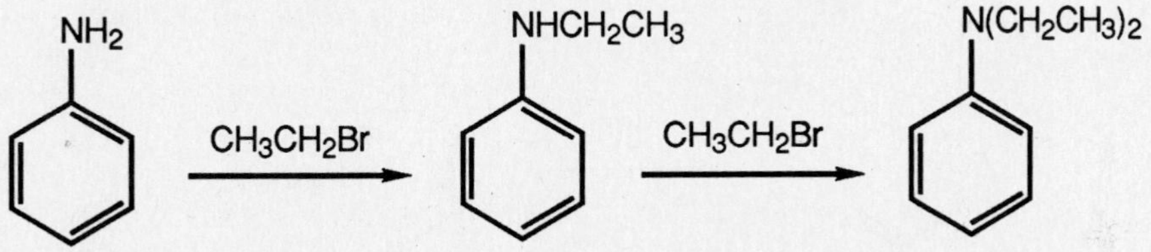

b. First nitrate, then chlorinate, to obtain the *meta* orientation.

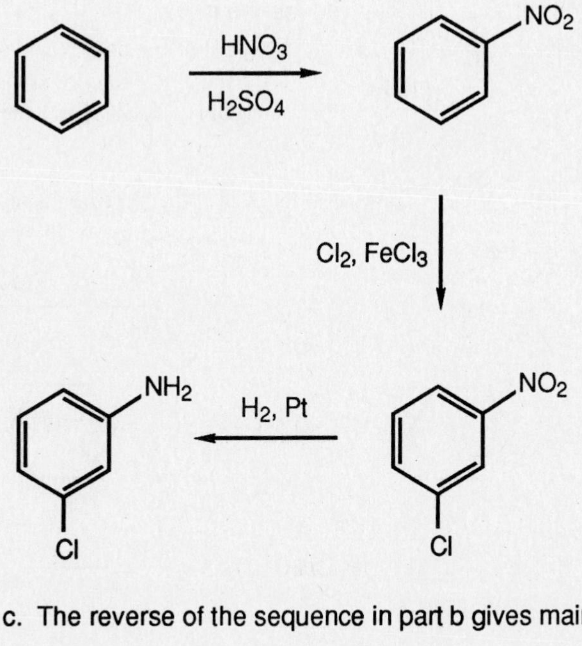

c. The reverse of the sequence in part b gives mainly *para* orientation.

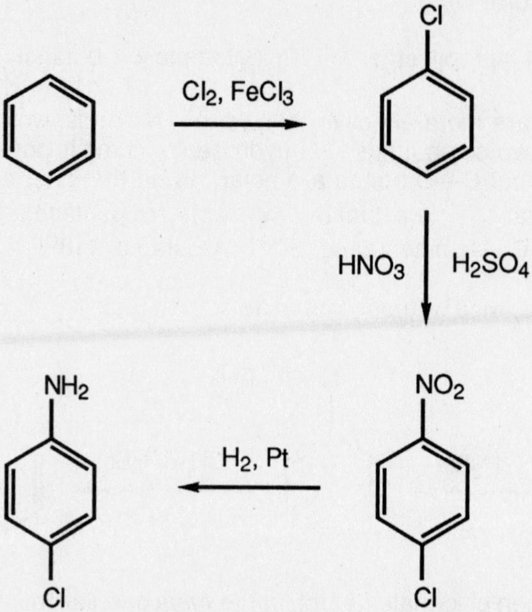

d. Displace the bromide and then reduce the nitrile.

$$CH_3CH_2CH_2CH_2Br \xrightarrow{NaCN} CH_3CH_2CH_2CH_2C{\equiv}N \xrightarrow{LiAlH_4}$$

$$CH_3CH_2CH_2CH_2CH_2NH_2$$

11.33 a.

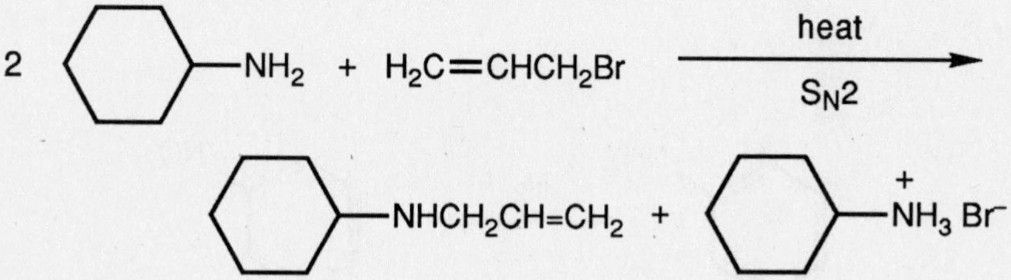

b.

$$CH_3-\overset{\overset{\displaystyle O}{\|}}{C}-NHCH_2CH(CH_3)_2 \quad \xrightarrow{\text{LiAlH}_4} \quad CH_3CH_2NHCH_2CH(CH_3)_2$$

c.

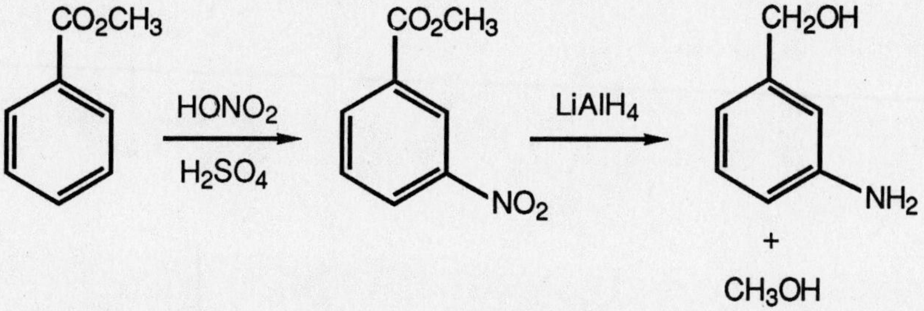

In the first step, the ester group is *meta*-directing. In the second step, both the nitro group and the ester group are reduced when excess LiAlH₄ is used.

d.

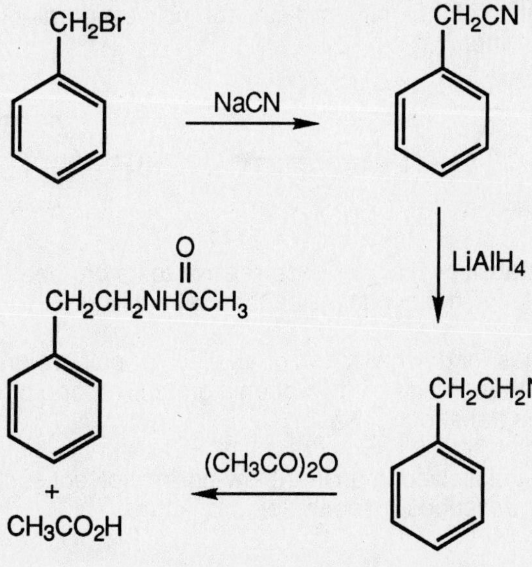

e.

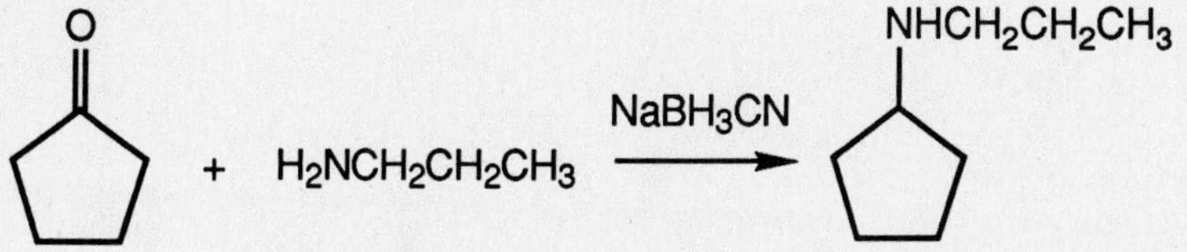

285

11.34

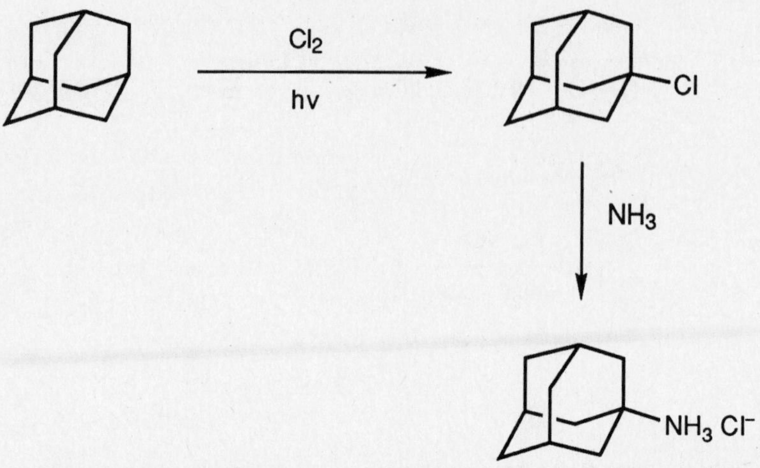

11.35 a. Aniline is the stronger base. The p-cyano group is electron-withdrawing and therefore decreases the basicity of aniline. Note that the possibilities for delocalization of the unshared electron pair are greater in p-cyanoaniline than in aniline.

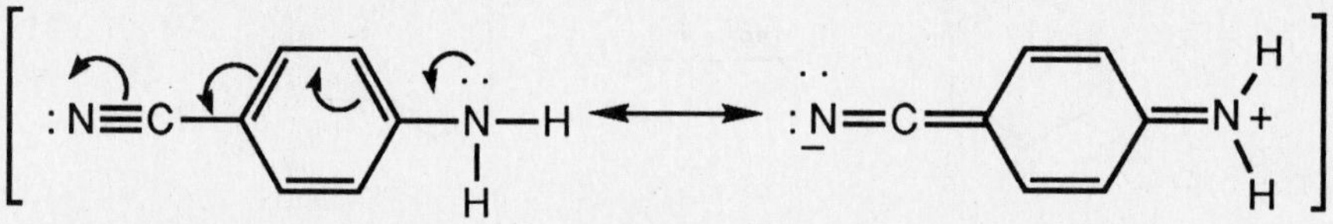

Resonance stabilizes the free base relative to its protonated form, and the effect is greater with p-cyanoaniline than with aniline.

b. The possibilities for delocalization of an electron pair are greater in diphenylamine than in aniline (two phenyl groups vs. one phenyl group). Thus aniline is the stronger base.

11.36 The mixture is first dissolved in an inert, low-boiling solvent such as ether. The following scheme describes a separation procedure:

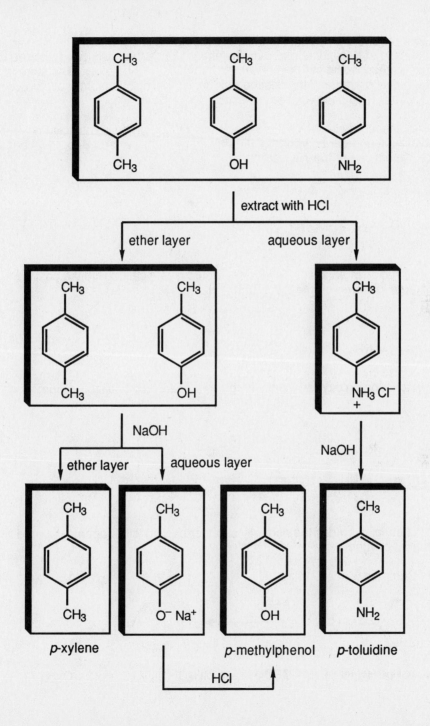

To recover the *p*-xylene, the ether is evaporated and the *p*-xylene distilled. In the case of *p*-toluidine and *p*-methylphenol, once the product is liberated from the corresponding salt, it is extracted from the water by ether. The ether is then evaporated and the desired product is distilled. The order of extraction-- acid first, then base--can be reversed.

11.37 The unshared electron pair on the amino group can be delocalized not only to the *ortho* and *para* carbons of the ring as with aniline (see Sec. 11.7) but also to the oxygen of the nitro group.

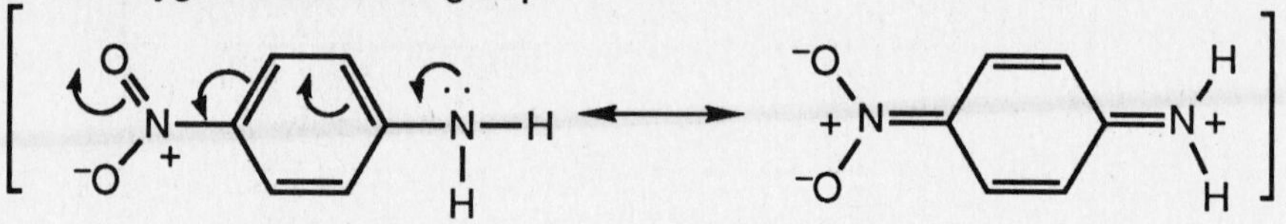

11.38 Protonation of an amide on nitrogen would give a cation with the positive charge localized on the nitrogen:

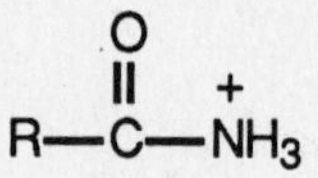

Protonation on oxygen, however, gives a charge-delocalized cation:

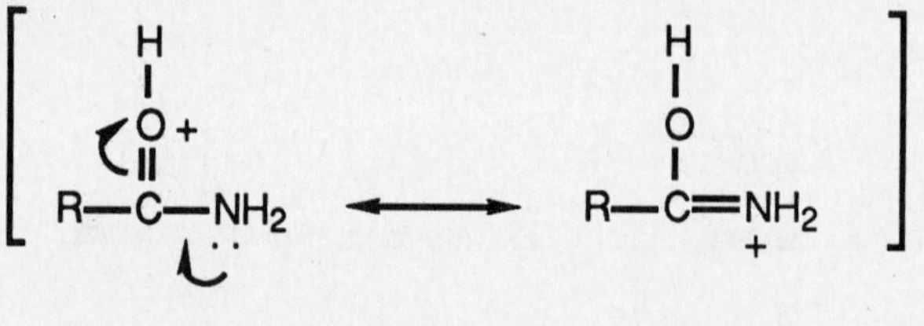

This effect overrides the normally stronger basicity of nitrogen compared to that of oxygen.

11.39 a. Compare with eq. 11.18.

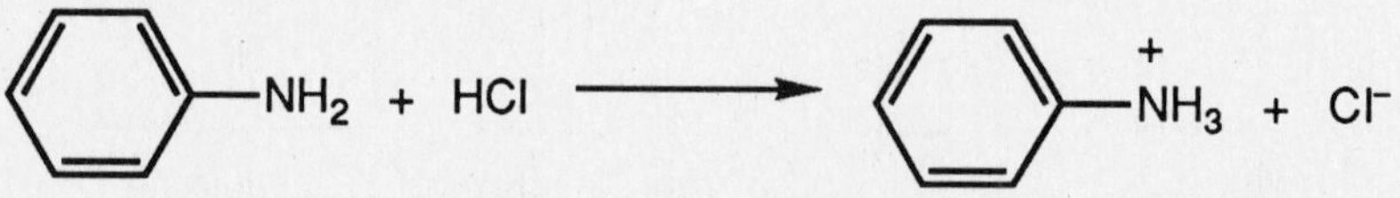

b. $(CH_3CH_2)_3N + H-OSO_3H \rightarrow (CH_3CH_2)_3NH^+ + {}^-OSO_3H$

c. $(CH_3CH_2)_2NH_2{}^+ Cl^- + Na^+OH^- \rightarrow (CH_3CH_2)_2NH + Na^+Cl^- + H_2O$

d.

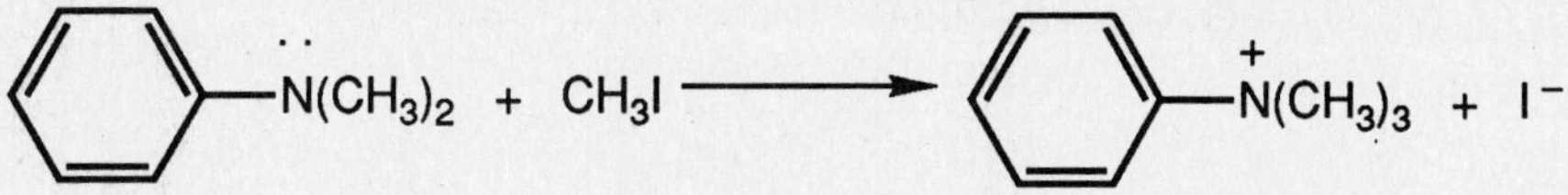

e.

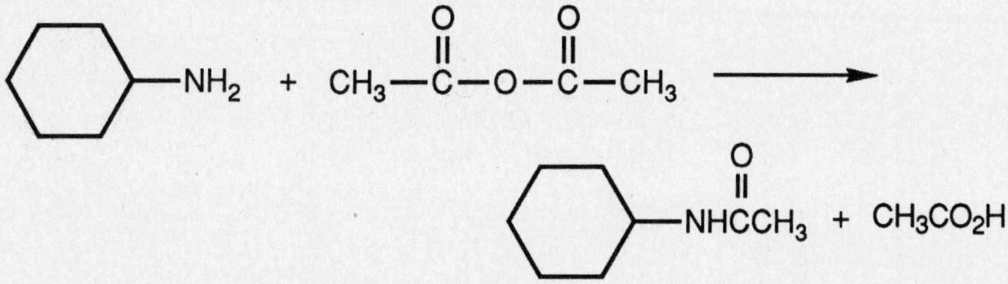

11.40 The reaction begins with nucleophilic attack by the amine on the carbonyl group of the anhydride.

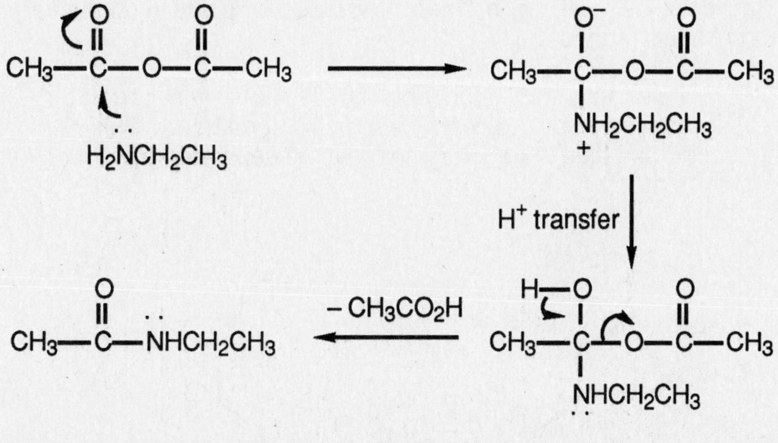

Even though the resulting amide has an unshared electron pair on nitrogen, it does not react with a second mole of acetic anhydride to become diacylated:

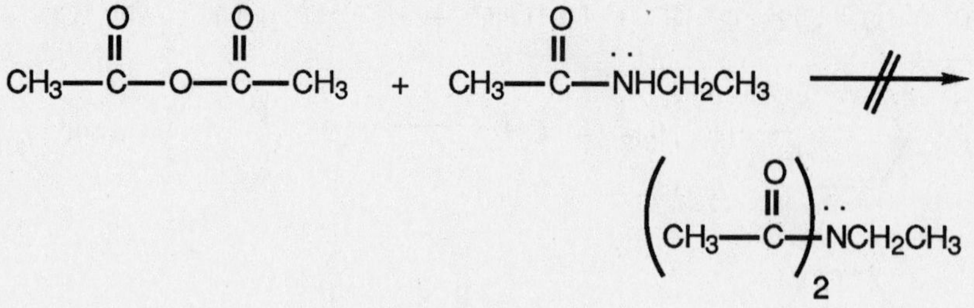

$$CH_3-\overset{\overset{O}{\|}}{C}-O-\overset{\overset{O}{\|}}{C}-CH_3 \quad + \quad CH_3-\overset{\overset{O}{\|}}{C}-\overset{..}{N}HCH_2CH_3 \quad \longrightarrow\!\!\!/\!\!\!\longrightarrow$$

$$\left(CH_3-\overset{\overset{O}{\|}}{C}\right)_2\overset{..}{N}CH_2CH_3$$

The reason is that amides are poor nucleophiles because the unshared electron pair on the nitrogen is delocalized through resonance:

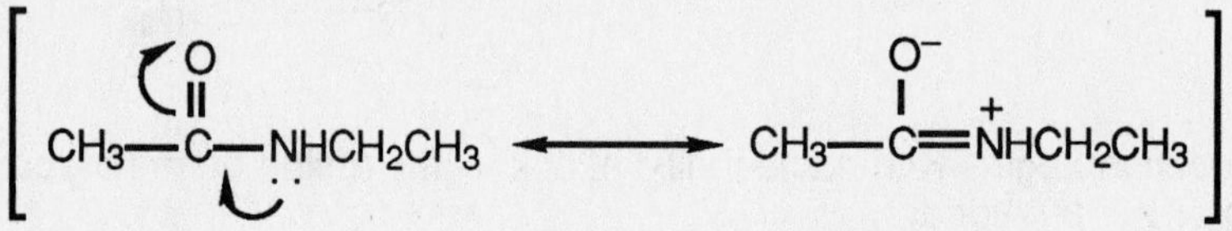

$$\left[\quad CH_3-\overset{\overset{O}{\|}}{C}-NHCH_2CH_3 \quad \longleftrightarrow \quad CH_3-\overset{\overset{O^-}{|}}{C}=\overset{+}{N}HCH_2CH_3 \quad \right]$$

The amide is ineffective with respect to nucleophilic attack on the carbonyl group of acetic anhydride.

11.41 The *R*- and *S*-enantiomers of ammonium salt **A** can interconvert only by processes that involve breaking a carbon–nitrogen bond. This does not occur easily, and thus the enantiomers can be separated by formation of diastereomeric salts.

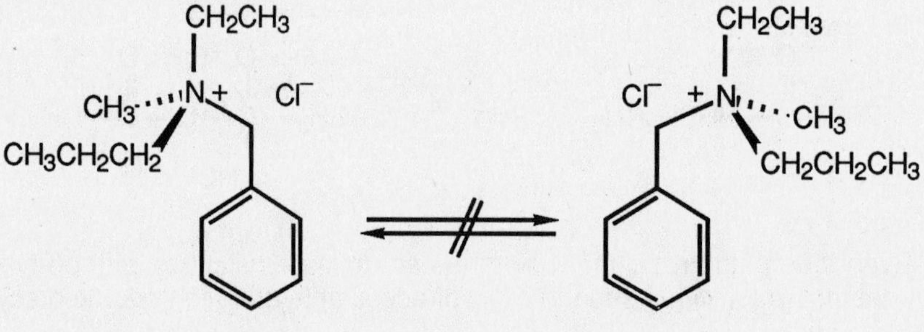

(*R*)-**A**   (*S*)-**A**

The enantiomers of amine **B** can easily interconvert by "inversion" of the nitrogen lone pair (see eq. 11.1) and thus cannot be separated.

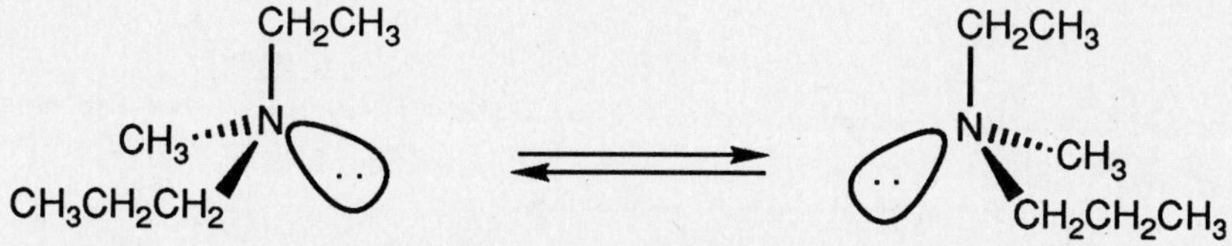

11.42 The priority order is $CH_2C_6H_5 > CH_2CH_2CH_3 > CH_2CH_3 > CH_3$.

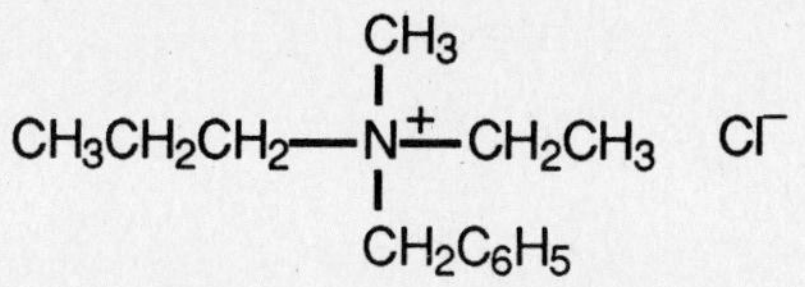

11.43

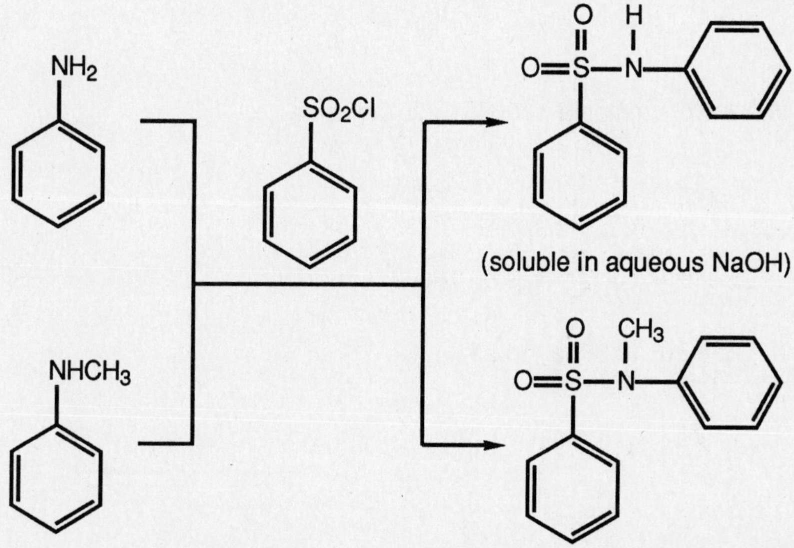

(soluble in aqueous NaOH)

(insoluble in aqueous NaOH)

N,N-Dimethylaniline will not react with benzenesulfonyl chloride.

11.44 a. The amino group of aniline would be protonated by chlorosulfonic acid. This would deactivate the aromatic ring toward electrophiles, and any electrophilic aromatic substitution that might occur would take place *meta* to the amino group.

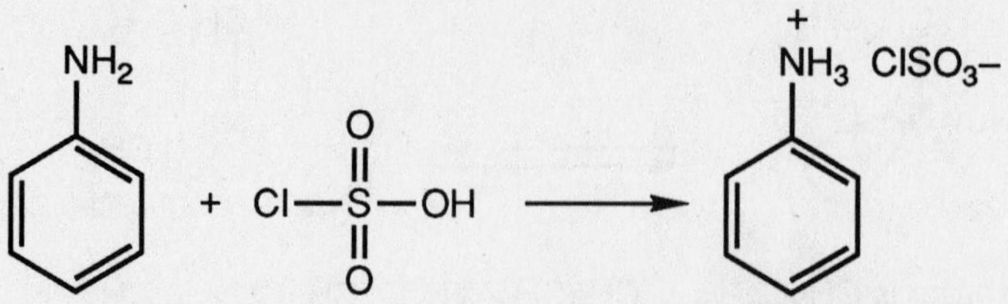

11. AMINES AND RELATED NITROGEN COMPOUNDS

The $-NH_3^+$ group is deactivating and *meta*-directing

b. The use of

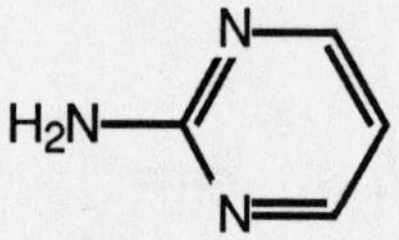

in place of ammonia would give sulfadiazine.

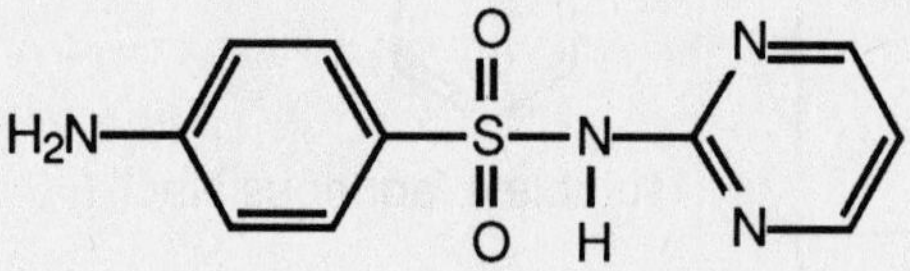

c.   See eq. 10.38 for an analogy.

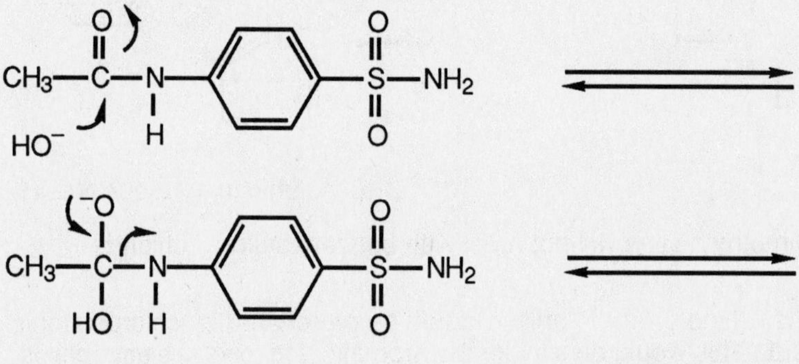

$$HN \text{:} \longrightarrow \text{—} SO_2 \text{—} NH_2 \longrightarrow H_2N \text{—} \longrightarrow SO_2 \text{—} NH_2$$

$$+$$

$$CH_3CO_2H \qquad\qquad\qquad CH_3CO_2^-$$

11.45 The reaction involves an S_N2 ring opening of ethylene oxide.

$$(CH_3)_3N\text{:} \qquad \xrightarrow{\text{base}} \qquad (CH_3)_3\overset{+}{N}CH_2CH_2O^-$$

$$H_2C\text{—}CH_2$$
$$\underset{O}{\diagdown}$$

$$\Big\updownarrow H_2O$$

$$[(CH_3)_3\overset{+}{N}CH_2CH_2OH] \; OH^-$$

choline

11.46

$$CH_3\text{—}\overset{\overset{O}{\|}}{C}\text{—}S\text{—}CoA \quad + \quad [(CH_3)_3\overset{+}{N}CH_2CH_2OH] \; OH^- \longrightarrow$$

$$CH_3\text{—}\overset{\overset{O}{\|}}{C}\text{—}O\text{—}CH_2CH_2\overset{+}{N}(CH_3)_3 \; OH^- \quad + \quad HS\text{—}CoA$$

11.47 Alkyl diazonium salts can lose nitrogen to give 1°, 2°, or 3° carbocations, depending on the nature of the alkyl group.

$$R\text{—}N_2^+ \longrightarrow R^+ \quad + \quad N_2$$

Aryl cations are less stable than 1°, 2°, or 3° carbocations, so aryl diazonium salts are more stable than alkyl diazonium salts.

11.48 These equations illustrate the reactions in Secs. 11.13 and 11.14.

a. $\quad CH_3\text{—}\bigcirc\text{—}N_2^+ \; HSO_4^- \quad + \quad KCN \quad \xrightarrow[\text{(see eq. 11.35)}]{Cu_2(CN)_2}$

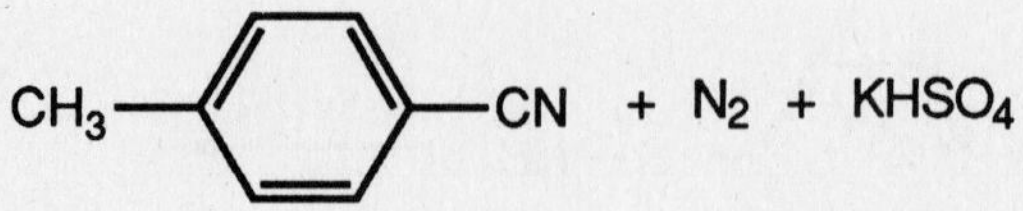

$$CH_3 - \!\!\!\bigcirc\!\!\!- CN \ + \ N_2 \ + \ KHSO_4$$

b. 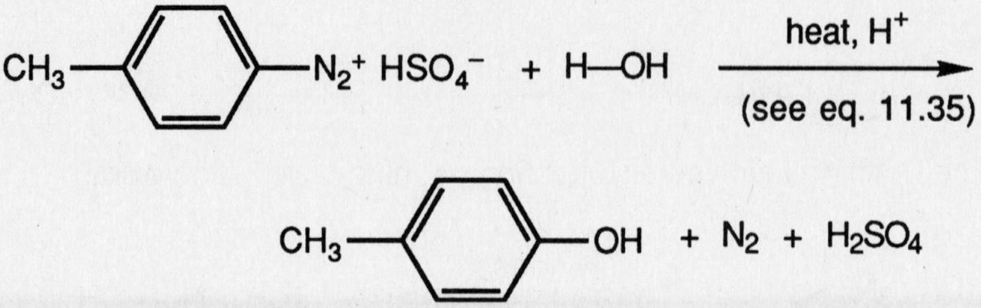 $CH_3 - \!\!\!\bigcirc\!\!\!- N_2^+ \ HSO_4^- \ + \ H\!-\!OH \xrightarrow[\text{(see eq. 11.35)}]{\text{heat, H}^+}$

$$CH_3 - \!\!\!\bigcirc\!\!\!- OH \ + \ N_2 \ + \ H_2SO_4$$

c. 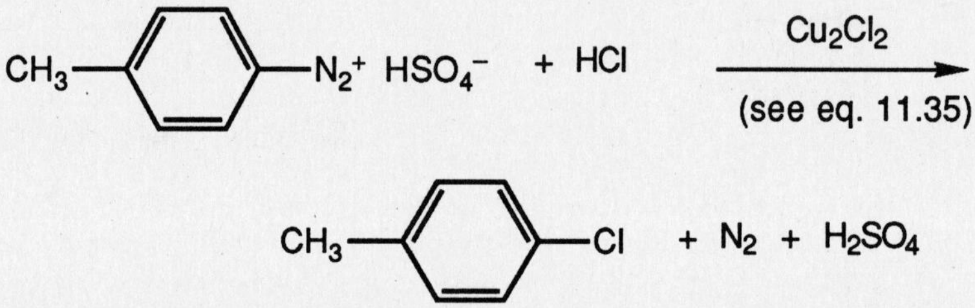 $CH_3 - \!\!\!\bigcirc\!\!\!- N_2^+ \ HSO_4^- \ + \ HCl \xrightarrow[\text{(see eq. 11.35)}]{Cu_2Cl_2}$

$$CH_3 - \!\!\!\bigcirc\!\!\!- Cl \ + \ N_2 \ + \ H_2SO_4$$

d. 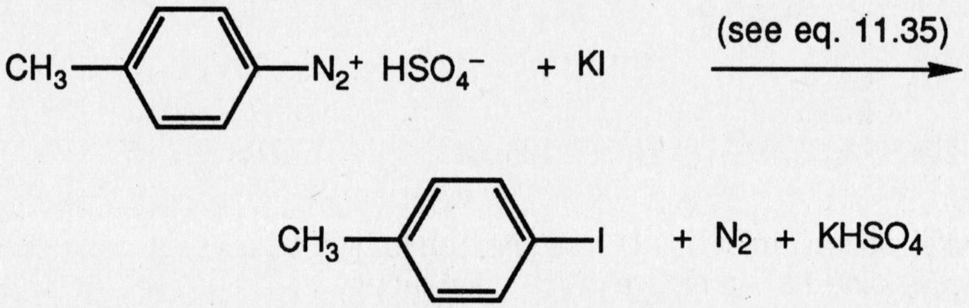 $CH_3 - \!\!\!\bigcirc\!\!\!- N_2^+ \ HSO_4^- \ + \ KI \xrightarrow{\text{(see eq. 11.35)}}$

$$CH_3 - \!\!\!\bigcirc\!\!\!- I \ + \ N_2 \ + \ KHSO_4$$

e. Since *para* coupling is blocked by the methyl substituent, ortho coupling occurs:

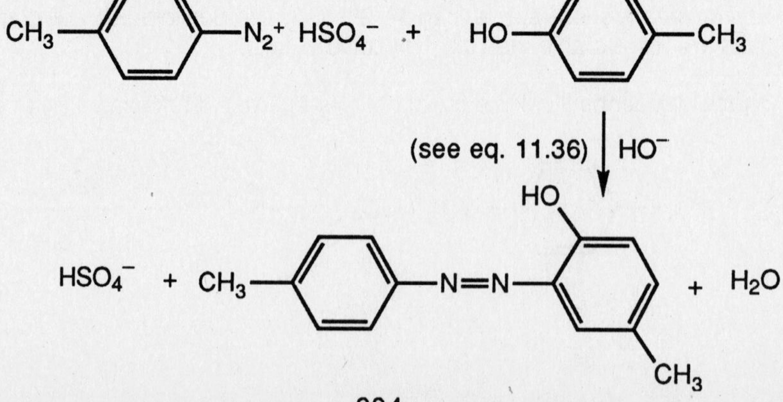

$$CH_3 - \!\!\!\bigcirc\!\!\!- N_2^+ \ HSO_4^- \ + \ HO - \!\!\!\bigcirc\!\!\!- CH_3$$

$$\xrightarrow[\text{(see eq. 11.36)}]{HO^-}$$

$$HSO_4^- \ + \ CH_3 - \!\!\!\bigcirc\!\!\!- N\!=\!N - \!\!\!\bigcirc\!\!\!- + \ H_2O$$

(product: ortho-coupled azo compound with HO and CH3 substituents)

f. CH_3—⟨benzene ring⟩—N_2^+ HSO_4^- + ⟨benzene ring⟩—$N(CH_3)_2$

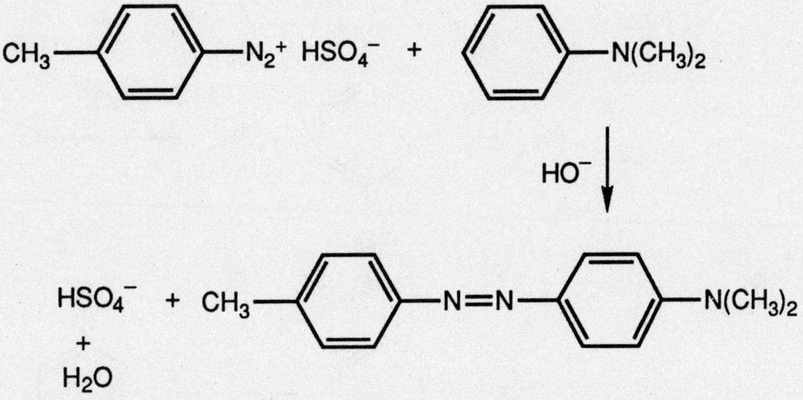

HO^- ↓

HSO_4^- + CH_3—⟨benzene ring⟩—$N=N$—⟨benzene ring⟩—$N(CH_3)_2$
+
H_2O

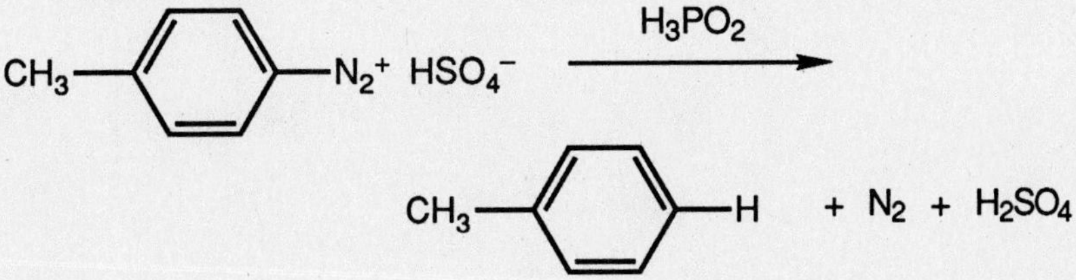

g. CH_3—⟨benzene ring⟩—N_2^+ HSO_4^- $\xrightarrow{H_3PO_2}$

CH_3—⟨benzene ring⟩—H + N_2 + H_2SO_4

h. CH_3—⟨benzene ring⟩—N_2^+ HSO_4^- $\xrightarrow[\text{2. heat}]{\text{1. } HBF_4}$

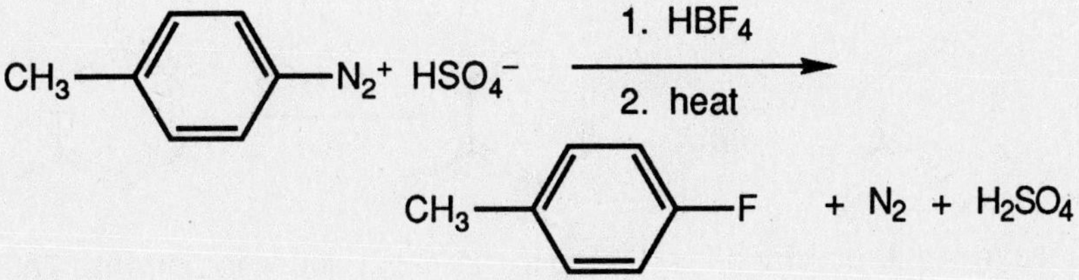

CH_3—⟨benzene ring⟩—F + N_2 + H_2SO_4

11.49 a. ⟨benzene ring with NH_2 top, Br bottom⟩ $\xrightarrow[\text{H}^+, 0°C]{HONO}$ ⟨benzene ring with N_2^+ top, Br bottom⟩ $\xrightarrow[\text{Cu}_2(CN)_2]{KCN}$

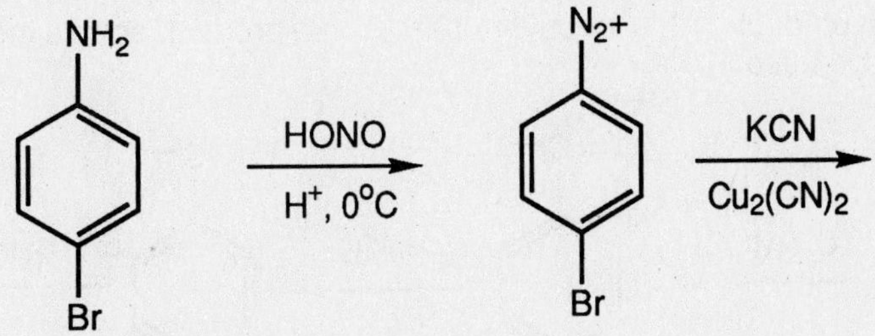

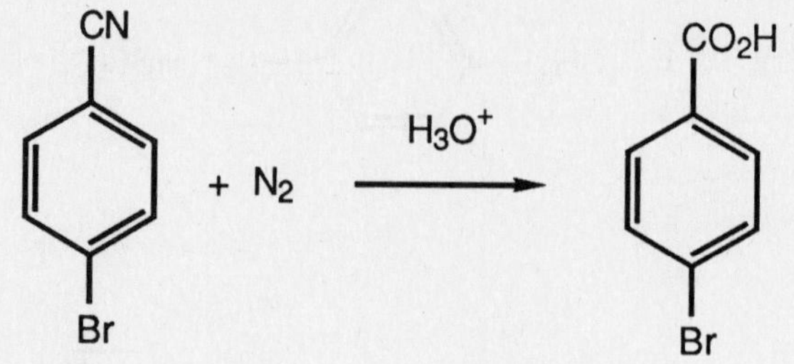

b.

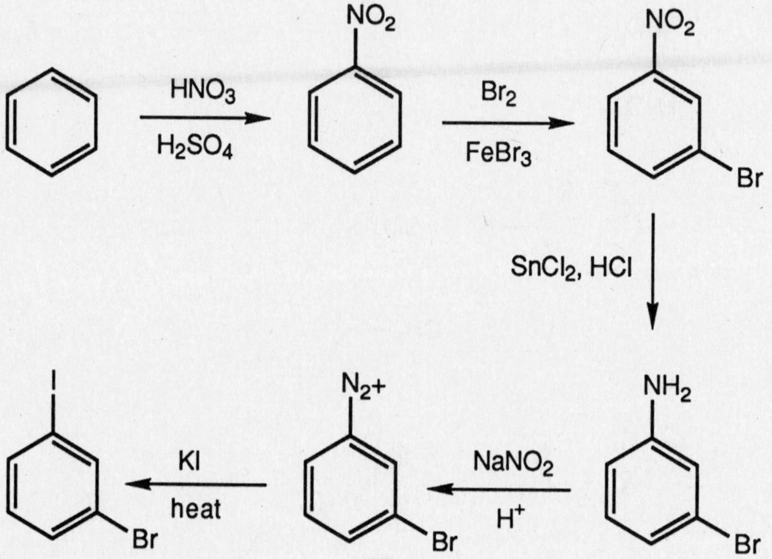

Note that the order of each step in the sequence is important. The benzene must be nitrated first and then brominated to attain *meta* orientation. Bromination of iodobenzene would not give *meta* product, so this indirect route must be used.

c.

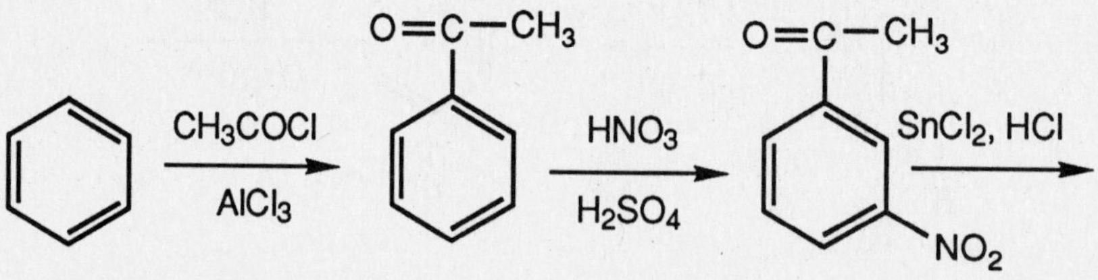

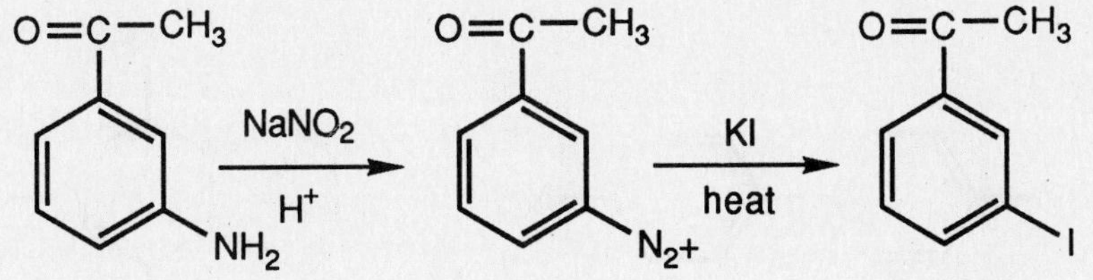

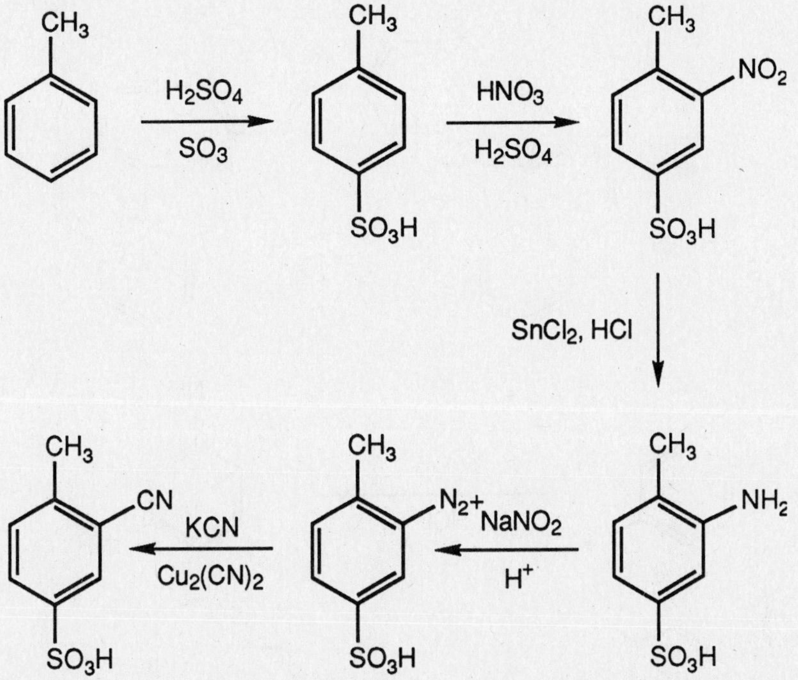

11.50 Benzidine can be diazotized at each amino group. It can then couple with two equivalents of the aminosulfonic acid. Coupling occurs *ortho* to the amino group since the *para* position is blocked by the sulfonic acid group.

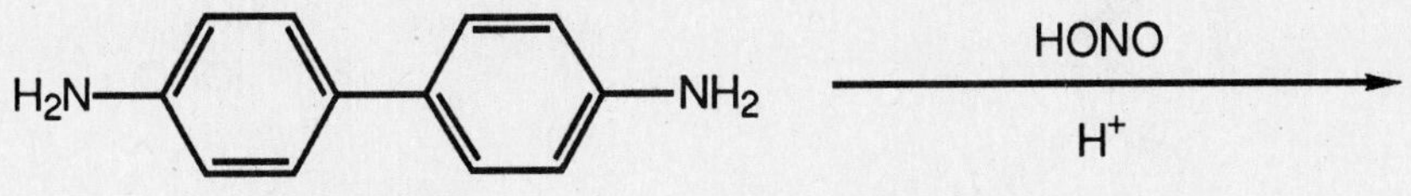

benzidine

11.51

11.52

CHAPTER TWELVE: SPECTROSCOPY AND STRUCTURE DETERMINATION

CHAPTER SUMMARY

Spectroscopic methods provide rapid, nondestructive ways to determine molecular structures. One of the most powerful of these methods is **nuclear magnetic resonance (NMR) spectroscopy,** which involves the excitation of nuclei from lower to higher energy spin states while they are placed between the poles of a powerful magnet. In organic chemistry, the most important nuclei measured are ^{1}H and ^{13}C.

Protons in different chemical environments have different **chemical shifts**, measured in δ (delta) units from the reference peak of tetramethylsilane [TMS, $(CH_3)_4Si$]. **Peak areas** are proportional to the number of protons. Peaks may be split (**spin-spin splitting**) depending on the number of nearby protons. Proton NMR gives at least three types of structural information: (1) the number of signals and their chemical shifts can be used to identify the kinds of chemically different protons in the molecules; (2) peak areas tell how many protons of each kind are present; (3) spin-spin splitting patterns identify the number of near-neighbor protons.

^{13}C NMR spectroscopy can tell how many different "kinds" of carbon atoms are present, and ^{1}H-^{13}C splitting can be used to determine the number of hydrogens on a given carbon.

NMR spectroscopy can be used to study problems of biological and medicinal importance, including analysis of bodily fluids for metabolites and monitoring of muscle tissue function. **Magnetic resonance imaging** (MRI) relies on the ^{1}H NMR spectrum of water and has been used in diagnostic medicine and to address problems in food science, agriculture, and the building industry.

Infrared spectroscopy is mainly used to tell what types of bonds are present in a molecule (using the **functional group region**, 1500-5000 cm^{-1}) and whether two substances are identical or different (using the **fingerprint region**, 700-1500 cm^{-1}).

Visible and ultraviolet spectroscopy employs radiation with wavelengths of 200-800 nm. This radiation corresponds to energies that are associated with **electronic transitions**, in which an electron "jumps" from a filled orbital to a vacant orbital with higher energy. Visible-ultraviolet spectra are most commonly used to

LEARNING OBJECTIVES

detect conjugation. In general, the greater the degree of the conjugation, the longer the wavelength of energy absorbed.

Mass spectra are used to determine molecular weights and molecular composition (from the **parent** or **molecular ion**) and to obtain structural information from the fragmentation of the molecular ion into **daughter ions.**

LEARNING OBJECTIVES

1. Know the meaning of: NMR, applied magnetic field, spin state, chemical shift, δ value, TMS (tetramethylsilane), peak area, spin-spin splitting, $n + 1$ rule, singlet, doublet, triplet, quartet, MRI.

2. Given a structure, tell how many different "kinds" of protons are present.

3. Given a simple structure, use the data in Table 12.2 to predict the appearance of its ^{1}H NMR spectrum.

4. Given a ^{1}H NMR spectrum and other data such as molecular formula, use Table 12.2 to deduce a possible structure.

5. Use spin-spin splitting patterns on a spectrum to help assign a structure.

6. Use ^{1}H and ^{13}C NMR spectra with other data or spectra to deduce a structure.

7. Know the meaning of the functional group region and fingerprint region of an infrared spectrum, and tell what kind of information can be obtained from each.

8. Use the infrared stretching frequencies in Table 12.4 to distinguish between different classes of organic compounds.

9. In connection with visible-ultraviolet spectroscopy, know the meaning of: nanometer, electronic transition, Beer's law, molar absorptivity or extinction coefficient.

10. Know the relationship between conjugation and visible-ultraviolet absorption, and use this relationship to distinguish between closely related structures.

11. Know the meaning of: mass spectrum, m/e ratio, molecular ion, parent ion, fragmentation, daughter ion.

12. Given the molecular formula of a compound, deduce the m/e ratio of the parent ion and predict the relative intensity of the parent + 1 and parent ions.

301

12. SPECTROSCOPY AND STRUCTURE DETERMINATION

13. Given the structure of a simple compound and its mass spectrum, deduce possible structures for the daughter ions.

14. Use all spectroscopic methods in conjunction to deduce a structure.

ANSWERS TO PROBLEMS

Problems Within the Chapter

12.1   All the protons in the structures in parts a and c are equivalent and appear as a sharp, single peak. One way to test whether protons are equivalent is to replace any one of them by some group X. If the same product is obtained regardless of which proton is replaced, then the protons must be equivalent. Try this test with the compounds in parts a and c. Note that diethyl ether, the compound in part b, has two sets of equivalent protons (CH_2 and CH_3).

12.2   a. CH_3OH 3:1

b. $CH_3CO_2CH_3$ 1:1 (or 3:3)

$$\underset{\underset{\displaystyle c.\ CH_3CH_2CCH_2CH_3}{\parallel}}{O}$$ 2:3 (or 4:6)

12.3   All the hydrogens in 1,2-dichloroethane are equivalent, and it will show only one peak in its NMR spectrum. The NMR spectrum of 1,1-dichloroethane will show two peaks in a 3:1 ratio.

12.4   a. CH_3CO_2H

There will be two peaks, at approximately δ 2.1–2.6 and δ 10–13, with relative areas 3:1.

b. $(CH_3)_2C=CH_2$

There will be two peaks, at approximately δ 1.6–1.9 and δ 4.6–5.0, with relative areas 6:2 (or 3:1).

12.5   The compound is

$$\underset{\substack{\delta\ 0.9 \\ 9\ H}}{(CH_3)_3C}-\overset{\overset{\displaystyle O}{\|}}{C}-\underset{\substack{\delta\ 3.6 \\ 3\ H}}{OCH_3}$$

The expected spectrum for its isomer is

$$\underset{\substack{\delta\ 2.1-2.6 \\ 3\ H}}{CH_3}-\overset{\overset{\displaystyle O}{\|}}{C}-\underset{\substack{\delta\ 0.9 \\ 9\ H}}{OC(CH_3)_3}$$

The only difference is the chemical shift of the methyl protons.

12.6   The greater the electronegativity of the atom attached to the methyl group (the more electron-withdrawing the atom), the further downfield the chemical shift.

CH_3-H	CH_3-I	CH_3-Br	CH_3-Cl
$\delta\ 0.23$	$\delta\ 2.16$	$\delta\ 2.68$	$\delta\ 3.05$
$E_N(H) = 2.1$	$E_N(I) = 2.5$	$E_N(Br) = 2.8$	$E_N(Cl) = 3.0$

12.7   One would expect peaks at approximately δ 1.0 (for the methyl groups) and δ 5.2–5.7 (for the vinylic protons), with relative areas of 18:2 (or 9:1). The experimental values are actually δ 0.97 (18 H) and δ 5.30 (2 H).

$$\underset{(CH_3)_3C}{\overset{H}{\diagdown}}C=C\underset{H}{\overset{C(CH_3)_3}{\diagup}}$$

12.8 $C\underline{H}_3CHCl_2$ doublet, δ 0.85–0.95, area = 3

       $CH_3C\underline{H}Cl_2$ quartet, δ 5.8–5.9, area = 1

12.9   a. $BrC\underline{H}_2CH_2Cl$ triplet, δ 3.4–3.6, area = 2

                $BrCH_2C\underline{H}_2Cl$ triplet, δ 3.6–3.8, area = 2

                Actually, the spectrum will be more complex than this because the chemical shifts of the two types of protons are close in value.

       b. $ClCH_2CH_2Cl$. The protons are all equivalent, and the spectrum will consist of a sharp singlet at δ 3.6–3.8.

12.10 $CH_3CH_2CH_2OH$

       The spectrum *without* ^1H coupling will appear as three peaks:

       C-1 $\delta \sim 65$
       C-2 $\delta \sim 32$
       C-3 $\delta \sim 11$

       In the ^1H-coupled spectra, the peaks for C-1 and C-2 would be triplets, and the C-3 peak would be a quartet.

12.11   a. One; all the carbons are equivalent.
       b. Two; the methyl carbons are equivalent.
       c. Three; the methyl carbons are equivalent.
       d. Four; the molecule has an internal plane of symmetry passing through carbon-2 and the C(4)-C(5) bond.

12.12   The IR spectrum of 1-hexyne will show a band in the $C\equiv C$ stretching frequency region (2100-2260 cm^{-1}) and in the $\equiv C$-H stretching frequency region (3200-3350 cm^{-1}). These bands will not appear in the IR spectrum of 1,3-hexadiene, which will show bands in the C=C (1600-1680 cm^{-1}) and =C-H (3030-3140 cm^{-1}) regions.

12.13   We use Beer's law: $A = \varepsilon\,c\,l$
       Rearranging, we get:

       $c = A\,/\,\varepsilon l = 2.2\,/\,12{,}600 \times 1 = 1.75 \times 10^{-4}$ mol/L

       Note that ultraviolet spectra are often obtained on very dilute solutions.

12.14   Conjugation is possible between the two rings in biphenyl, but in diphenylmethane the —CH_2— group interrupts this conjugation. Thus for comparable electronic transitions, biphenyl is expected to absorb at longer wavelengths.

12.15 Azulene. The blue color indicates that azulene absorbs light in the visible region of the spectrum (400–800 nm). Naphthalene undergoes electronic transitions in the ultraviolet region of the spectrum (314 nm). The longer the wavelength (nm), the lower the energy of the π-electronic transition.

12.16 Alkanes have the molecular formula C_nH_{2n+2}. Therefore, for any alkane

n(atomic wt. of C) + (2n+2) (atomic wt. of H) = m/e ratio of parent ion (M$^{+\cdot}$).

In this case,

n(12) + (2n+2)(1) = 142

Rearranging,

14n = 140   or n = 10

and the molecular formula of the alkane is $C_{10}H_{22}$. The intensity of the peak at m/e 143 should be (1.1)(10) or 11% of the intensity of the peak at m/e 142.

12.17 Using the approach described in Prob. 12.16, the molecular formula of the compound is C_3H_3Cl. The molecular formula indicates that the compound has two π-bonds or one π-bond and one ring. Possible structures are:

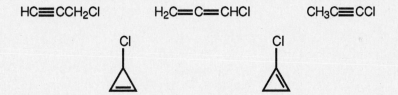

HC≡CCH$_2$Cl   H$_2$C=C=CHCl   CH$_3$C≡CCl

12.18 The molecular ion peak ($C_7H_{14}O^+$) will appear at m/e = 114 (14 mass units, or one —CH$_2$— group less than for 4-octanone). Since the ketone is symmetrical, it should fragment to yield a $C_3H_7CO^+$ peak (m/e = 71) and a $C_3H_7^+$ peak (m/e = 43). Only one set of daughter ions (instead of the two seen in Figure 12.10) will be observed.

Additional Problems

12.19 Possible solutions are:

a.

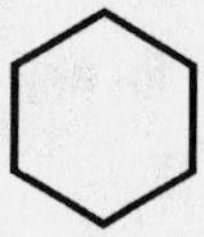

b.

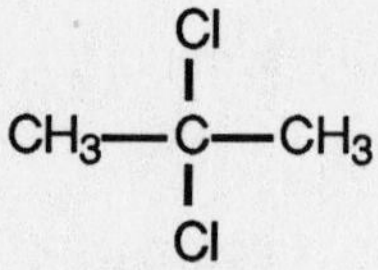

$$CH_3-\underset{\underset{Cl}{|}}{\overset{\overset{Cl}{|}}{C}}-CH_3$$

c. $CH_3C\equiv CCH_3$

d.

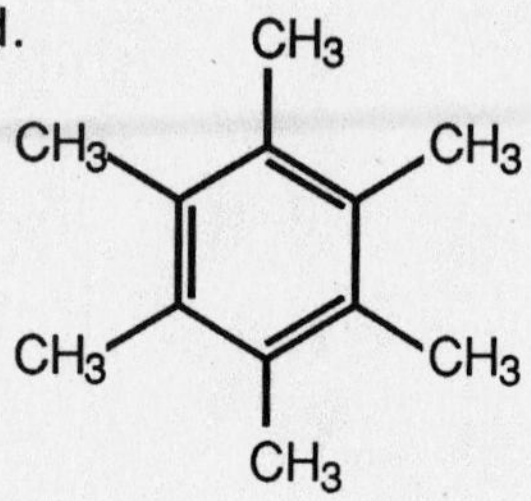

e. CH_3OCH_3

f.

$$CH_3-\underset{\underset{CH_3}{|}}{\overset{\overset{CH_3}{|}}{C}}-CH_3$$

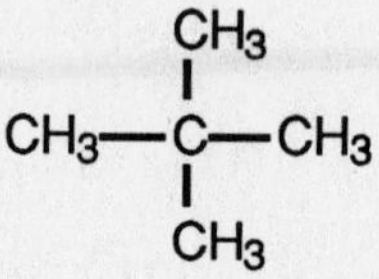

12.20 a. There are four different types of protons:

$$\underset{a}{CH_3}\diagdown\underset{\underset{a}{CH_3}\diagup}{CH}-\overset{c}{CH_2}-\overset{d}{CH_3}$$

b

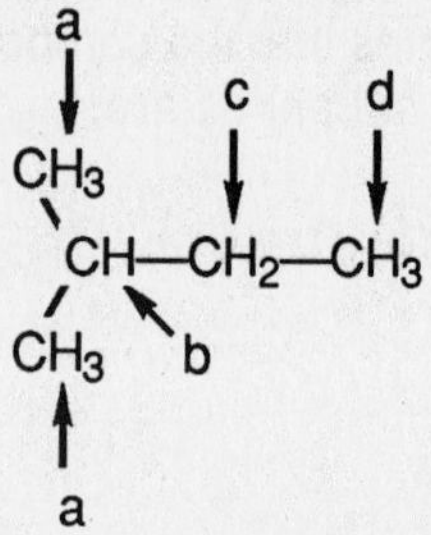

b. There are three different types of protons:

$$\underset{a}{CH_3}\diagdown\underset{\underset{a}{CH_3}\diagup}{N}-\overset{b}{CH_2}-\overset{c}{CH_3}$$

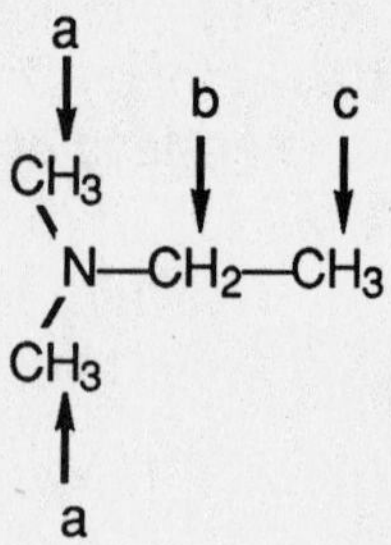

c. There are three different types of protons:

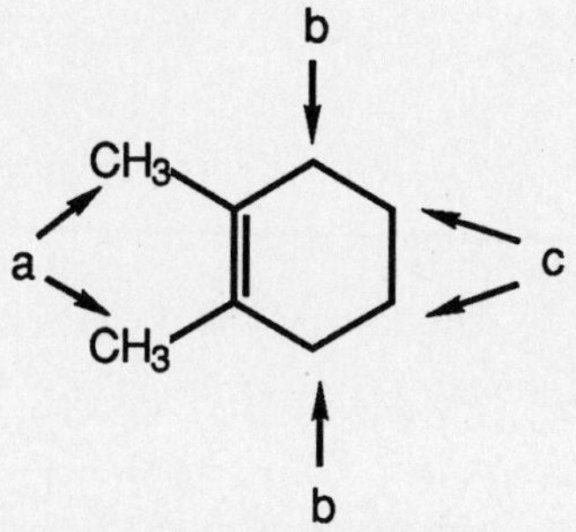

d. There are three different types of protons:

$$\underset{a}{CH_3} - \underset{b}{CH_2} - \underset{c}{OH}$$

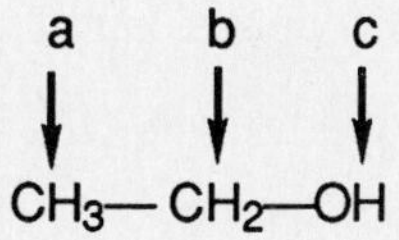

12.21 The first compound must have nine equivalent hydrogens. The only possible structure is *t*-butyl bromide:

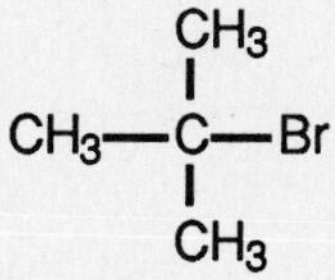

$$CH_3 - \underset{\underset{CH_3}{|}}{\overset{\overset{CH_3}{|}}{C}} - Br$$

Its isomer has three different types of hydrogens, two of one kind, one unique, and six equivalent. The compound must be isobutyl bromide. The chemical shifts and spin-spin splitting pattern fit this structure:

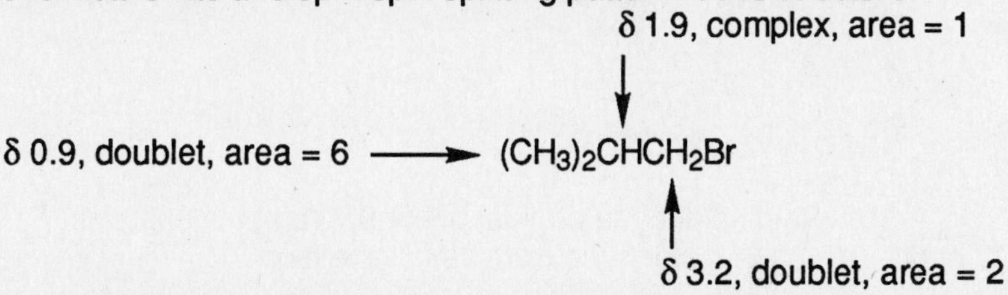

δ 1.9, complex, area = 1

δ 0.9, doublet, area = 6 ⟶ (CH₃)₂CHCH₂Br

δ 3.2, doublet, area = 2

(It is notable that a singlet or doublet that integrates for 6 H is almost always two CH₃'s rather than three CH₂'s)

12.22 The chemical shifts suggest that carbon is more electronegative (more electron-withdrawing) than silicon. Electronegativity generally decreases as one goes down a column in the periodic table. The Pauling electronegativities of carbon and silicon are 2.5 and 1.8, respectively.

12.23 a. CH_3CCl_3 All the protons will appear as a singlet.

$CH_2ClCHCl_2$ There will be two sets of peaks, with an area ratio of 2:1. The former will be a doublet, the latter a triplet.

b. $CH_3CH_2CH_2OH$ In addition to the O—H peak, there will be three sets of proton peaks, with area ratio 3:2:2.

$(CH_3)_2CHOH$ In addition to the O—H peak, there will be only two sets of proton peaks, with area ratio 6:1.

c.

$$\underset{\displaystyle CH_3CH_2-\overset{\textstyle O}{\overset{\|}{C}}-OCH_3}{}$$

The spectrum will show a singlet at about δ 3.5–3.8 for the O—CH$_3$ protons and a quartet and triplet near δ 2.6 and 0.95, respectively.

$$\underset{\displaystyle CH_3-\overset{\textstyle O}{\overset{\|}{C}}-OCH_2CH_3}{}$$

The spectrum will show a singlet at about δ 2.1–2.6 for the CH$_3$C(O)– and a quartet and triplet near δ 3.8 and 0.95, respectively.

d.

A one-proton aldehyde peak at δ 9.5–9.7 (a triplet) will easily distinguish this aldehyde from its ketone isomer.

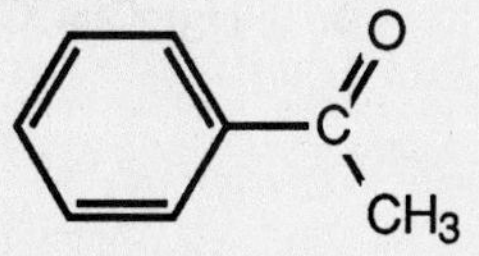

The aliphatic protons will give a sharp three-proton singlet at about δ 2.1–2.6.

12.24 At very low temperatures, the interconversion of one chair form to another can be frozen out. If this happens, there will be two sets of protons, equatorial and axial, with different chemical shifts. At room temperature, the interconversion is so fast that the spectrum shows only one signal, an average between the two types.

12.25 The peaks are assigned as follows:

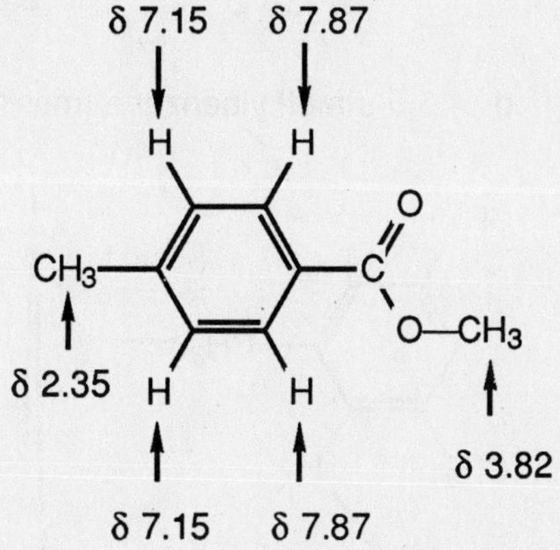

In general, aromatic protons show up at lower field strengths if they are adjacent to electron-withdrawing substitutents. This is the basis for distinguishing between the two sets of aromatic protons.

12.26 a. CH_3CHO b. $(CH_3)_2CHOCH(CH_3)_2$

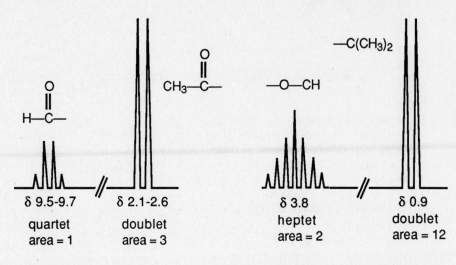

c. $CH_3CH=CCl_2$ d. 1,3,5-trimethylbenzene (mesitylene)

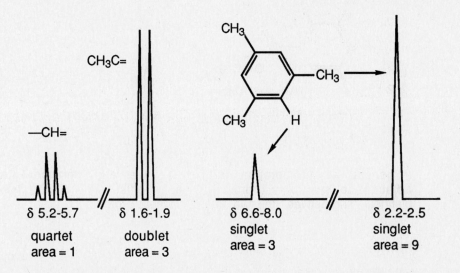

12.27 Replacement of the five chlorines with hydrogens gives C_3H_8. Thus there are no double bonds or rings. To split each other, the protons must be on adjacent carbon atoms. Two possibilities are:

$ClCH_2$—$CHCl$—CCl_3 and Cl_2CH—CH_2—CCl_3

However, neither of these alternatives fits the chemical shift data. For the first structure, the two-proton signal should be at δ 3.5 or a little lower but not at δ 6.0. For the second structure, the CH_2 protons should be at higher field than the CH proton, whereas the opposite is true. We must therefore seek another possibility, and this one comes to mind:

δ 4.5, triplet

$$Cl_2CH-\overset{\overset{\text{H}}{|}}{\underset{\underset{\text{Cl}}{|}}{C}}-CHCl_2$$

δ 6.0, doublet

The two protons on C-1 and C-3 are equivalent and appear at low field because of the two chlorines.

12.28 There are three possibilities. The number of different types of carbons is shown under each structure.

9   9   7

The compound must be the *para* isomer. For its proton NMR spectrum, see the answer to Prob. 12.25.

12.29 No. (R)-2-butanol and (S)-2-butanol are enantiomers. They rotate plane-polarized light in opposite directions and interact differently with chiral substances but exhibit identical ^1H and ^{13}C NMR spectra when taken in achiral solvents such as carbon tetrachloride or chloroform.

12.30 Yes. *Meso-* and (2R, 3R)-2,3-butanediol are diastereomers and have different physical properties, including ^1H and ^{13}C NMR spectra.

12.31 The absence of a band at 3500 cm^{-1} indicates that there is no hydroxyl group. The absence of a band at 1720 cm^{-1} indicates that the compound is not an aldehyde or ketone. This suggests that the oxygen function is probably an ether. Possible structures are:

$$CH_3OCH\!=\!CH_2 \qquad \begin{matrix} H_2C\!-\!O \\ |\quad\ | \\ H_2C\!-\!CH_2 \end{matrix} \qquad \begin{matrix} O \\ / \ \backslash \\ H_2C\!-\!CH\!-\!CH_3 \end{matrix}$$

An NMR spectrum would readily distinguish between these possibilities. The presence of a band at about 1650 cm^{-1} in the IR spectrum (C=C stretch) would indicate that the compound is the vinyl ether (the first of the three compounds) and not the cyclic ethers.

12.32 The lower the wave number (cm^{-1}), the easier it is to stretch the bond.

$$\begin{matrix} | \\ -\!C\!-\!H \\ | \end{matrix} \qquad\qquad =\!C\!\!\begin{matrix}H\\ \\ \backslash\end{matrix} \qquad\qquad \equiv\!C\!-\!H$$

$$2850\text{–}3000 \text{ cm}^{-1} \qquad 3030\text{–}3140 \text{ cm}^{-1} \qquad 3200\text{–}3350 \text{ cm}^{-1}$$

$$C(sp^3)\!-\!H \qquad\qquad C(sp^2)\!-\!H \qquad\qquad C(sp)\!-\!H$$

The longer the bond, the weaker the bond, and the easier it will be to stretch the bond. C—H bond lengths increase as one goes from sp-hybridized carbon to sp^3-hybridized carbon, consistent with the observed trend in stretching frequency.

12.33 These infrared data provide direct evidence for hydrogen bonding in alcohols. In dilute solution, the alcohol molecules are isolated, being surrounded by inert solvent molecules. The sharp band at 3580 cm^{-1} is caused by the O—H stretching frequency in an isolated ethanol molecule. As the concentration of ethanol is increased, alcohol molecules can come in contact with one another and form hydrogen bonds. Hydrogen-bonded O—H is less "tight" than an isolated O—H and has a variable length (as the proton is transferred back and forth between oxygen atoms). Consequently, hydrogen-bonded O—H absorbs at a lower frequency and with a broader range (3250–3350 cm^{-1}) than the isolated O—H group.

12.34 Refer to Table 12.4 when answering this problem.

a. The first compound, a ketone, will have an intense band in the 1650-1780 cm^{-1} region. The ether will not absorb in this region.

b. The distinguishing feature is the C=C bond in the second com̱ ̱und which should absorb in the 1600-1680 cm^{-1} region.

c. The second compound will show an O—H stretching band near 3500 cm^{-1}.

d. Both compounds will show C=O stretching bands, but only the acid will also show a broad O—H band in the 2500-3000 cm^{-1} region.

12.35 Both compounds will be similar in the functional group region of the spectrum with bands at 3500 cm^{-1} for the O—H stretch and 1700 cm^{-1} for the C=O stretch. But their fingerprint regions (1500-700 cm^{-1}) are expected to differ from one another.

12.36 a. The first compound, a ketone, would show a carbonyl stretching band at about 1720 cm^{-1}. The alcohol would not show this band and would show bands for the hydroxyl and alkene groups at 3200-3700 cm^{-1} and 1600-1680 cm^{-1}, respectively.

b. The first compound, an aldehyde, would show a carbonyl stretching band at about 1720 cm^{-1}. This band would be absent in the IR spectrum of the vinyl ether, which would show a C=C stretching band at 1600-1680 cm^{-1}.

c. The secondary amine [$(CH_3CH_2)_2NH$] will show an N-H stretching band at 3200-3600 cm^{-1}. The tertiary amine (triethylamine) has no N-H bonds and this band will be absent.

12.37 The band at 1725 cm^{-1} in the infrared spectrum is due to a carbonyl group, probably a ketone. The quartet-triplet pattern in the NMR spectrum suggests an ethyl group. The compound is 3-pentanone:

δ 2.7, quartet, area = 2 (or 4)

$$CH_3—CH_2—\overset{\overset{\textstyle O}{\|}}{C}—CH_2—CH_3$$

δ 0.9, triplet, area = 3 (or 6)

12.38 The quartet-triplet pattern suggests that the ten protons are present as two ethyl groups. This gives a partial structure of $(CH_3CH_2)_2CO_3$. The chemical shift of the CH_2 groups (δ 4.15) suggests that they are attached to the oxygen atoms. Finally, the infrared band at 1745 cm^{-1} suggests a carbonyl function. The structure is diethyl carbonate:

$$CH_3-CH_2-O-\overset{\overset{\displaystyle O}{\|}}{C}-O-CH_2-CH_3$$

12.39 You could monitor disappearance of the O—H stretching band in cyclohexanol and the appearance of the carbonyl stretching band in cyclohexanone to follow the progress of the reaction and determine the purity of the product.

12.40 Compounds a, c, and e have no unsaturation and will not absorb in the ultraviolet region of the spectrum.

12.41 See Sec. 12.6 in the text. As the number of double bonds increases, so does the extent of conjugation. Thus the absorption maximum moves to longer and longer wavelengths.

12.42 In *trans*-1,2-diphenylethylene, both phenyl groups can be conjugated with the double bond, and all π-bonds can lie in one plane. In *cis*-1,2-diphenylethylene, only one phenyl group at a time can be conjugated with the double bond for steric reasons. The more extended conjugation in the *trans* compound leads to a lower energy (longer wavelength) λ_{max} than in the *cis* compound.

λ_{max} = 295 nm

λ_{max} = 280 nm

ANSWERS TO PROBLEMS

12.43 We must use Beer's law (eq. 12.5) to solve this problem.

$A = \varepsilon c l,$ or $c = A / \varepsilon l$

$c = 0.43 / 215 \times 1 = 2 \times 10^{-3}$ mol/L

To go further, 1 L cyclohexane will contain $2 \times 10^{-3} \times 78 = 0.156$ g benzene as a contaminant. As you can see, ultraviolet spectroscopy can be a very sensitive tool for detecting impurities.

12.44 Follow the example in eq. 12.7.

$[CH_3CH_2\overset{\displaystyle .}{\underset{\displaystyle ..}{O}}\text{---}H]^+$

12.45 The molecular ion of 1-pentanol will be:

$[CH_3CH_2CH_2CH_2CH_2\overset{\displaystyle ..}{O}H]^+$

Fragmentation between C-1 and C-2 would give a daughter ion with $m/e = 31$ (compare with eq. 12.8). A possible mechanism for this cleavage is:

$m/e = 31$

12.46 The formula $C_5H_{12}O = C_5H_{11}OH$ tells us that the alcohols are saturated. The peak at $m/e = 59$ cannot be caused by a four-carbon fragment ($C_4 = 4 \times 12 = 48$; this would require 11 hydrogens, too many for four carbons). Thus the peak must contain one oxygen, leaving $59-16 = 43$ for carbon and hydrogen. A satisfactory composition for the peak at 59 is $C_3H_7O^+$ (or $C_3H_6OH^+$). Similarly, the $m/e = 45$ peak corresponds to $C_2H_5O^+$ (or $C_2H_4OH^+$). Fragmentation of alcohols often occurs between the hydroxyl-bearing carbon and an attached hydrogen or carbon (eq. 12.8).

Possible structures for the first alcohol are

$$CH_3CH_2 \overset{\overset{\displaystyle H}{|}}{\underset{\underset{\displaystyle CH_2CH_3}{|}}{C}} OH \qquad\qquad CH_3 \overset{\overset{\displaystyle CH_3}{|}}{\underset{\underset{\displaystyle CH_2CH_3}{|}}{C}} OH$$

3-pentanol   2-methyl-2-butanol

Possible structures of its isomer are

$$CH_3 \overset{\overset{\displaystyle H}{|}}{\underset{\underset{\displaystyle CH_2CH_2CH_3}{|}}{C}} OH \qquad\qquad CH_3 \overset{\overset{\displaystyle H}{|}}{\underset{\underset{\displaystyle CH(CH_3)_2}{|}}{C}} OH$$

2-pentanol   3-methyl-2-butanol

The correct structure for the first isomer could be deduced by NMR spectroscopy as follows:

$$CH_3 - CH_2 - \overset{\overset{\displaystyle H}{|}}{\underset{\underset{\displaystyle OH}{|}}{C}} - CH_2 - CH_3$$

H ← δ 4, 1 H, quintet

δ 1, 6 H, triplet   δ 1.3, 4 H, quintet or multiplet

three ^{13}C peaks

$$CH_3 - \overset{\overset{\displaystyle CH_3}{|}}{\underset{\underset{\displaystyle OH}{|}}{C}} - CH_2 - CH_3$$

δ 1.3, 2H, quartet

δ 1, 6 H, singlet   δ 1, 3 H, triplet

four ^{13}C peaks

The correct structure for the second isomer also could be deduced by NMR spectroscopy. However, both proton spectra are likely to be quite complex because of similar chemical shifts and a great deal of spin-spin splitting. The ^{13}C spectra are simpler and diagnostic:

$$CH_3CHCH_2CH_2CH_3$$
$$|$$
$$OH$$

six types of protons

five ^{13}C peaks

$$CH_3CHCH(CH_3)_2$$
$$|$$
$$OH$$

five types of protons

four ^{13}C peaks

12.47 The NMR peak at δ 7.4 with an area of 5 suggests that the compound may have a phenyl group, C_6H_5—. If so, this accounts for 77 of the 102 mass units. This leaves only 25 mass units, one of which must be a hydrogen (for the NMR peak at δ 3.08). The other 24 units must be two carbon atoms, since the compound is a hydrocarbon (no other elements present except C and H). Phenylacetylene fits all the data:

12.48 Bromine consists of a 50:50 mixture of ^{79}Br and ^{81}Br. Therefore mono-bromides show two parent ions of equal intensity (see Example 12.8). Dibromides will show three parent ion peaks with a relative intensity of 1:2:1. The peak at m/e 198 is due to a dibromide that contains two ^{79}Br's. This leaves 40 mass units to be divided between carbon and hydrogen, and C_3H_4 is most reasonable. Therefore the molecular formula of the compound is $C_3H_4Br_2$.

CHAPTER THIRTEEN: HETEROCYCLIC COMPOUNDS

CHAPTER SUMMARY

Atoms other than carbon and hydrogen that appear in organic compounds are called **heteroatoms**. Cyclic organic compounds that contain one or more heteroatoms are called **heterocycles**. Heterocyclic compounds are the largest class of organic compounds and can be either **aromatic** (such as pyridine, pyrrole, and furan) or **nonaromatic** (such as piperidine, pyrrolidine, and tetrahydrofuran).

Pyridine is a six-membered ring heterocycle that has a structure **isoelectronic** with the aromatic hydrocarbon benzene. In pyridine, one of the –(CH)= units in benzene is replace by an sp^2-hybridized nitrogen [–(N:)=], and the nitrogen contributes one electron to the aromatic ring. Pyridine undergoes **electrophilic aromatic substitution** reactions at the 3-position, but at reaction rates much slower than benzene, partly because of the electron withdrawing and deactivating effect of the nitrogen. Pyridine undergoes **nucleophilic aromatic substitution** upon treatment with strong nucleophiles like sodium amide and sodium methoxide. The reaction mechanism involves addition of the nucleophile to the electron-deficient pyridine followed by elimination of a leaving group. The nitrogen of pyridine is basic, and the nonbonded lone pair is protonated by mineral acids to give **pyridinium salts**. Catalytic hydrogenation of pyridine gives the nonaromatic six-membered ring heterocycle **piperidine**. **Nicotine** (from tobacco) and **pyridoxine** (vitamin B_6) are two naturally occurring substituted pyridines.

Polycyclic aromatic heterocycles that contain pyridine rings fused with benzene rings include **quinoline** and **isoquinoline**. **Quinine**, used to treat malaria, is an example of a naturally occurring quinoline.

The **diazines (pyridazine, pyrimidine,** and **pyrazine)** are six-membered ring aromatic heterocycles that have two nitrogens in the ring. **Cytosine, thymine,** and **uracil** are derivatives of pyrimidine that are important bases in nucleic acids (DNA and RNA). Heterocyclic analogs of the aromatic hydrocarbon naphthalene include **pteridines**, which have four nitrogens in the rings. Naturally occurring pteridine derivatives include **xanthopterin** (a pigment) and **folic acid** (a vitamin). **Methotrexate** is a pteridine used in cancer chemotherapy.

REACTION SUMMARY

Pyrylium ions are six-membered ring heterocycles in which a positively charged sp^2-hybridized oxygen replaces the nitrogen in pyridine. The pyrylium ring appears in many naturally occurring flower **pigments**.

Pyrrole, **furan,** and **thiophene** are five-membered ring aromatic heterocycles with one heteroatom. In pyrrole, the nitrogen is sp^2-hybridized and contributes two electrons to the 6π aromatic ring. Furan and thiophene are isoelectronic with pyrrole, the [–(HN:)–] unit being replaced by –(:O:)– and –(:S:)– units, respectively. Pyrrole, furan, and thiophene are electron-rich (there are six π electrons distributed over five atoms) and undergo electrophilic aromatic substitution at the 2-position with reaction rates much faster than benzene. Pyrrole rings form the building blocks of biologically important pigments called **porphyrins**. **Hemoglobin** and **myoglobin** (important in oxygen transport) and **chlorophyll** (important in photosynthesis) are examples of naturally occurring porphyrins. Furans are obtained commercially from **furfural** (furan-2-carboxaldehyde), which is produced by heating corn cobs with strong acid.

Polycyclic aromatic heterocycles that contain the 2- and 3-positions of pyrrole fused to a benzene ring are called **indoles**. The indole ring occurs in many medicinally important natural products, such as the neurotransmitter serotonin.

The **azoles** (**oxazole**, **imidazole,** and **thiazole**) are five-membered ring aromatic heterocycles that have two heteroatoms in the ring. One of the heteroatoms in each of these heterocycles is an sp^2-hybridized nitrogen that contributes one electron to the 6π aromatic system and has a basic nonbonded lone pair. The other heteroatom (oxygen, nitrogen, or sulfur) contributes two electrons to the 6π system. The imidazole skeleton is present in the amino acid **histidine**. The thiazole ring occurs in **thiamin** (vitamin B_1).

The **purines** are an important class of heterocycles that contain an imidazole ring fused to a pyrimidine ring. **Uric acid** (the main product of nitrogen metabolism in birds and reptiles), **caffeine** (present in coffee), and **adenine** and **guanine** (nitrogen bases present in the nucleic acids DNA and RNA) are examples of naturally occurring purines.

REACTION SUMMARY
Reactions of Pyridine and Related Six-Membered Ring Aromatic Heterocycles

1. Protonation

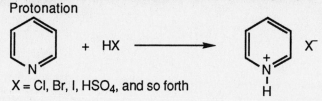

X = Cl, Br, I, HSO$_4$, and so forth

13. HETEROCYCLIC COMPOUNDS

2.   Electrophilic Aromatic Substitution

$$\text{pyridine} + E^+ \longrightarrow \text{3-E-pyridine} + H^+$$

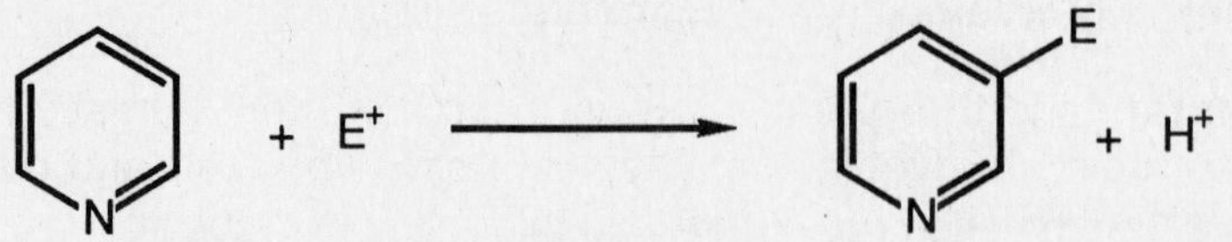

3.   Nucleophilic Aromatic Substitution

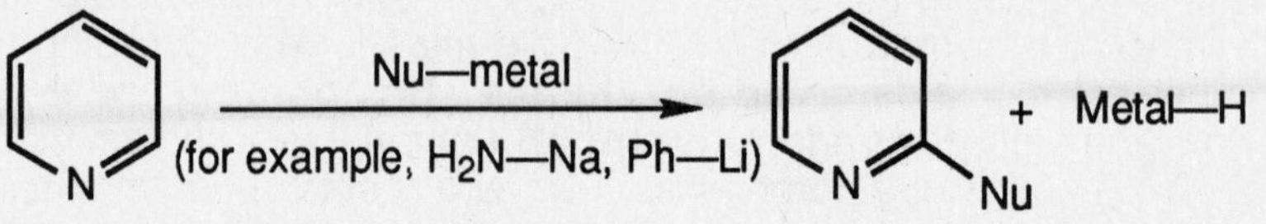

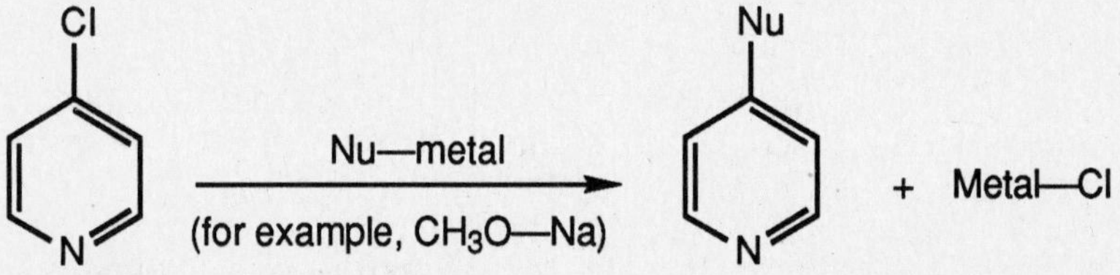

pyridine $\xrightarrow{\text{Nu—metal}}$ (for example, H_2N—Na, Ph—Li) → 2-Nu-pyridine + Metal—H

4-Cl-pyridine $\xrightarrow{\text{Nu—metal}}$ (for example, CH_3O—Na) → 4-Nu-pyridine + Metal—Cl

4.   Oxidation

3-CH₃-pyridine $\xrightarrow{KMnO_4}$ 3-CO₂H-pyridine

5.   Reduction

pyridine $\xrightarrow[\text{Pt}]{H_2}$ piperidine

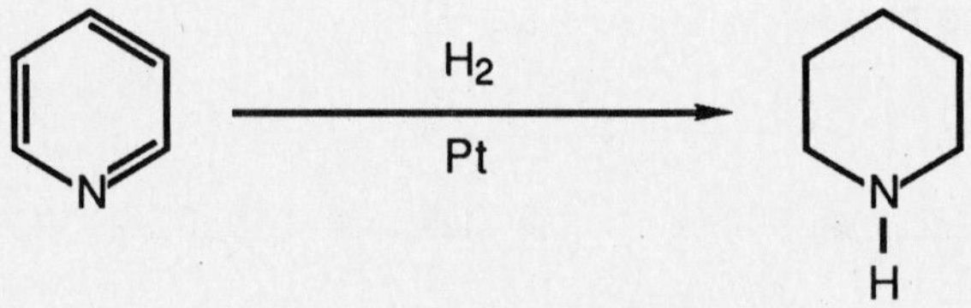

MECHANISM SUMMARY

Electrophilic Aromatic Substitution Reactions of Five-Membered Ring Aromatic Heterocycles

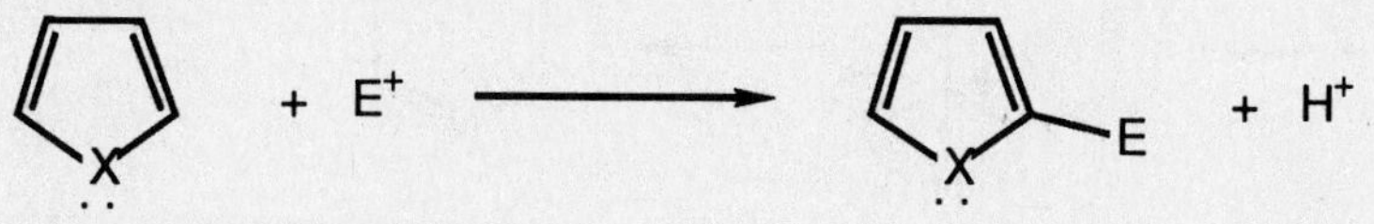

$$X = O:, S:, N-H$$

MECHANISM SUMMARY

Electrophilic Aromatic Substitution of Pyridine

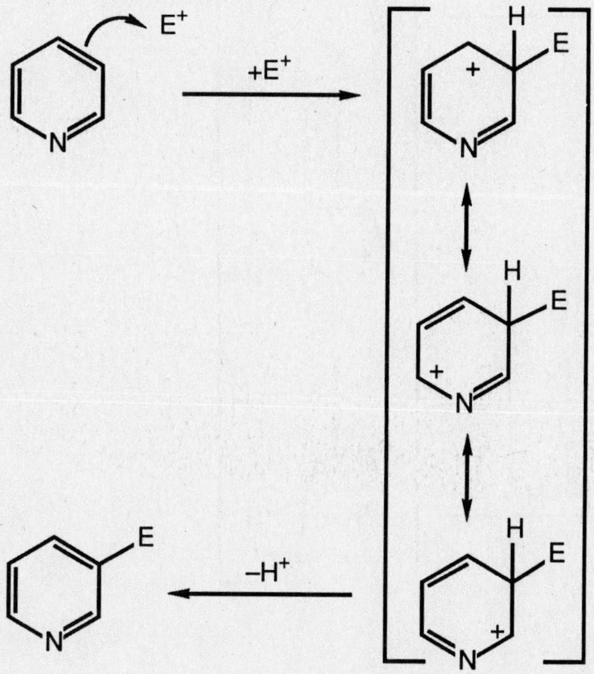

13. HETEROCYCLIC COMPOUNDS

Nucleophilic Aromatic Substitutions of Pyridines

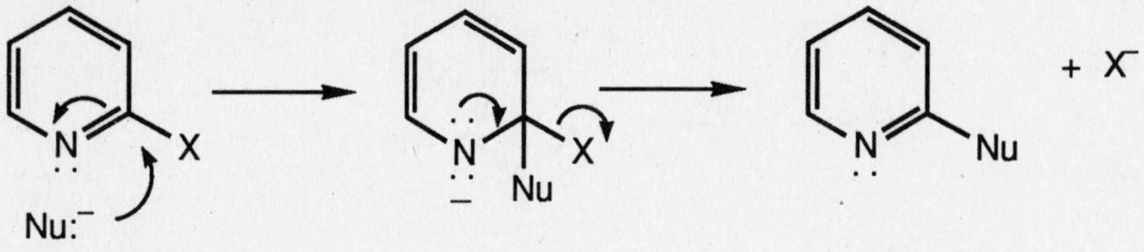

X = leaving group

Electrophilic Aromatic Substitution of Five-Membered Ring Heterocycles

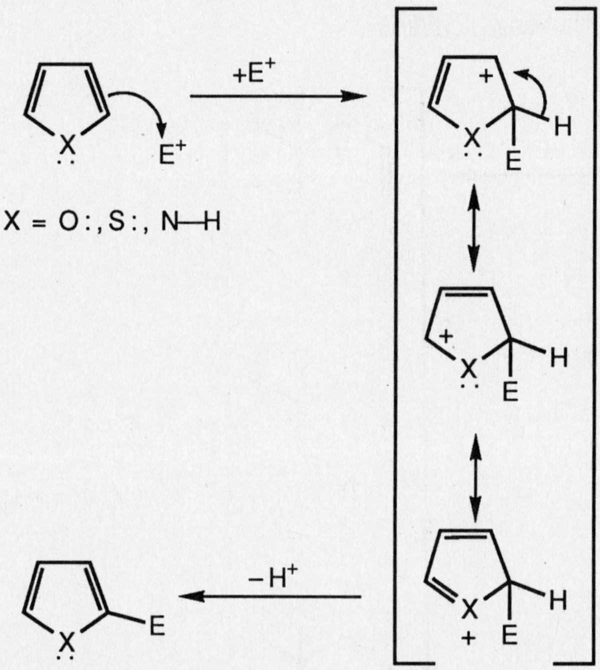

X = O:, S:, N—H

LEARNING OBJECTIVES

1. Know the meaning of: heteroatom, aromatic heterocycle, nonaromatic heterocycle, diazine, azole.

2. Draw the structure of: pyridine, pyrrolidine, nicotine, quinoline, isoquinoline, pyrimidine, pteridine, the pyrylium cation.

LEARNING OBJECTIVES

3. Draw the structure of: pyrrole, furan, thiophene, furfural, indole, oxazole, imidazole, thiazole, purine.

4. Determine how many electrons are contributed to the aromatic π system by each heteroatom in the aforementioned heterocycles.

5. Determine the position of basic nonbonded lone pairs in the aforementioned heterocycles.

6. Predict the product of electrophilic aromatic substitution reactions of pyridine and quinoline.

7. Write the mechanism of electrophilic aromatic substitution reactions of pyridine.

8. Predict the product of nucleophilic aromatic substitution reactions of pyridine and substituted pyridines.

9. Write the mechanism of nucleophilic aromatic substitution reactions of pyridine.

10. Predict the product expected from hydrogenation of aromatic heterocycles.

11. Predict the product expected from potassium permanganate oxidation of a methylated pyridine.

12. Predict the product expected from electrophilic aromatic substitution reactions of pyrrole, furan, and thiophene.

13. Write the mechanism of electrophilic aromatic substitution reactions of pyrrole, furan, and thiophene.

14. Explain why pyridine is less reactive than benzene in electrophilic aromatic substitution reactions.

15. Explain why pyrrole, furan, and thiophene are more reactive than benzene or pyridine in electrophilic aromatic substitution reactions.

16. Predict the products expected from treatment of pyridine and imidazole with Brønsted-Lowry acids and electrophiles such as iodomethane.

ANSWERS TO PROBLEMS

Problems Within the Chapter

13.1   a.

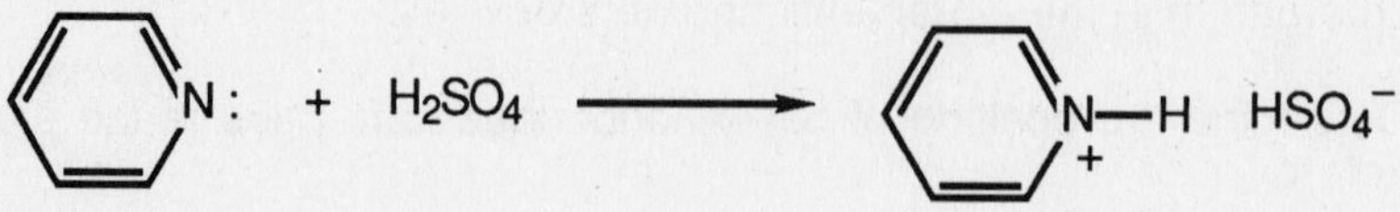

b.

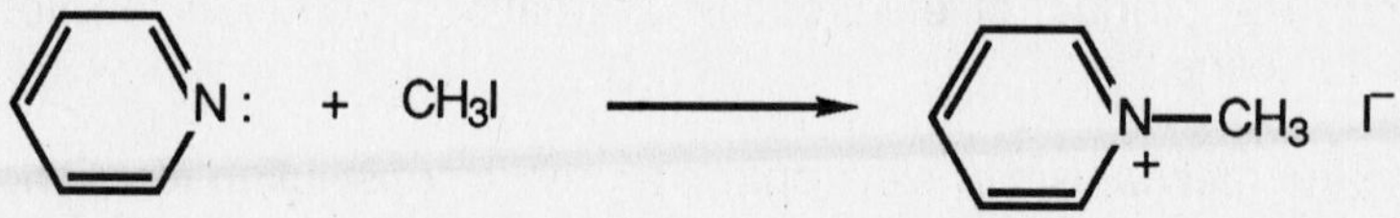

13.2   The contributors to the resonance hybrid derived from electrophilic attack at C-2 are:

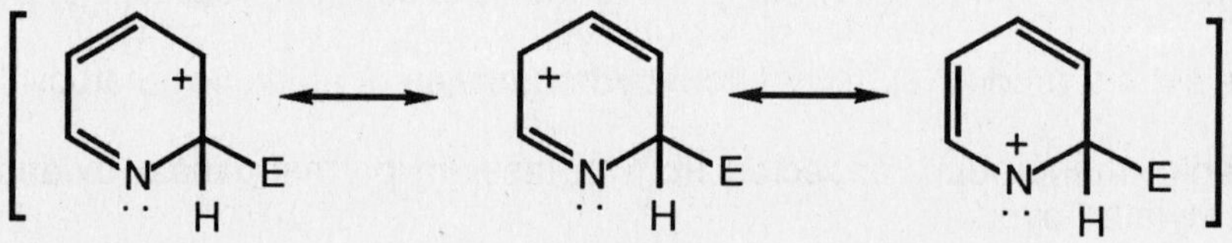

The contributors to the structure derived from attack at C-4 are:

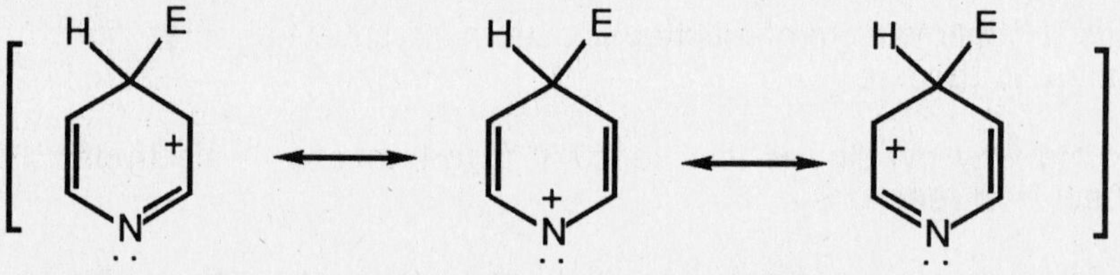

In each case, positive charge is placed on the nitrogen atom in one resonance contributor. This is a destabilizing effect due to the electronegativity of nitrogen. Electrophilic attack at C-3 gives a cation in which the positive charge does not reside on nitrogen (see Example 13.1). Therefore attack at C-3 is favored relative to attack at C-2 and C-4.

13.3 Methoxide adds to C-4 followed by elimination of chloride ion.

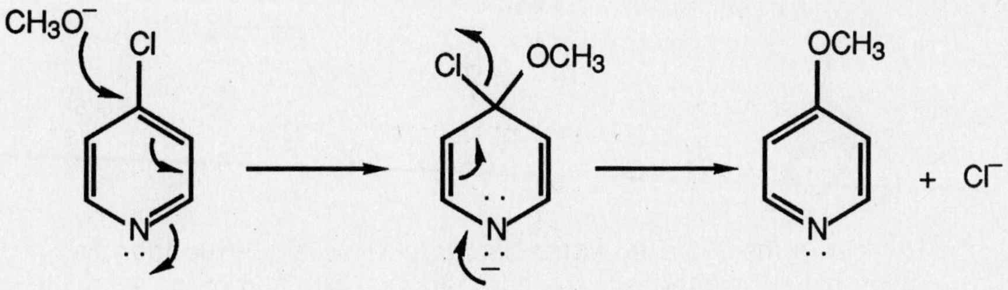

13.4 The configuration of (+)-coniine is *R* (review Sec. 5.4).

13.5

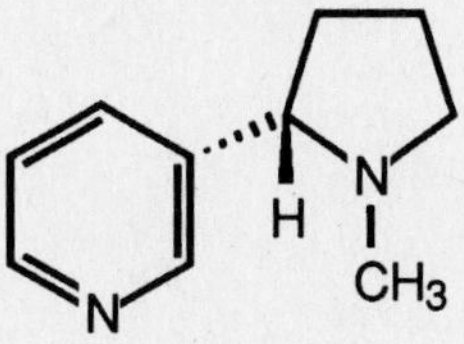

(*S*)–(–)–nicotine

13.6 The first equivalent of HCl reacts with the more basic, *sp*^{3}-hybridized nitrogen to give:

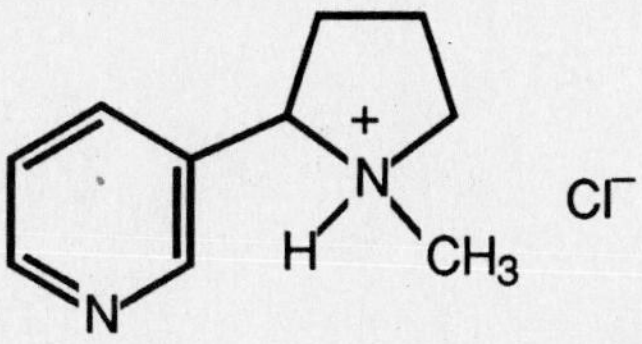

Cl$^-$

The second equivalent of HCl reacts with the *sp*^{2}-hybridized pyridine nitrogen:

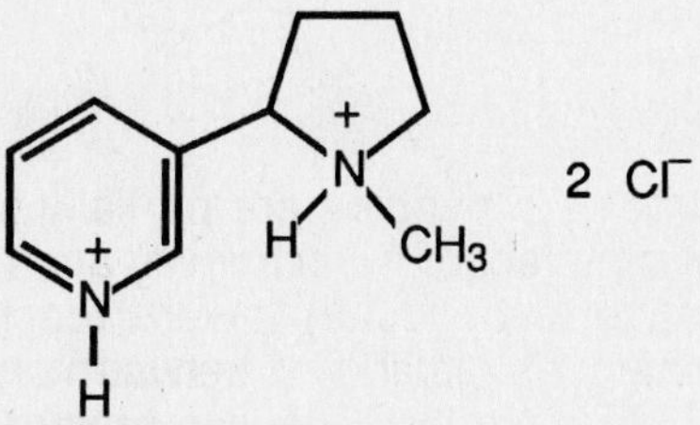

2 Cl$^-$

13. HETEROCYCLIC COMPOUNDS

13.7   Pyrrolidine is miscible because the basic, nonbonded lone pair electrons can hydrogen bond to water as follows:

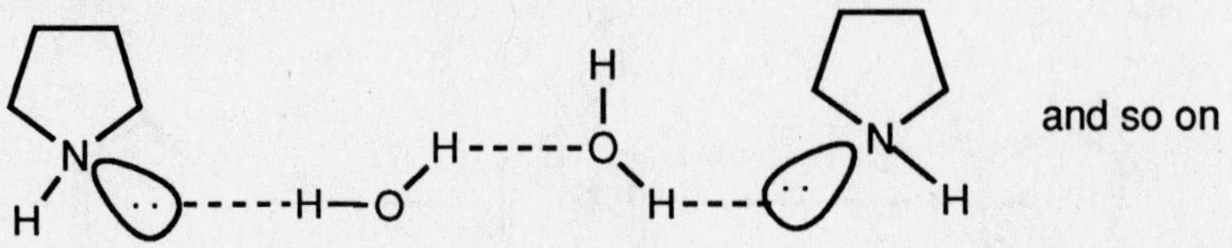

and so on

Pyrrole is insoluble in water because there are no longer any non-bonded electrons associated with the nitrogen; they are part of the aromatic π system.

13.8

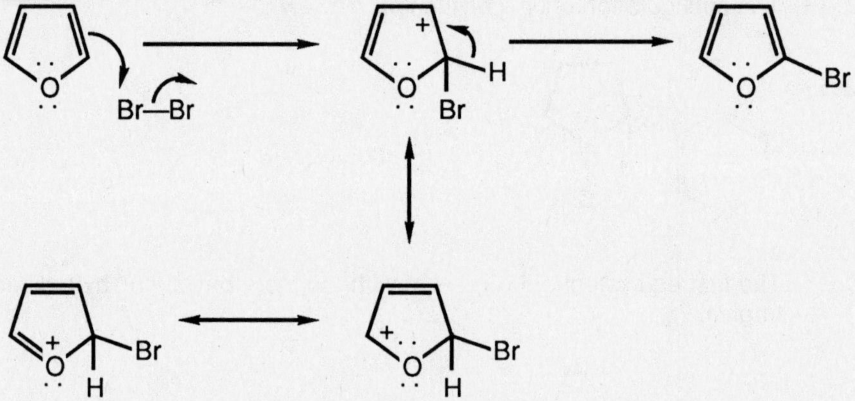

Additional Problems

13.9   Three minor dipolar contributors are:

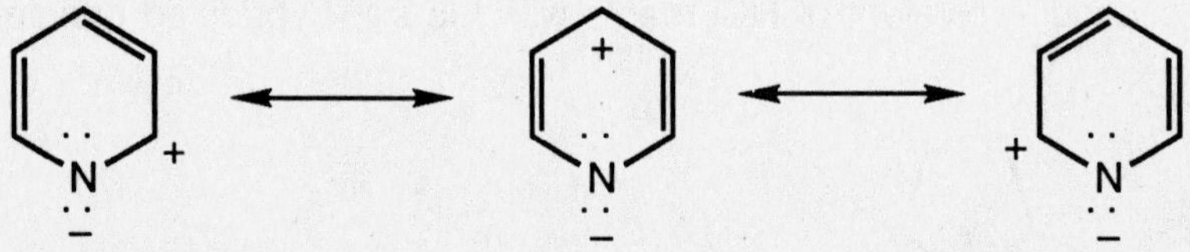

These structures suggest that the carbons in pyridine are partially positively charged (due to the electron-withdrawing effect of the nitrogen) and therefore are expected to be deactivated (relative to benzene) toward reaction with electrophiles. Note that the positive charge is distributed between carbons 2, 4, and 6. Therefore these carbons should be less reactive toward electrophiles than carbon 3 (or 5).

13.10

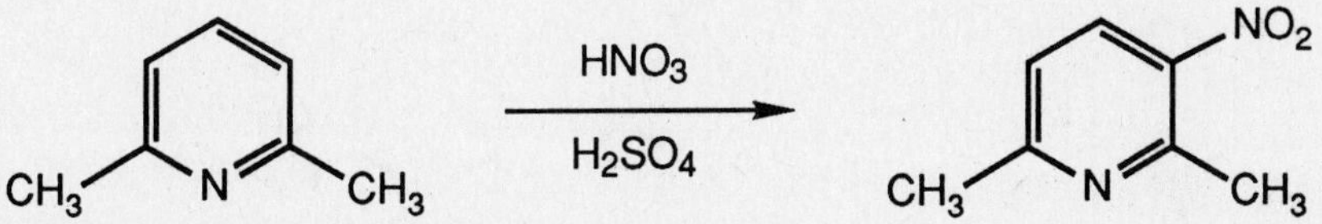

The methyl groups are electron-donating and activate aromatic compounds toward electrophilic aromatic substitution (see Sec. 4.11). Therefore 2,6-dimethylpyridine undergoes nitration under much milder conditions than required by pyridine.

13.11 Compare with eq. 13.3 and Example 13.2.

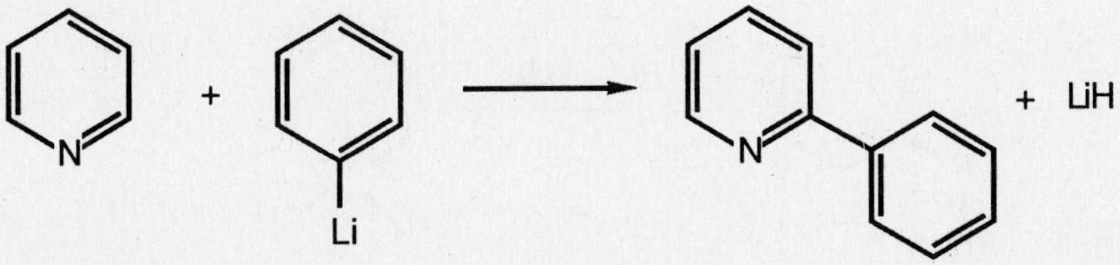

Mechanism

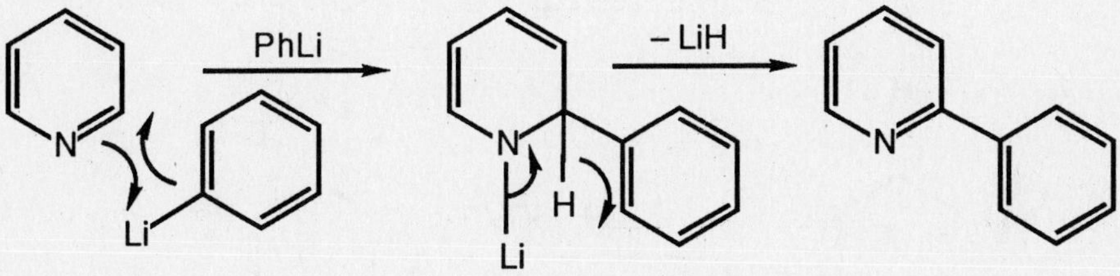

13.12

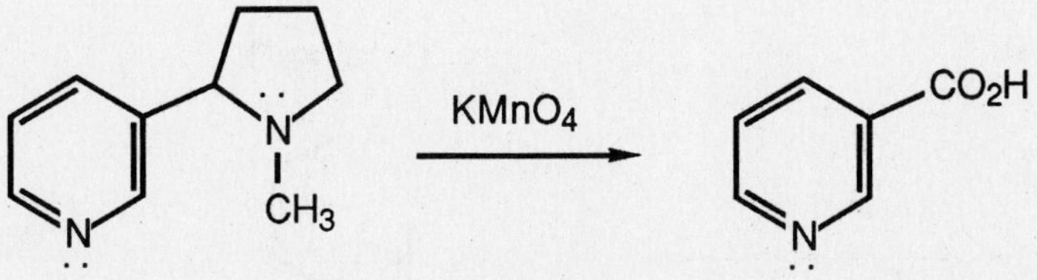

327

13.13 a.

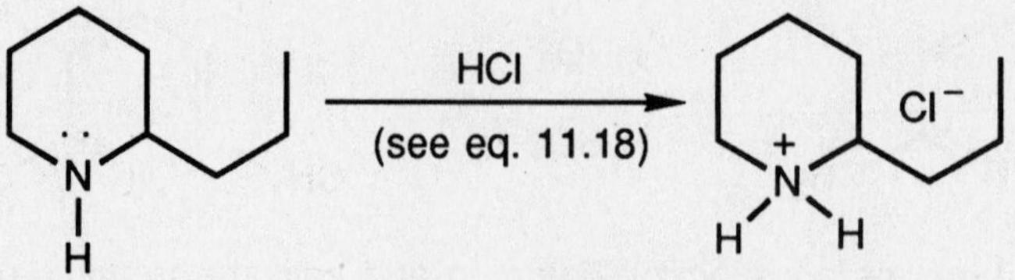
HCl
(see eq. 11.18)
Cl⁻

b.

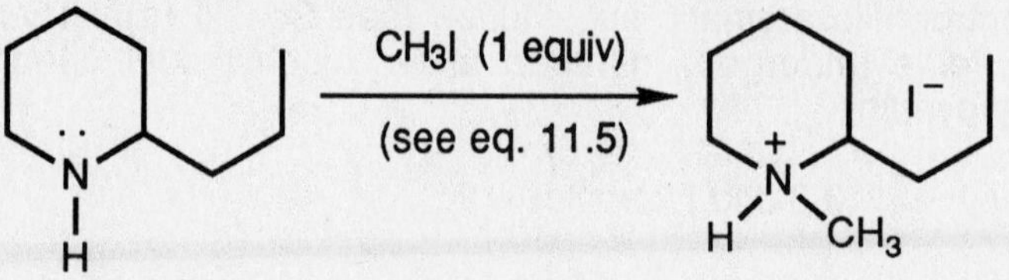
CH3I (1 equiv)
(see eq. 11.5)
I⁻
CH3

c.

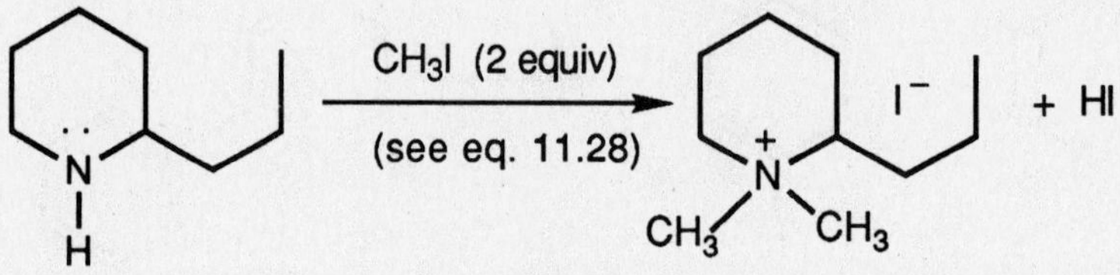
CH3I (2 equiv)
(see eq. 11.28)
I⁻
+ HI
CH3
CH3

d.

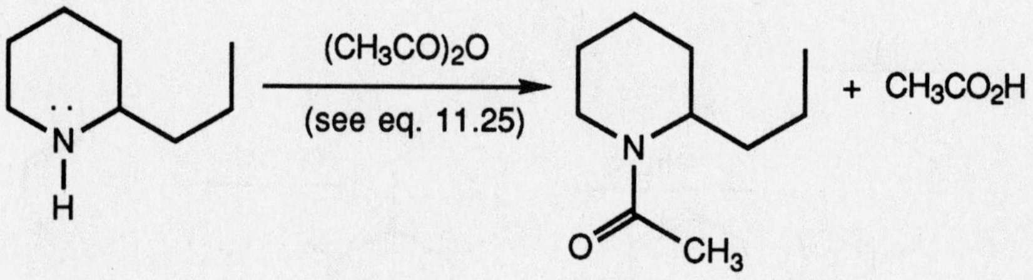
(CH3CO)2O
(see eq. 11.25)
+ CH3CO2H
CH3

e.

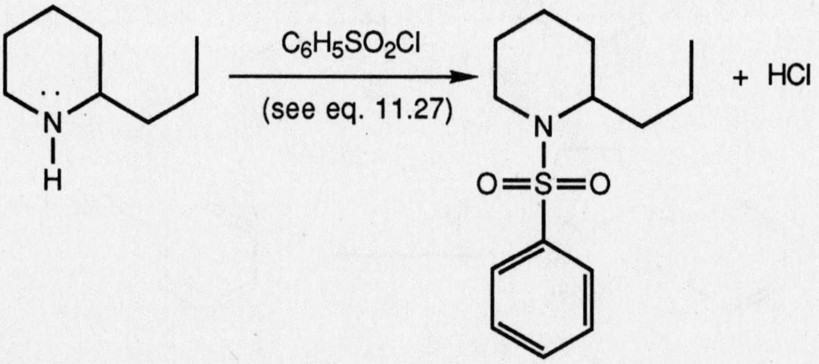
C6H5SO2Cl
(see eq. 11.27)
+ HCl

13.14 Substitution occurs in the more electron-rich "benzene" ring. When substitution occurs at C-5 or C-8, two resonance structures that retain an aromatic pyridine ring are possible.

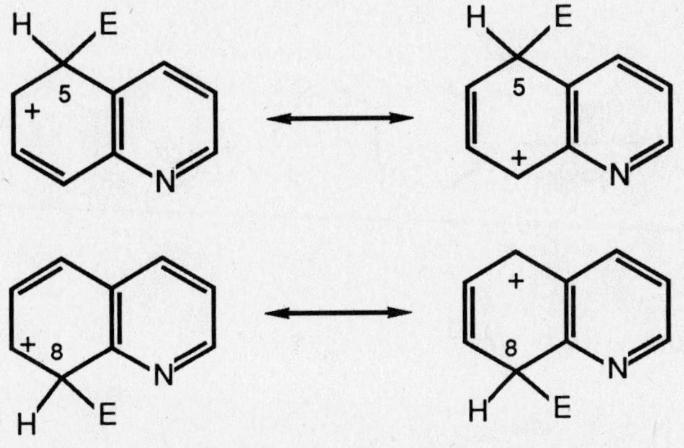

When substitution occurs at C-6 or C-7, only one resonance structure that retains an aromatic pyridine ring is possible.

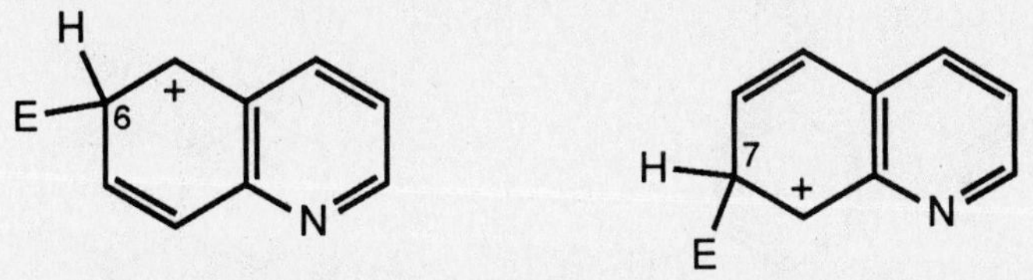

Since the intermediate carbocations derived from attack at C-5 and C-8 are more stable than those derived from attack at C-6 and C-7, substitution occurs at those positions. It is notable that the behavior of quinoline toward electrophiles is similar to the behavior of naphthalene (see eq. 4.37, Example 4.3, and Prob. 4.16).

13.15 a. Quinoline will behave like pyridine (see eq. 13.1).

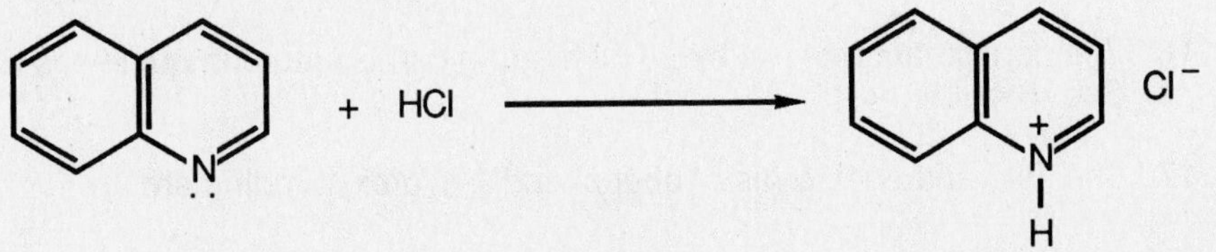

b. Substitution will occur in the most electron-rich aromatic ring by way of the most stabilized carbocation intermediates (see Prob. 13.14).

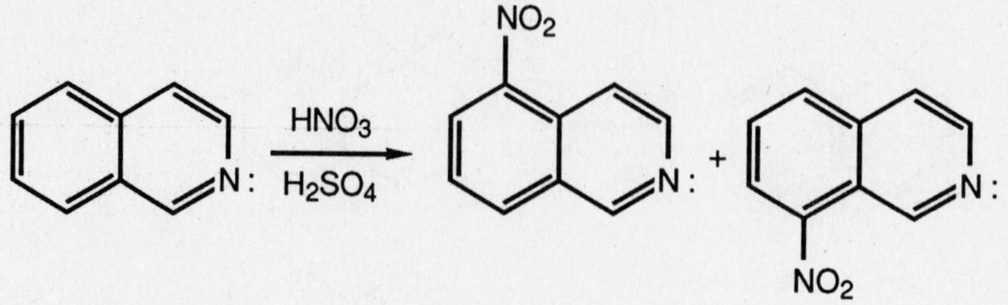

NO₂ becomes NO_2

$$\text{isoquinoline} \xrightarrow[\text{H}_2\text{SO}_4]{\text{HNO}_3} \text{5-nitro} + \text{8-nitro}$$

c. Quinoline will behave like pyridine (see eq. 13.3).

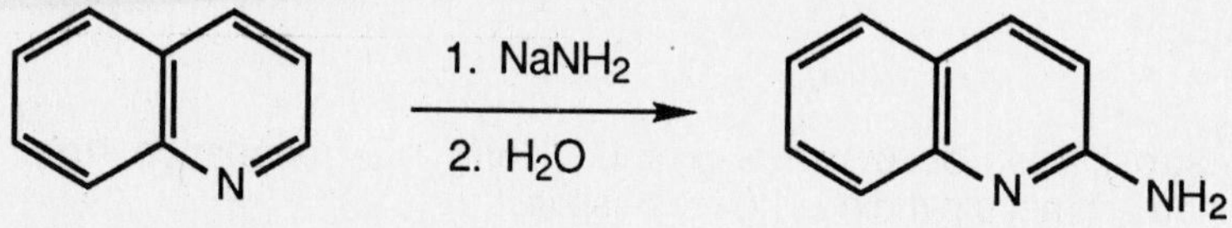

$$\text{quinoline} \xrightarrow[\text{2. H}_2\text{O}]{\text{1. NaNH}_2} \text{2-aminoquinoline}$$

d. Substitution occurs at C-1:

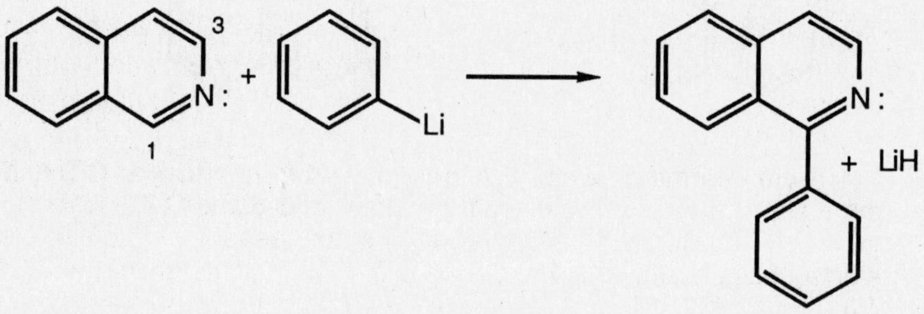

$$\text{isoquinoline} + \text{C}_6\text{H}_5\text{Li} \longrightarrow \text{1-phenylisoquinoline} + \text{LiH}$$

Nucleophilic attack at C-3 would give an intermediate lacking aromaticity in both rings and therefore is disfavored.

14.16 The configuration of the hydroxyl-bearing carbon atom in quinine is S (review Sec. 5.4).

14.17 The keto and enol forms of phenol and 2-hydroxypyridine are:

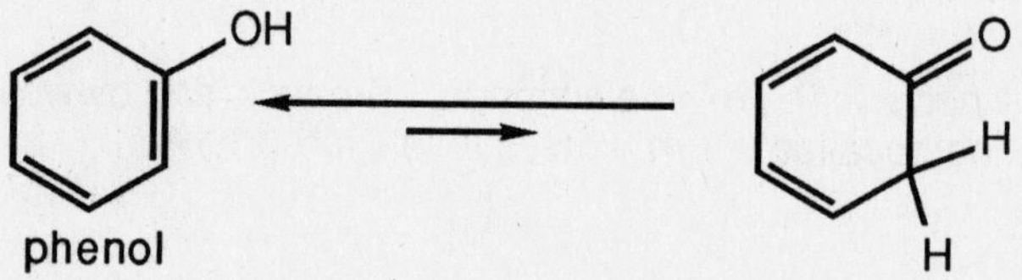

phenol

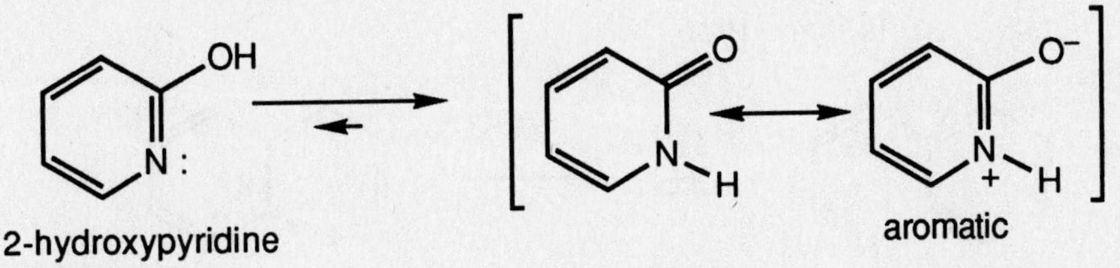

2-hydroxypyridine   aromatic

The keto tautomer of phenol does not retain any of the resonance energy associated with the aromatic ring, while the keto tautomer of 2-hydroxypyridine does.

13.18   a. See eq. 11.25.

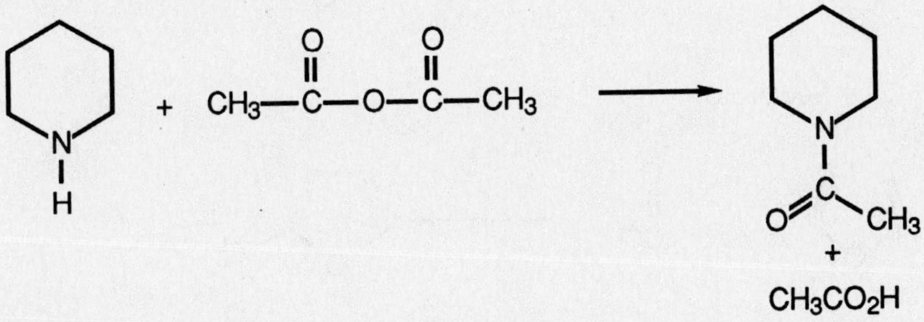

b. See eq. 11.18.

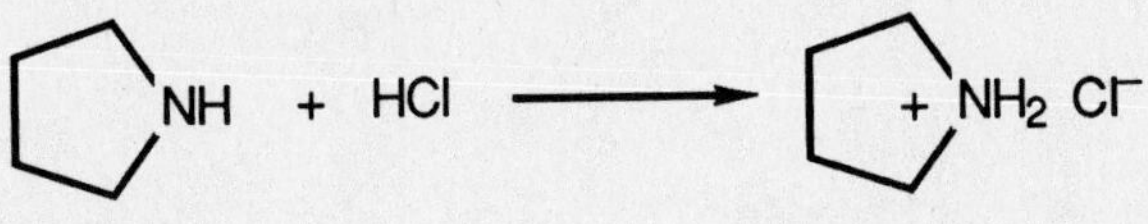

c. See eq. 13.1.

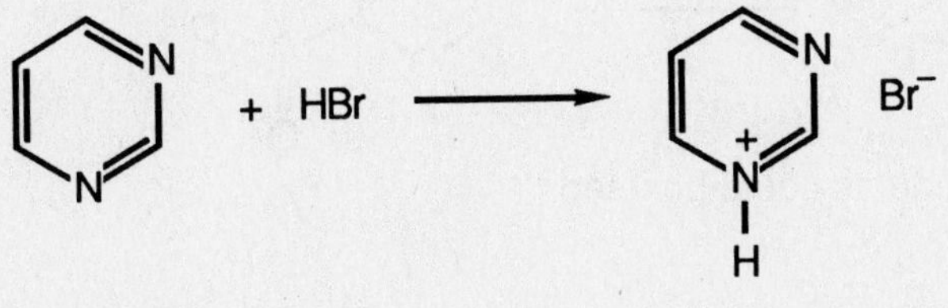

13. HETEROCYCLIC COMPOUNDS

d. See eq. 11.6 and Prob. 13.1

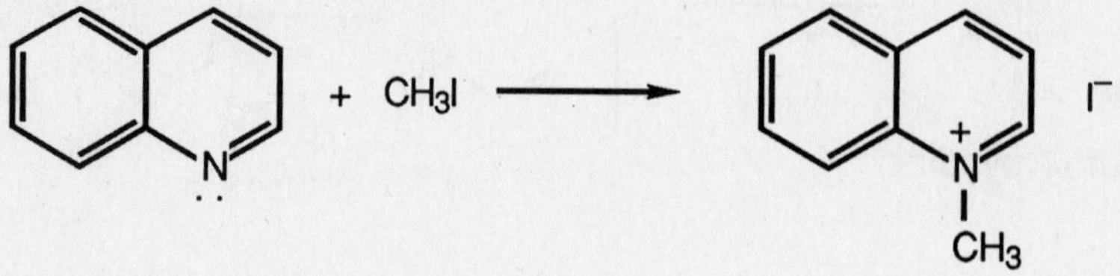

e. See Sec. 13.6.

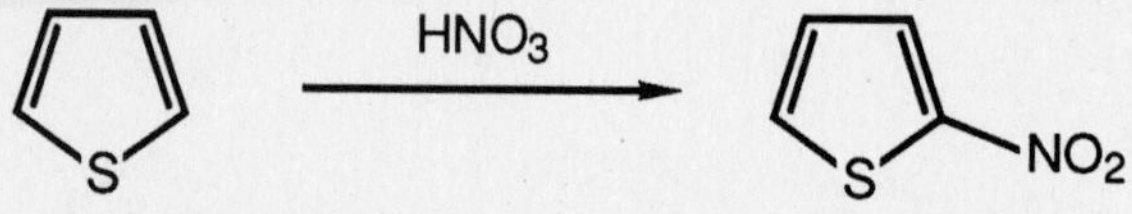

f. See Sec. 13.6.

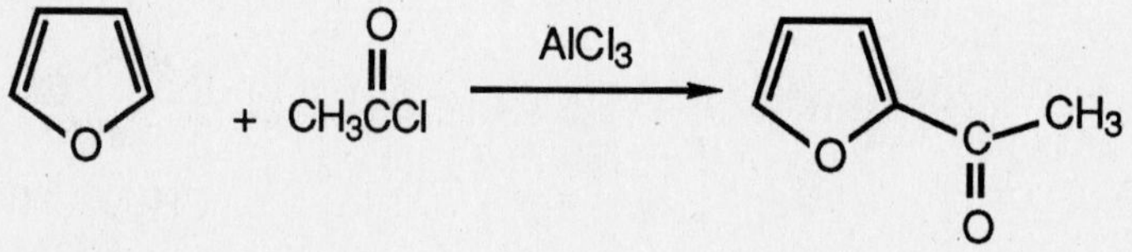

13.19 a.

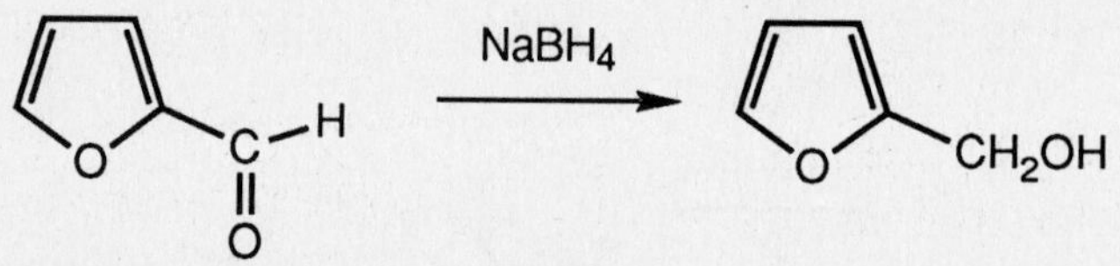

Other reducing agents could be used.

b.

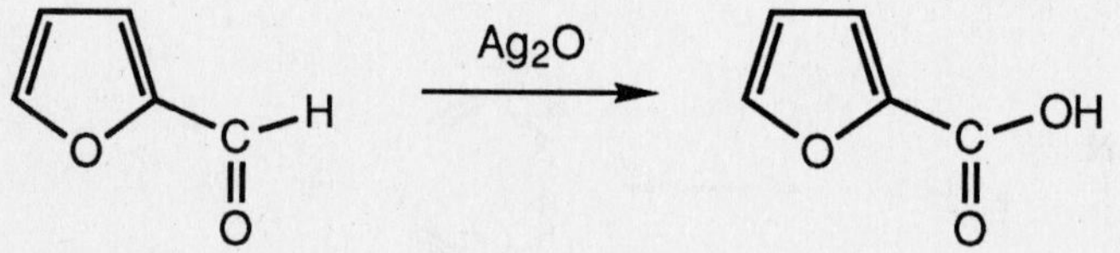

Other oxidizing agents could be used.

13.20 Examine the intermediates derived from electrophilic attack at each position .

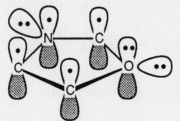

Attack at C-3 gives a carbocation that is resonance stabilized by a nonbonded pair of electrons *on the adjacent nitrogen*. Attack at C-2 gives a cation in which the positive charge is not adjacent to an atom bearing nonbonded electrons. Any additional resonance contributors disrupt the benzenoid structure in the "left" ring. Therefore electrophilic aromatic substitution reactions of indole occur primarily at C-3.

13.21

The nitrogen has a nonbonded electron pair in an sp^2 orbital at right angles to the cyclic array of p orbitals. Therefore it will be more basic than pyrrole.

13. HETEROCYCLIC COMPOUNDS

13.22 a.

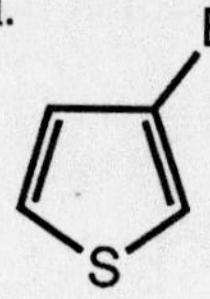

b.

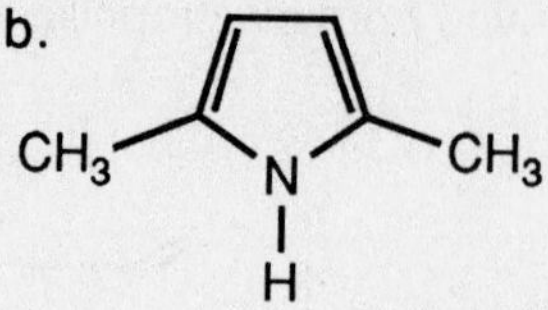

c.

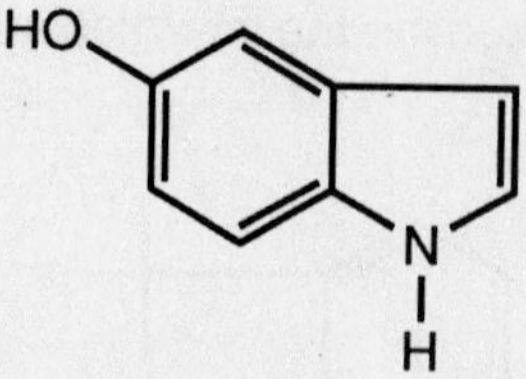

d.

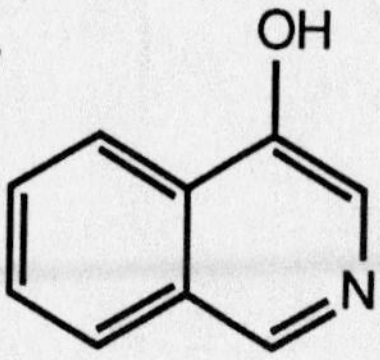

e.

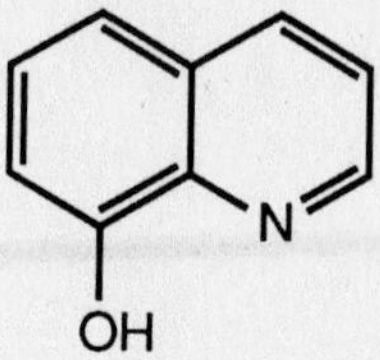

f.

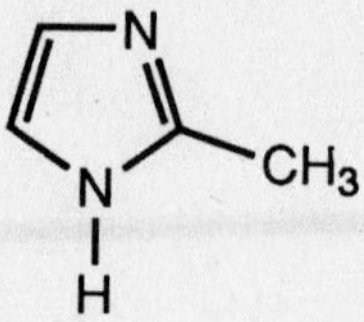

g.

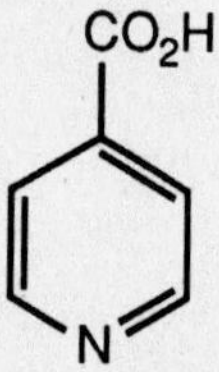

h.

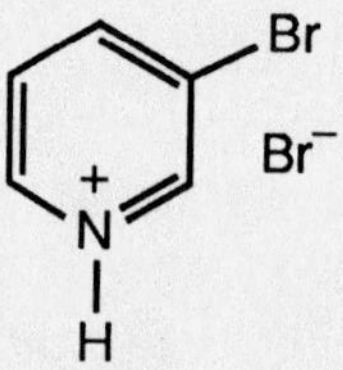

i.

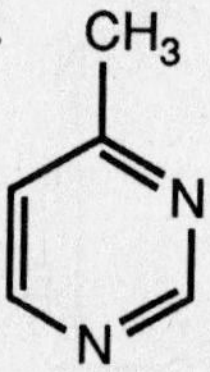

j.

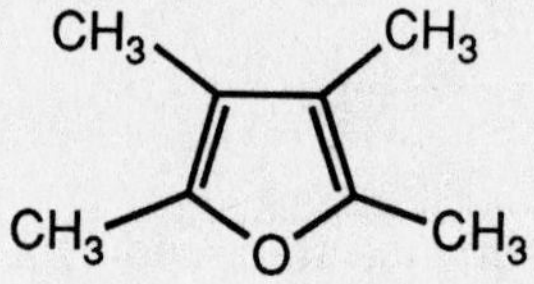

13.23 a.

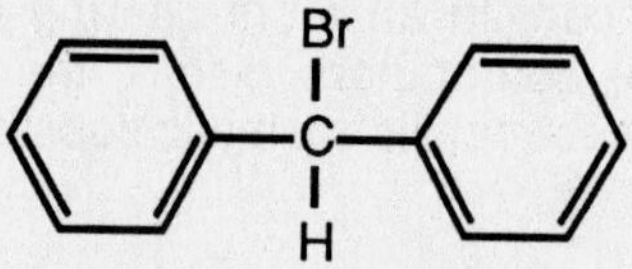

b. $(CH_3)_2NH$ + H_2C-CH_2 (epoxide, O) $\longrightarrow$ $(CH_3)_2NCH_2CH_2OH$

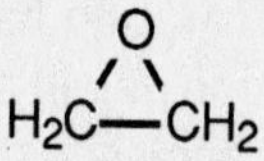

c. The first step is a free radical halogenation (see Sec. 2.16).

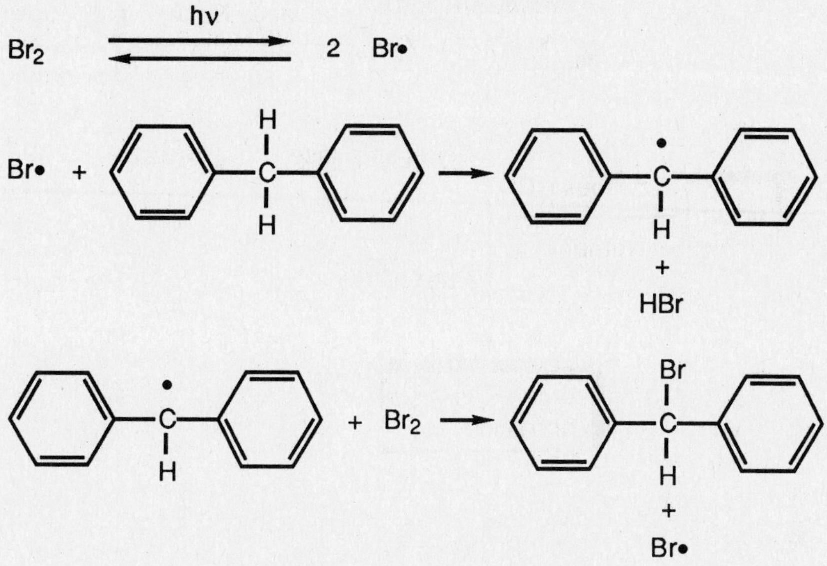

$$Br_2 \; \underset{\longleftarrow}{\overset{h\nu}{\longrightarrow}} \; 2 \;\; Br\bullet$$

Br• + [diphenylmethane] → [diphenylmethyl radical]

+

HBr

[diphenylmethyl radical] + Br₂ → [bromodiphenylmethane]

+

Br•

The second step is an S$_N$1 reaction (see Sec. 6.6).

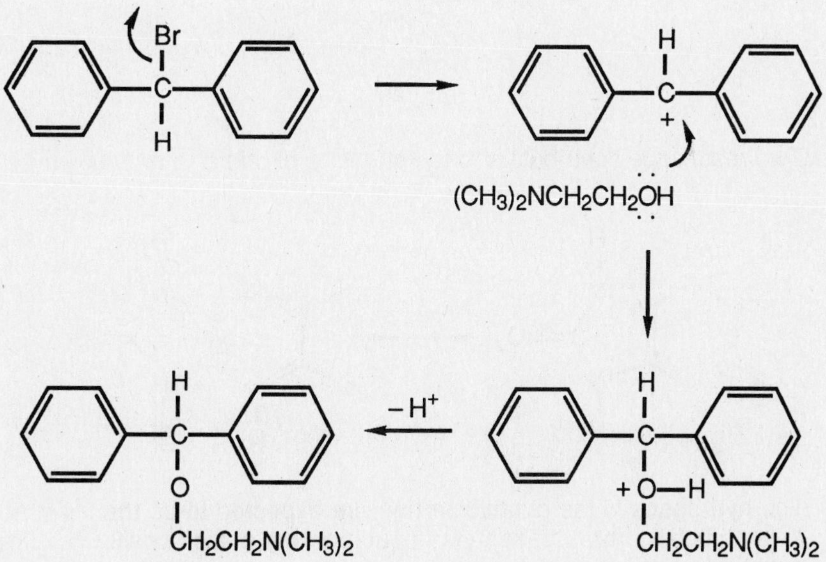

[benzhydryl bromide with Br leaving] → [benzhydryl cation]

(CH₃)₂NCH₂CH₂OH

[benzhydryl ether with +O—H]

CH₂CH₂N(CH₃)₂

−H⁺ →

[benadryl product]

O

CH₂CH₂N(CH₃)₂

benadryl

13.24

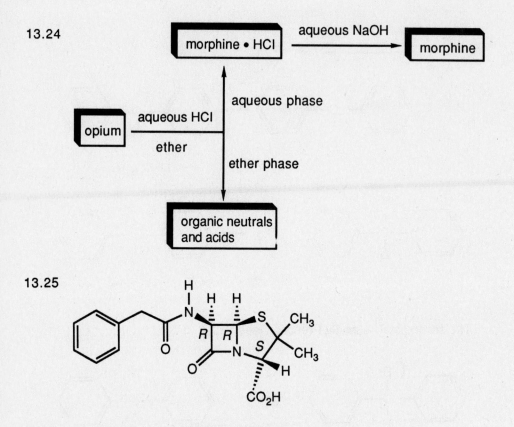

13.25

13.26 Two resonance contributors to the structure of uric acid are shown below:

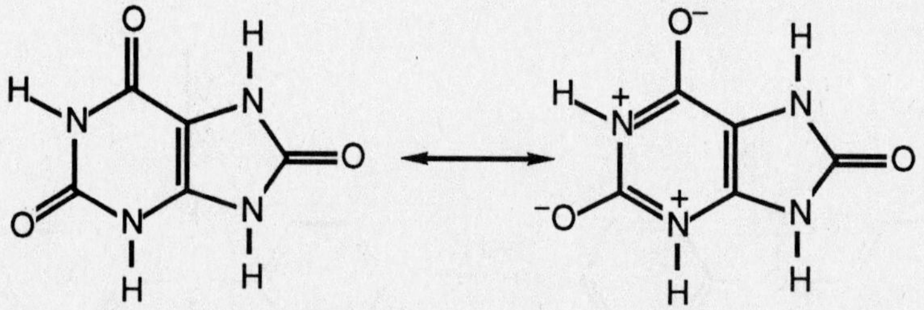

The hydrogens in the pyrimidine ring are expected to be the most acidic due to contribution from a resonance structure that places positive charge on the pyrimidine ring nitrogens.

CHAPTER FOURTEEN: SYNTHETIC POLYMERS

CHAPTER SUMMARY

Polymers are **macromolecules** built of smaller units called **monomers**. The process by which they are formed is called **polymerization**. They may be synthetic (nylon, Teflon, and Plexiglas) or natural (such as the biopolymers starch, cellulose, proteins, DNA, RNA). **Homopolymers** are made from a single monomer. **Copolymers** are made from two or more monomers. Polymers may be **linear**, **branched**, or **cross-linked**, depending on how the monomer units are arranged. These details of structure affect polymers' properties.

Chain-growth, or **addition, polymers** are made by adding one monomer unit at a time to the growing polymer chain. The reaction requires initiation to produce some sort of reactive intermediate, which may be a **free radical**, a **cation**, or an **anion**. The intermediate adds to the monomer, giving a new intermediate, and the process continues until the chain is terminated in some way. **Polystyrene** is a typical free-radical chain-growth polymer.

Chiral centers can be generated when a substituted vinyl compound is polymerized. The resulting polymers are classified as **atactic**, **isotactic**, or **syndiotactic**, depending on whether the chiral centers are **random, identical,** or **alternating** in configuration as one proceeds down the polymer chain. Ziegler–Natta catalysts (one example is a mixture of trialkylaluminum and titanium tetrachloride) usually produce stereoregular polymers, whereas free-radical catalysts generally give stereorandom polymers.

Step-growth, or **condensation, polymers** are usually formed in a reaction between two monomers, each of which is at least difunctional. Polyesters and polyamides are typical examples of step-growth polymers. These polymers grow by steps or leaps rather than one monomer unit at a time.

REACTION SUMMARY

Free-Radical Chain-Growth Polymerization

$$\text{Initiation: Initiator} \xrightarrow{\text{heat or light}} 2 \text{ R}\bullet$$

Propagation: $R\bullet$ + $H_2C{=}CH$ $\longrightarrow$ $R{-}CH_2\overset{\bullet}{C}H$ $\xrightarrow{\underset{X}{\overset{H_2C=CH}{|}}}$ and so on

$\quad\quad\quad\quad\quad\quad\quad\quad\quad\quad$ $\underset{X}{|}$ $\quad\quad\quad\quad\quad\quad\quad$ $\underset{X}{|}$

Termination: $\sim\!\!\!\sim\!\overset{\bullet}{C}H$ + $\overset{\bullet}{C}H\!\!\sim\!\!\!\sim$ $\longrightarrow$ $\sim\!\!\!\sim\!HC{-}CH\!\sim\!\!\!\sim$

$\quad\quad\quad\quad\quad\quad\underset{X}{|}\quad\quad\underset{X}{|}\quad$ (radical coupling) $\quad\underset{X}{|}\;\underset{X}{|}$

$\quad\quad\quad\quad\quad\quad\quad\quad\quad\quad\quad\quad\quad\quad\quad$ disproportionation

$\quad\quad\quad\sim\!\!\!\sim\!CH_2\overset{\bullet}{C}H$ + $\overset{\bullet}{C}HCH_2\!\sim\!\!\!\sim$ $\longrightarrow$

$\quad\quad\quad\quad\quad\quad\underset{X}{|}\quad\quad\underset{X}{|}$

$\quad\quad\quad\quad\quad\quad\quad\quad\quad\quad\quad\sim\!\!\!\sim\!CH_2CH_2$ + $CH{=}CH\!\sim\!\!\!\sim$

$\quad\quad\quad\quad\quad\quad\quad\quad\quad\quad\quad\quad\quad\underset{X}{|}\quad\quad\underset{X}{|}$

Chain Transfer (Hydrogen Abstraction)

$\sim\!\!\!\sim\!\overset{\bullet}{H}C$ + $\sim\!\!\!\sim\!\overset{\overset{H}{|}}{C}\!\sim\!\!\!\sim$ $\longrightarrow$ $\sim\!\!\!\sim\!H_2C$ + $\sim\!\!\!\sim\!\overset{\bullet}{C}\!\sim\!\!\!\sim$

$\quad\underset{X}{|}\quad\quad\quad\underset{X}{|}\quad\quad\quad\quad\quad\quad\quad\quad\underset{X}{|}\quad\quad\quad\underset{X}{|}$

Cationic Chain-Growth Polymerization

R^+ + $H_2C{=}CH$ $\longrightarrow$ $RCH_2{-}\overset{+}{C}H$ $\xrightarrow{\underset{X}{\overset{H_2C=CH}{|}}}$ and so on

$\quad\quad\quad\quad\underset{X}{|}\quad\quad\quad\quad\quad\quad\quad\underset{X}{|}$

Anionic Chain-Growth Polymerization

R^- + $H_2C{=}CH$ $\longrightarrow$ $RCH_2{-}\overset{\cdot\cdot}{C}H$ $\xrightarrow{\underset{X}{\overset{H_2C=CH}{|}}}$ and so on

$\quad\quad\quad\quad\underset{X}{|}\quad\quad\quad\quad\quad\quad\quad\underset{X}{|}$

Step-Growth Polymerization (illustrated for polyesters)

LEARNING OBJECTIVES

1. Know the meaning of: monomer, polymer, macromolecule, polymerization, average molecular weight, degree of polymerization.

2. Know the meaning of: homopolymer, copolymer, linear, branched, and cross-linked polymer. For copolymers, know the meaning of: alternating, random, block, and graft.

3. Know the meaning of thermoplastic and thermosetting polymers.

4. Know the meaning of and illustrate the difference between chain-growth (addition) and step-growth (condensation) polymerization.

5. Write the mechanism for an addition polymerization via a radical, cationic, or anionic intermediate. In each case, predict the direction of addition to the monomer if it is an unsymmetrical alkene.

6. Write resonance structures for the reactive intermediate in addition polymerization, to show how it is stabilized by the alkene substituent.

7. Know the meaning of: radical coupling, radical disproportionation, and chain transfer, and illustrate each with examples.

8. Know the meaning of and illustrate with examples: Ziegler–Natta polymerization, atactic, isotactic, and syndiotactic polymers.

9. Write a mechanism for a step-growth polymerization, as in the formation of a polyester, polyamide, polyurethane, or phenol–formaldehyde polymer.

10. Write the structure of the repeating unit of a chain-growth polymer derived from a given alkene monomer.

11. Write the structure of the repeating unit of a step-growth polymer derived from two given monomers.

12. Write the structure of a polyether derived from polymerization of a given epoxide.

ANSWERS TO PROBLEMS

Problems Within the Chapter

14.1　The odd electron can be delocalized to the *ortho* and *para* carbons of the phenyl group.

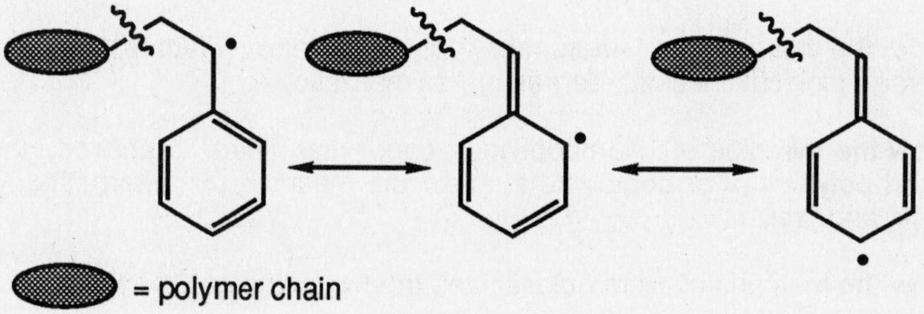

= polymer chain

14.2　Sulfonation of polystyrene occurs by an electrophilic aromatic substitution mechanism (see Sec. 4.10c). Since the phenyl groups in polystyrene are substituted with alkyl groups, which are *ortho,para* directing (see Sec. 4.12), sulfonation occurs at the *para* position (the *ortho* position is disfavored due to steric hindrance).

14.3

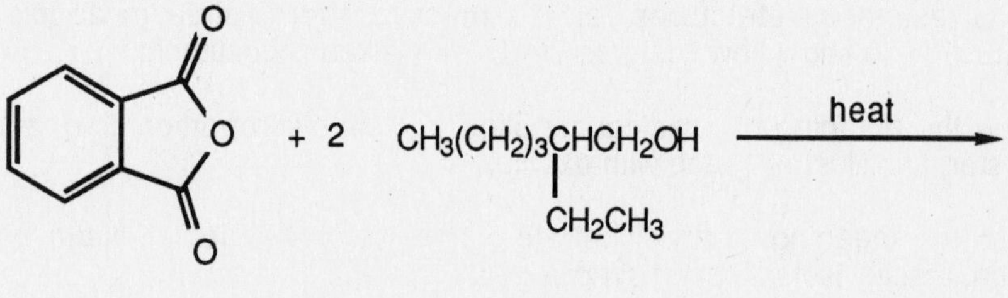

$+ 2$　$CH_3(CH_2)_3CHCH_2OH$　$\xrightarrow{\text{heat}}$
　　　　　　　　　　　　　$|$
　　　　　　　　　　　　CH_2CH_3

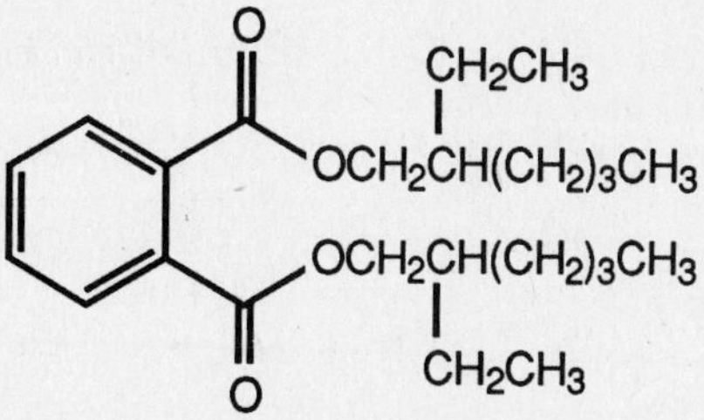

Phthalate diester structure (benzene ring with two ester groups):

$$\text{o-C}_6\text{H}_4(\text{C(=O)OCH}_2\text{CH(CH}_2\text{CH}_3)(\text{CH}_2)_3\text{CH}_3)_2$$

—C(=O)OCH₂CH(CH₂CH₃)(CH₂)₃CH₃ (two such ester groups on adjacent ring positions)

14.4 a. 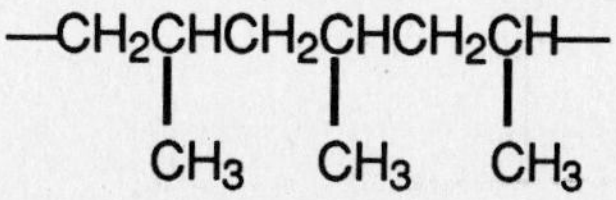 $-\text{CH}_2\text{CHCH}_2\text{CHCH}_2\text{CH}-$
           | | |
          CH₃   CH₃   CH₃

b. 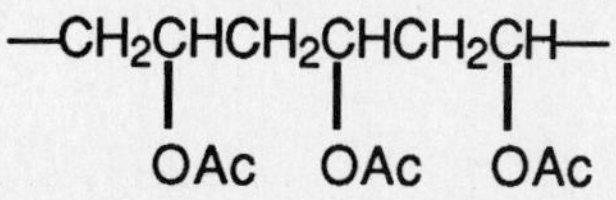 $-\text{CH}_2\text{CHCH}_2\text{CHCH}_2\text{CH}-$
           | | |
          OAc   OAc   OAc

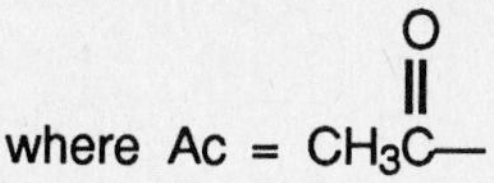

where Ac = $\text{CH}_3\overset{\text{O}}{\overset{\|}{\text{C}}}-$

c.
           E   E   E
           | | |
$-\text{CH}_2\text{CCH}_2\text{CCH}_2\text{C}-$
           | | |
          CH₃   CH₃   CH₃

where E = $\text{CH}_3\text{O}\overset{\text{O}}{\overset{\|}{\text{C}}}-$

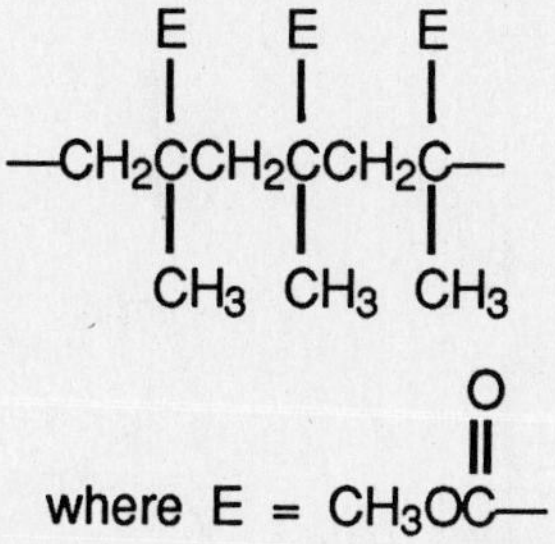

d. 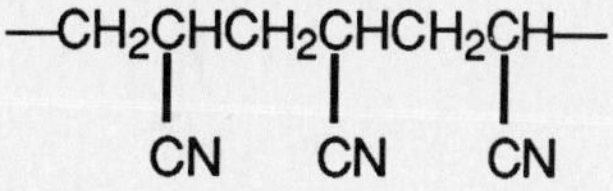 $-\text{CH}_2\text{CHCH}_2\text{CHCH}_2\text{CH}-$
           | | |
          CN   CN   CN

14.5 a. 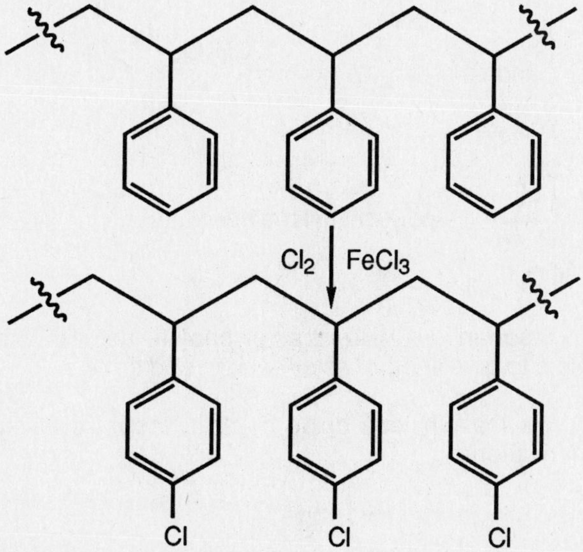 (polystyrene chain with three phenyl groups)

$\xrightarrow[\text{FeCl}_3]{\text{Cl}_2}$

(poly(4-chlorostyrene) chain with three para-chlorophenyl groups, Cl substituents)

b.

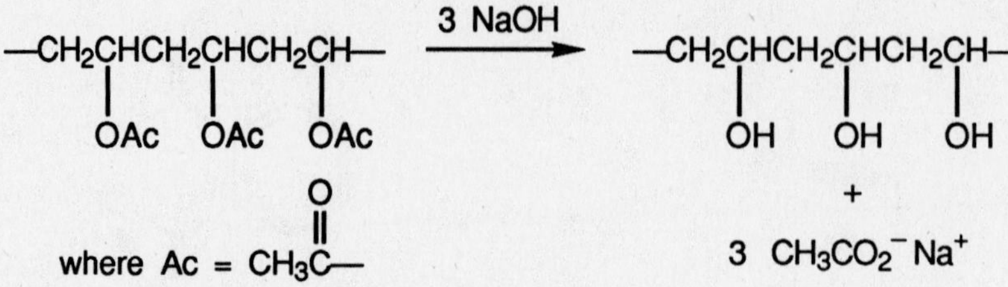

$$-CH_2CHCH_2CHCH_2CH- \xrightarrow{\text{3 NaOH}} -CH_2CHCH_2CHCH_2CH-$$

with OAc groups on the left and OH groups on the right

$$\text{where Ac} = CH_3C-$$

$$+$$

$$3 \ CH_3CO_2^- Na^+$$

14.6   1,4-Addition of *n*-butyllithium to the unsaturated ester initiates polymerization:

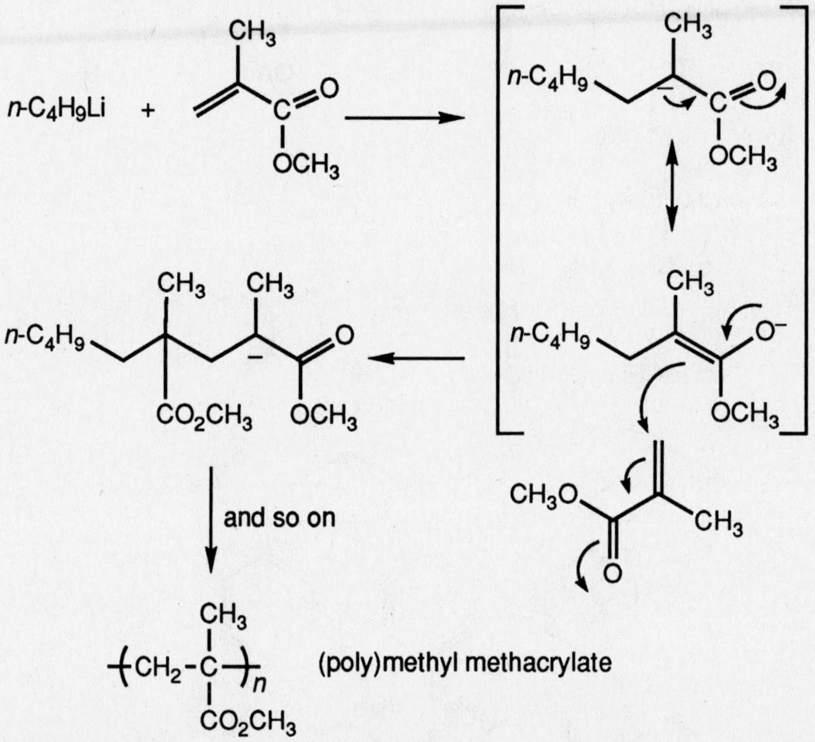

$$-\left(CH_2-\underset{\underset{CO_2CH_3}{|}}{\overset{\overset{CH_3}{|}}{C}}\right)_n$$ (poly)methyl methacrylate

The intermediate resonance-stabilized ester enolate adds to a molecule of methyl methacrylate to give another ester enolate and so on.

14.7   Hydroxide ion opens the ethylene oxide by a nucleophilic displacement to initiate the polymerization.

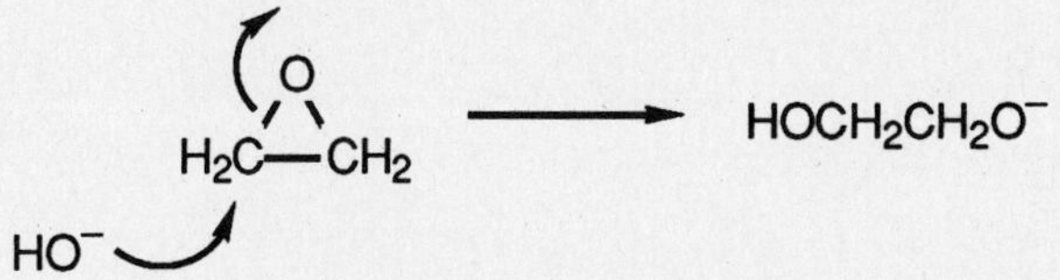

ANSWERS TO PROBLEMS

$$H_2C\!-\!CH_2 \xrightarrow{\hspace{2cm}} HOCH_2CH_2O^-$$

HO^-

The resulting alkoxide reacts with another molecule of ethylene oxide.

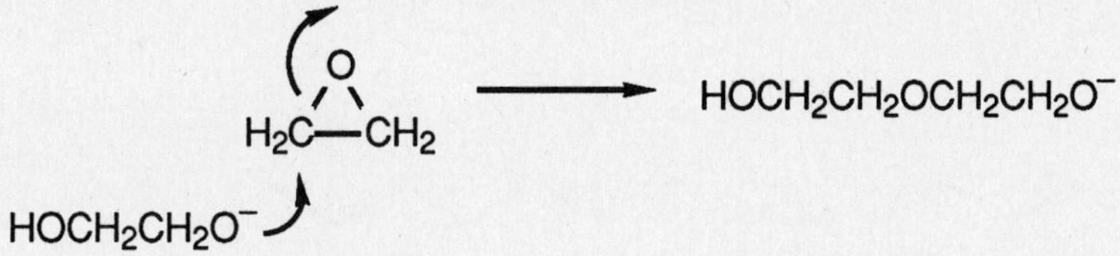

$$H_2C\!-\!CH_2 \xrightarrow{\hspace{2cm}} HOCH_2CH_2OCH_2CH_2O^-$$

$HOCH_2CH_2O^-$

and so on to give carbowax.

$$HOCH_2CH_2O\!-\!(CH_2CH_2O)_n\!-$$

carbowax

14.8   a.

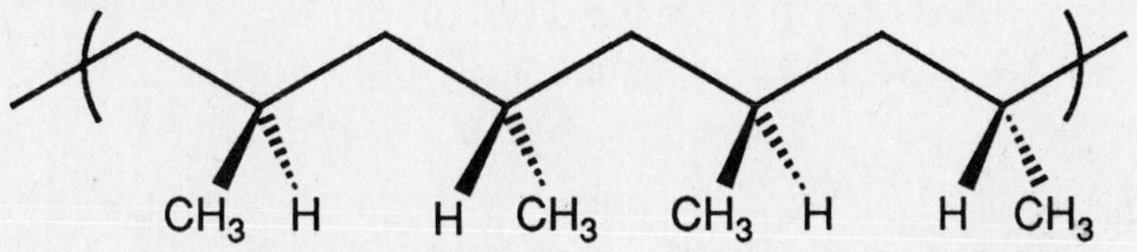

CH_3 H H CH_3 CH_3 H H CH_3

syndiotactic polypropylene: asymmetric carbons alternate configuration

b.

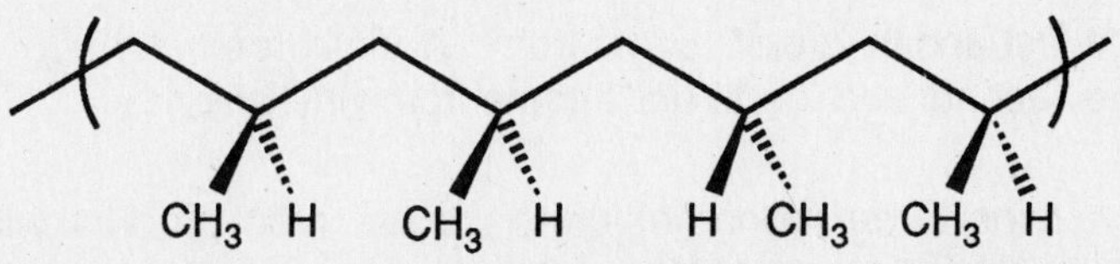

CH_3 H CH_3 H H CH_3 CH_3 H

atactic polypropylene: asymmetric carbons have random configurations

343

14.9

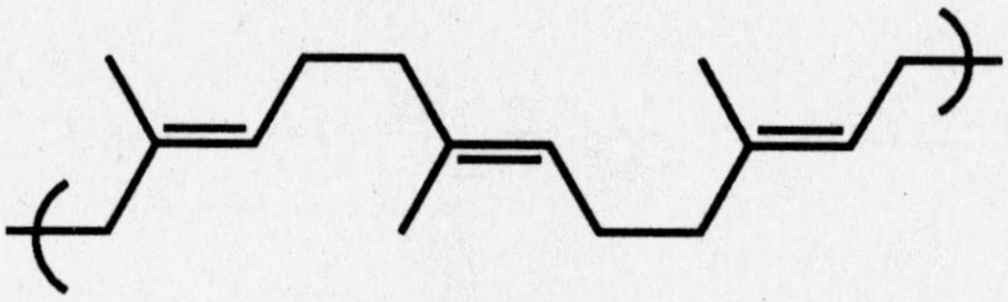

gutta-percha

14.10 a.

cis-poly(1,3-butadiene)

b.

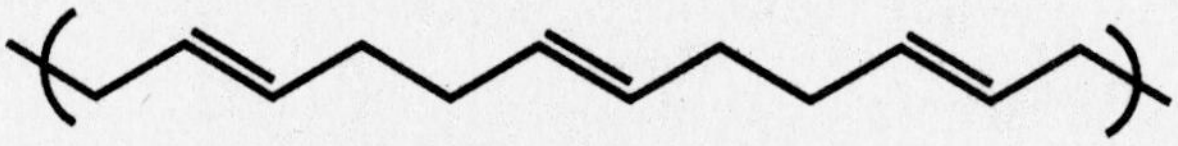

trans-poly(1,3-butadiene)

c.

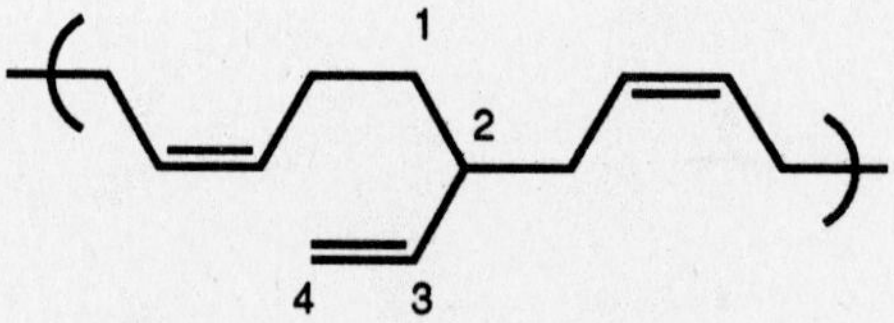

The origin of the middle unit is
indicated by the numbers.

14.11

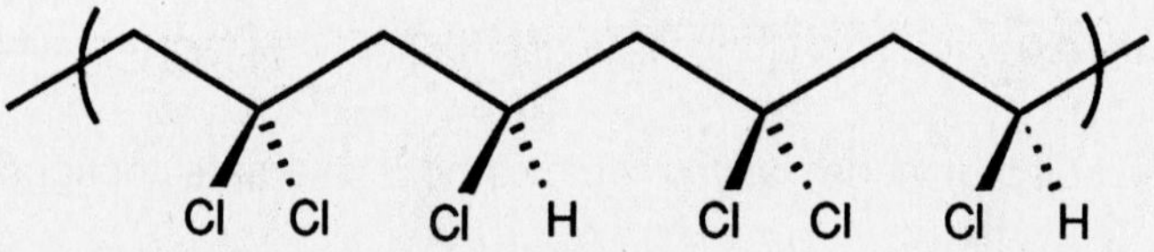

The first and third units come from 1,1-dichloroethene.
The second and fourth units come from vinyl chloride.

14.12 Each contains three monomer units, so the next product will contain six
monomer units. Its structure is:

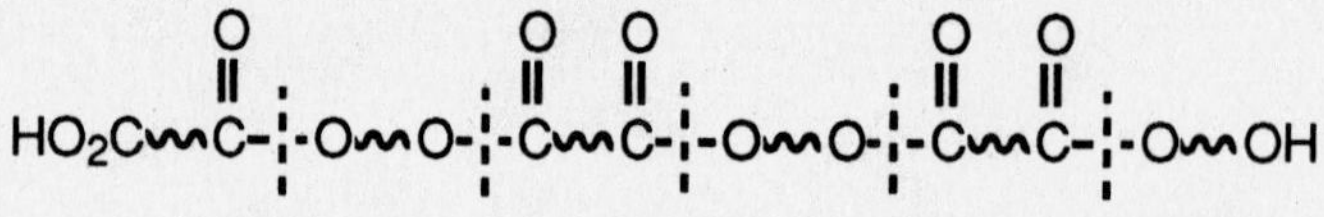

The monomer units are marked off by dashed lines. If the diester-diol reacted only with monomeric diacid, or if the diester-diacid reacted only with monomeric diol, the product would contain only four monomer units. But by reacting with each other, the two compounds engage in step growth to six monomer units.

14.13 $HO_2C\sim CO_2H$ + $HO\sim OH$ $\longrightarrow$ $HO\sim O-\overset{\overset{O}{\|}}{C}\sim\overset{\overset{O}{\|}}{C}-O\sim OH$

(large excess)   diester-diol

$HO\sim OH$ + $HO\sim O-\overset{\overset{O}{\|}}{C}\sim\overset{\overset{O}{\|}}{C}-O\sim OH$ $\longrightarrow$

(large excess)

$HO-\overset{\overset{O}{\|}}{C}\sim\overset{\overset{O}{\|}}{C}-O\sim O-\overset{\overset{O}{\|}}{C}\sim\overset{\overset{O}{\|}}{C}-OH$

diester-diacid

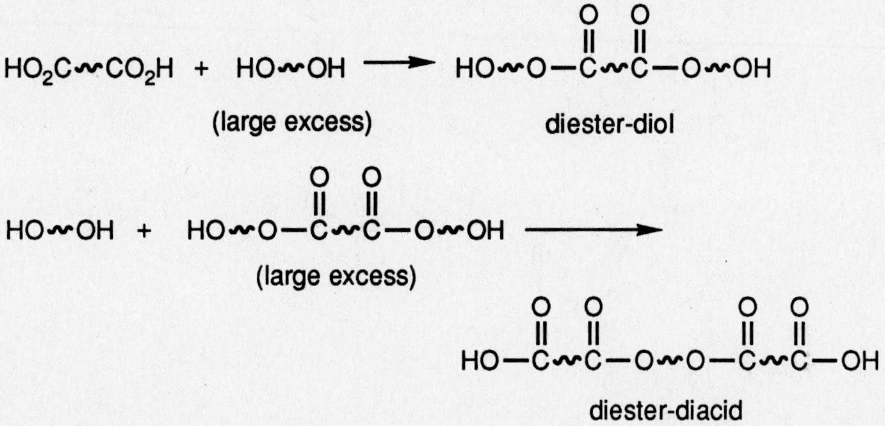

14.14 The monomers are:

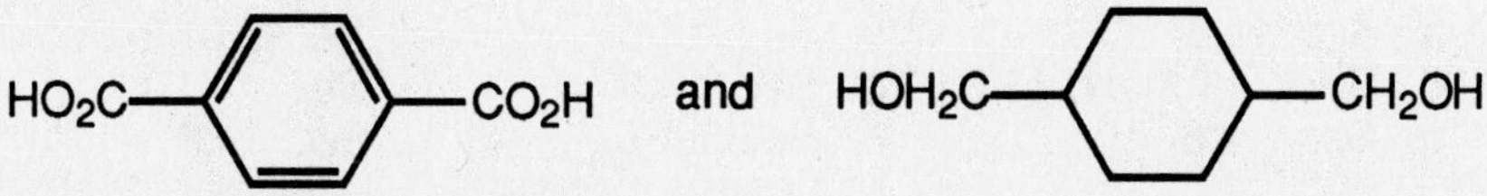

HO_2C-⟨benzene ring⟩$-CO_2H$ and HOH_2C-⟨cyclohexane ring⟩$-CH_2OH$

Note that the diol used for Kodel can be made by complete reduction of the dicarboxylic acid (terephthalic acid).

14.15

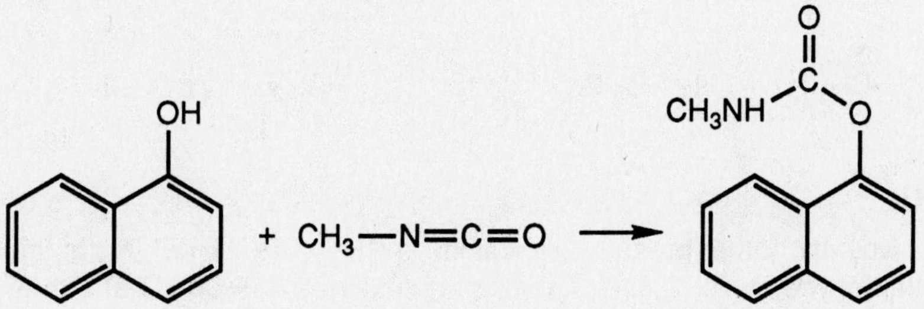

⟨1-naphthol with OH⟩ + $CH_3-N=C=O$ $\longrightarrow$ ⟨naphthalene with $CH_3NH-\overset{\overset{O}{\|}}{C}-O$ substituent⟩

14.16

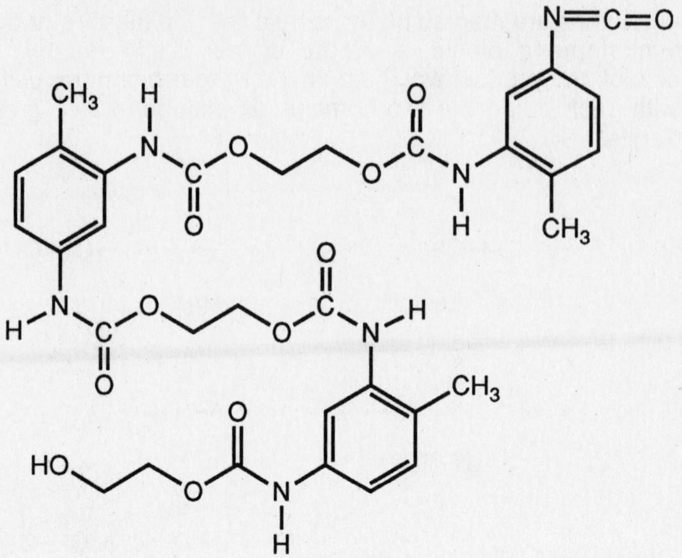

$$CH_3 \quad H \qquad\qquad O \qquad\qquad\qquad N\!=\!C\!=\!O$$

(structure: a diisocyanate-derived carbamate network)

CH₃–aryl–N(H)–C(=O)–O–CH₂CH₂–O–C(=O)–N(H)–aryl(CH₃)(N=C=O)

H–N–C(=O)–O–CH₂CH₂–O–C(=O)–N–H ... aryl(CH₃)

HO–CH₂CH₂–O–C(=O)–N(H)–aryl(CH₃)

Additional Problems

14.17 For the definitions and examples, see the indicated sections in the text:

a. Sec. 14.8	b. Sec. 14.8	c. Secs. 14.2-14.5
d. Secs. 14.3 and 14.7	e. Sec. 14.3	f. Sec. 14.10
g. Sec. 14.6	h. Sec. 14.6	i. Sec. 14.3

14.18 Follow eqs. 14.5-14.6, with L = Cl.

$$R\bullet \; + \; H_2C\!=\!CHCl \longrightarrow RCH_2\overset{\bullet}{C}HCl$$

$$RCH_2\overset{\bullet}{C}HCl \; + \; H_2C\!=\!CHCl \longrightarrow RCH_2\underset{\underset{Cl}{|}}{C}HCH_2\overset{\bullet}{C}HCl$$

14.19 The vinyl monomer presumably would be $CH_2\!=\!CHOH$, but this is the *enol* of acetaldehyde, which exists almost completely as $CH_3CH\!=\!O$ and thus cannot be a vinyl monomer. Polyvinyl acetate can, however, serve as the precursor to polyvinyl alcohol:

$$-CH_2CH-(CH_2CH)_n-CH_2CH- \quad \text{and so on}$$
$$|||$$
$$OHOHOH$$

14.20 The growing polymer chain can abstract a hydrogen atom from the methyl group of a propylene monomer:

$$\overset{\bullet}{\text{w}CH_2\text{-}CH} + CH_3\text{—}CH\text{=}CH_2 \longrightarrow \text{w}CH_2\text{-}CH_2 + H_2\overset{\bullet}{C}\text{—}CH\text{=}CH_2$$
$$\phantom{\text{w}CH_2\text{-}}|\phantom{CH + CH_3\text{—}CH\text{=}CH_2 \longrightarrow \text{w}CH_2\text{-}CH_2}|$$
$$\phantom{\text{w}CH_2\text{-}}CH_3\phantom{+ CH_3\text{—}CH\text{=}CH_2 \longrightarrow \text{w}CH_2\text{-}CH_2}CH_3$$

This process terminates the chain growth. It is an energetically favorable process because the resulting radical is allylic and resonance-stabilized:

$$\left[\quad H_2\overset{\bullet}{C}\text{—}CH\text{=}CH_2 \longleftrightarrow H_2C\text{=}CH\text{—}\overset{\bullet}{C}H_2 \quad \right]$$

14.21. The structure of a polymer derived from 1 benzoyloxy radical and 1000 ethylene units is:

$$\overset{\displaystyle O}{\underset{\displaystyle }{C_6H_5\text{—}\overset{\|}{C}\text{—}O\text{—}(CH_2CH_2)_{1000}\text{—}}}$$

mw = 121   mw = 28,000

The percentage of the molecular weight due to the initiator is:

$$\frac{121}{28,121} \times 100 = 0.43\,\%$$

The polymer derived from 1 benzoyloxy radical and 1000 styrene units is:

$$\overset{\displaystyle O}{\underset{\displaystyle }{C_6H_5\text{—}\overset{\|}{C}\text{—}O\text{—}(CH_2CHC_6H_5)_{1000}\text{—}}}$$

mw = 121   mw = 104,000

and the percentage of initiator is:

$$\frac{121}{104,121} \times 100 = 0.11\%$$

14.22 Chain termination by ion combination is not possible because the ions have like charges. In cationic polymerizations, termination by proton loss is possible:

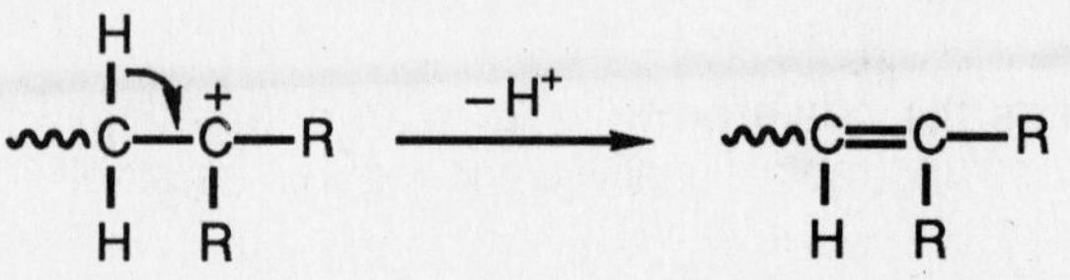

In anionic polymerizations, termination by proton abstraction (carbanions are strong bases) is one possibility:

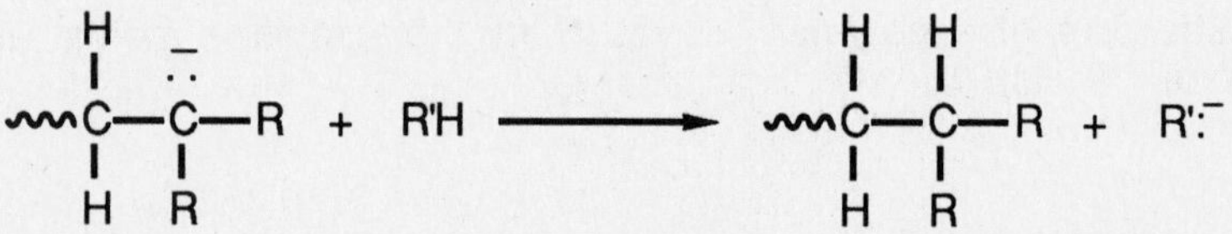

14.23 The organometallic reagent might simply add to the ester carbonyl group (Sec. 10.16):

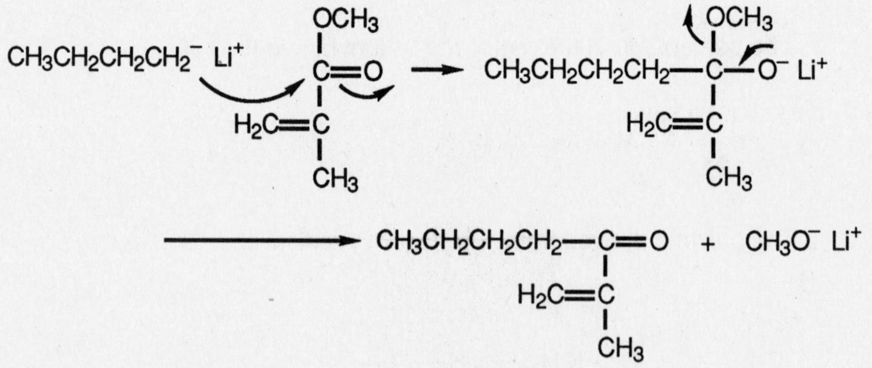

The reaction could continue further, to give a tertiary alcohol.

ANSWERS TO PROBLEMS

14.24 CH$_3$CHCH$_2$CHCH$_2$CHCH=CHCH$_3$
           | | |
          CH$_3$ CH$_3$ CH$_3$

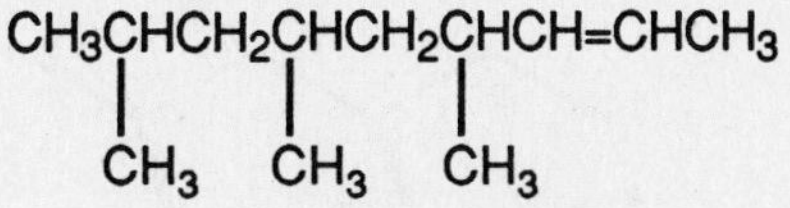

atactic polypropylene

14.25 The polymer has the following structure:

$$\left(O-CH_2-\underset{\underset{CH_3}{|}}{CH}\right)_n$$

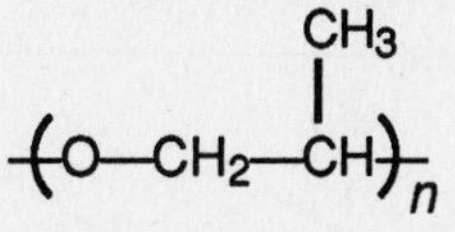

which is formed by S$_N$2 displacements on the primary carbon of the epoxide:

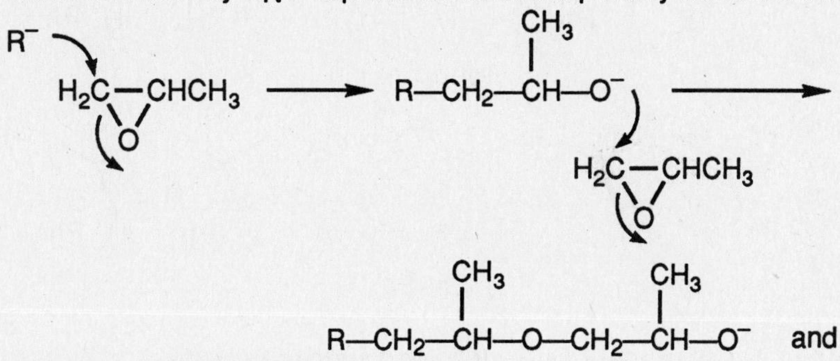

R$^-$ H$_2$C—CHCH$_3$ ⟶ R—CH$_2$—CH—O$^-$ ⟶

O

H$_2$C—CHCH$_3$

O

R—CH$_2$—CH—O—CH$_2$—CH—O$^-$ and so on.

CH$_3$ CH$_3$

14.26 The polymer has the following structure:

$$\left(CH_2-\underset{\underset{CO_2CH_3}{|}}{\overset{\overset{CN}{|}}{C}}\right)_n$$

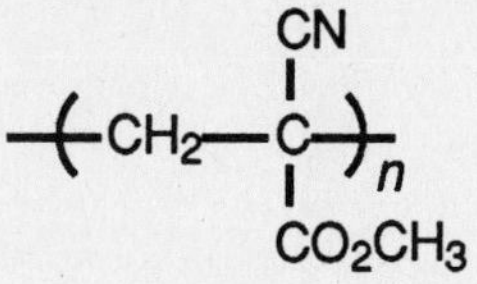

The monomer is susceptible to anionic polymerization because the intermediate anion is stabilized by two electron-withdrawing groups.

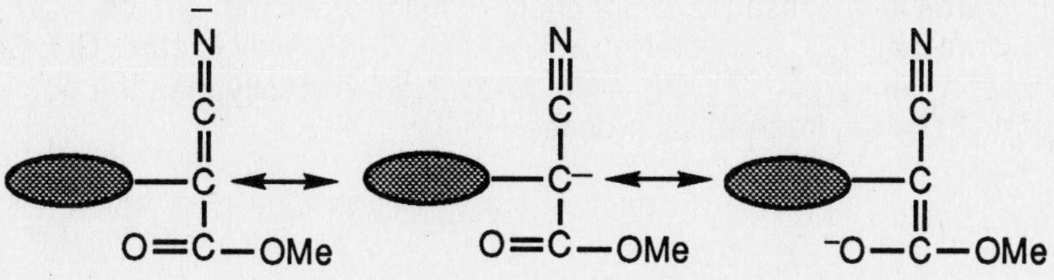

N≡C / N≡C / N≡C structures with C and O=C—OMe resonance forms

349

14.27

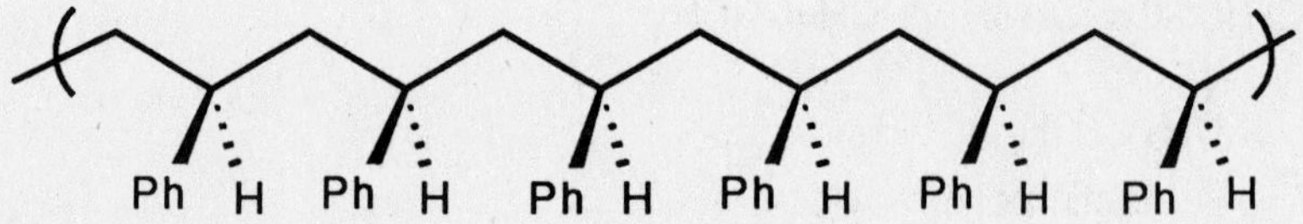

isotactic

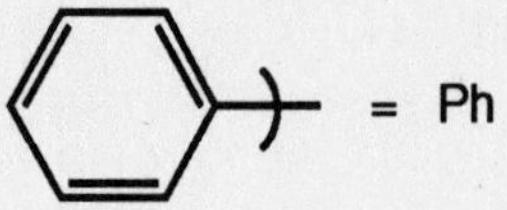

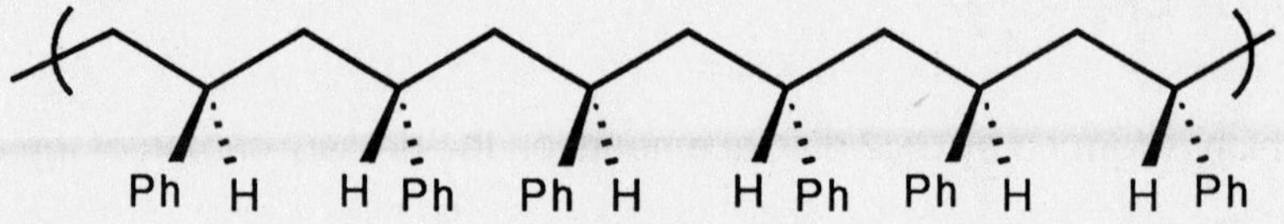

= Ph

syndiotactic

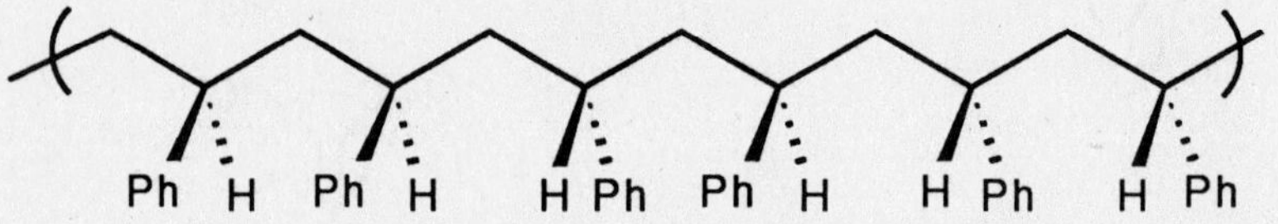

atactic

14.28 No. Poly(isobutylene) has the following structure:

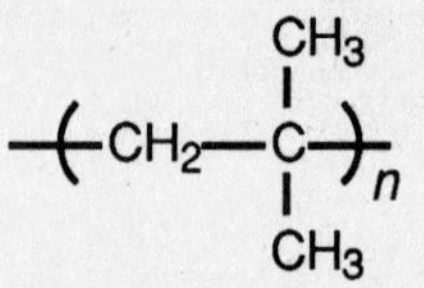

$$\left(\!\!CH_2\!\!-\!\!\underset{\underset{CH_3}{|}}{\overset{\overset{CH_3}{|}}{C}}\!\!\right)_{\!n}$$

It has no asymmetric carbons and thus does not exist in stereoisomeric forms.

14.29 Polyethylenes obtained by free-radical polymerization have highly branched structures as a consequence of chain-transfer reactions (see eq. 3.37 and the structure below it). Ziegler–Natta polyethylene is mainly linear: $(CH_2CH_2)_n$. It has a higher degree of crystallinity and a higher density than the polyethylene obtained by the free-radical process.

14.30 The structure of natural rubber is:

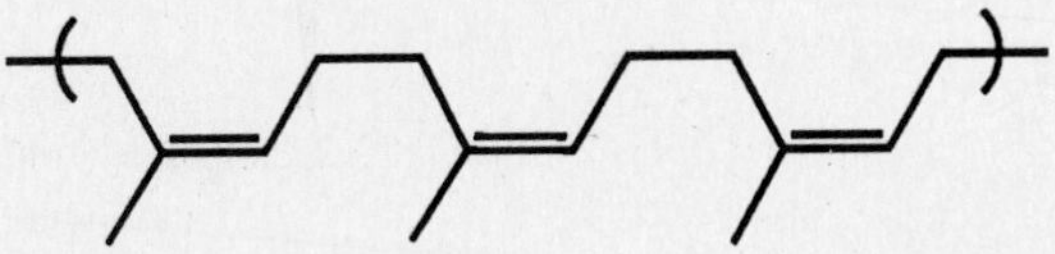

It will be cleaved by ozone at each double bond. The product of ozonolysis is therefore

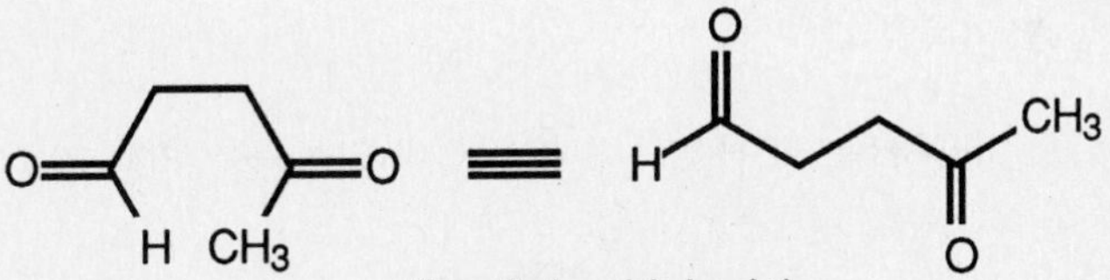

(levulinic aldehyde)

14.31 The reaction may be initiated by addition of a free radical from the catalyst to either butadiene or styrene:

$$R\cdot \ + \ H_2C=CHCH=CH_2 \ \longrightarrow \ \left[\begin{array}{c} R\overset{\bullet}{C}H_2CH-CH=CH_2 \\ \updownarrow \\ RCH_2CH=CH\overset{\bullet}{C}H_2 \end{array} \right]$$

$$R\cdot \ + \ H_2C=CHC_6H_5 \ \longrightarrow \ RCH_2\overset{\bullet}{C}HC_6H_5$$

These radicals may then add to either butadiene or styrene. The allylic radical from butadiene may add in either a 1,2- or a 1,4-manner (shown on the next page is only one of several alternative sequences of addition):

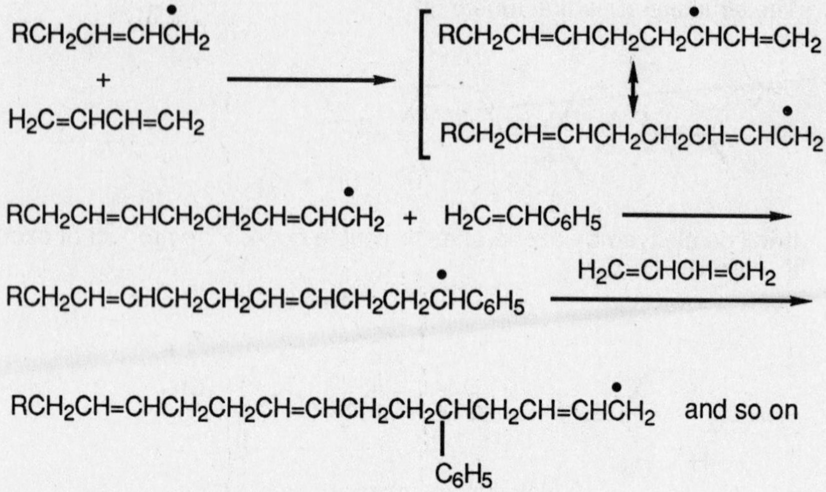

$$RCH_2CH=CH\overset{\bullet}{C}H_2$$

$$+$$

$$H_2C=CHCH=CH_2 \longrightarrow \left[\begin{array}{c} RCH_2CH=CHCH_2CH_2\overset{\bullet}{C}HCH=CH_2 \\ \updownarrow \\ RCH_2CH=CHCH_2CH_2CH=CH\overset{\bullet}{C}H_2 \end{array} \right]$$

$$RCH_2CH=CHCH_2CH_2CH=CH\overset{\bullet}{C}H_2 \ + \ H_2C=CHC_6H_5 \longrightarrow$$

$$RCH_2CH=CHCH_2CH_2CH=CHCH_2CH_2\overset{\bullet}{C}HC_6H_5 \xrightarrow{\quad H_2C=CHCH=CH_2 \quad}$$

$$RCH_2CH=CHCH_2CH_2CH=CHCH_2CH_2\underset{\underset{C_6H_5}{|}}{C}HCH_2CH=CH\overset{\bullet}{C}H_2 \quad \text{and so on}$$

14.32 The structure is analogous to that of natural rubber, except that the methyl group in each isoprene unit is replaced by a chlorine atom.

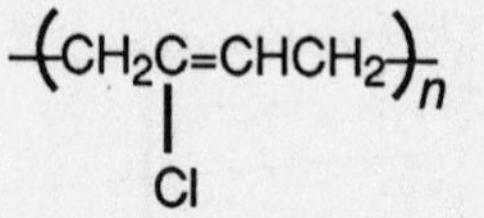

$$\left(CH_2\underset{\underset{Cl}{|}}{C}=CHCH_2\right)_n$$

14.33 The structure of a styrene–methyl methacrylate copolymer is:

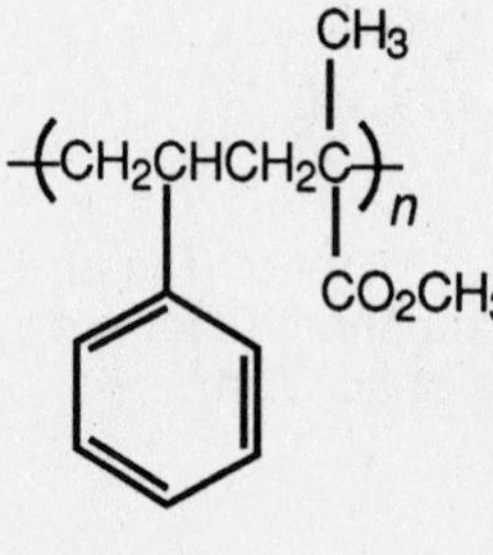

$$\left(CH_2CHCH_2\underset{\underset{CO_2CH_3}{|}}{\overset{\overset{CH_3}{|}}{C}}\right)_n$$

14.34 a.

$$-\!\!\left(\overset{\displaystyle O}{\overset{\|}{C}}(CH_2)_6\overset{\displaystyle O}{\overset{\|}{C}}NH(CH_2)_6NH\right)\!\!_n$$

b.

$$-\!\!\left(\overset{\displaystyle O}{\overset{\|}{C}}-N\!\!\underset{H}{\overset{H}{|}}\!-\!\!\bigcirc\!\!-\overset{H}{\underset{H}{\overset{|}{C}}}\!-\!\!\bigcirc\!\!-N\!\!\underset{H}{\overset{\displaystyle O}{|}}\!-\overset{\displaystyle O}{\overset{\|}{C}}\;\;OCH_2CH_2O\right)\!\!_n$$

c.

$$-\!\!\left(\overset{\displaystyle O}{\overset{\|}{C}}(CH_2)_4\overset{\displaystyle O}{\overset{\|}{C}}OCH_2CH_2O\right)\!\!_n$$

14.35

$$-\!\!\left(\overset{\displaystyle O}{\overset{\|}{C}}-O-\!\!\bigcirc\!\!-\overset{CH_3}{\underset{CH_3}{\overset{|}{\underset{|}{C}}}}\!-\!\!\bigcirc\!\!-O\right)\!\!_n$$

The other product of the polymerization is phenol.

14.36 The "unzipping" reaction, which gives formaldehyde, is expressed by the following curved arrows:

$$H\!-\!O\!-\!CH_2\!-\!O\!-\!CH_2\!-\!O\!-\!CH_2 \;\ldots\; \longrightarrow \; H^+ \;+\; n\; O\!=\!CH_2$$

This reaction is made possible by dissociation of a proton from a terminal hydroxyl group (an alcohol; see eq. 7.11). If the polymer is treated with acetic anhydride, the terminal hydroxyl groups are esterified, and unzipping is no longer possible:

$$CH_3\!-\!\overset{\displaystyle O}{\overset{\|}{C}}\!-\!O\!-\!CH_2\!-\!(OCH_2)_n\!-\!O\!-\!CH_2\!-\!O\!-\!\overset{\displaystyle O}{\overset{\|}{C}}\!-\!CH_3$$

14.37

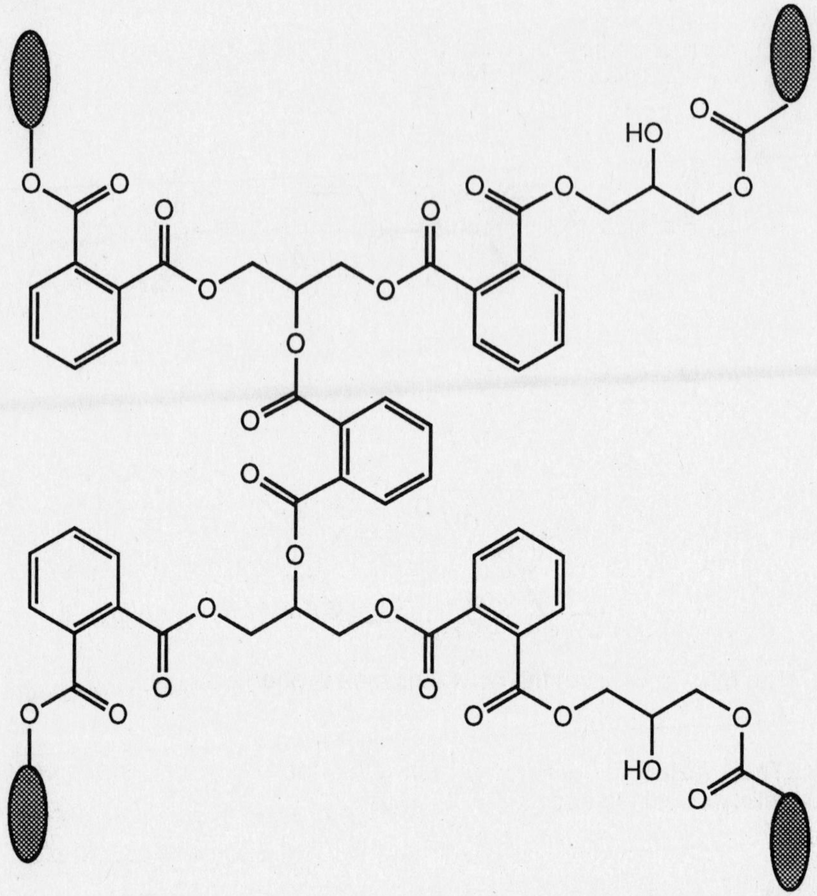

14.38 With the *para* position "blocked" by a methyl substituent, condensation can only occur *ortho* to the phenolic hydroxyl, leading to a linear (*not* a cross-linked) polymer.

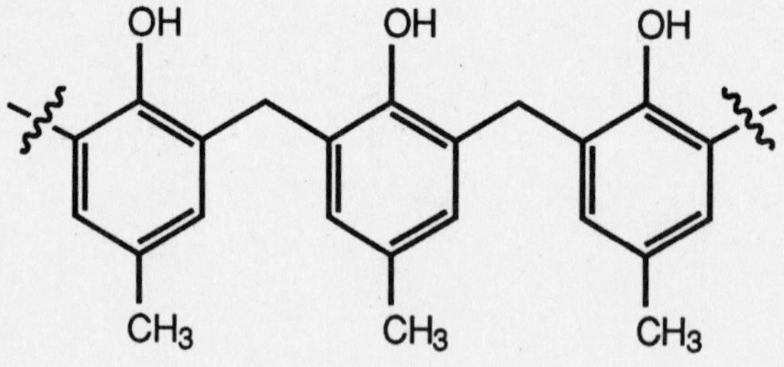

14.39 $H_2C=CHCH=CH_2$ $\xrightarrow{\text{Cl}_2}$ $Cl-CH_2CH=CHCH_2-Cl$

$\downarrow$ NaCN

$H_2N-(CH_2)_6-NH_2$ $\xleftarrow[\text{Ni}]{H_2}$ $NC-CH_2CH=CHCH_2-CN$

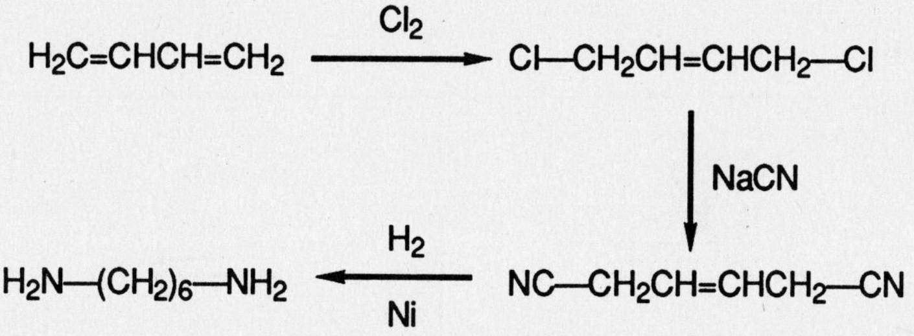

14.40

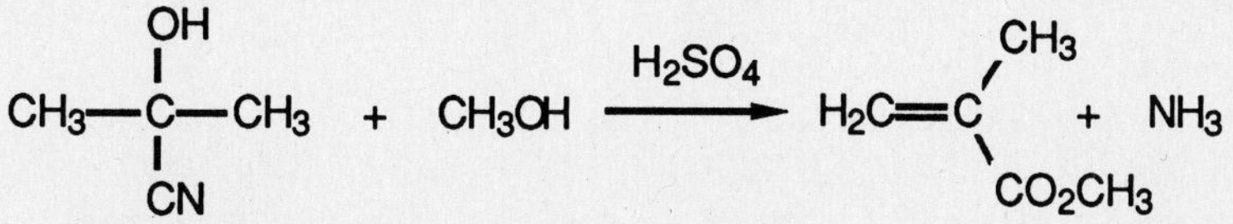

$$\underset{\substack{| \\ CN}}{\overset{\substack{OH \\ |}}{CH_3-C-CH_3}} + CH_3OH \xrightarrow{H_2SO_4} H_2C=\underset{CO_2CH_3}{\overset{CH_3}{C}} + NH_3$$

Mechanism:

$$\underset{\substack{| \\ CN}}{\overset{\substack{OH \\ |}}{CH_3-C-CH_3}} \rightleftharpoons \underset{\substack{| \quad | \\ H \quad CN}}{\overset{\substack{+ \\ OH_2 \\ |}}{H_2C-C-CH_3}} \longrightarrow \underset{\substack{| \\ C\equiv N}}{\overset{CH_3}{H_2C=C}}$$

$\downarrow\uparrow$ H$^+$

$$\underset{\substack{| \\ CH_3-O}}{\overset{\substack{CH_3 \\ |}}{H_2O:\curvearrowright \ C\overset{+}{=}NH_2}} \quad H_2C=C \xleftarrow{\text{H}^+ \text{ transfer}} \underset{\substack{| \\ CH_3-\overset{+}{O}H}}{\overset{\substack{CH_3 \\ |}}{H_2C=C}} \underset{C=NH}{\rightleftharpoons} \underset{\substack{| \\ CH_3\ddot{O}H}}{\overset{\substack{CH_3 \\ |}}{H_2C=C}} \underset{\substack{| \\ C\equiv \overset{+}{N}H}}{}$$

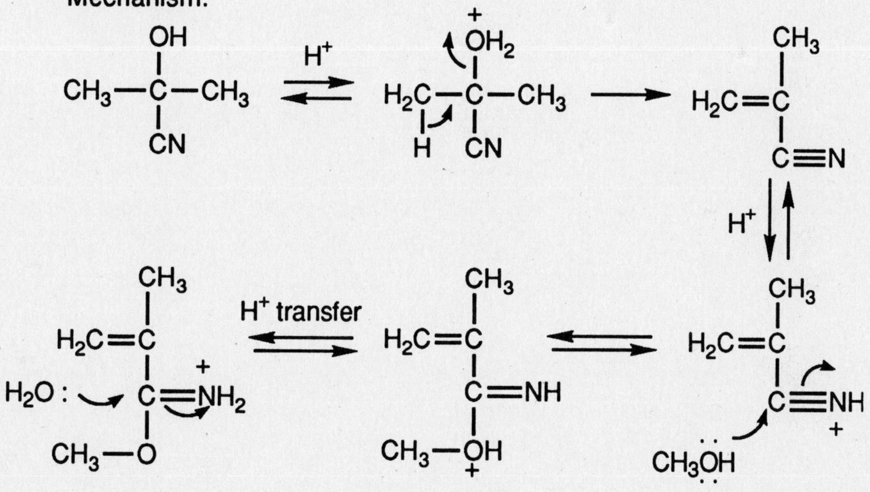

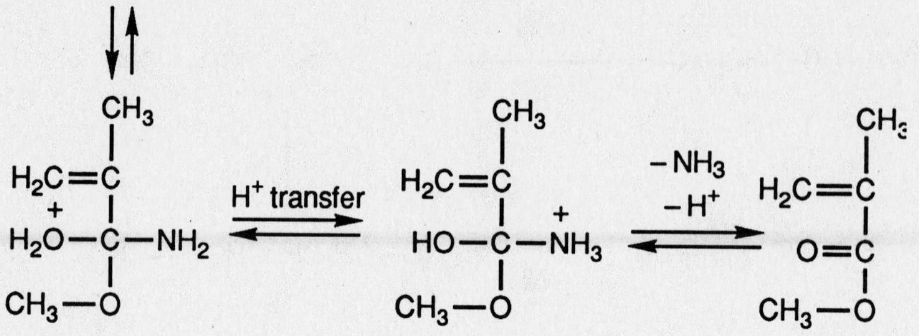

$$\text{CH}_3$$
$$\updownarrow$$

$$
\begin{array}{c}
\text{CH}_3 \\
| \\
\text{H}_2\text{C}=\text{C} \\
\overset{+}{} \quad | \\
\text{H}_2\text{O}-\text{C}-\text{NH}_2 \\
| \\
\text{CH}_3-\text{O}
\end{array}
\quad \xrightarrow[\longleftarrow]{\text{H}^+ \text{ transfer}} \quad
\begin{array}{c}
\text{CH}_3 \\
| \\
\text{H}_2\text{C}=\text{C} \\
\quad | \quad \overset{+}{} \\
\text{HO}-\text{C}-\text{NH}_3 \\
| \\
\text{CH}_3-\text{O}
\end{array}
\quad
\begin{array}{c}
-\text{NH}_3 \\
-\text{H}^+ \\
\longleftarrow
\end{array}
\quad
\begin{array}{c}
\text{CH}_3 \\
| \\
\text{H}_2\text{C}=\text{C} \\
| \\
\text{O}=\text{C} \\
| \\
\text{CH}_3-\text{O}
\end{array}
$$

CHAPTER FIFTEEN: LIPIDS AND DETERGENTS

CHAPTER SUMMARY

Lipids are constituents of plants and animals that are characteristically insoluble in water. **Fats** and **oils** are lipids that are triesters formed by reaction of the triol **glycerol** with long-chain saturated or unsaturated acids called **fatty acids**. The common acids in fats and oils have an even number of carbons (for example, **stearic acid** has 18 carbons), and if unsaturated (for example, **oleic acid**), they have the Z configuration (Table 15.1 lists the fatty acids, with their common names). **Hydrogenation** of oils, which have a high percentage of unsaturated acids, converts them to solid fats in a process called **hardening**.

Saponification of fats and oils by boiling with strong base gives glycerol and the sodium salts of fatty acids. The latter are **soaps**. Soaps contain a long carbon chain that is **lipophilic** and a terminal polar group that is **hydrophilic**. Soap molecules aggregate in water to form **micelles**, which help emulsify droplets of oil or grease.

Two disadvantages of ordinary soaps are their alkalinity and tendency to form insoluble salts with the Ca^{2+} or Mg^{2+} ions present in "hard" water. **Synthetic detergents** or **syndets** overcome these disadvantages. The most widely used syndets at present are **straight-chain alkylbenzenesulfonates**, obtained by successive alkylation and sulfonation of benzene. The polar portion of alkylbenzenesulfonates is anionic, and the straight chains are necessary for biodegradability. Syndets, in which the polar group is cationic, neutral, or dipolar, are also known. The manufacture of a detergent is a complex process, and the **surfactant** is usually only one portion of the commercial product. Detergents may also contain builders (to remove calcium and magnesium ions), bleaches, fabric softeners, enzymes (for stain removal), antiredeposition agents (to prevent soil deposition), optical brighters, antistatic agents, fragrances, and perfumes.

Lipids play a number of important biological roles. **Phospholipids** are triesters of glycerol in which one ester is derived from a phosphatidylamine. They are an important structural unit in cell membranes. **Prostaglandins** are 20-carbon cyclopentane derivatives of **arachidonic acid** that have profound biological effects even in minute quantities. **Waxes** are monoesters of long-chain acids and alcohols.

Terpenes are natural products usually obtained from the essential oils of plants. They contain multiples of five-carbon atoms (5, 10, 15, and so on). Each five-carbon arrangement is called an **isoprene unit**, a four-carbon chain with a one-carbon branch at C-2. Terpenes are frequently used in fragrances and perfumes. **Steroids** are lipids that contain a unique four-ring structure and are biosynthetically related to terpenes. Important examples of steroids include **cholesterol**, the **bile acids**, and the **sex hormones**.

REACTION SUMMARY

Saponification of a Triglyceride

$$
\begin{array}{l}
\text{H}_2\text{C}-\text{O}-\overset{\displaystyle \text{O}}{\overset{\|}{\text{C}}}-\text{R} \\[6pt]
\text{HC}-\text{O}-\overset{\displaystyle \text{O}}{\overset{\|}{\text{C}}}-\text{R} \;+\; 3\ \text{NaOH} \longrightarrow \\[6pt]
\text{H}_2\text{C}-\text{O}-\overset{\text{C}}{}-\text{R} \\[-2pt]
\qquad\qquad\overset{\|}{\text{O}}
\end{array}
\qquad
\begin{array}{l}
\text{H}_2\text{C}-\text{OH} \\[4pt]
\text{HC}-\text{OH} \quad +\; 3\ \text{Na}^{+\,-}\text{O}-\overset{\displaystyle \text{O}}{\overset{\|}{\text{C}}}-\text{R} \\[4pt]
\text{H}_2\text{C}-\text{OH}
\end{array}
$$

Hydrogenation of a Triglyceride (Hardening)

$$
\begin{array}{l}
\text{H}_2\text{C}-\text{O}-\overset{\displaystyle \text{O}}{\overset{\|}{\text{C}}}-(\text{CH}_2)_n\text{CH}=\text{CH}(\text{CH}_2)_m\text{CH}_3 \\[6pt]
\text{HC}-\text{O}-\overset{\displaystyle \text{O}}{\overset{\|}{\text{C}}}-(\text{CH}_2)_n\text{CH}=\text{CH}(\text{CH}_2)_m\text{CH}_3 \\[6pt]
\text{H}_2\text{C}-\text{O}-\overset{\text{C}}{}-(\text{CH}_2)_n\text{CH}=\text{CH}(\text{CH}_2)_m\text{CH}_3 \\[-2pt]
\qquad\qquad\overset{\|}{\text{O}}
\end{array}
\quad \xrightarrow[\text{Ni, heat}]{3\ \text{H}_2}
$$

$$
\begin{array}{l}
\text{H}_2\text{C}-\text{O}-\overset{\displaystyle \text{O}}{\overset{\|}{\text{C}}}-(\text{CH}_2)_n\text{CH}_2\text{CH}_2(\text{CH}_2)_m\text{CH}_3 \\[6pt]
\text{HC}-\text{O}-\overset{\displaystyle \text{O}}{\overset{\|}{\text{C}}}-(\text{CH}_2)_n\text{CH}_2\text{CH}_2(\text{CH}_2)_m\text{CH}_3 \\[6pt]
\text{H}_2\text{C}-\text{O}-\overset{\text{C}}{}-(\text{CH}_2)_n\text{CH}_2\text{CH}_2(\text{CH}_2)_m\text{CH}_3 \\[-2pt]
\qquad\qquad\overset{\|}{\text{O}}
\end{array}
$$

LEARNING OBJECTIVES

Hydrogenolysis of a Triglyceride

$$\text{H}_2\text{C}-\text{O}-\overset{\overset{\text{O}}{\|}}{\text{C}}-\text{R}$$
$$\text{HC}-\text{O}-\overset{\overset{\text{O}}{\|}}{\text{C}}-\text{R} \quad \xrightarrow[\text{zinc chromite}]{6\ \text{H}_2} \quad \begin{array}{l}\text{H}_2\text{C}-\text{OH}\\ \text{HC}-\text{OH}\\ \text{H}_2\text{C}-\text{OH}\end{array} + 3\ \text{HOH}_2\text{C}-\text{R}$$
$$\text{H}_2\text{C}-\text{O}-\overset{\underset{\text{O}}{\|}}{\text{C}}-\text{R}$$

LEARNING OBJECTIVES

1. Know the meaning of: triglyceride, fatty acid, fat, oil, hardening of a vegetable oil, hydrogenolysis, soap, saponification.

2. Know the structures and common names of the acids listed in Table 15.1.

3. Given the name of a glyceride, write its structure.

4. Given the name or structure of a carboxylic acid, write the formula for the corresponding glyceride.

5. Given the structure of a glyceride, write the equation for its saponification.

6. Given the structure of an unsaturated glyceride, write equations (including catalysts) for its hydrogenation and hydrogenolysis.

7. Explain the difference between the structure of a fat and that of a vegetable oil.

8. Describe the structural features essential for a good soap or detergent.

9. Explain, with the aid of a diagram, how a soap emulsifies fats and oils.

10. Explain, with the aid of equations, what happens when an ordinary soap is used in hard water and how synthetic detergents overcome this difficulty.

11. Know the maining of: lipophilic, hydrophilic, sodium alkyl sulfate, alkylben-zenesulfonate.

15. LIPIDS AND DETERGENTS

12. Explain the difference between anionic, cationic, neutral, and amphoteric detergents.

13. Know the meaning of: lipid, phospholipid, wax, prostaglandin.

14. Know the meaning of: terpene, isoprene unit, monoterpene, sesquiterpene, diterpene, triterpene, steroid, sex hormone, estrogen, androgen.

15. Identify the isoprene units in a given terpene.

ANSWERS TO PROBLEMS

Problems Within the Chapter

15.1

$CH_3(CH_2)_4$ ⟋═⟍⟋═⟍ $(CH_2)_7CO_2H$

15.2   a.

$H_2C-O-\overset{\overset{O}{\|}}{C}-(CH_2)_{14}CH_3$

$HC-O-\overset{\overset{O}{\|}}{C}-(CH_2)_{14}CH_3$

$H_2C-O-\underset{\underset{O}{\|}}{C}-(CH_2)_{14}CH_3$

b.

$H_2C-O-\overset{\overset{O}{\|}}{C}-(CH_2)_{14}CH_3$

$HC-O-\overset{\overset{O}{\|}}{C}-(CH_2)_7CH=CH(CH_2)_7CH_3$

$H_2C-O-\underset{\underset{O}{\|}}{C}-(CH_2)_{16}CH_3$

For mixed triglycerides, the name indicates the order in which the fatty acids are arranged. For example, glyceryl palmitooleostearate (Prob. 15.2b), glyceryl stearopalmitooleate (Example 15.2 on p. 406) and glyceryl palmitostearooleate (p. 406) represent three isomeric mixed triglycerides.

15.3   a. glycerol and sodium palmitate, $CH_3(CH_2)_{14}CO_2^-\ Na^+$

     b. glycerol and equimolar amounts of the sodium salts of the three carboxylic acids:

      $CH_3(CH_2)_{14}CO_2^-\ Na^+$
      sodium palmitate

      $CH_3(CH_2)_7CH=CH(CH_2)_7CO_2^-\ Na^+$
      sodium oleate

$CH_3(CH_2)_{16}CO_2^- Na^+$
sodium stearate

The triglyceride shown in Example 15.2 would give the same saponification products.

15.4   In general, the ratio of unsaturated to saturated acids is greater in vegetable oils than in animal fat (palm oil appears to be an exception).

15.5

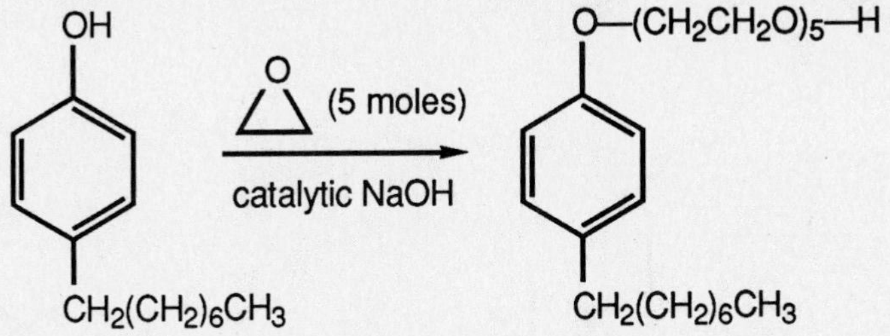

15.6 $CH_3(CH_2)_{10}CH_2N(CH_3)_2 \longrightarrow CH_3(CH_2)_{10}CH_2-\overset{\underset{\displaystyle CH_3}{|}}{\underset{\underset{\displaystyle CH_3}{|}}{N}}{}^+-CH_2CO_2H \ Br^-$

          $+$

          $BrCH_2CO_2H$

$\downarrow NaOH$

$H_2O \ + \ NaBr \ + \ CH_3(CH_2)_{10}CH_2-\overset{\underset{\displaystyle CH_3}{|}}{\underset{\underset{\displaystyle CH_3}{|}}{N}}{}^+-CH_2CO_2^-$

15.7

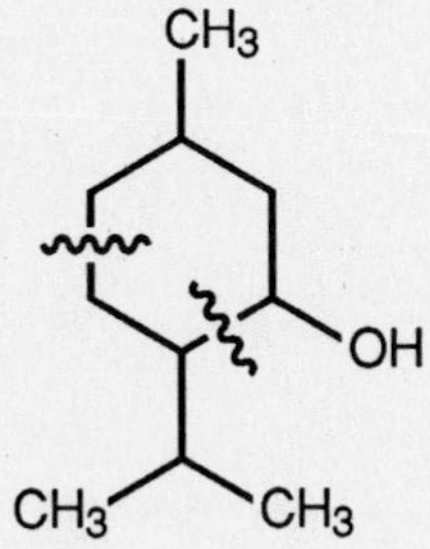

15.8

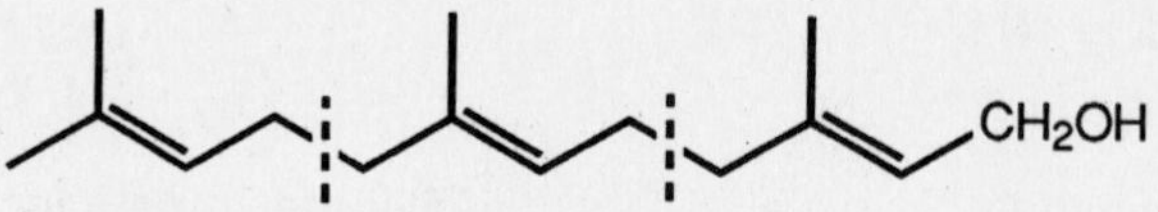

farnesol

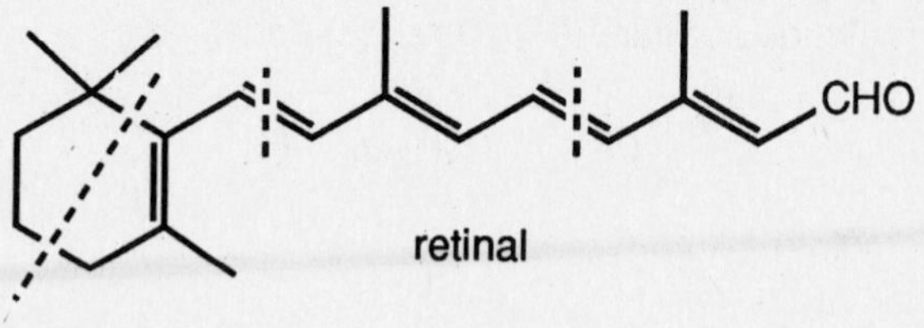

retinal

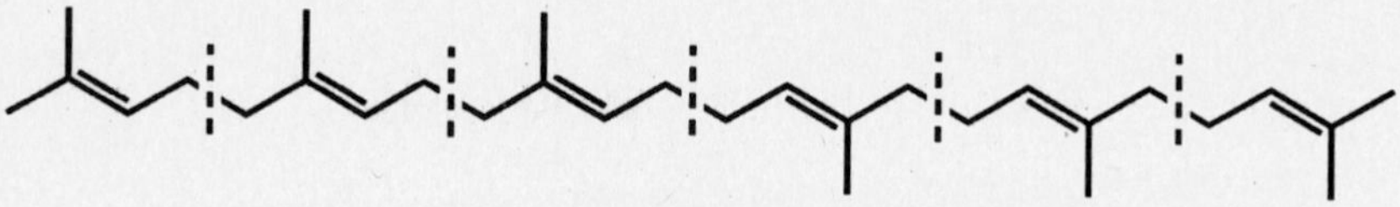

squalene

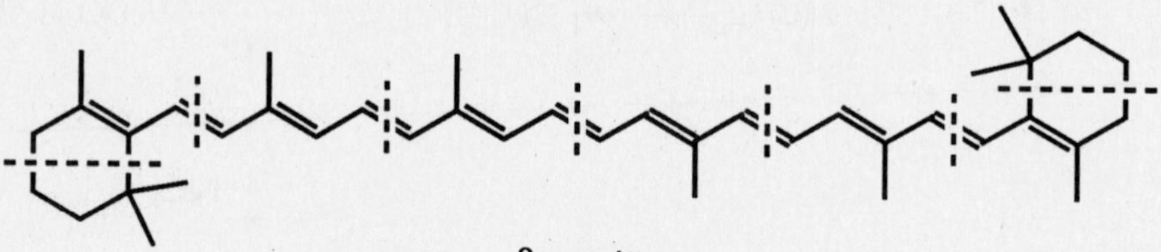

β-carotene

15.9 There are no asymmetric centers in squalene, while there are seven in lanosterol, as indicated below with asterisks.

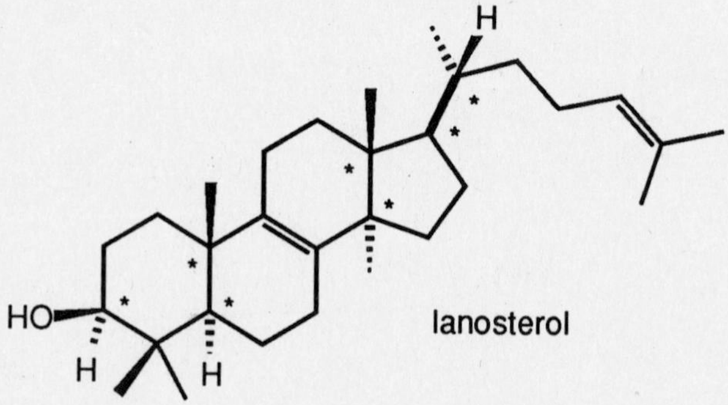

lanosterol

ANSWERS TO PROBLEMS

Additional Problems

15.10   a. $CH_3(CH_2)_{14}CO_2^- K^+$

b. $[CH_3(CH_2)_7CH{=}CH(CH_2)_7CO_2^-]_2\ Mg^{2+}$

c.

$$
\begin{array}{l}
\quad\quad\quad\quad\ \ \overset{\displaystyle O}{\overset{\displaystyle \|}{}} \\
H_2C-O-C-(CH_2)_{10}CH_3 \\
\quad\quad\quad\quad\ \ \overset{\displaystyle O}{\overset{\displaystyle \|}{}} \\
HC-O-C-(CH_2)_{10}CH_3 \\
H_2C-O-C-(CH_2)_{10}CH_3 \\
\quad\quad\quad\quad\ \ \underset{\displaystyle O}{\underset{\displaystyle \|}{}}
\end{array}
$$

d.

$$
\begin{array}{l}
\quad\quad\quad\quad\ \ \overset{\displaystyle O}{\overset{\displaystyle \|}{}} \\
H_2C-O-C-CH_2CH_2CH_3 \\
\quad\quad\quad\quad\ \ \overset{\displaystyle O}{\overset{\displaystyle \|}{}} \\
HC-O-C-(CH_2)_{14}CH_3 \\
H_2C-O-C-(CH_2)_7CH{=}CH(CH_2)_7CH_3 \\
\quad\quad\quad\quad\ \ \underset{\displaystyle O}{\underset{\displaystyle \|}{}}
\end{array}
$$

e. $CH_3(CH_2)_4CH{=}CHCH_2CH{=}CH(CH_2)_7CO_2(CH_2)_{13}CH_3$

f. $CH_3(CH_2)_{18}CO_2CH_2CH_3$

15.11 Saponification:

$$
\begin{array}{l}
\quad\quad\quad\quad\ \ \overset{\displaystyle O}{\overset{\displaystyle \|}{}} \\
H_2C-O-C-(CH_2)_7CH{=}CHCH_2CH{=}CHCH_2CH{=}CHCH_2CH_3 \\
\quad\quad\quad\quad\ \ \overset{\displaystyle O}{\overset{\displaystyle \|}{}} \\
HC-O-C-(CH_2)_7CH{=}CHCH_2CH{=}CHCH_2CH{=}CHCH_2CH_3 \\
H_2C-O-C-(CH_2)_7CH{=}CHCH_2CH{=}CHCH_2CH{=}CHCH_2CH_3 \\
\quad\quad\quad\quad\ \ \underset{\displaystyle O}{\underset{\displaystyle \|}{}}
\end{array}
$$

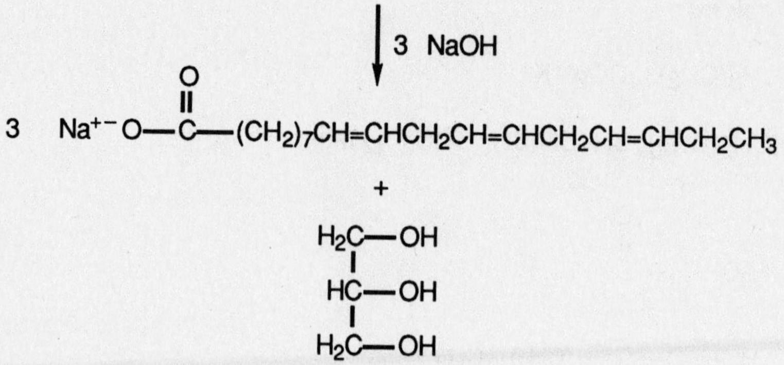

$$3 \text{ NaOH} \downarrow$$

$$3 \quad \text{Na}^{+-}\text{O}-\overset{\overset{\textstyle O}{\|}}{C}-(CH_2)_7CH=CHCH_2CH=CHCH_2CH=CHCH_2CH_3$$

$$+$$

$$
\begin{array}{l}
H_2C-OH \\
| \\
HC-OH \\
| \\
H_2C-OH
\end{array}
$$

Hydrogenation:

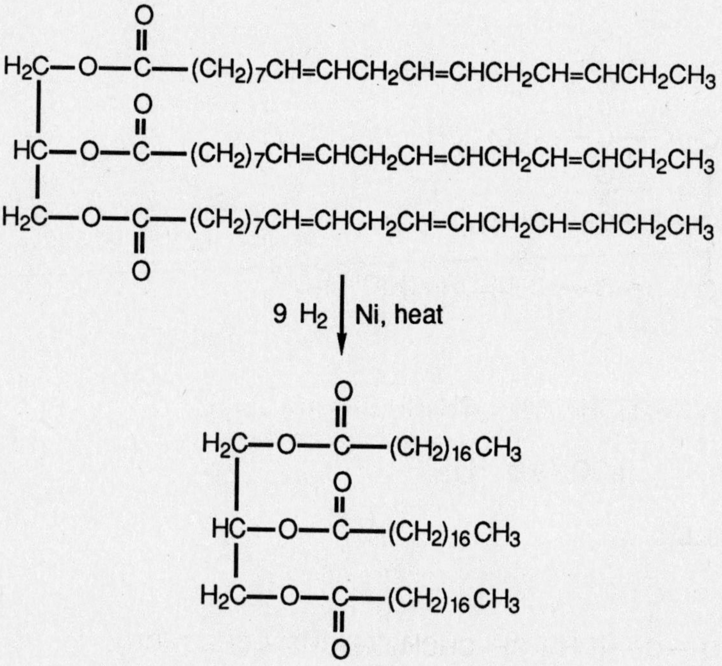

$$
\begin{array}{l}
H_2C-O-\overset{\overset{\textstyle O}{\|}}{C}-(CH_2)_7CH=CHCH_2CH=CHCH_2CH=CHCH_2CH_3 \\
| \\
HC-O-\overset{\overset{\textstyle O}{\|}}{C}-(CH_2)_7CH=CHCH_2CH=CHCH_2CH=CHCH_2CH_3 \\
| \\
H_2C-O-\underset{\underset{\textstyle O}{\|}}{C}-(CH_2)_7CH=CHCH_2CH=CHCH_2CH=CHCH_2CH_3
\end{array}
$$

$$9 \text{ H}_2 \downarrow \text{ Ni, heat}$$

$$
\begin{array}{l}
H_2C-O-\overset{\overset{\textstyle O}{\|}}{C}-(CH_2)_{16}CH_3 \\
| \\
HC-O-\overset{\overset{\textstyle O}{\|}}{C}-(CH_2)_{16}CH_3 \\
| \\
H_2C-O-\underset{\underset{\textstyle O}{\|}}{C}-(CH_2)_{16}CH_3
\end{array}
$$

Hydrogenolysis:

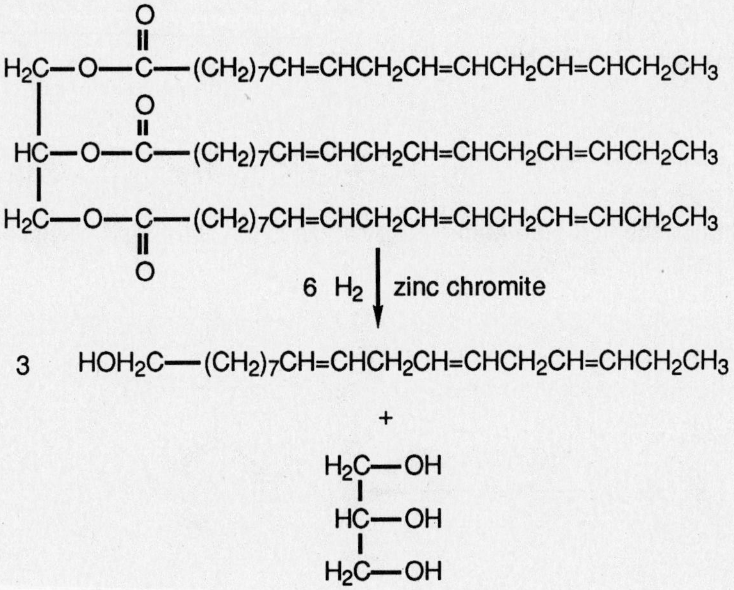

$$H_2C-O-\overset{\overset{O}{\|}}{C}-(CH_2)_7CH=CHCH_2CH=CHCH_2CH=CHCH_2CH_3$$

$$HC-O-\overset{\overset{O}{\|}}{C}-(CH_2)_7CH=CHCH_2CH=CHCH_2CH=CHCH_2CH_3$$

$$H_2C-O-\underset{\underset{O}{\|}}{C}-(CH_2)_7CH=CHCH_2CH=CHCH_2CH=CHCH_2CH_3$$

6 H$_2$ | zinc chromite

3   $HOH_2C-(CH_2)_7CH=CHCH_2CH=CHCH_2CH=CHCH_2CH_3$

+

$$H_2C-OH$$
$$HC-OH$$
$$H_2C-OH$$

15.12

$$H_2C-O-\overset{\overset{O}{\|}}{C}-(CH_2)_7CH=CHCH_2CH(OH)(CH_2)_5CH_3$$

$$HC-O-\overset{\overset{O}{\|}}{C}-(CH_2)_7CH=CHCH_2CH(OH)(CH_2)_5CH_3$$

$$H_2C-O-\underset{\underset{O}{\|}}{C}-(CH_2)_7CH=CHCH_2CH(OH)(CH_2)_5CH_3$$

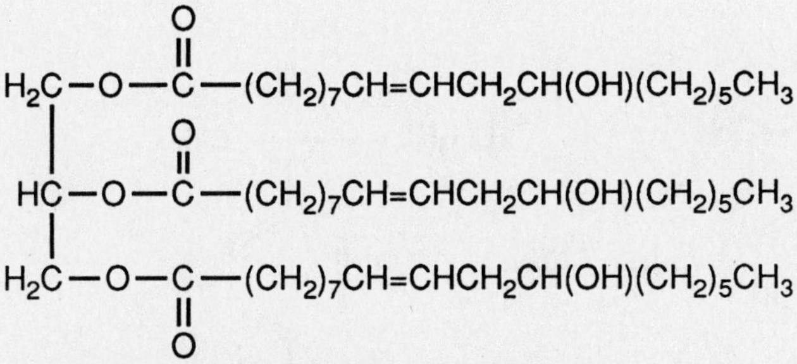

All the double bonds have *cis* geometry.

15.13   a. Compare with eq. 15.5.

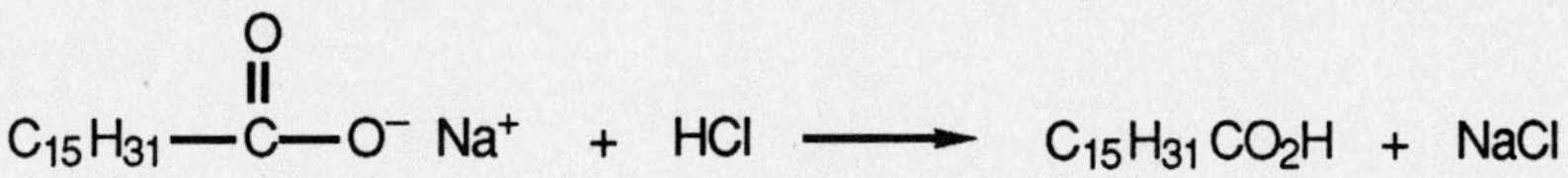

$$C_{15}H_{31}-\overset{\overset{O}{\|}}{C}-O^- Na^+ \;+\; HCl \longrightarrow \; C_{15}H_{31}CO_2H \;+\; NaCl$$

b. Compare with eq. 15.6.

$$2 \quad C_{15}H_{31}-\overset{\overset{\displaystyle O}{\|}}{C}-O^- \ Na^+ \ + \ Mg^{2+} \longrightarrow$$

$$(C_{15}H_{31}-\overset{\overset{\displaystyle O}{\|}}{C}-O^-)_2{-}Mg^{2+} \ (\downarrow) \quad + \quad 2 \ Na^+$$

15.14 Note that in the first step Markovnikov's rule is followed, and alkylation occurs via a secondary carbocation.

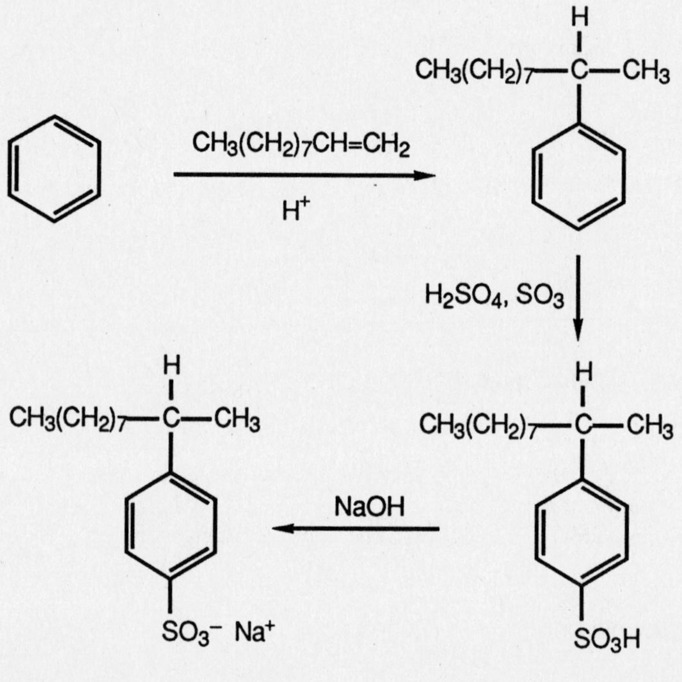

15.15 The first steps are analogous to eq. 8.20, and the last steps are analogous to eq. 15.8.

$$CH_3(CH_2)_{10}CH_2OH \xrightarrow[H^+]{\overset{O}{\underset{H_2C-CH_2}{\triangle}}} CH_3(CH_2)_{10}CH_2OCH_2CH_2OH$$

repeat twice with ethylene oxide

$$CH_3(CH_2)_{10}CH_2(OCH_2CH_2)_3OSO_3^-Na^+$$

$$\xleftarrow[\text{2. NaOH}]{\text{1. cold } H_2SO_4} CH_3(CH_2)_{10}CH_2(OCH_2CH_2)_3OH$$

15.16 A good syndet should have a lipophilic portion with a hydrocarbon chain of appropriate length to dissolve oil droplets. The hydrocarbon chain should be biodegradable and therefore should not have any branching. A good syndet should also have a polar portion that creates a micelle surface that is attractive to (or soluble in) water. If the polar group is anionic, it should not form insoluble salts with trace metals commonly found in water (Mg^{2+}, Ca^{2+}).

15.17 Polar portions of opposite charge might be attracted to one another. This would produce "reverse" micelles that would not be capable of emulsifying oil droplets in water or even be soluble in water.

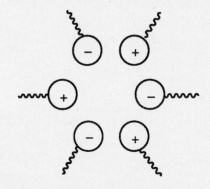

15.18 Many answers are possible. Definitions or examples can be found in the indicated sections of the text.

a. Sec. 15.2   b. Secs. 15.2 and 15.3
c. Sec. 15.9   d. Sec. 15.4

15. LIPIDS AND DETERGENTS

15.19 The priority order of groups attached to the chiral center is:

$$R'CO_2\rightarrow -CH_2OP > -CH_2OC(=O)- > -H$$

The structure of the lecithin is therefore

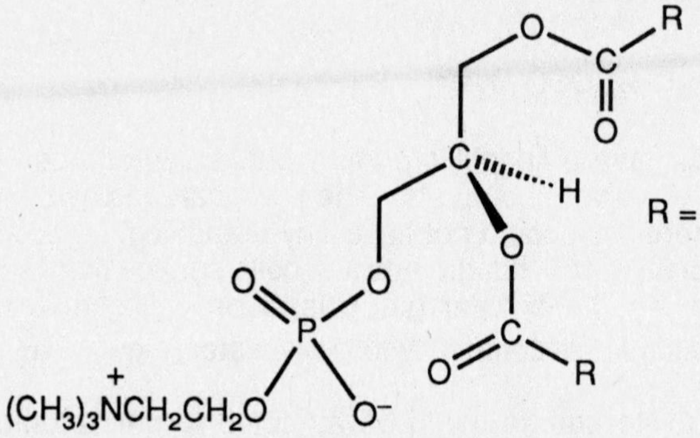

R = long chain hydrocarbon
(see Sec. 15.7)

15.20 All the double bonds of arachidonic acid have Z geometry.

15.21 a. There are four chiral centers in PGE$_2$.

b. 8R, 11R, 12R, 15S.

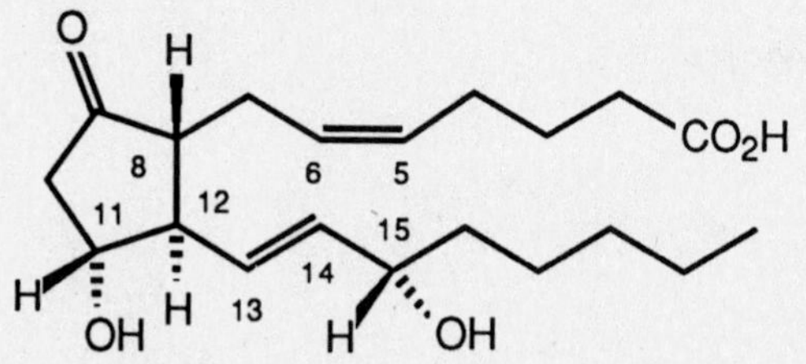

c. 5Z, 13E.

d. The side chains are *trans* to one another.

15.22 See Sec. 15.9.

$$CH_3(CH_2)_{14}-\overset{\overset{\displaystyle O}{\|}}{C}-O(CH_2)_{15}CH_3 \xrightarrow{\text{NaOH}} CH_3(CH_2)_{14}CO_2^-\,Na^+$$

$$+$$

$$CH_3(CH_2)_{15}OH$$

15.23 Saponification of fats and oils with strong alkali, such as sodium hydroxide, gives sodium salts of fatty acids and glycerol, both of which are soluble in water. On the other hand, hydrolysis of waxes gives sodium salts of fatty acids and long chain alcohols. The long chain alcohols are too "hydrocarbonlike" to be soluble in water and do not dissolve (consult Table 7.1).

15.24 a.   b.

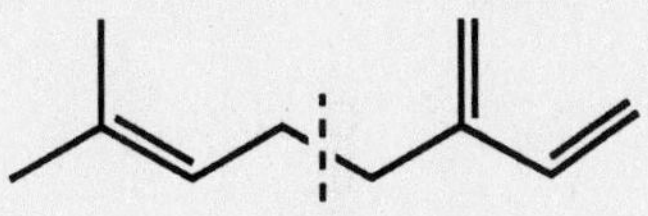

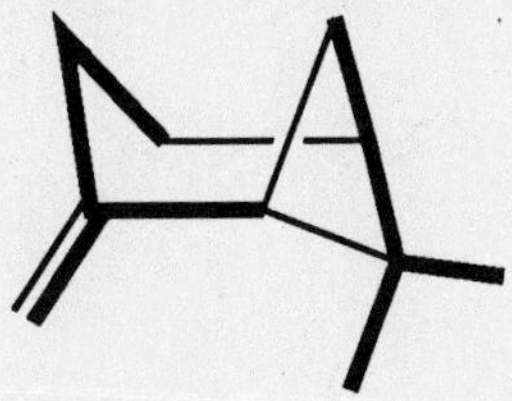

The two isoprene units
are outlined in bold.

15.25 The head-to-tail arrangement of isoprene units breaks down in squalene at the position indicated below:

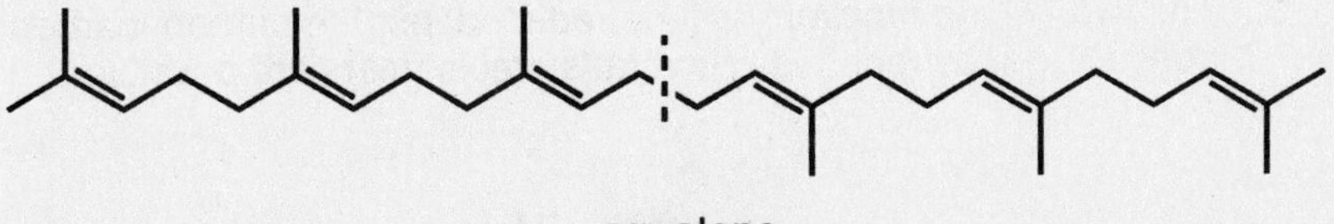

squalene

This suggests that squalene is biosynthesized by the head-to-head connection of two 15-carbon pieces rather than, say, stepwise connection of six isoprene units in a linear fashion. β-Carotene may be biosynthesized in a similar manner by the head-to-head connection of two 20-carbon units at the head-to-tail break point indicated at the top of the next page:

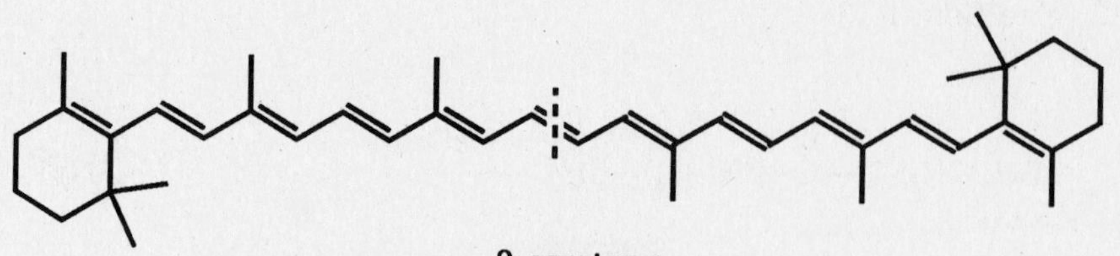

β-carotene

15.26 a. The two hydroxyl groups will be esterified.

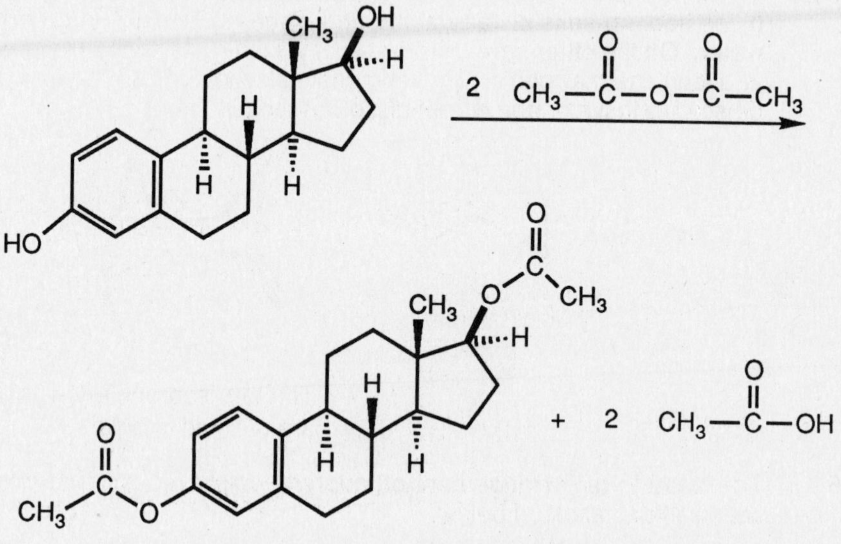

b. The two ketone functions will be reduced, but the carbon–carbon double bond will not be reduced. Several stereoisomers are possible.

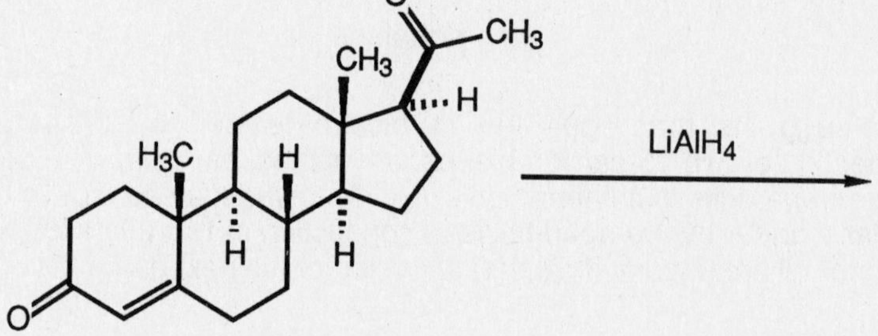

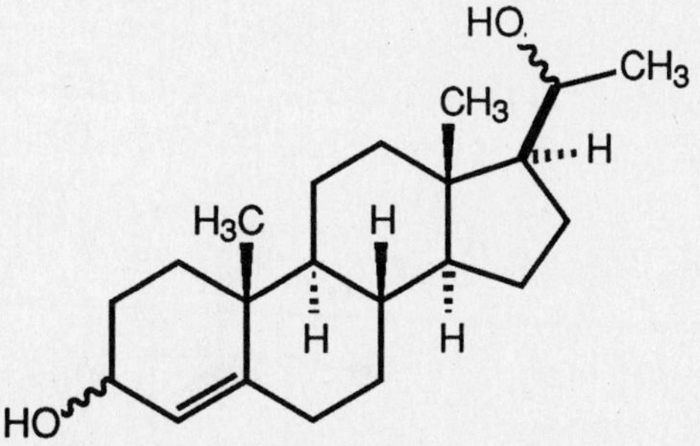

c. The alkene will be converted to an epoxide. Two stereoisomers are possible.

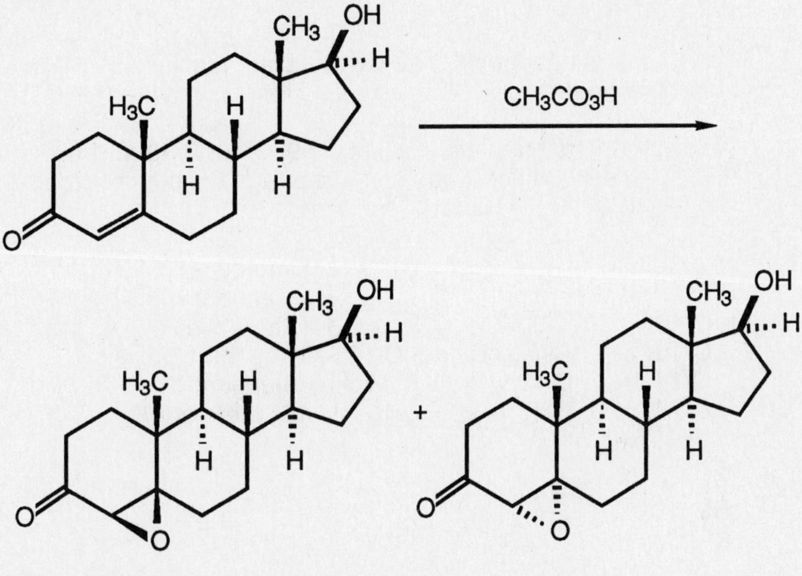

d. The secondary alcohol will be oxidized to a ketone.

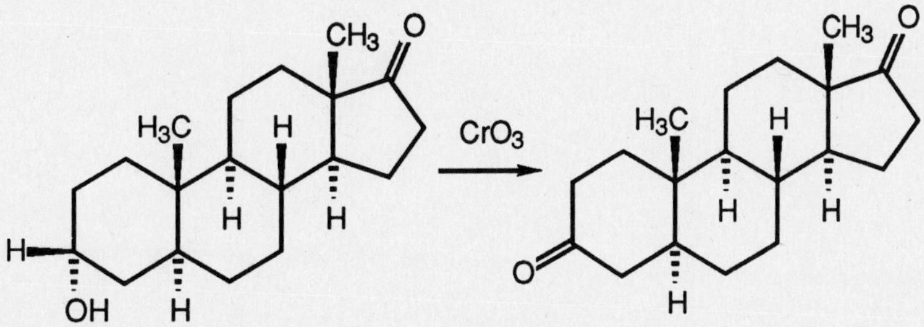

15.27 a.   Use the steroid numbering system shown on page 420 of the text.

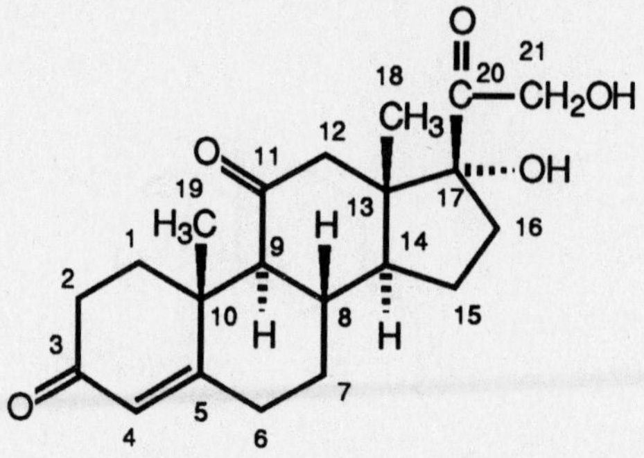

Note that the "angular" methyl groups are numbered 18 and 19 and that numbering then continues with the side chain attached to C-17.

b.   Chiral centers are present at C-8, C-9, C-10, C-13, C-14, and C-17. To assign configuration, you must first assign the priority order at each chiral center.

C-8:   C-9 > C-14 > C-7 > H, therefore *S*
C-9:   C-11 > C-10 > C-8 > H, therefore *S*
C-10:   C-5 > C-9 > C-1 > C-19, therefore *R*
C-13:   C-17 > C-14 > C-12 > C-18, therefore *S*
C-14:   C-13 > C-8 > C-15 > H, therefore *S*
C-17:   OH > C-20 > C-13 > C-16, therefore *R*

CHAPTER SIXTEEN: CARBOHYDRATES

CHAPTER SUMMARY

Carbohydrates are polyhydroxy aldehydes or ketones, or substances that give such compounds on hydrolysis. They are classified as polysaccharides, oligosaccharides, or monosaccharides.

Monosaccharides, also called simple sugars, are classified by the number of carbon atoms (triose, tetrose, pentose, etc.) and by the nature of the carbonyl group (aldose or ketose).

R-(+)-Glyceraldehyde is an aldotriose designated by the following **Fischer projection formula:**

This configuration is designated D, whereas the enantiomer with the H and OH positions reversed is L. In larger monosaccharides, the letters D and L are used to designate the configuration of the chiral center with the *highest* number, the chiral carbon most remote from the carbonyl group. The D-aldoses through the hexoses are listed in Figure 16.1.

Epimers are stereoisomers that differ in configuration at *only one* chiral center.

Monosaccharides with more than four carbons usually exist in a cyclic hemiacetal form, in which a hydroxyl group on C-4 or C-5 reacts with the carbonyl group (at C-1 in aldoses) to form a hemiacetal. In this way, C-1 also becomes chiral and is called the **anomeric carbon. Anomers** differ in configuration only at this chiral center and are designated α or β. The common ring sizes for the cyclic hemiacetals are six-membered (called **pyranoses**) or five-membered (called **furanoses**). The rings contain one oxygen atom and five or four carbon atoms, respectively.

Anomers usually interconvert in solution, resulting in a gradual change in optical rotation from that of the pure anomer to an equilibrium value for the mixture. This rotational change is called **mutarotation**.

Haworth formulas are a useful way of representing the cyclic forms of monosaccharides. The rings are depicted as flat, with hydroxyl groups or other substituents above or below the ring plane.

Monosaccharides can be oxidized at the aldehyde carbon to give carboxylic acids called **aldonic acids.** Oxidation at both ends of the carbon chain gives **aldaric acids.** Reduction of the carbonyl group to an alcohol gives polyols called **alditols.** The —OH groups in sugars, like those in simpler alcohols, can be esterified or etherified.

Monosaccharides react with alcohols (H^+ catalyst) to give **glycosides**. The —OH group at the anomeric carbon is replaced by an —OR group. The product is an acetal. Alcohols and phenols often occur in nature combined with sugars as glycosides, which renders them water soluble.

Disaccharides consist of two monosaccharides linked by a glycosidic bond between the anomeric carbon of one unit and a hydroxyl group (often on C-4) of the other unit. Examples include **maltose** and **cellobiose** (formed from two glucose units joined by a 1,4-linkage and differing only in configuration at the anomeric carbon, being α and β, respectively), **lactose** (from a galactose and glucose unit linked 1,4 and β), and **sucrose**, or cane sugar (from a fructose and glucose unit, linked at the anomeric carbon of each, or 1,2).

Sugars such as fructose, glucose, and sucrose are sweet, but others (for example, lactose and galactose) are not. Some noncarbohydrates, such as **saccharin** and **aspartame**, also taste sweet.

Polysaccharides have many monosaccharide units linked by glycosidic bonds. **Starch** and **glycogen** are polymers of D-glucose, mainly linked 1,4 or 1,6 and α. **Cellulose** consists of D-glucose units linked 1,4 and β.

Monosaccharides with modified structures are often biologically important. Examples include **sugar phosphates, deoxy sugars, amino sugars,** and **ascorbic acid** (vitamin C).

REACTION SUMMARY

Hydrolysis

$$\text{polysaccharide} \xrightarrow{H_3O^+} \text{oligosaccharide} \xrightarrow{H_3O^+} \text{monosaccharide}$$

REACTION SUMMARY

Acyclic and Cyclic Equilibration

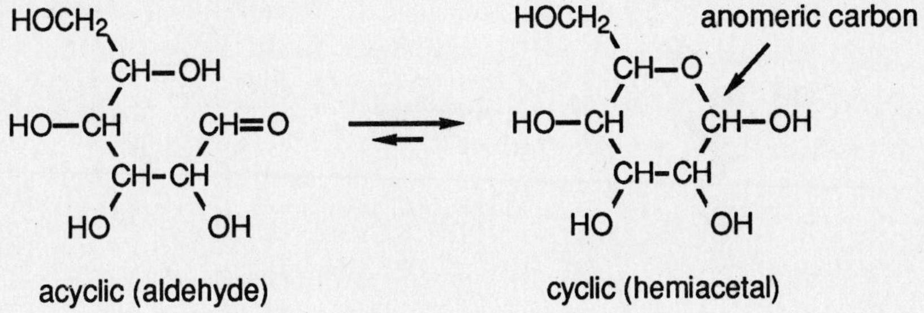

acyclic (aldehyde)   cyclic (hemiacetal)

Oxidation

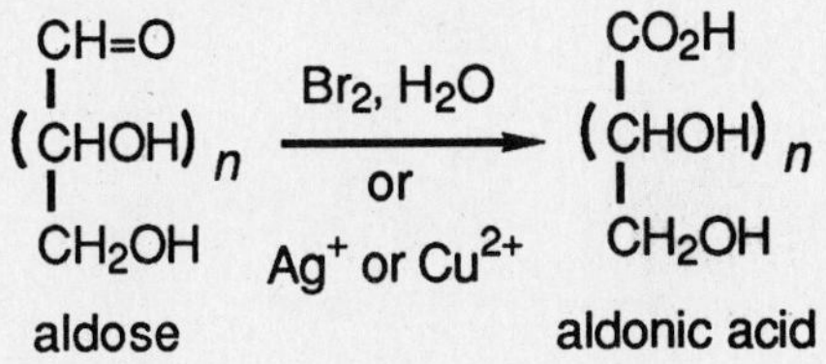

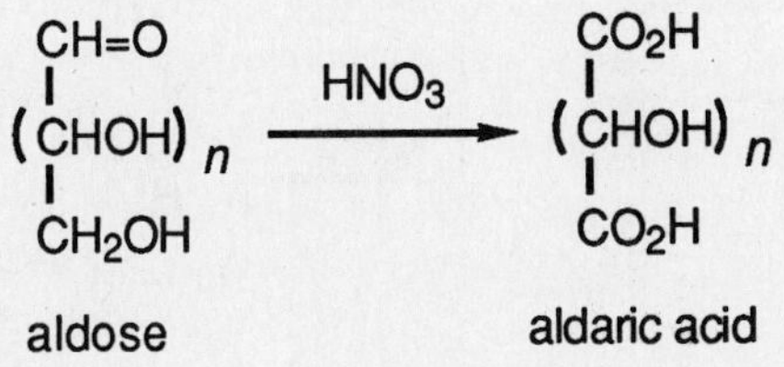

Reduction

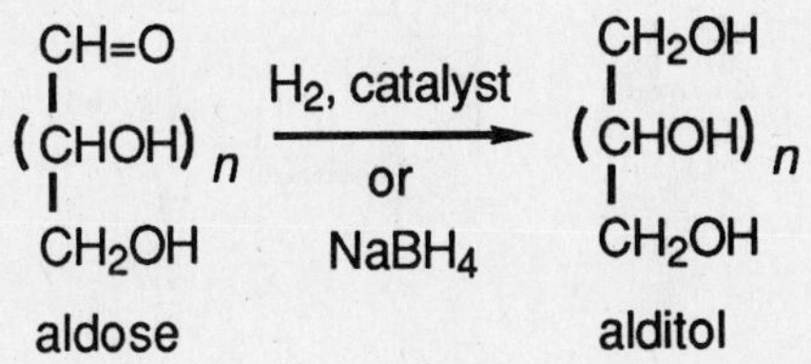

16. CARBOHYDRATES

Esterification and Etherification

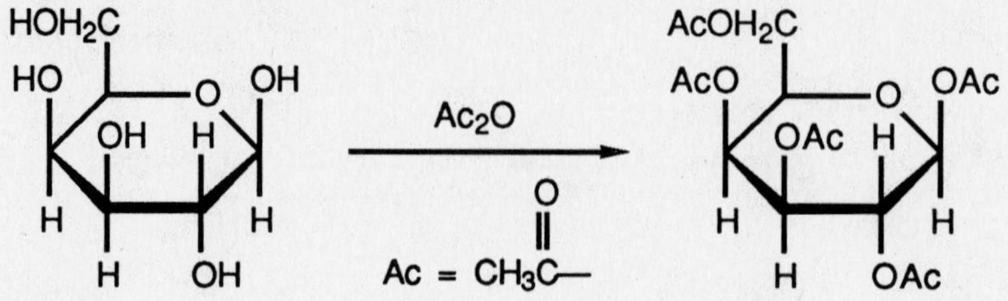

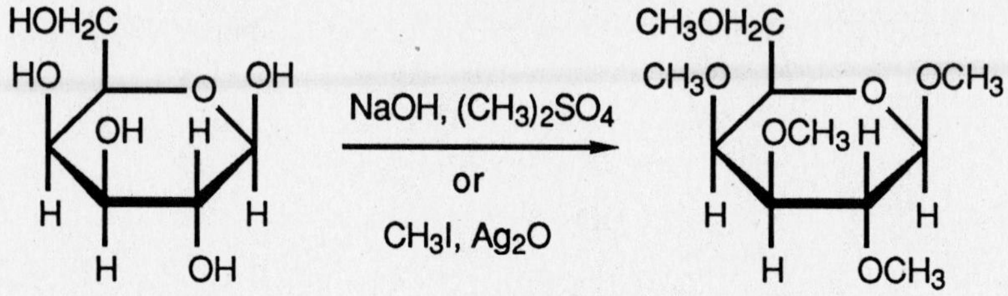

$$Ac_2O$$

$$Ac = CH_3\overset{\overset{O}{\|}}{C}-$$

$$NaOH, (CH_3)_2SO_4$$
$$or$$
$$CH_3I, Ag_2O$$

Formation of Glycosides

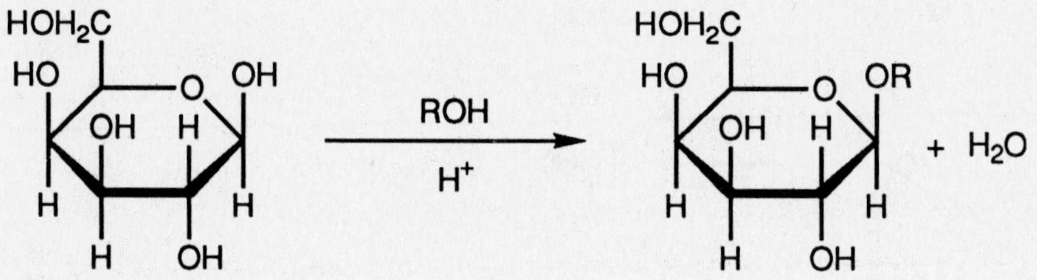

$$ROH$$
$$H^+$$

$$+ \ H_2O$$

Hydrolysis of Glycosides

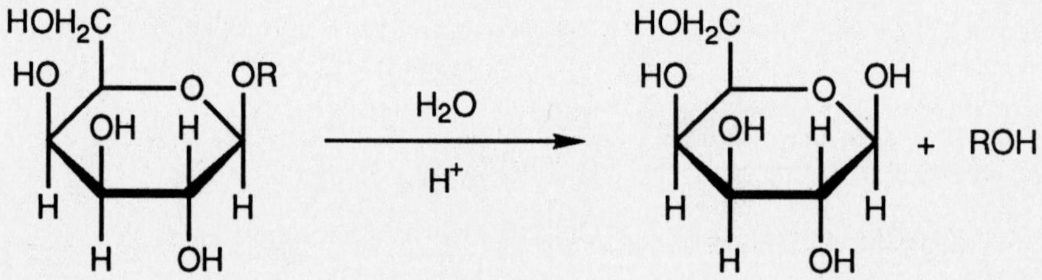

$$H_2O$$
$$H^+$$

$$+ \ ROH$$

LEARNING OBJECTIVES

1. Know the meaning of: carbohydrate, monosaccharide, oligosaccharide, polysaccharide, disaccharide, trisaccharide.

LEARNING OBJECTIVES

2. Know the meaning of: aldose, ketose, triose, tetrose, pentose, hexose, glyceraldehyde, dihydroxyacetone.

3. Know the meaning of: D- or L-sugar, Fischer projection formula, Haworth formula, epimer.

4. Know the meaning of: α and β configurations, anomer, furanose and pyranose forms, mutarotation.

5. Know the meaning of: glycosidic bond, reducing and nonreducing sugar, aldaric acid, aldonic acid, alditol.

6. Learn the formulas for some common monosaccharides, especially the D forms of glucose, mannose, galactose, and fructose.

7. Draw the Fischer projection formula for a simple monosaccharide.

8. Convert the Fischer projection formula for a tetrose to a sawhorse or Newman projection formula, and vice versa.

9. Tell whether two structures are epimers or anomers.

10. Given the acyclic formula for a monosaccharide, draw its cyclic structure in either the pyranose or furanose form and either α or β configuration.

11. Given the rotations of two pure anomers and their equilibrium mixture, calculate the percentage of each anomer present at equilibrium.

12. Given the formula for a monosaccharide, draw the formula for its glycoside with a given alcohol or with a given additional monosaccharide.

13. Draw the cyclic structures (Haworth projection and conformational structure) for α-D- and β-D-glucose and the corresponding methyl glucosides.

14. Write all the steps in the mechanism for the formation of a glycoside from a given sugar and alcohol.

15. Write all the steps in the mechanism for the hydrolysis of a given glycoside to the corresponding sugar and alcohol.

16. Write all the steps in the mechanism for the hydrolysis of a given disaccharide to the component monosaccharides.

16. CARBOHYDRATES

17. Given the structure of a sugar, write equations for its reaction with each of the following reagents: acetic anhydride, bromine water, nitric acid, sodium borohydride, and Tollens' or Fehling's reagent.

18. Know the structures of: maltose, cellobiose, lactose, sucrose.

19. Write the structures for the repeating units in starch and cellulose.

20. Know the meaning of: sugar phosphate, deoxy sugar, amino sugar, ascorbic acid (vitamin C).

ANSWERS TO PROBLEMS

Problems Within the Chapter

16.1 The L isomer is the mirror image of the D isomer. The configuration at *every* chiral center is reversed.

a.
$$
\begin{array}{c}
CH{=}O \\
H{-\!\!-\!\!-}OH \\
HO{-\!\!-\!\!-}H \\
CH_2OH
\end{array}
$$

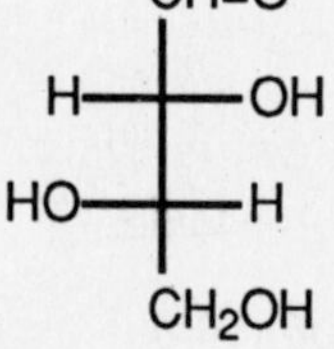

b.
$$
\begin{array}{c}
CH{=}O \\
HO{-\!\!-\!\!-}H \\
H{-\!\!-\!\!-}OH \\
HO{-\!\!-\!\!-}H \\
HO{-\!\!-\!\!-}H \\
CH_2OH
\end{array}
$$

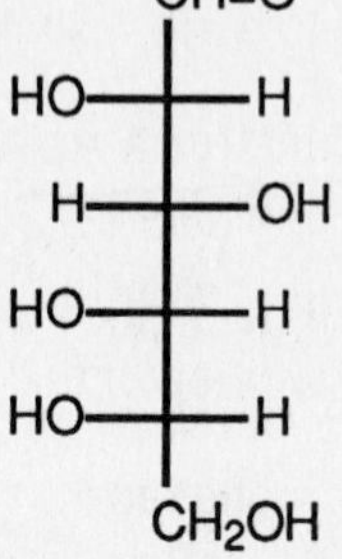

16.2 Follow Example 16.2 as a guide.

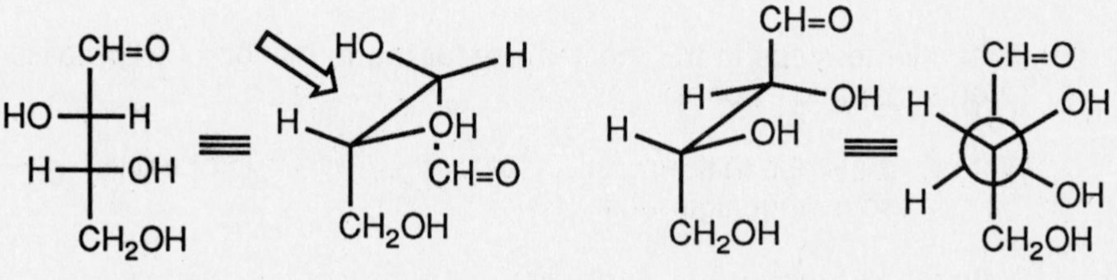

378

If you view the second formula from the top you will see that it is just a three-dimensional representation of the Fischer projection. Horizontal groups at each chiral center come up toward you, and vertical groups recede away from you. The second formula represents an eclipsed conformation of D-threose. The third and fourth formulas represent sawhorse and Newman projections of a staggered conformation of D-threose.

16.3 For each D-aldohexose there will be, as one more chiral center is added, two aldoheptoses. Altogether, there are sixteen D-aldoheptoses.

16.4 D-Ribose and D-xylose are identical except for the configuration at C-3. So too are D-arabinose and D-lyxose.

16.5 D-Galactose is the C-4 epimer of D-glucose (see Figure 16.1). Therefore the Haworth projection will be identical to that for D-glucose, except the C-4 hydroxyl group will be up rather than down.

D-galactose

16.6 D-Erythrose, a tetrose, has only four carbons. Therefore it cannot form a pyranose that would require the presence of at least five carbons. Bonding between the C-4 hydroxyl group and the carbonyl carbon, however, does lead to two anomeric furanoses.

β-anomer   and   ∝-anomer

16. CARBOHYDRATES

16.7 Use eq. 16.6 as a guide. Note that β-D-galactopyranose is the C-4 epimer of β-D-glucopyranose.

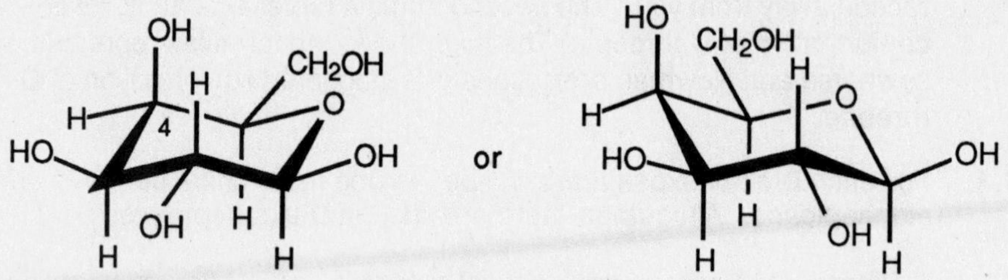

β-D-galactopyranose

16.8

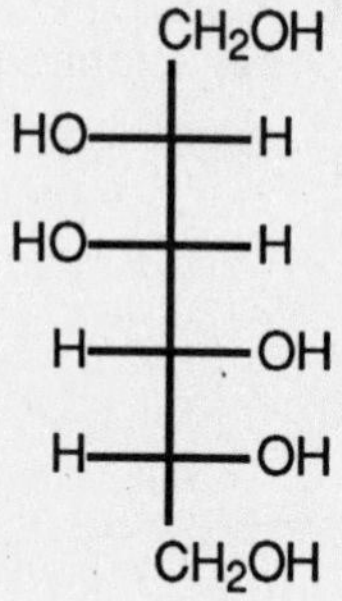

16.9

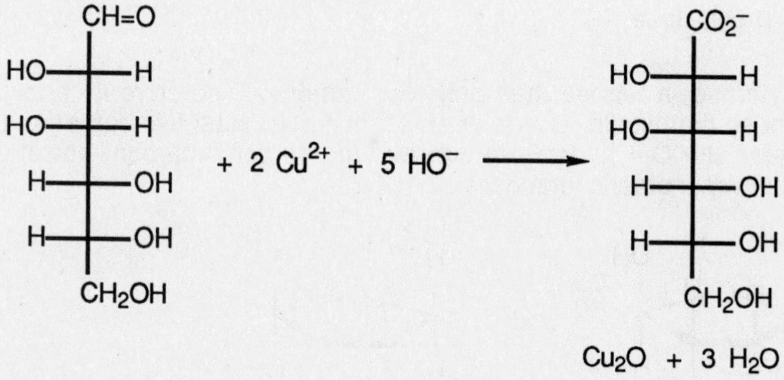

$$+ \; 2 \; Cu^{2+} \; + \; 5 \; HO^{-} \longrightarrow$$

Cu$_2$O + 3 H$_2$O

16.10

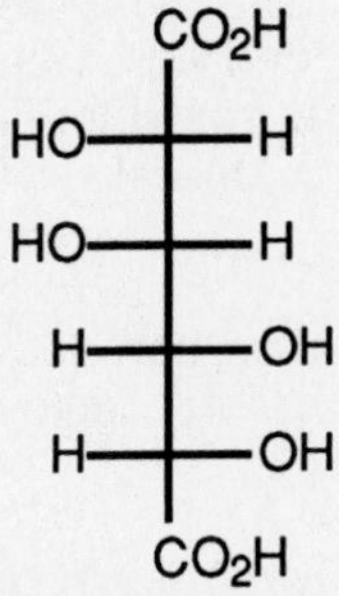

$$\begin{array}{c}
\text{CO}_2\text{H} \\
\text{HO} \overline{} \text{H} \\
\text{HO} \overline{} \text{H} \\
\text{H} \overline{} \text{OH} \\
\text{H} \overline{} \text{OH} \\
\text{CO}_2\text{H}
\end{array}$$

16.11 Using Haworth formulas, we have the following:

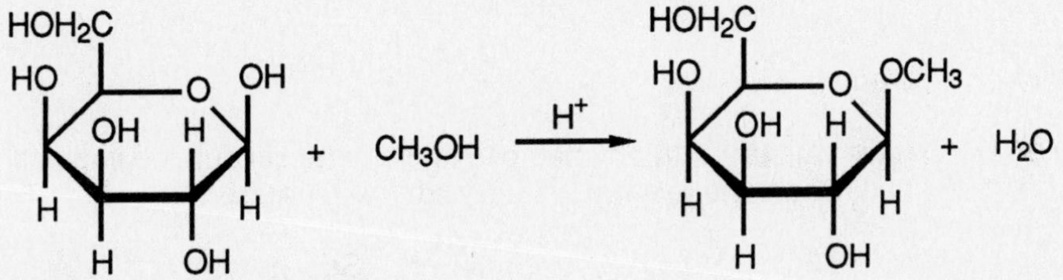

Although we show only the methyl β-D-galactoside as a product, some of the α epimer will also be formed because the carbocation intermediate can be attacked at either face by the methanol.

16.12 Let us examine carbocation formation at C-2.

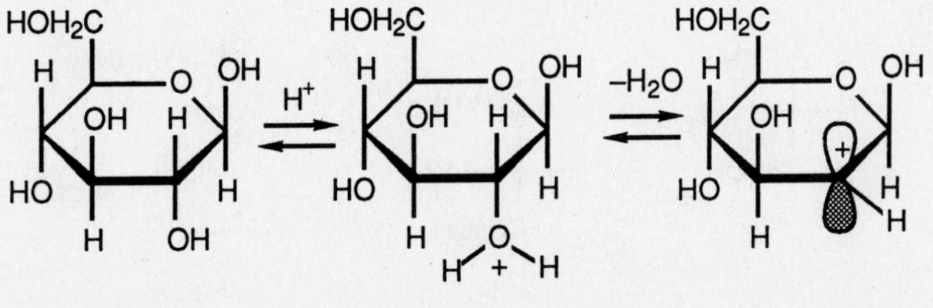

The resulting carbocation is secondary because two carbons and a hydrogen are attached to the positive carbon. The same would be true of a carbocation with the positive charge on C-3 or C-4. The carbocation with the positive charge on C-6 would be primary. In contrast, the C-1 carbocation (see Example 16.7) has an *oxygen*, a carbon, and a hydrogen attached to the positive carbon.

16. CARBOHYDRATES

16.13 Carbon-1 of the second (right-hand) glucose unit in maltose is a hemiacetal carbon. Thus the α and β forms at this carbon atom can equilibrate via the open-chain aldehyde form. Mutarotation is therefore possible. Note that carbon-1 of the first (left-hand) glucose unit is an acetal carbon. Its configuration is therefore fixed (as α).

16.14 Carbon-1 of the glucose unit in lactose is a hemiacetal carbon and will be in equilibrium with the open-chain aldehyde form. Therefore, lactose will be oxidized by Fehling's solution and will mutarotate.

16.15 Carbon-1 of methyl β-D-glucopyranoside is an acetal carbon. In solution it is not in equilibrium with an open-chain aldehyde. Therefore it cannot reduce Ag^+ or Cu^{2+} and does not give positive tests with Tollens', Fehling's, or Benedict's reagent.

Additional Problems

16.16 If you have difficulty with any part of this problem, consult the indicated section in the text, where the term is defined and/or illustrated.

a. Sec. 16.3

b. Sec. 16.3

c. Secs. 16.2 and 16.3

d. Secs. 16.2 and 16.13

e. Secs. 16.2 and 16.14

f. Sec. 16.7

g. Sec. 16.7

h. Sec. 16.12

i. Sec. 16.6

16.17 D-Sugars have the same configuration at the carbon atom adjacent to the primary alcohol function as D-(+)-glyceraldehyde (which is R).

D-(+)-glyceraldehyde   a D-aldose   a D-ketose

L-Sugars have the opposite configuration at the chiral center adjacent to the primary alcohol function.

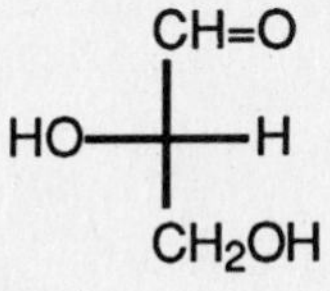

CH=O
HO———H
CH$_2$OH

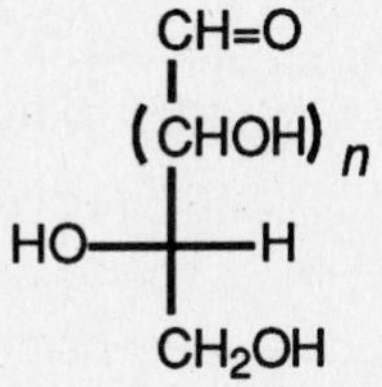

CH=O
(CHOH)$_n$
HO———H
CH$_2$OH

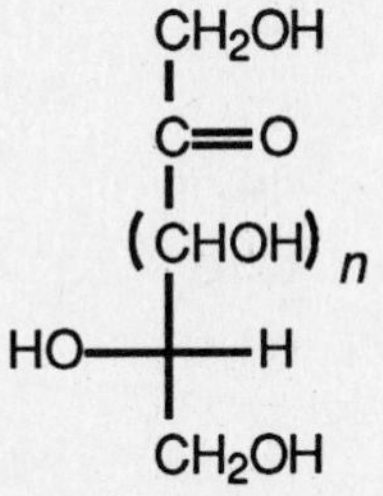

CH$_2$OH
C=O
(CHOH)$_n$
HO———H
CH$_2$OH

L-(–)-glyceraldehyde   an L-aldose   an L-ketose

16.18 Although the hydroxyl groups at carbons 2,3, and 4 of D-talose are on the left (Figure 16.1), the hydroxyl group on *the highest numbered asymmetric carbon (C-5) is on the right.* Only the stereochemistry at that carbon determines whether a sugar belongs to the D- or L-series.

16.19 Consult Sec. 5.4.

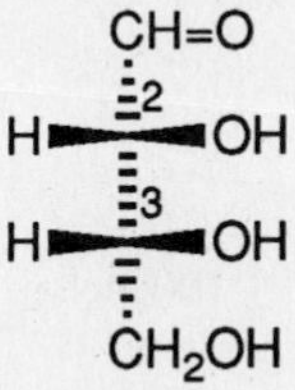

CH=O
2
H———OH
3
H———OH
CH$_2$OH

D-erythrose

At C-2, the priority order of groups is OH > CH=O > CH(OH)CH$_2$OH > H. The configuration is *R*.

At C-3, the priority order of groups is OH > CH(OH)CHO > CH$_2$OH > H. The configuration is *R*.

16.20 Use the answer to Prob. 16.19 to help you set up the priority orders.

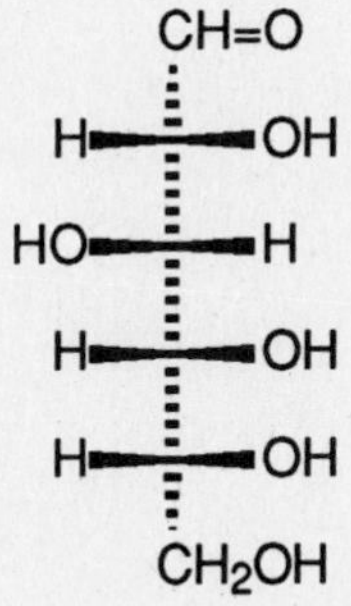

CH=O
H———OH
HO———H
H———OH
H———OH
CH$_2$OH

C1: *R*
C2: *R*
C3: *S*
C4: *R*
C5: *R*

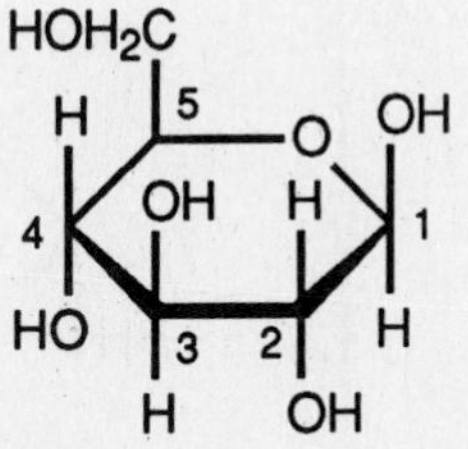

HOH$_2$C
H   5   O   OH
OH   H
4   1
HO   3   2   H
H   OH

16.21 D-Gulose and D-idose differ only in the configuration at C-2. They are epimers at C-2.

16. CARBOHYDRATES

16.22 The D-ketoses:

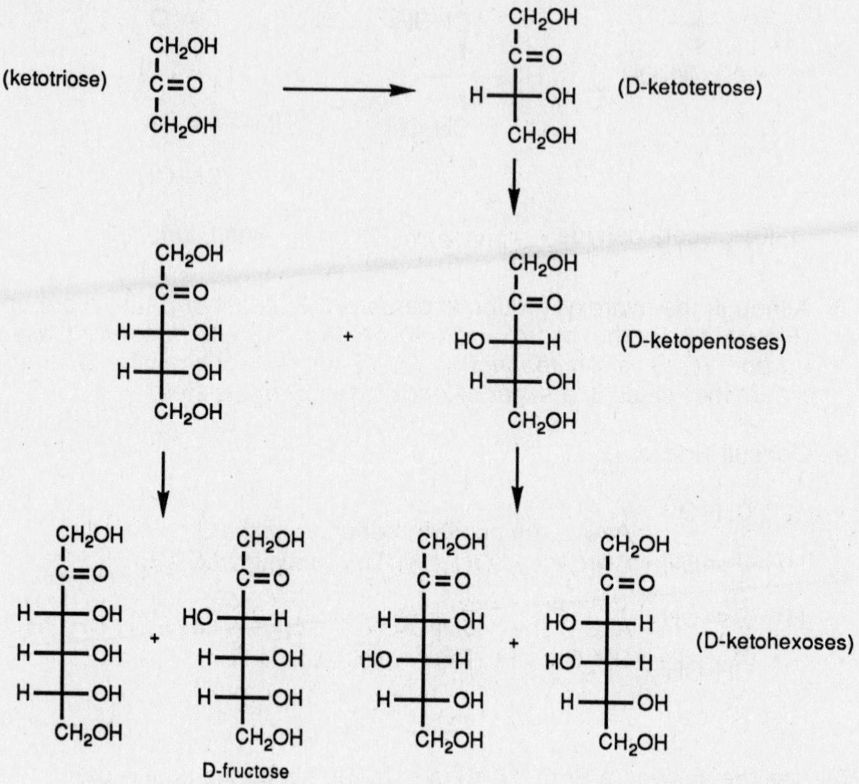

(ketotriose)

CH_2OH
$C=O$
CH_2OH

→

CH_2OH
$C=O$
H—OH
CH_2OH

(D-ketotetrose)

CH_2OH
$C=O$
H—OH
H—OH
CH_2OH

+

CH_2OH
$C=O$
HO—H
H—OH
CH_2OH

(D-ketopentoses)

CH_2OH
$C=O$
H—OH
H—OH
H—OH
CH_2OH

+

CH_2OH
$C=O$
HO—H
H—OH
H—OH
CH_2OH

D-fructose

CH_2OH
$C=O$
H—OH
HO—H
H—OH
CH_2OH

+

CH_2OH
$C=O$
HO—H
HO—H
H—OH
CH_2OH

(D-ketohexoses)

16.23 a.

H—OCH_3
H—OH
HO—H
H—OH
H—
CH_2OH

HOH_2C
H
OH H
HO
H OH
O
H
OCH_3

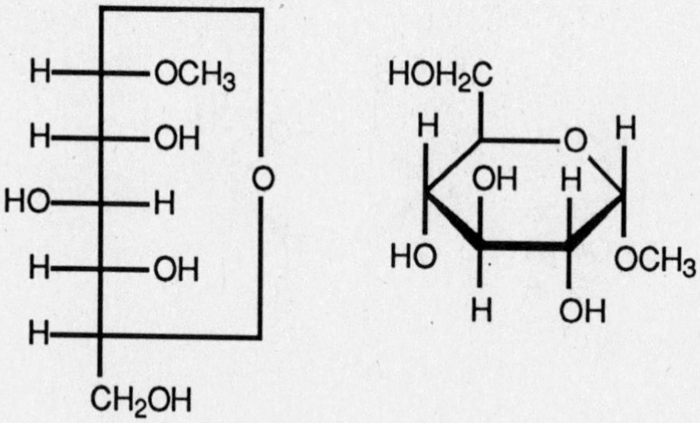

b.

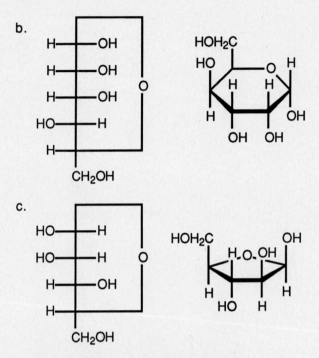

c.

d. The structure is the enantiomer of the
   structure shown in part a above.

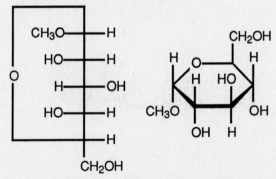

16.24 a. Consult Figure 16.1 for the Fischer projection formula of D-mannose. The
           L isomer is enantiomeric with the D isomer.

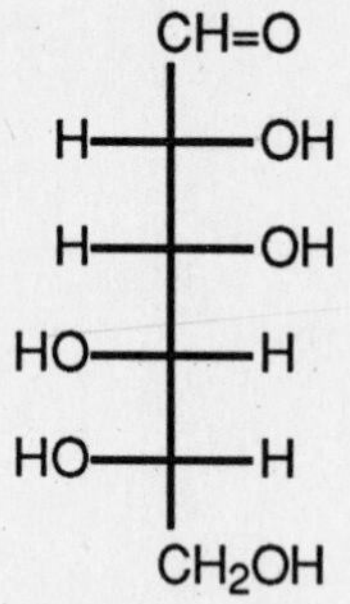

$$
\begin{array}{c}
\text{CH=O} \\
\text{H}\!-\!\!-\!\!-\!\text{OH} \\
\text{H}\!-\!\!-\!\!-\!\text{OH} \\
\text{HO}\!-\!\!-\!\!-\!\text{H} \\
\text{HO}\!-\!\!-\!\!-\!\text{H} \\
\text{CH}_2\text{OH}
\end{array}
$$

L-mannose

b. See the answer to Prob. 16.22 for the Fischer projection formula of D-fructose.

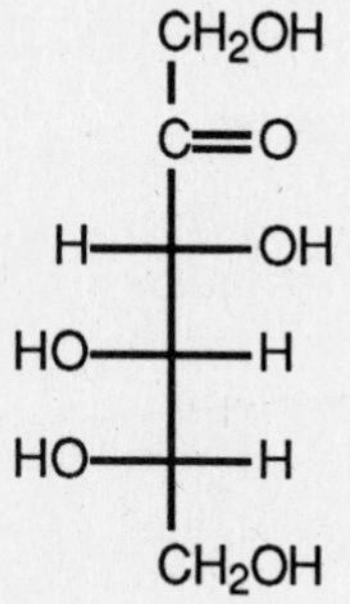

$$
\begin{array}{c}
\text{CH}_2\text{OH} \\
\text{C=O} \\
\text{H}\!-\!\!-\!\!-\!\text{OH} \\
\text{HO}\!-\!\!-\!\!-\!\text{H} \\
\text{HO}\!-\!\!-\!\!-\!\text{H} \\
\text{CH}_2\text{OH}
\end{array}
$$

L-fructose

16.25

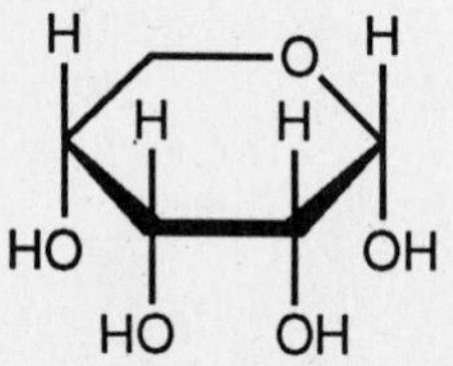

α-pyranose 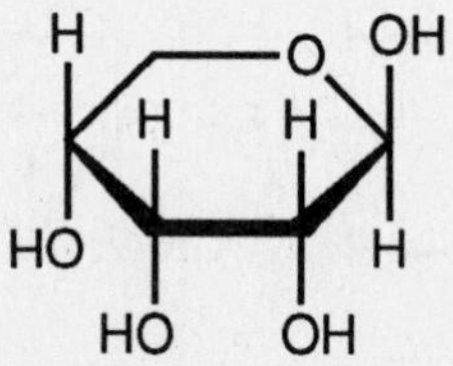   β-pyranose

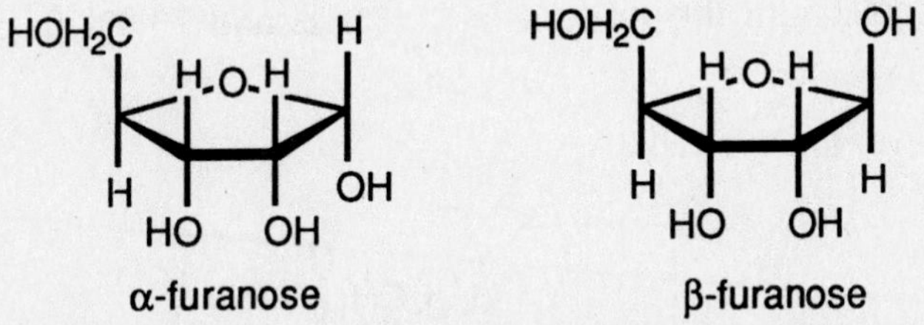

α-furanose β-furanose

16.26 See Figure 16.1 for the Fischer projection of the acyclic form.

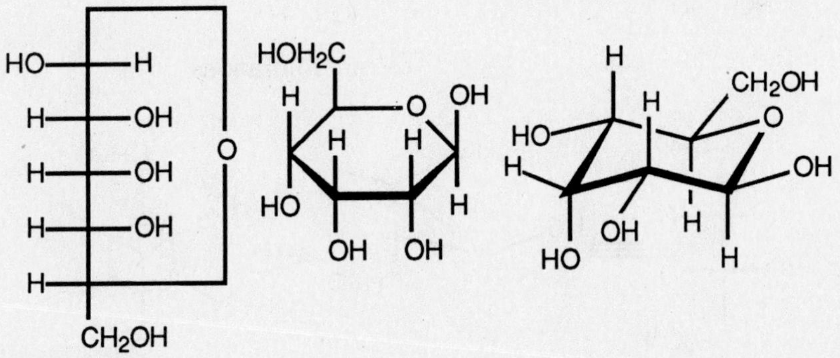

16.27 The compounds are diastereomers. That is, they are stereoisomers but not mirror images. They are therefore expected to differ in all properties, including water solubility.

16.28 $[\alpha]_e = [\alpha]_\alpha$ (mole fraction α) + $[\alpha]_\beta$ (mole fraction β)

where $[\alpha]_e$ = rotation at equilibrium = −92

$[\alpha]_\alpha$ = rotation of pure α anomer = +21

$[\alpha]_\beta$ = rotation of pure β anomer = −133

Since (mole fraction β) = (1 − mole fraction α), and letting (mole fraction α) = X, we can write:

$[\alpha]_e = [\alpha]_\alpha X + [\alpha]_\beta (1 - X) = [\alpha]_\beta + X ([\alpha]_\alpha - [\alpha]_\beta)$ or

$X = [\alpha]_e - [\alpha]_\beta / [\alpha]_\alpha - [\alpha]_\beta = (-92) - (-133) / (21) - (-133) = 41 / 154 = 0.266$

Therefore at equilibrium the mixture is 26.6% α and 73.4% β (also see Example 16.4).

16.29 Consult the answer to Prob. 16.6.

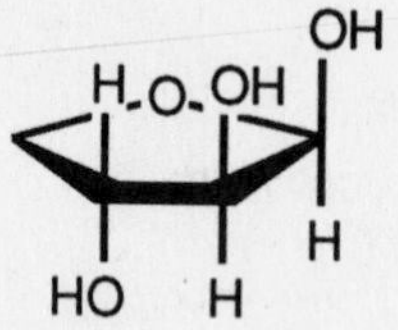

β-D-threofuranose 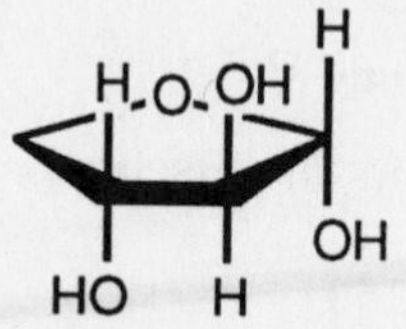   α-D-threofuranose

16.30

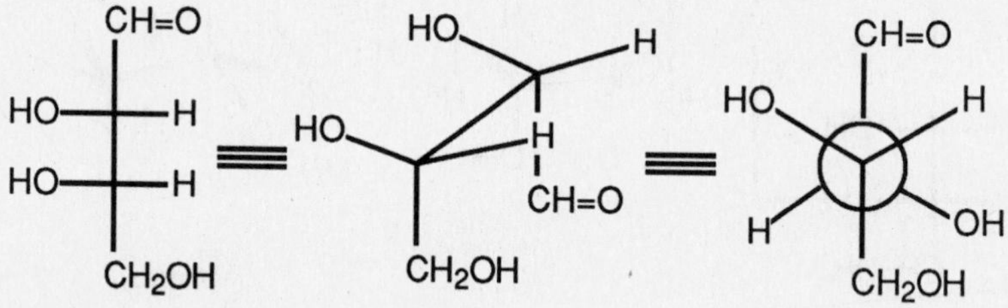

L-erythrose

16.31

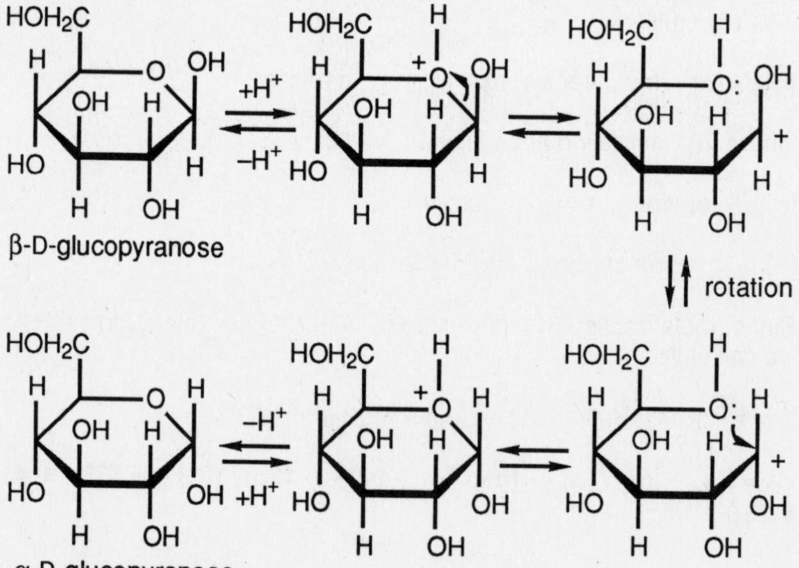

β-D-glucopyranose

rotation

α-D-glucopyranose

16.32 The structures are shown in Figure 16.1. Oxidation of D-erythrose gives optically inactive *meso*-tartaric acid:

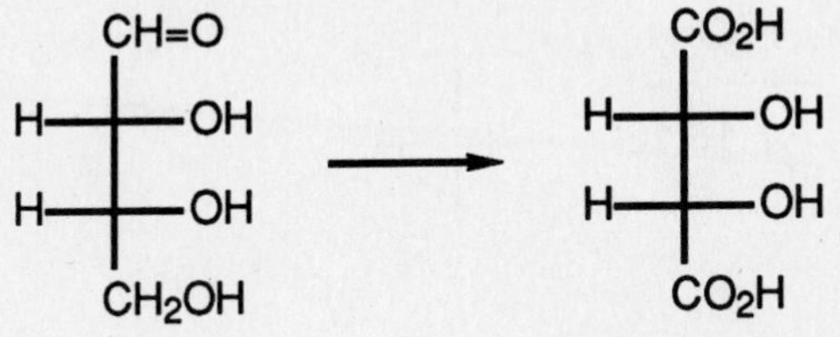

```
   CH=O                      CO2H
H ─┼─ OH                  H ─┼─ OH
H ─┼─ OH       ──►        H ─┼─ OH
   CH2OH                     CO2H

D-erythrose              meso-tartaric acid
                         (optically inactive)
```

Analogous oxidation of D-threose gives an optically active tartaric acid:

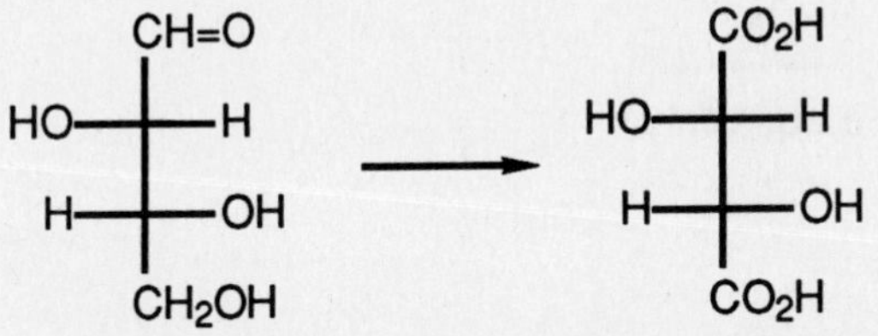

```
    CH=O                      CO2H
HO ─┼─ H                  HO ─┼─ H
 H ─┼─ OH      ──►          H ─┼─ OH
    CH2OH                     CO2H

D-threose                S,S-tartaric acid
                         (optically active)
```

In this way, we can readily assign structures to the two tetroses.

16.33 Consult Sec. 16.11 as a guide.

```
a.     CO2H              b.      CO2H
   H ─┼─ OH                  H ─┼─ OH
  HO ─┼─ H                  HO ─┼─ H
  HO ─┼─ H                  HO ─┼─ H
   H ─┼─ OH                  H ─┼─ OH
      CH2OH                     CO2H
```

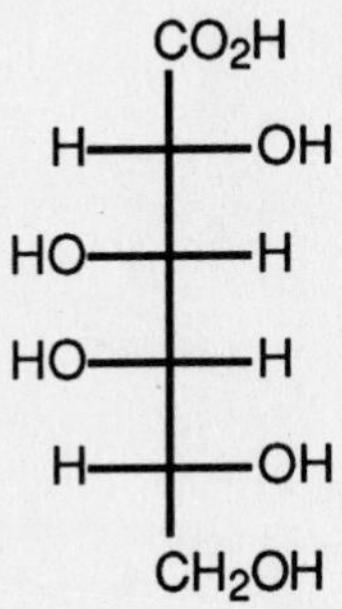

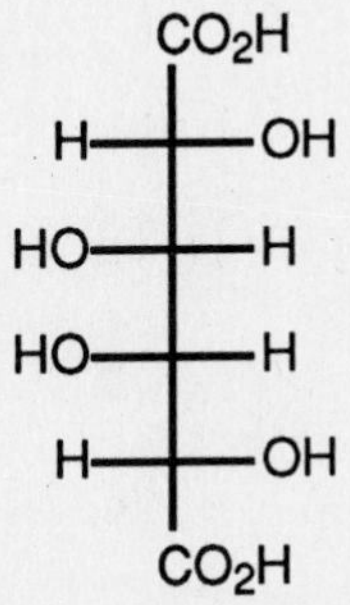

16.34 a.

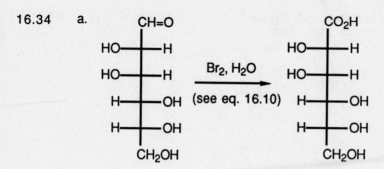

D-mannonic acid

b.

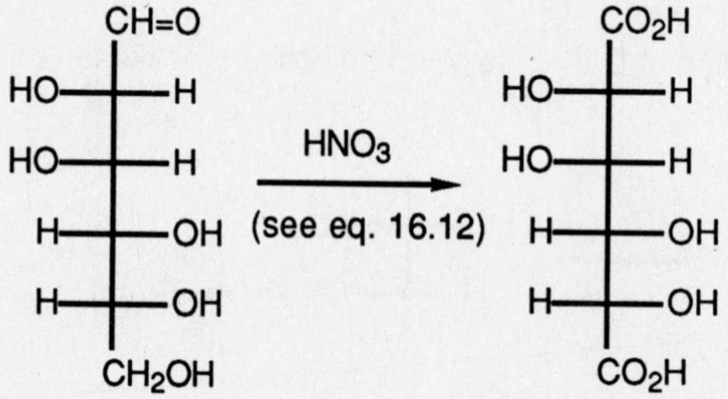

D-mannaric acid

c.

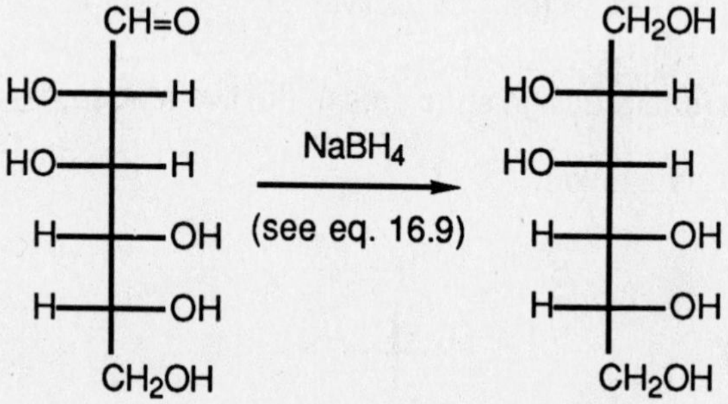

D-mannitol

ANSWERS TO PROBLEMS

d.

β-D-mannose $\quad Ac = CH_3\overset{\overset{O}{\|}}{C}-\quad$ β-D-mannose pentaacetate

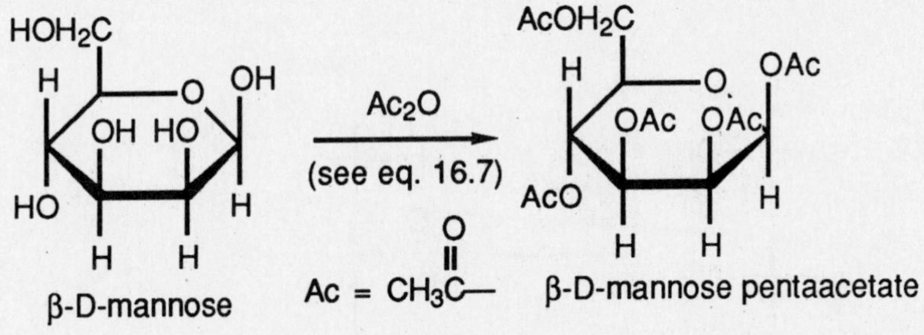

16.35 Using Fischer projection formulas, we can write the structures of D-glucitol (eq. 16.9) and D-mannitol (see the answer to Prob. 16.8).

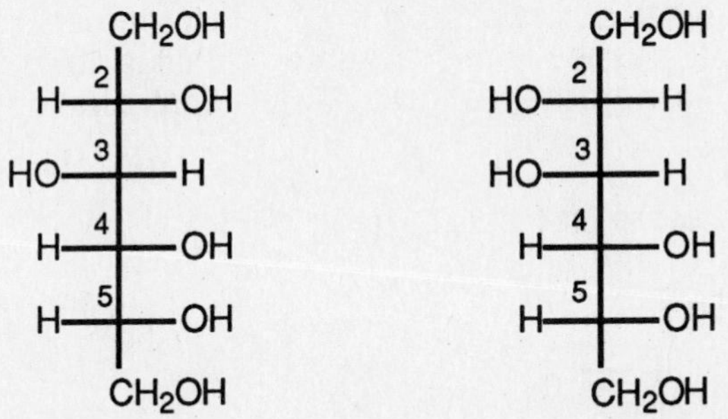

D-glucitol

D-mannitol

Note that the configurations are identical at C-3, C-4, and C-5. The fact that D-fructose gives both of these polyols on reduction tells us that it must also have the same configuration as they do at C-3, C-4, and C-5. Thus the keto group of D-fructose must be at C-2.

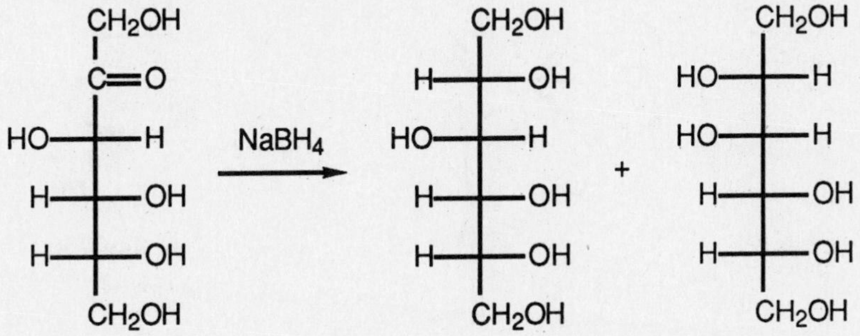

16.36 See eq. 16.12

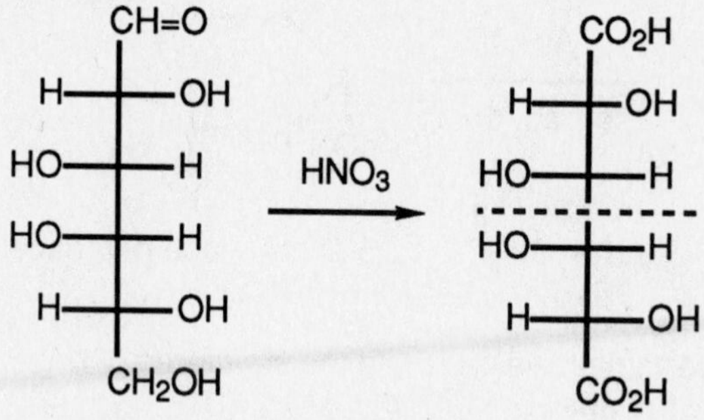

The dicarboxylic acid obtained as the product is a *meso* form, with a plane of symmetry shown by the dashed line. Therefore, this acid is achiral. (Experiments such as this were helpful in assigning configurations to the various monosaccharides).

16.37 a. For the structure of maltose, see Sec. 16.13a.

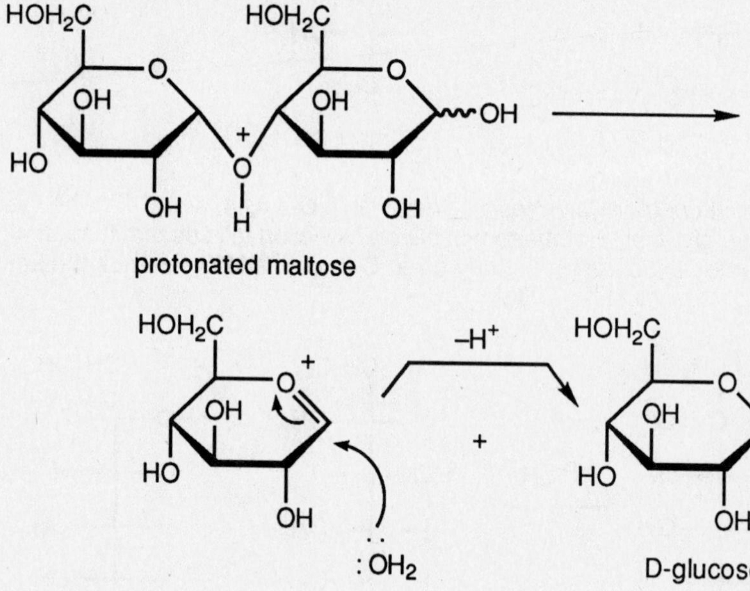

protonated maltose

D-glucose

b. For the structure of lactose, see Sec. 16.13c.

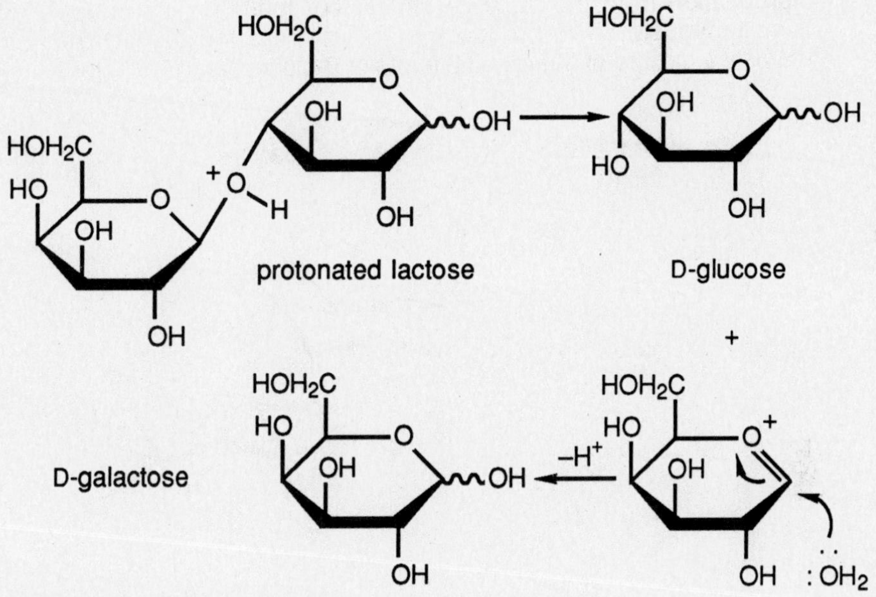

protonated lactose

D-glucose

D-galactose

+

c. For the structure of sucrose, see Sec. 16.13d.

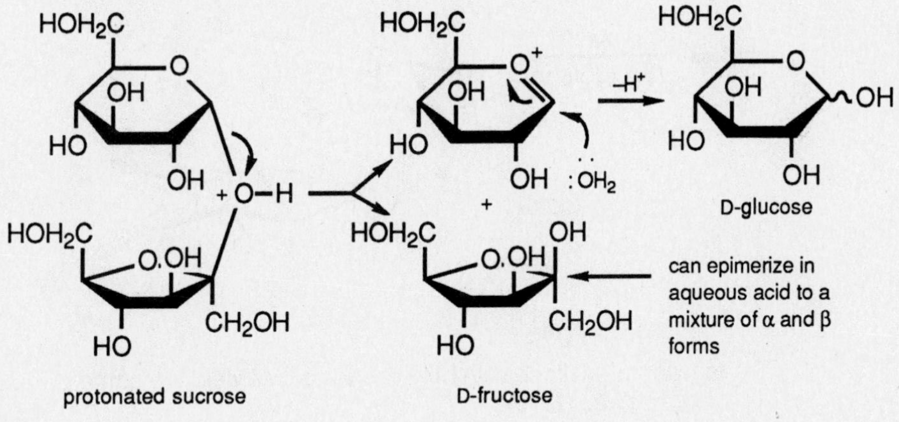

protonated sucrose

D-fructose

D-glucose

can epimerize in aqueous acid to a mixture of α and β forms

The mechanism shown is one of two that are possible. The alternative mechanism would break the other glycoside bond to give glucose and the carbocation from the fructose unit. Both mechanisms undoubtedly occur simultaneously. The products, of course, are the same from both paths, namely a mixture of the α and β forms of D-glucose and D-fructose.

16.38   The formula for maltose is given in Sec. 16.13a.

a.   maltose $\xrightarrow[\text{H}^+]{\text{CH}_3\text{OH}}$ (see eq .16.13)

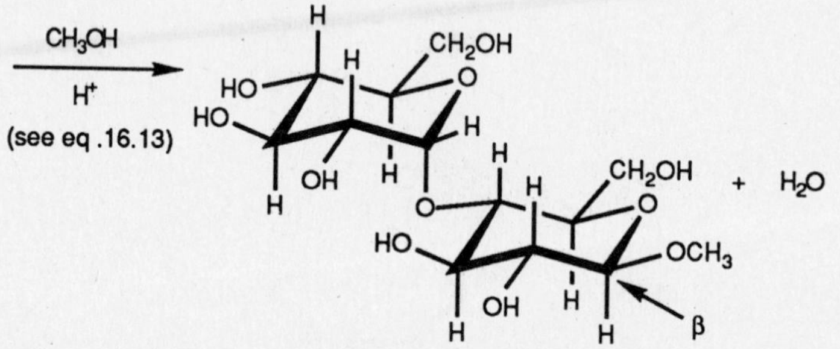

Only the β isomer is shown, but the α isomer will also be formed.

b.   maltose $\xrightarrow[\text{(see eq .16.10)}]{\text{Ag}^+}$

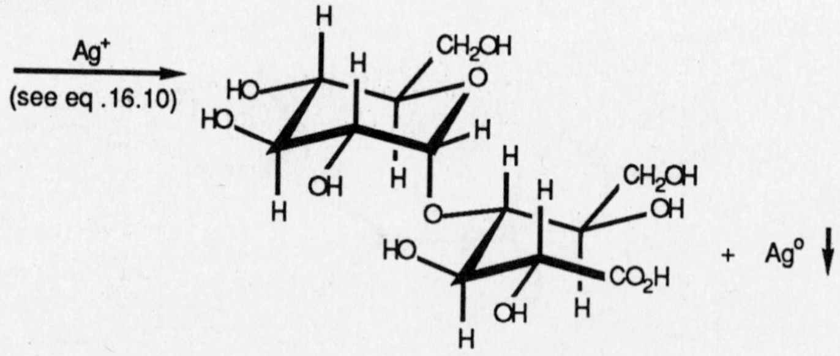

c.   The reaction of maltose with bromine water will yield the same product as in part b (see eq. 16.10).

d. maltose $\xrightarrow[\text{(see eq. 16.7)}]{Ac_2O}$

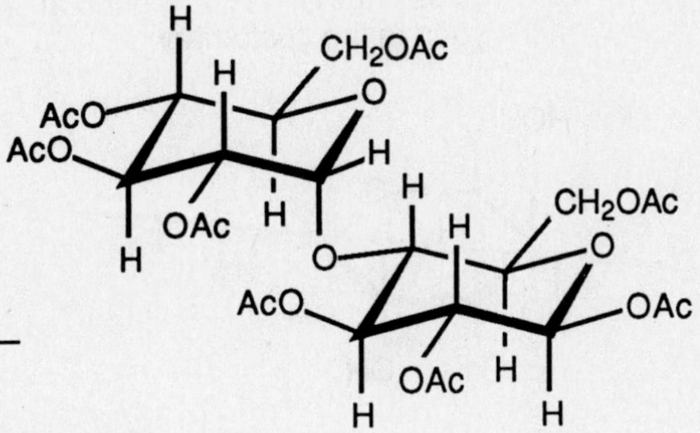

$$Ac = CH_3\overset{\overset{\textstyle O}{\|}}{C}-$$

Eight equivalents of acetic anhydride are required to acetylate all eight hydroxyl groups of maltose.

16.39 a. Hydrolysis of trehalose gives two moles of D-glucose. From the "left" portion, we get:

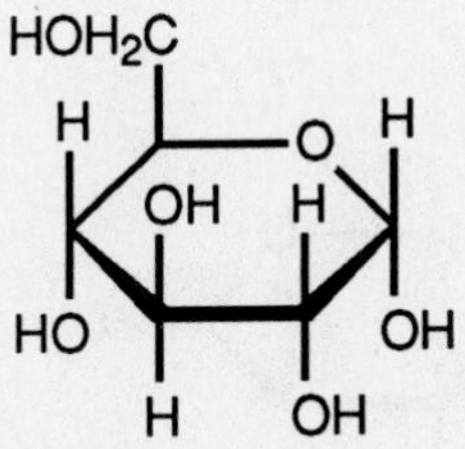

which is easily recognizable as D-glucose. From the "right" portion, we get

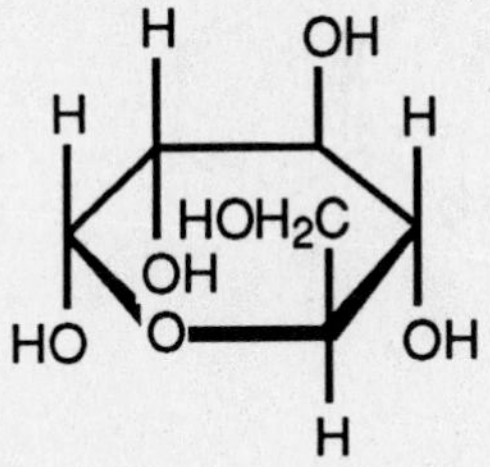

Turning this structure 180° around an axis in the plane of the paper to put the ring oxygen in the customary position gives

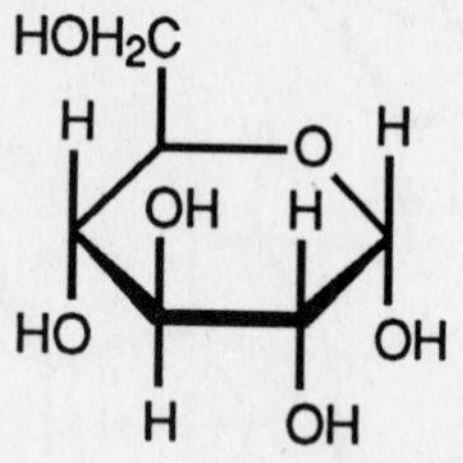

which is now also recognizable as D-glucose.

b. Since both anomeric carbons in trehalose are tied up in the glucosidic bond, no hemiacetal group remains in the structure. Therefore, equilibration with the aldehyde form and oxidation by Fehling's reagent are not possible. The test, as with sucrose, will be negative.

16.40

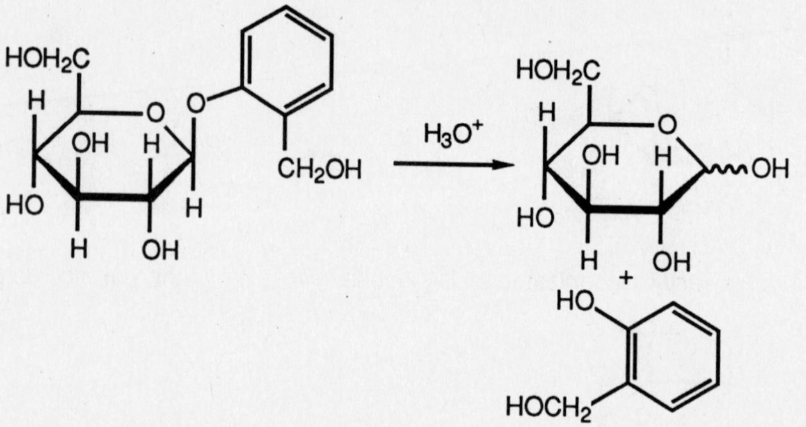

ANSWERS TO PROBLEMS

16.41 a. For the structure, see Sec. 16.13c. The α form is shown. The β form is identical with it, except for the configuration at C-1 in the "right-hand" unit, where the OH group is equatorial instead of axial.

   b. Follow Example 16.4 or Prob. 16.28 as guides for the calculation.

$$\% \text{ of } \beta = \frac{92.6 - 52}{92.6 - 34} \times 100 = \frac{40.6}{58.6} = 69.3 \%$$

   The % of α is 100 − 69.3 = 30.7%.

16.42 a. See eq. 16.7.

   b. See eq. 16.10.

   c. See eq. 16.9.

   d. Compare with Table 9.1.

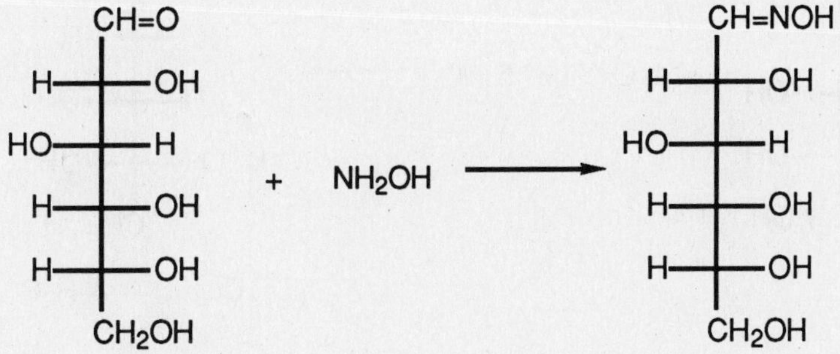

   e. See eq. 16.13.

f.   Compare with eq. 9.25.

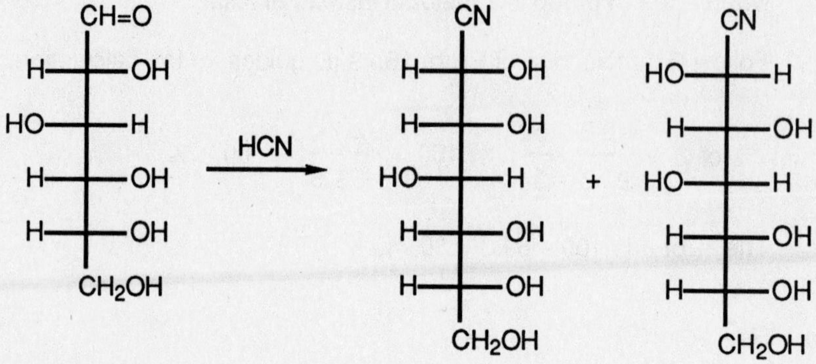

HCN

g.   Compare with eq. 16.11.

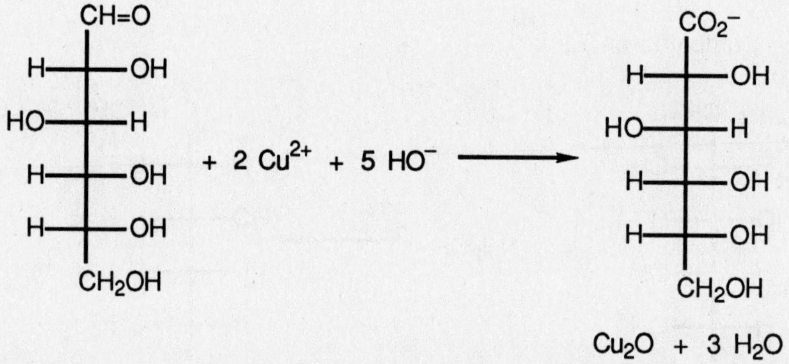

$$+ \ 2 \ Cu^{2+} \ + \ 5 \ HO^{-} \longrightarrow$$

$$Cu_2O \ + \ 3 \ H_2O$$

16.43 In sucrose, both anomeric carbons are involved in the glycosidic bond. No hemiacetal function is present. Both anomeric carbons are in the acetal (ketal) form. Therefore, equilibration with an acyclic aldehyde form is not possible, and the sugar cannot reduce Tollens', Fehling's, or Benedict's reagent. In maltose (Sec. 16.13a), on the other hand, carbon-1 of the right-hand glucose unit is a hemiacetal carbon, in equilibrium with the open-chain aldehyde form, which can reduce those same reagents.

16.44 Step 1 is an electrophilic aromatic substitution (See Sec. 4.9 and Word About 24):

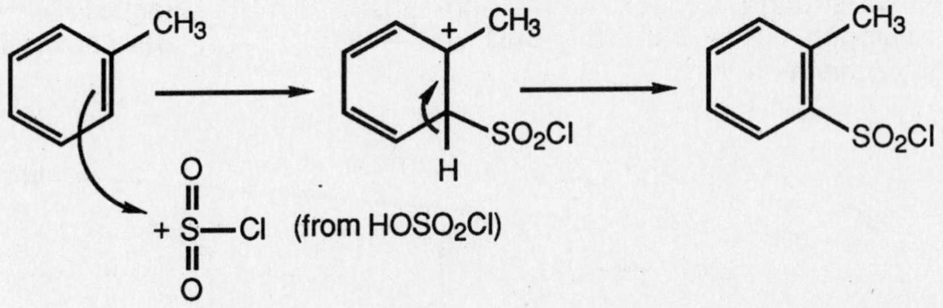

ANSWERS TO PROBLEMS

$+ \; \underset{O}{\overset{O}{\underset{\parallel}{S}}} - Cl \quad$ (from $HOSO_2Cl$)

Step 2 involves sulfonamide formation (consult Sec. 11.11) and resembles the first step in the Hinsburg test (eqs. 11.26 and 11.27):

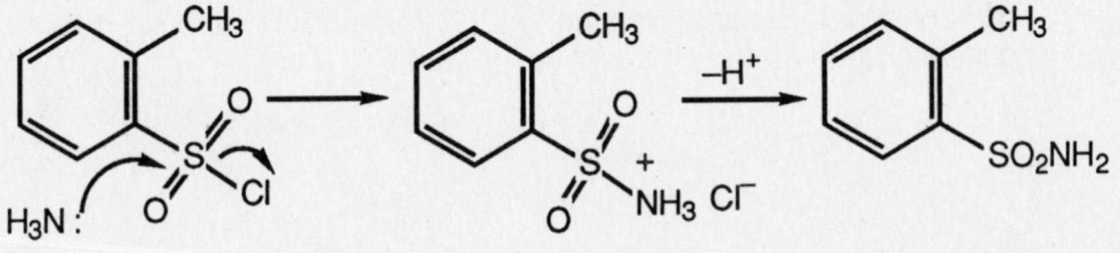

Step 3 involves oxidation of the aryl side-chain methyl group (consult eq. 10.7). Step 4 involves cyclic amide formation:

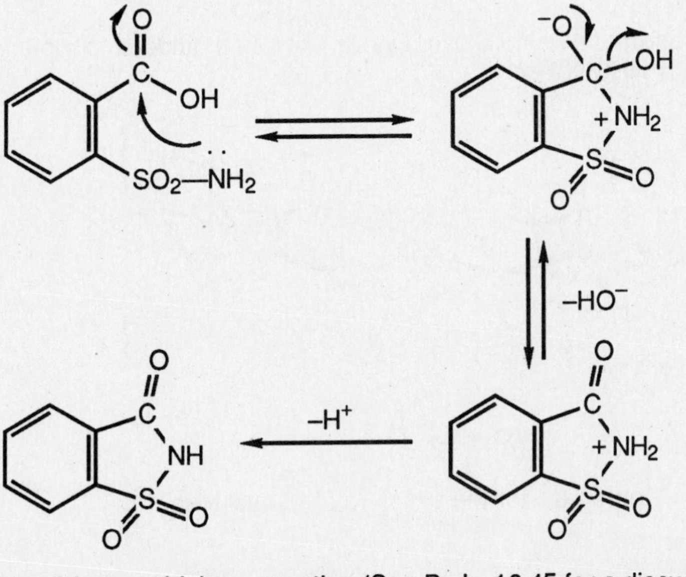

Step 5 is an acid–base reaction (See Prob. 16.45 for a discussion).

399

16.45 Two strongly electron-withdrawing substituents are attached to the amide nitrogen. They are the —C=O and —SO$_2$ groups. Both act to enhance saccharin's acidity.

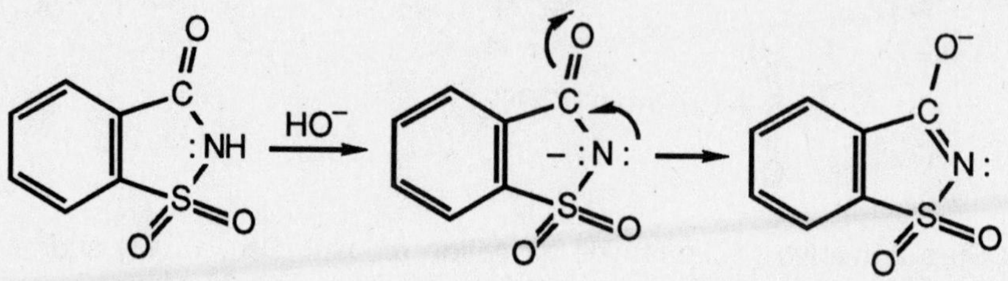

The negative charge is not only delocalized to the carbonyl oxygen (as with ordinary amide anions; see eq. 11.17) but is also stabilized by the partial positive charge on the adjacent sulfur atom.

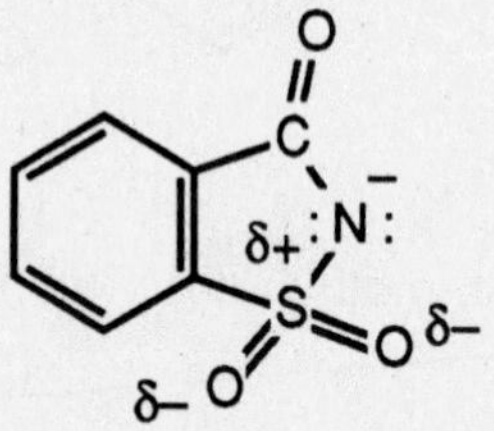

16.46 a. Use the formula for cellulose (Figure 16.6) as a guide, and replace the OH at C-2 by NHCOCH$_3$:

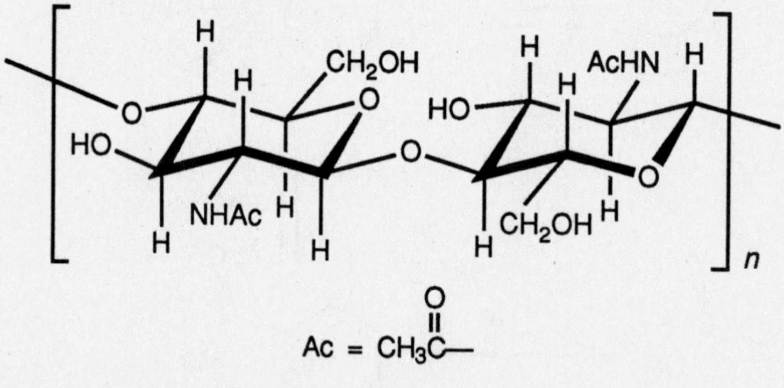

$$Ac = CH_3C—$$

b. Consult Figure 16.1 for the structure of D-galactose.

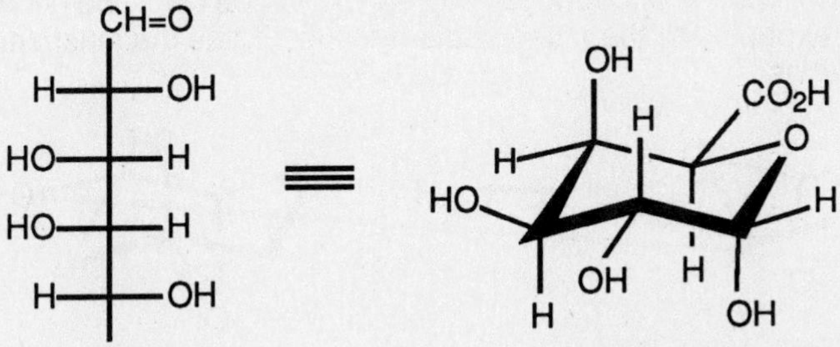

α-D-galacturonic acid

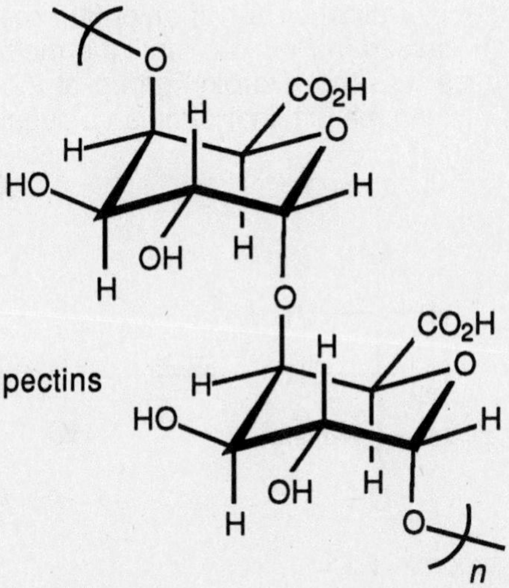

pectins

16.47 The formula of L-galactose can be obtained from that of D-galactose, which is shown in Figure 16.1. Deoxy sugars are discussed in Sec. 16.16.

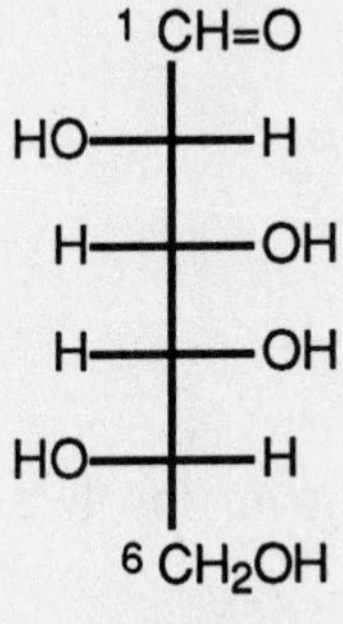

L-galactose

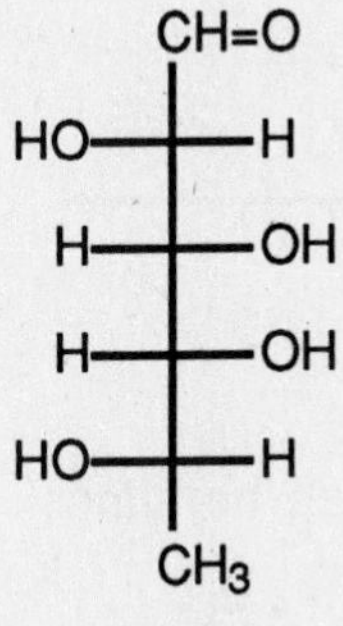

L-fucose

(6-deoxy-L-galactose)

16.48   Two chair conformations of daunosamine are shown below. The conformation on the right is expected to the most stable because it has the least severe 1,3-diaxial interactions.

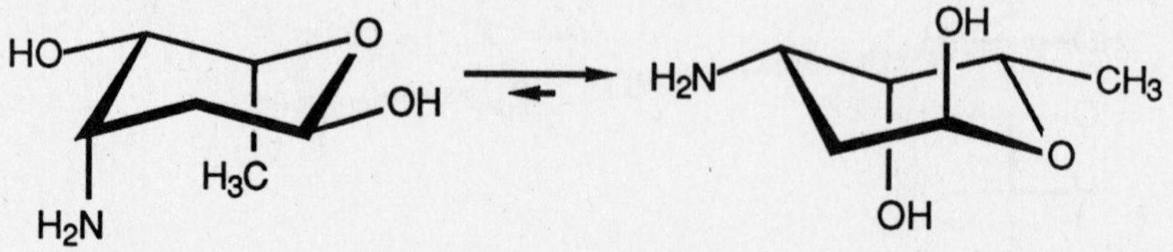

Daunosamine is an L-sugar. First draw the Fischer projection of the open-chain aldehyde derived from daunosamine. Then manipulate the groups attached to C-5, using the rules discussed in Sec. 5.8, until the methyl group occupies a vertical position. Notice that the hydroxyl group at the highest numbered asymmetric carbon (C-5) is on the left, indicating an L-sugar.

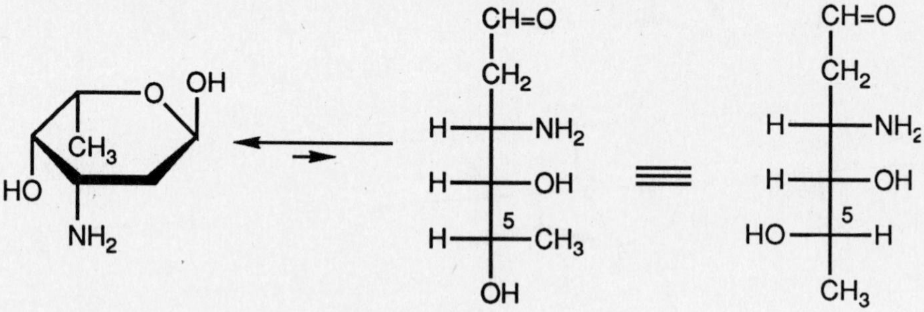

16.49   See eq. 16.14 for the formula of ascorbic acid.

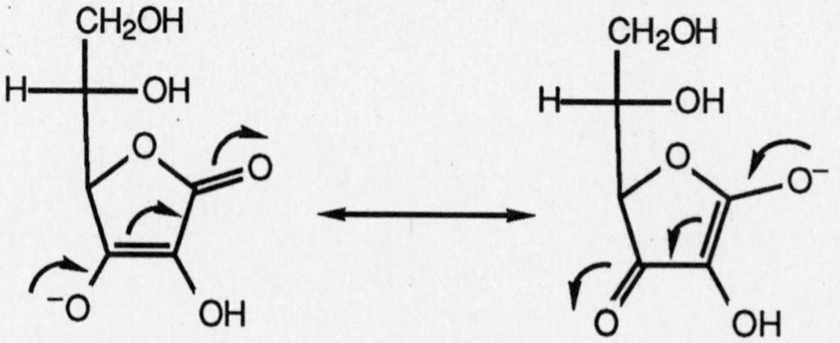

The negative charge can be almost equally spread over the two oxygen atoms.

16.50 The formula for β-D-xylopyranose, derived from the Fischer projection of D-xylose in Figure 16.1 , is

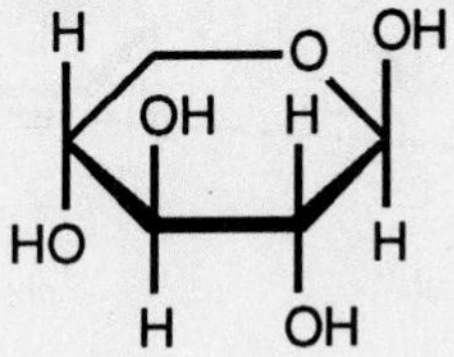

Since the xylans have these units linked 1,4, their structure is:

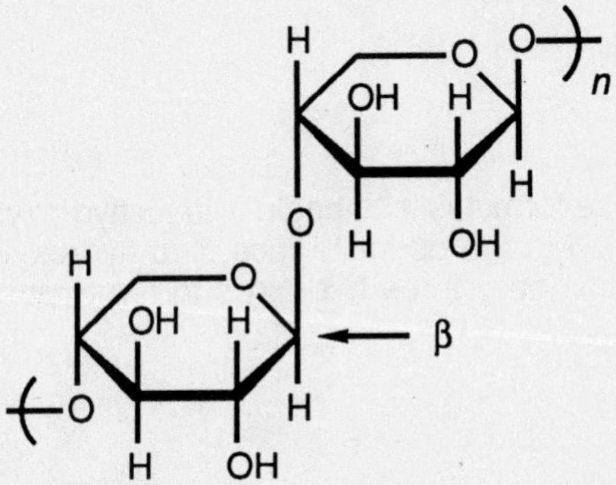

16.51 Be systematic in deducing all the structures.

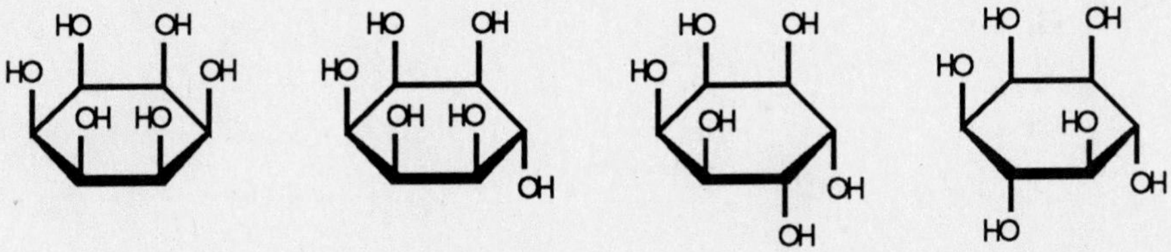

achiral   achiral   achiral   achiral

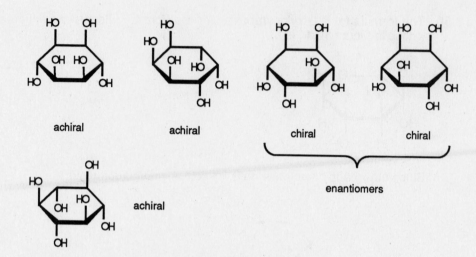

The system to use in writing these formulas is to begin with all hydroxyl groups *cis*. Then we move one hydroxyl to the opposite face, two hydroxyls to the opposite face (1,2; 1,3; 1,4), and finally three hydroxyls to the opposite face (1,2,3; 1,2,4; 1,3,5).

CHAPTER SEVENTEEN: AMINO ACIDS, PEPTIDES, AND PROTEINS

CHAPTER SUMMARY

Proteins are natural polymers composed of α-**amino acids** linked by **amide (peptide) bonds.** Except for **glycine** (aminoacetic acid), protein-derived amino acids are chiral and have the L-configuration. Table 17.1 lists the names, three-letter abbreviations, and structures of the twenty common amino acids. Of these, eight (the **essential amino acids**) cannot be synthesized in the bodies of adult humans and must be ingested in food. A measurement of the gradual change in optical activity of amino acids that remain in human or animal bones and teeth after death can be used in archaeological dating.

Amino acids with one amino and one carboxyl group exist as **dipolar ions**. Amino acids are **amphoteric:** in strong acid, they are protonated and become positively charged (ammonium carboxylic acids); in base, they lose a proton and become negatively charged (amine carboxylates). If placed in an electric field, amino acids migrate toward the cathode (negative electrode) at low pH and toward the anode (positive electrode) at high pH. The intermediate pH at which they do not migrate toward either electrode is the **isoelectric point**. (The isoelectric points for the twenty common amino acids are listed in Table 17.1). For amino acids with one amino group and one carboxyl group, the isoelectric pH is about 6. Amino acids with two carboxyl groups and one amino group have isoelectric points at a low pH (about 3), whereas amino acids with two amino groups and one carboxyl group have isoelectric points at a high pH (about 9). **Electrophoresis** is a process that makes use of the dependence of charge on pH to separate amino acids and proteins.

Amino acids undergo reactions that correspond to each functional group. In addition to exhibiting both acidic and basic behavior, for example, the carboxyl group can be esterified and the amino group can be acylated.

Amino acids react with **ninhydrin** to give a violet dye. This reaction is useful for detection and quantitative analysis of amino acids.

An amide bond between the carboxyl group of one amino acid and the amino group of another is called a **peptide bond,** which links amino acids to one another in peptides and proteins. The amino acid (often abbreviated **aa**) at one end of the peptide chain will have a free amino group (the **N-terminal amino acid**), and the amino acid at the other end of the chain (the **C-terminal amino acid**) will have a free carboxyl group. By convention, we write these structures from left to right starting at the N-terminal end.

Another type of covalent bond in proteins is the **disulfide bond**, formed by oxidative coupling of the —SH groups in **cysteine**. **Oxytocin** is an example of a cyclic peptide with an S—S bond.

By the **primary structure** of a peptide or protein, we mean its **amino acid sequence**. Complete hydrolysis gives the amino acid content. The N-terminal amino acid can be identified by the **Sanger method**, using **2,4-dinitrofluorobenzene**. The **Edman degradation** uses **phenyl isothiocyanate** to clip off one amino acid at a time from the N-terminal. Other reagents selectively cleave peptide chains at certain amino acid links. A combination of these methods can "sequence" a protein, and the methods have been automated.

Polypeptides are usually synthesized by the **Merrifield solid-phase technique.** An N-protected amino acid is linked by an ester bond to a benzyl chloride-type unit in a polystyrene. The protecting group is then removed, and the next N-protected amino acid is linked to the polymer-bound one. The cycle is repeated until the peptide is assembled, after which it is detached from the polymer. The most common N-protecting group is the *t*-**butoxycarbonyl (Boc) group**: $(CH_3)_3COC(=O)-$, which is connected to the amino group using di-*t*-butyldicarbonate and can be removed from the amino group using mild acid. The reagent used to attach each N-protected amino acid to the growing peptide chain is **dicyclohexyl-carbodiimide.** Detachment of the peptide chain from the polymer is accomplished using HF.

Two features that affect **secondary protein structure** (molecular shape) include the rigid, planar geometry and restricted rotation of the peptide bond, and interchain or intrachain hydrogen bonding of the type C=O···H—N. The α **helix** and the **pleated sheet** are common protein shapes.

Proteins may be **fibrous** or **globular**. The structure and polarity of the particular amino acid R groups and their sequence affect the solubility properties and **tertiary structure** of proteins. **Quaternary structure** refers to the aggregation of similar protein subunits.

REACTION SUMMARY

REACTION SUMMARY

Dissociation of Amino Acids

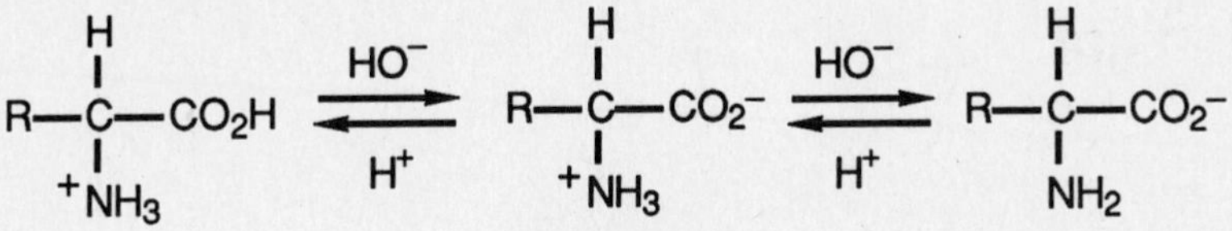

$$R-\underset{\underset{+NH_3}{|}}{\overset{\overset{H}{|}}{C}}-CO_2H \;\underset{H^+}{\overset{HO^-}{\rightleftharpoons}}\; R-\underset{\underset{+NH_3}{|}}{\overset{\overset{H}{|}}{C}}-CO_2^- \;\underset{H^+}{\overset{HO^-}{\rightleftharpoons}}\; R-\underset{\underset{NH_2}{|}}{\overset{\overset{H}{|}}{C}}-CO_2^-$$

dipolar ion

Esterification

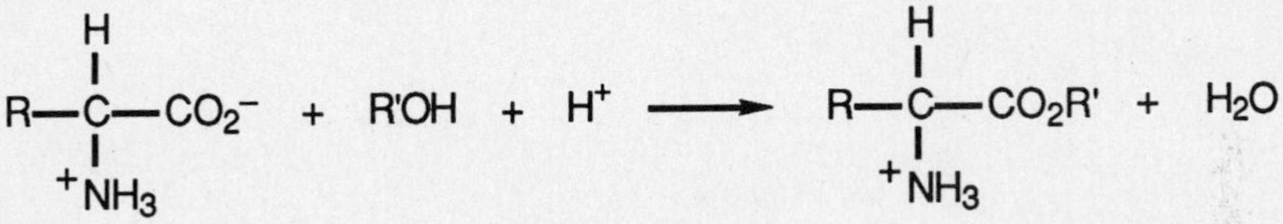

$$R-\underset{\underset{+NH_3}{|}}{\overset{\overset{H}{|}}{C}}-CO_2^- \;+\; R'OH \;+\; H^+ \longrightarrow R-\underset{\underset{+NH_3}{|}}{\overset{\overset{H}{|}}{C}}-CO_2R' \;+\; H_2O$$

Acylation

$$R-\underset{\underset{+NH_3}{|}}{\overset{\overset{H}{|}}{C}}-CO_2^- \;+\; R'-\underset{\underset{O}{\|}}{C}-Cl \;\xrightarrow{2\,HO^-}\; R-\underset{\underset{R'-\underset{\underset{O}{\|}}{C}-NH}{|}}{\overset{\overset{H}{|}}{C}}-CO_2^- \;+\; 2\,H_2O \;+\; Cl^-$$

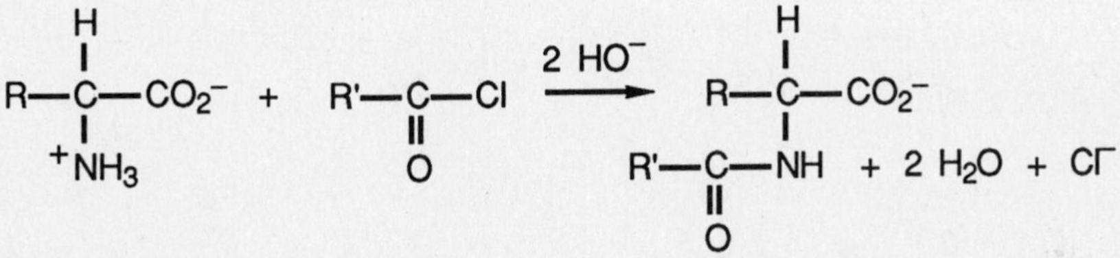

Ninhydrin Reaction

$$2\; \text{(indane-1,3-dione-2,2-diol)} \;+\; R-\underset{\underset{+NH_3}{|}}{\overset{\overset{H}{|}}{C}}-CO_2^- \longrightarrow$$

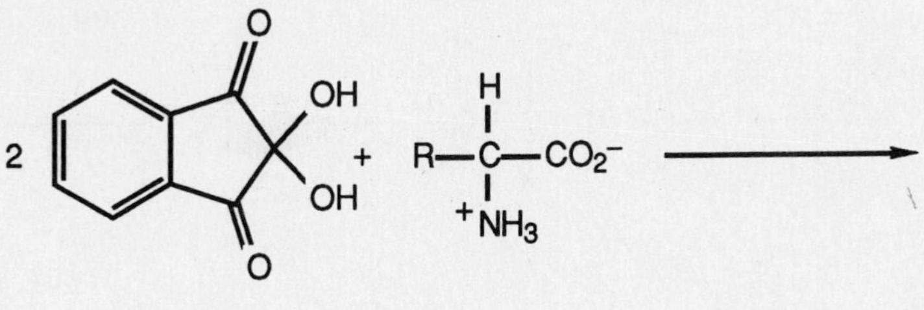

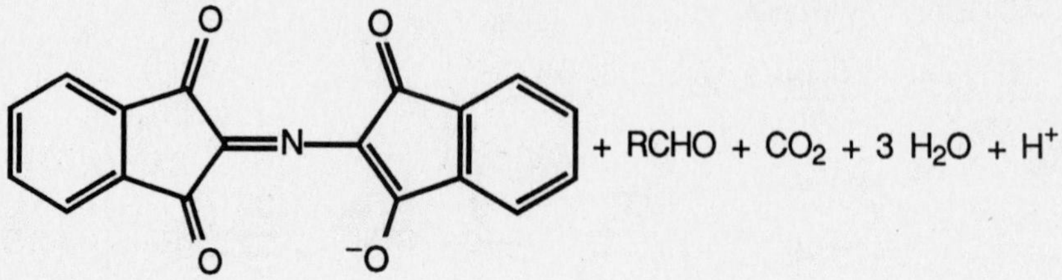

$$+ \; RCHO \; + \; CO_2 \; + \; 3 \; H_2O \; + \; H^+$$

Sanger's Reagent

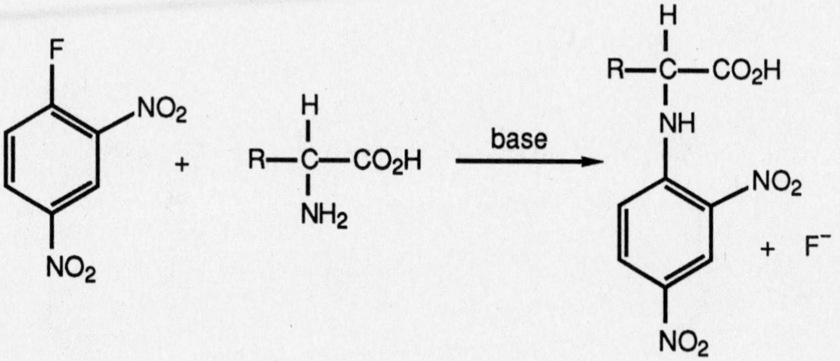

base $\longrightarrow$

$+ \; F^-$

Edman Degradation

$\longrightarrow$

$\xrightarrow[\text{H}_2\text{O}]{\text{HCl}}$

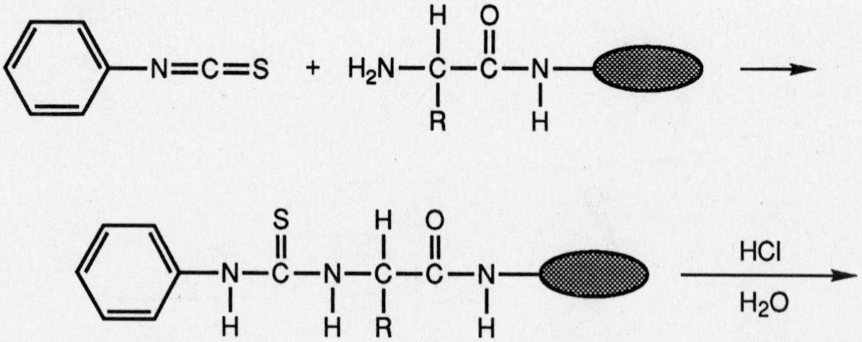

REACTION SUMMARY

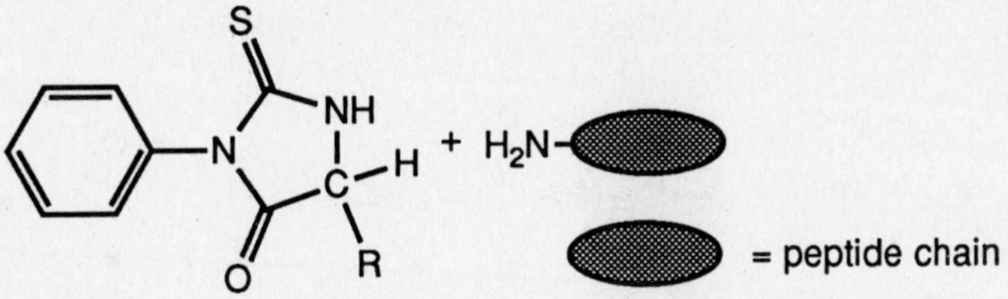

= peptide chain

Peptide Synthesis

a. N-protection:

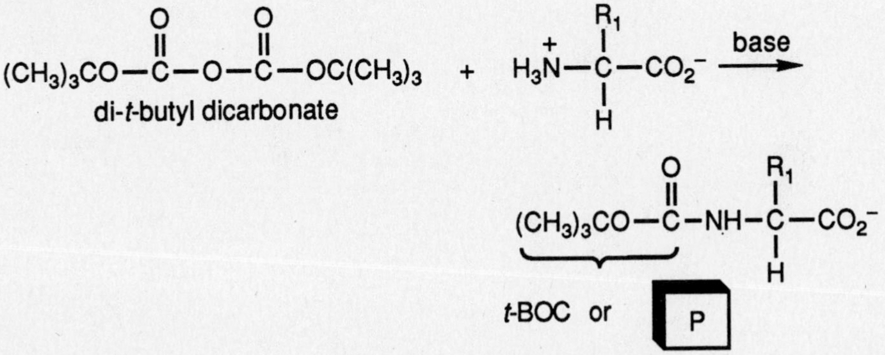

di-*t*-butyl dicarbonate

t-BOC or P

b. Polymer attachment:

= polymer

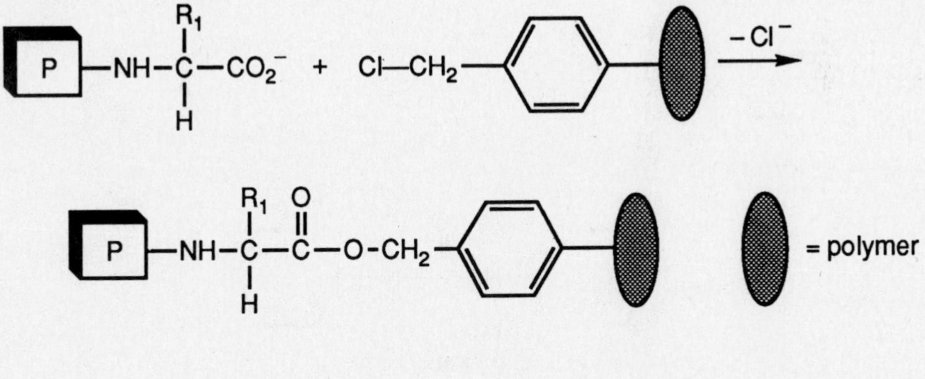

c. Deprotection (removal of the protecting group):

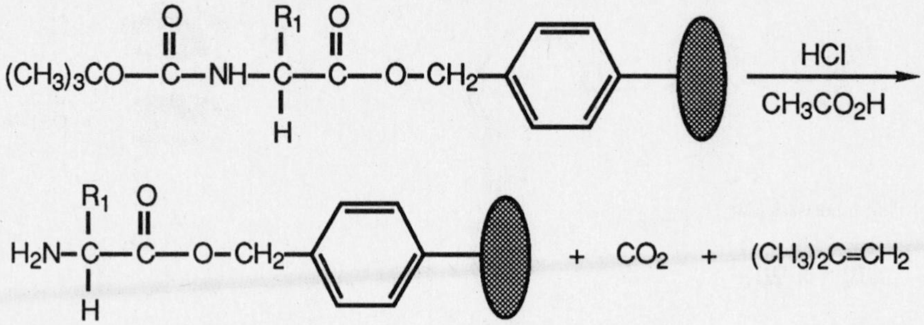

$(CH_3)_3CO-C(=O)-NH-CH(R_1)-C(=O)-O-CH_2$—⟨benzene⟩—⟨resin⟩ $\xrightarrow[CH_3CO_2H]{HCl}$

$H_2N-CH(R_1)-C(=O)-O-CH_2$—⟨benzene⟩—⟨resin⟩ $+$ CO_2 $+$ $(CH_3)_2C=CH_2$

d. Amino acid coupling with DDC:

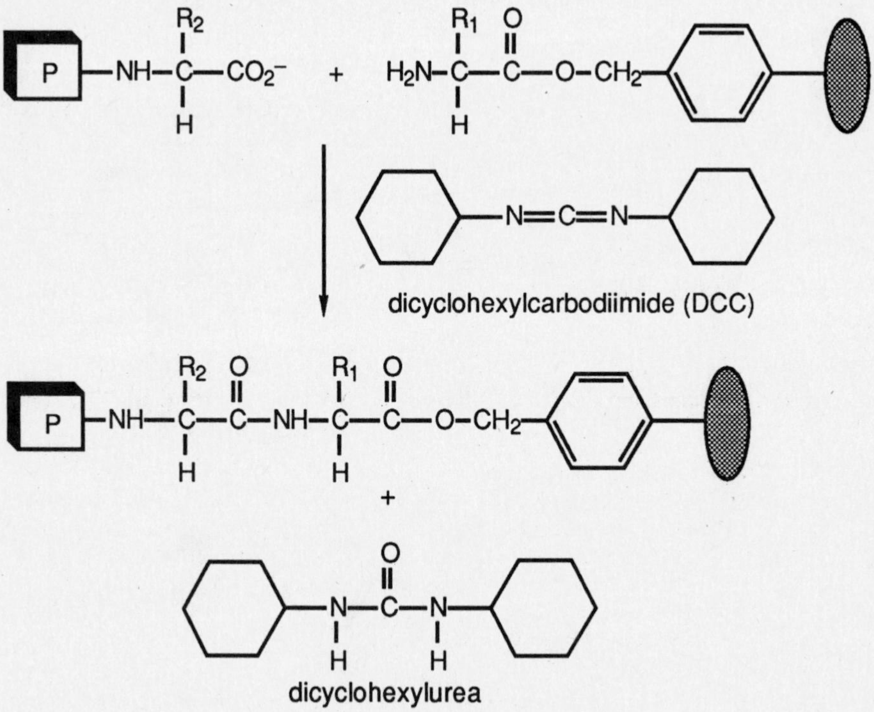

$\boxed{P}-NH-CH(R_2)-CO_2^-$ $+$ $H_2N-CH(R_1)-C(=O)-O-CH_2$—⟨benzene⟩—⟨resin⟩

⟨cyclohexyl⟩$-N=C=N-$⟨cyclohexyl⟩

dicyclohexylcarbodiimide (DCC)

$\boxed{P}-NH-CH(R_2)-C(=O)-NH-CH(R_1)-C(=O)-O-CH_2$—⟨benzene⟩—⟨resin⟩

$+$

⟨cyclohexyl⟩$-NH-C(=O)-NH-$⟨cyclohexyl⟩

dicyclohexylurea

ANSWERS TO PROBLEMS

e. Detachment from the polymer:

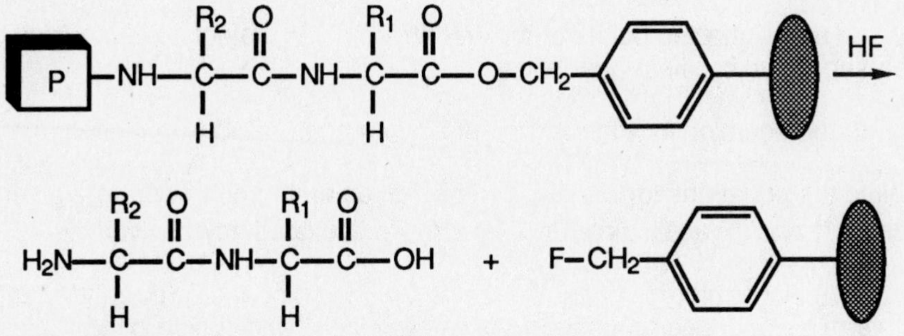

MECHANISM SUMMARY

Nucleophilic Aromatic Substitution (Addition–Elimination)

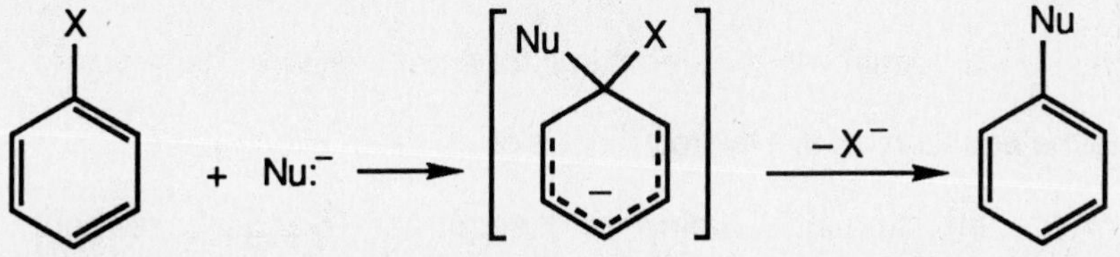

Exemplified by Sanger's reaction (eq. 17.11) and facilitated by electron-withdrawing (carbanion-stabilizing) groups *ortho* or *para* to the leaving group X.

LEARNING OBJECTIVES

1. Know the meaning of: α-amino acid, essential amino acid, dipolar ion, amphoteric, L configuration, isoelectric point (pI), electrophoresis, pK_a.

2. Learn the names, structures, and abbreviation of the amino acids listed in Table 17.1.

3. Write the structure of a given amino acid as a function of the pH of the solution.

4. Write an equation for the reaction of an amino acid (dipolar form) with strong acid or strong base.

5. Write the form of a given amino acid that is likely to predominate at high or low pH and at the isoelectric pH.

6. Given the isoelectric points of several amino acids, predict the directions of their migration during electrophoresis at a given pH.

7. Given the isoelectric points of several amino acids, select a pH at which they can be separated by electrophoresis.

8. Write the equation for the reaction of a given amino acid with ninhydrin reagent.

9. Write the equations for the reaction of a given amino acid with (a) a given alcohol and H^+, and (b) acetic anhydride or another activated acyl derivative.

10. Know the meaning of: peptide bond, peptide, dipeptide, tripeptide, and so on; N-terminal and C-terminal amino acid; cysteine unit; disulfide bond.

11. Given the structures of the component amino acids and the name of a di-, tri-, or polypeptide, draw its structure.

12. Identify the N-terminal and C-terminal amino acid for a given peptide.

13. Given the three-letter abbreviated name for a peptide, write its structure.

14. Write the equation for the hydrolysis of a given di- or polypeptide.

15. Know the meaning of: amino acid sequence, Sanger's reagent, Edman degradation, selective peptide cleavage.

16. Write the equation for the reaction of an amino acid with 2,4-dinitro-fluorobenzene.

17. Write the equations for the Edman degradation of a given tri- or polypeptide.

18. Given information on selective peptide cleavage and the sequences of fragment peptides, deduce the sequence of the original polypeptide.

19. Know the meaning of: protecting group, solid-phase technique, t-butoxycar-bonyl (Boc) group, dicyclohexylcarbodiimide (DCC).

20. Write an equation for the protection of a given amino acid using di-t-butyl dicarbonate and for its deprotection using acid.

21. Write an equation for the linking of two protected amino acids with dicyclohexylcarbodiimide (DCC).

ANSWERS TO PROBLEMS

22. Write an equation for the linking of the N-protected C-terminal amino acid to a polymer for peptide synthesis, and write an equation for detachment of the peptide from the polymer.

23. Describe the geometry of the peptide bond.

24. Know the meaning of: primary, secondary, tertiary, and quaternary protein structure; α helix; pleated sheet; fibrous and globular protein.

ANSWERS TO PROBLEMS

Problems Within the Chapter

17.1 These reactions occur if acid is added to the product of eq. 17.3. That is, they are essentially the reverse of the process shown in eqs. 17.3 and 17.2.

$$CH_3-\underset{\underset{NH_2}{|}}{\overset{\overset{H}{|}}{C}}-CO_2^-\ Na^+\ +\ HCl\ \longrightarrow\ CH_3-\underset{\underset{+NH_3}{|}}{\overset{\overset{H}{|}}{C}}-CO_2^-\ +\ NaCl$$

$$CH_3-\underset{\underset{+NH_3}{|}}{\overset{\overset{H}{|}}{C}}-CO_2^-\ +\ HCl\ \longrightarrow\ CH_3-\underset{\underset{+NH_3}{|}}{\overset{\overset{H}{|}}{C}}-CO_2H\ +\ Cl^-$$

17.2 The $-CO_2H$ group is more acidic than the $-NH_3^+$ group. Thus, on treatment with base, a proton is removed from the $-CO_2H$ group (see eq. 17.2). Then, with a second equivalent of base, a proton is removed from the $-NH_3^+$ group (see eq. 17.3).

17.3 The $-NH_2$ group is more basic than the $-CO_2^-$ group. Thus, on treatment with acid (as in Prob.17.1), a proton adds first to the $-NH_2$ group. Then, with a second equivalent of acid, a proton adds to the $-CO_2^-$ group.

17.4 a. The dipolar ion will not migrate toward either electrode.

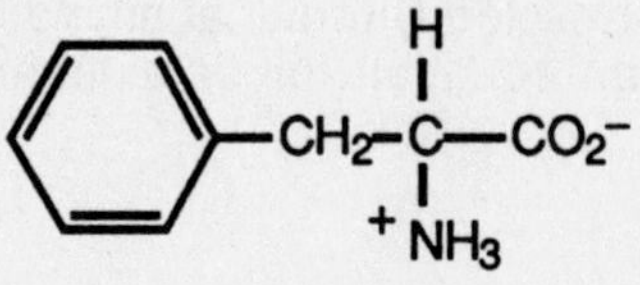

b. The ion will move toward the cathode (negative electrode).

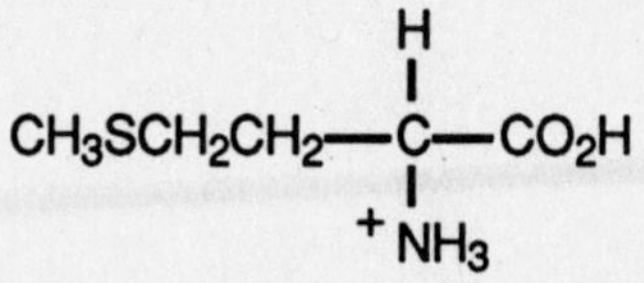

c. The ion will migrate toward the anode (positive electrode).

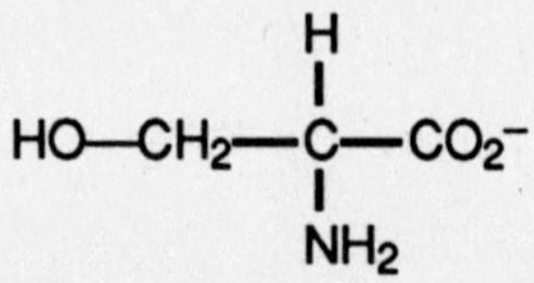

17.5 The least acidic group in aspartic acid is the ammonium group, as shown in eq. 17.5. Carboxyl groups are more acidic than ammonium groups.

17.6 The equilibria for arginine are:

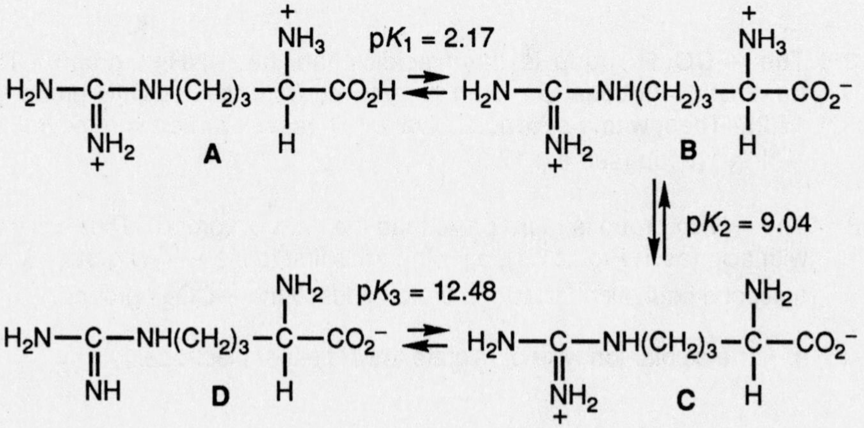

The dipolar ion has structure **C** and the isoelectric point (pI) will be the average of the pK_a's of **B** and **C**. Specifically,

pI $=$ [pK_2 + pK_3]/2 $=$ [9.04 + 12.48]/2 $=$ 10.76

17.7　a.　At pH 7.0, glycine (isoelectric point = 6.0) is present mainly in the form H_3N^+—CH_2—CO_2^- with some H_2N—CH_2—CO_2^- and will migrate toward the anode (positive electrode), whereas lysine (isoelectric point = 9.7) is present as:

$$\overset{+}{H_3N}CH_2CH_2CH_2CH_2-\overset{\overset{\displaystyle H}{|}}{\underset{\underset{\displaystyle +NH_3}{|}}{C}}-CO_2H \quad \text{and} \quad \overset{+}{H_3N}CH_2CH_2CH_2CH_2-\overset{\overset{\displaystyle H}{|}}{\underset{\underset{\displaystyle +NH_3}{|}}{C}}-CO_2^-$$

and will migrate toward the cathode (negative electrode).

b.　At pH 6.0, phenylalanine (isoelectric point = 5.5) will migrate toward the anode. Its structure is

$$\text{C}_6\text{H}_5-CH_2-\overset{\overset{\displaystyle H}{|}}{\underset{\underset{\displaystyle NH_2}{|}}{C}}-CO_2^-$$

Leucine (isoelectric point = 6.0) will not migtrate. Its structure is

$$(CH_3)_2CHCH_2-\overset{\overset{\displaystyle H}{|}}{\underset{\underset{\displaystyle +NH_3}{|}}{C}}-CO_2^-$$

Proline (isoelectric point = 6.3) will migrate toward the cathode. Its structure is

To generalize, if the solution pH is less than the isoelectric pH, the net charge on the amino acid will be positive and it will migrate toward the cathode. If the solution pH is greater than the isoelectric pH, the net charge on the amino acid will be negative and it will migrate toward the anode.

17.8   a.

$$\text{C}_6\text{H}_5-\text{CH}_2-\underset{\overset{|}{+\text{NH}_3}}{\overset{\text{H}}{\underset{|}{\text{C}}}}-\text{CO}_2^- \ + \ \text{CH}_3\text{OH} \ + \ \text{HCl} \ \longrightarrow$$

$$\text{C}_6\text{H}_5-\text{CH}_2-\underset{\overset{|}{+\text{NH}_3}}{\overset{\text{H}}{\underset{|}{\text{C}}}}-\text{CO}_2\text{CH}_3 \ + \ \text{H}_2\text{O} \ + \ \text{Cl}^-$$

b.

$$(\text{CH}_3)_2\text{CH}-\underset{\overset{|}{+\text{NH}_3}}{\overset{\text{H}}{\underset{|}{\text{C}}}}-\text{CO}_2^- \ + \ \text{C}_6\text{H}_5-\overset{\overset{\text{O}}{\|}}{\text{C}}-\text{Cl} \ + \ 2 \ \text{HO}^- \ \longrightarrow$$

$$(\text{CH}_3)_2\text{CH}-\underset{\overset{|}{\text{HN}-\overset{\overset{|}{}}{\underset{\underset{\|}{\text{O}}}{\text{C}}}-\text{C}_6\text{H}_5}}{\overset{\text{H}}{\underset{|}{\text{C}}}}-\text{CO}_2^- \ + \ 2 \ \text{H}_2\text{O} \ + \ \text{Cl}^-$$

c.

$$\text{H}-\underset{\overset{|}{+\text{NH}_3}}{\overset{\text{H}}{\underset{|}{\text{C}}}}-\text{CO}_2^- \ + \ \text{CH}_3-\overset{\overset{\text{O}}{\|}}{\text{C}}-\text{O}-\overset{\overset{\text{O}}{\|}}{\text{C}}-\text{CH}_3 \ \longrightarrow$$

$$\text{CH}_3-\overset{\overset{\text{O}}{\|}}{\text{C}}-\text{NHCH}_2\text{CO}_2\text{H} \ + \ \text{CH}_3\text{CO}_2\text{H}$$

17.9   Rewrite eq. 17.9, with R = CH$_3$.

17.10 Both Gly—Ala and Ala—Gly have one acidic and one basic group. Using Table

17.2 as a guide, the expected pI for these dipeptides is approximately 6. Gly—Ala is expected to be positively charged at pH 3:

$$\overset{+}{H_3}NCH_2—\overset{\overset{\displaystyle O}{\|}}{C}-NH\text{-}\underset{\underset{\displaystyle CH_3}{|}}{CH}-CO_2H \longleftarrow pK_a \approx 2.3$$

and predominantly negatively charged at pH 9:

$$H_2NCH_2—\overset{\overset{\displaystyle O}{\|}}{C}-NH-\overset{\overset{\displaystyle H}{|}}{\underset{\underset{\displaystyle CH_3}{|}}{C}}-CO_2^-$$

$pK_a \approx 9.4$

17.11   a.

$$(CH_3)_2CH—\overset{\overset{\displaystyle H}{|}}{\underset{\underset{\displaystyle +NH_3}{|}}{C}}—\overset{\overset{\displaystyle O}{\|}}{C}-NH-\overset{\overset{\displaystyle CH_3}{|}}{\underset{\underset{\displaystyle H}{|}}{C}}-CO_2^-$$

b.

$$CH_3—\overset{\overset{\displaystyle H}{|}}{\underset{\underset{\displaystyle +NH_3}{|}}{C}}—\overset{\overset{\displaystyle O}{\|}}{C}-NH-\overset{\overset{\displaystyle CH(CH_3)_2}{|}}{\underset{\underset{\displaystyle H}{|}}{C}}-CO_2^-$$

17.12

$$H_2N—CH_2—\overset{\overset{\displaystyle O}{\|}}{C}-NH-\overset{\overset{\displaystyle H}{|}}{\underset{\underset{\displaystyle CH_3}{|}}{C}}—\overset{\overset{\displaystyle O}{\|}}{C}-NH-\overset{\overset{\displaystyle H}{|}}{\underset{\underset{\displaystyle CH_2OH}{|}}{C}}-CO_2H$$

Note that, by convention, in writing peptide structures, each peptide bond is written in the following direction:

$$—\overset{\overset{\displaystyle O}{\|}}{C}-NH—$$

17.13 Gly—Ser—Ala
Ala—Gly—Ser
Ala—Ser—Gly
Ser—Gly—Ala
Ser—Ala—Gly

There are six isomeric structures altogether, including Gly—Ala—Ser.

17.14 For alanylglycine, the reactions are as follows:

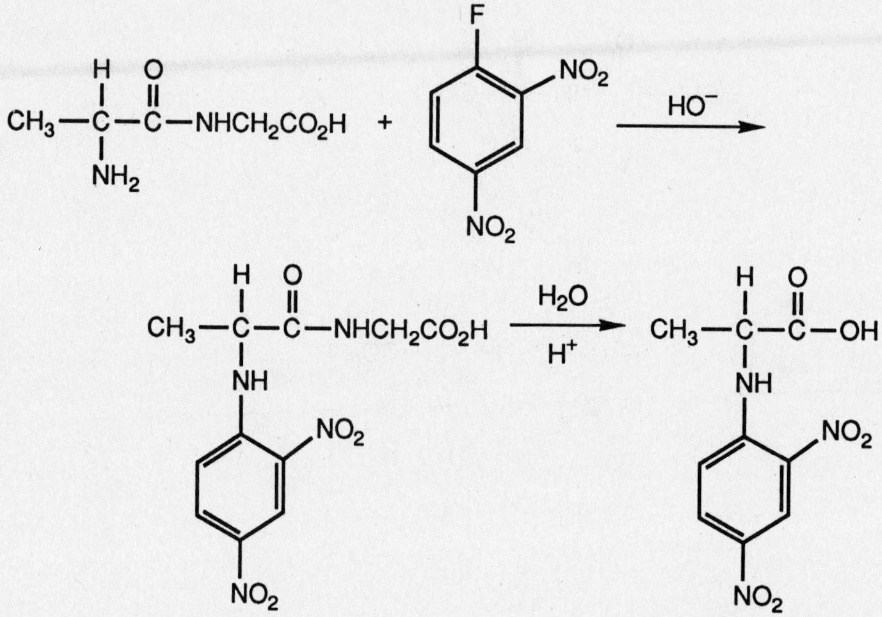

(DNP-alanine)

+

$H_2NCH_2CO_2H$

(glycine)

For glycylalanine, we have the following:

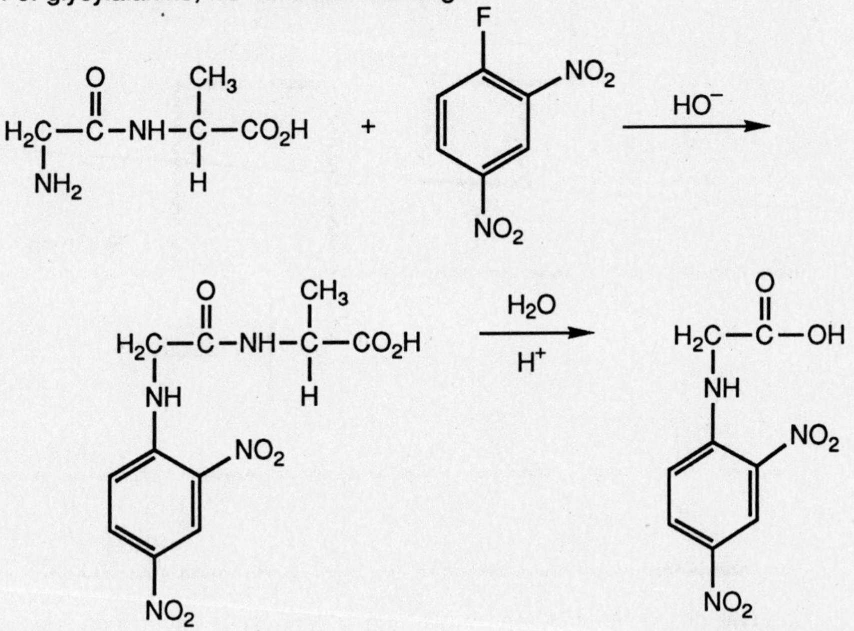

(DNP-glycine)

+

$H_2NCH(CH_3)CO_2H$

(alanine)

17.15 a. Trypsin cleaves peptides on the *carboxyl* side of lysine and arginine. Bradykinin contains no lysines and two arginines. Cleavage will occur after the first arginine (carboxyl side) but *not* in front of the second arginine (amino side). The cleavage products will be

Arg and Pro—Pro—Gly—Phe—Ser—Pro—Phe—Arg

b. Chymotrypsin will cleave bradykinin on the *carboxyl* sides of Phe to give three products:

Arg—Pro—Pro—Gly—Phe and Ser—Pro—Phe and Arg

17.16 Reduction of the disulfide bonds will chemically separate the A and B chains (see eq. 17.10).

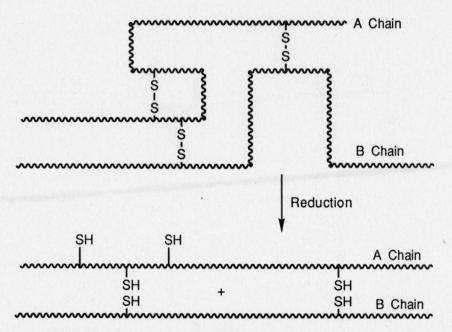

The two chains could then be physically separated using a chromatographic technique, for example, electrophoresis (Figure 17.3).

17.17 The base is needed to convert the —NH$_3^+$ group to a nucleophilic —NH$_2$ group.

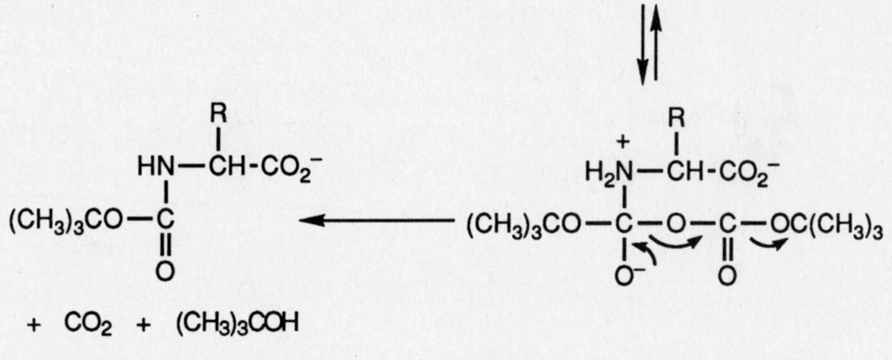

$$+ \quad CO_2 \quad + \quad (CH_3)_3COH$$

The free carboxyl is obtained upon neutralizing the basic reaction mixture.

17.18   First protect the amino group of the C-terminal amino acid (Phe) as described in eq. 17.15:

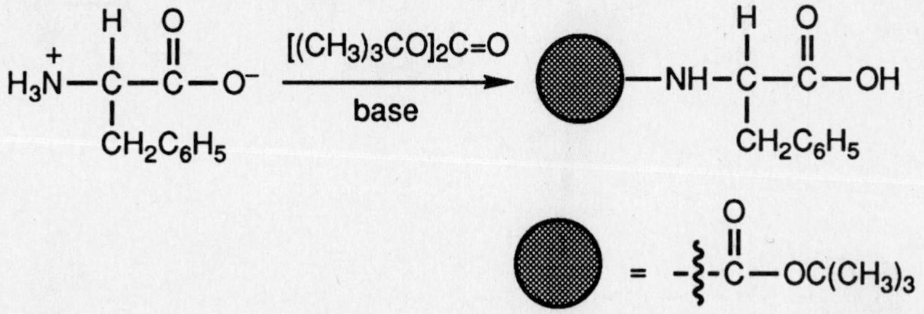

Next attach the N-protected phenylalanine to the chloromethylated polystyrene:

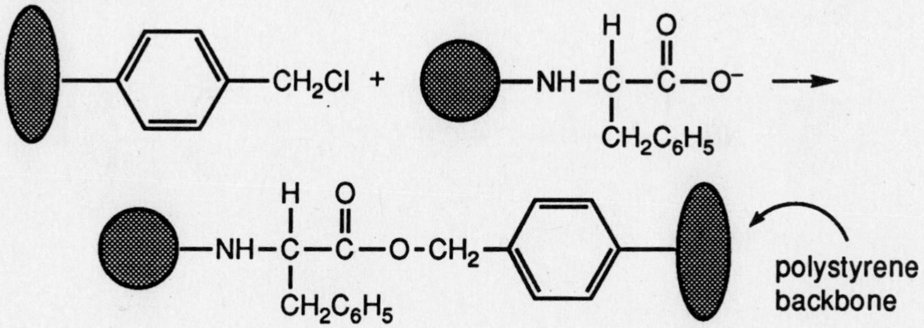

polystyrene
backbone

Then deprotect the amino group of the Phe:

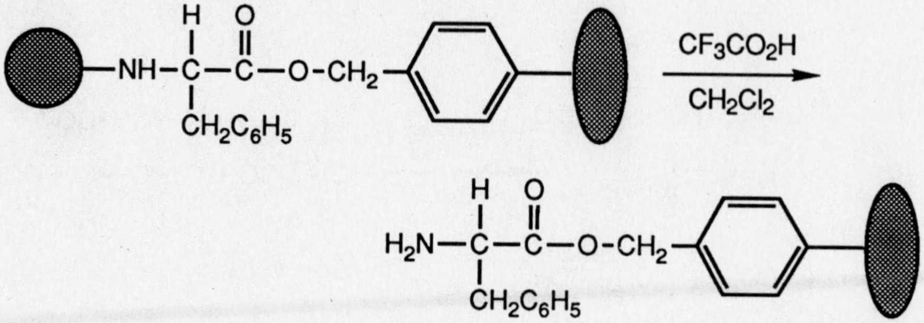

Next couple the polymer bound phenylalanine with N-protected alanine (Ala):

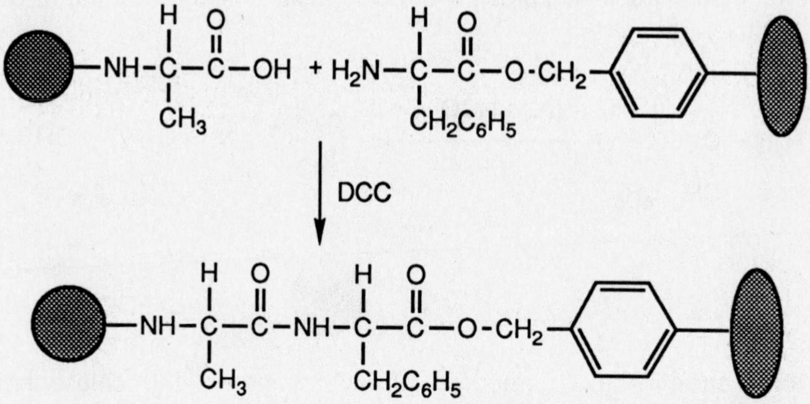

Then sequentially deblock the Ala, couple with N-blocked glycine (Gly), and deblock the Gly:

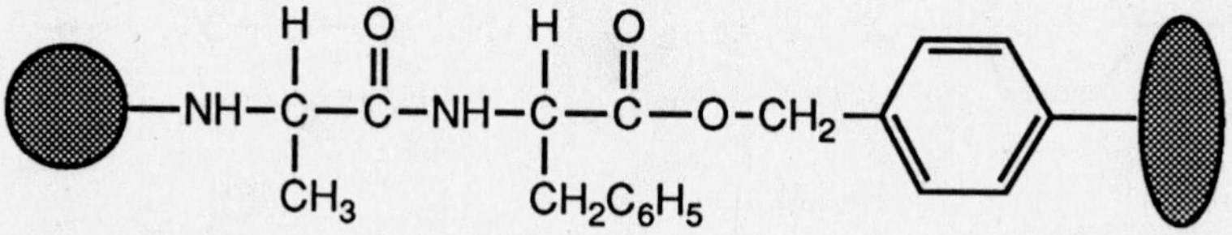

1. CF_3CO_2H, CH_2Cl_2

2. ⬤—NH—CH_2CO_2H

3. CF_3CO_2H, CH_2Cl_2

$$H_2N-CH_2-\overset{\overset{\displaystyle O}{\|}}{C}-NH-\overset{\overset{\displaystyle H}{|}}{\underset{\underset{\displaystyle CH_3}{|}}{C}}-\overset{\overset{\displaystyle O}{\|}}{C}-NH-\overset{\overset{\displaystyle H}{|}}{\underset{\underset{\displaystyle CH_2C_6H_5}{|}}{C}}-\overset{\overset{\displaystyle O}{\|}}{C}-O-CH_2-C_6H_4-⬤$$

Finally, detach the tripeptide from the polymer.

$$H_2N-CH_2-\overset{\overset{\displaystyle O}{\|}}{C}-NH-\overset{\overset{\displaystyle H}{|}}{\underset{\underset{\displaystyle CH_3}{|}}{C}}-\overset{\overset{\displaystyle O}{\|}}{C}-NH-\overset{\overset{\displaystyle H}{|}}{\underset{\underset{\displaystyle CH_2C_6H_5}{|}}{C}}-\overset{\overset{\displaystyle O}{\|}}{C}-O-CH_2-C_6H_4-⬤$$

HF

$$H_2N-CH_2-\overset{\overset{\displaystyle O}{\|}}{C}-NH-\overset{\overset{\displaystyle H}{|}}{\underset{\underset{\displaystyle CH_3}{|}}{C}}-\overset{\overset{\displaystyle O}{\|}}{C}-NH-\overset{\overset{\displaystyle H}{|}}{\underset{\underset{\displaystyle CH_2C_6H_5}{|}}{C}}-\overset{\overset{\displaystyle O}{\|}}{C}-OH$$

(H_2N—Gly—Ala—Phe—CO_2H)

Notice that the peptide is built up from the C-terminal amino acid toward the N-terminal amino acid.

17.19 Hydrocarbon-like side chains are nonpolar:

Gly, Ala, Val, Leu, Ile, Phe, and even Met (a thioether)

Alcohol, thiol, amine, amide, carboxylic acid, guanidine, and imidazole side chains are polar:

Ser, Thr, Cys, Pro, Asp, Glu, Asn, Gln, Lys, Arg, Trp, Tyr and His

Aromatic side chains are relatively flat:

Phe, Tyr, Trp, His

Additional Problems

17.20 Definitions and/or examples are given in the text, where indicated.

a.	Sec. 17.8	b.	Sec. 17.3
c.	Sec. 17.8	d.	Sec. 17.2
e.	Sec. 17.2	f.	Table 17.1, entries 1–5, 9–11
g.	Table 17.1, entries 6–8, 12–20	h.	Sec. 17.3
i.	Sec. 17.3	j.	Sec. 17.7

17.21 The configuration is S.

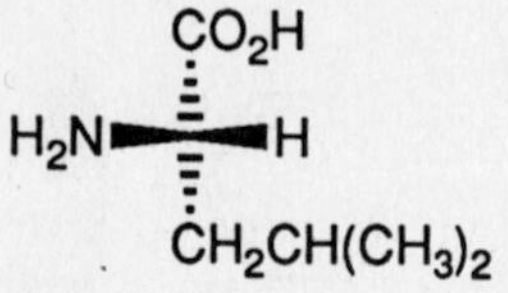

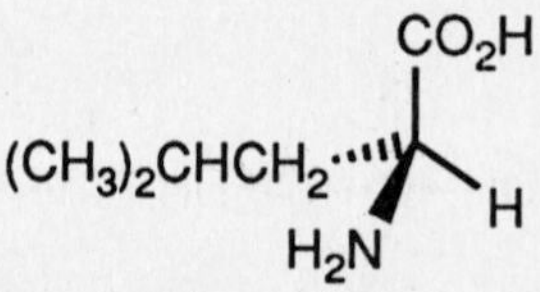

L-leucine

priority order: $NH_2 > CO_2H > CH_2CH(CH_3)_2 > H$

17.22 Use Figure 17.1 and Table 17.1 as guides.

a.

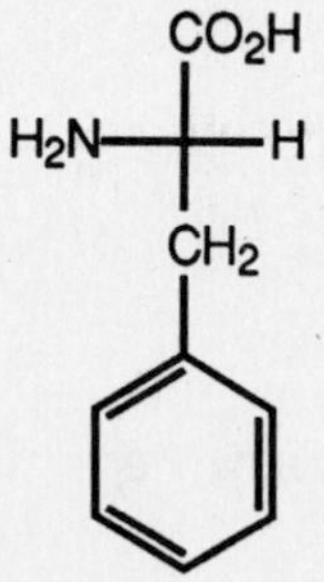

b.

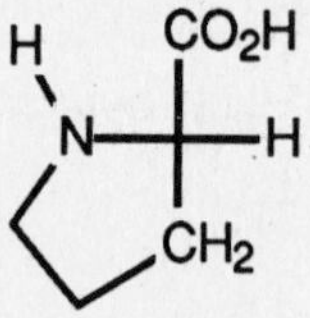

17.23 a.

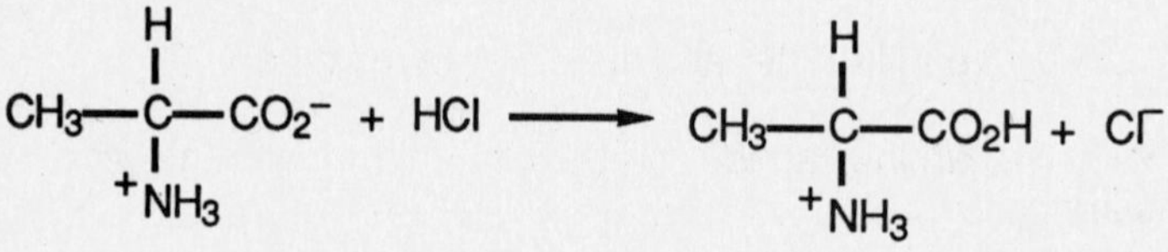

b.

$$CH_3-\overset{\overset{\displaystyle H}{|}}{\underset{\underset{\displaystyle +NH_3}{|}}{C}}-CO_2^- \;+\; NaOH \;\longrightarrow\; CH_3-\overset{\overset{\displaystyle H}{|}}{\underset{\underset{\displaystyle NH_2}{|}}{C}}-CO_2^-\,Na^+ \;+\; H_2O$$

17.24   a.

$$(CH_3)_2CH-\overset{\overset{\displaystyle H}{|}}{\underset{\underset{\displaystyle +NH_3}{|}}{C}}-CO_2^-$$

b.

$$HOCH_2-\overset{\overset{\displaystyle H}{|}}{\underset{\underset{\displaystyle +NH_3}{|}}{C}}-CO_2^-$$

c.

$$H_2C-\overset{\overset{\displaystyle H}{|}}{\underset{\underset{\displaystyle NH_2}{|}}{C}}-CO_2^-$$

d.

$$HO-\!\!\bigcirc\!\!-CH_2-\overset{\overset{\displaystyle H}{|}}{\underset{\underset{\displaystyle +NH_3}{|}}{C}}-CO_2^-$$

17.25   a.   The most acidic proton is on the carboxyl group nearest the ammonium ion. The product of deprotonation is

$$HO_2CCH_2CH_2-\overset{\overset{\displaystyle H}{|}}{\underset{\underset{\displaystyle +NH_3}{|}}{C}}-CO_2^-$$

b.   The ammonium ion is more acidic than the alcohol function. The product of deprotonation is

$$HOCH_2-\overset{\overset{\displaystyle H}{|}}{\underset{\underset{\displaystyle NH_2}{|}}{C}}-CO_2^-$$

c.   The carboxyl group is a stronger acid than the ammonium group. The product of deprotonation is

$$(CH_3)_2CH-\overset{\displaystyle H}{\underset{\displaystyle +NH_3}{C}}-CO_2^-$$

d. The proton on the positive nitrogen is the only appreciably acidic proton. The product of deprotonation is

$$HN\overset{\displaystyle }{=}\underset{\displaystyle NH_2}{C}-NH(CH_2)_3-\overset{\displaystyle H}{\underset{\displaystyle NH_2}{C}}-CO_2^-$$

17.26 a. The most basic group is CO_2^-. The product of protonation is

$$CH_3-\overset{\displaystyle H}{\underset{\displaystyle HO}{C}}-\overset{\displaystyle H}{\underset{\displaystyle +NH_3}{C}}-CO_2H$$

b. The carboxylate ion remote from the ammonium group is most basic. The product of protonation is

$$HO_2CCH_2-\overset{\displaystyle H}{\underset{\displaystyle +NH_3}{C}}-CO_2^-$$

17.27 Because of its positive charge, the ammonium group is an electron-attracting substituent. As such, it enhances the acidity of the nearby carboxyl group (in protonated alanine).

17.28

$$HO_2CCH_2CH_2-\overset{\displaystyle +NH_3}{\underset{\displaystyle H}{C}}-CO_2H \underset{\displaystyle }{\overset{\displaystyle pH = 2.19}{\rightleftharpoons}} HO_2CCH_2CH_2-\overset{\displaystyle +NH_3}{\underset{\displaystyle H}{C}}-CO_2^-$$

pH = 4.25

$^-O_2CCH_2CH_2{-}\underset{\underset{H}{|}}{\overset{\overset{NH_2}{|}}{C}}{-}CO_2^-$ ⇌ (pH = 9.67) $^-O_2CCH_2CH_2{-}\underset{\underset{H}{|}}{\overset{\overset{\overset{+}{NH_3}}{|}}{C}}{-}CO_2^-$

17.29 The order of acidities is COOH > $^{\oplus}NH_3$ > ArOH (a phenol).

$HO{-}\text{⬡}{-}CH_2{-}\underset{\underset{H}{|}}{\overset{\overset{\overset{+}{NH_3}}{|}}{C}}{-}CO_2H$ ⇌ (pH = 2.20) $HO{-}\text{⬡}{-}CH_2{-}\underset{\underset{H}{|}}{\overset{\overset{\overset{+}{NH_3}}{|}}{C}}{-}CO_2^-$

pH = 9.11

$^-O{-}\text{⬡}{-}CH_2{-}\underset{\underset{H}{|}}{\overset{\overset{NH_2}{|}}{C}}{-}CO_2^-$ ⇌ (pH = 10.07) $HO{-}\text{⬡}{-}CH_2{-}\underset{\underset{H}{|}}{\overset{\overset{NH_2}{|}}{C}}{-}CO_2^-$

17.30

$H_2N{-}\underset{\underset{NH}{\|}}{C}{-}NH(CH_2)_3{-}\underset{\underset{H}{|}}{\overset{\overset{NH_2}{|}}{C}}{-}CO_2^-$ ⇌ (pH = 12.48) $H_2N{-}\underset{\underset{\overset{NH_2}{+}}{\|}}{C}{-}NH(CH_2)_3{-}\underset{\underset{H}{|}}{\overset{\overset{NH_2}{|}}{C}}{-}CO_2^-$

pH = 9.04

$H_2N{-}\underset{\underset{\overset{NH_2}{+}}{\|}}{C}{-}NH(CH_2)_3{-}\underset{\underset{H}{|}}{\overset{\overset{\overset{+}{NH_3}}{|}}{C}}{-}CO_2H$ ⇌ (pH = 2.17) $H_2N{-}\underset{\underset{\overset{NH_2}{+}}{\|}}{C}{-}NH(CH_2)_3{-}\underset{\underset{H}{|}}{\overset{\overset{\overset{+}{NH_3}}{|}}{C}}{-}CO_2^-$

The order of basicities (and therefore the sequence of protonation of groups) is guanidine > —NH$_2$ > —CO$_2^-$.

17.31 In strong acid (pH 1), histidine has two resonance contributors:

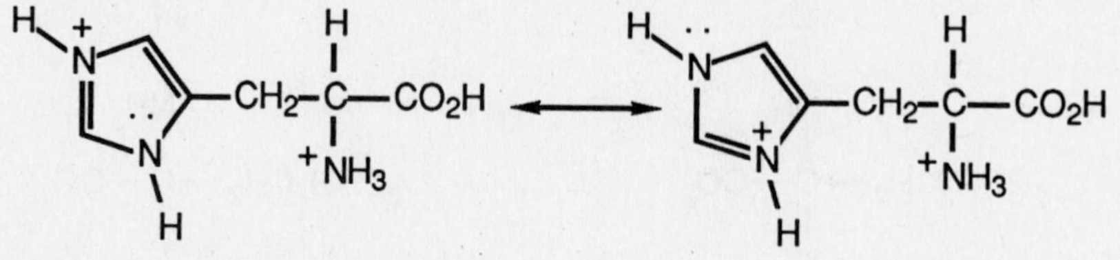

17.32 When an amino acid is in aqueous solution at a pH greater than its isoelectric point (pI), it will be negatively charged and will migrate toward the anode (positive electrode) of an electrophoresis apparatus. If the pH is less than the pI, the amino acid will migrate toward the cathode (negative electrode). The pI's of Asn, His, and Asp are 5.4, 7.6, and 3.0, respectively (consult Table 17.1). Therefore at pH 6, His will migrate toward the cathode, while Asp and Asn will migrate toward the anode. Asp will migrate faster than Asn because it will be more negatively charged at pH 6.

17.33 a. Use eq. 17.7 with R = CH_3 and R' = CH_3CH_2—.

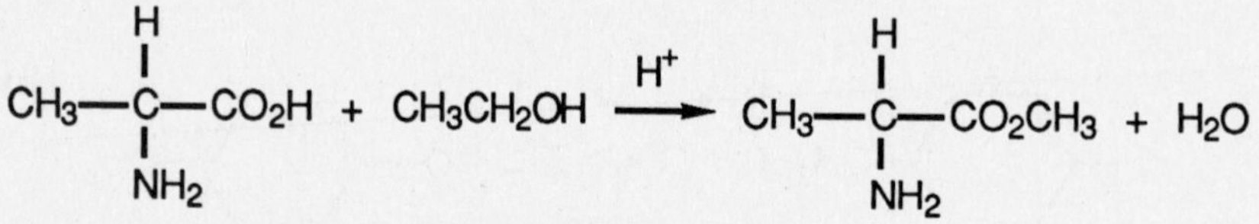

b. Use eq. 17.8 with R = CH_3 and R' = C_6H_5.

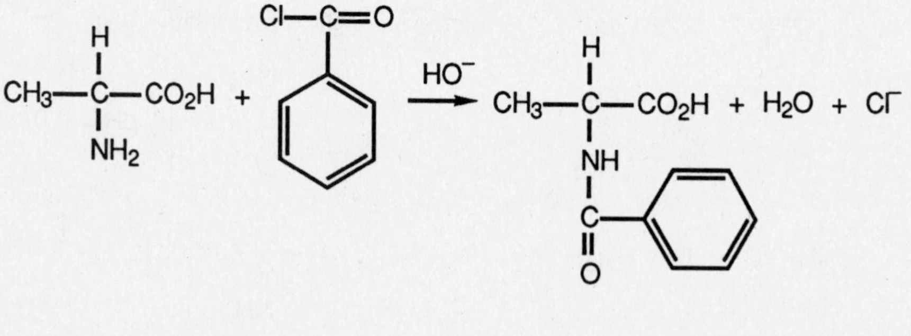

c. Use eq. 17.15 with R = CH$_3$ and substitute (CH$_3$)$_3$CO with CH$_3$.

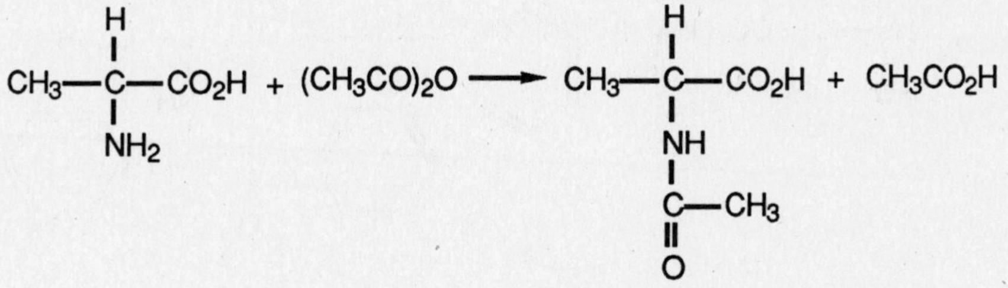

$$CH_3-\underset{\underset{NH_2}{|}}{\overset{\overset{H}{|}}{C}}-CO_2H \ + \ (CH_3CO)_2O \longrightarrow CH_3-\underset{\underset{\underset{\overset{||}{O}}{\underset{C-CH_3}{|}}}{\underset{NH}{|}}}{\overset{\overset{H}{|}}{C}}-CO_2H \ + \ CH_3CO_2H$$

17.34 a. The amino group and the hydroxyl group react with acetic anhydride.

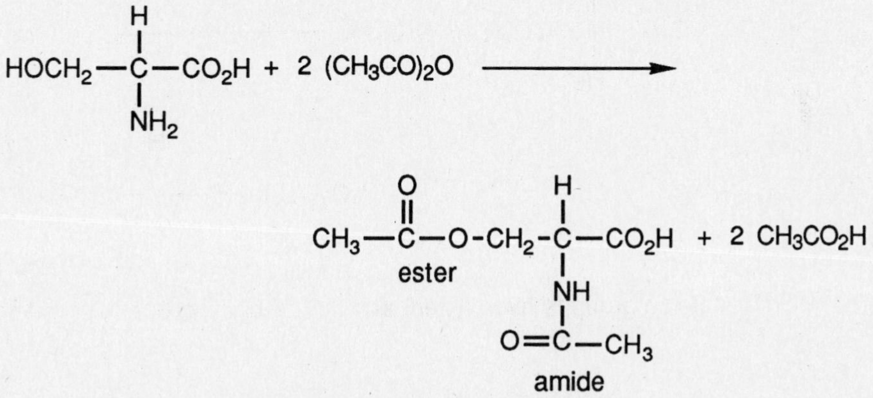

$$HOCH_2-\underset{\underset{NH_2}{|}}{\overset{\overset{H}{|}}{C}}-CO_2H \ + \ 2 \ (CH_3CO)_2O \longrightarrow$$

$$\underset{ester}{CH_3-\overset{\overset{O}{||}}{C}-O-CH_2}-\underset{\underset{\underset{O=C-CH_3}{|}}{\underset{NH}{|}}}{\overset{\overset{H}{|}}{C}}-CO_2H \ + \ 2 \ CH_3CO_2H$$
amide

b. The phenolic ring is easily brominated (see Sec. 7.15, eq. 7.40).

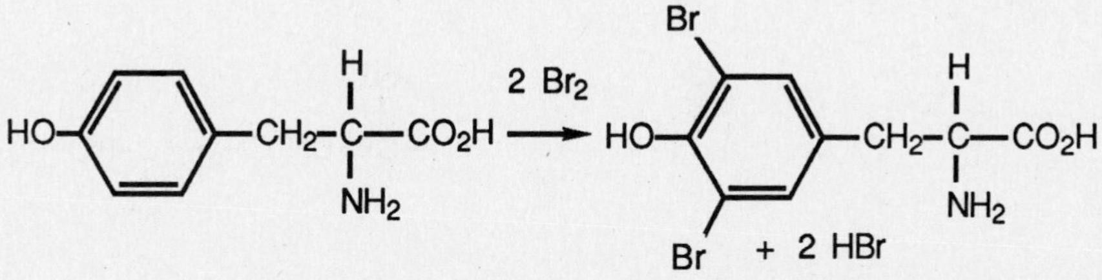

c. Both the hydroxyl and amino groups react with benzoyl chloride.

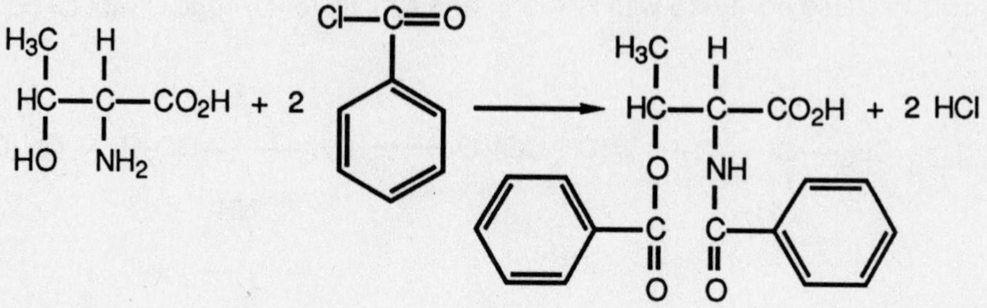

$$H_3C \quad H$$
$$HC-C-CO_2H + 2 \quad Cl-C=O \quad \longrightarrow \quad HC-C-CO_2H + 2 \; HCl$$
$$HO \quad NH_2$$

d. Both carboxyl groups are esterified.

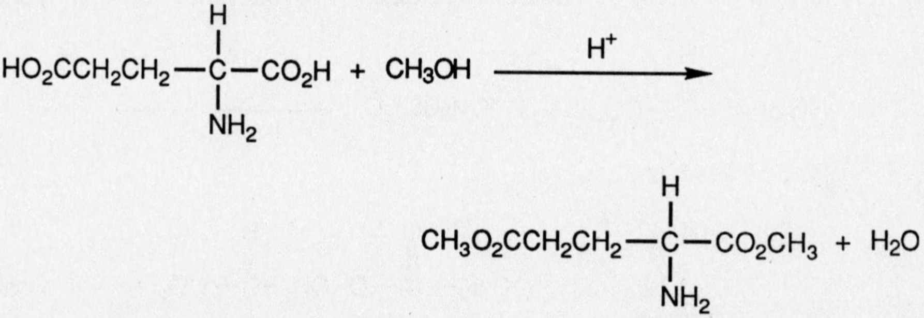

$$HO_2CCH_2CH_2-\overset{H}{\underset{NH_2}{C}}-CO_2H + CH_3OH \xrightarrow{\;H^+\;}$$

$$CH_3O_2CCH_2CH_2-\overset{H}{\underset{NH_2}{C}}-CO_2CH_3 + H_2O$$

e. The amide group is hydrolyzed, and the carboxyl group is converted to its salt.

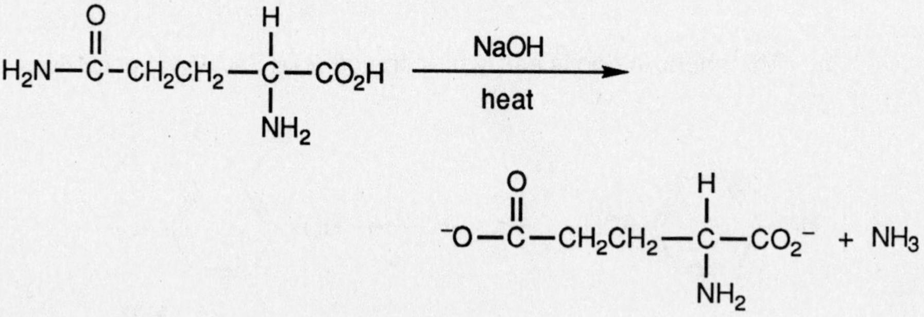

$$H_2N-\overset{O}{\overset{\|}{C}}-CH_2CH_2-\overset{H}{\underset{NH_2}{C}}-CO_2H \xrightarrow[\text{heat}]{\text{NaOH}}$$

$$^-O-\overset{O}{\overset{\|}{C}}-CH_2CH_2-\overset{H}{\underset{NH_2}{C}}-CO_2^- + NH_3$$

17.35 Follow eq. 17.9, with R = benzyl ($C_6H_5CH_2$—).

17.36 Review Sec. 17.8 if necessary, and use Table 17.1 for the structures of the amino acids. Write the structures in neutral form, recognizing of course that dipolar ion structures are possible and that the exact form and degree of ionization depend on the pH of the solution .

ANSWERS TO PROBLEMS

a.

$$H_2N-\overset{\overset{\displaystyle H}{|}}{C}-\overset{\overset{\displaystyle O}{||}}{C}-NH-\overset{\overset{\displaystyle H}{|}}{C}-CO_2H$$

with CH_3 on the first carbon and CH_3 on the second carbon.

b.

$$H_2N-\overset{\overset{\displaystyle H}{|}}{C}-\overset{\overset{\displaystyle O}{||}}{C}-NH-\overset{\overset{\displaystyle H}{|}}{C}-CO_2H$$

with $(CH_3)_2HC$ on the first carbon and CH_2—(indole ring) on the second carbon.

c.

$$H_2N-\overset{\overset{\displaystyle H}{|}}{C}-\overset{\overset{\displaystyle O}{||}}{C}-NH-\overset{\overset{\displaystyle H}{|}}{C}-CO_2H$$

with CH_2—(indole ring) on the first carbon and $CH(CH_3)_2$ on the second carbon.

d.

$$H_2N-CH_2-\overset{\overset{\displaystyle O}{||}}{C}-NH-\overset{\overset{\displaystyle H}{|}}{C}-\overset{\overset{\displaystyle O}{||}}{C}-NHCH_2CO_2H$$

with CH_3 on the central carbon.

e.

$$H_2N-\overset{\overset{\displaystyle H}{|}}{C}-\overset{\overset{\displaystyle O}{||}}{C}-NH-\overset{\overset{\displaystyle H}{|}}{C}-\overset{\overset{\displaystyle O}{||}}{C}-NH-\overset{\overset{\displaystyle H}{|}}{C}-\overset{\overset{\displaystyle O}{||}}{C}-OH$$

with CH_2OH on the first carbon, CH_2—$CH(CH_3)_2$ on the second carbon, and $CH_2CH_2CH_2-NH-\overset{\overset{\displaystyle ||}{}}{C}-NH_2$ (with NH) on the third carbon.

431

f.

$$H_2N-\overset{\overset{\displaystyle H}{|}}{C}-\overset{\overset{\displaystyle O}{\|}}{C}-NH\cdot CH_2-\overset{\overset{\displaystyle O}{\|}}{C}-NH\cdot CH_2-\overset{\overset{\displaystyle O}{\|}}{C}-NH-\overset{\overset{\displaystyle H}{|}}{C}-\overset{\overset{\displaystyle O}{\|}}{C}-OH$$

with side chains CH_2 (bearing imidazole ring: HN–N) and $CH_2CH_2CO_2H$

13.37 a.

$$H_2N-\overset{\overset{\displaystyle H}{|}}{C}-\overset{\overset{\displaystyle O}{\|}}{C}-NH-\overset{\overset{\displaystyle H}{|}}{C}-\overset{\overset{\displaystyle O}{\|}}{C}-OH \quad \xrightarrow[H^+]{H_2O}$$

with side chains CH_2, $CH(CH_3)_2$ and CH_2OH

$$H_2N-\overset{\overset{\displaystyle H}{|}}{C}-\overset{\overset{\displaystyle O}{\|}}{C}-OH \quad + \quad H_2N-\overset{\overset{\displaystyle H}{|}}{C}-\overset{\overset{\displaystyle O}{\|}}{C}-OH$$

with side chains CH_2, $CH(CH_3)_2$ Leu   and CH_2OH Ser

b.

$$H_2N-\overset{\overset{\displaystyle H}{|}}{C}-\overset{\overset{\displaystyle O}{\|}}{C}-NH-\overset{\overset{\displaystyle H}{|}}{C}-\overset{\overset{\displaystyle O}{\|}}{C}-OH \quad \xrightarrow[H^+]{H_2O}$$

with side chains CH_2OH and $CH_2CH(CH_3)_2$

$$H_2N-\overset{\overset{\displaystyle H}{|}}{C}-\overset{\overset{\displaystyle O}{\|}}{C}-OH \quad + \quad H_2N-\overset{\overset{\displaystyle H}{|}}{C}-\overset{\overset{\displaystyle O}{\|}}{C}-OH$$

with side chains CH_2, $CH(CH_3)_2$ Leu   and CH_2OH Ser

The hydrolysis products in parts a and b are identical.

c.

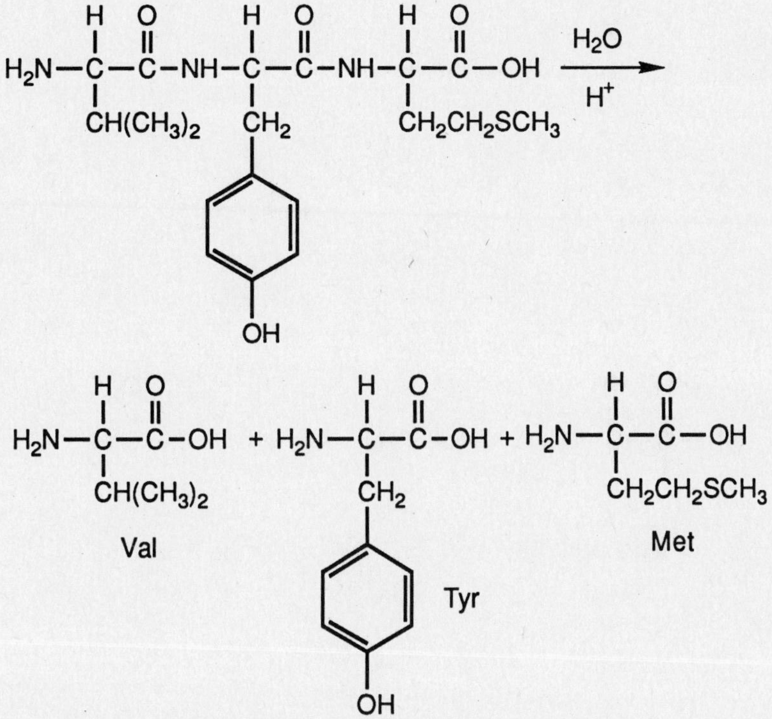

$$H_2N-\overset{\overset{\displaystyle H}{|}}{C}-\overset{\overset{\displaystyle O}{\|}}{C}-NH-\overset{\overset{\displaystyle H}{|}}{C}-\overset{\overset{\displaystyle O}{\|}}{C}-NH-\overset{\overset{\displaystyle H}{|}}{C}-\overset{\overset{\displaystyle O}{\|}}{C}-OH \quad \xrightarrow[H^+]{H_2O}$$

with side chains $CH(CH_3)_2$, CH_2 (to phenol ring with OH), $CH_2CH_2SCH_3$

$$H_2N-\overset{\overset{\displaystyle H}{|}}{C}-\overset{\overset{\displaystyle O}{\|}}{C}-OH \;+\; H_2N-\overset{\overset{\displaystyle H}{|}}{C}-\overset{\overset{\displaystyle O}{\|}}{C}-OH \;+\; H_2N-\overset{\overset{\displaystyle H}{|}}{C}-\overset{\overset{\displaystyle O}{\|}}{C}-OH$$

$CH(CH_3)_2$ — Val

CH_2 (phenol ring with OH) — Tyr

$CH_2CH_2SCH_3$ — Met

17.38

$$H_2N-\overset{\overset{\displaystyle H}{|}}{C}-\overset{\overset{\displaystyle O}{\|}}{C}-NH-\overset{\overset{\displaystyle H}{|}}{C}-\overset{\overset{\displaystyle O}{\|}}{C}-OCH_3 \quad \xrightarrow[\substack{H^+ \\ heat}]{H_2O}$$

with side chains CH_2–CO_2H and CH_2–(phenyl ring)

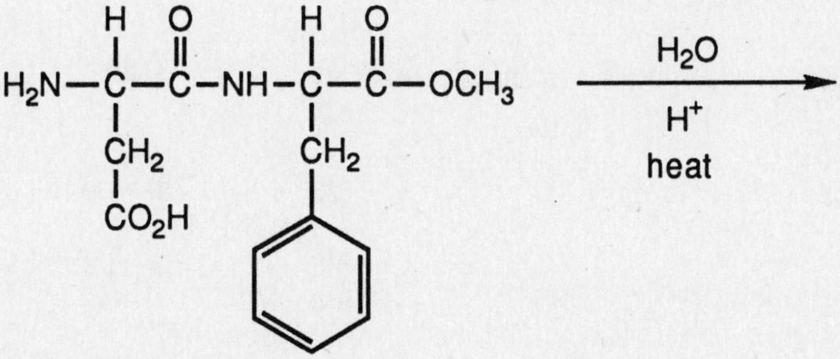

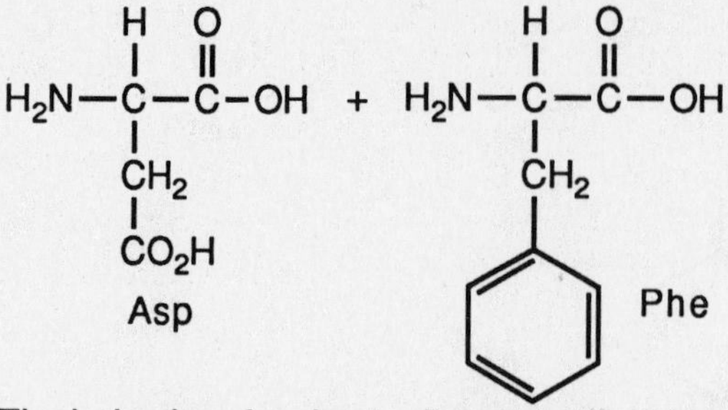

Asp + Phe

17.39 The behavior of a simple dipeptide like this, with no acidic or basic functions in the R group of each unit, is very much like that of a simple amino acid.

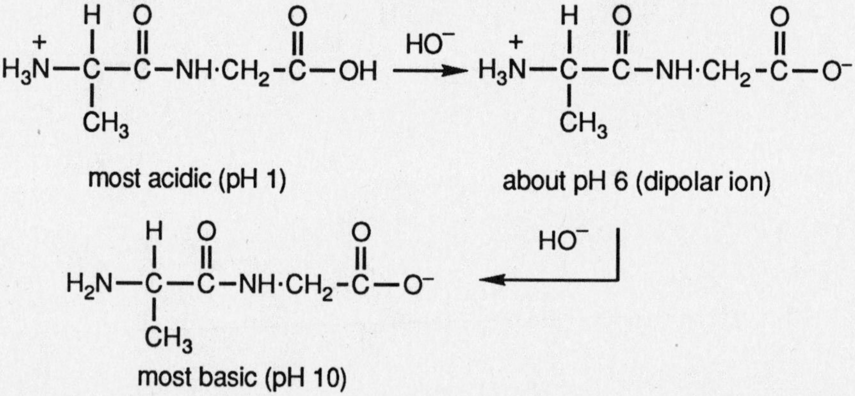

most acidic (pH 1)   about pH 6 (dipolar ion)

most basic (pH 10)

17.40 With four different amino acids there are 4!, or $4 \times 3 \times 2 \times 1 = 24$, possible structures. They are as follows:

Gly—Ala—Val—Leu	Ala—Gly—Val—Leu
Gly—Ala—Leu—Val	Ala—Gly—Leu—Val
Gly—Val—Ala—Leu	Ala—Val—Gly—Leu
Gly—Val—Leu—Ala	Ala—Val—Leu—Gly
Gly—Leu—Ala—Val	Ala—Leu—Gly—Val
Gly—Leu—Val—Ala	Ala—Leu—Val—Gly
Val—Gly—Ala—Leu	Leu—Gly—Ala—Val
Val—Gly—Leu—Ala	Leu—Gly—Val—Ala
Val—Ala—Gly—Leu	Leu—Ala—Gly—Val
Val—Ala—Leu—Gly	Leu—Ala—Val—Gly
Val—Leu—Gly—Ala	Leu—Val—Gly—Ala
Val—Leu—Ala—Gly	Leu—Val—Ala—Gly

17.41 Hydrogen peroxide oxidizes thiols to disulfides (Sec. 7.17, eq. 7.45). See also eq. 17.10.

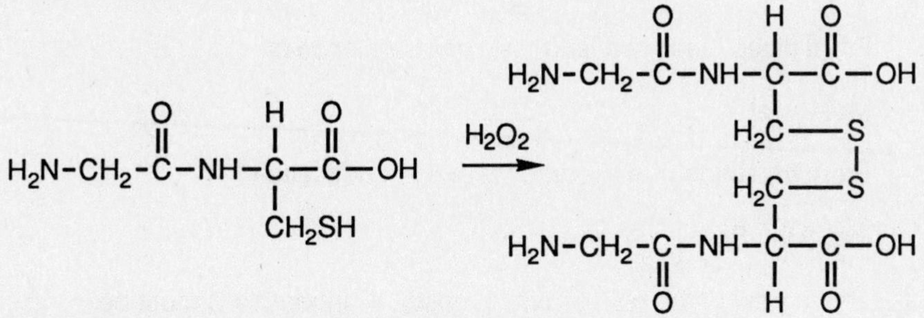

17.42 a. Use eq. 17.11, with R_1 = H and ... = OH.

b. Lysine contains two primary amino groups. Each group can react with 2,4-dinitrofluorobenzene:

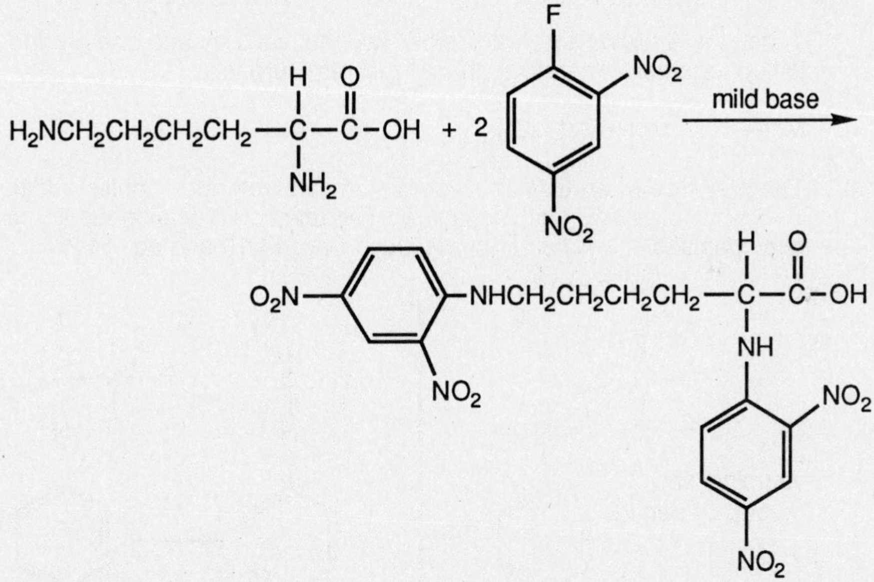

17.43 The first result tells us that the N-terminal amino acid is methionine. The structure at this point is

Met (2 Met, Ser, Gly)

From dipeptide D, we learn that one sequence is

Ser—Met

From dipeptide C, we learn that one sequence is

Met—Met

From tripeptide B, we learn that one sequence is

Met (Met, Ser)

and in view of the result from dipeptide D, tripeptide B must be

Met—Ser—Met

Since two methionines must be adjacent (dipeptide C), the possibilities are

Met—Met—Ser—Met—Gly and Met—Ser—Met—Met—Gly

Tripeptide A allows a decision since two methionines and one glycine must be joined together. Therefore, the correct structure is

Met—Ser—Met—Met—Gly

17.44 The first step in an Edman degradation involves nucleophilic addition to the C=S bond of phenyl isothiocyanate. The reaction is analogous to the addition of alcohols to isocyanates discussed in Sec. 14.10 and eq. 14.28.

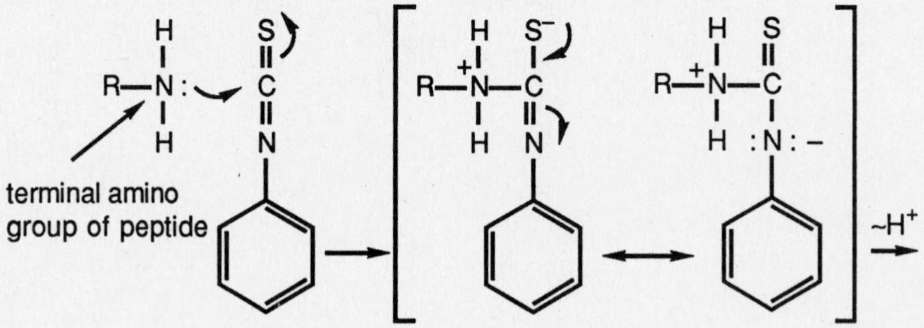

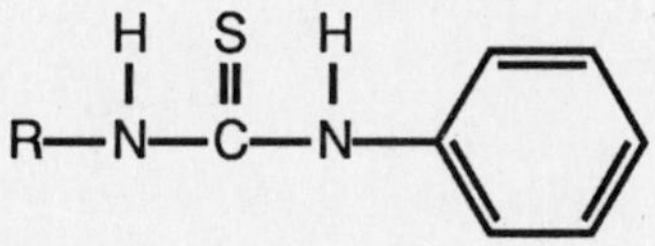

The product is a thiourea (compare its structure with the structure of urea on page 401 of the text).

17.45 Follow the scheme in Figure 17.6, with R_1 = CH$_3$, R_2 = H, and R_3 = (CH$_3$)$_2$CH, and with an —OH attached to the carbonyl group at the right. The final products are the phenylthiohydantoin of alanine and the dipeptide glycylvaline.

17.46

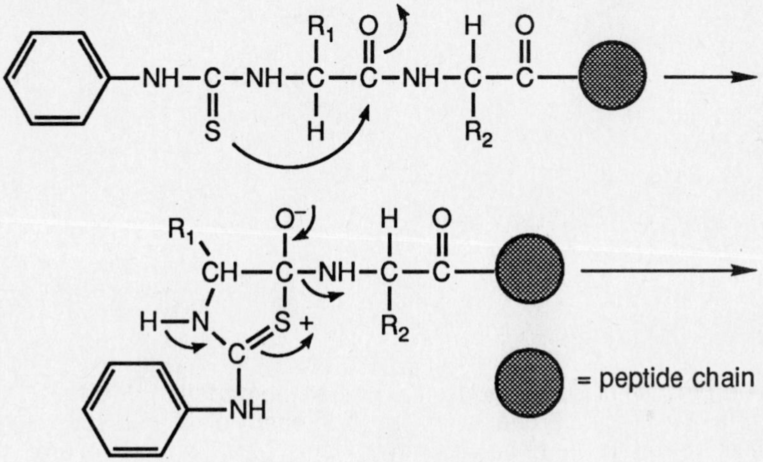

= peptide chain

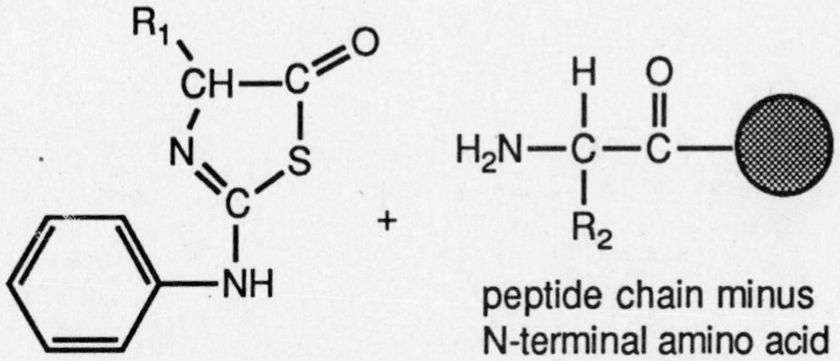

a thiazolinone of the
N-terminal amino acid

peptide chain minus
N-terminal amino acid

One possible mechanism for rearrangement to the hydantoin is the following:

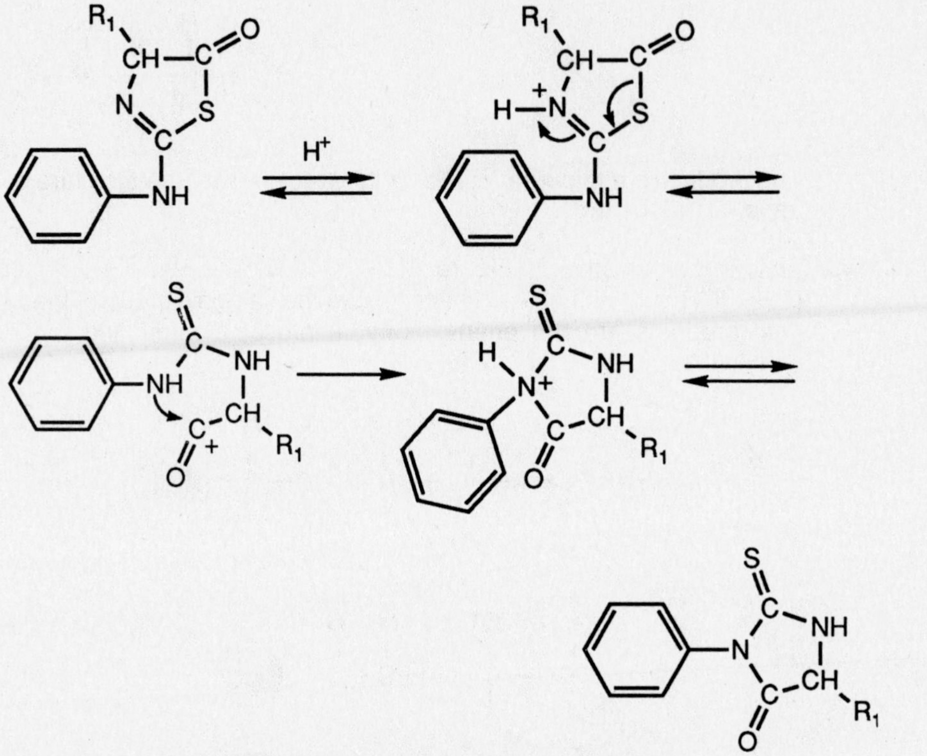

17.47 When insulin is subjected to the Edman degradation, the N-terminal amino acids of the A and B chains (glycine and phenylalanine, respectively) will both be converted to their phenylthiohydantoins. Their structures are:

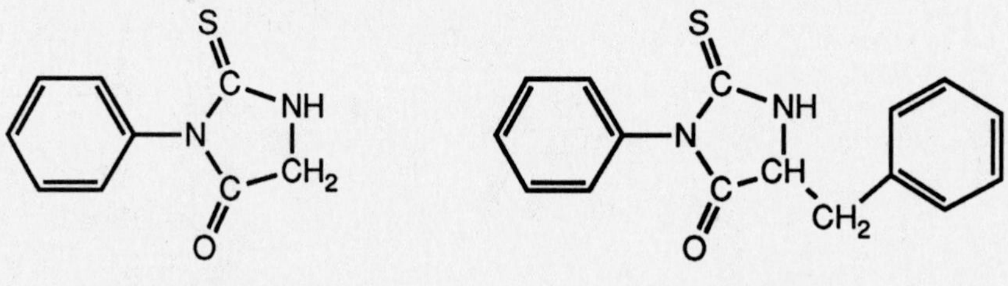

(from Gly)   (from Phe)

17.48 The hydrolysis products must overlap as shown below:

```
Ala—Gly
   Gly—Val
   Gly—Val—Tyr
        Val—Tyr—Cys
           Tyr—Cys—Phe
               Cys—Phe—Leu
                    Phe—Leu—Try
```

The peptide must be Ala—Gly—Val—Tyr—Cys—Phe—Leu—Try, an octapeptide. The N-terminal amino acid is alanine, the C-terminal one is tryptophan, and the name is alanylglycylvalyltyrosylcystylphenylalanyl-leucyltryptophan.

17.49 The point of this problem is to illustrate that it is deceptively easy to use the three-letter abbreviations in designating peptide structures. Although Try—Gly—Gly—Phe—Met appears quite simple, the full structure is not. With a molecular formula of $C_{29}H_{36}O_6N_6S$, this pentapeptide has a molecular weight of 596 and is quite a complex polyfunctional molecule.

17.50 The complete hydrolysis tells us what eight amino acids are in angiotensin II. Reaction with Sanger's reagent indicates that Asp is the N-terminal amino acid. Partial digestion with carboxypeptidase indicates that Phe is the C-terminal amino acid. We can now write the following partial structure:

Asp—(Arg, His, Ile, Pro, Tyr, Val)—Phe

Trypsin cleaves peptides at the carboxyl side of Arg and chymotrypsin cleaves peptides at the carboxyl side of Tyr and Phe. Since we already know that Phe is the C-terminal amino acid, the chymotrypsin treatment tells us that Tyr is the fourth amino acid from the N-terminus, and we can write the following partial structure:

Asp—(Arg, Val)—Tyr—Ile—His—Pro—Phe

The trypsin treatment tells us that Arg is the second amino acid from the N-terminus. Therefore angiotensin II is:

Asp—Arg—Val—Tyr—Ile—His—Pro—Phe

17.51 Trypsin cleaves peptides at the carboxyl side of Lys and Arg. The only fragment that does not end in Lys is the dipeptide. Therefore, it must come at the C-terminal end of endorphin, and the C-terminal amino acid is Gln.

Cyanogen bromide cleaves peptides at the carboxyl side of Met. There is only one Met in the structure (see the fifth trypsin fragment), and the cyanogen bromide cleavage gives a hexapeptide consisting of the first six amino acids of this fragment. The N-terminal amino acid must be Tyr. We now know the first ten and the last two amino acids of β-endorphin.

Chymotrypsin cleaves peptides at the carboxyl side of Phe, Tyr, and Trp. Working backward in the 15-unit fragment, we see that it starts at the ninth amino acid in the last fragment from the trypsin digestion. We see that the first amino acid in this fragment (Ser) must be connected to the Lys at the end of the fifth fragment from the trypsin cleavage. Since this fragment begins with the N-terminal amino acid, we now know the sequence of the first twenty and the last two amino acids. The partial structure of β-endorphin must be

Tyr–Gly–Gly–Phe–Leu–Met–Thr–Ser–Glu–Lys–Ser–Gln–Thr–Pro–Leu–Val

Thr–Leu–Phe–Lys–(Lys)(Asn–Ala–His–Lys)(Asn–Ala–Ile–Val–Lys)–Gly–Gln

We cannot deduce, from the data given, the sequence of the amino acids in parentheses. To put them in order we must seek peptides from partial hydrolysis that contain overlapping sequences of these amino acids.

17.52 The nucleophile is the carboxylate anion; the leaving group is chloride ion from the benzyl chloride. Benzyl halides react rapidly in S_N2 displacements.

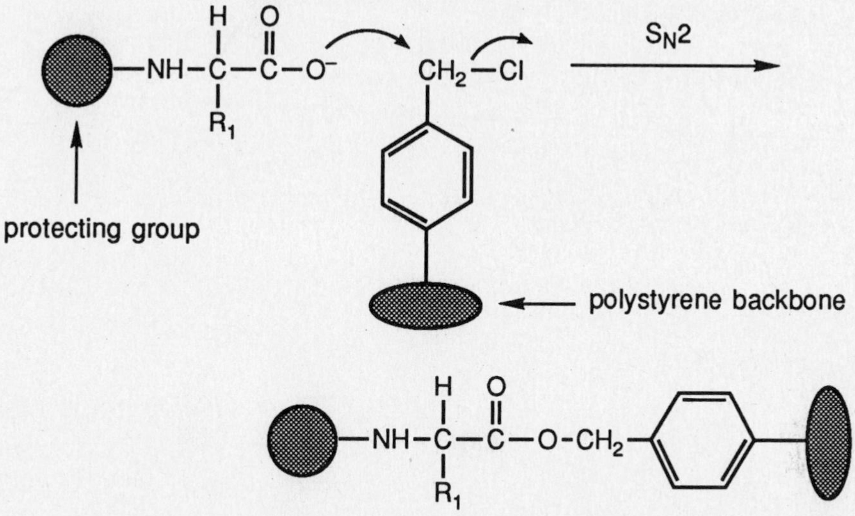

H
O
NH
C
C
O⁻
R1
CH2
Cl
SN2
protecting group
polystyrene backbone

17.53 The protonated peptide is the leaving group. The nucleophile is fluoride ion.

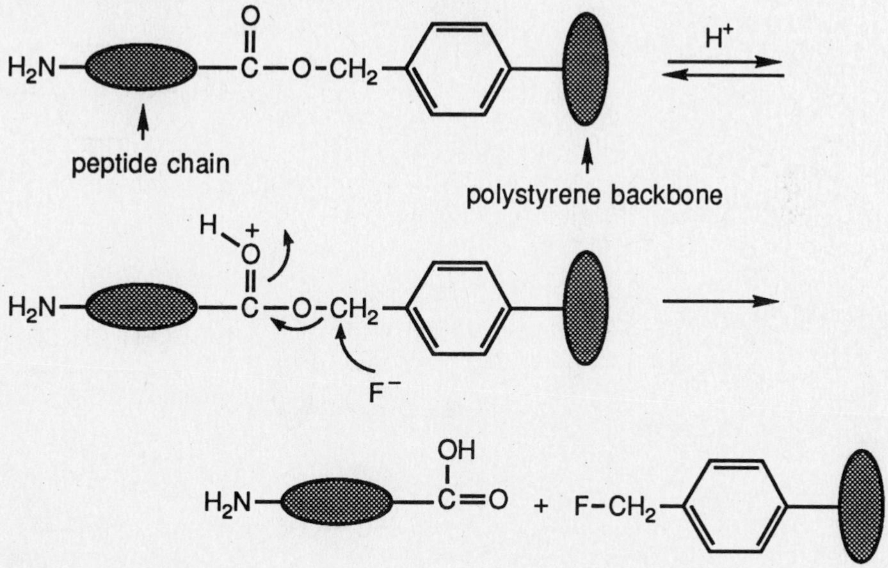

H2N
C
O
O-CH2
H+
peptide chain
polystyrene backbone
H
F⁻
OH
C=O
+
F-CH2

17.54 First, protect the C-terminal amino acid (proline):

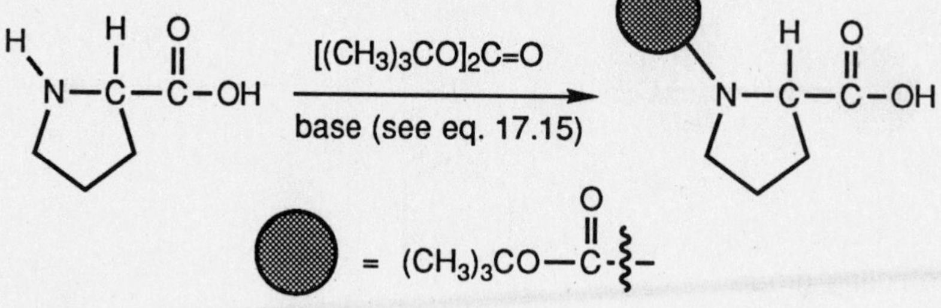
[(CH3)3CO]2C=O
base (see eq. 17.15)
= (CH3)3CO—C-

Then attach it to the polymer:

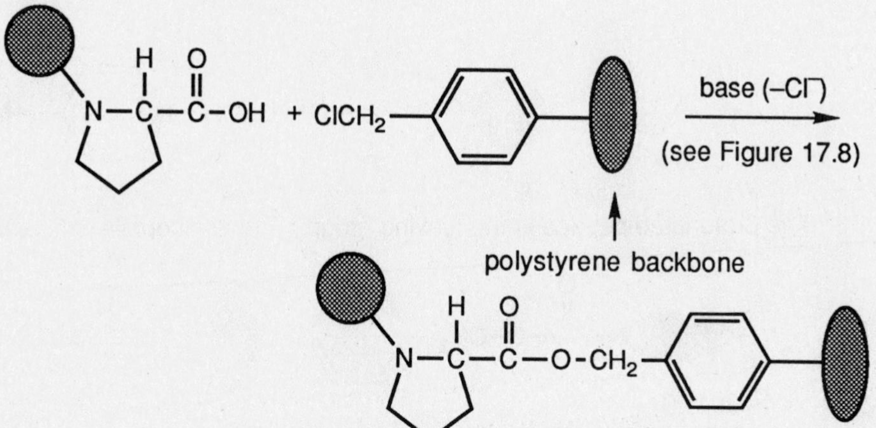
+ ClCH2
base (–Cl⁻)
(see Figure 17.8)
polystyrene backbone

Next, deprotect the attached proline:

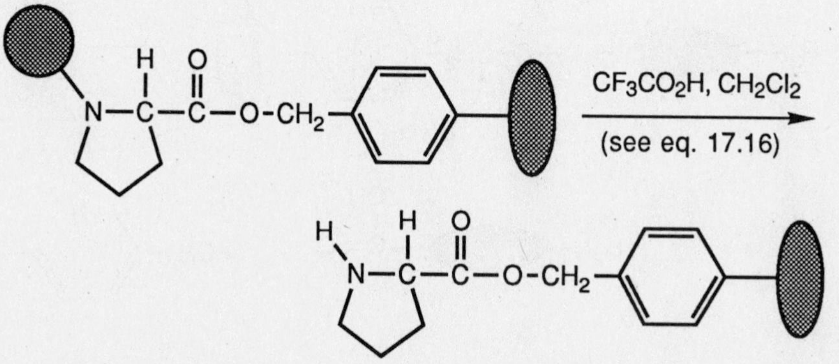
CF3CO2H, CH2Cl2
(see eq. 17.16)

Couple protected leucine to the deprotected, polymer-bound proline:

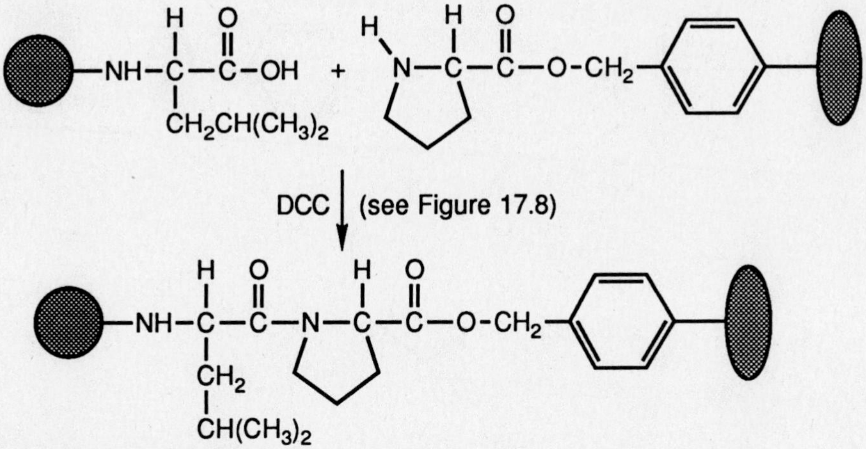

DCC (see Figure 17.8)

Then deprotect again.

CF_3CO_2H

CH_2Cl_2

(see eq. 17.16)

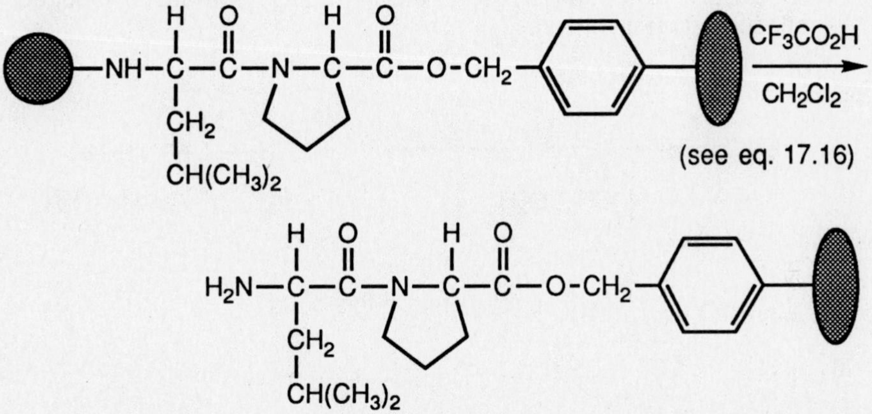

At this stage, the dipeptide is constructed, but it must be detached from the polymer:

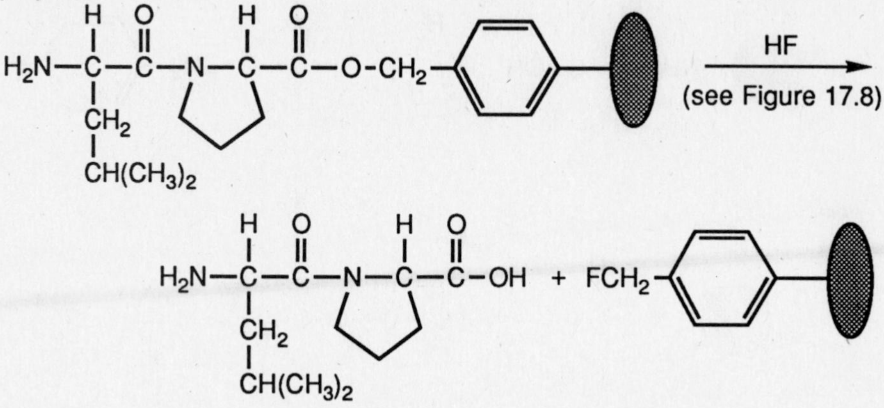

$$\xrightarrow{\text{HF}}$$

(see Figure 17.8)

(Leu—Pro)

This problem, which involves only the construction of a simple dipeptide, should give you some sense of the number of operations required in solid-phase peptide syntheses.

17.55

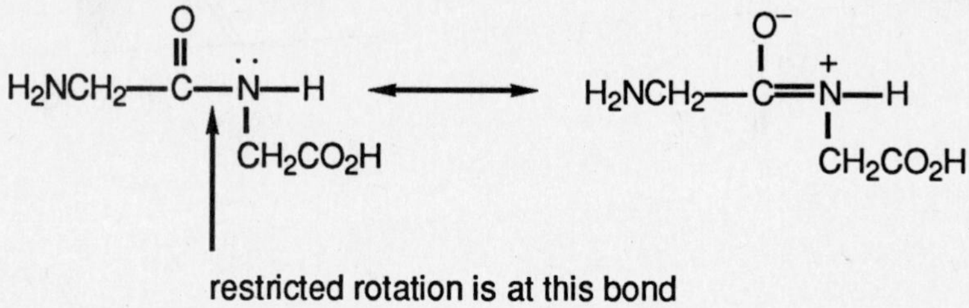

restricted rotation is at this bond

17.56 Use Table 17.3 as a guide.

a. With trypsin, we expect four fragments:

His—Ser—Glu—Gly—Thr—Phe—Thr—Ser—Asp—Tyr—Ser—Lys
Tyr—Leu—Asp—Ser—Arg
Arg
Ala—Gln—Asp—Phe—Val—Gln—Trp—Leu—Met—Asn—Thr

b.   With chymotrypsin, we expect six fragments:

His—Ser—Glu—Gly—Thr—Phe
Thr—Ser—Asp—Tyr
Ser—Lys—Tyr
Leu—Asp—Ser—Arg—Arg—Ala—Gln—Asp—Phe
Val—Gln—Trp
Leu—Met—Asn—Thr

17.57   Amino acids with nonpolar R groups will have those groups pointing toward the center of a globular protein. These are phenylalanine (b) and isoleucine (c). The other amino acids listed (a, d, e, f) have polar R groups, which will point toward the surface and hydrogen-bond with water, thus helping to solubilize the protein.

CHAPTER EIGHTEEN: NUCLEOTIDES AND NUCLEIC ACIDS

CHAPTER SUMMARY

Nucleic acids, the carriers of genetic information, are macromolecules that are composed of and can be hydrolyzed to **nucleotide units**. Hydrolysis of a nucleotide gives one equivalent each of a **nucleoside** and phosphoric acid. Further hydrolysis of a nucleoside gives one equivalent each of a sugar and a heterocyclic base.

The DNA sugar is **2-deoxy-D-ribose**. The four heterocyclic bases in DNA are **cytosine, thymine, adenine,** and **guanine**. The first two bases are **pyrimidines,** and the latter two are **purines**. In nucleosides, the bases are attached to the anomeric carbon (C-1) of the sugar as β-**N-glycosides**. In nucleotides, the —OH at C-3 or C-5 of the sugar is present as a phosphate ester.

The primary structure of DNA consists of nucleotides linked by a phosphodiester bond between the 5'—OH of one unit and the 3'—OH of the next unit. To fully describe a DNA molecule, the **base sequence** must be known. Methods for sequencing are rapidly being developed, and at present, over 150 bases can be sequenced per day. The counterpart of sequencing, the synthesis of oligonucleotides having known base sequences, is also rapidly progressing.

The secondary structure of DNA is a **double helix**. Two helical poly-nucleotide chains coil around a common axis. In **B-DNA**, the predominant form, each helix is right-handed, and the two run in opposite directions with respect to their 3' and 5' ends. The bases are located inside the double helix, in planes perpendicular to the helix axis. They are paired (A—T and G—C) by hydrogen bonds, which hold the two chains together. The sugar–phosphate backbones form the exterior surface of the double helix. Genetic information is passed on when the double helix uncoils and each strand acts as a template for binding and linking nucleotides to form the next generation.

Other forms of DNA include **A-DNA** (a right-handed double helix in which base pairs are tilted to the helical axis) and **Z-DNA** (a left-handed double helix).

RNA differs from DNA in three ways: The sugar is D-ribose, the pyrimidine **uracil** replaces thymine (the other three bases are the same), and the molecules are mainly single-stranded. The three principal types of RNA are **messenger RNA**

LEARNING OBJECTIVES

(involved in transcribing the genetic code), **transfer RNA** (which carries a specific amino acid to the site of protein synthesis), and **ribosomal RNA**.

The **genetic code** involves sequences of three bases called **codons,** each of which translates to a specific amino acid. The code is degenerate (that is, there is more than one codon per amino acid), and some codons are "stop" signals that terminate synthesis. **Protein biosynthesis** is the process by which the message carried in the base sequence is transformed into an amino acid sequence in a peptide or protein.

In addition to their role in genetics, nucleotides play other important roles in biochemistry. Key enzymes and coenzymes such as **nicotinamide adenine dinucleotide** (NAD), **flavin adenine dinucleotide** (FAD), and **vitamin B$_{12}$** also include nucleotides as part of their structures. Also, the major component of **viruses** is DNA.

REACTION SUMMARY

Hydrolysis of Nucleic Acids

LEARNING OBJECTIVES

1. Know the meaning of: nucleic acid, nucleotide, nucleoside.

2. Know the structures of: cytosine, thymine, adenine, guanine, uracil, 2-deoxy-D-ribose, D-ribose.

3. Know the meaning of: DNA, B-DNA, A-DNA, Z-DNA, RNA, N-glycoside, pyrimidine base, purine base.

18. NUCLEOTIDES AND NUCLEIC ACIDS

4. Given the name, draw the structure of a specific nucleoside.

5. Write an equation for the hydrolysis of a specific nucleoside by aqueous acid. Write the steps in the reaction mechanism.

6. Given the name, draw the structure of a specific nucleotide.

7. Write an equation for the hydrolysis of a specific nucleotide by aqueous base.

8. Draw the structure of an N-glycoside and an O-glycoside.

9. Given the name or abbreviation for a DNA or RNA nucleotide or nucleoside, draw its structure.

10. Draw the primary structure of a segment of an RNA or DNA chain.

11. Explain why only pyrimidine–purine base pairing is permissible in the double helix structure.

12. Describe the main features of the secondary structure of DNA.

13. Explain, with the aid of structures, the role of hydrogen bonding in nucleic acid structures.

14. Describe the main features of DNA replication.

15. Given the base sequence in one strand of a DNA molecule, write the base sequence in the other strand, or in the derived mRNA. Conversely, given a base sequence for mRNA, write the base sequence in one strand of the corresponding DNA.

16. Given a synthetic polyribonucleotide and the peptide sequence in the resulting polypeptide, deduce the codons for the amino acids.

17. Describe the main features of protein biosynthesis.

18. Explain the different functions of messenger, ribosomal, and transfer RNA.

19. Know the meaning of: codon, anticodon, genetic code, transcription.

20. Draw the structure of: AMP, ADP, ATP, and cAMP.

21. Describe the main features of DNA profiling.

ANSWERS TO PROBLEMS

Problems Within the Chapter

18.1

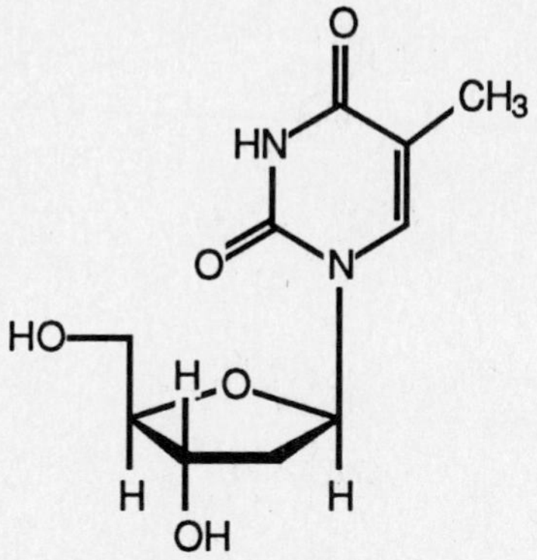

2'-deoxythymidine

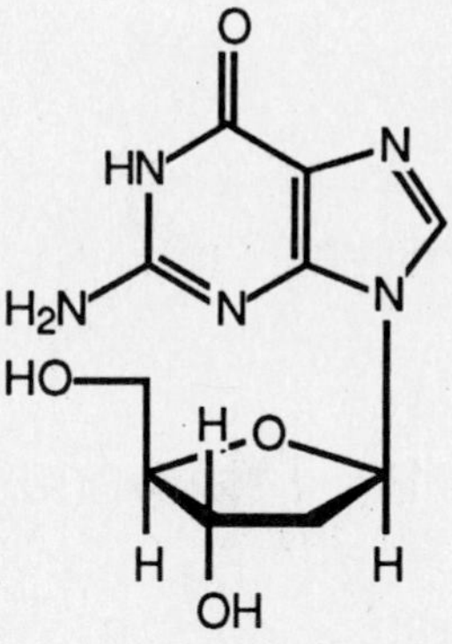

2'-deoxyguanosine

18.2

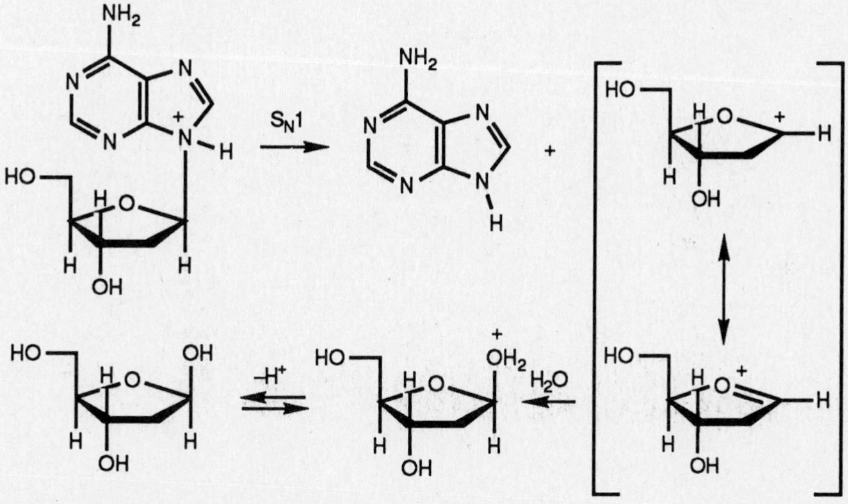

After protonation of the nitrogen, the bond between the anomeric carbon of the sugar and the nitrogen of the heterocyclic base is cleaved by an S_N1 mechanism. The resulting carbocation is stabilized by resonance as shown. Reaction with water as a nucleophile (followed by proton loss) then gives 2-deoxyribose, either the β form as shown or the α form. That is, the water can attack from the top or bottom face of the furanose ring. This mechanism is exactly analogous to that of the hydrolysis of an acetal (the reverse of eq. 9.12; see also the answer to Prob. 9.11).

18.3   a. 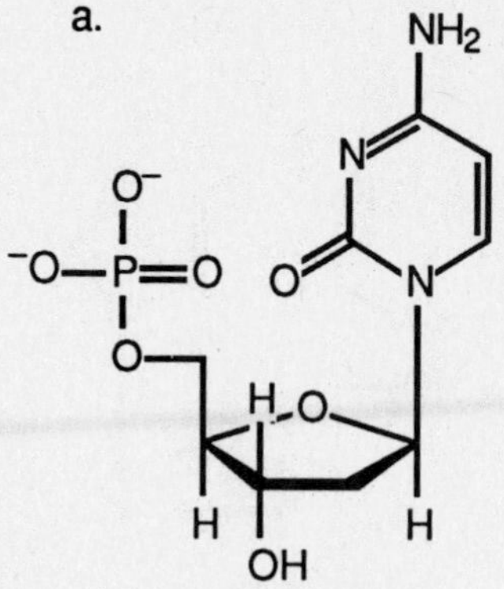   b.

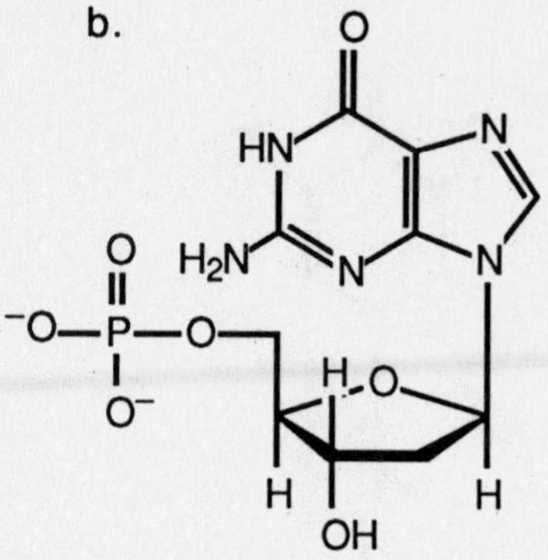

18.4   Consult eqs. 18.3 and 18.2.

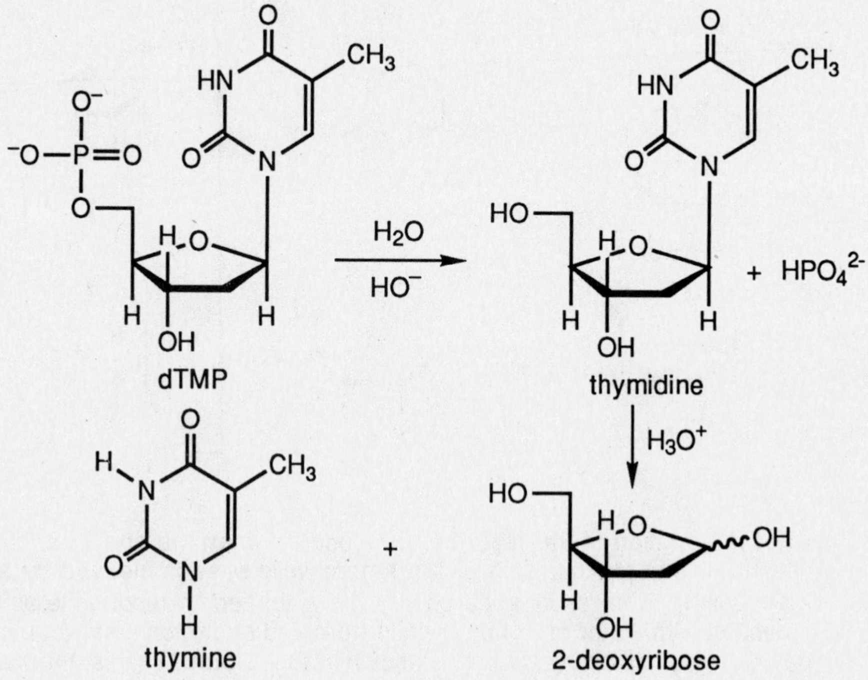

18.5

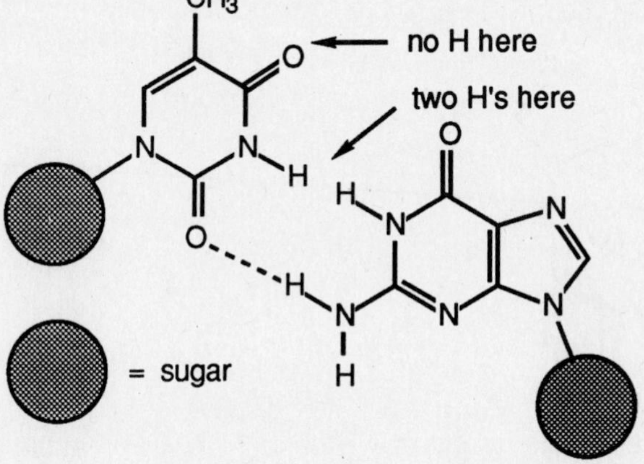

no H here

two H's here

This structure has only one hydrogen bond.

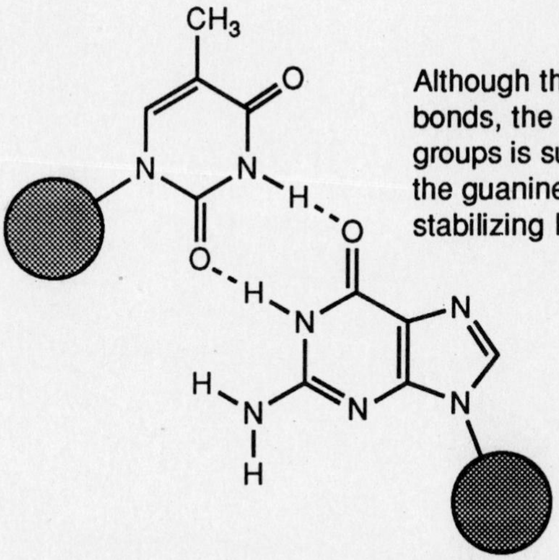

= sugar

Although this structure has two hydrogen bonds, the distance between the sugar groups is substantially increased, and the guanine amino group is left with no stabilizing hydrogen bond.

18.6 —TCGGTACA— (written from the 3' end to the 5' end).

18.7 a. Note that C-2' has a hydroxyl group. AMP is an RNA mononucleotide.

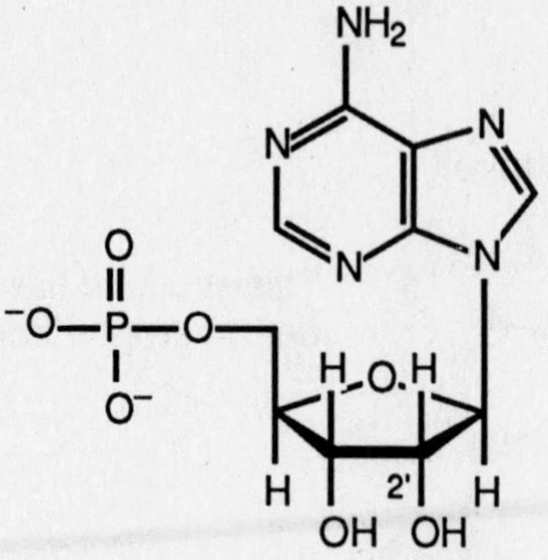

b. Note that there is a hydroxyl group at C-2' of each unit and that the 5' and 3' ends do not have phosphate groups attached.

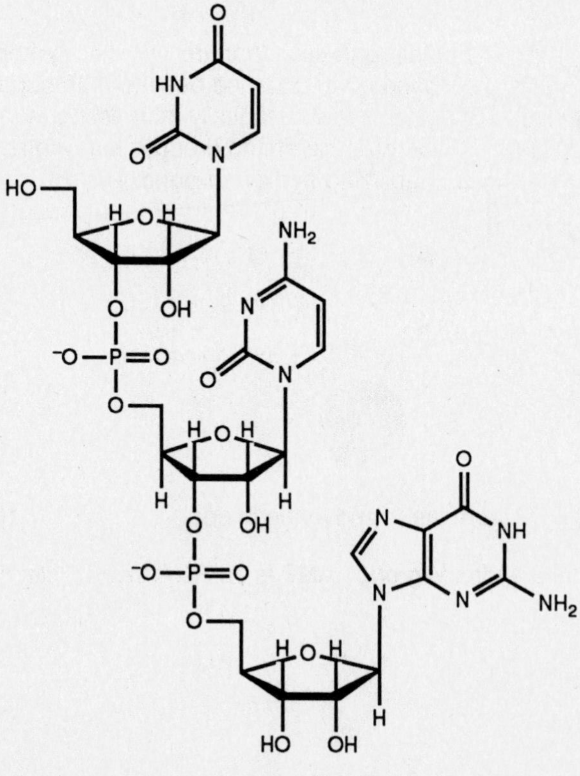

18.8   A polynucleotide made from UA will have the codons UAU and AUA:

UAUAUAUA ...

UAU is the codon for Tyr, and AUA is a codon for Ile. Note that this codon differs from the codon in Example 18.4 only in the last letter. Such differences are common. The first two letters of codons are frequently more important than the third letter.

18.9   Consult Table 18.2. The codon UUU translates to the amino acid phenyl-alanine. If a UUU sequence were to be mutated to UCU (the codon for serine), the protein produced would have a Ser residue in place of the Phe. Since UCU and UCC both code for Ser, a UCU → UCC mutation would not lead to a change in the protein sequence. Thus, the advantage of a redundant code is that not all mutations cause disadvantageous changes in protein structure.

Additional Problems

18.10 a. See the formulas for cytosine and thymine (Figure 18.1) and uracil (Sec. 18.11).
       b. See the formulas for adenine and guanine (Figure 18.1).
       c. See Sec. 18.4.
       d. See Sec. 18.5.

18.11

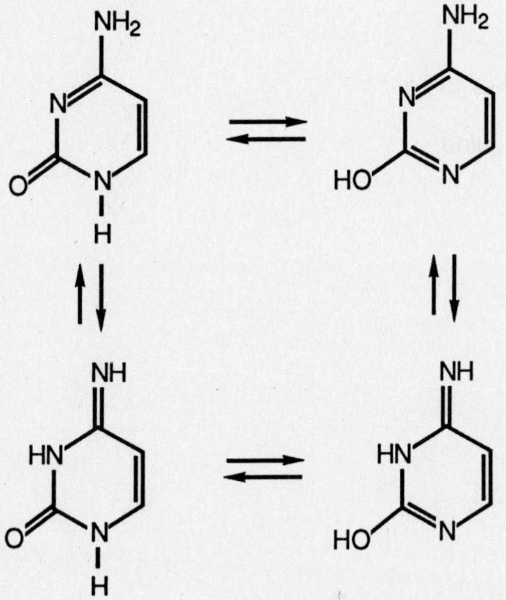

18.12 All the ring atoms in adenine but one (the N–H nitrogen in the five-membered ring) are sp^2-hybridized and planar. Since both rings are aromatic, we expect the molecule to be planar or nearly so. In guanine, the ring atoms are also sp^2-hybridized (except for one NH nitrogen in each ring). The amide group in the six–membered ring (N-1, C-6) is also planar. Therefore, once again we expect the molecule to be planar, or very nearly so. In the pyrimidine bases cytosine and thymine, the rings will also be essentially planar, for the same reasons.

18.13 a. Use Figure 18.2 as a guide.

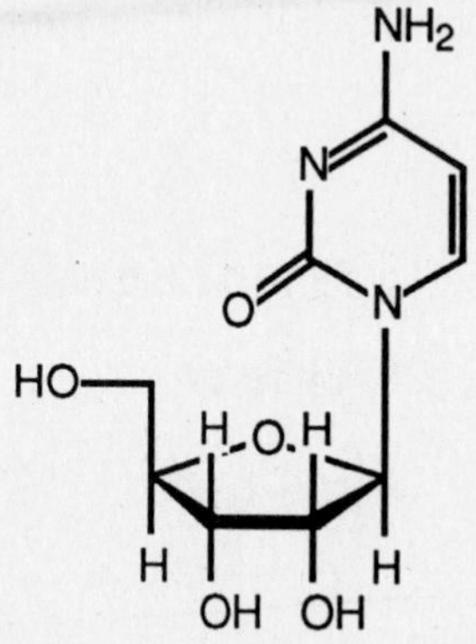

b. See the starting material in eq. 18.2.

c. See Sec. 18.11 as a guide.

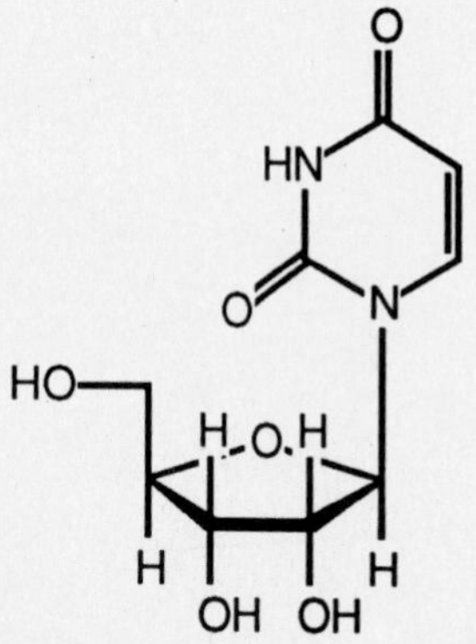

d. See the answer to Prob. 18.1.

18.14 The products are adenine, ribose, and inorganic phosphate.

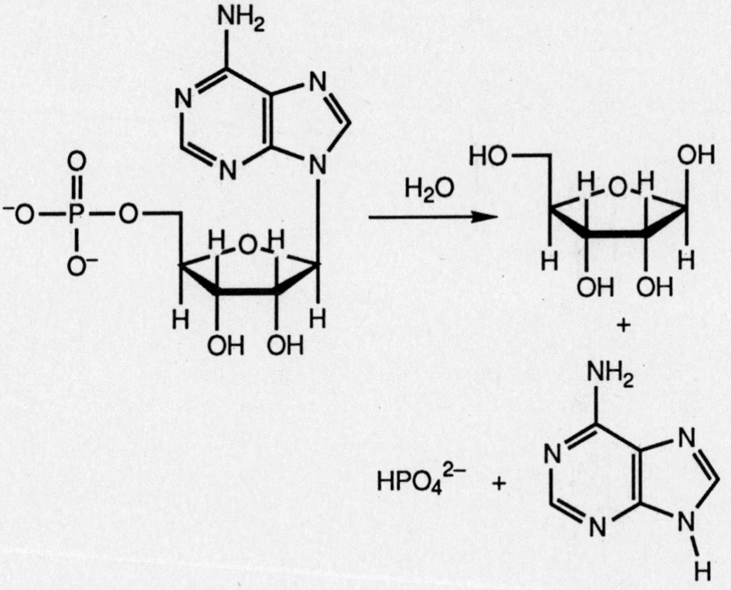

18.15 a. The sugar part of the molecule has a hydroxyl group at C-2' (D-ribose).

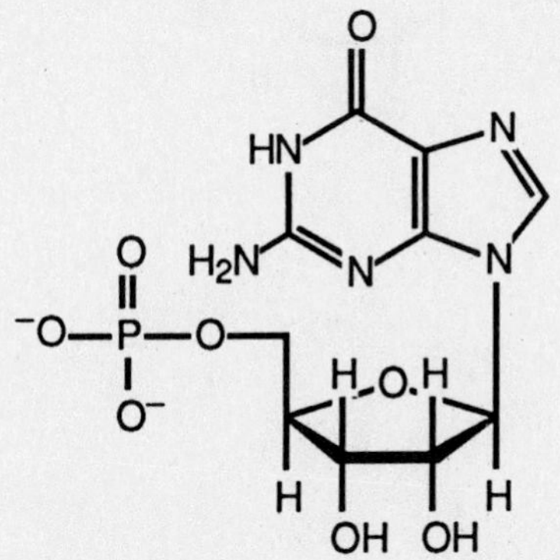

b. The sugar part of the molecule lacks a hydroxyl group at C-2' (2-deoxy-D-ribose).

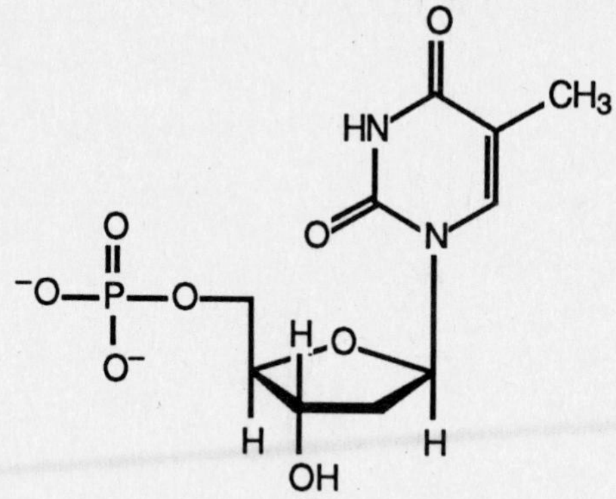

18.16 Hydroxide ion attacks the primary C-5' carbon atom in an S_N2 displacement. The leaving group is phosphate ion.

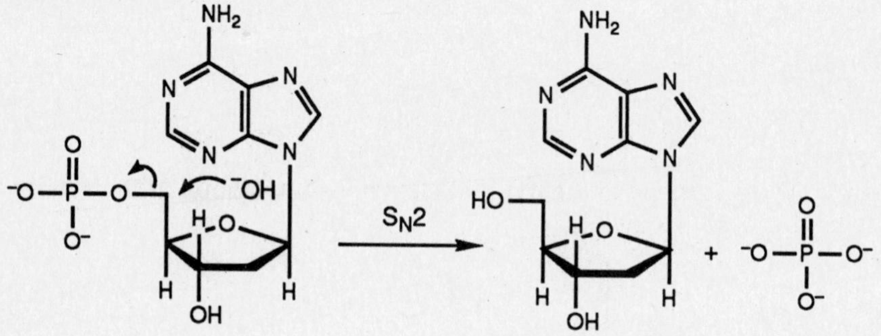

18.17   a.

b.

c.

18.18   a.

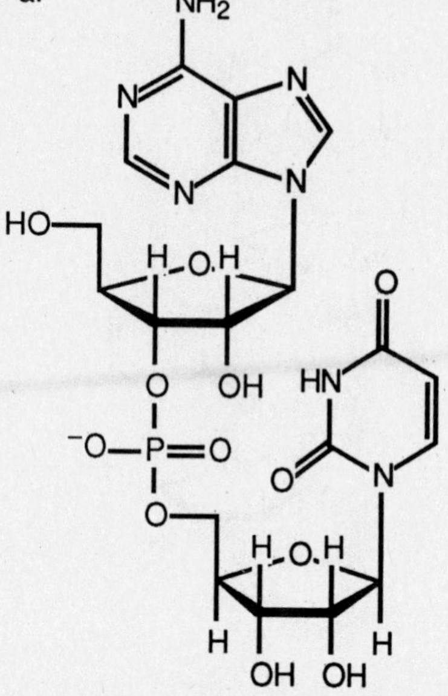

b.

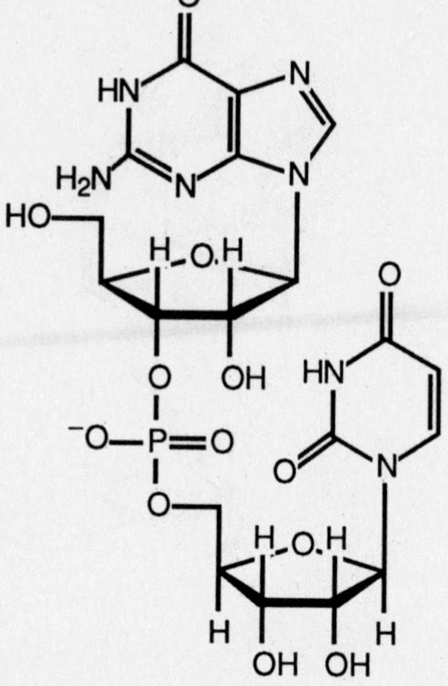

c.

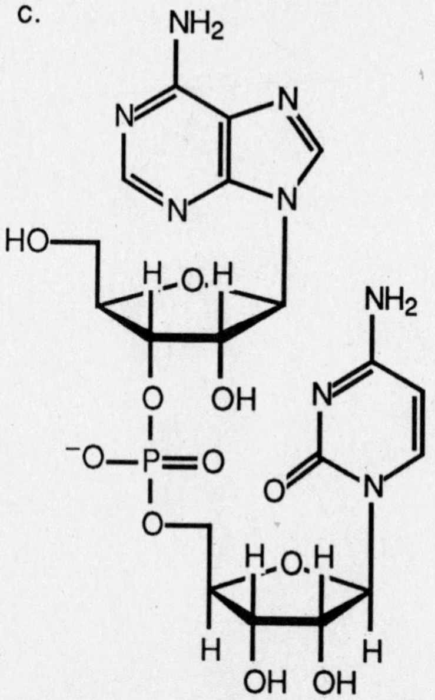

18.19 a. The products are two equivalents of 2'-deoxyadenosine-3'-monophos-
phate, one equivalent of 2'-deoxythymidine-3'-monophosphate, and one
equivalent of 2'-deoxycytidine.

b. The products are two equivalents of adenine, one equivalent of thymine, one equivalent of cytosine, three equivalents of inorganic phosphate, and four equivalents of 2-deoxy-D-ribose.

18.20   a.   b.

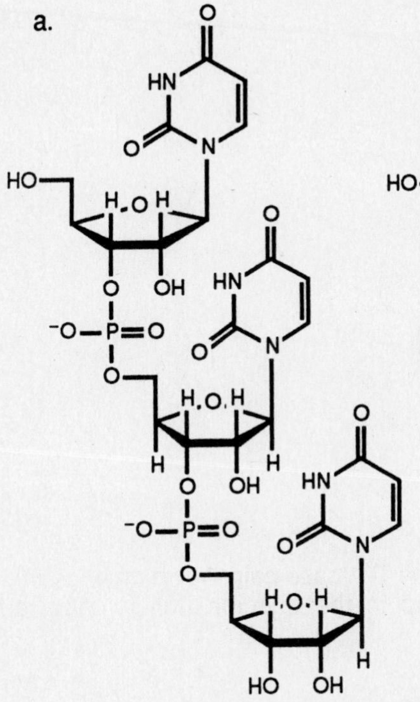

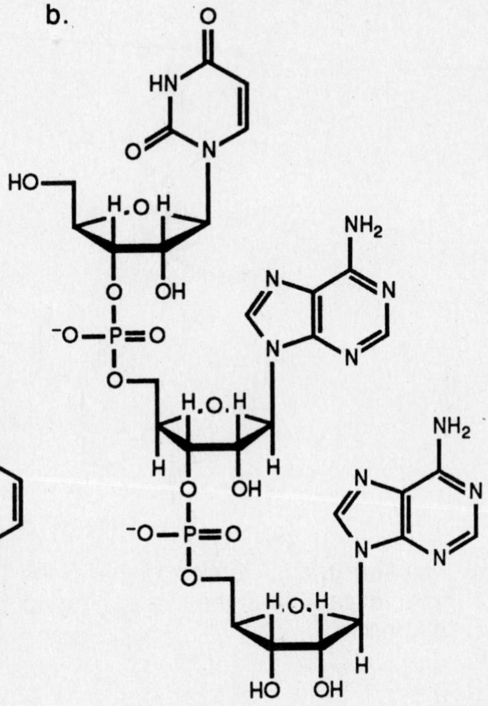

c.

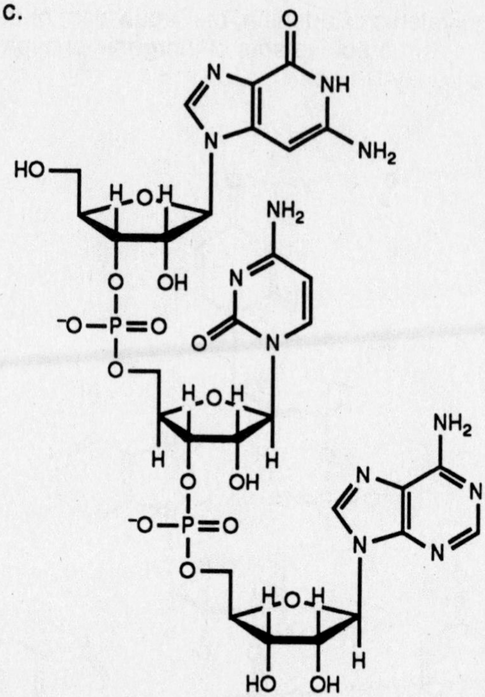

18.21 The structure is identical to that of the T-A base pair shown on page 504 of the text, except that the methyl group in the thymine unit is replaced by a hydrogen.

18.22 3' T—T—C—G—A—C—A—T—G 5'

18.23 3' U—U—C—G—A—C—A—U—G 5'

The only difference between this sequence and the answer to Prob. 18.22 is that each T is replaced by U.

18.24 Given *m*RNA sequence: 5' A—G—C—U—G—C—U—C—A 3'

DNA strand from which *m*RNA was transcribed:

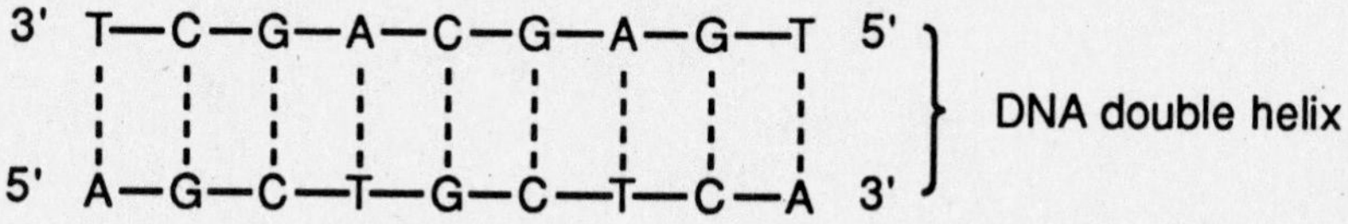

3' T—C—G—A—C—G—A—G—T 5'
     | | | | | | | | | } DNA double helix
5' A—G—C—T—G—C—T—C—A 3'

Note that the DNA strand which was *not* transcribed (the last segment shown above) is identical with the given *m*RNA segment, except that each U is replaced by T.

18.25 For each T there must be an A. For each G there must be a C. As a consequence of this base pairing, the mole percentages of T in any sample of DNA will be equal to the mole percentages of A, and the same will be true for G and C. It is *not* necessary, however, that there be any special relationship between the percentages of the two pyrimidines (T and C) or of the two purines (A and G).

18.26 The code reads from the 5' to the 3' end of *m*RNA.

5' C—A—U 3' *m*RNA
3' G—T—A 5' DNA-transcribed chain
5' C—A—T 3' DNA complement

18.27 A purine → purine mutation in the third base of a codon will only result in a change of biosynthesized protein in the following instances:

UGA → UGG, which will result in a stop command being replaced by a Trp, and
AUA → AUG, which will cause an Ile to be changed to a Met or start command.

Pyrimidine → pyrimidine mutations in the third base pair will not alter protein biosynthesis.

18.28 Mutations in the first and second base of a codon are far more serious. For example UUU, UCU, UAU, and UGU code for four different amino acids, as do UUU, CUU, AUU, and GUU.

18.29 The AUG and UAG codons would start and stop peptide synthesis:

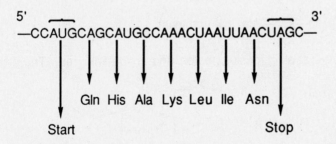

Thus, the *m*RNA strand codes for synthesis of the heptapeptide Gln—His—Ala—Lys—Leu—Ile—Asn.

18. NUCLEOTIDES AND NUCLEIC ACIDS

18.30 The AUG and UAA codons would now start and stop peptide synthesis:

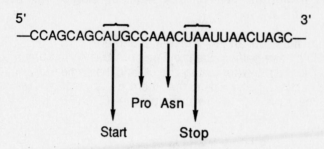

Thus, the mRNA would code for the synthesis of the dipeptide Pro—Asn.

18.31

5' —TTACCGTCTGCTGCCCCCCAT— 3' (DNA)

transcription

3' —AAUGGCAGACGACGGGGGGUA— 5' (mRNA)

translation

Ala Asp Asp Gly Gly

The AUG codon would start peptide synthesis and Ala would be the C-terminal amino acid. Peptide synthesis would continue until a stop codon (UAA, UAG, or UGA) appeared. The peptide fragment from the DNA strand would be:

—Gly—Gly—Asp—Asp—Ala (C–terminus)

18.32 Hydrolysis occurs at all glycosidic, ester, and amide linkages. The products are as follows:

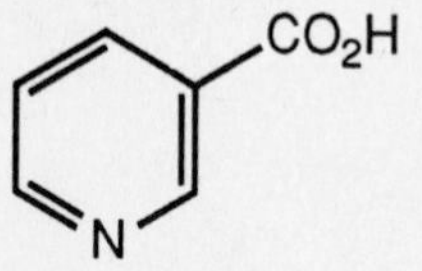

nicotinic acid   ammonia   ribose (2 mol)

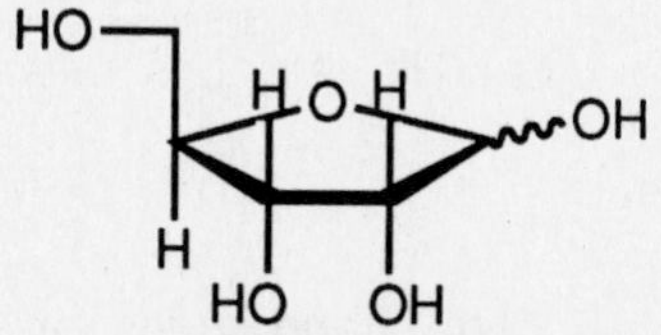

ANSWERS TO PROBLEMS

adenine

H_3PO_4

phosphoric acid
(2 mol)

18.33 Use the figure on page 513 as a guide.

18.34 The formula for uridine monophosphate is shown in Sec. 18.11 of the text, and the formula for α-D-glucose is shown in eq. 16.3. The structure of UDP-glucose is

18.35 There are no N—H bonds in caffeine. Therefore (unlike adenine and guanine), it cannot form N-glycosides.

18.36

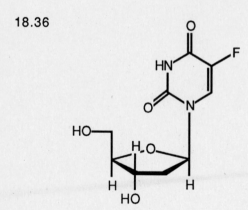

18.37

SUMMARY OF SYNTHETIC METHODS

In the text, methods for preparing the particular classes of compound are presented in two ways. In certain chapters, methods for synthesizing the class of compound with which the chapter deals are enumerated and discussed. More frequently, however, syntheses are presented less formally as a consequence of various reactions.

In this summary of the main synthetic methods for each important class of compound, a general equation for each reaction is given. Also, section numbers, given in parentheses, refer you to the place or places in the book where the reaction is described. References to "A Word About" sections are abbreviated AWA.

1. Alkanes and Cycloalkanes

    a. Alkenes + H_2 (3.8; 4.8; 15.3)

$$\begin{array}{c}\diagdown\\ \diagup\end{array}C = C\begin{array}{c}\diagup\\ \diagdown\end{array} \quad + \quad H_2 \quad \xrightarrow[\text{catalyst}]{\text{Ni or Pt}} \quad \begin{array}{cc}| & |\\ -C - C-\\ | & |\\ H & H\end{array}$$

    b. Alkynes + H_2 (3.1)

$$-C \equiv C- \quad + \quad 2 \quad H_2 \quad \xrightarrow[\text{catalyst}]{\text{Ni or Pt}} \quad \begin{array}{cc}H & H\\ | & |\\ -C - C-\\ | & |\\ H & H\end{array}$$

    c. Cyclohexanes from aromatic compounds + H_2 (4.8; AWA 6, Ch. 4)

$$\bighexagon \quad + \quad 3 \ H_2 \quad \xrightarrow[\text{catalyst}]{\text{Ni or Pt}} \quad \bighexagon$$

SUMMARY OF SYNTHETIC METHODS

    d. Grignard reagent + H_2O (or D_2O) (8.5)

$$R-MgX + H-OH \longrightarrow R-H + Mg(OH)X$$

2.   <u>Alkenes and Cycloalkenes</u>

    a. Alkynes + H_2 (3.19)

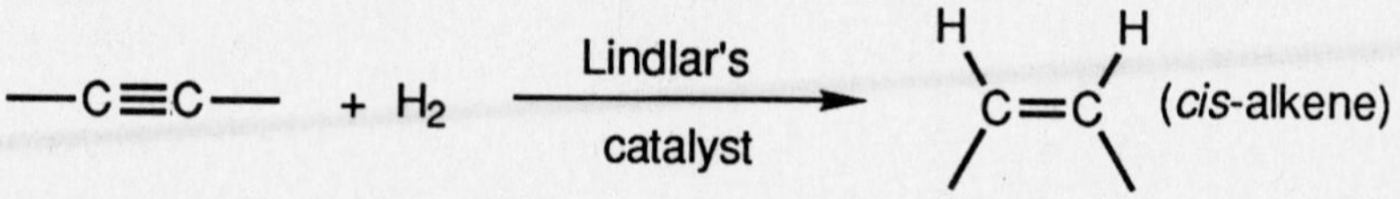

$$-C\equiv C- \ + \ H_2 \xrightarrow[\text{catalyst}]{\text{Lindlar's}} \overset{H}{\underset{}{C}}=\overset{H}{\underset{}{C} } \quad (\textit{cis}\text{-alkene})$$

    b. Dehydration of alcohols (7.9)

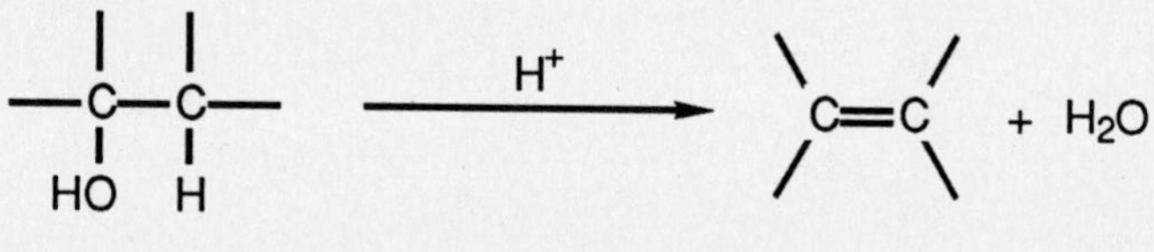

$$-\overset{|}{\underset{HO}{C}}-\overset{|}{\underset{H}{C}}- \xrightarrow{H^+} \ \ C=C \ + \ H_2O$$

    c. Elimination reaction; alkyl halide + strong base (6.8; 6.9)

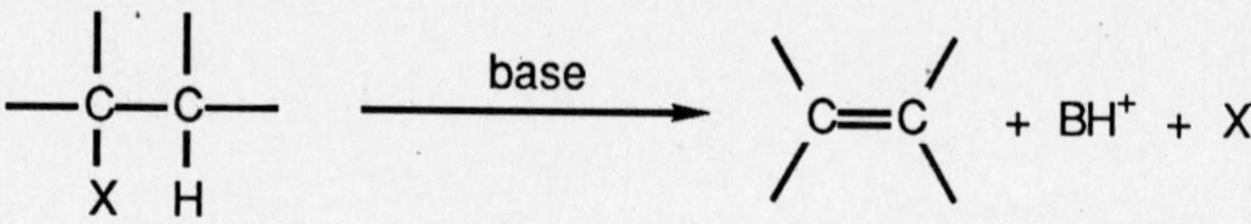

$$-\overset{|}{\underset{X}{C}}-\overset{|}{\underset{H}{C}}- \xrightarrow{\text{base}} \ \ C=C \ + \ BH^+ + \ X^-$$

    d. *cis-trans* Isomerism (AWA 3, Ch. 3)

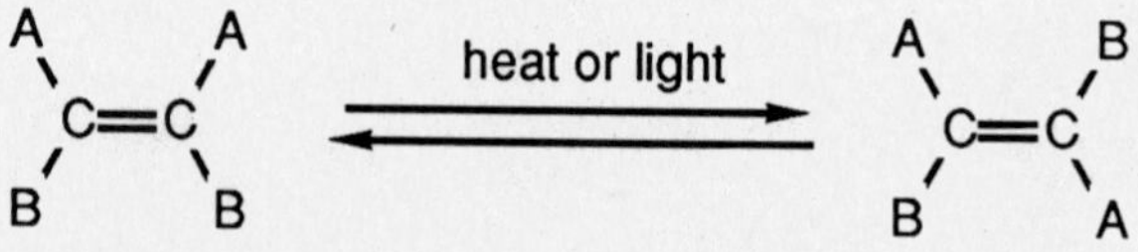

$$\overset{A}{\underset{B}{}}C=C\overset{A}{\underset{B}{}} \ \xrightleftharpoons{\text{heat or light}} \ \overset{A}{\underset{B}{}}C=C\overset{B}{\underset{A}{}}$$

    e. Diels-Alder reaction (3.14b)

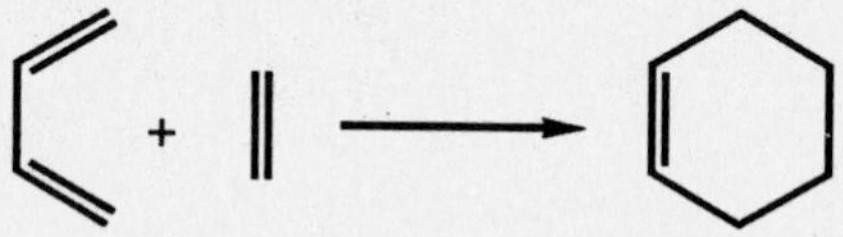

SUMMARY OF SYNTHETIC METHODS

f. Cracking of alkanes (AWA 4 and AWA 5, Ch. 3)

$$CH_3CH_3 \xrightarrow{700\text{--}900^\circ C} H_2C{=}CH_2 + H_2$$

$$C_nH_{2n+2} \xrightarrow{catalyst} C_mH_{2m} + C_pH_{2p+2} \quad (m+p=n)$$

3. Alkynes

a. From acetylides and alkyl halides (3.50; 6.3)

$$R{-}C{\equiv}C{-}H \xrightarrow[NH_3]{NaNH_2} R{-}C{\equiv}C^- Na^+ \xrightarrow{R'X} R{-}C{\equiv}C{-}R'$$

(best for R' = primary)

b. Pyrolysis of methane (AWA 4, Ch. 3)

$$2\ CH_4 \xrightarrow{1500^\circ C} R{-}C{\equiv}C{-}H + 3\ H_2$$

4. Aromatic Compounds

a. Alkylbenzenes from Friedel-Crafts reactions (4.9; 4.10d; AWA 6, Ch. 4; 15.6)

$$R{-}X + Ar{-}H \xrightarrow{AlCl_3} R{-}Ar + HX$$

$$\underset{/}{\overset{\backslash}{C}}{=}\underset{\backslash}{\overset{/}{C}} + Ar{-}H \xrightarrow{H^+} Ar{-}\overset{|}{\underset{|}{C}}{-}\overset{|}{\underset{|}{C}}{-}H$$

b. Acylbenzenes from Friedel-Crafts acylations (4.9; 4.10d)

$$Ar{-}H + R{-}\overset{O}{\overset{\|}{C}}{-}Cl \xrightarrow{AlCl_3} R{-}\overset{O}{\overset{\|}{C}}{-}Ar + HCl$$

467

c. Aromatic nitro compounds, by nitration (4.9; 4.10b; 4.12; 4.13; 7.15; 13.3)

$$Ar-H + HONO_2 \xrightarrow{H^+} Ar-NO_2 + H_2O$$

d. Aromatic sulfonic acids, by sulfonation (4.9; 4.10c; 15.6)

$$Ar-H + HOSO_3H \xrightarrow{H^+} Ar-SO_3H + H_2O$$

e. Aromatic halogen compounds, by halogenation (4.2; 4.9; 4.10a; 4.12; 4.13; 7.15; 13.3)

$$Ar-H + X_2 \xrightarrow{FeX_3} Ar-X + HX \quad (X = Cl, Br)$$

f. Aromatic halogen compounds from diazonium salts (11.13)

$$Ar-N_2^+ + HX \xrightarrow{Cu_2X_2} Ar-X + N_2 \quad (X = Cl, Br)$$

$$Ar-N_2^+ + KI \longrightarrow Ar-I + N_2 + K^+$$

g. Alkylbenzenes from alkanes (AWA 5, Ch. 3)

h. Pyridines (13.3; 13.4)

i. Quinolines (13.4)

j. Pyrroles (13.6)

k. Furans (13.5; 13.6)

l. Thiophenes (13.6)

SUMMARY OF SYNTHETIC METHODS

5.   Alcohols

   a. Hydration of alkenes (3.9b; 3.11)

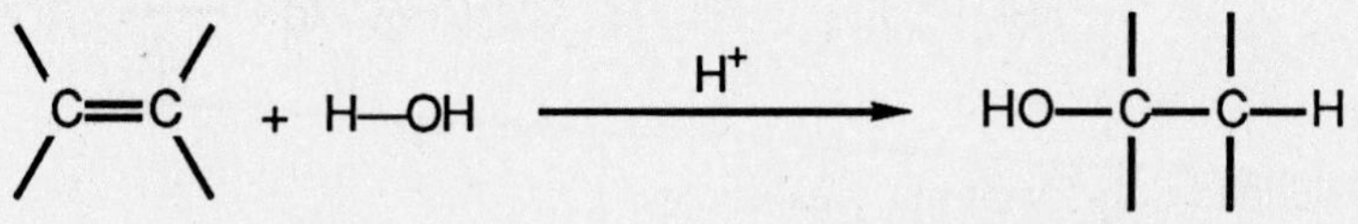

$$\text{C=C} + \text{H—OH} \xrightarrow{H^+} \text{HO—C—C—H}$$

(follow Markovnikov's rule)

   b. Hydroboration-oxidation of alkenes (3.13)

$$RCH=CH_2 \xrightarrow[\text{2. } H_2O_2,\ HO^-]{\text{1. } BH_3} RCH_2CH_2OH$$

   c. Alkyl halides + aqueous base (6.2; 6.3; 6.6; 6.7)

$$R\text{—}X + HO^- \longrightarrow R\text{—}OH + X^- \quad \text{(best for R = primary)}$$

   d. Grignard reagent + carbonyl compound (9.10)

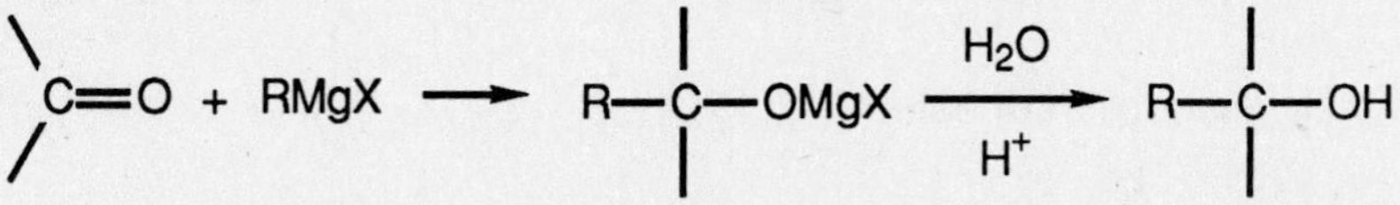

$$\text{C=O} + RMgX \longrightarrow R\text{—C—OMgX} \xrightarrow[H^+]{H_2O} R\text{—C—OH}$$

For primary alcohols, use formaldehyde. For secondary alcohols, use other aldehydes. For tertiary alcohols, use a ketone.

   e. Reduction of aldehydes or ketones (9.13; 9.20; 16.10)

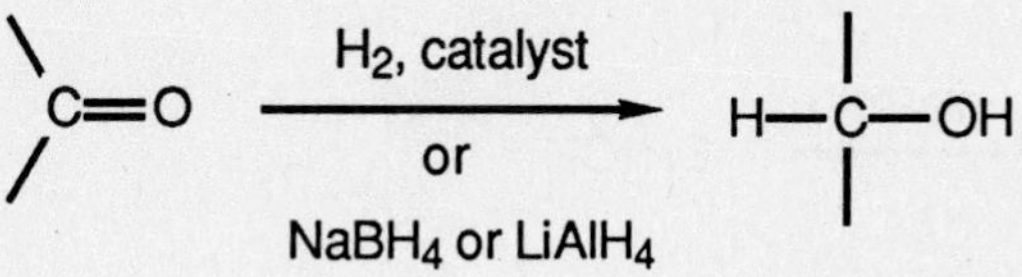

$$\text{C=O} \xrightarrow[\substack{\text{or} \\ NaBH_4 \text{ or } LiAlH_4}]{H_2,\ \text{catalyst}} H\text{—C—OH}$$

f. Grignard reagent + ethylene oxide (8.9)

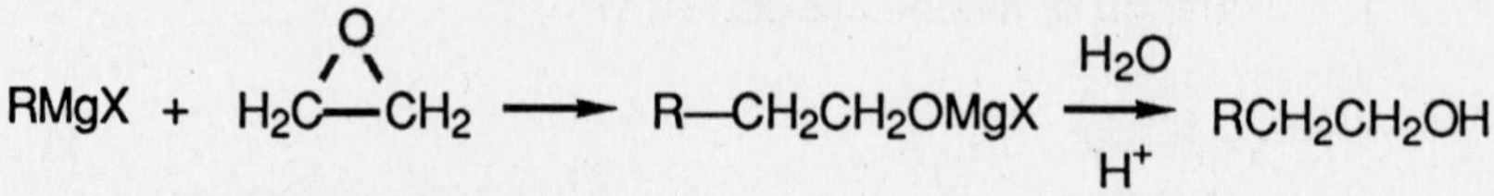

$$\text{RMgX} + \underset{\underset{\displaystyle H_2C—CH_2}{}}{\overset{\overset{\displaystyle O}{}}{}} \longrightarrow R—CH_2CH_2OMgX \xrightarrow[H^+]{H_2O} RCH_2CH_2OH$$

Useful only for primary alcohols.

g. Grignard reagent (excess) + ester (10.16)

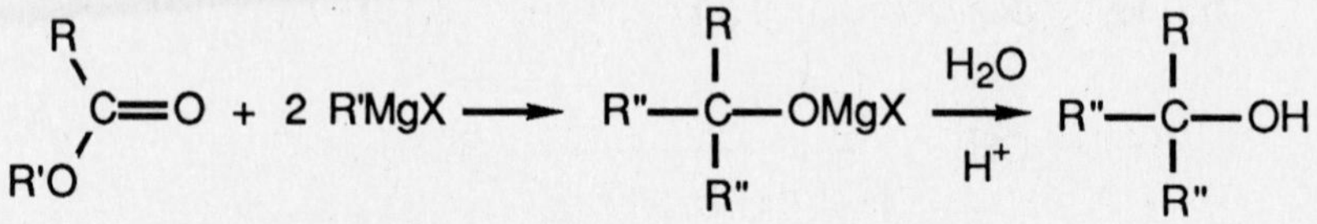

$$\underset{\displaystyle R'O}{\overset{\displaystyle R}{}}C{=}O + 2\ R'MgX \longrightarrow \underset{\displaystyle R''}{\overset{\displaystyle R}{}}R''{—}C{—}OMgX \xrightarrow[H^+]{H_2O} \underset{\displaystyle R''}{\overset{\displaystyle R}{}}R''{—}C{—}OH$$

Useful for tertiary alcohols with at least two identical R groups.

h. Saponification of esters (10.14)

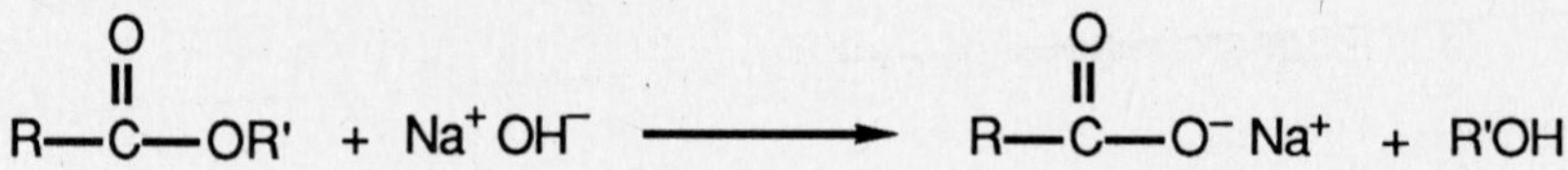

$$\overset{\overset{\displaystyle O}{\|}}{R—C—OR'} + Na^+OH^- \longrightarrow \overset{\overset{\displaystyle O}{\|}}{R—C—O^-Na^+} + R'OH$$

i. Reduction of esters (10.17; 15.6)

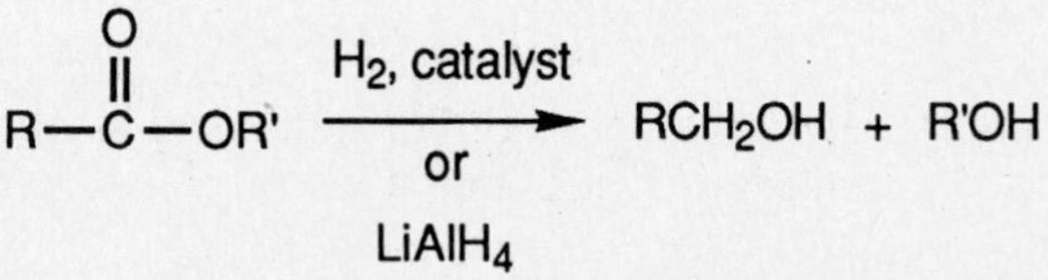

$$\overset{\overset{\displaystyle O}{\|}}{R—C—OR'} \xrightarrow[\substack{or\\ LiAlH_4}]{H_2,\ catalyst} RCH_2OH + R'OH$$

j. Methanol from carbon monoxide and hydrogen (AWA 14, Ch. 7)

$$CO + 2\ H_2 \xrightarrow{catalyst} CH_3OH$$

SUMMARY OF SYNTHETIC METHODS

6.   Phenols

    a.   From diazonium salts and base (11.13)

$$Ar\!-\!N_2^+X^- \ + \ NaOH \ \xrightarrow[\text{warm}]{H_2O} \ Ar\!-\!OH \ + \ N_2 \ + \ Na^+X^-$$

    b.   From phenoxides and acid (7.7)

$$Ar\!-\!O^-Na^+ \ + \ H^+X^- \ \longrightarrow \ Ar\!-\!OH \ + \ Na^+X^-$$

    c.   Phenol from isopropylbenzene (AWA 6, Ch. 4)

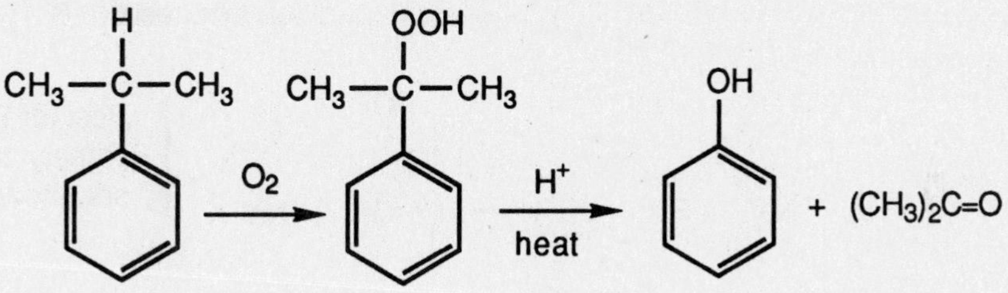

    d.   Reduction of quinones (AWA 21, Ch. 9)

7.   Glycols

    a.   Ring opening of epoxides (8.9)

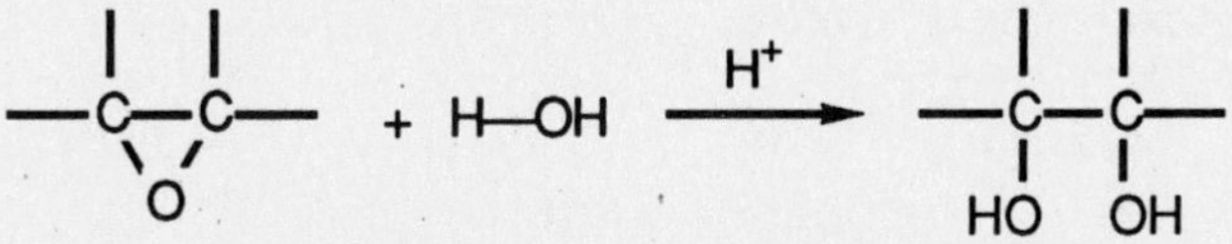

    b.   Oxidation of alkenes (3.16)

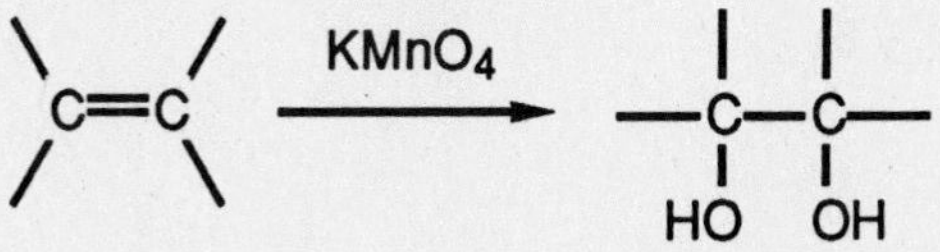

    c.   Hydrolysis of a fat or oil to give glycerol (15.2; 15.4)

$$H_2C-O-\overset{\overset{\displaystyle O}{\|}}{C}-R$$
$$HC-O-\overset{\overset{\displaystyle O}{\|}}{C}-R \xrightarrow[\text{heat}]{3 \ NaOH}$$
$$H_2C-O-\underset{\underset{\displaystyle O}{\|}}{C}-R$$

$$H_2C-OH$$
$$HC-OH + 3 \ Na^{+-}O-\overset{\overset{\displaystyle O}{\|}}{C}-R$$
$$H_2C-OH$$

8. <u>Ethers and Epoxides</u>

   a. From alkoxides and alkyl halides; Williamson synthesis (8.6; 16.9)

$$RO^-Na^+ + R'X \longrightarrow ROR + Na^+X^-$$

$$ArO^-Na^+ + R'X \longrightarrow ArOR + Na^+X^-$$

   Best for R = primary or secondary

   b. Dehydration of alcohols (8.6)

$$2 \ ROH \xrightarrow{H^+} ROR + H_2O$$

   Most useful for symmetric ethers.

   c. Ethylene oxide from ethylene and air (8.8)

$$H_2C{=}CH_2 + O_2 \xrightarrow[\text{catalyst}]{Ag} H-\underset{\underset{\displaystyle O}{\diagdown \diagup}}{\overset{\overset{\displaystyle H}{|}}{C}}-\overset{\overset{\displaystyle H}{|}}{C}-H$$

   d. Alkenes and peracids (8.8)

$$\diagup\hspace{-0.5em}\diagdown C{=}C\diagup\hspace{-0.5em}\diagdown + R-\overset{\overset{\displaystyle O}{\|}}{C}-O-OH \longrightarrow -\underset{\underset{\displaystyle O}{\diagdown \diagup}}{\overset{\overset{\displaystyle |}{C}}{}}-\overset{\overset{\displaystyle |}{C}}{} + R-\overset{\overset{\displaystyle O}{\|}}{C}-OH$$

R is usually CH_3—, C_6H_5—, or m-ClC_6H_4—

e. Ring opening of epoxides with alcohols (8.9)

$$-\overset{|}{\underset{\underset{O}{\diagdown\diagup}}{C}}-\overset{|}{C}- \quad + \quad R-OH \quad \xrightarrow{\;H^+\;} \quad -\overset{|}{\underset{HO}{C}}-\overset{|}{\underset{OR}{C}}-$$

9. Alkyl Halides

a. Halogenation of alkanes (2.15)

$$-\overset{|}{\underset{|}{C}}-H \quad + \quad X_2 \quad \xrightarrow{\;heat\ or\ light\;} \quad -\overset{|}{\underset{|}{C}}-X \quad + \quad H-X$$

b. Alkenes (or dienes) + hydrogen halides (3.9c; 3.14a; 5.12)

$$\overset{\diagdown}{\underset{\diagup}{C}}=\overset{\diagup}{\underset{\diagdown}{C}} \quad \xrightarrow{\;H-X\;} \quad -\overset{|}{\underset{H}{C}}-\overset{|}{\underset{X}{C}}-$$

c. Vinyl halides from alkynes + hydrogen halides (3.19)

$$-C\equiv C- \quad + \quad H-X \quad \longrightarrow \quad \overset{H}{\underset{\diagup}{\overset{\diagdown}{C}}}=\overset{\diagup}{\underset{\diagdown X}{C}}$$

d. Alcohols + hydrogen halides (7.10)

$$R-OH \quad + \quad H-X \quad \longrightarrow \quad R-X \quad + \quad H_2O$$

(Catalysts such as ZnX_2 are required when R is primary.)

e. Alcohols + thionyl chloride or phosphorus halides (7.11)

$$R{-}OH + SOCl_2 \longrightarrow R{-}Cl + SO_2 + HCl$$

$$3\ R{-}OH + PX_3 \longrightarrow 3\ R{-}X + H_3PO_3 \quad (X = Cl, Br)$$

f. Cleavage of ethers with hydrogen halides (8.7)

$$R{-}O{-}R' + 2\ H{-}X \longrightarrow R{-}X + R'{-}X + H_2O$$

g. Alkyl iodides from alkyl chlorides (6.3)

$$R{-}Cl + NaI \longrightarrow R{-}I + NaCl$$

10. <u>Polyhalogen Compounds</u>

a. Halogenation of alkanes (2.15; 6.10)

b. Addition of halogen to alkenes, dienes, and alkynes (3.9; 3.19)

11. <u>Aldehydes and Ketones</u>

a. Oxidation of alcohols (7.13; 9.4)

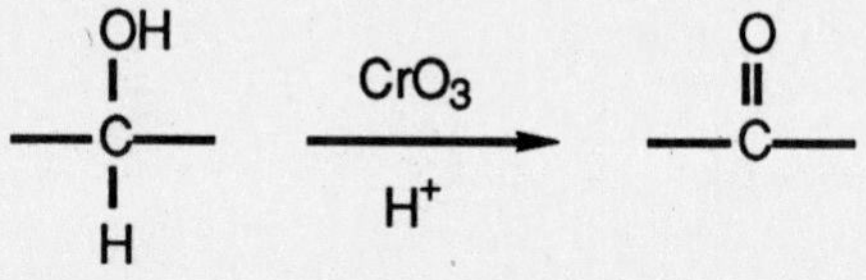

Primary alcohols give aldehydes. The best oxidant is pyridinium chloro-chromate (PCC). Secondary alcohols give ketones; the reagent is usually CrO_3, H^+.

b. Hydration of alkynes (3.19; 9.4)

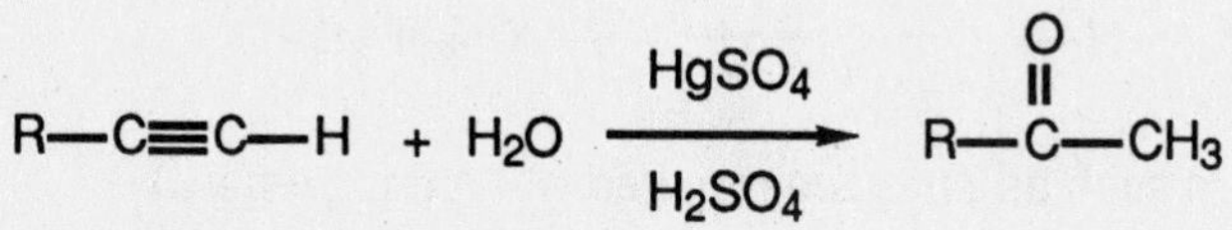

c. Ozonolysis of alkenes (3.16b)

$$\text{C}=\text{C} + \text{O}_3 \longrightarrow \text{C}=\text{O} + \text{O}=\text{C}$$

d. Hydrolysis of acetals and ketals (9.8)

$$\begin{array}{c} \text{OR} \\ | \\ -\text{C}- \\ | \\ \text{OR} \end{array} \quad \xrightarrow[\text{H}^+]{\text{H}_2\text{O}} \quad \begin{array}{c} \text{O} \\ || \\ -\text{C}- \end{array} + \ 2\ \text{ROH}$$

e. Acylation of benzenes (4.9; 4.10d; 9.4)

$$\text{benzene} \quad \xrightarrow[\text{AlCl}_3]{\overset{\overset{\text{O}}{||}}{\text{R}-\text{C}-\text{Cl}}} \quad \text{C}_6\text{H}_5-\text{C}(=\text{O})-\text{R}$$

f. Acetaldehyde from ethylene (9.3)

$$2\ \text{H}_2\text{C}=\text{CH}_2 + \text{O}_2 \quad \xrightarrow[100-130^\circ\text{C}]{\text{Pd-Cu}} \quad 2\ \text{CH}_3\text{CH}=\text{O}$$

g. Formaldehyde from methanol (9.3)

$$\text{CH}_3\text{OH} \quad \xrightarrow[600-700^\circ\text{C}]{\text{Ag}} \quad \text{H}_2\text{C}=\text{O} + \text{H}_2$$

h. Deuterated aldehydes or ketones (9.17)

$$\begin{array}{cc} \text{H} & \text{O} \\ | & || \\ -\text{C}-\text{C}- \\ | \end{array} \quad \xrightarrow[\text{CH}_3\text{OD}]{\text{CH}_3\text{O}^-\ \text{Na}^+} \quad \begin{array}{cc} \text{D} & \text{O} \\ | & || \\ -\text{C}-\text{C}- \\ | \end{array}$$

Only α-hydrogens exchange. Acid catalysis is also used, especially with aldehydes.

i. β-Hydroxy carbonyl compounds and α,β-unsaturated carbonyl compounds by way of the aldol condensation (9.18; 9.19; 9.20)

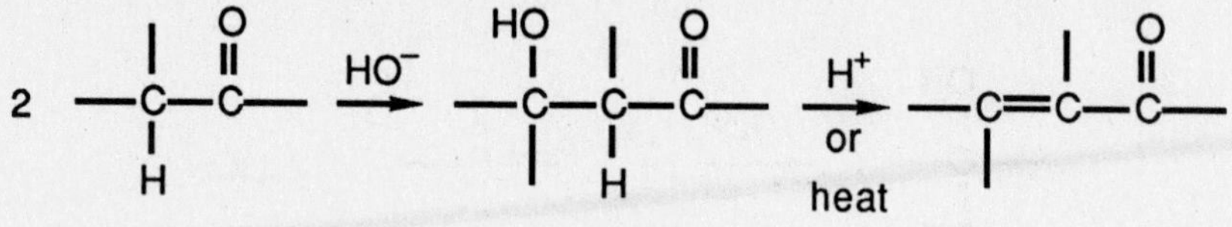

$$2 \quad \underset{H}{\overset{O}{-C-C-}} \quad \xrightarrow{HO^-} \quad \underset{H}{\overset{HO \quad \quad O}{-C-C-C-}} \quad \xrightarrow[heat]{H^+ \ or} \quad \overset{O}{-C=C-C-}$$

an aldol

12. Carboxylic Acids

a. Hydrolysis of nitriles (cyanides) (10.8d)

$$R-C{\equiv}N \quad \xrightarrow[H^+ \ or \ HO^-]{H_2O} \quad R-\overset{O}{C}-OH \quad (+ \ NH_3 \ or \ NH_4^+)$$

b. Grignard reagents + carbon dioxide (10.8c)

$$R-MgX + O{=}C{=}O \longrightarrow R-\overset{O}{C}-OMgX \xrightarrow[H^+]{H_2O} R-\overset{O}{C}-OH$$

c. Oxidation of aromatic side chains (10.8b)

$$Ar-CH_3 \xrightarrow{KMnO_4} Ar-\overset{O}{C}-OH$$

d. Oxidation of aldehydes (9.14; 10.8a; 16.11)

$$R-CH{=}O \xrightarrow{Ag \ or \ other \ oxidant} R-\overset{O}{C}-OH$$

e. Oxidation of primary alcohols (10.8a; 16.11)

$$RCH_2OH \xrightarrow[H^+]{K_2Cr_2O_7} RCO_2H$$

f. Saponification of esters (10.14)

$$R-\overset{\displaystyle O}{\overset{\|}{C}}-OR' + NaOH \longrightarrow R'OH + R-\overset{\displaystyle O}{\overset{\|}{C}}-O^- Na^+$$

$$\downarrow H^+$$

$$R-\overset{\displaystyle O}{\overset{\|}{C}}-OH$$

g. Hydrolysis of acid derivatives (10.19; 10.20; 10.21)

$$R-\overset{\displaystyle O}{\overset{\|}{C}}-Cl + H_2O \longrightarrow R-\overset{\displaystyle O}{\overset{\|}{C}}-OH + HCl$$

$$R-\overset{\displaystyle O}{\overset{\|}{C}}-O-\overset{\displaystyle O}{\overset{\|}{C}}-R + H_2O \longrightarrow 2\ R-\overset{\displaystyle O}{\overset{\|}{C}}-OH$$

$$R-\overset{\displaystyle O}{\overset{\|}{C}}-NH_2 + H_2O \xrightarrow[\substack{or \\ HO^-}]{H^+} R-\overset{\displaystyle O}{\overset{\|}{C}}-OH + NH_3$$

h. Dicarboxylic acids from cyclic ketones (9.14)

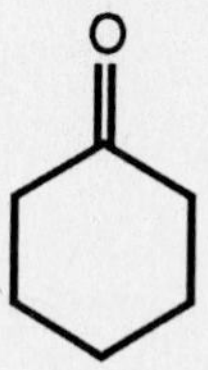

$+ HNO_3 \xrightarrow{V_2O_5} HOOC(CH_2)_4COOH$

i. Keto acids from hydroxy acids (5.7)

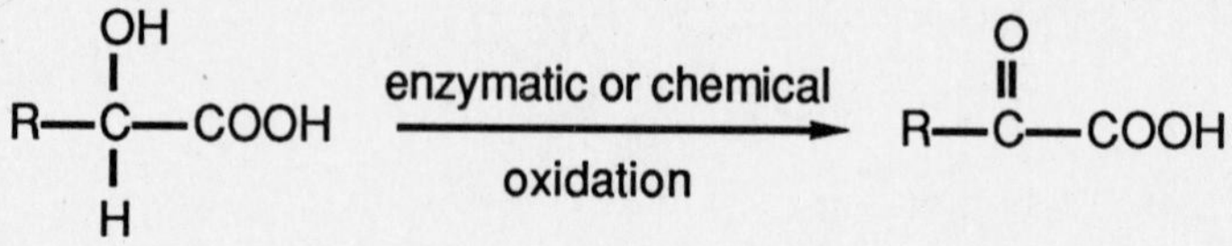

$$\underset{\overset{|}{H}}{\overset{\overset{OH}{|}}{R-C-COOH}} \quad \xrightarrow[\text{oxidation}]{\text{enzymatic or chemical}} \quad \overset{\overset{O}{\parallel}}{R-C-COOH}$$

13. <u>Esters</u>

a. From an alcohol and an acid (10.11; 10.12; 14.9; 17.6)

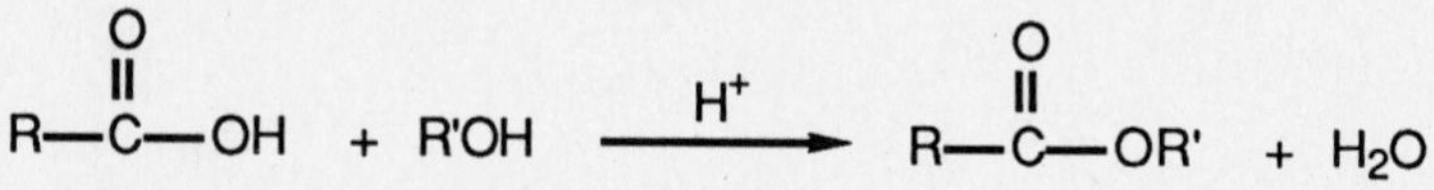

$$\overset{\overset{O}{\parallel}}{R-C-OH} \ + \ R'OH \quad \xrightarrow{H^+} \quad \overset{\overset{O}{\parallel}}{R-C-OR'} \ + \ H_2O$$

b. From an alcohol and an acid derivative (10.19; 10.20; 10.22; 16.9; 16.14b)

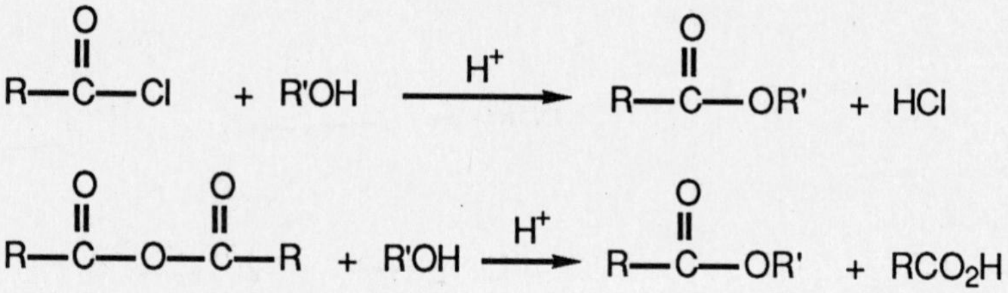

$$\overset{\overset{O}{\parallel}}{R-C-Cl} \ + \ R'OH \quad \xrightarrow{H^+} \quad \overset{\overset{O}{\parallel}}{R-C-OR'} \ + \ HCl$$

$$\overset{\overset{O}{\parallel}}{R-C}-O-\overset{\overset{O}{\parallel}}{C-R} \ + \ R'OH \quad \xrightarrow{H^+} \quad \overset{\overset{O}{\parallel}}{R-C-OR'} \ + \ RCO_2H$$

c. Salt + alkyl halide (6.3; 17.12)

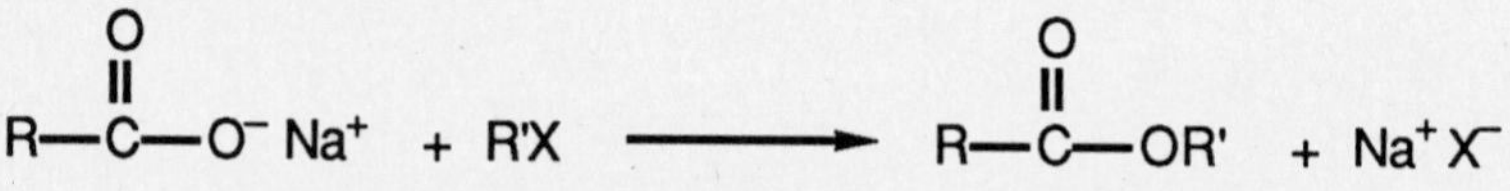

$$\overset{\overset{O}{\parallel}}{R-C}-O^-\,Na^+ \ + \ R'X \quad \longrightarrow \quad \overset{\overset{O}{\parallel}}{R-C-OR'} \ + \ Na^+X^-$$

d. Lactones from hydroxy acids (10.13)

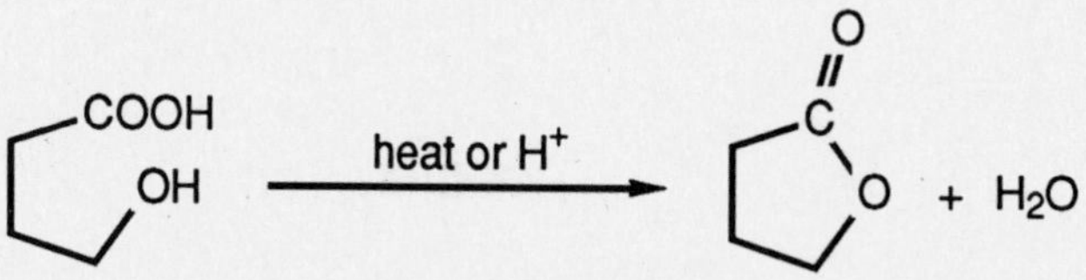

e. β-Keto esters by way of the Claisen condensation (10.23)

$$2 \quad RCH_2\!-\!\overset{\overset{\displaystyle O}{\|}}{C}\!-\!OR' \quad\xrightarrow{\text{base}}\quad RCH_2\!-\!\overset{\overset{\displaystyle O}{\|}}{C}\!-\!\underset{\underset{\displaystyle R}{|}}{CH}\!-\!\overset{\overset{\displaystyle O}{\|}}{C}\!-\!OR' \;+\; R'OH$$

14. Amides

a. Acyl halides + ammonia (or primary or secondary amines) (10.19; 11.11; 17.6)

$$R\!-\!\overset{\overset{\displaystyle O}{\|}}{C}\!-\!Cl \;+\; \overset{\overset{\displaystyle H}{|}}{\underset{\underset{\displaystyle R''}{|}}{N}}\!-\!R' \;\longrightarrow\; R\!-\!\overset{\overset{\displaystyle O}{\|}}{C}\!-\!\underset{\underset{\displaystyle R''}{|}}{N}\!-\!R' \;+\; HCl$$

(R' and R" = H, alkyl, or aryl)

b. Acid anhydrides and ammonia (or primary or secondary amines) (10.20; 11.11)

$$R\!-\!\overset{\overset{\displaystyle O}{\|}}{C}\!-\!O\!-\!\overset{\overset{\displaystyle O}{\|}}{C}\!-\!R \;+\; \overset{\overset{\displaystyle H}{|}}{\underset{\underset{\displaystyle R''}{|}}{N}}\!-\!R' \;\longrightarrow\; R\!-\!\overset{\overset{\displaystyle O}{\|}}{C}\!-\!\underset{\underset{\displaystyle R''}{|}}{N}\!-\!R' \;+\; RCO_2H$$

c. Esters and ammonia (or primary or secondary amines) (10.15)

$$R\!-\!\overset{\overset{\displaystyle O}{\|}}{C}\!-\!OR' \;+\; \overset{\overset{\displaystyle H}{|}}{\underset{\underset{\displaystyle R''}{|}}{N}}\!-\!R''' \;\longrightarrow\; R\!-\!\overset{\overset{\displaystyle O}{\|}}{C}\!-\!\underset{\underset{\displaystyle R''}{|}}{N}\!-\!R''' \;+\; R'OH$$

d. From acids and ammonia (or primary or secondary amines) (10.21)

$$R\!-\!\overset{\overset{\displaystyle O}{\|}}{C}\!-\!OH \;+\; \overset{\overset{\displaystyle H}{|}}{\underset{\underset{\displaystyle R'}{|}}{N}}\!-\!R' \quad\xrightarrow{\text{heat}}\quad R\!-\!\overset{\overset{\displaystyle O}{\|}}{C}\!-\!\underset{\underset{\displaystyle R'}{|}}{N}\!-\!R' \;+\; H_2O$$

(R' = H or alkyl)

e. Polyamides from diamines and dicarboxylic acids or their derivatives (14.2; AWA 28, Ch. 14)

$$H_2N\text{-----}NH_2 \quad + \quad HOOC\text{-----}COOH \longrightarrow$$

$$\text{---}\overset{\overset{\textstyle O}{\|}}{C}\text{--N}\text{-----}N\text{--}\overset{\overset{\textstyle O}{\|}}{C}\text{---}\Big)_n$$
$$\qquad\qquad |\qquad\quad|$$
$$\qquad\qquad H\qquad\ H$$

g. Peptides (17.12)

15. <u>Other Carboxylic Acid Derivatives</u>

a. Salts from acids and bases (10.7)

$$R\text{--}\overset{\overset{\textstyle O}{\|}}{C}\text{--OH} \; + \; NaOH \longrightarrow R\text{--}\overset{\overset{\textstyle O}{\|}}{C}\text{--}O^-\,Na^+ \; + \; H_2O$$

b. Salts (soaps) by saponification of esters (10.14; 15.4)

$$R\text{--}\overset{\overset{\textstyle O}{\|}}{C}\text{--OR'} \; + \; NaOH \xrightarrow{\text{heat}} R\text{--}\overset{\overset{\textstyle O}{\|}}{C}\text{--}O^-\,Na^+ \; + \; R'OH$$

c. Acyl halides from acids (10.19)

$$R\text{--}\overset{\overset{\textstyle O}{\|}}{C}\text{--OH} \xrightarrow{\text{SOCl}_2 \text{ or PCl}_3} R\text{--}\overset{\overset{\textstyle O}{\|}}{C}\text{--Cl}$$

d. Anhydrides from diacids (10.20)

heat or Ac$_2$O

16. <u>Nitriles (Cyanides)</u>

   a. Nitriles from alkyl halides and inorganic cyanides (6.3; 10.8d)

$$R-X + Na^+ CN^- \longrightarrow R-CN + Na^+ X^-$$

   (R = primary or secondary)

   b. Nitriles from diazonium ions (11.13)

$$Ar-N_2^+ + KCN \xrightarrow{Cu_2(CN)_2} Ar-CN + N_2 + Cu_2X_2$$

17. <u>Amines and Related Compounds</u>

   a. Alkylation of ammonia or amines (6.3; 11.5)

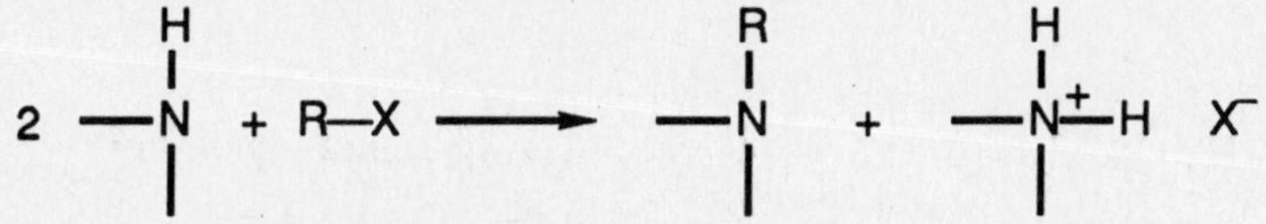

$$2 \ \overset{H}{\underset{|}{-N-}} + R-X \longrightarrow \overset{R}{\underset{|}{-N-}} + \overset{H}{\underset{|}{-N^+-H}} \ X^-$$

   b. Reduction of nitriles (11.6)

$$R-C\equiv N + 2 H_2 \xrightarrow{\text{LiAlH}_4 \text{ or } H_2, \text{ Ni catalyst}} RCH_2NH_2$$

   c. Reduction of nitro compounds (11.6)

$$Ar-NO_2 \xrightarrow{\text{SnCl}_2, \text{ HCl or } H_2, \text{ catalyst}} Ar-NH_2 \ (+ \ 2 \ H_2O)$$

   d. Reduction of amides (10.21; 11.6)

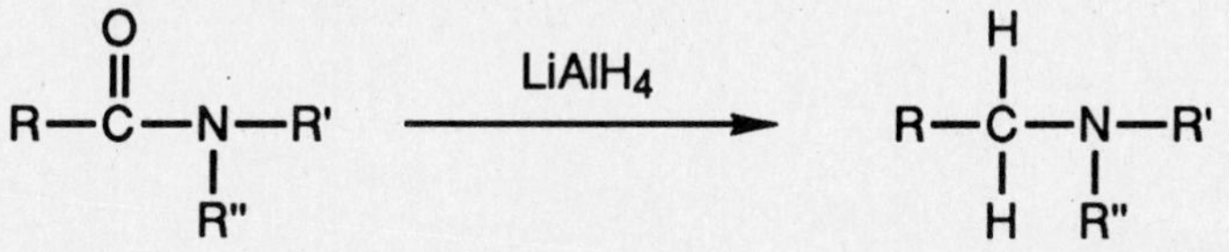

$$\overset{O}{\overset{\|}{R-C-\underset{\underset{R''}{|}}{N}-R'}} \xrightarrow{\text{LiAlH}_4} \overset{H}{\underset{\underset{H}{|}}{R-C}}-\underset{\underset{R''}{|}}{N}-R'$$

e. Hydrolysis of amides (10.21)

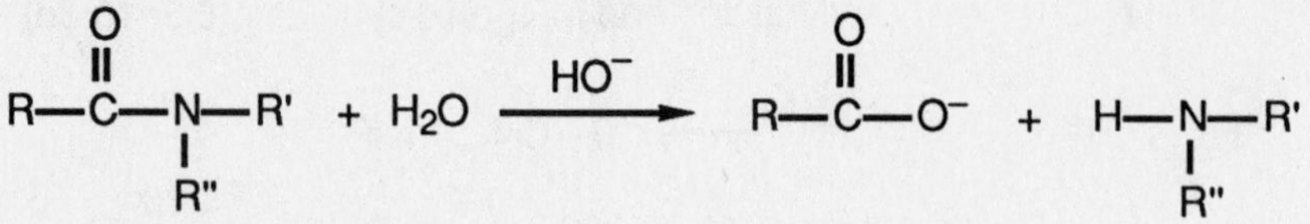

$$\underset{\substack{|\\R''}}{R-\overset{\overset{\displaystyle O}{\|}}{C}-N-R'} + H_2O \xrightarrow{HO^-} R-\overset{\overset{\displaystyle O}{\|}}{C}-O^- + \underset{\substack{|\\R''}}{H-N-R'}$$

(R' and R" = alkyl or aryl)

f. Amine salts from amines (7.13; 11.9; 13.2)

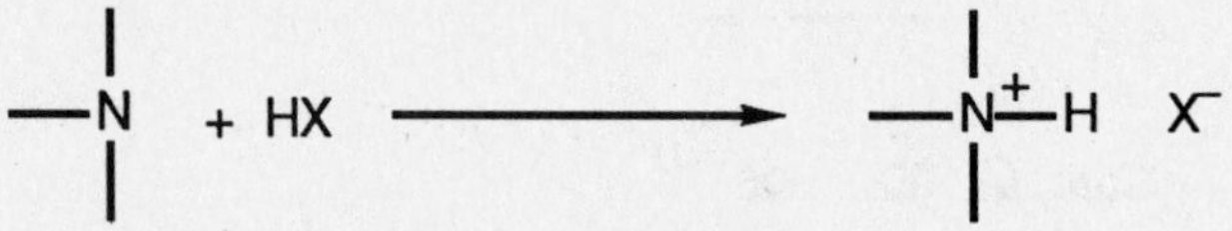

$$\underset{|}{\overset{|}{-N}} + HX \longrightarrow \underset{|}{\overset{|}{-N^+\!-H}}\ \ X^-$$

g. Quaternary ammonium salts (6.3; 11.12)

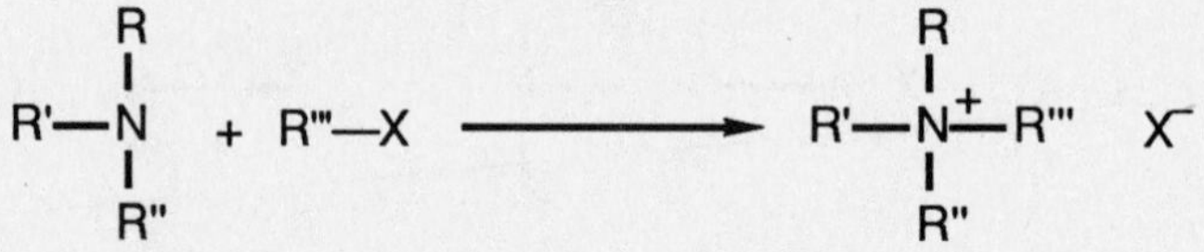

$$\underset{\substack{|\\R''}}{\overset{\substack{R\\|}}{R'-N}} + R'''-X \longrightarrow \underset{\substack{|\\R''}}{\overset{\substack{R\\|}}{R'-N^+\!-R'''}}\ \ X^-$$

(R''' = primary or secondary)

18. Miscellaneous Nitrogen Compounds

a. Oximes, hydrazones, and imines from carbonyl compounds and ammonia derivatives (9.12)

$$\overset{\diagdown}{\underset{\diagup}{C}}{=}O + H_2N-R \longrightarrow \overset{\diagdown}{\underset{\diagup}{C}}{=}N-R + H_2O$$

R = alkyl, aryl, OH, NH$_2$

b. Cyanohydrins from carbonyl compounds and hydrogen cyanide (9.11)

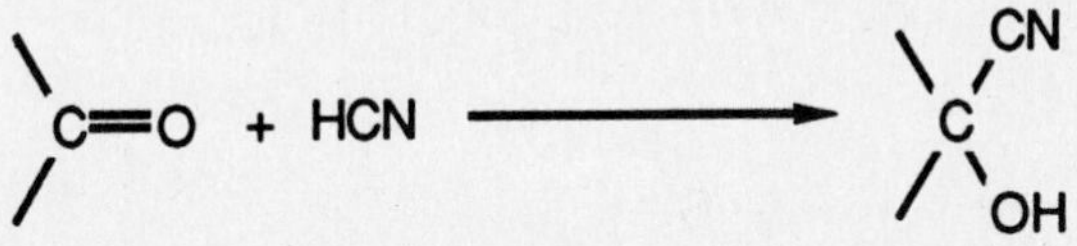

$$\text{\textbackslash}C{=}O \; + \; HCN \; \longrightarrow \; \underset{/}{\overset{\textbackslash}{C}}\underset{OH}{\overset{CN}{}}$$

c. Diazonium compounds from primary aromatic amines and nitrous acid (11.13)

$$Ar{-}NH_2 \; + \; HONO \; + \; HX \; \longrightarrow \; Ar{-}N_2^+X^- \; + \; 2\,H_2O$$

d. Azo compounds, via coupling reactions (11.14)

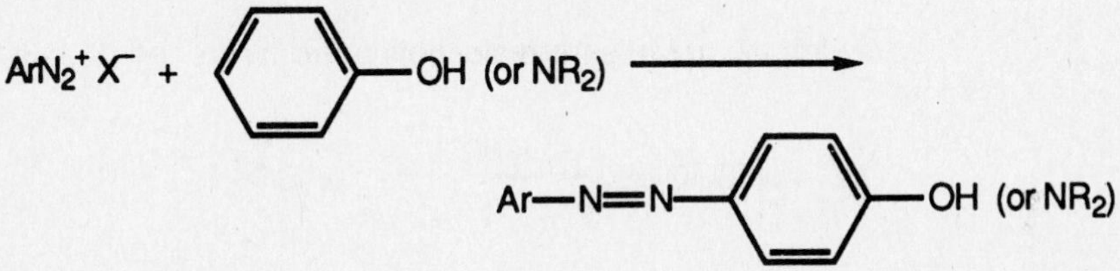

$$ArN_2^+X^- \; + \; \text{(benzene ring)}{-}OH \; (\text{or } NR_2) \; \longrightarrow$$

$$Ar{-}N{=}N{-}\text{(benzene ring)}{-}OH \; (\text{or } NR_2)$$

e. Alkyl nitrates (7.14; 16.14b)

$$ROH \; + \; HONO_2 \; \longrightarrow \; RONO_2 \; + \; H_2O$$

f. Ureas (14.10)

g. Carbamates (urethanes) (14.10)

h. Piperidine (13.3)

19. <u>Organic Sulfur Compounds</u>

a. Thiols from alkyl halides and sodium hydrosulfide (6.3; 7.17)

$$R{-}X \; + \; Na^+SH^- \; \longrightarrow \; R{-}SH \; + \; Na^+X^- \quad (\text{best when R is primary})$$

b. Thioethers from alkyl halides and sodium mercaptides (6.3)

$$R{-}X \; + \; Na^+SR^- \; \longrightarrow \; R{-}S{-}R' \; + \; Na^+X^- \quad (\text{best when R is primary})$$

c. Disulfides from thiols (7.17; AWA 16, p. Ch. 7; 17.9)

$$2 \text{ R—SH} \xrightarrow{\text{H}_2\text{O}_2} \text{R—S—S—R}$$

d. Sulfonium salts (6.3; AWA 10, Ch. 6)

$$
\begin{array}{c}
\text{R'} \\
| \\
:\text{S}: \\
| \\
\text{R}
\end{array}
+ \text{ R''—X} \longrightarrow
\begin{array}{c}
\text{R'} \\
|\ + \\
\text{R''—S}: \\
| \\
\text{R}
\end{array}
\text{ X}^-
$$

e. Alkyl hydrogen sulfates from alcohols or from alkenes (3.9; 3.11; 15.6)

$$\text{R—OH} + \text{HOSO}_3\text{H} \xrightarrow{\text{cold}} \text{ROSO}_3\text{H} + \text{H}_2\text{O}$$

$$
\begin{array}{c}
\backslash \quad / \\
\text{C}=\text{C} \\
/ \quad \backslash
\end{array}
+ \text{HOSO}_3\text{H} \xrightarrow{\text{cold}}
\begin{array}{c}
|\quad| \\
\text{—C—C—} \\
|\quad| \\
\text{H} \quad \text{OSO}_3\text{H}
\end{array}
$$

f. Sulfonic acids by sulfonation (4.9; 4.10c; 15.6)

g. Sulfonamides (AWA 24, Ch. 11; 11.11)

20. Miscellaneous Classes of Compounds

a. Grignard reagents (8.5)

$$\text{R—X} + \text{Mg} \xrightarrow{\text{ether}} \text{RMgX}$$

b. Organolithium compounds (8.5)

$$\text{R—X} + 2 \text{ Li} \longrightarrow \text{RLi} + \text{LiX}$$

c. Acetylides (3.50)

$$R-C\equiv C-H + NaNH_2 \xrightarrow{NH_3} R-C\equiv C^- Na^+ + NH_3$$

d. Alkoxides and phenoxides (7.7)

$$2 \ ROH + 2 \ Na \longrightarrow 2 \ RO^- Na^+ + H_2$$

$$Ar-OH + Na^+ OH^- \longrightarrow Ar-O^- Na^+ + H_2$$

e. Hemiacetals and acetals (9.8; 16.5; 16.6; 16.12)

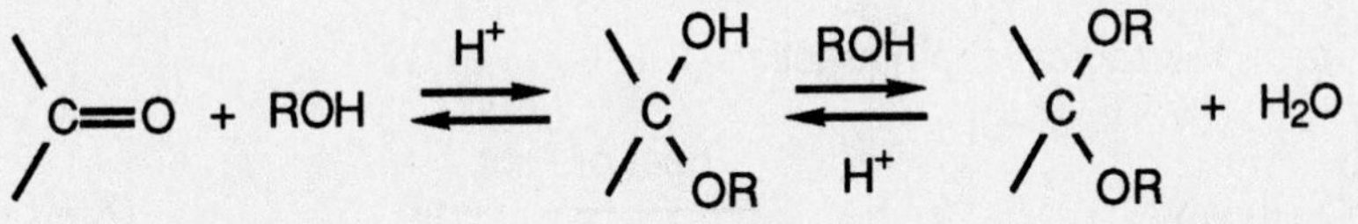

f. Quinones (7.16)

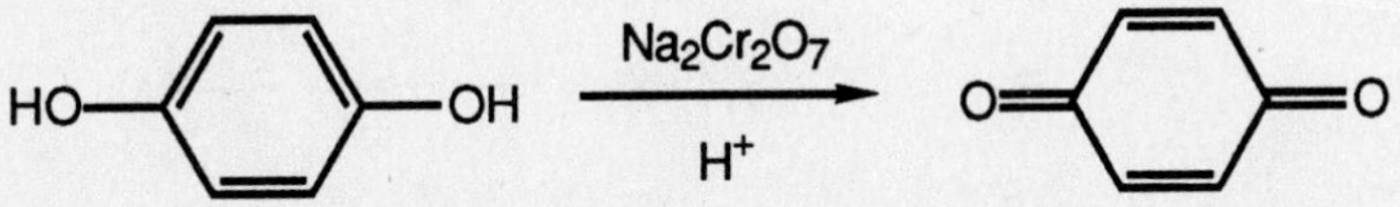

g. Vinyl polymers (3.15; 6.10; 14.2; 14.3; 14.4; 14.5; 14.6)

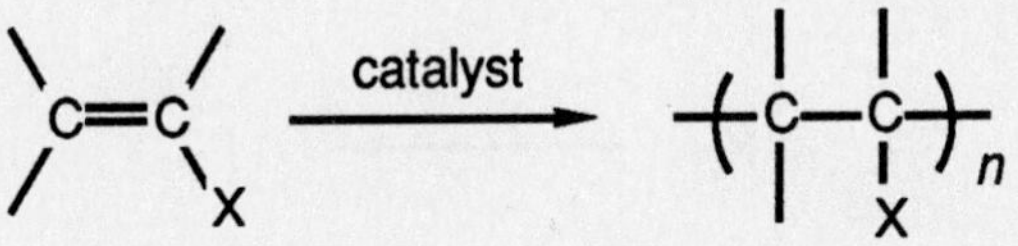

SUMMARY OF REACTION MECHANISMS

Although a substantial number of reactions are described in the text, they belong to a relatively modest number of mechanistic types. The preparation of alkyl halides from alcohols and HX, the cleavage of ethers, the preparation of amines from alkyl halides and ammonia (and many other reactions) all, for example, occur by a nucleophilic substitution mechanism. The following is a brief review of the main mechanistic pathways discussed in the text.

1. Substitution Reactions

    a. Free-radical chain reaction (2.16)

    Initiation: $:\ddot{X}-\ddot{X}: \xrightarrow{\text{heat or light}} 2 \; :\ddot{X}\cdot$ (X = Cl, Br)

    Propagation: $R-H + :\ddot{X}\cdot \longrightarrow R\cdot + H-\ddot{X}:$

                   $R\cdot + :\ddot{X}-\ddot{X}: \longrightarrow R-\ddot{X}: + :\ddot{X}\cdot$

    Termination: $2 :\ddot{X}\cdot \longrightarrow :\ddot{X}-\ddot{X}:$

                   $2 R\cdot \longrightarrow R-R$

                   $R\cdot + :\ddot{X}\cdot \longrightarrow R-\ddot{X}:$

The sum of the propagation steps gives the overall reaction.

    b. Electrophilic aromatic substitution (4.10)

    Reaction occurs in two steps: addition of an electrophile to the aromatic π electron system followed by loss of a ring proton.

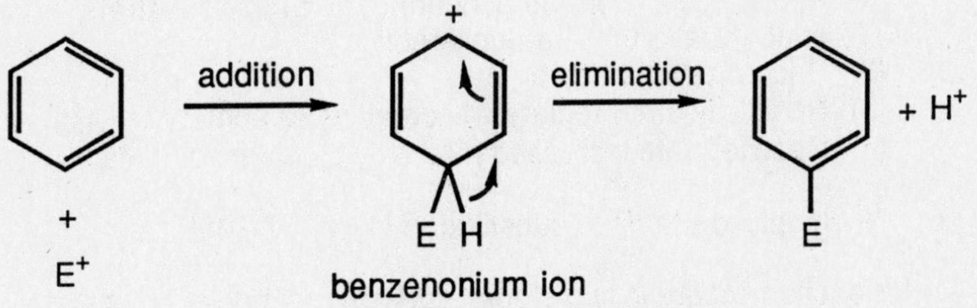

benzenonium ion

Substituents already on the aromatic ring affect the reaction rate and the orientation of subsequent substitutions (4.11; 4.12).

c. Nucleophilic aliphatic substitution (6.4; 6.5; 6.6; 6.7)

S_N2 (substitution, nucleophilic, bimolecular):

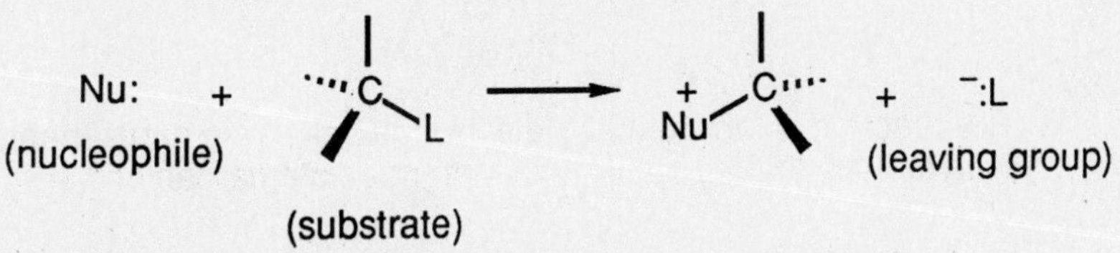

(1) Rate depends on the concentration of both reactants, that is, the nucleophile and the substrate.
(2) Inversion of configuration at carbon
(3) Reactivity order CH_3 > primary > secondary >> tertiary
(4) Rate only mildly dependent on solvent polarity

S_N1 (substitution, nucleophilic, unimolecular):

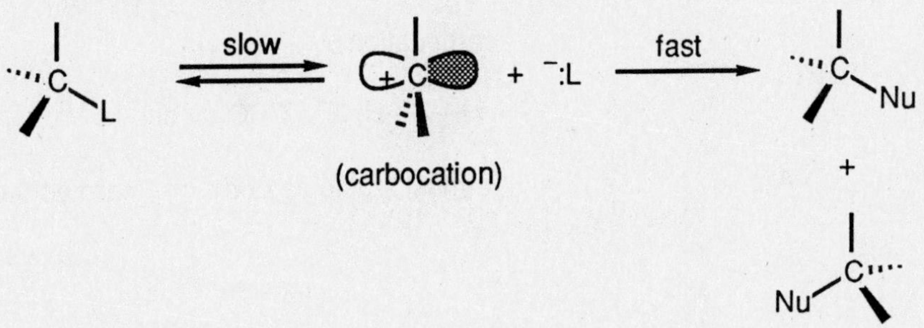

(carbocation)

      (1) Rate depends on the concentration of the substrate but *not* on the concentration of the nucleophile.
      (2) Racemization at carbon
      (3) Reactivity order tertiary > secondary >> primary or CH_3
      (4) Reaction rate increased markedly by polar solvents

d. Nucleophilic aromatic substitution (13.3; 17.10b)

Reaction occurs in two steps: addition of a nucleophile to the aromatic ring followed by loss of a leaving group, often an anion.

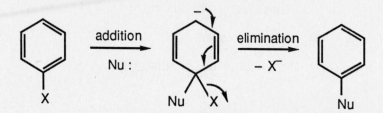

The reaction is facilitated by electron-withdrawing substituents *ortho* or *para* to X since they stabilize the intermediate carbanion.

2. <u>Addition Reactions</u>

a. Electrophilic additions to C=C and C≡C

<u>Addition of acids (3.11):</u>

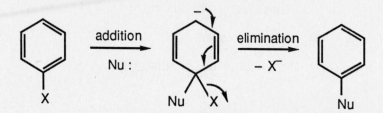

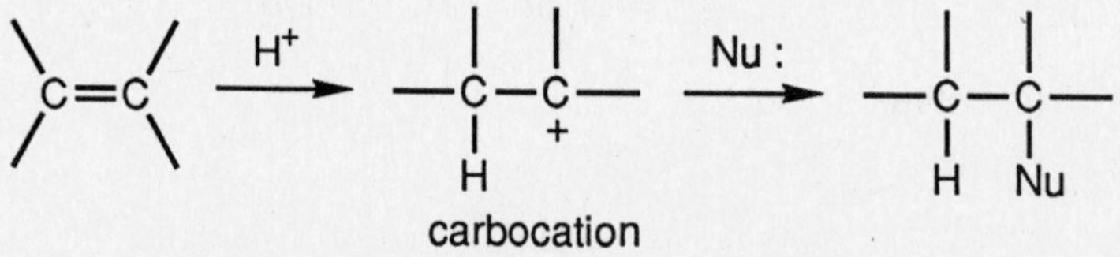

carbocation

Nu : = R—Ö—H, H—Ö—H, X⁻, ⁻OSO_3H, and so on

Addition proceeds via the most stable carbocation intermediate (Markovnikov's rule) (3.10; 3.12).

SUMMARY OF REACTION MECHANISMS

Addition to alkynes (3.19):

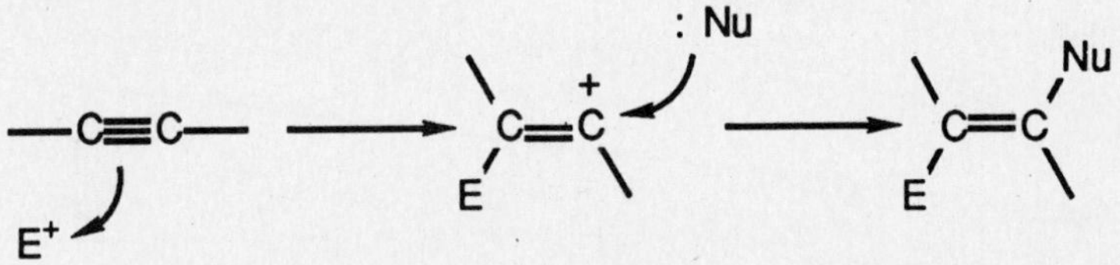

1,4-Addition to conjugated dienes (3.14a):

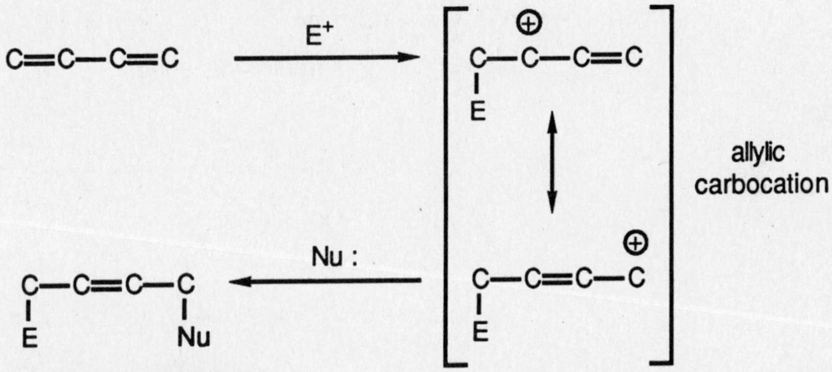

allylic carbocation

(+ normal 1,2-addition)

cycloaddition (3.14b):

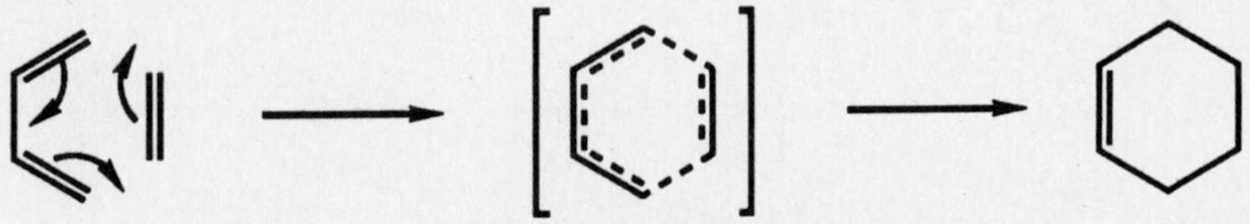

transition state

The reaction involves six π electrons in a cyclic transition state. The dotted bonds in the transition state represent partially broken or partially formed bonds.

b. Free-radical addition (3.15; 14.3).

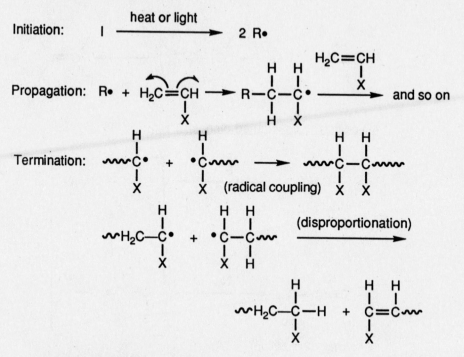

Another reaction type associated with such reactions is chain-transfer (or hydrogen abstraction) (14.3).

Cationic and anionic chain-growth polymerizations occur by chain reactions similar to those for free-radical polymerizations but involving charged intermediates (14.4; 14.5).

c. Hydroboration via a cyclic transition state (3.13).

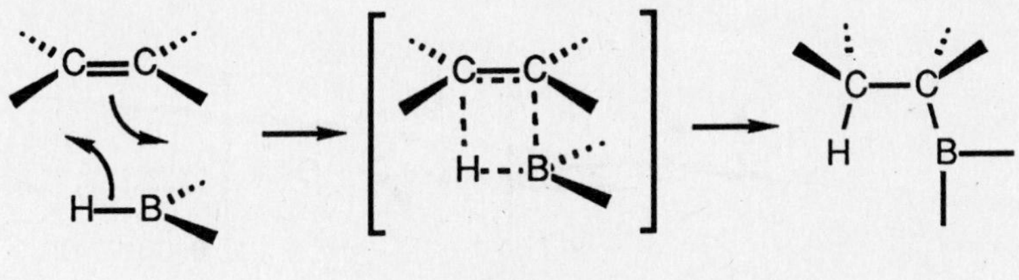

transition state

The boron adds to the least crowded carbon; both the boron and hydrogen add to the same "face" of the double bond. Subsequent oxidation with H_2O_2 and OH^- places a hydroxyl group in the exact position previously occupied by boron:

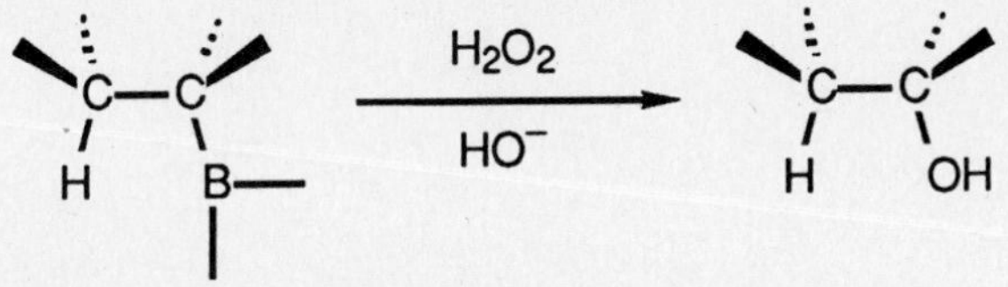

d. Nucleophilic addition to C=O (9.7; 10.12).

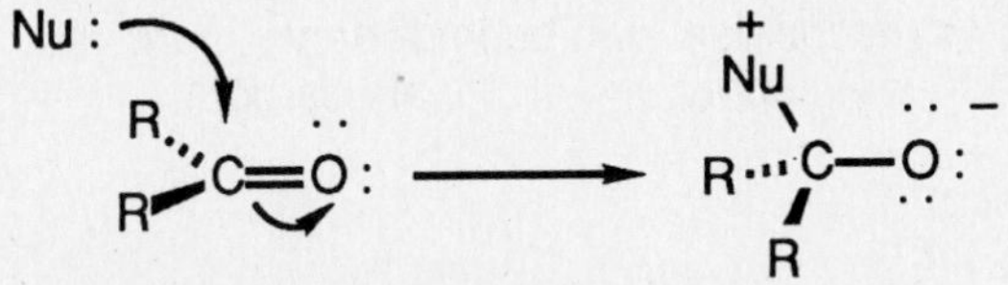

The reaction is often followed by protonation of the oxygen. The C=O group may be present in an aldehyde, ketone, ester, anhydride, acyl halide, amide, and so forth.

Acid catalysts may be necessary with weak nucleophiles. The acid protonates the oxygen, making the carbon more positive and thus more electrophilic.

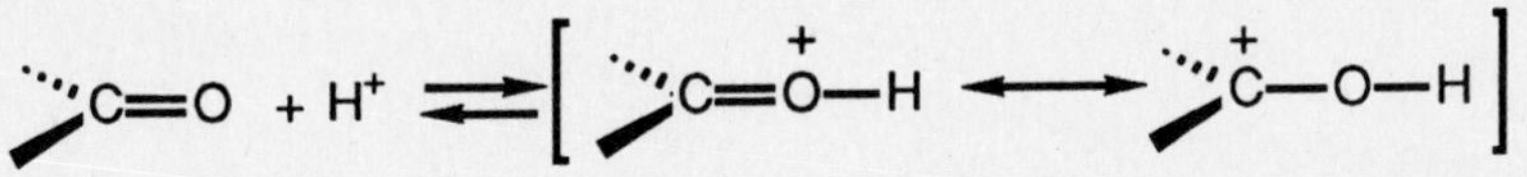

Nucleophilic *substitution* at a carbonyl group (10.12) takes place by an addition–elimination mechanism (10.18):

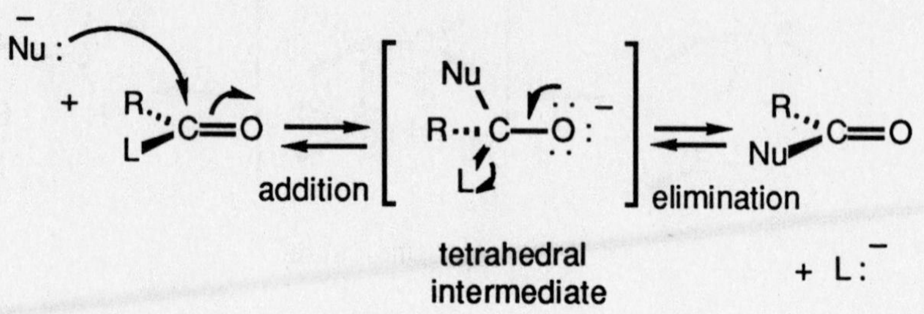

3. Elimination Reactions

a. The E2 mechanism (6.8).

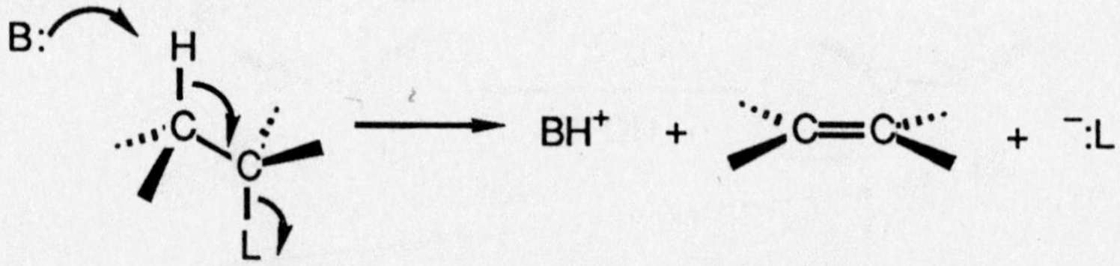

A planar, *anti* arrangement of the eliminated groups (H and L) is preferred. The reaction rate depends on the concentration of both the base (B) and the substrate. This mechanism can be important regardless of whether L is attached to a primary, secondary, or tertiary carbon.

b. The E1 mechanism (6.8).

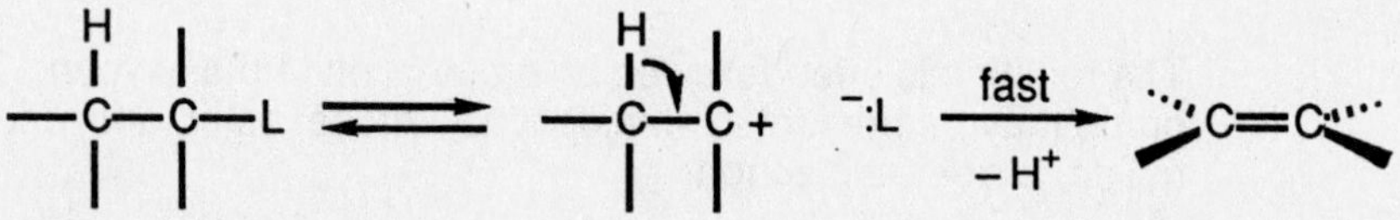

The first step is the same as the first step of the S_N1 mechanism. The reaction rate depends only on the substrate concentration. This mechanism is most important when L is attached to a tertiary carbon.

REVIEW PROBLEMS ON SYNTHESIS

The following problems are designed to give you practice in multistep syntheses and should be helpful in preparing for quizzes or examinations. To add to the challenge and make this preparation more realistic , no answers are given. If you have difficulty with some of the problems, consult your textbook first and then your instructor.

The best technique for solving a multistep synthesis problem is to work backward from the final goal. Keep in mind the structure of the available starting material, however, so that eventually you can link the starting material with the product. The other constraint on synthesis problems is that you *must* use combinations of known reactions to achieve your ultimate goal. Although research chemists do try to invent and develop new reactions, you cannot afford that luxury until you have already mastered known reactions. All the following problems can be solved using reactions in your text.

1. Show how each of the following can be prepared from propene:

   a. propane
   b. 2-bromopropane
   c. 1,2-dichloropropane
   d. 2-propanol
   e. 1-propanol

2. Each of the following conversions requires two reactions, in the proper sequence. Write equations for each conversion.

   a. *n*-propyl bromide to propene to 1,2-dibromopropane
   b. isopropyl alcohol to propene to 2-iodopropane
   c. 1-bromobutane to 2-chlorobutane
   d. 2-butanol to butane
   e. bromocyclopentane to 1,2-dibromocyclopentane
   f. *t*-butyl chloride to isobutyl alcohol

3. Starting with acetylene, write equations for the preparation of:

   a. ethane
   c. 1-butyne
   b. ethyl iodide
   d. 1,1-diiodoethane

e. 2,2-dibromobutane   f. 3-hexyne
g. 1,1,2,2-tetrabromoethane   h. *cis*-3-hexene

4. Write equations for each of the following conversions:

   a. 2-butene to 1,3-butadiene (two steps)
   b. 2-propanol to 1-propanol (three steps)
   c. 1-bromopropane to 1-aminobutane (two steps)
   d. 1,3-butadiene to 1,4-dibromobutane (two steps)

In solving problems of this type, carefully examine the structures of the starting material and the final product. Seek out similarities and differences. Note the types of bonds that must be made or broken to go from one structure to the other. Sometimes it is profitable to work backward from the product and forward from the starting material simultaneously, with the goal of arriving at a common intermediate.

5. Using benzene or toluene as the only organic starting material, devise a synthesis for each of the following:

   a. *m*-chlorobenzenesulfonic acid   b. 2,4,6-tribromotoluene

   c.

   d.

   e.

   f.

6. Write equations for the preparation of:

   a. 1-phenylethanol from styrene
   b. 2-phenylethanol from styrene
   c. 1-butanol from 1-bromobutane
   d. sodium 2-butoxide from 1-butene
   e. 1-butanethiol from 1-butanol
   f. 2,4,6-tribromobenzoic acid from toluene
   g. ethyl cyclohexyl ether from ethanol and phenol
   h. di-*n*-butyl ether from 1-butanol

7. Write equations that show how 2-propanol can be converted to each of the following:

    a. 2-chloropropane
    c. 2-methoxypropane

    b. 1,2-bromopropane
    d. isopropylbenzene

8. 2-Bromobutane can be obtained in one step from each of the following precursors: butane, 2-butanol, 1-butene, and 2-butene. Write an equation for each method and describe the advantages or disadvantages of each.

9. Starting with an unsaturated hydrocarbon, show how each of the following can be prepared:

    a. 1,2-dibromobutane
    c. 1,2,3,4-tetrabromobutane
    e. 1,4-dibromo-2-butene
    g. 1-bromo-1-phenylethane

    b. 1,1-dichloroethane
    d. cyclohexyl iodide
    f. 1,1,2,2-tetrachloropropane
    h. 1,2,5,6-tetrabromocyclooctane

10. Give equations for the preparation of the following carbonyl compounds:

    a. 2-pentanone from an alcohol
    c. cyclohexanone from phenol
       (two steps)

    b. pentanal from an alcohol
    d. acetone from propyne

11. Complete each of the following equations, giving the structures and names of the main organic products:

    a. benzoic acid + ethylene glycol + H^+
    c. n-propylamine + acetic anhydride

    e. n-propylbenzene + $K_2Cr_2O_7$ + H^+

    g. pentanedioic acid + thionyl chloride
    i. methyl 3-butenoate + $LiAlH_4$

    b. $C_6H_5CH_2MgBr$ + CO_2, followed by H_3O^+
    d. p-hydroxybenzoic acid + acetic anhydride
    f. phthalic anhydride + methanol + H^+
    h. cyclopropanecarboxylic acid + NH_4OH, then heat
    j. ethyl propanoate + NaOH, heat, then acidify

12. Show how each of the following conversions can be accomplished:

    a. butanoyl chloride to methyl butanoate
    c. butanoic acid to 1-butanol

    b. propanoic anhydride to propanamide
    d. 1-pentanol to pentanoic acid

495

e. propanoyl bromide to *N*-ethyl propanamide

f. oxalic acid to diethyl oxalate

g. urea to ammonia and CO_2

h. benzoyl chloride to *N*-methylbenzamide

13. Show how each of the following compounds can be prepared from the appropriate acid:

a. propanoyl bromide
c. *n*-butanamide
e. calcium oxalate
g. isopropyl benzoate

b. ethyl pentanoate
d. phthalic anhydride
f. phenylacetamide
h. *m*-nitrobenzoyl chloride

14. Write equations to describe how each of the following conversions might be accomplished:

a. *n*-butyl chloride to *n*-butyltrimethylammonium chloride
b. *o*-toluidine to *o*-toluic acid
c. *o*-toluidine to *o*-bromobenzoic acid
d. 1-butene to 2-methyl-1-aminobutane

15. Starting with benzene, toluene, or any alcohol with four carbon atoms or fewer, and any essential inorganic reagents, outline steps for the synthesis of the following compounds:

a. *n*-butylamine
c. 1-aminopentane
e. *m*-aminobenzoic acid
g. 1,4-diaminobutane

b. *p*-toluidine
d. *N*-ethylaniline
f. tri-*n*-butylamine
h. ethyl cyclohexanecarboxylate

16. Show how each of the following compounds can be prepared using a Diels-Alder reaction:

a.

b.

c.

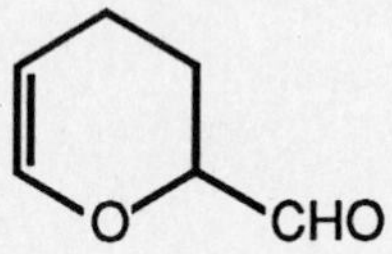

d.

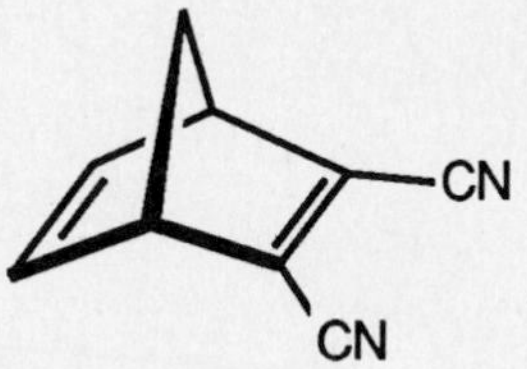

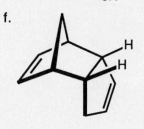

e.

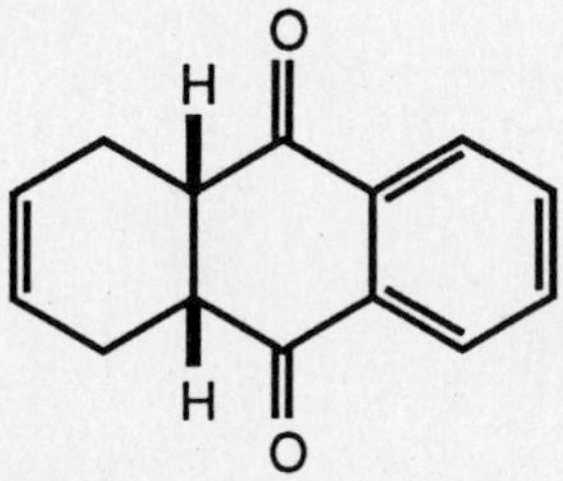

f.

17. Plan a synthesis of each of the following compounds, using starting materials with five or fewer carbon atoms:

a. 1-hexanol
b. 2-hexanol
c. 3-hexanol
d. 1-heptanol
e. ethylcyclopentane
f. 1-ethynylcyclopentene
g. *cis*-3-hexene
h. 2-ethyl-2-hexenal

18. Show how you could accomplish the following conversions:

a. 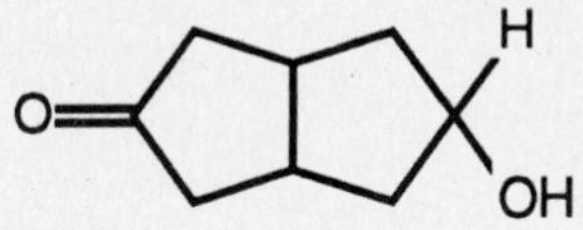   O= ... H, OH   to   CH₃, HO ...

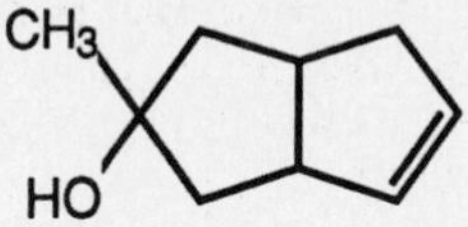

b. 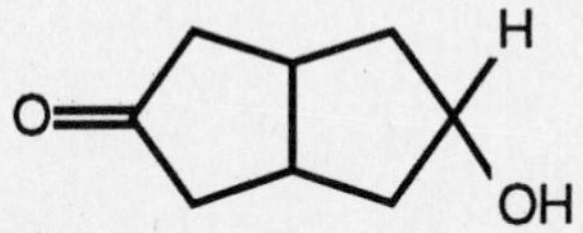   O= ... H, OH   to   O= ... CH₃, OH

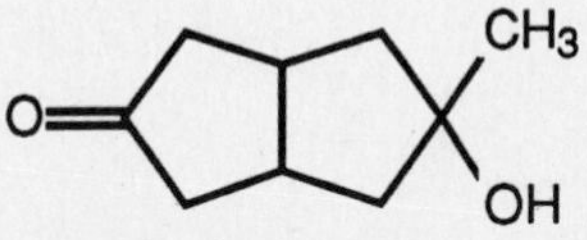

19. Show how each of the following compounds could be made starting with benzene:

REVIEW PROBLEMS ON SYNTHESIS

a.

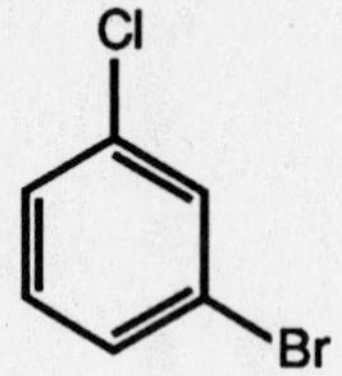

b.

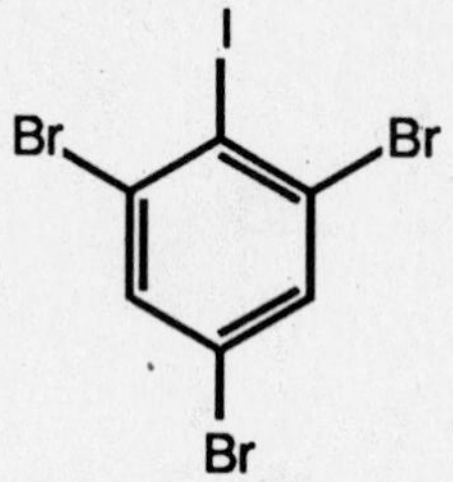

c.

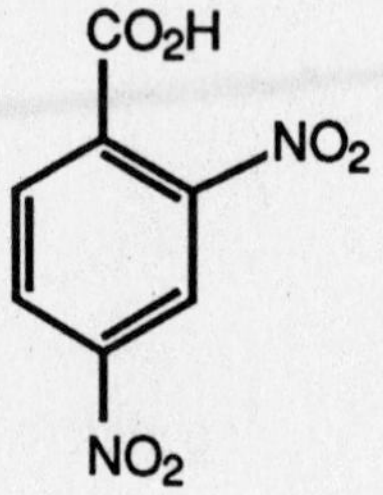

d.

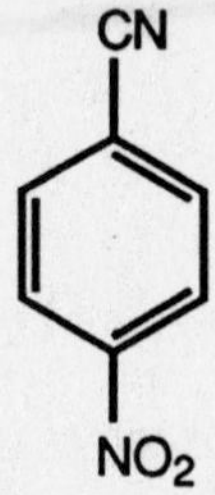

20. Give the structure of the expected product when each of the following is treated with strong aqueous base:

a. CH_3CH_2O— —CHO + CH_3CHO

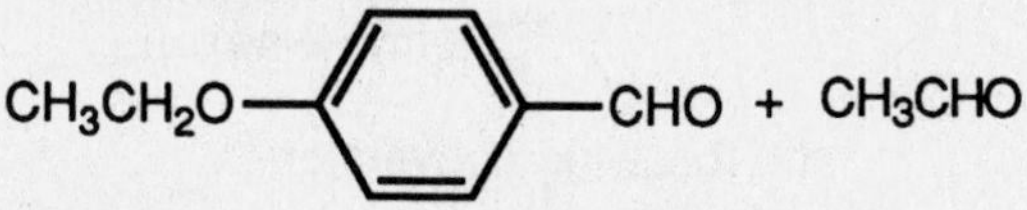

b. + CH_3CH_2CHO

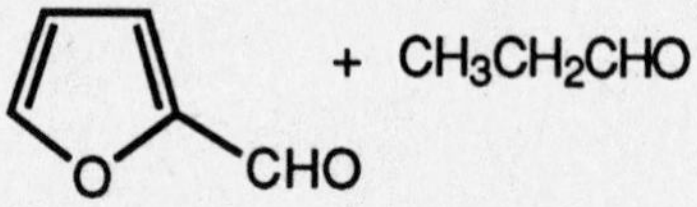

c. + $(CH_3)_3CCHO$

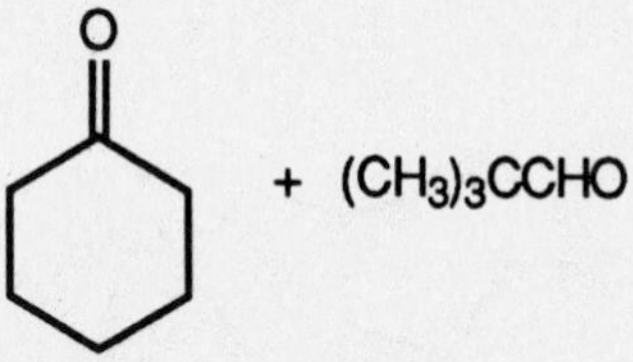

d.

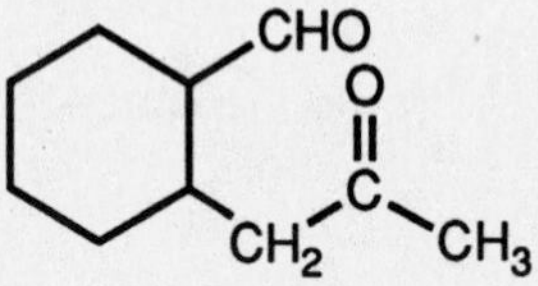

21. Starting with 4-methylpyridine, write equations for the preparation of:

    a.   2-amino-4-methylpyridine
    b.   3-amino-4-methylpyridine
    c.   pyridine-4-carboxylic acid
    d.   4-methylpiperidine

22. Plan a synthesis of the following polymers from appropriate monomers:

    a. $\left(CH_2CH\right)_n$ with CN substituent

    b. $\left(CH_2CH\right)_n$ with OH substituent

    c. $\left(OCH_2CH\right)_n$ with CH_3 substituent

    d. $\left(NHCH_2CH_2CH_2CH_2CH_2\overset{\displaystyle O}{\overset{\|}{C}}\right)_n$

    e. $\left(OCH_2CH_2O\overset{\displaystyle O}{\overset{\|}{C}}CH_2CH_2CH_2CH_2\overset{\displaystyle O}{\overset{\|}{C}}\right)_n$

    f. 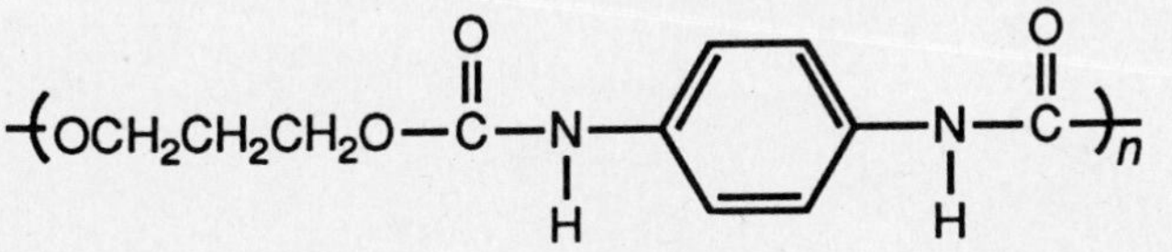 $\left(OCH_2CH_2CH_2O-\overset{\displaystyle O}{\overset{\|}{C}}-\underset{H}{N}-\text{C}_6\text{H}_4-\underset{H}{N}-\overset{\displaystyle O}{\overset{\|}{C}}\right)_n$

These questions, taken from mid-term and final examinations, are presented here for your practice. Although they do not cover all of the subject matter in the text, these questions provide a fair review of the material and may give you some idea of what to expect if (because of large classes) your instructor uses multiple-choice exams. Do not just guess at the correct answer. Take the time to write out the structures or equations before you make a choice.

1. Which of the following compounds may exist as *cis-trans* isomers?

    a. 1-butene   b. 2-butene   c. 2-pentene

    1. a   2. b   3. a and b   4. b and c   5. a and c

2. What is the molecular formula of the following compound?

    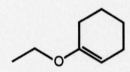

    1. $C_7H_{15}O$ 2. $C_8H_{13}O$ 3. $C_8H_{14}O$ 4. $C_8H_{15}O$ 5. $C_9H_{14}O$

3. How many σ (sigma) bonds are there in $CH_2=CH-CH=CH_2$?

    1. 3   2. 9   3. 10   4. 11   5. 12

4.   The preferred conformation of butane is:

1. 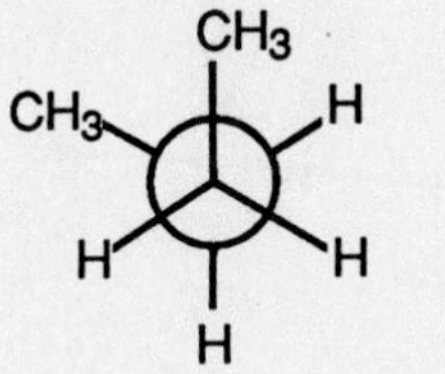   2. 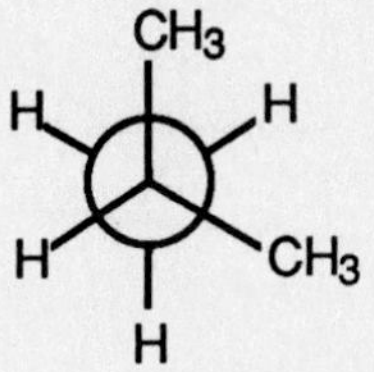   3.

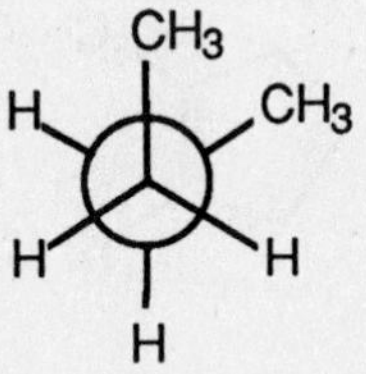

4. 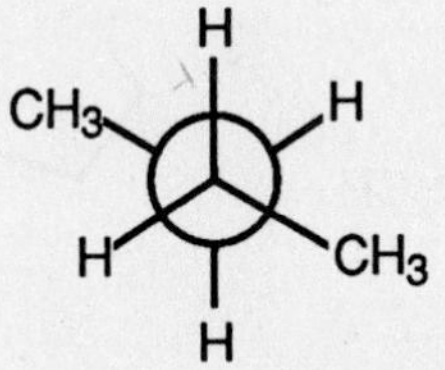   5. none of these

5.   The preferred conformation of cis-1,3-dimethylcyclohexane is:

1. 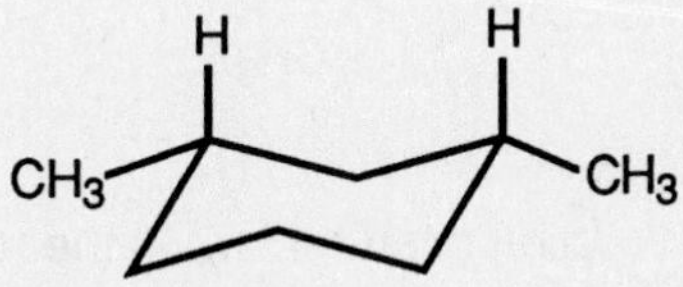   2.

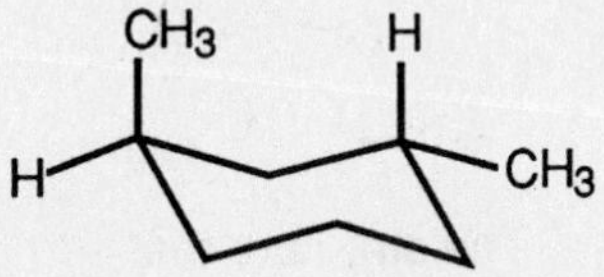

3. 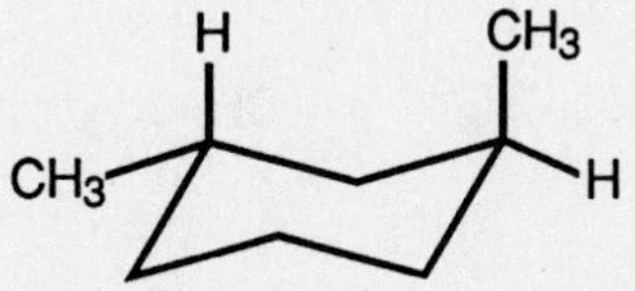   4.

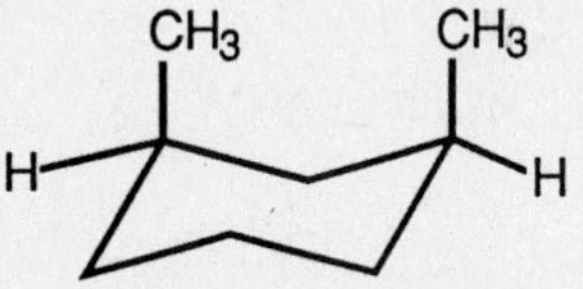

5. none of these

6. The major reaction product of

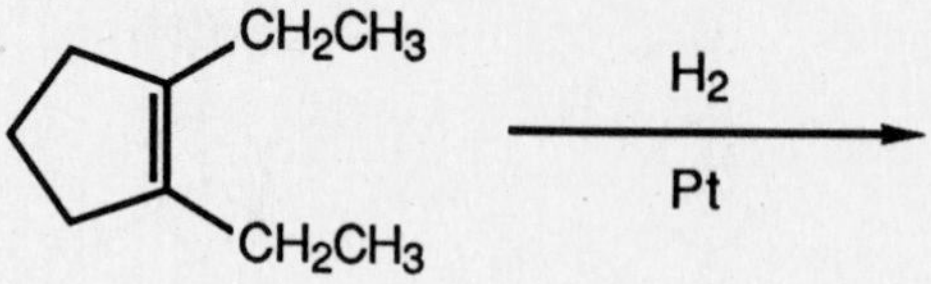

is:

1. 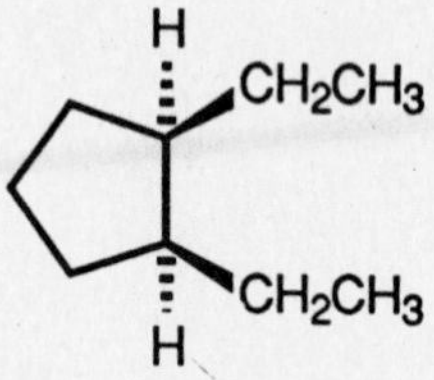
2. 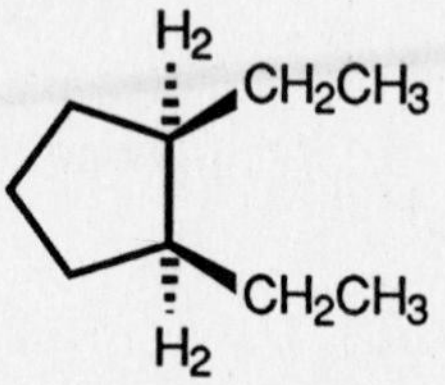
3.

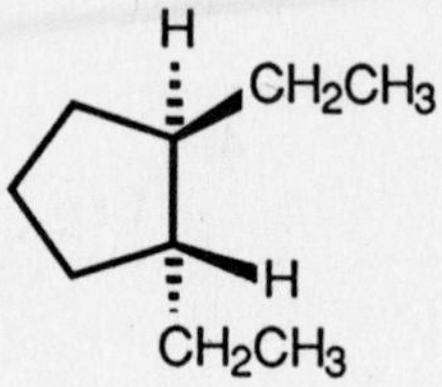

4. 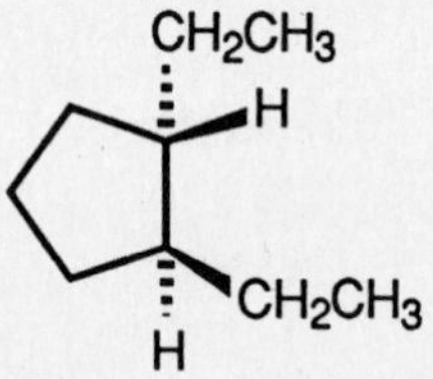
5.

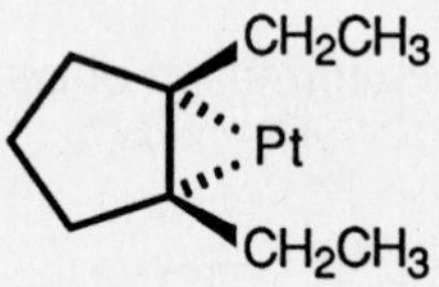

7. The structure that corresponds to a reaction intermediate in the reaction

$$CH_3—CH_3 + Br_2 \xrightarrow{heat} \text{is:}$$

1. $CH_3CH_2^+$ 2. $CH_3CH_2\bullet$ 3. $CH_3CH_2^-$ 4. $CH_2CH_3^+$

5. $\bullet CH_2CH_2\bullet$

8. The best way to prepare $BrCH_2CH_2Br$ is to start with:

1. $HC\equiv CH + Br_2$ (excess)   2. $CH_2=CH_2 + HBr$

3. $HC\equiv CH + HBr$ (excess)   4. $ClCH_2CH_2Cl + Br_2$

5. $CH_2=CH_2 + Br_2$

SAMPLE MULTIPLE-CHOICE QUESTIONS

9.   The IUPAC name (*E*)-2-methyl-3-hexene corresponds to:

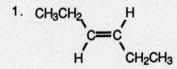

1. CH₃CH₂   H
         \\   /
          C=C
         /   \\
        H   CH₂CH₃

2. CH₃CH₂   CH(CH₃)₂
         \\   /
          C=C
         /   \\
        H   H

3. CH₃CH₂   H
         \\   /
          C=C
         /   \\
        H   CH(CH₃)₂

4. CH₃   CH₂CH₃
       \\   /
        C=C
       /   \\
     CH₃   H

5. CH₃   H
       \\   /
        C=C
       /   \\
     CH₃   CH₂CH₃

10.   The major product of the reaction

CH₃   CH₃
   \\   /   H₂O
    C=C   ──────────→   is:
   /   \\   H₂SO₄
  H   H

         OH OH
          | |
1. CH₃—C—C—CH₃
          | |
          H   H

         OH OSO₃H
          | |
2. CH₃—C—C—CH₃
          | |
          H   H

                 OSO₃H
                   |
3. CH₃-CH₂-C-CH₃
                   |
                   H

4. CH₃-CH₂-CH₂-CH₃

             OH
              |
5. CH₃-CH₂-C-CH₃
              |
              H

SAMPLE MULTIPLE-CHOICE QUESTIONS

11. The most likely structure for an unknown that yields only $CH_3CH_2CH=O$ on ozonolysis is:

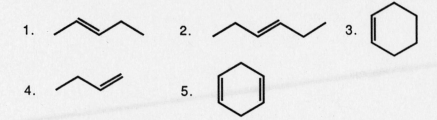

1.   2.   3.

4.   5.

12. Considering the hybridization of carbon orbitals, which of the following structures is *least* likely to exist?

1.    2.    3.    4. 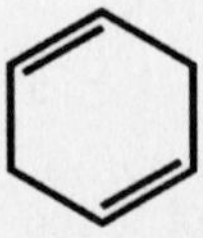   5.

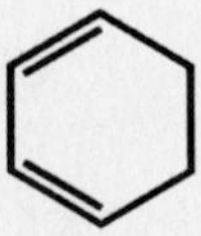

13. What is the correct name for the following structure?

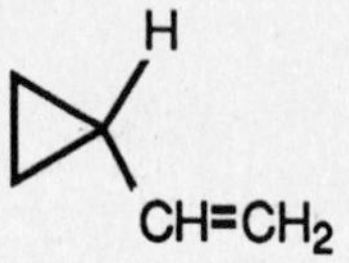

1. cyclopentene   2. cyclopropylpropene   3. vinylcyclopropane
4. vinylcyclobutane 5. 1-cyclopropyl-2-ene

14. H$^+$ often reacts as a(an):

1. electrophile   2. nucleophile   3. radical
4. electron   5. carbocation

15.   The major reaction product of

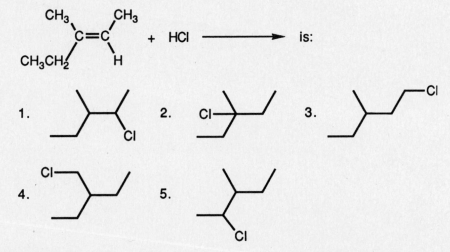

16.   Which of the following classes of compounds is unreactive toward sulfuric acid?

    a. alkanes   b. alcohols   c. alkenes   d. alkynes

    1. only a   b. only b   3. a and b   4. only d   5. a, b and d

17.   Which of the following structures represents a *trans*-dibromo compound?

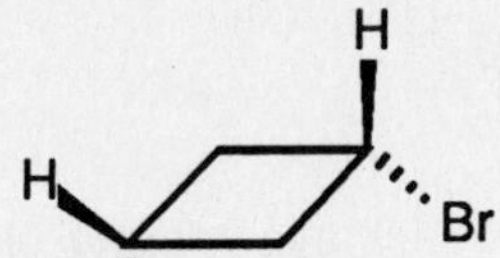

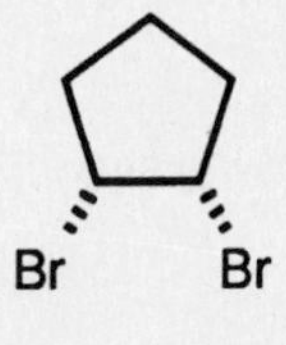

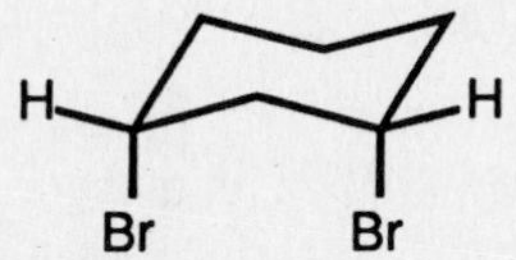

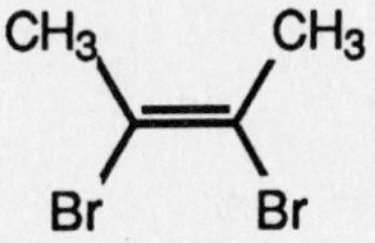

5. none of these

18. How many *mono* chlorination products are possible in the reaction of 2,2-dimethylbutane with Cl_2 and light?

    1. 2   2. 3   3. 4   4. 5   5. 6

19. Teflon is represented by the structure $—CF_2CF_2(CF_2CF_2)_nCF_2CF_2—$. Which of the following monomers is used to make teflon?

    1. $CF_2—CF=F$ 2. $CF_2=CF—CF=CF_2$ 3. $CF_2=CFCF_3$

    4. $CHF_2—CHF_2$ 5. $CF_2=CF_2$

20. Which of the following compounds contain one or more polar covalent bonds?

    a. CH_3CH_2OH b. $CH_3CH_2CH_3$ c. $CHCl=CHCl$ d. Cl_2

    1. a and b   2. a and c   3. b and c   4. only d   5. a and d

21. What is the relationship between

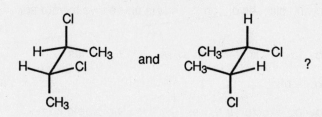

    1. structural isomers   2. geometric isomers
    3. conformational isomers   4. identical
    5. none of these

22. In which of the following structures does nitrogen have a formal charge of +1?

    a. 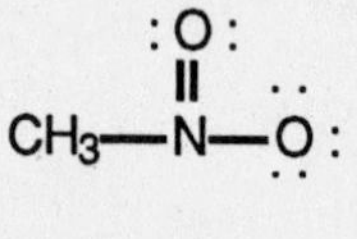 $CH_3—\overset{\overset{\displaystyle :O:}{\|}}{N}—\overset{..}{\underset{..}{O}}:$ b. 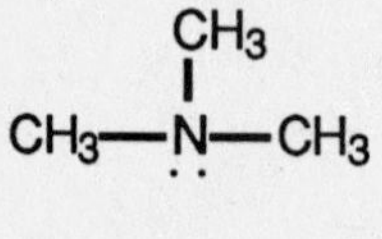 $CH_3—\overset{\overset{\displaystyle CH_3}{|}}{\underset{..}{N}}—CH_3$ c. NH_4^+

    1. a and c   2. a and b   3. b and c   4. a, b, and c 5. only a

23. Which of the following statements is(are) true of $CH_3CH_2CH_3$?

    a. all bond angles are about 109.5°
    b. each carbon is sp^3-hybridized
    c. the compound is combustible

    1. only a   2. only b   3. a and b   4. a and c   5. a, b, and c

24. Which reagent can accomplish this conversion?

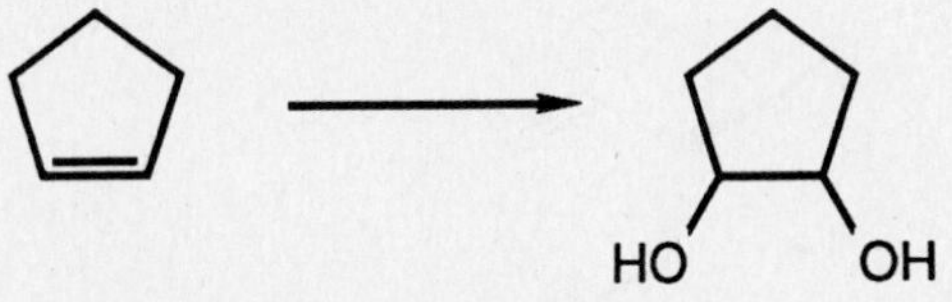

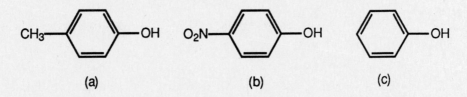

    1. $KMnO_4$ 2. H_2O, H^+ 3. O_3, then Zn, H^+ 4. OH^-
    5. H_2O, heat

25. How many elimination products are possible for the following reaction?

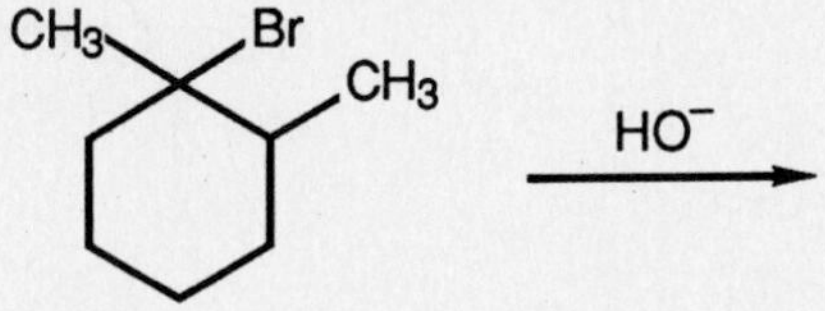

    1. 0   2. 1   3. 2   4. 3   5. 4

26. Arrange the following compounds in order of increasing acidity:

    (a)   (b)   (c)

    1. a < b < c   2. c < b < a   3. b < a < c   4. a < c < b   5. b < c < a

SAMPLE MULTIPLE-CHOICE QUESTIONS

27. Which structure corresponds to *R*-1-chloroethylbenzene?

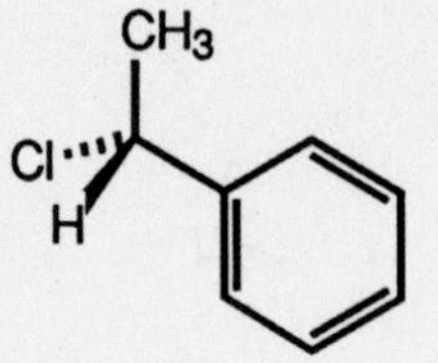

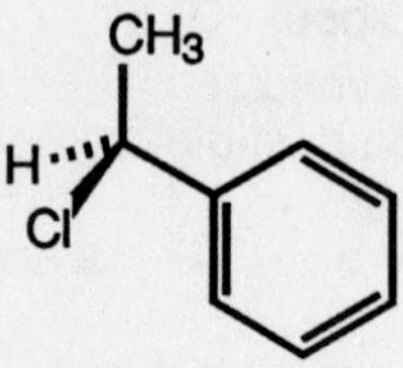

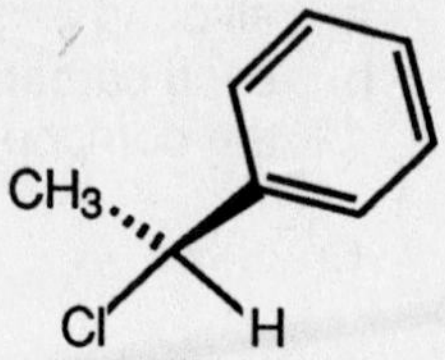

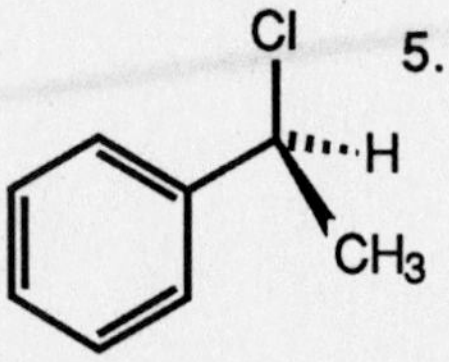

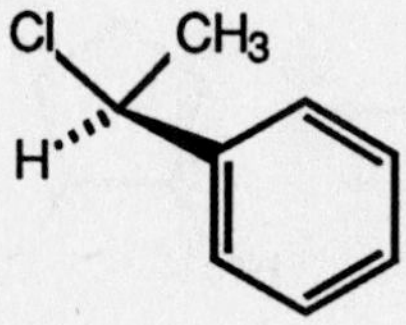

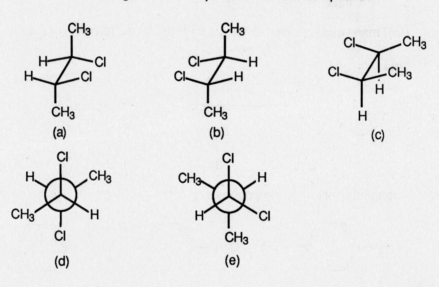

28. Which of the following structures represents a *meso* compound?

1. only e   2. a and b   3. c and d   4. d and e   5. a, b, and c

SAMPLE MULTIPLE-CHOICE QUESTIONS

29. What is the correct name for this compound?

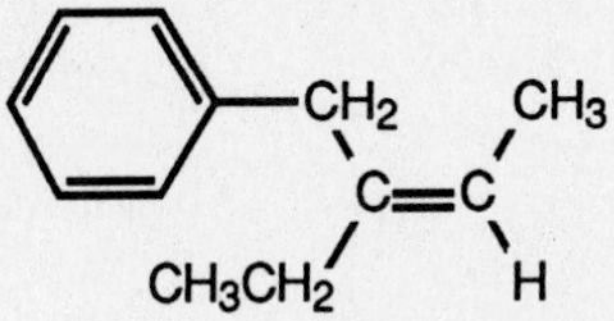

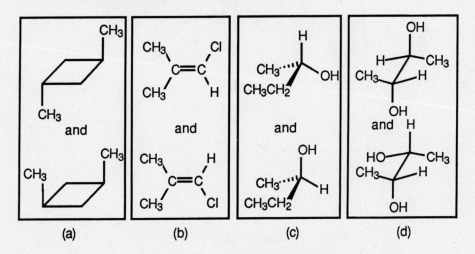

1. *cis*-3-phenyl-2-pentene
2. *trans*-3-phenyl-2-pentene
3. (*Z*)-3-benzyl-2-pentene
4. (*E*)-3-benzyl-2-pentene
5. (*Z*)-3-benzyl-3-pentene

30. Which of the following pairs of compounds can be separated by physical methods (for example, crystallization or distillation)?

(a)   (b)   (c)   (d)

1. all of them   2. a and d   3. b and c   4. only   5. only c

SAMPLE MULTIPLE-CHOICE QUESTIONS

31.   To which Fischer projection formula does the following drawing correspond?

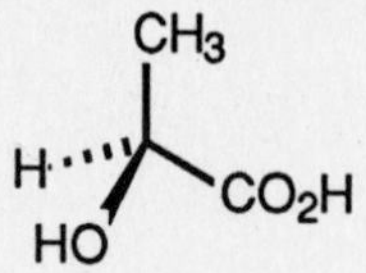

1.

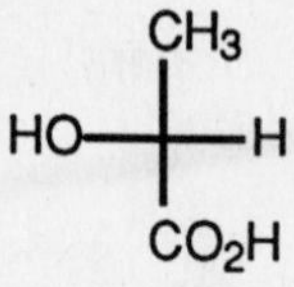

2.

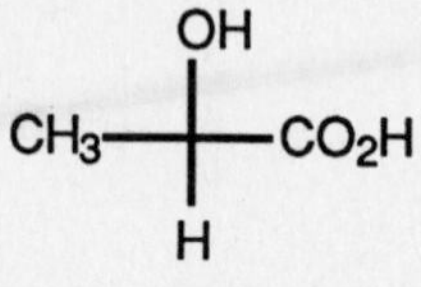

3.

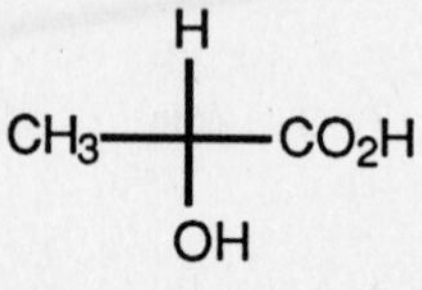

4.

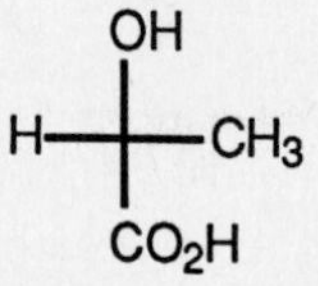

5.

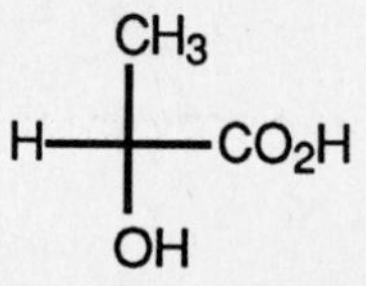

32.   What is the correct name for the following compound?

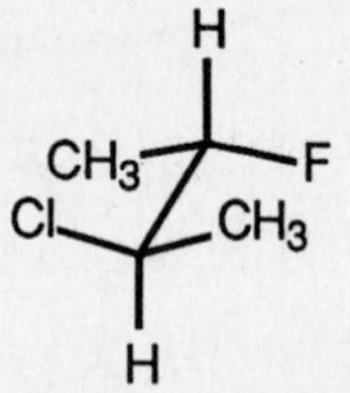

1. (2R,3R)-2-chloro-3-fluorobutane
2. (2R,3S)-2-chloro-3-fluorobutane
3. (2S,3R)-2-chloro-3-fluorobutane
4. (2S,3S)-2-chloro-3-fluorobutane
5. none of the above

33. Which of the following compounds reacts least rapidly in electrophilic substitutions?

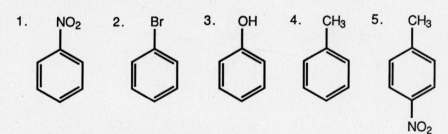

34. Which of the following compounds gives off the *least* amount of heat during hydrogenation?

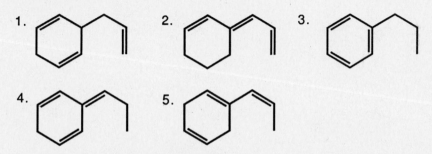

35. The resonance contributor that is most important in the intermediate for *para* electrophilic substitution in toluene is:

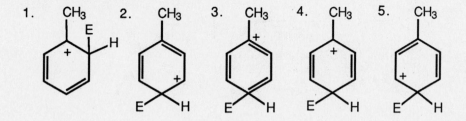

SAMPLE MULTIPLE-CHOICE QUESTIONS

36. How are the following structures related?

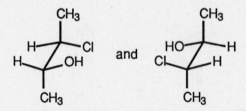

1. enantiomers   2. diastereomers   3. *meso* compounds
4. identical   5. achiral compounds

37. Which reagent would be most useful for carrying out the following transformation?

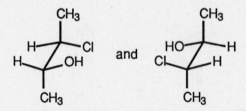

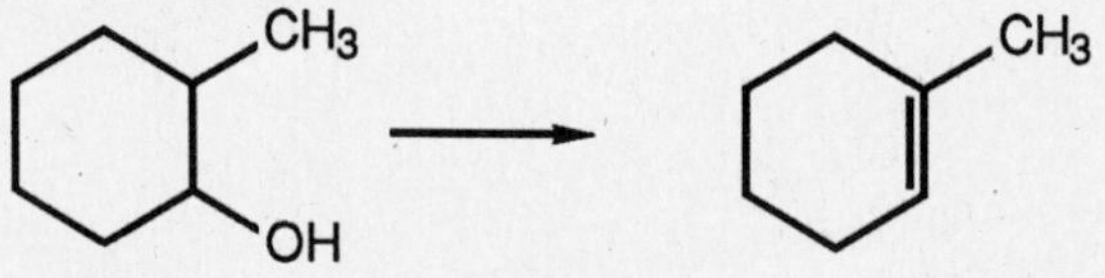

1. LiAlH$_4$ 2. conc. H$_2$SO$_4$ 3. H$_2$/Pd
4. CrO$_3$ 5. NaOH

38. By which of the indicated mechanisms would the following reaction proceed?

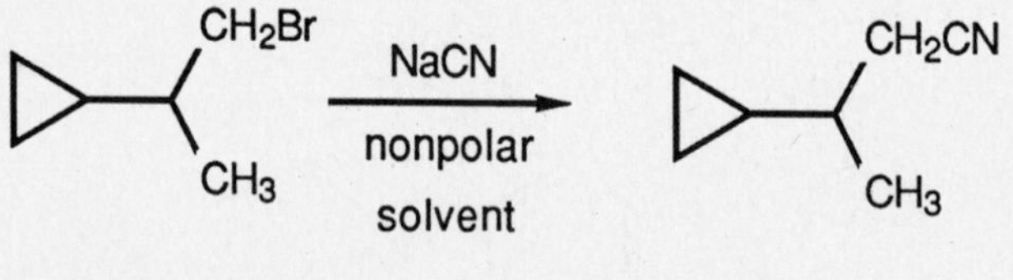

1. S$_N$1   2. S$_N$2   3. E1   4. E2
5. free-radical

39. What is the most likely product of the following reaction?

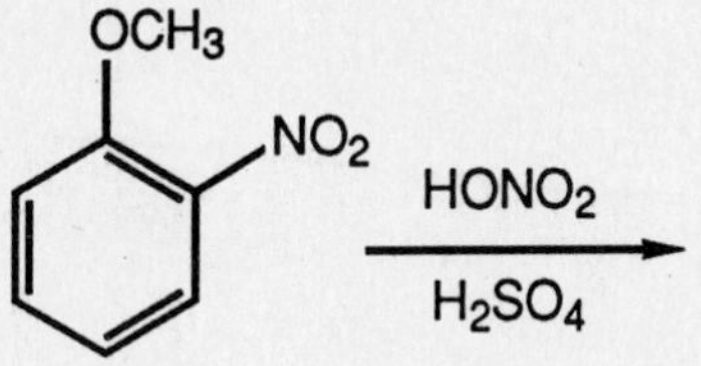

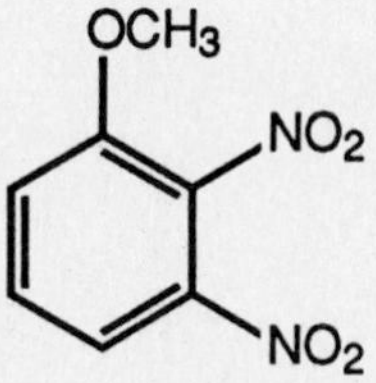

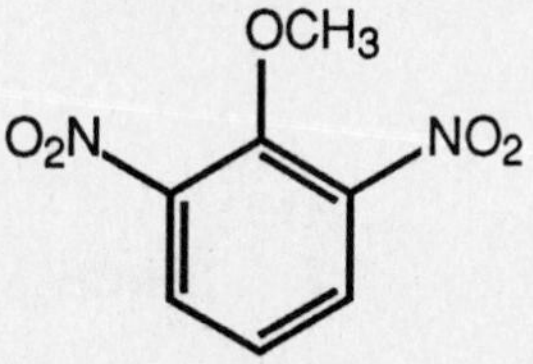

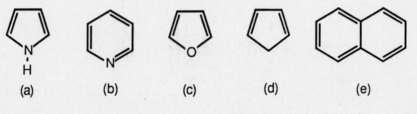

40. Which of the following compound(s) is(are) not aromatic?

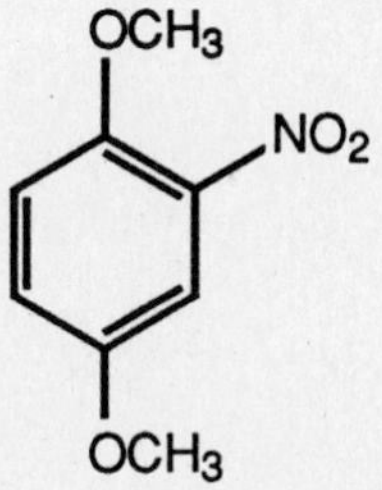

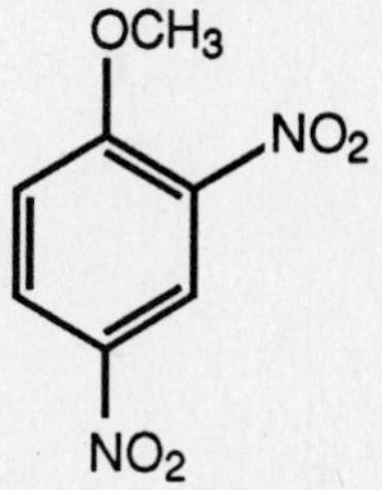

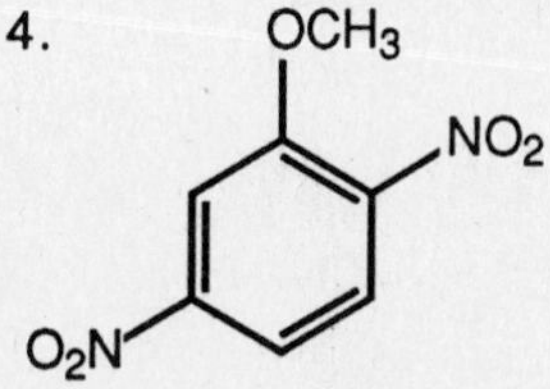

(a)   (b)   (c)   (d)   (e)

1. only d   2. only e   3. d and e   4. a and c   5. b and d

41.   What are the appropriate reagents to accomplish the following transformations?

$$CH_3CH_2OCH_2CH_3 \xrightarrow{A} CH_3CH_2Br \xrightarrow{B} CH_3CH_3$$

1. Br_2 2. HBr   3. Na   4. conc. H_2SO_4
5. Mg, then   6. CrO_3
   H_2O

A = _____   B = _____

42.   Which of the following compounds would react most rapidly in an S_N2 reaction?

1. 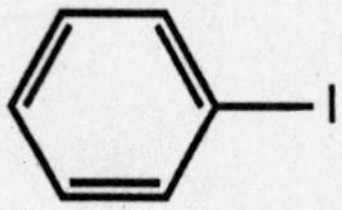 [benzene ring with —I]   2. CH_3CH_2I 3. $H_2C=CH—I$

4. $(CH_3)_2CH—I$ 5. $(CH_3)_3C—I$

43.   Which of the following compounds is a secondary alcohol?

1. 3-hexanol   2. 1-hexanol   3. 2-methyl-2-hexanol
4. 1-ethyl-1-cyclohexanol   5. cyclohexylmethanol

44.   The conversion $CH_3CO_2CH_3 \rightarrow CH_3CH_2OH$ can be called:

1. oxidation   2. reduction   3. hydrolysis
4. saponification   5. dehydration

45. Which of the following compounds can be used as a detergent?

1. NaOH   2.

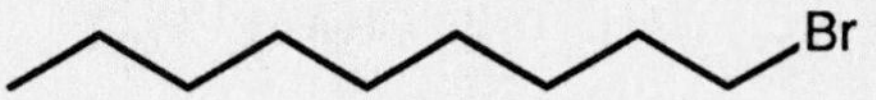

3.

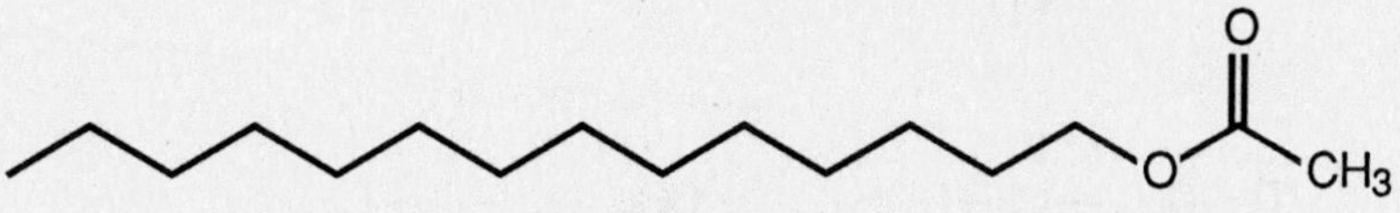

4.

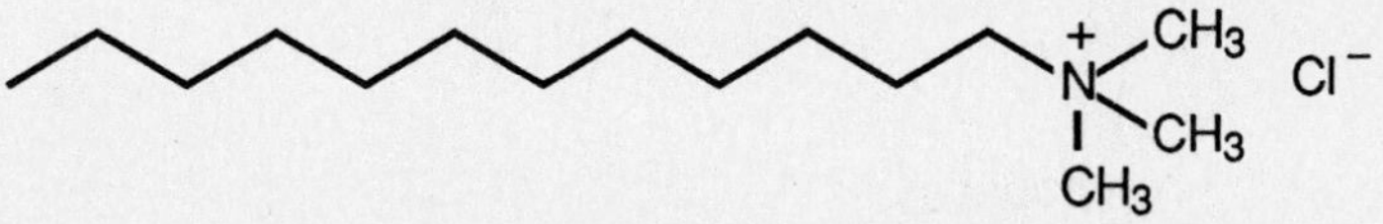

5.

46. Which of the following compounds is a tautomer of phenol?

1. 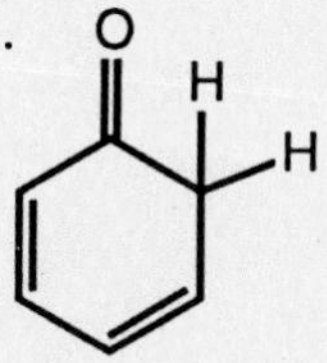   2. 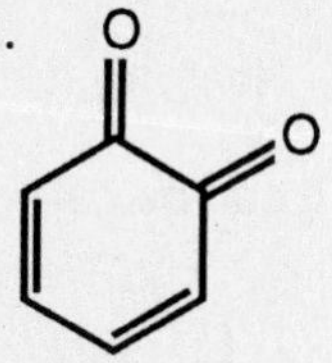   3.

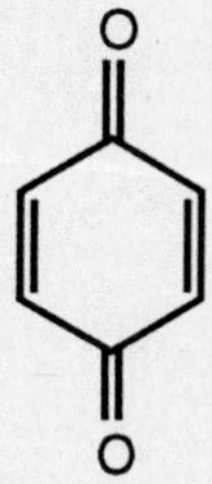

4. 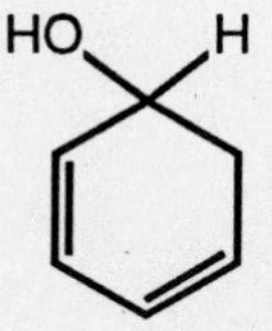   5.

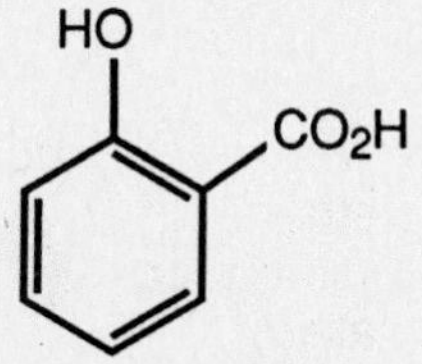

47. Which compound is the major product of the following reaction?

$$CH_3CH_2CH{=}O \ + \ CH_3MgBr \xrightarrow{\quad} \xrightarrow[H^+]{H_2O}$$

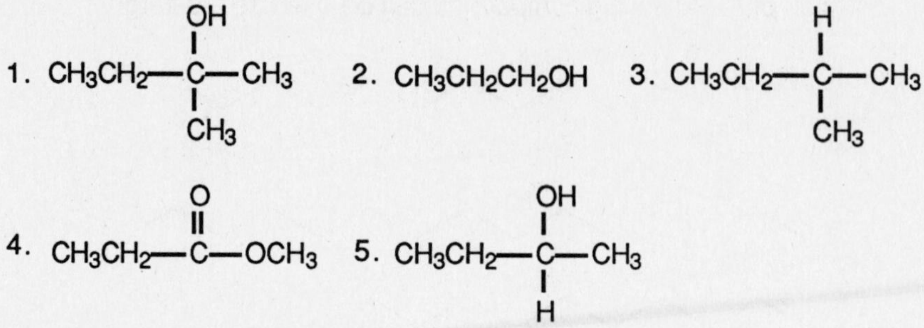

1. $CH_3CH_2-\overset{\displaystyle OH}{\underset{\displaystyle CH_3}{\overset{|}{\underset{|}{C}}}}-CH_3$ 2. $CH_3CH_2CH_2OH$ 3. $CH_3CH_2-\overset{\displaystyle H}{\underset{\displaystyle CH_3}{\overset{|}{\underset{|}{C}}}}-CH_3$

4. $CH_3CH_2-\overset{\displaystyle O}{\overset{\|}{C}}-OCH_3$ 5. $CH_3CH_2-\overset{\displaystyle OH}{\underset{\displaystyle H}{\overset{|}{\underset{|}{C}}}}-CH_3$

48.   The following compound is a:

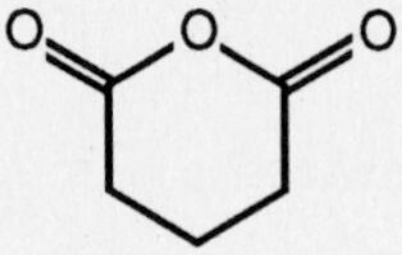

1. diketone   2. keto ether   3. lactone   4. anhydride   5. ketal

49.   Nylon has the following structure. It can be called a:

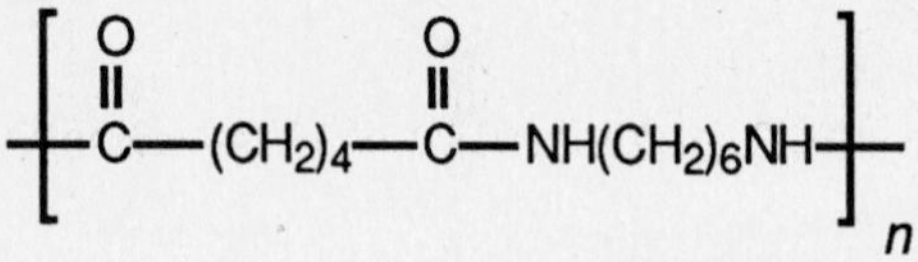

1. polyamide   2. polyamine   3. polyester
4. vinyl polymer   5. polyurethane

50.   The following reaction can be called:

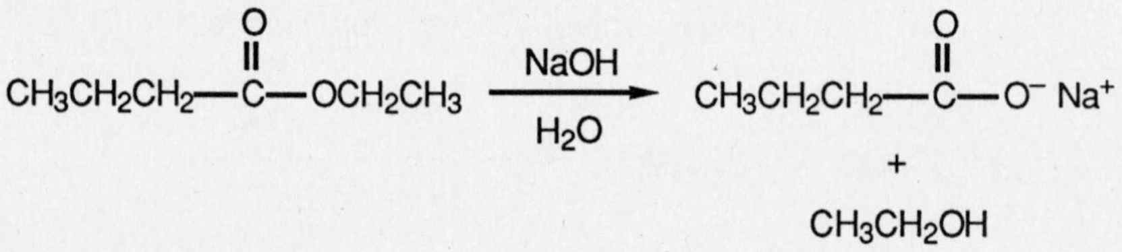

$$CH_3CH_2CH_2-\overset{O}{\overset{\|}{C}}-OCH_2CH_3 \quad \xrightarrow[\text{H}_2\text{O}]{\text{NaOH}} \quad CH_3CH_2CH_2-\overset{O}{\overset{\|}{C}}-O^- \, Na^+$$

$$+$$

$$CH_3CH_2OH$$

1. enolization   2. elimination   3. condensation
4. esterification   5. saponification

SAMPLE MULTIPLE-CHOICE QUESTIONS

51. The enol of 2-butanone is:

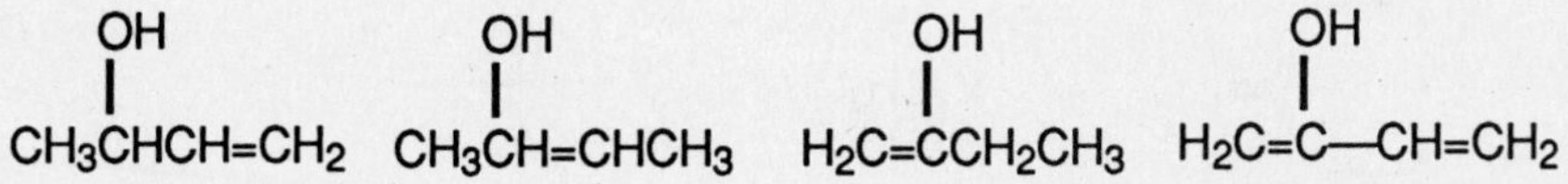

$$\underset{\text{(a)}}{\overset{\overset{\displaystyle OH}{|}}{CH_3CHCH=CH_2}} \qquad \underset{\text{(b)}}{\overset{\overset{\displaystyle OH}{|}}{CH_3CH=CHCH_3}} \qquad \underset{\text{(c)}}{\overset{\overset{\displaystyle OH}{|}}{H_2C=CCH_2CH_3}} \qquad \underset{\text{(d)}}{\overset{\overset{\displaystyle OH}{|}}{H_2C=C-CH=CH_2}}$$

1. only a   2. a and b   3. b and c   4. a and d   5. a, b and c

52. What is the major product of the following reaction?

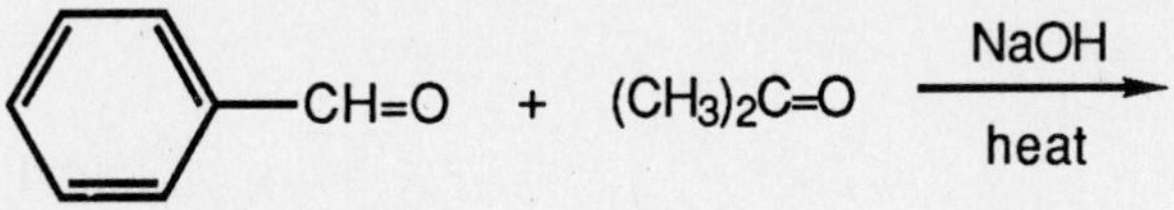

$$\text{C}_6\text{H}_5-CH=O \;+\; (CH_3)_2C=O \;\xrightarrow[\text{heat}]{\text{NaOH}}$$

1. 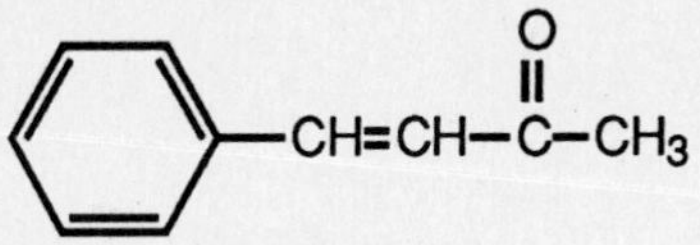 $C_6H_5-CH=CH-\overset{\overset{\displaystyle O}{\|}}{C}-CH_3$

2. 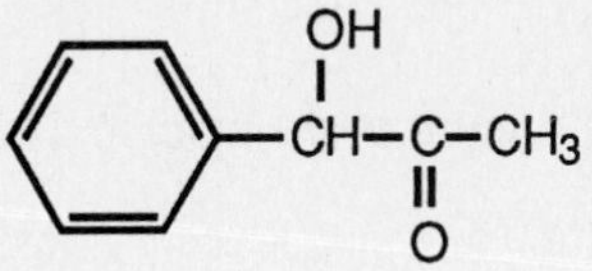 $C_6H_5-\overset{\overset{\displaystyle OH}{|}}{CH}-\underset{\underset{\displaystyle O}{\|}}{C}-CH_3$

3. $(CH_3)_2\overset{\overset{\displaystyle OH}{|}}{C}CH_2\overset{\overset{\displaystyle O}{\|}}{C}CH_3$

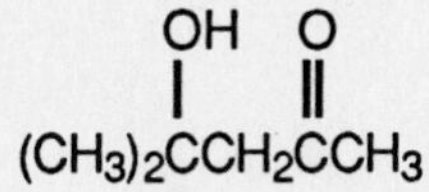

4. 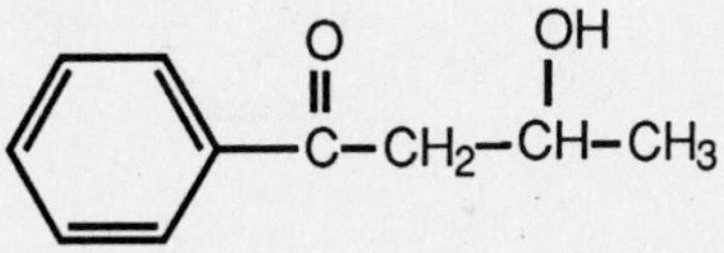 $C_6H_5-\overset{\overset{\displaystyle O}{\|}}{C}-CH_2-\overset{\overset{\displaystyle OH}{|}}{CH}-CH_3$

5. $C_6H_5-\overset{\overset{\displaystyle O}{\|}}{C}-\underset{\underset{\displaystyle OH}{|}}{\overset{\overset{\displaystyle CH_3}{|}}{C}}-CH_3$

517

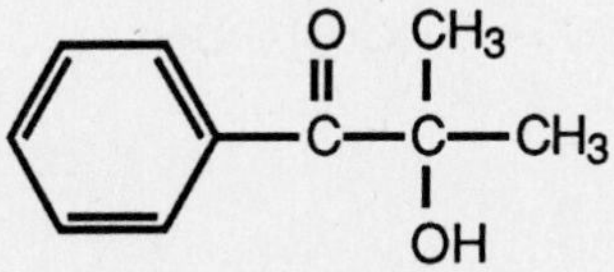

53.   What is the major product of the following reaction?

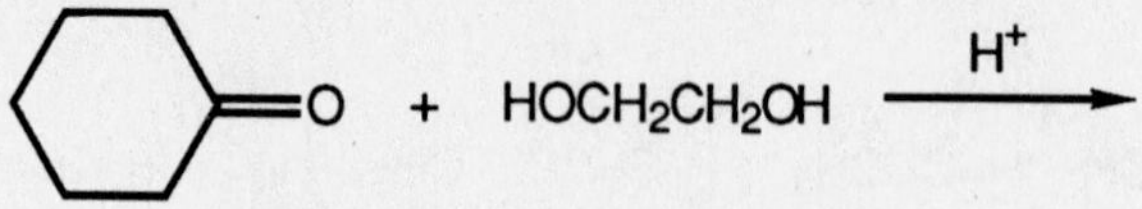

1. 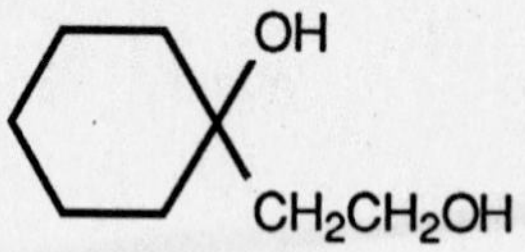
2. a polyester   3.

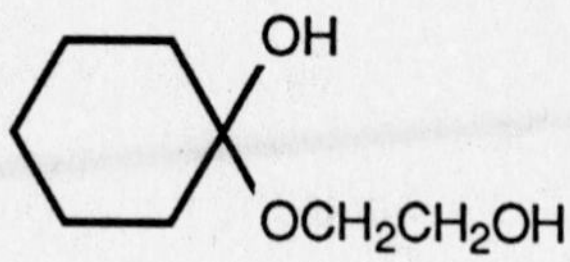

4. 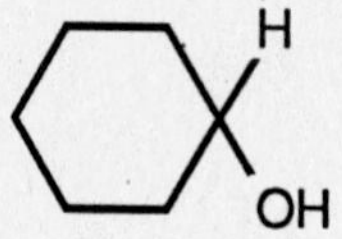
5.

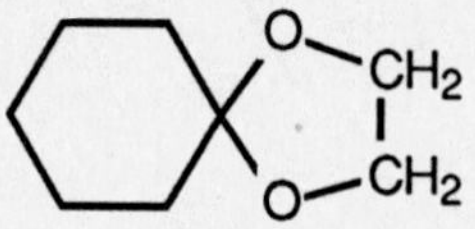

54.   The structure of malonic acid is:

1. 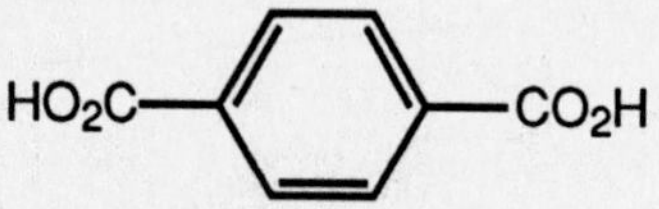 HO_2C—(benzene ring)—CO_2H

2. $HO—\overset{\overset{O}{\|}}{C}—\overset{\overset{O}{\|}}{C}—OH$

3. $HO—\overset{\overset{O}{\|}}{C}-CH_2-\overset{\overset{O}{\|}}{C}—OH$

4. $HO—\overset{\overset{O}{\|}}{C}-(CH_2)_4-\overset{\overset{O}{\|}}{C}—OH$

5. $CH_3—\overset{\overset{O}{\|}}{C}—\overset{\overset{O}{\|}}{C}—OH$

55. Which hybrid structure represents the most stable enolate anion formed in the following reaction?

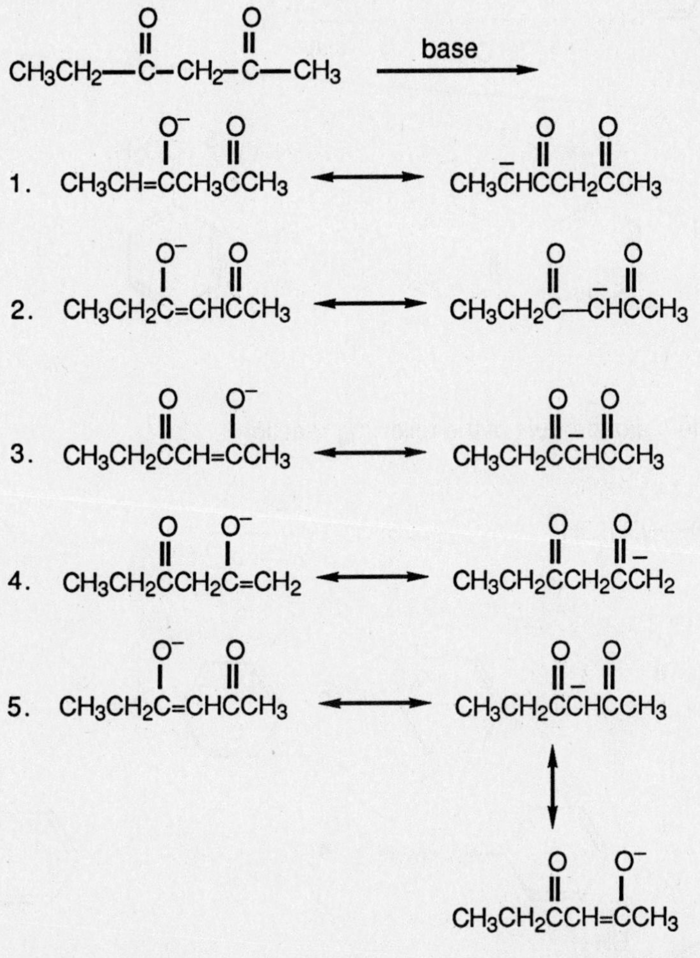

$$CH_3CH_2-\overset{\overset{O}{\|}}{C}-CH_2-\overset{\overset{O}{\|}}{C}-CH_3 \xrightarrow{\text{base}}$$

1. $CH_3CH=\overset{\overset{O^-}{|}}{C}CH_3\overset{\overset{O}{\|}}{C}CH_3 \longleftrightarrow CH_3\overset{-}{C}H\overset{\overset{O}{\|}}{C}CH_2\overset{\overset{O}{\|}}{C}CH_3$

2. $CH_3CH_2\overset{\overset{O^-}{|}}{C}=CH\overset{\overset{O}{\|}}{C}CH_3 \longleftrightarrow CH_3CH_2\overset{\overset{O}{\|}}{C}-\overset{-}{C}H\overset{\overset{O}{\|}}{C}CH_3$

3. $CH_3CH_2\overset{\overset{O}{\|}}{C}CH=\overset{\overset{O^-}{|}}{C}CH_3 \longleftrightarrow CH_3CH_2\overset{\overset{O}{\|}}{C}\overset{-}{C}H\overset{\overset{O}{\|}}{C}CH_3$

4. $CH_3CH_2\overset{\overset{O}{\|}}{C}CH_2\overset{\overset{O^-}{|}}{C}=CH_2 \longleftrightarrow CH_3CH_2\overset{\overset{O}{\|}}{C}CH_2\overset{\overset{O}{\|}}{C}\overset{-}{C}H_2$

5. $CH_3CH_2\overset{\overset{O^-}{|}}{C}=CH\overset{\overset{O}{\|}}{C}CH_3 \longleftrightarrow CH_3CH_2\overset{\overset{O}{\|}}{C}\overset{-}{C}H\overset{\overset{O}{\|}}{C}CH_3$

$\updownarrow$

$CH_3CH_2\overset{\overset{O}{\|}}{C}CH=\overset{\overset{O^-}{|}}{C}CH_3$

56. Arrange the following compounds in order of decreasing basicity:

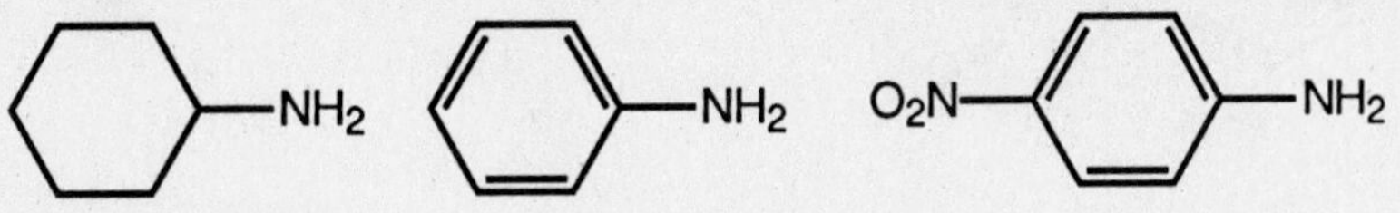

(a)   (b)   (c)

1. a > b > c   2. c > b > a   3. b > a > c   4. c > a > b   5. a > c > b

SAMPLE MULTIPLE-CHOICE QUESTIONS

57.   What is the major product of the following reaction?

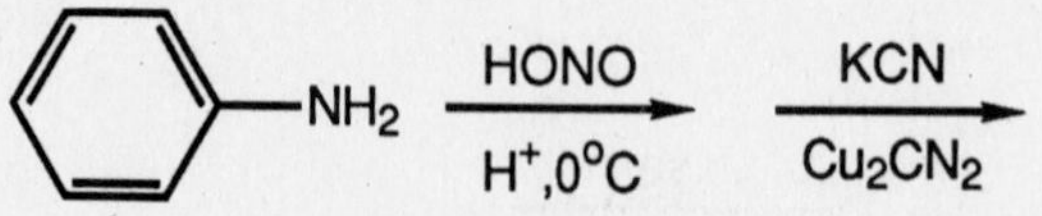

1. 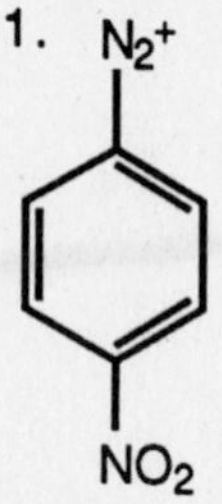 N_2^+ 2. 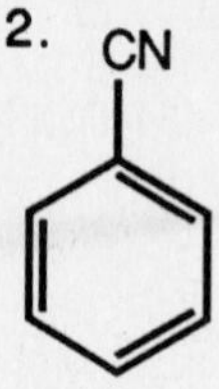 CN   3. 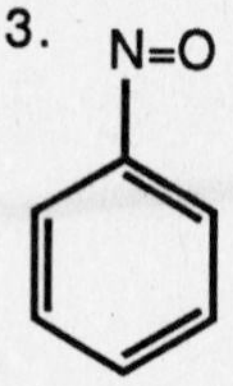 N=O   4. 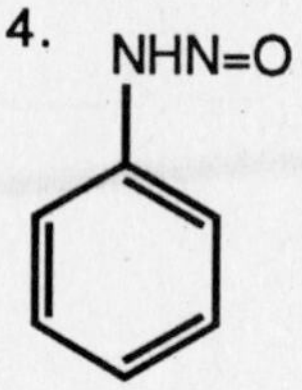 NHN=O   5. 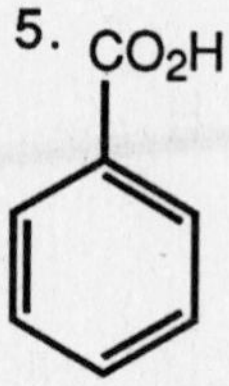 CO_2H

NO_2

58.   What is the major product of the following reaction?

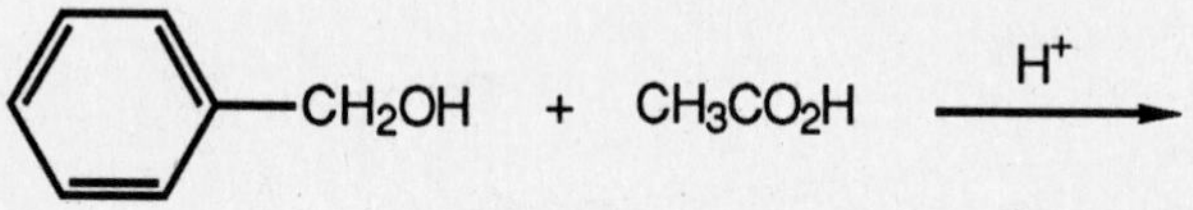

$\text{—CH}_2\text{OH}$ + CH_3CO_2H $\xrightarrow{H^+}$

1. 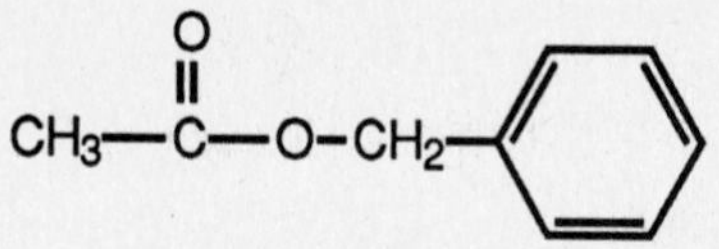 $CH_3\text{—}\overset{O}{\overset{\|}{C}}\text{—O–}CH_2\text{—}$ 2. $\text{—}CH_2\text{–}\overset{O}{\overset{\|}{C}}\text{—}OCH_3$

3. $CH_3\text{—}\overset{O}{\overset{\|}{C}}\text{—}\text{—}CH_2OH$ 4. 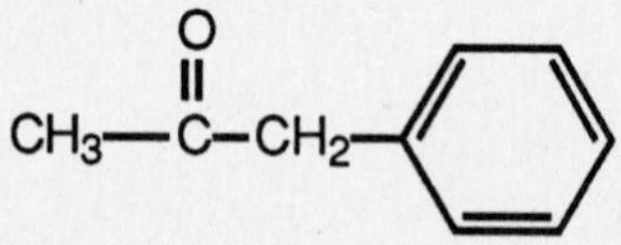 $CH_3\text{—}\overset{O}{\overset{\|}{C}}\text{–}CH_2\text{—}$

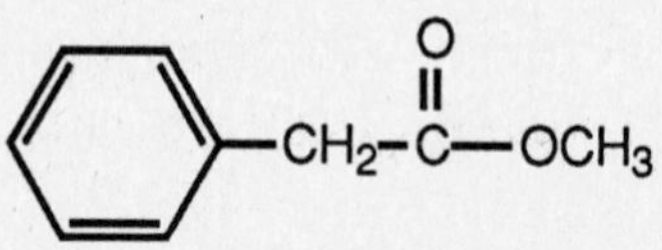

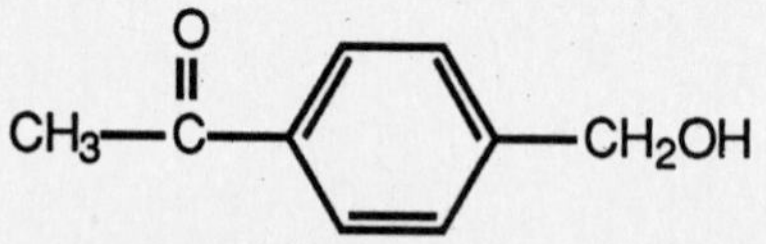

5.

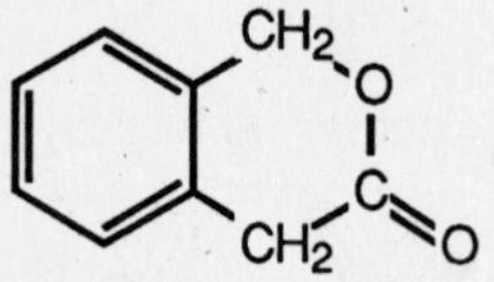

59. What is the major product of the following reaction?

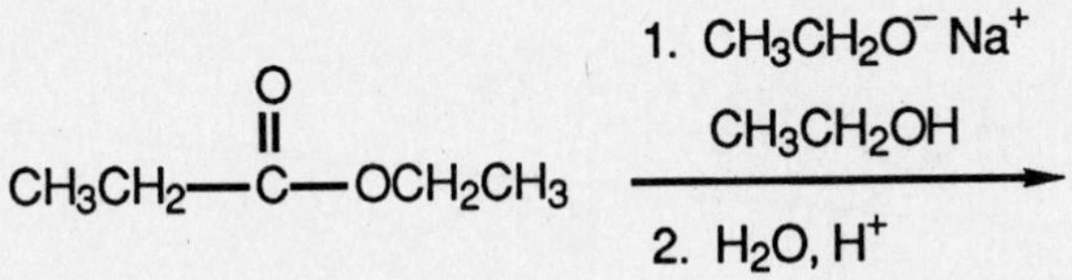

$$CH_3CH_2—\overset{\overset{\displaystyle O}{\|}}{C}—OCH_2CH_3 \quad \xrightarrow[\text{2. } H_2O,\ H^+]{\begin{array}{c}\text{1. } CH_3CH_2O^-\ Na^+ \\ CH_3CH_2OH \end{array}}$$

1. $CH_3CH_2\overset{\overset{\displaystyle O}{\|}}{C}—O—\overset{\overset{\displaystyle O}{\|}}{C}CH_2CH_3$ 2. $CH_3CH_2\overset{\overset{\displaystyle O}{\|}}{C}—CH_2CH_2—\overset{\overset{\displaystyle O}{\|}}{C}OCH_2CH_3$

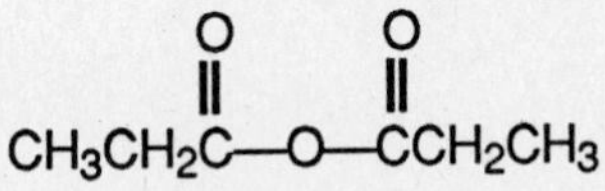

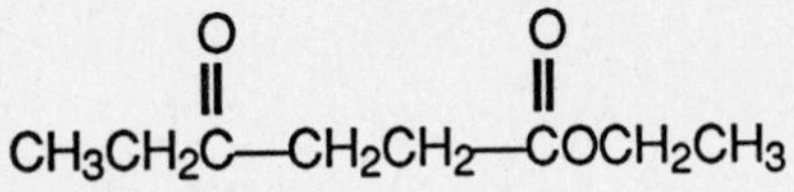

3. $CH_3CH_2\overset{\overset{\displaystyle OH}{|}}{C}H\overset{\overset{\displaystyle O}{\|}}{C}H\overset{\displaystyle C}{}OCH_2CH_3$ 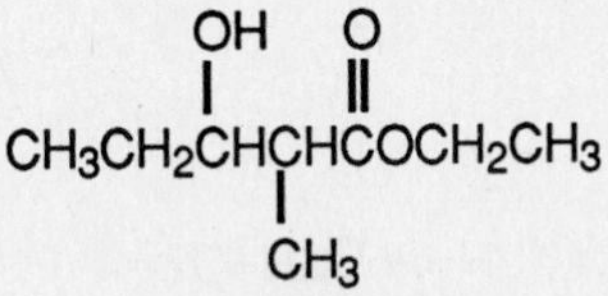   4. 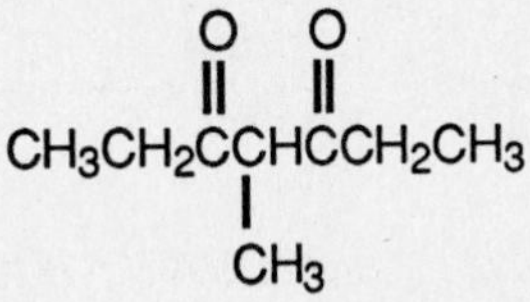 $CH_3CH_2\overset{\overset{\displaystyle O}{\|}}{C}\overset{\overset{\displaystyle O}{\|}}{C}HC\overset{\displaystyle }{}CH_2CH_3$
          $\overset{\displaystyle }{\underset{\displaystyle CH_3}{|}}$ $\overset{\displaystyle }{\underset{\displaystyle CH_3}{|}}$

5. 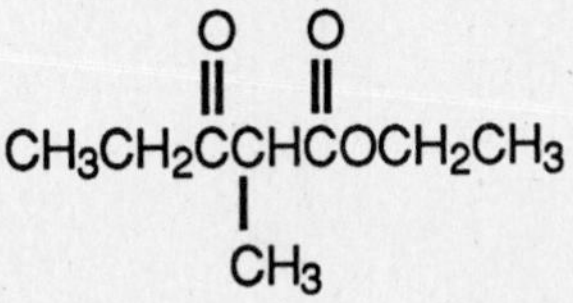 $CH_3CH_2\overset{\overset{\displaystyle O}{\|}}{C}\overset{\overset{\displaystyle O}{\|}}{C}HC\overset{\displaystyle }{}OCH_2CH_3$
          $\overset{\displaystyle }{\underset{\displaystyle CH_3}{|}}$

60. Which of the following compounds is a secondary amine?

1. 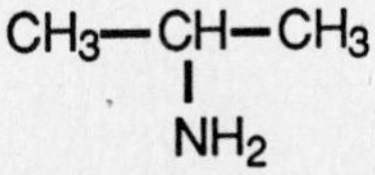 $CH_3—\overset{\overset{\displaystyle }{}}{\underset{\overset{\displaystyle |}{NH_2}}{C}}H—CH_3$ 2. $CH_3CH_2NHCH_3$ 3. $CH_3—\overset{\overset{\displaystyle O}{\|}}{C}—\overset{\overset{\displaystyle }{}}{\underset{\overset{\displaystyle |}{CH_3}}{N}}—CH_3$

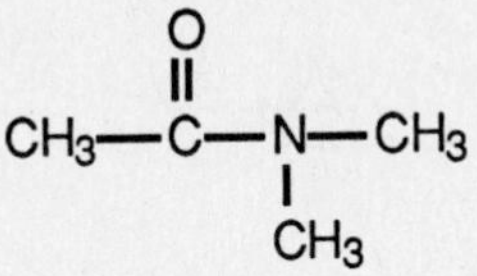

4. $H_2NCH_2CH_2NH_2$ 5. $(CH_3)_3N$

61. The resonance contributor that is most important in the intermediate anion formed upon reaction of 4-chloropyridine with sodium methoxide is:

1. 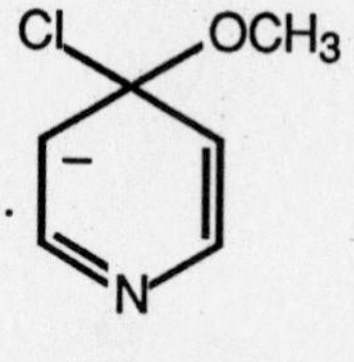
2. 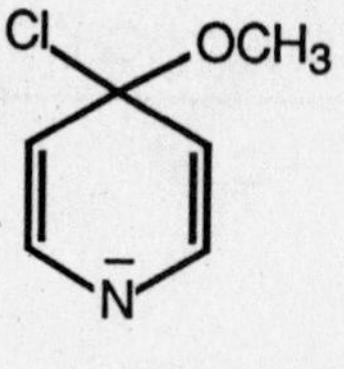
3. 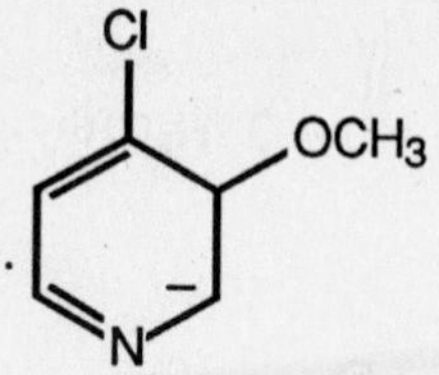
4. 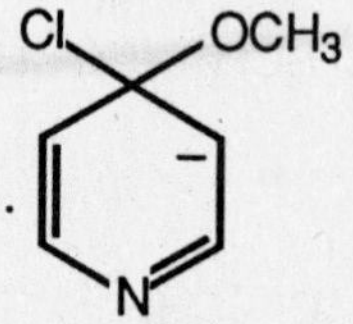
5. 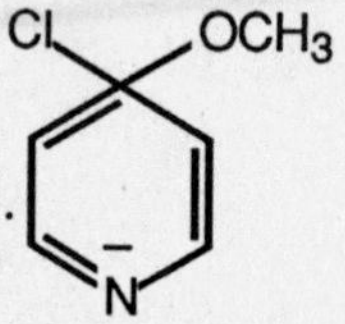

62. What is the correct IUPAC name for $BrCH_2CH_2CH_2CH{=}O$?

1. γ-bromobutyraldehyde
2. 1-bromo-3-carbonylpropane
3. 1-bromo-4-butanal
4. 4-bromo-1-butanone
5. 4-bromobutanal

63. What is the product of the following reaction?

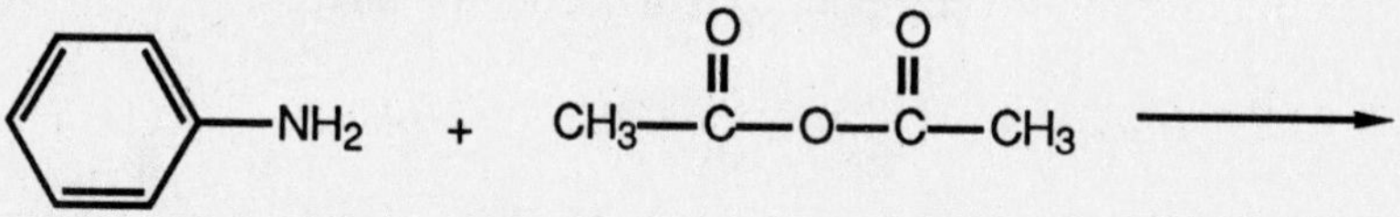

1. 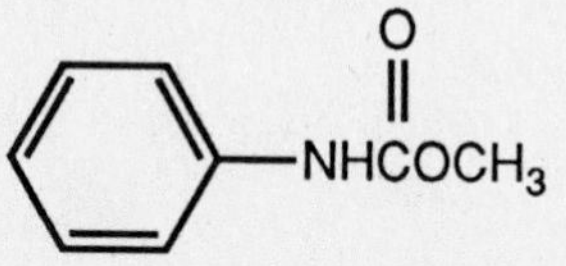
2. 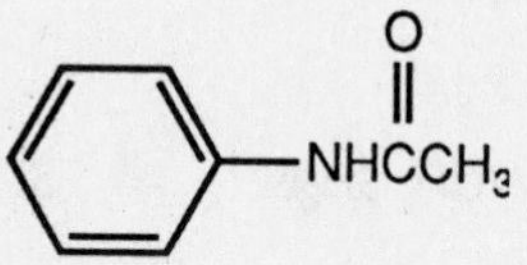
3. 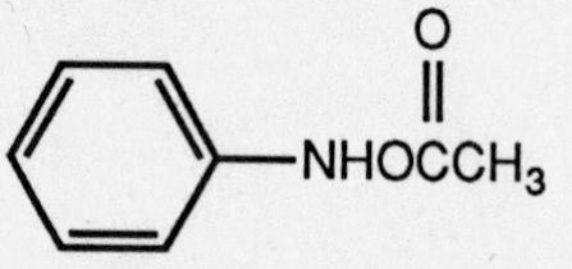
4. 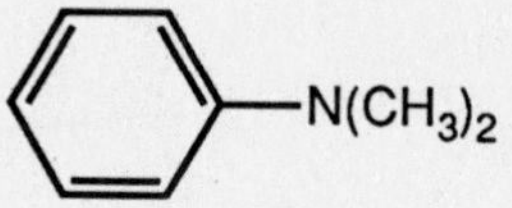
5. 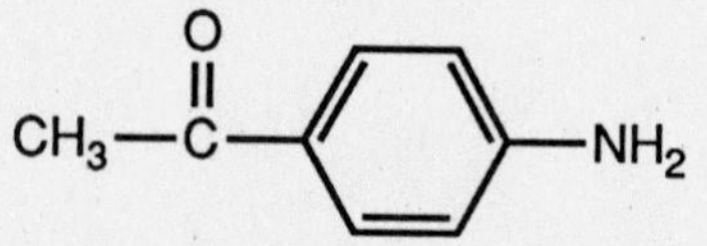

64. Which reaction is most appropriate for the preparation of $(CH_3)_2CHCH_2CH_2CO_2H$?

1. $(CH_3)_2CHCH_2Br$ $\xrightarrow{\text{1. Mg} \atop \text{2. CO}_2 \atop \text{3. H}_3O^+}$

2. $(CH_3)_2CHCH_2CH=O$ $\xrightarrow{Ag(NH_3)_2^+}$

3. $(CH_3)_2CHCH_2CH_2Br$ $\xrightarrow{\text{1. KCN} \atop \text{2. H}_3O^+}$

4. $(CH_3)_2CHMgBr$ $\xrightarrow{H_2C-CH_2}$ (epoxide)

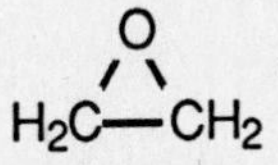

5. $(CH_3)_2CHBr$ + $BrCH_2CH_2CO_2H$ $\xrightarrow{Na}$

65. Arrange the following compounds in order of decreasing acidity:

(a) $BrCH_2CH_2CO_2H$ $CH_3CHCOOH$ $CH_3CHCOOH$

(a)   Br (b)   F (c)

1. a > b > c   2. c > b > a   3. c > a > b   4. b > a > c   5. b > c > a

66. Which structure corresponds to cis-3-pentenoic acid?

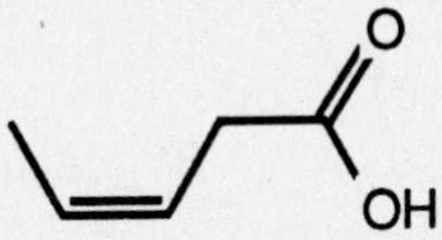

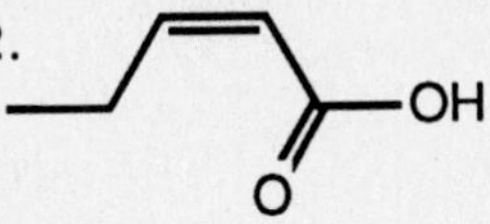

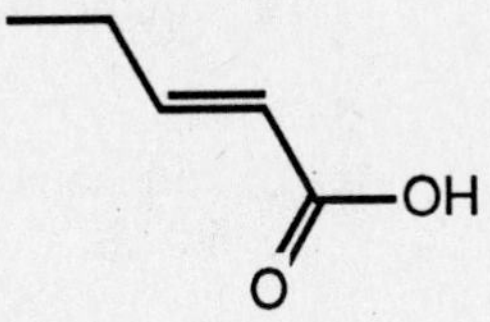

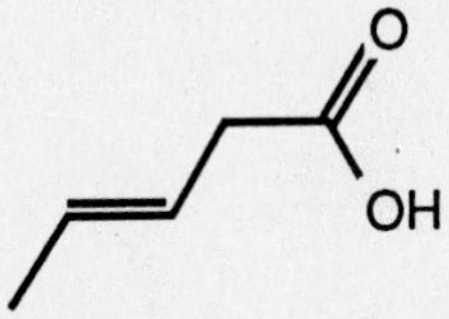

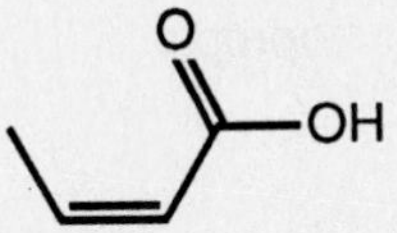

67. What is(are) the product(s) of the following reaction?

$$CH_3CH_2CH(OCH_2CH_3)_2 \xrightarrow[H^+]{H_2O}$$

a. CH_3CH_2OH b. $CH_3CH_2\overset{O}{\overset{\|}{C}}OH$ c. $CH_3CH_2\underset{OH}{\overset{OH}{\underset{|}{\overset{|}{C}}}}OCH_2CH_3$ d. $CH_3CH_2CH{=}O$

    1. a and b   2. a and d   3. a and c   4. only c   5. b and d

68. Cyclic ethers with a three-membered ring

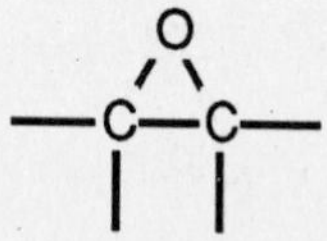

are called:

    1. epoxy resins   2. lactones   3. oxiranes
    4. lactams   5. alkoxides

69. The appearance of a silver mirror in a Tollen's test indicates the presence of:

    1. a monosaccharide   2. an aldehyde   3. cellulose
    4. cellulose   5. glucose

70. The name benzyl α-bromopropionate refers to:

1. 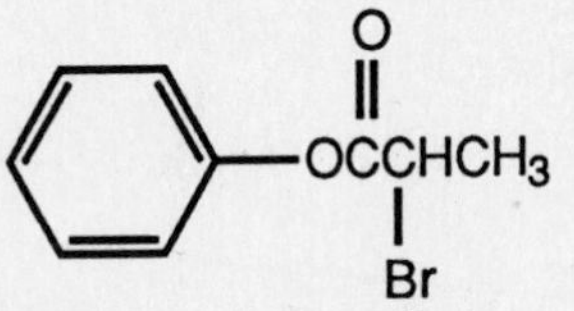
$$\text{C}_6\text{H}_5\text{—OCCHCH}_3$$
with C=O (O double bond) and Br substituent

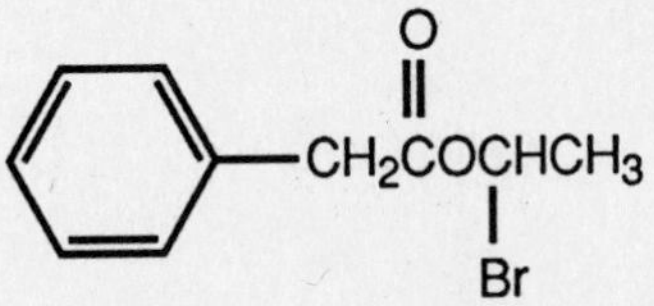

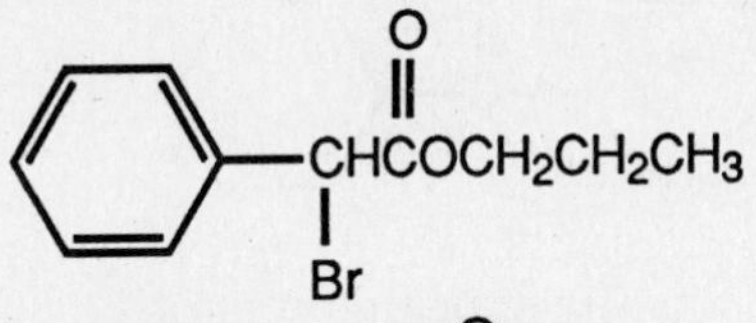

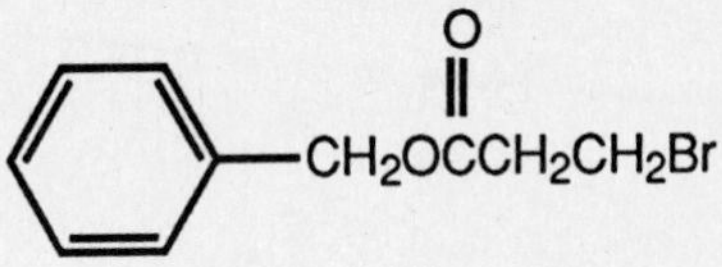

2.
$$\text{C}_6\text{H}_5\text{—CH}_2\text{COCHCH}_3$$
with C=O (O double bond) and Br substituent

3.
$$\text{C}_6\text{H}_5\text{—CHCOCH}_2\text{CH}_2\text{CH}_3$$
with C=O (O double bond) and Br substituent

4.
$$\text{C}_6\text{H}_5\text{—CH}_2\text{OCCH}_2\text{CH}_2\text{Br}$$
with C=O (O double bond)

5.
$$\text{C}_6\text{H}_5\text{—CH}_2\text{OCCHCH}_3$$
with C=O (O double bond) and Br substituent

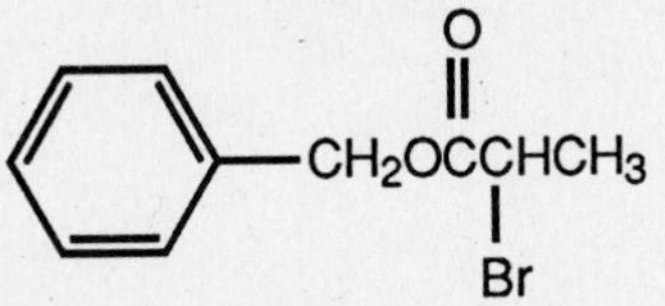

71. Compound X gives the following test results:

$$X + Ag(NH_3)_2^+ \longrightarrow \text{no silver mirror}$$

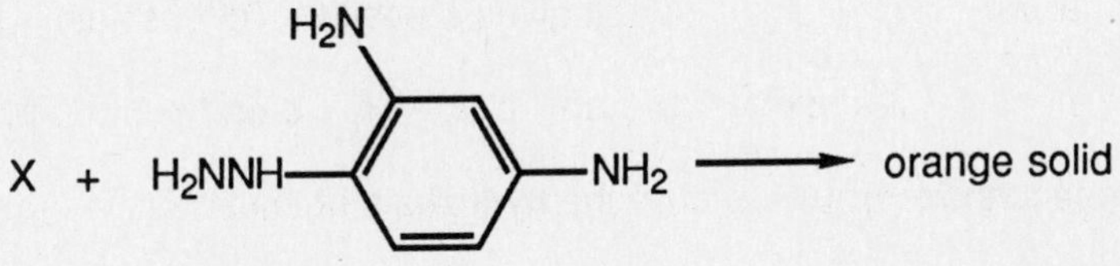

$$X + H_2NNH\text{—}\underset{\underset{NH_2}{}}{\overset{H_2N}{\bigcirc}}\text{—NH}_2 \longrightarrow \text{orange solid}$$

$$X \xrightarrow[\text{CH}_3\text{OD}]{\text{CH}_3\text{O}^-\text{Na}^+} \text{exchanges } \textit{four} \text{ hydrogens for deuterium}$$

Which of the following compounds is a possible structure for X?

SAMPLE MULTIPLE-CHOICE QUESTIONS

1. $CH_3COC(CH_3)_3$ 2. $CH_3CH_2CH_2CCH_2CH_3$ 3. 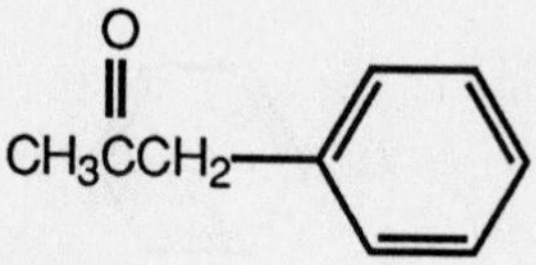 CH_3CCH_2-

4. $O=CHCH_2CH_2CH=O$ 5.

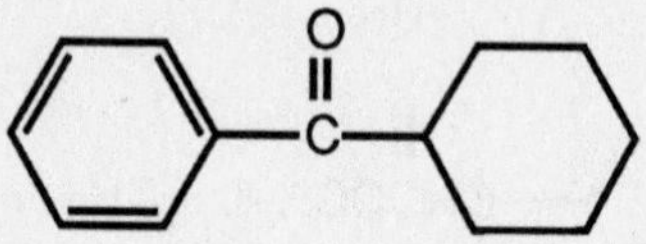

72.   Which statements regarding the following structure are true?

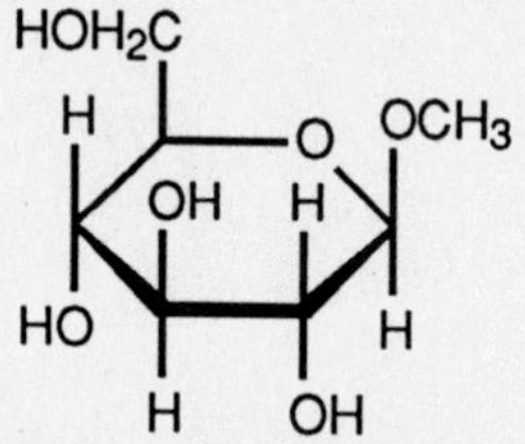

a.   It is a hemiacetal.
b.   It is an acetal.

c.   It can be hydrolyzed by H_3O^+.
d.   It gives a positive Tollens' test.

1. b and d   2. a and d   3. only b   4. b and c   5. b, c, and d

73.   The monosaccharide obtained from the hydrolysis of starch is:

1.   maltose   2.   D-ribose   3.   2-deoxy-D-ribose
4.   D-galactose   5.   D-glucose

74.   The products of the complete hydrolysis of DNA are:

1.   D-ribose, phosphoric acid, and four heterocyclic bases
2.   nucleosides
3.   nucleotides
4.   glucose and fructose
5.   2-deoxy-D-ribose, phosphoric acid, and four heterocyclic bases

75.   Uracil is:

1.   a constituent of urine   2.   a pyrimidine   3.   a purine
4.   a heterocyclic base   5.   a potent carcinogen
     present in DNA

526

76. The primary structure of a protein refers to:

1. the amino acid sequence in the polypeptide chain
2. the presence or absence of an α helix
3. the orientation of the amino acid side chains in space
4. interchain cross-links with disulfide bonds
5. whether the protein is fibrous or globular

77. Which of the following compounds is a fat?

1. $CH_3(CH_2)_{16}\overset{\displaystyle O}{\overset{\|}{C}}O(CH_2)_{15}CH_3$

2.

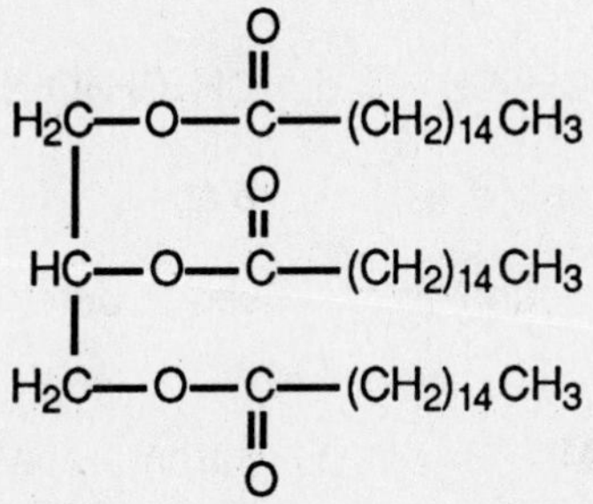

3.

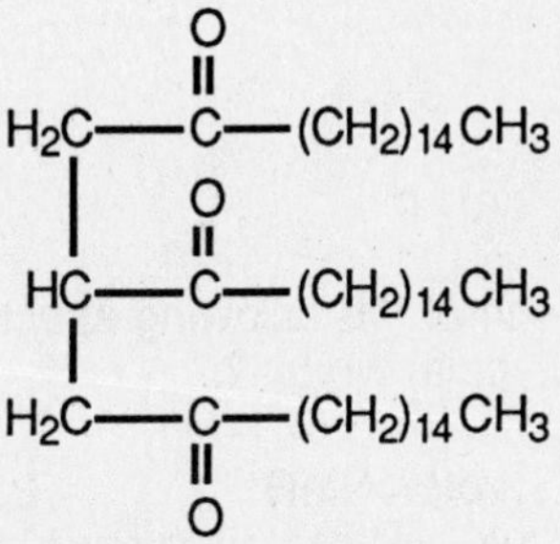

4.

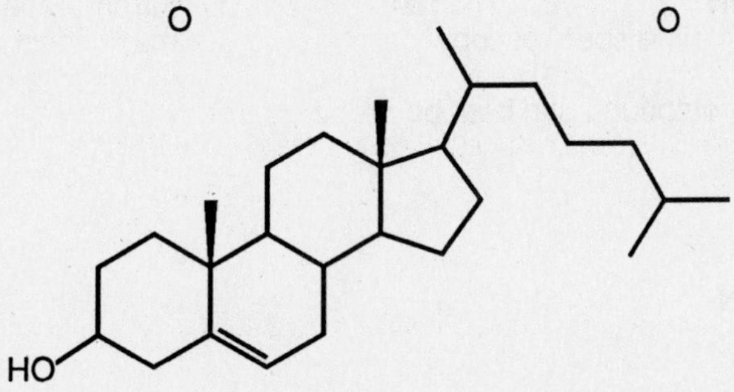

5.

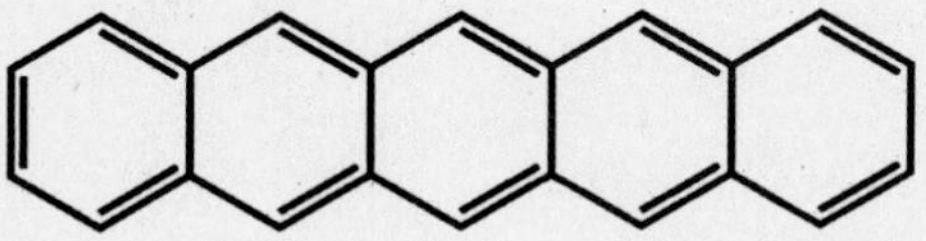

78. Which of the following structures is a zwitterion?

1. $^-OCCH_2CH_2CO^-$ (with two C=O)   2. Ca^{2+} 3. $H_3\overset{+}{N}CH_2CO^-$ (with C=O)

4. $CH_3(CH_2)_{16}CO^- Na^+$ (with C=O)   5. $R-C\overset{O}{\underset{O^-}{<}} \longleftrightarrow R-C\overset{O^-}{\underset{O}{<}}$

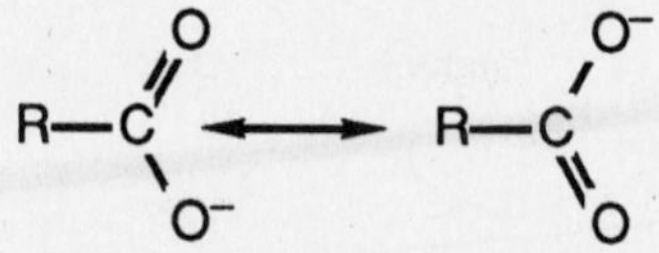

79. Which of the following compounds would show only a single peak in a proton NMR spectrum?

   a. $(CH_3)_4Si$ b. CCl_4 c. CH_2Cl_2 d. $CH_3CH_2OCH_2CH_3$

   1. only b   2. a and c   3. only d   4. b and c   5. a, b, and c

80. Which of the following spectroscopic techniques is used to determine molecular weight?

   1. proton NMR   2. ^{13}C NMR   3. infrared spectroscopy
   4. ultraviolet-visible spectroscopy   5. mass spectrometry

81. The following compound can best be prepared from:

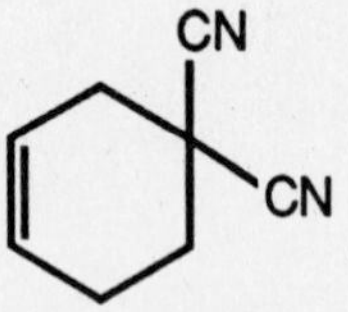

1. 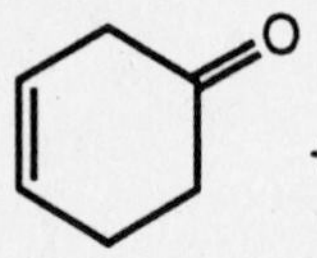 + HCN

2.

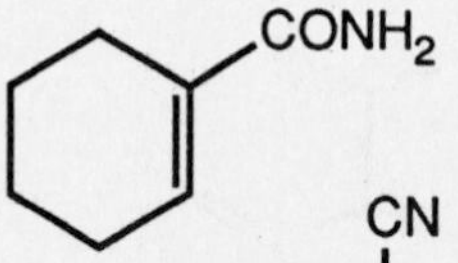

3. $H_2C=CH-CH=CH_2 + H_2C=C(CN)_2$ 4.

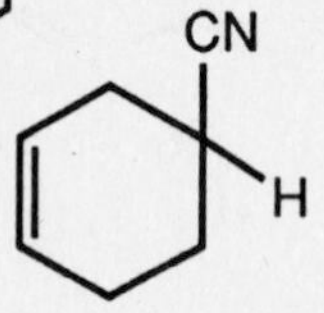

5. $H_2C=CH-CH=CH_2 + NC-C\equiv C-CN$

SAMPLE MULTIPLE-CHOICE QUESTIONS

82. Which of the following terms can be used to describe this pair of compounds?

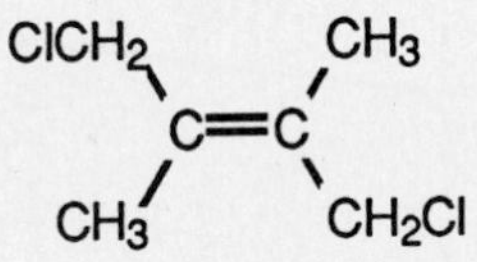

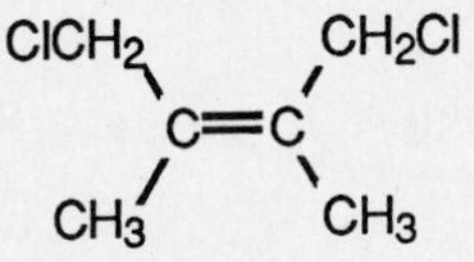

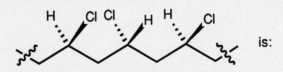

a. enantiomers   b. diastereomers   c. constitutional isomers
d. stereoisomers e. chiral   f. achiral

1. a, b, and d   2. a, c, and e   3. b, d, and e
4. a, d, and e   5. b, d, and f

83. Which of the following best describes $C_6H_9Br_3$?

1. no rings, no double bonds   2. one ring, no double bonds
3. two double bonds   4. one triple bond
5. one ring and one double bond   6. two rings

84. The polymer

is:

1. syndiotactic   2. amorphous   3. atactic
4. block   5. glassy   6. isotactic

529

85.   Which of the following is a step-growth polymer?

1. polystyrene   2. nylon   3. Teflon   4. polyvinyl
5. synthetic rubber

86.   How many chiral centers are present in cortisone?

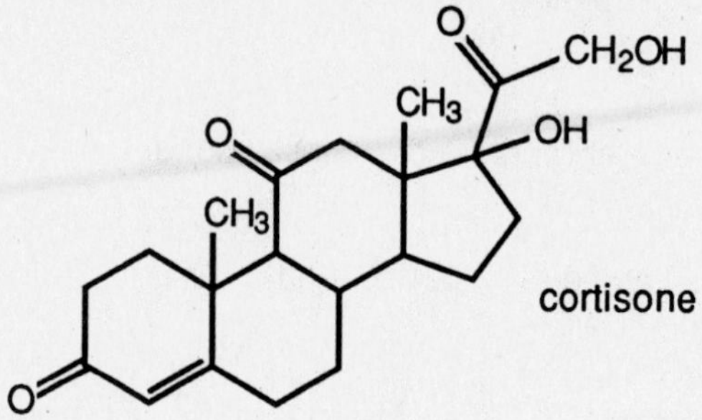

cortisone

1. 0   2. 3   3. 4   4. 6   5. 8

87.   The following Fischer projections

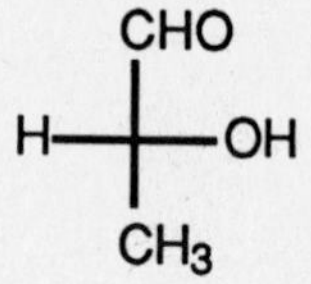

and 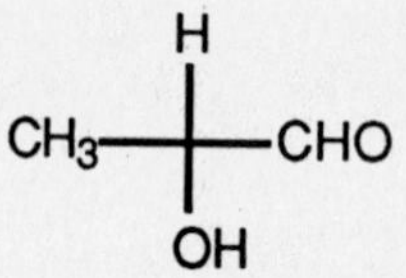   represent:

1. the same compound   2. enantiomers   3. diastereomers
4. *meso* forms   5. rotamers

88. The parent heterocyclic ring present in vitamin B_6 is:

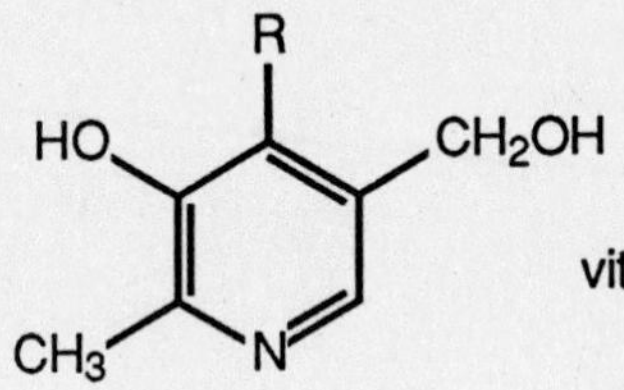

vitamin B_6

1. pyridine   2. pyrrole   3. pyrimidine
4. purine   5. imidazole

89. Oxidation of ethanol can give:

1. $CH_2=O$ 2. $HOCH_2CH_2OH$ and $CH_3CH(OH)_2$ 3. $CO_2 + H_2$

4. $CH_3CHO + CH_3CO_2H$ 5. $CH_3OCH_3 + CH_3CO_2H$

90. The structure of purine is:

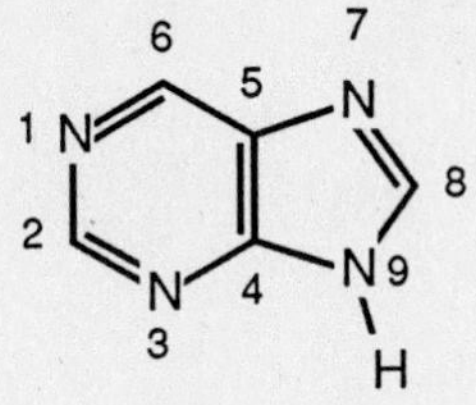

The least basic nitrogen present is:

1. N-1   2. N-3   3. N-7   4. N-9
5. They all are approximately equally basic.

91.   The formulas for the amino acids glycine and alanine are as follows:

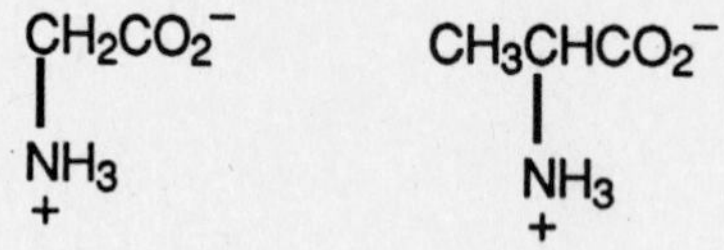

$$CH_2CO_2^-$$
|
$$NH_3$$
$$+$$

glycine

$$CH_3CHCO_2^-$$
|
$$NH_3$$
$$+$$

alanine

The correct formula for alanylglycine is:

1. 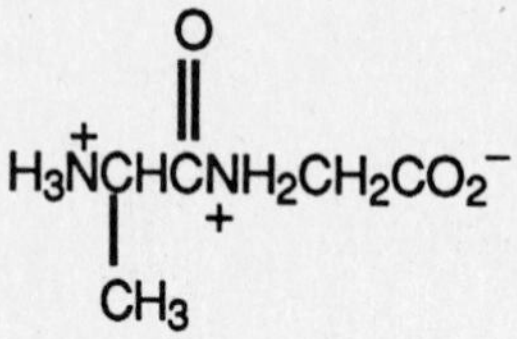 $H_3\overset{+}{N}CHC\overset{O}{\overset{||}{}}NH_2CH_2CO_2^-$
    |
    CH_3 $+$

2. $H_3\overset{+}{N}CH_2C\overset{O}{\overset{||}{}}NHCHCO_2^-$
    |
    CH_3

3. $H_3\overset{+}{N}CHC\overset{O}{\overset{||}{}}CHCO_2^-$
    | |
    CH_3 NH_3 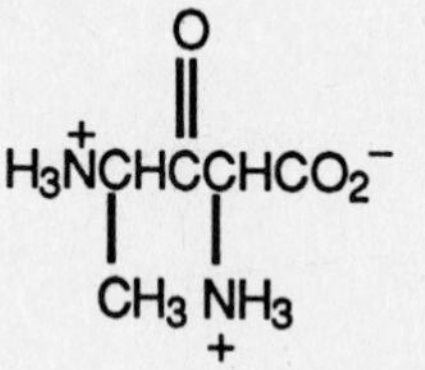
             $+$

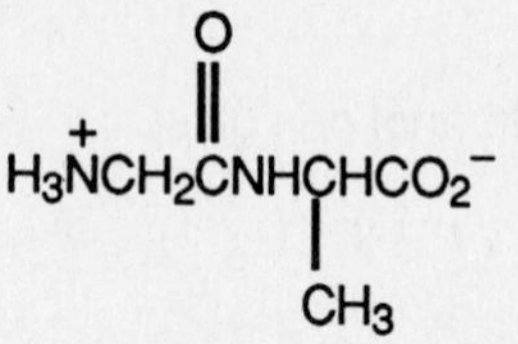

4. 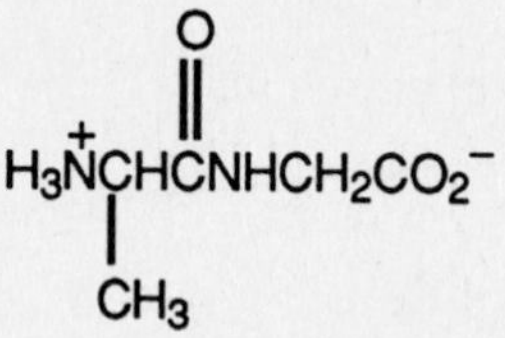 $H_3\overset{+}{N}CHC\overset{O}{\overset{||}{}}NHCH_2CO_2^-$
    |
    CH_3

5. 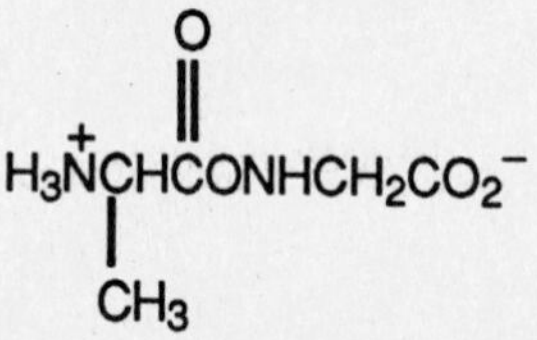 $H_3\overset{+}{N}CHC\overset{O}{\overset{||}{}}ONHCH_2CO_2^-$
    |
    CH_3

92.   Which of the following can act as a nucleophile?

   a. H_2O b. OH^- c. H^+ d. CN^- e. Na^+

   1. a, b, and d   2. a and b   3. c and e   4. only d   5. b and d

SAMPLE MULTIPLE-CHOICE QUESTIONS

93.   Penicillin is a β-lactam. Its formula is:

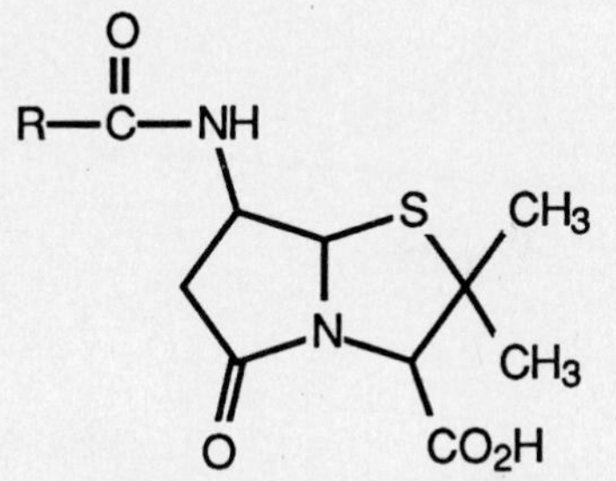

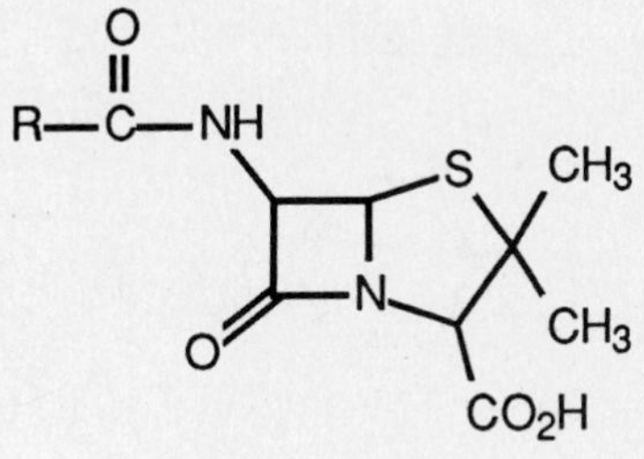

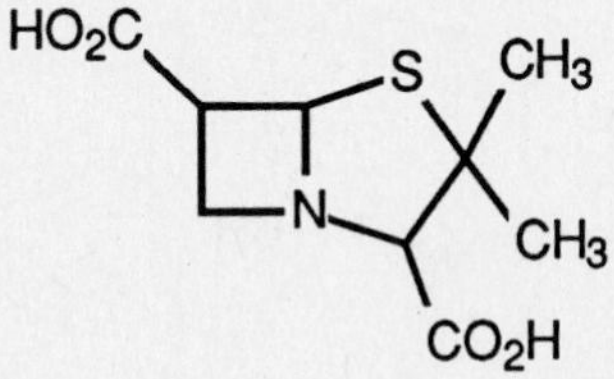

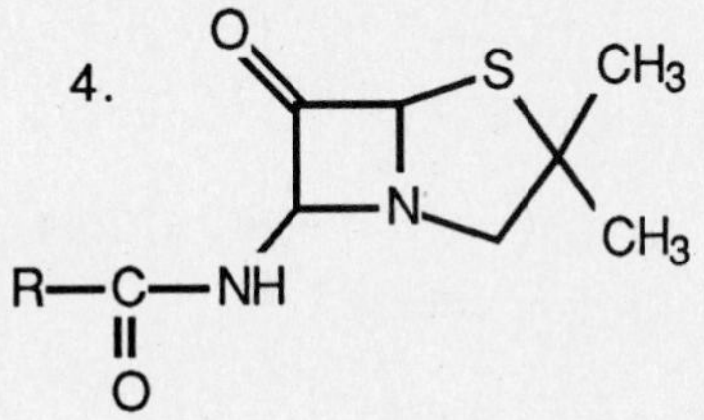

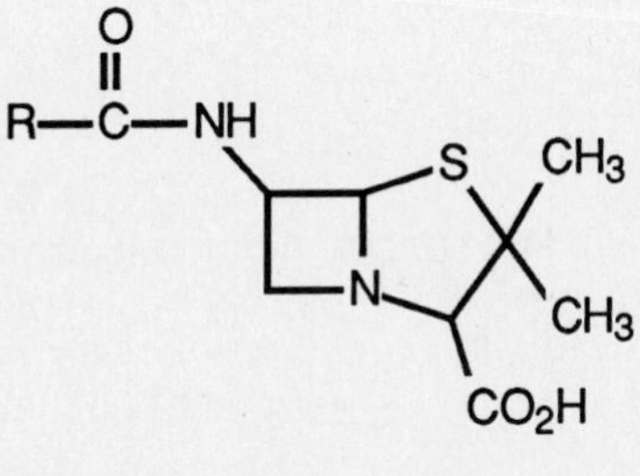

94.   The antihistamine diphenhydramine (Benadryl) has the formula:

On heating with aqueous acid, it gives:

1. (phenyl)$_2$CHOH + HOCH$_2$CH$_2$N(CH$_3$)$_2$

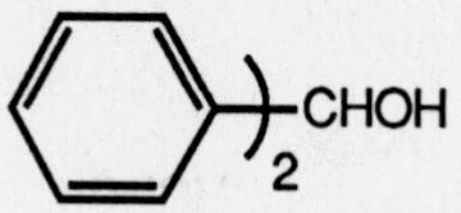

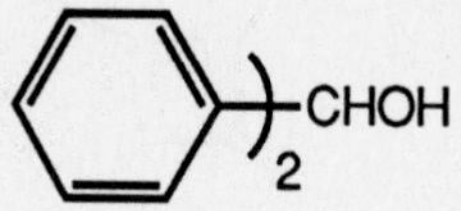

2. (phenyl)$_2$CHOH + CH$_3$CH$_2$N(CH$_3$)$_2$

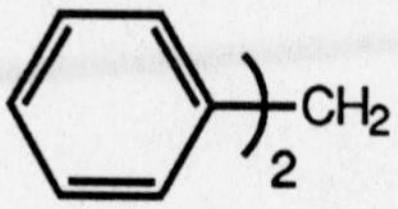

3. (phenyl)$_2$CH$_2$ + HOCH$_2$CH$_2$N(CH$_3$)$_2$

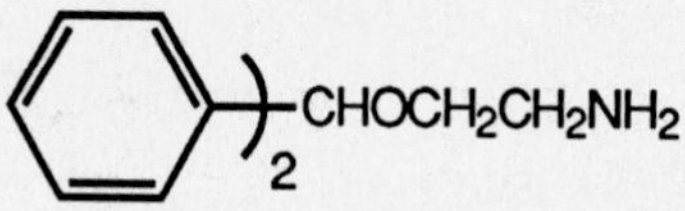

4. (phenyl)$_2$CHOCH$_2$CH$_2$NH$_2$ + CH$_3$—CH$_3$

5. (phenyl)$_2$CHOCH$_2$CH$_2$N(CH$_3$)$_3$$^+$

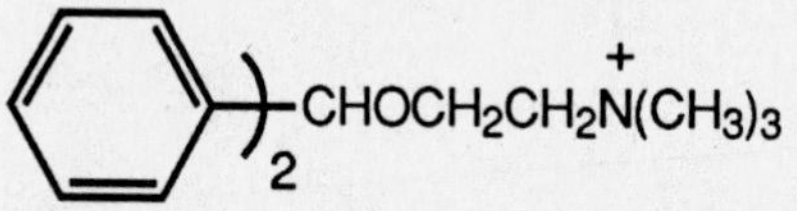

95. How many peaks appear in the ^{1}H NMR spectrum of the following compound?

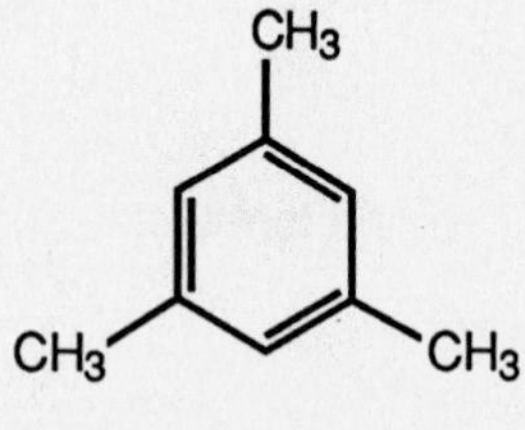

CH$_3$

(benzene ring with CH$_3$, CH$_3$, CH$_3$ substituents)

1. 1   2. 2   3. 3   4. 6   5. 9

96. How many peaks appear in the ^{13}C NMR spectrum of the compound in Question 95?

1. 1   2. 2   3. 3   4. 6   5. 9

97. The ^{1}H NMR spectrum of one isomer of $C_3H_3Cl_5$ consists of a triplet at δ 4.5 and a doublet at δ 6.0, with relative areas 1:2. The structure of the isomer is

1. $CH_2ClCHClCCl_3$ 2. $CH_2ClCCl_2CHCl_2$ 3. $CHCl_2CH_2CCl_3$

4. $CHCl_2CHClCHCl_2$ 5. $CH_3CCl_2CCl_3$

98. Cocaine is an alkaloid. Its correct structure is

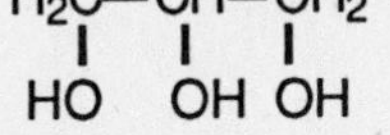

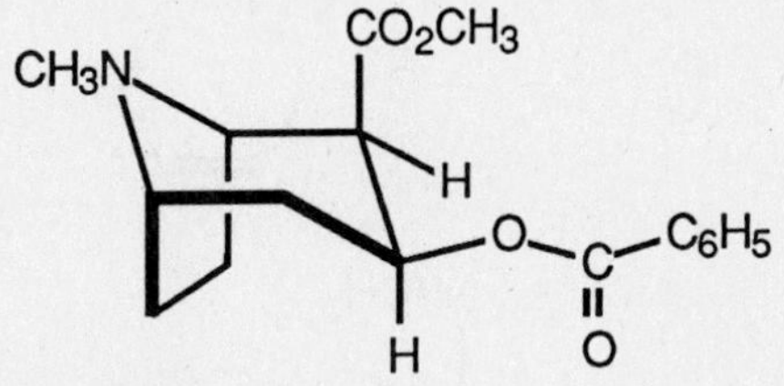

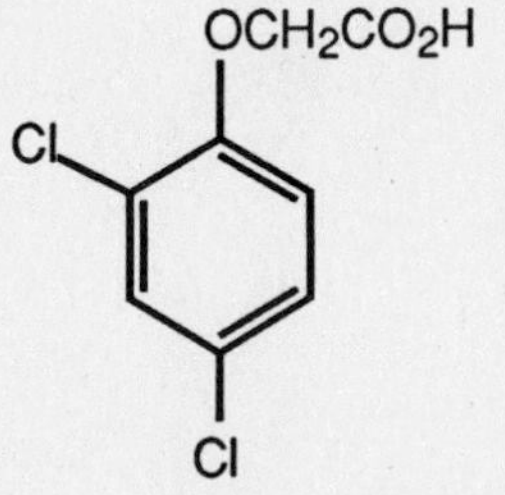

99. Which of the following statements is *not* true?

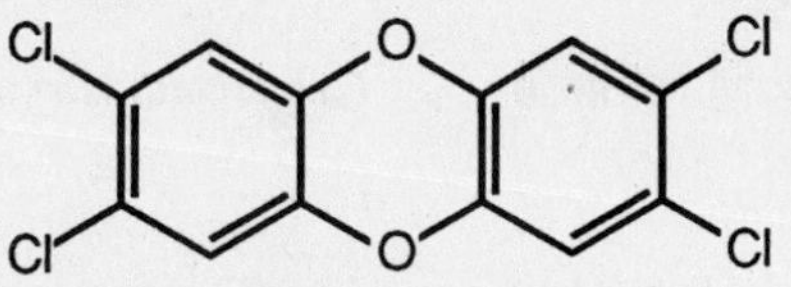

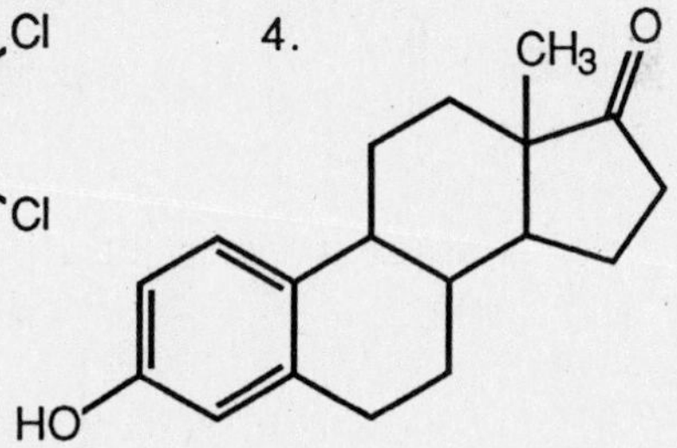

1. Natural rubber is a hydrocarbon.
2. Natural rubber is made of isoprene units.
3. Natural rubber is a polymer of 1,3-butadiene.
4. Natural rubber has *cis* double bonds.
5. Natural rubber can be vulcanized.

100. What is(are) the product(s) of the following reaction?

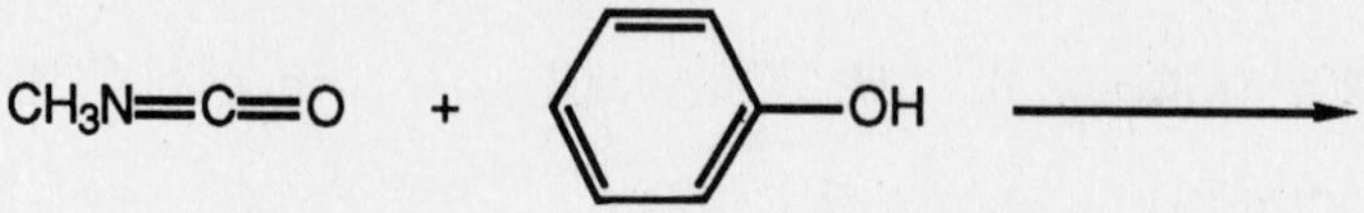

$$CH_3N=C=O \quad + \quad$$ 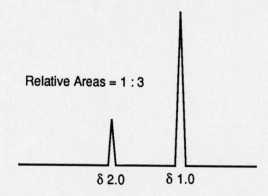 —OH $\longrightarrow$

1. 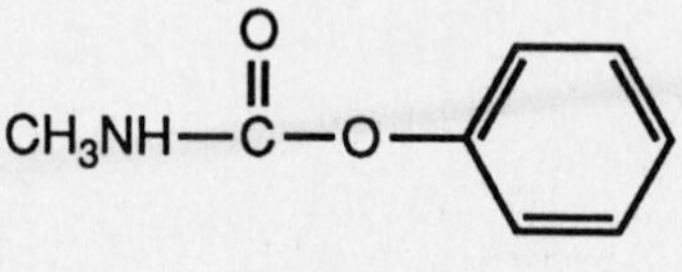 $CH_3NH-\overset{\overset{\textstyle O}{\|}}{C}-O-$⟨⟩

2. 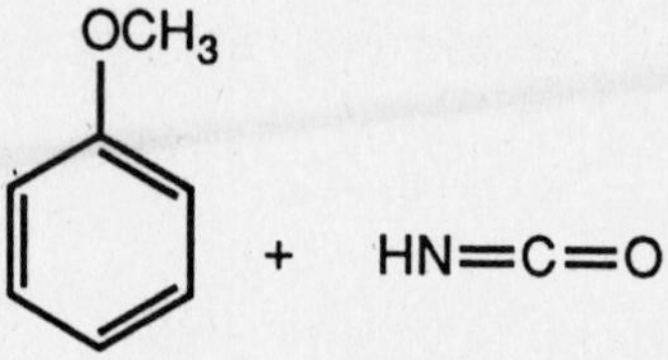 ⟨⟩ with OCH₃ $\quad + \quad HN=C=O$

3. 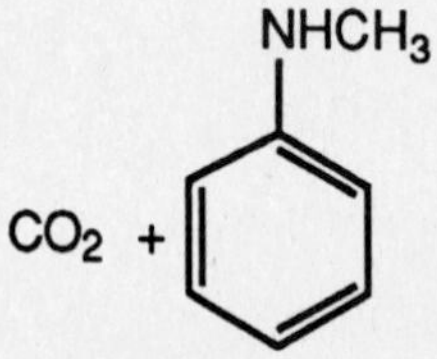 $CO_2 \; +$ ⟨⟩ with NHCH₃

4. a polyurethane

5. 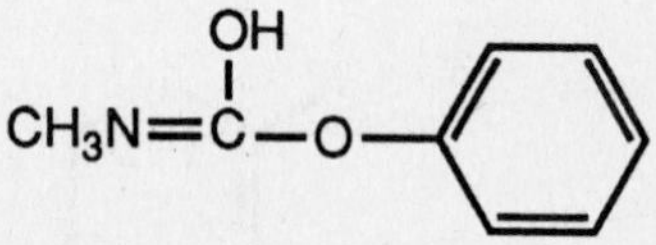 $CH_3N=\overset{\overset{\textstyle OH}{|}}{C}-O-$⟨⟩

101. Which structure is consistent with the following ^1H NMR spectrum?

Relative Areas = 1 : 3

δ 2.0   δ 1.0

1. $(CH_3)_3CO-\overset{\overset{\textstyle O}{\|}}{C}-CH_3$

2. $(CH_3)_3CO-\overset{\overset{\textstyle O}{\|}}{C}-OCH_3$

3. $(CH_3)_3C-\overset{\overset{\textstyle O}{\|}}{C}-OCH_3$

4. $CH_3-\overset{\overset{\textstyle O}{\|}}{C}-OCH_3$

5. $CH_3-\overset{\overset{\textstyle O}{\|}}{C}-CH_3$

102. The infrared spectrum of a compound shows absorptions at 1700 cm^{-1} and 3600 cm^{-1}. A possible structure is:

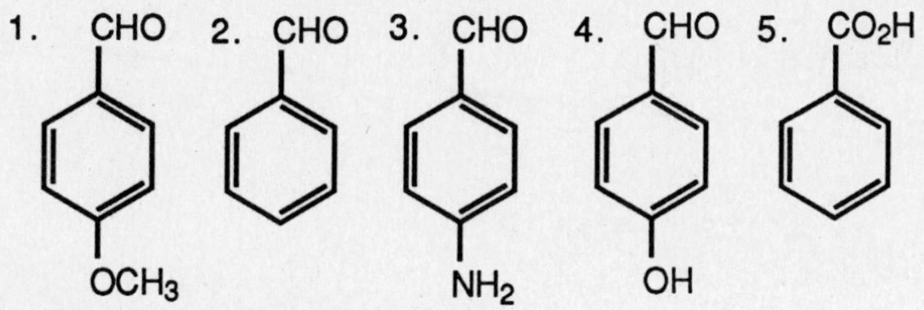

1. CHO   2. CHO   3. CHO   4. CHO   5. CO$_2$H

(structures 1–5 shown)

OCH$_3$ NH$_2$ OH

103. The parent ion in the mass spectrum of 2-propanone has the following structure:

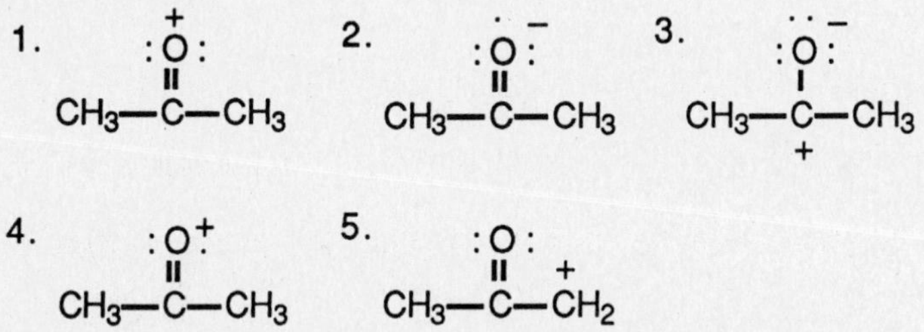

1.   :O: (+)
      ‖
   CH$_3$—C—CH$_3$

2.   :O: (• —)
      ‖
   CH$_3$—C—CH$_3$

3.   :O: (•• —)
      |
   CH$_3$—C—CH$_3$
           (+)

4.   :O:(+)
      ‖
   CH$_3$—C—CH$_3$

5.   :O:
      ‖   (+)
   CH$_3$—C—CH$_2$

104. The major product of the following reaction is:

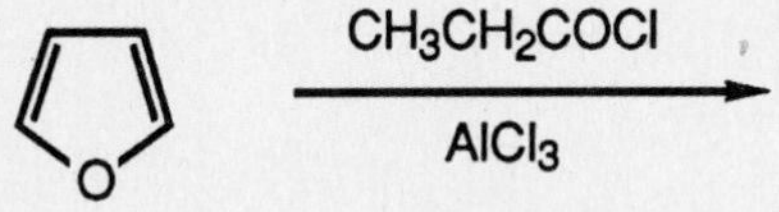

(furan) $\xrightarrow[\text{AlCl}_3]{\text{CH}_3\text{CH}_2\text{COCl}}$

1.

2.

3.

4.

5.

— AlCl₃

105. Which reaction sequence can be used to accomplish the following transformation:

1. (a) LiAlH₄ (b) CH₃OH, H⁺ (c) H₂O, H⁺
2. (a) CH₃OH, H⁺ (b) H₂O, H⁺ (c) LiAlH₄
3. (a) CH₃OH, H⁺ (b) LiAlH₄ (c) H₂O, H⁺
4. (a) CH₃OH, H⁺ (b) PCC (c) H₂O, H⁺
5. (a) NaBH₄ (b) LiAlH₄ (c) PCC

106. Arrange the following compounds in order of increasing electrophilic aromatic substitution rates.

SAMPLE MULTIPLE-CHOICE QUESTIONS

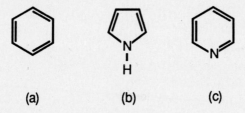

(a)   (b)   (c)

1. a < b < c   2. c < b < a   3. b < c < a   4. c < a < b   5. b < a < c

107.   The following compound is a:

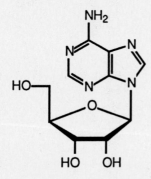

1. nucleotide   2. glycoside   3. nucleoside
4. codon   5. pyrimidine

108.   The DNA strand that is complementary to the following sequence of bases is:

5'—TTACCG—3'

1. 3'—AATGGC—5'   2. 3'—UUTGGC—5'   3. 5'—AATGGC—3'
4. 3'—CCGTTA—5'   5. 5'—CCGTTA—3'

109.   The mRNA strand produced upon transcription of the following DNA strand is:

5'—TTACCG—3'

1. 5'—AAUGGC—3'   2. 3'—AAUGGC—5'   3. 5'—AATGGC—3'
4. 3'—AATGGC—5'   5. 3'—GCCATT—5'

110.   Translation is the process by which:

1. DNA replication occurs
2. DNA is transcribed into *m*RNA
3. the sequence of nucleotides in a DNA strand is determined
4. nucleic acids are synthesized in the laboratory
5. genetic messages are decoded and used to build proteins

Key to the Sample Multiple-Choice Questions

1. 4	26. 4	51. 3	76. 1	101. 1
2. 3	27. 1	52. 1	77. 2	102. 4
3. 2	28. 3	53. 5	78. 3	103. 4
4. 4	29. 3	54. 3	79. 2	104. 3
5. 1	30. 2	55. 5	80. 5	105. 3
6. 1	31. 3	56. 1	81. 3	106. 4
7. 2	32. 2	57. 2	82. 5	107. 3
8. 5	33. 1	58. 1	83. 2	108. 1
9. 3	34. 3	59. 5	84. 1	109. 2
10. 5	35. 4	60. 2	85. 2	110. 5
11. 2	36. 1	61. 2	86. 4	
12. 2	37. 2	62. 5	87. 2	
13. 3	38. 2	63. 2	88. 1	
14. 1	39. 3	64. 3	89. 4	
15. 2	40. 1	65. 2	90. 4	
16. 1	41. 2, 5	66. 1	91. 4	
17. 5	42. 2	67. 2	92. 1	
18. 2	43. 1	68. 3	93. 2	
19. 5	44. 2	69. 2	94. 1	
20. 2	45. 4	70. 5	95. 2	
21. 4	46. 1	71. 2	96. 3	
22. 1	47. 5	72. 4	97. 4	
23. 5	48. 4	73. 5	98. 2	
24. 1	49. 1	74. 5	99. 3	
25. 4	50. 5	75. 2	100. 1	